Resveratrol in Health and Disease

OXIDATIVE STRESS AND DISEASE

Series Editors

LESTER PACKER, PH.D.
ENRIQUE CADENAS, M.D., PH.D.
University of Southern California School of Pharmacy
Los Angeles, California

1. Oxidative Stress in Cancer, AIDS, and Neurodegenerative Diseases, *edited by Luc Montagnier, René Olivier, and Catherine Pasquier*
2. Understanding the Process of Aging: The Roles of Mitochondria, Free Radicals, and Antioxidants, *edited by Enrique Cadenas and Lester Packer*
3. Redox Regulation of Cell Signaling and Its Clinical Application, *edited by Lester Packer and Junji Yodoi*
4. Antioxidants in Diabetes Management, *edited by Lester Packer, Peter Rösen, Hans J. Tritschler, George L. King, and Angelo Azzi*
5. Free Radicals in Brain Pathophysiology, *edited by Giuseppe Poli, Enrique Cadenas, and Lester Packer*
6. Nutraceuticals in Health and Disease Prevention, *edited by Klaus Krämer, Peter-Paul Hoppe, and Lester Packer*
7. Environmental Stressors in Health and Disease, *edited by Jürgen Fuchs and Lester Packer*
8. Handbook of Antioxidants: Second Edition, Revised and Expanded, *edited by Enrique Cadenas and Lester Packer*
9. Flavonoids in Health and Disease: Second Edition, Revised and Expanded, *edited by Catherine A. Rice-Evans and Lester Packer*
10. Redox–Genome Interactions in Health and Disease, *edited by Jürgen Fuchs, Maurizio Podda, and Lester Packer*
11. Thiamine: Catalytic Mechanisms in Normal and Disease States, *edited by Frank Jordan and Mulchand S. Patel*
12. Phytochemicals in Health and Disease, *edited by Yongping Bao and Roger Fenwick*
13. Carotenoids in Health and Disease, *edited by Norman I. Krinsky, Susan T. Mayne, and Helmut Sies*

14. Herbal and Traditional Medicine: Molecular Aspects of Health, *edited by Lester Packer, Choon Nam Ong, and Barry Halliwell*
15. Nutrients and Cell Signaling, *edited by Janos Zempleni and Krishnamurti Dakshinamurti*
16. Mitochondria in Health and Disease, *edited by Carolyn D. Berdanier*
17. Nutrigenomics, *edited by Gerald Rimbach, Jürgen Fuchs, and Lester Packer*
18. Oxidative Stress, Inflammation, and Health, *edited by Young-Joon Surh and Lester Packer*
19. Nitric Oxide, Cell Signaling, and Gene Expression, *edited by Santiago Lamas and Enrique Cadenas*
20. Resveratrol in Health and Disease, *edited by Bharat B. Aggarwal and Shishir Shishodia*

Resveratrol in Health and Disease

edited by

Bharat B. Aggarwal
Shishir Shishodia

A CRC title, part of the Taylor & Francis imprint, a member of the Taylor & Francis Group, the academic division of T&F Informa plc.

Published in 2006 by
CRC Press
Taylor & Francis Group
6000 Broken Sound Parkway NW, Suite 300
Boca Raton, FL 33487-2742

Printed in the United States of America on acid-free paper
10 9 8 7 6 5 4 3 2 1

International Standard Book Number-10: 0-8493-3371-7 (Hardcover)
International Standard Book Number-13: 978-0-8493-3371-2 (Hardcover)

Library of Congress Cataloging-in-Publication Data

Catalog record is available from the Library of Congress

Taylor & Francis Group
is the Academic Division of Informa plc.

Visit the Taylor & Francis Web site at
http://www.taylorandfrancis.com

and the CRC Press Web site at
http://www.crcpress.com

Series Introduction

Oxygen is a dangerous friend. Through evolution, oxygen — itself a free radical — was chosen as the terminal electron acceptor for respiration. The two unpaired electrons of oxygen spin in the same direction; thus, oxygen is a biradical. Other oxygen-derived free radicals, such as superoxide anions or hydroxyl radicals, formed during metabolism or by ionizing radiation are stronger oxidants, i.e., endowed with a higher chemical reactivity. Oxygen-derived free radicals are generated during oxidative metabolism and energy production in the body and are involved in regulation of signal transduction and gene expression, activation of receptors and nuclear transcription factors, oxidative damage to cell components, the antimicrobial and cytotoxic action of immune system cells, neutrophils, and macrophages, as well as in aging and age-related degenerative diseases. Overwhelming evidence indicates that oxidative stress can lead to cell and tissue injury. However, the same free radicals that are generated during oxidative stress are produced during normal metabolism and, as a corollary, are involved in both human health and disease.

In addition to reactive oxygen species, research on reactive nitrogen species has been gathering momentum to develop an area of enormous importance in biology and medicine. Nitric oxide or nitrogen monoxide (NO) is a free radical generated by nitric oxide synthase (NOS). This enzyme modulates physiological responses in the circulation such as vasodilation (eNOS) or signaling in the brain (nNOS). However, during inflammation, a third isoenzyme is induced, iNOS, resulting in the overproduction of NO and causing damage to targeted infectious organisms and to healthy tissues in the vicinity. More worrisome, however, is the fact that NO can react with the superoxide anion to yield a strong oxidant, peroxynitrite. Oxidation of lipids, proteins, and DNA by peroxynitrite increases the likelihood of tissue injury.

Both reactive oxygen and nitrogen species are involved in the redox regulation of cell functions. Oxidative stress is increasingly viewed as a major upstream component in the signaling cascade involved in inflammatory responses and stimulation of adhesion molecule and chemoattractant production. Hydrogen peroxide decomposes in the presence of transition metals to the highly reactive hydroxyl radical, which by two major reactions — hydrogen abstraction and addition — accounts for most of the oxidative damage to proteins, lipids, sugars, and nucleic acids. Hydrogen peroxide is also an important signaling molecule that, among others, can activate NF-κB, an important transcription factor involved in inflammatory responses. At low concentrations hydrogen peroxide regulates cell signaling and stimulates cell proliferation; at higher concentrations it triggers apoptosis and, at even higher levels, necrosis.

Virtually all diseases thus far examined involve free radicals. In most cases, free radicals are secondary to the disease process, but in some instances free radicals are causal. Thus, there is a delicate balance between oxidants and antioxidants in health and disease. Their proper balance is essential for ensuring healthy aging.

The term oxidative stress indicates that the antioxidant status of cells and tissues is altered by exposure to oxidants. The redox status is thus dependent on the degree to which a cell's components are in the oxidized state. In general, the reducing environment inside cells helps to prevent oxidative damage. In this reducing environment, disulfide bonds (S–S) do not spontaneously form because sulfhydryl groups are maintained in the reduced state (SH), thus preventing protein misfolding or aggregation. This reducing environment is maintained by oxidative metabolism and by the action of antioxidant enzymes and substances, such as glutathione, thioredoxin, vitamins E and C, and enzymes such as superoxide dismutases, catalase, and the selenium-dependent glutathione reductase and glutathione and thioredoxin hydroperoxidases, which serve to remove reactive oxygen species (hydroperoxides).

Changes in the redox status and depletion of antioxidants occur during oxidative stress. The thiol redox status is a useful index of oxidative stress mainly because metabolism and NADPH-dependent enzymes maintain cell glutathione (GSH) almost completely in its reduced state. Oxidized glutathione (glutathione disulfide, GSSG) accumulates under conditions of oxidant exposure and this changes the GSSG/GSH ratio; an increased ratio is usually taken as indicating oxidative stress. Other oxidative stress indicators are ratios of redox couples such as NADPH/NADP, NADH/NAD, $\text{thioredoxin}_{\text{reduced}}/\text{thioredoxin}_{\text{oxidized}}$, dihydrolipoic acid/$\alpha$-lipoic acid, and lactate/pyruvate. Changes in these ratios affect the energy status of the cell, largely determined by the ratio ATP/ADP + AMP. Many tissues contain large amounts of glutathione, 2 to 4 mM in erythrocytes or neural tissues and up to 8 mM in hepatic tissues. Reactive oxygen and nitrogen species can oxidize glutathione, thus lowering the levels of the most abundant nonprotein thiol, sometimes designated as the cell's primary preventative antioxidant.

Current hypotheses favor the idea that lowering oxidative stress can have a health benefit. Free radicals can be overproduced or the natural antioxidant system defenses weakened, first resulting in oxidative stress, and then leading to oxidative injury and disease. Examples of this process include heart disease, cancer, and neurodegenerative disorders. Oxidation of human low-density lipoproteins is considered an early step in the progression and eventual development of atherosclerosis, thus leading to cardiovascular disease. Oxidative DNA damage may initiate carcinogenesis. Environmental sources of reactive oxygen species are also important in relation to oxidative stress and disease. For example, ultraviolet radiation, ozone, cigarette smoke, and others are significant sources of oxidative stress.

Compelling support for the involvement of free radicals in disease development originates from epidemiological studies showing that an enhanced antioxidant status is associated with reduced risk of several diseases. Vitamins C and E and prevention of cardiovascular disease are a notable example. Elevated antioxidant status is also associated with decreased incidence of cataracts, cancer, and neurodegenerative disorders. Some recent reports have suggested an inverse correlation between antioxidant status and the occurrence of rheumatoid arthritis and diabetes mellitus. Indeed, the number of indications in which antioxidants may be useful in the prevention and/or the treatment of disease is increasing.

Oxidative stress, rather than being the primary cause of disease, is more often a secondary complication in many disorders. Oxidative stress diseases include inflammatory bowel diseases, retinal ischemia, cardiovascular disease and restenosis, AIDS, adult respiratory distress syndrome, and neurodegenerative diseases such as stroke, Parkinson's disease, and Alzheimer's disease. Such indications may prove amenable to antioxidant treatment (in combination with conventional therapies) because there is a clear involvement of oxidative injury in these disorders.

In this series of books, the importance of oxidative stress and disease associated with organ systems of the body is highlighted by exploring the scientific evidence and the medical applications of this knowledge. The series also highlights the major natural antioxidant enzymes and antioxidant substances such as vitamins E, A, and C, flavonoids, polyphenols, carotenoids, lipoic acid, coenzyme Q_{10}, carnitine, and other micronutrients present in food and beverages. Oxidative stress is an underlying factor in health and disease. More and more evidence indicates that a proper balance between oxidants and antioxidants is involved in maintaining health and longevity and that altering this balance in favor of oxidants may result in pathophysiological responses causing functional disorders and disease. This series is intended for researchers in the basic biomedical sciences and clinicians. The potential of such knowledge for healthy aging and disease prevention warrants further knowledge about how oxidants and antioxidants modulate cell and tissue function.

Lester Packer
Enrique Cadenas

Foreword

During the past half century, major advances have been made in our understanding of the biology of cancer and other diseases. Recent knowledge of genomics, proteomics, and bioinformatics will further add to our knowledge underlying the molecular anatomy of diseases in the future. Knowledge of the treatment of most of these diseases, however, is lagging behind. It is a general belief that it is easier to prevent a disease than to treat it. Numerous epidemiological studies have shown that consumption of fruits and vegetables can minimize the incidence of various diseases including cancer. How fruits and vegetables prevent disease is not well understood. What components of the fruits and vegetables mediate these activities is mostly unclear. Knowledge about the genes and cell signaling pathways that are modulated by the active principle derived from fruits and vegetables is also incomplete.

This book addresses some of the problems outlined above. The book shows that resveratrol is a component of various fruits and vegetables, including red grapes, peanuts, mulberry, cranberry, blueberry, and jackfruit. This book also describes various genes and the cell signaling pathways that are affected by resveratrol. The book further elucidates various preclinical studies, both *in vitro* and *in vivo*, demonstrating the potential of resveratrol as regards cancer, cardiovascular diseases, and others.

The goal of preventive medicine is to find a drug that is pharmacologically safe and efficacious. The question of safety becomes even more critical if an agent is to be administered to normal populations over long periods of time for prevention of a disease. Agents such as resveratrol, which are likely to be safe, are critically needed for prevention of cancer and other diseases. However, only systematic clinical trials will provide the kind of proof that is needed. Nevertheless, this book is an excellent source of information for investigators in this field. I am sure both basic and clinical researchers will find it highly useful.

Waun Ki Hong, M.D.
Samsung Distinguished University Chair
Head, Division of Cancer Medicine
The University of Texas M.D. Anderson Cancer Center
Houston, Texas

Preface

Modern medicine is referred to as "evidence-based medicine," but even millennia ago indigenous peoples used plants as medicine based on empirical evidence, chronicled in books or folklore, that they were effective against particular diseases. Almost 70% of all drugs currently approved by the U.S. Food and Drug Administration have originated from plants. Hippocrates remarked almost 25 centuries ago: "Let food be thy medicine and medicine be thy food." This admonition parallels the common American saying, "you are what you eat," and the recommendation from the U.S. National Institutes of Health to consume as many as "eight servings of fruits and vegetables a day" to prevent common diseases. This story of food as medicine raises a scientific question of what components in food are responsible for prevention of disease and how they act. Recent evidence shows that red wine has salutary effects on health and that the red wine constituent resveratrol may be responsible. The use of resveratrol, a component of numerous plants, was perhaps first described in Ayurveda almost 5000 years ago (referred to as "draksha" which is fermented grape juice) as a cardiotonic. Besides grapes, we now know that resveratrol is also present in peanuts, cranberry, blueberry, mulberry, jackfruit, and other fruits and vegetables. The relatives of resveratrol, how resveratrol and its relatives affect disease, and what their molecular targets are is the focus of this book. Through this book, we prove that, like modern medicine, ancient medicine was also evidence-based but based on technology different from that of today. Products that are safe and yet efficacious are needed today more than ever before. Considering the cost of modern medicine, unaffordable by more than 80% of the world's population, compounds like resveratrol are in even greater need. Overall, we hope that the information provided in this book will be found useful by scientists, clinicians, herbalogists, naturopaths, and above all, the people who use such products.

We would like to thank all the contributors who made this book possible.

Bharat B. Aggarwal, Ph.D.
Shishir Shishodia, Ph.D.

Dedicated to:

Our gurus and parents whose guidance continues to inspire us!

"*Gururbrahma Gururvishnu Gururdevo Maheshwrah, Guru Sakshat Parm Brahma Tasme Srigurve Namaha*"

Yatkaromi Yatashnami Yajjuhomi Dadami Yat
Yatpsyami Mahadeva Tatkromi Tavarpanam
(modified from Gita 9-27)

Editors

Bharat B. Aggarwal, Ph.D. received his Ph.D. in biochemistry from the University of California, Berkeley, completed his postdoctoral fellowship in endocrinology from the University of California Medical Center, San Francisco, and then worked in a biotechnology company (Genentech Inc.) where he discovered TNF, an essential component of the immune system. In 1989 Dr. Aggarwal accepted a position as a professor and chief of the Cytokine Research Section at the University of Texas M. D. Anderson Cancer Center in Houston. He currently holds the Ransom Horne Jr. Endowed Professorship in Cancer Research. He has published over 350 original peer-reviewed articles and reviews, edited eight books, and been granted almost 35 patents. Since 2001, Dr. Aggarwal has been listed as one of the world's most highly cited scientists by the Institute of Scientific Information.

Shishir Shishodia, Ph.D. is currently a postdoctoral fellow at the University of Texas M. D. Anderson Cancer Center, Houston. He received his M.Sc. and Ph.D. from Banaras Hindu University (BHU), Varanasi, India. He was a junior research fellow and a senior research fellow (1994–1996) at the School of Biotechnology, BHU. He served as a lecturer in the Department of Zoology at Patna University, India (1996–2001).

Dr. Shishodia's research interests include cytokine signaling, the role of transcription factors in tumorigenesis, and regulation of transcription by natural products. He has published over 30 peer-reviewed papers.

Dr. Shishodia is a recipient of the BHU Gold Medal and the Theodore N. Law Award for outstanding scientific achievements.

Contributors

Marielle Adrian
Institut Universitaire de la Vigne et du Vin
Université de Bourgogne
Dijon, France

Bharat B. Aggarwal
Cytokine Research Laboratory
Department of Experimental Therapeutics
University of Texas M.D. Anderson Cancer Center
Houston, Texas

Abdullah Shafique Ahmad
Johns Hopkins University
School of Medicine
Baltimore, Maryland

Nihal Ahmad
Department of Dermatology
University of Wisconsin
Madison, Wisconsin

Catalina Alarcón de la Lastra
Department of Pharmacology
University of Seville, Spain

Stéphane Bastianetto
Department of Psychiatry
Douglas Hospital Research Centre
McGill University
Montreal, Quebec, Canada

Alberto A. E. Bertelli
Department of Human Morphology
University of Milan, Italy

Benjamin Bonavida
Department of Microbiology, Immunology, and Molecular Genetics
Jonsson Comprehensive Cancer Center
David Geffen School of Medicine at UCLA
Los Angeles, California

Adriana Borriello
Department of Biochemistry and Biophysics
"F. Cedrangolo" Medical School
Second University of Naples, Italy

Ebba Bråkenhielm
Department of Urology
David Geffen School of Medicine at UCLA
Los Angeles, California

Sukesh Burjonroppa
Division of Cardiology
Department of Internal Medicine and Research Center for Cardiovascular Diseases
Brown Foundation Institute of Molecular Medicine for the Prevention of Human Diseases
University of Texas Health Science Center
Houston, Texas

Renhai Cao
Laboratory of Angiogenesis Research
Microbiology and Tumor Biology Center
Karolinska Institute
Stockholm, Sweden

Yihai Cao
Laboratory of Angiogenesis Research
Microbiology and Tumor Biology Center
Karolinska Institute
Stockholm, Sweden

Feng Chu
Department of Cancer Biology
University of Texas M.D. Anderson Cancer Center
Houston, Texas

Valeria Cucciolla
Department of Biochemistry and Biophysics
"F. Cedrangolo" Medical School
Second University of Naples, Italy

Dipak K. Das
Cardiovascular Research Center
University of Connecticut School of Medicine
Farmington, Connecticut

Klaus-Michael Debatin
University Children's Hospital
Ulm, Germany

Sylvain Doré
Department of Anesthesiology/Critical Care Medicine
Neuro Research Division
Johns Hopkins University
School of Medicine
Baltimore, Maryland

Scott A. Dulchavsky
Department of Surgery
Henry Ford Health System
Detroit, Michigan

Ken Fujise
Division of Cardiology
Department of Internal Medicine and Research Center for
Cardiovascular Diseases
Brown Foundation Institute of Molecular Medicine for
the Prevention of Human Diseases
University of Texas Health Science Center
Houston, Texas

Simone Fulda
University Children's Hospital
Ulm, Germany

Xiaohua Gao
Department of Surgery
Henry Ford Health System
Detroit, Michigan

Subhash C. Gautam
Department of Surgery
Henry Ford Health System
Detroit, Michigan

Barry D. Gehm
Science Division
Lyon College
Batesville, Arkansas

Riccardo Ghidoni
Laboratory of Biochemistry and Molecular Biology
San Paolo University Hospital
School of Medicine
University of Milan, Italy

Ying-Shan Han
Department of Psychiatry
Douglas Hospital Research Centre
McGill University
Montreal, Quebec, Canada

Andrea Lisa Holme
National University Medical Institutes
Clinical Research Center
National University of Singapore
Singapore

Ali R. Jazirehi
Department of Microbiology, Immunology, and Molecular Genetics
Jonsson Comprehensive Cancer Center
David Geffen School of Medicine at UCLA
Los Angeles, California

Philippe Jeandet
Laboratoire d'Oenologie et Chimie Appliquée
Université de Reims, France

Vishal V. Kulkarni
Department of Chemistry
University of Pune
Ganeshkhind, Pune, India

Joydeb Kumar Kundu
National Research Laboratory of Molecular Carcinogenesis
and Chemoprevention
College of Pharmacy
Seoul National University
Seoul, Korea

Anait S. Levenson
Department of Orthopedic Surgery,
Comprehensive Cancer Center
Northwestern University's Feinberg School of Medicine
Chicago, Illinois

Jen-Kun Lin
National Taiwan University
Taipei, Taiwan

Gail B. Mahady
Department of Pharmacy Practice and
Department of Medicinal Chemistry and Pharmacognosy
College of Pharmacy
University of Illinois at Chicago, Illinois

Antonio Ramón Martín
Department of Pharmacology
University of Seville, Spain

Bhagavathi A. Narayanan
New York University School of Medicine
Tuxedo, New York

Narayanan K. Narayanan
New York University School of Medicine
Tuxedo, New York

Catherine A. O'Brian
Department of Cancer Biology
University of Texas M.D. Anderson Cancer Center
Houston, Texas

Francisco Orallo
Departamento de Farmacologia
Facultad de Farmacia
Universidad de Santiago de Compostela
Santiago de Compostela, Spain

Subhash Padhye
Department of Chemistry
University of Pune
Ganeshkhind, Pune, India

Trevor M. Penning
Department of Pharmacology
University of Pennsylvania School of Medicine
Philadelphia, Pennsylvania

Shazib Pervaiz
National University Medical Institutes
Graduate School of Integrative Sciences and Engineering
Department of Physiology
National University of Singapore
Singapore

John M. Pezzuto
Purdue University
College of Pharmacy
Nursing and Health Sciences
West Lafayette, Indiana

Rémi Quirion
Department of Psychiatry
Douglas Hospital Research Centre
McGill University
Montreal, Quebec, Canada

Fulvio Della Ragione
Department of Biochemistry and Biophysics
"F. Cedrangolo" Medical School
Second University of Naples, Italy

Francis Raul
Laboratory of Nutritional Oncology
University Louis Pasteur
Strasbourg, France

Shannon Reagan-Shaw
Department of Dermatology
University of Wisconsin
Madison, Wisconsin

Sofiyan Saleem
Johns Hopkins University
School of Medicine
Baltimore, Maryland

Navindra P. Seeram
Center for Human Nutrition
David Geffen School of Medicine at UCLA
Los Angeles, California

Shishir Shishodia
Cytokine Research Laboratory
Department of Experimental Therapeutics
University of Texas M. D. Anderson Cancer Center
Houston, Texas

Yogeshwer Shukla
Department of Dermatology
University of Wisconsin
Madison, Wisconsin

Paola Signorelli
Laboratory of Biochemistry and Molecular Biology
San Paolo University Hospital
School of Medicine
University of Milan, Italy

Young-Joon Surh
National Research Laboratory of Molecular Carcinogenesis
and Chemoprevention
College of Pharmacy
Seoul National University
Seoul, Korea

Lawrence M. Szewczuk
Department of Biochemistry and Physics
University of Pennsylvania School of Medicine
Philadelphia, Pennsylvania

Isabel Villegas
Department of Pharmacology
University of Seville, Spain

Vincenzo Zappia
Department of Biochemistry and Biophysics
"F. Cedrangolo" Medical School
Second University of Naples, Italy

Contents

1 Resveratrol: A Polyphenol for All Seasons

Shishir Shishodia and Bharat B. Aggarwal

CONTENTS

Extensive research in the last few years has revealed that most diseases are caused by dysregulation of multiple genes. Thus drugs targeted to a single gene are not likely to cure a disease even when the gene's association with the particular disease is well established. As an alternative, certain foods and food-derived phytochemicals have been used as treatments. One such phytochemical is resveratrol, a component of red grapes, peanuts, berries, and several other food plants. Although once considered a disadvantage, resveratrol's ability to modulate multiple cellular targets makes it suitable for the prevention and treatment of a wide variety of diseases. Among a large number of resveratrol's targets, cyclooxygenase-2 is ideal, and resveratrol presumably lacks the toxicity recently assigned to Vioxx, Celebrex, and Bextra. However, the lack of a patent is a disadvantage for most pharmaceutical companies that might want to develop the drug. We hope that federal agencies will consider its development, given the spectrum of its therapeutic activities described in this book.

INTRODUCTION

The National Institutes of Health, after extensive research over the last 50 years, has recently stated that a diet rich in fruits and vegetables can reduce the risk of many diseases including cancer and cardiovascular disease [1]. One may ask whether that recommendation is any different from what Hippocrates (460–377 BC), a Greek physician and the acknowledged "Father of Medicine," remarked thousands of years ago: "Let food be thy medicine and medicine be thy food." For example, Hippocrates reported that chewing on the bark of the willow tree relieved aches and fevers. This observation led to the discovery of aspirin in 1890. More than a century later, aspirin is recommended not only for aches and fevers but also to counteract arthritis, cardiovascular diseases, and even cancer. A healthy diet can act like medicine, boosting the immune system so that it can ward off illnesses such as cancer and heart disease. Indeed, the right foods can have a much broader effect than modern medications on a variety of health problems.

DISCOVERY OF RESVERATROL

The history of resveratrol, the active ingredient in red grapes, peanuts, berries, and several other food plants, indirectly dates back to the Ayurveda, the ancient Indian treatise on the science of longevity. Perhaps the first known use of grape extracts for human health can be dated to 2500 BC or earlier. Grape (*Vitis vinifera* L.) juice was the chief component of "darakchasava" (fermented juice of red grapes), a well-known Indian herbal preparation prescribed as a cardiotonic and also given for other disorders [2] (see Figure 1.1). Using high-performance liquid chromatography analysis, it has now been shown that the main components of darakchasava are the polyphenols resveratrol and pterostilbene, which account for its numerous medicinal properties.

Resveratrol was first identified in 1940 as a constituent of the roots of white hellebore (*Veratrum grandiflorum* O. Loes), and later in the dried roots of *Polygonum cuspidatum*, called Ko-jo-kon in Japanese, used in traditional Chinese and Japanese medicine to treat suppurative dermatitis,

FIGURE 1.1 (See color insert following page 546.) Plant sources of resveratrol.

gonorrhea favus, athlete's foot (tinea pedis), and hyperlipidemia [3–6]. Resveratrol is now recognized as a naturally occurring phytoalexin produced by a wide variety of plants other than grapes such as peanuts and mulberries in response to stress, injury, ultraviolet (UV) irradiation, and fungal (*Botrytis cinerea*) infection as part of their defense mechanism. In 1976 resveratrol was also detected in the leaf epidermis and the skin of grape berries but not in the flesh [7–9]. Fresh grape skins contain 50 to 100 mg resveratrol per gram, and the concentration in wine may range from 0.2 to 7.7 mg per liter. The epidemiologic finding of an inverse relationship between consumption of red wine and incidence of cardiovascular disease has led to the "French paradox," which is consistent with its known activity [10,11].

SALIENT FEATURES OF RESVERATROL

Inflammation plays a major role in the pathogenesis of a wide variety of diseases including cardiovascular diseases, cancer, diabetes, Alzheimer's disease, and autoimmune diseases. Agents that can suppress inflammation thus have a potential in mitigating the symptoms of the disease. Resveratrol exhibits antioxidant and antiinflammatory activities and thus may have potential in the treatment of these diseases (see Figure 1.2). The numerous targets that have been identified for resveratrol are listed in Table 1.1. Microarray analysis has also shown that resveratrol differentially modulates the expression of many genes [12–15] in multiple cell-signaling pathways.

RESVERATROL AS A COX-2 INHIBITOR

Activation of cyclooxygenase (COX)-2 leads to the production of prostaglandin E2 (PGE2), which causes inflammation. Although COX-2-specific

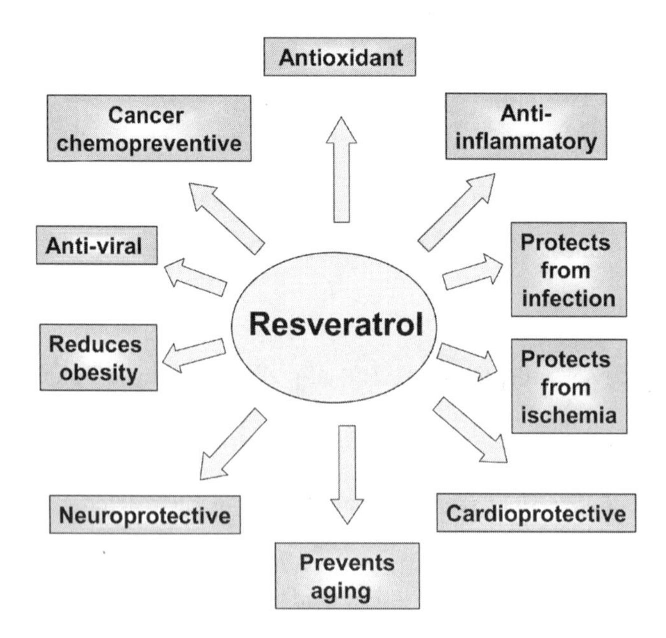

FIGURE 1.2 Health benefits of resveratrol.

TABLE 1.1
Molecular Targets of Resveratrol

Target	Effect
Cytokines	
Transforming growth factor β2 (TGFβ 2)	Upregulated
Transforming growth factor (TGF)-α	Downregulated
Epidermal growth factor (EGF)	Downregulated
Tumor necrosis factor (TNF)	Downregulated
FasL	Upregulated
Interleukin (IL)-1β	Downregulated
Interleukin (IL)-6	Downregulated
Vascular epithelial growth factor (VEGF)	Downregulated
Insulin-like growth factor 1 receptor (IGF-1R)	Downregulated
Transcription Factors	
Activator protein-1 (AP-1)	Downregulated
Nuclear factor-kappa B (NF- κB)	Downregulated
Beta catenin (β-catenin)	Downregulated
Early growth response (egr)-1	Upregulated
Androgen receptor (AR)	Downregulated
Cell Cycle Proteins	
Cyclin D1	Downregulated
Retinoblastoma (Rb)	Downregulated
Cyclin A	Downregulated
Cyclin-dependent kinase (cdk)-2	Upregulated
Cyclin B1	Downregulated
$p21^{Cip1/WAF1}$	Upregulated
$p27^{kip1}$	Upregulated
Invasion and Metastasis	
Cyclooxygenase (COX)-2	Downregulated
5-Lipoxygenase (5-LOX)	Downregulated
Inducible nitric oxide synthase (iNOS)	Downregulated
Vascular cell adhesion molecules (VCAM-1)	Downregulated
Intracellular adhesion molecule (ICAM-1)	Downregulated
Tissue Factor	Downregulated
NADPH:quinone oxidoreductase (NQO)-1	Upregulated
Apoptosis	
Bax	Upregulated
Bcl-2	Downregulated
Survivin	Downregulated
p53	Upregulated
Kinases	
Protein kinase C (PKC)	Downregulated
Syk	Downregulated
Protein kinase D (PKD)	Downregulated

(*continued*)

TABLE 1.1
Continued

Target	Effect
Caesin kinase II (CKII)	Downregulated
Extracellular signal-regulated kinase (ERK) 1/2	Downregulated
Others	
Ribonucleotide reductase	Downregulated
DNA polymerase	Downregulated
CYP1A1	Downregulated
Nonsteroidal antiinflammatory drug-activated gene (NAG-1)	Upregulated

Note: For references, see Aggarwal BB, Bhardwaj A, Aggarwal RS, Seeram NP, Shishodia S and Takada Y, *Anticancer Res* 24, 2783–2840, 2004.

inhibitors have been approved for the treatment of pain, fever, arthritis, and cancer, they have been found to be cardiotoxic (see Table 1.2). One attractive alternative for treating proinflammatory diseases is resveratrol, which has been found to suppress COX-2 expression [16]. It has not yet been extensively tested in patients, however. Therefore its toxicity profile has not been definitively established.

Murias et al. found that resveratrol is a nonselective inhibitor of COX-1 and COX-2 [17]. To produce more selective COX-2 inhibitors, they synthesized a series of methoxylated and hydroxylated derivatives and evaluated their activity against COX-1 and COX-2 using *in vitro* PGE2 assays. Hydroxylated but not methoxylated resveratrol derivatives showed a high rate of inhibition. The most potent resveratrol compounds were *trans*-3,3′,4′,5-tetrahydroxystilbene (COX-1: IC(50) = 4.713 μM, COX-2: IC(50) = 0.0113 μM, selectivity index = 417.08) and *trans*-3,3′,4,4′,5,5′-hexahydroxystilbene (COX-1: IC(50) = 0.748 μM, COX-2: IC(50) = 0.00104 μM, selectivity index = 719.23). Their selectivity index was in part higher than celecoxib, a selective COX-2 inhibitor already established on the market (COX-1: IC(50) = 19.026 μM, COX-2: IC(50) = 0.03482 μM, selectivity index = 546.41). The effect of structural parameters on COX-2 inhibition was evaluated by quantitative structure–activity relationship analysis. A high correlation was found with the topological surface area ($r = 0.93$). Docking studies on both COX-1 and COX-2 protein structures also revealed that hydroxylated but not methoxylated resveratrol analogs bind to the previously identified binding sites of the enzymes. Hydroxylated resveratrol analogs therefore represent a novel class of highly selective COX-2 inhibitors and promising candidates for *in vivo* studies.

TABLE 1.2
Adverse Effects of Recent Food and Drug Administration-Approved Drugs

Drug	Target	Use	Year approved	Year withdrawn/ warning label	Reason
Vioxx (rofecoxib)	COX-2	Rheumatoid arthritis	1999	2004	Heart attack
Celebrex[a] (celcoxib)	COX-2	Rheumatoid arthritis	1998	2004	Heart attack
Bextra[a] (valdecoxib)	COX-2	Rheumatoid arthritis	2001	2004	Heart attack
Iressa (gefitinib)	EGFR tyrosine kinase	Nonsmall-cell lung carcinoma	2002	2004	No clinical benefits
Remicade (anti-TNF antibody)	TNF	Crohn's disease, rheumatoid arthritis	1998	2001[b]	High risk of lymphoma
Prozac (fluoxetine)	Serotonin	Antidepressant	1986	2004[b]	Suicidal thoughts and behavior
Strattera (atomoxetine hydrochloride)	Norepinephrine reuptake inhibitor	Attention deficit hyperactivity disorder	2002	2004[b]	Hepatotoxic
Fen-phen (phentermine/ fenfluramine)	Serotonin/catecholamines	Antiobesity	1959/1973	1997	Valvular heart disease, PPH
Rezulin (troglitazone)	PPARγ	Diabetes	1999	2000	Hepatotoxic
Avandia (rosiglitazone maleate)	PPARγ	Diabetes	1999	2001[b]	Hepatotoxic
Propulsid (cisapride)	Potassium channels	Heart burn	1993	2000	Cardiotoxic
PPA	ANS	Cold, cough		2000	Hemorrhagic stroke
Baycol (cerivastatin)	HMG CoA reductase	Cholesterol lowering	2000	2003	Rhabdomyolysis
Arava (leflunomide)	Dihydroorotate dehydrogenase	Rheumatoid arthritis	1997		Peripheral neuropathy
Lotronex (alosetron hydrochloride)	5-HT3 receptor antagonist	Irritable bowel	2000	2000	Ischemic colitis
Serzone (nefazodone hydrochloride)	Serotonin type 2 receptor	Antidepressant		2004	Liver injury

Note: COX, cyclooxygenase; EGFR, epidermal growth factor receptor; TNF, tumor necrosis factor; PPH, primary pulmonary hypertension; PPAR, peroxisome proliferators-activated receptor; PPA, phenylpropanolamine; ANS, autonomic nervous system; HMG-CoA, 3-hydroxy-3-methylglutaryl coenzyme A; rhabdomyolysis, muscle breakdown; 5-HT3R, serotonin 5-HT3A receptor.

[a] Although Celebrex has been linked with cardiotoxicity, no black label warning has yet been issued.

[b] Black box warning.

Chemosensitization by Resveratrol

Most chemotherapeutic agents are highly toxic not only to tumor cells but also to normal cells. Additionally, tumors eventually develop resistance to chemotherapeutic agents. Thus, agents that can sensitize tumors to chemotherapeutic agents are needed. Bcl-2 and Bcl-xl have been implicated in chemoresistance. Opipari et al. found that the cells transfected to express high levels of the antiapoptotic proteins Bcl-xL and Bcl-2 were just as sensitive as control cells to resveratrol [18]. This mechanism may give resveratrol and its derivatives a distinct advantage in the treatment of ovarian cancer that is chemoresistant on the basis of ineffective apoptosis.

Wu et al. investigated the ability of resveratrol to enhance the sensitivity to chemotherapeutic agents *in vivo*. They used the transplantable murine hepatoma22 model to evaluate the antitumor activity of resveratrol alone or in combination with 5-fluorouracil (5-FU) *in vivo* [19]. They found that 10 mg/kg or 15 mg/kg resveratrol for ten days inhibited the growth of murine hepatoma22 by 36.3% ($n = 10$) and 49.3% ($n = 9$), respectively, which increased obviously compared with that in the control group (85 ± 22 vs. 68 ± 17, $P < 0.01$). The enhanced inhibition of tumor growth by 5-FU was also observed in hepatoma22-bearing mice when 5-FU was administered in combination with 10 mg/kg resveratrol. The inhibition rates for 20 or 10 mg/kg 5-FU in combination with 10 mg/kg resveratrol were 77.4 and 72.4%, respectively, compared with the group given the same doses of 5-FU alone, in which the inhibition rates were 53.4 and 43.8%, respectively ($n = 8$). Thus resveratrol enhanced the antitumor effect of 5-FU on murine hepatoma22 and antagonized its toxicity markedly. These results suggest that resveratrol, as a biochemical modulator of the therapeutic effects of 5-FU, may be useful in cancer chemotherapy.

While enhancing the toxicity against tumor cells, resveratrol seems to protect normal cells from chemotherapeutic agents. Olas et al. found that resveratrol protects hematopoietic cells from chemotherapy-induced toxicity [20]. They found that resveratrol protected blood against changes in platelet thiols induced by platinum compounds. Cisplatin is especially useful in the treatment of epithelial malignancies, but its use is accompanied by several toxicities including hematological toxicity. Unlike cisplatin, the selenium-cisplatin conjugate ((NH(3))(2)Pt(SeO(3)); Se-Pt) has only a slight toxic effect on blood platelet function. They found that platinum compounds caused the reduction of platelet protein thiols. Resveratrol (after 30 minutes) at a concentration of 25 μg/ml partly reduced the decrease in platelet thiols caused by platinum compounds, particularly the thiols in the acid-soluble fraction.

Radiosensitization by Resveratrol

A search for new agents that can sensitize cancer cells to ionizing radiation led Baatout et al. to discover that resveratrol can radiosensitize cancer cells [21]. Exposure of the human cancer cell lines HeLa (cervix carcinoma), K-562 (chronic myeloid leukemia), and IM-9 (multiple myeloma) to x-rays (doses from 0 to 8 Gy) and resveratrol (concentrations ranging from 0 to 200 μM) produced a synergistic killing effect at the highest dose of 200 μM. These results show that resveratrol can act as a potential radiation sensitizer at high concentrations.

Because prostaglandin has been implicated in the cytotoxic and cytoprotective responses of tumor cells to ionizing radiation, Zoberi et al. hypothesized that tumor cells may exhibit changes in the cellular response to ionizing radiation following exposure to resveratrol [22]. Clonogenic cell survival assays indicated that irradiated HeLa and SiHa cells pretreated with resveratrol exhibited enhanced tumor cell killing. These results suggest that resveratrol alters both cell cycle progression and the cytotoxic response to ionizing radiation in tumor cell lines.

Resveratrol as Antiviral Agent

Docherty et al. found that resveratrol inhibits herpes simplex virus types 1 and 2 (HSV-1 and HSV-2) replication [23]. The observed reduction in virus yield was not caused by the direct inactivation of HSV by resveratrol or inhibition of virus attachment to the cell. The chemical did, however, target an early event in the virus replication cycle since it was most effective when added within 1 hour of cell infection, less effective if added 6 hours after infection, and not effective if added 9 hours after infection. Resveratrol also delayed the cell cycle at the S-G2-M interphase, inhibited reactivation of virus from latently infected neurons, and reduced the amount of ICP-4, a major immediate early viral regulatory protein. These results suggest that a critical early event in the viral replication cycle that has a compensatory cellular counterpart is being adversely affected.

In a later study, the same investigators determined the effect of resveratrol on HSV infection *in vivo* [24]. The abraded epidermis of SKH1 mice was infected with HSV-1 and topically treated with 12.5 or 25% resveratrol cream or cream only. Initial studies demonstrated that: (1) 25% resveratrol cream topically applied two, three, or five times a day effectively suppressed lesion development, whereas 12.5% resveratrol cream effectively suppressed lesion formation only when applied five times a day starting 1 hour after infection; (2) both 12.5 and 25% resveratrol were effective at 1 and 6 hours after infection, but not if applied 12 hours after infection. Comparative studies between resveratrol cream, 10% docosanol cream (Abreva), and

5% acyclovir ointment (Zovirax) were also carried out. When treatment was begun 1 hour after infection and repeated every 3 hours 5 times a day for 5 days, 12.5 and 25% resveratrol significantly inhibited the development of HSV-1-induced skin lesions. Acyclovir was as effective as resveratrol. Animals that were topically treated with docosanol were not protected and developed lesions indistinguishable from those of cream-only controls. These studies were repeated with an HSV-1 acyclovir-resistant virus. As before, 12.5 and 25% resveratrol cream effectively suppressed lesion formation. The skin of resveratrol-treated animals showed no apparent dermal toxicity such as erythema, scaling, crusting, lichenification, or excoriation. These studies demonstrate that topically applied resveratrol inhibits HSV lesion formation in the skin of mice.

BIOAVAILABILITY OF RESVERATROL

Although resveratrol has been shown to have chemopreventive activity against cardiovascular disease and a variety of cancers in model systems, it is not clear whether the drug reaches the proposed sites of action *in vivo* after oral ingestion, especially in humans. Walle et al. examined the absorption, bioavailability, and metabolism of ^{14}C-resveratrol after oral and i.v. doses in six human volunteers [25]. The absorption of a dietarily relevant 25 mg oral dose was at least 70%, with peak plasma levels of resveratrol and metabolites reaching 491 ± 90 ng/ml (about 2 μM) and a plasma half-life of 9.2 ± 0.6 hours. However, only trace amounts of unchanged resveratrol (<5 ng/ml) could be detected in plasma. Most of the oral dose was recovered in urine, and liquid chromatography/mass spectrometry analysis identified three metabolic pathways, i.e., sulfate and glucuronic acid conjugation of the phenolic groups and, interestingly, hydrogenation of the aliphatic double bond, the latter likely produced by intestinal microflora. Extremely rapid sulfate conjugation by the intestine/liver appeared to be the rate-limiting step in resveratrol's bioavailability. Although the systemic bioavailability of resveratrol is very low, accumulation of resveratrol in epithelial cells along the aerodigestive tract and potentially active resveratrol metabolites may still produce cancer-preventive and other effects.

CLINICAL TRIALS WITH RESVERATROL

Resveratrol is now commonly available (see Figure 1.3). It is estimated that development of a typical drug requires 10 to 15 years of testing and as much as a billion dollars. Will that amount of resources be invested in resveratrol for which a strong patent position is questionable? In spite of the

Research grade

http://www.caymanchem.com/neptune/servlet/neptune/catalog/70675/template/Product.vm
http://www.alexis-corp.com/
http://www.sigmaaldrich.com/Area_of_Interest/The_Americas/United_States.html
http://www.emdbiosciences.com/product/554325
http://www.alexis-corp.com/other_inhibitors_of_cox-LKT-R1776/opfa.1.1.LKT-R1776.893.4.1.html

Dietary supplement

http://www.gettingwell.com/drug_info/nmdrugprofiles/nutsupdrugs/res_0224.shtml
15 mg, 50 mg, 200 mg capsules; 10 mg tablets.
http://www.vrp.com/affiliate?vNAV=YES&vMBR=238900&vPORT=ECART&vITM=5512%2C5171%2C1930
100 mg capsules
http://www.lef.org/newshop/items/item00655.html?source=Google&key=Resveratrol&WT.srch=1
20 mg capsules
http://www.vitacost.com/science/hn/Supp/Resveratrol.htm
40 mg tablets
http://www.longevinex.com/sdm.asp?pg=default
capsules
http://www.iherb.com/resveratrol2.html
15 mg, 50 mg, 200 mg capsules; 10 mg tablets.
http://www.harvesthealth.com/sore60ca75.html
75 mg capsules
http://www.allvita.net/resverat.htm
Resveratrol Forte® Grape Skin Extract & Grape seed Extract with Resveratrol (min 400 mcg/gm), Proanthocyanidins, Polyphenols, Anthocyanidinis, Catechine, Bioflavonoids (100 mg)
http://www.fslabs.com/resveratrol.html
50 mg capsules
http://www.activeplaza.com/buy_supplements/resveratrol-store6018.html
40 mg tablets
http://www.novanat.com/en/index/index.asp
http://www.health-marketplace.com/Resveratrol.htm
20 mg capsules
http://www.naturalpharmacy.com/supplement-category/Resveratrol
200 mg capsules

Clinical grade

www.resverine.com

FIGURE 1.3 Industrial sources of resveratrol.

financial risks, Royalmount Pharma (Montreal, Canada), a privately held development pharmaceutical company, has initiated a phase II trial for herpes infection using "Resverin" a patented synthetic compound. The company received an approval from Health Canada to initiate the trial. It will be conducted at six sites across Canada and will evaluate the drug's efficacy in approximately 120 patients with recurrent HSV-1 infection, commonly known as cold sores. These studies are also sponsored by the U.S. National Cancer Institute.

CONCLUSION

Although numerous studies (over 1000 citations) have indicated that resveratrol has a great potential for the treatment of a wide variety of diseases, only clinical trials can reveal its true potential. Whether it is preferable to consume resveratrol in a purified form or the plant part of which it is a constituent is an important question. Clinical trials with green tea-derived catechins [26], carrot-derived beta-carotenes [27], and tomato-derived lycopenes [28] have raised doubts on the use of these phytochemicals in excess for a variety of reasons. First, toxicity is associated with excessive use of a phytochemical; second, to use it in the absence of other components of the plants may decrease bioavailability; third, other components may also have other activities. For instance, Szewczuk et al. found that besides resveratrol red wine contains other constituents, namely the catechins and epicatechins, which inhibit COX-1 but not COX-2 [29]. Diwadkar-Navsariwala found that the absorption of lycopene was not directly proportional to the dose administered [30]. Cohn et al. found that the systemic availability of synthetic lycopene from a tablet formulation is comparable to that observed from processed tomatoes (soup from tomato paste) and superior to that from tomato juice [28]. No differences were observed in disposition kinetics of natural and synthetic lycopene. Whether resveratrol, synthetic or natural, proves to be as effective as red grapes/wine remains to be determined.

ACKNOWLEDGMENTS

The authors would like to thank Walter Pagel for a careful review of the manuscript. Dr. Aggarwal is a Ransom Horne Jr. Distinguished Professor of Cancer Research. This work was supported in part by the Odyssey Program and the Theodore N. Law Award for Scientific Achievement at the University of Texas M. D. Anderson Cancer Center (to S.S.), the Clayton Foundation for Research (to B.B.A.), Department of Defense U.S. Army Breast Cancer Research Program Grant (BC010610, to B.B.A.), PO1 grant (CA91844) from the National Institutes of Health (to B.B.A.), and Specialized Program of Research Excellence grant from the National Institutes of Health (to B.B.A.).

REFERENCES

1. Heber D, Vegetables, fruits and phytoestrogens in the prevention of diseases, *J Postgrad Med* 50, 145–149, 2004.
2. Paul B, Masih I, Deopujari J, and Charpentier C, Occurrence of resveratrol and pterostilbene in age-old darakchasava, an Ayurvedic medicine from India, *J Ethnopharmacol* 68, 71–76, 1999.
3. Takaoka MJ, *J Faculty Sci Hokkaido Imperial Univ* 3, 1–16, 1940.
4. Vastano BC, Chen Y, Zhu N, Ho CT, Zhou Z, and Rosen RT, Isolation and identification of stilbenes in two varieties of *Polygonum cuspidatum*, *J Agric Food Chem* 48, 253–256, 2000.
5. Lee SK, Mbwambo ZH, Chung H, Luyengi L, Gamez EJ, Mehta RG, Kinghorn AD, and Pezzuto JM, Evaluation of the antioxidant potential of natural products, *Comb. Chem. High Throughput Screen* 1, 35–46, 1998.
6. Cichewicz RH and Kouzi SA, Resveratrol oligomers: structure, chemistry, and biological activity, *J Nat. Prod. Chem* 26, 507–579, 2002.
7. Langcake P and Pryce RJ, The production of resveratrol by *Vitis vinifera* and other members of the *Vitaseae* as a response to infection or injury, *Physiol Plant Pathol* 9, 77–86, 1976.
8. Langcake P and McCarthy W, The relationship of resveratrol production to infection of grapevine leaves by *Botrytis cinerea*, *Vitis* 18, 244–253, 1979.
9. Creasy LL and Coffee M, Phytoalexin production potential of grape berries, *J Am Soc Hortic Sci* 113, 230–234, 1988.
10. Kopp P, Resveratrol, a phytoestrogen found in red wine. A possible explanation for the conundrum of the "French paradox"?, *Eur J Endocrinol* 138, 619–620, 1998.
11. Sun AY, Simonyi A, and Sun GY, The "French Paradox" and beyond: neuroprotective effects of polyphenols, *Free Radical Biol Med* 32, 314–318, 2002.
12. Levenson AS, Svoboda KM, Pease KM, Kaiser SA, Chen B, Simons LA, Jovanovic BD, Dyck PA, and Jordan VC, Gene expression profiles with activation of the estrogen receptor alpha-selective estrogen receptor modulator complex in breast cancer cells expressing wild-type estrogen receptor, *Cancer Res* 62, 4419–4426, 2002.
13. Narayanan BA, Narayanan NK, Re GG, and Nixon DW, Differential expression of genes induced by resveratrol in LNCaP cells: P53-mediated molecular targets, *Int J Cancer* 104, 204–212, 2003.
14. Yang SH, Kim JS, Oh TJ, Kim MS, Lee SW, Woo SK, Cho HS, Choi YH, Kim YH, Rha SY, Chung HC, and An SW, Genome-scale analysis of resveratrol-induced gene expression profile in human ovarian cancer cells using a cDNA microarray, *Int J Oncol* 22, 741–750, 2003.

15. Shi T, Liou LS, Sadhukhan P, Duan ZH, Novick AC, Hissong JG, Almasan A, and DiDonato JA, Effects of resveratrol on gene expression in renal cell carcinoma, *Cancer Biol Ther* 3, 2004.
16. Subbaramaiah K, Chung WJ, Michaluart P, Telang N, Tanabe T, Inoue H, Jang M, Pezzuto JM, and Dannenberg AJ, Resveratrol inhibits cyclooxygenase-2 transcription and activity in phorbol ester-treated human mammary epithelial cells, *J Biol Chem* 273, 21875–21882, 1998.
17. Murias M, Handler N, Erker T, Pleban K, Ecker G, Saiko P, Szekeres T, and Jager W, Resveratrol analogues as selective cyclooxygenase-2 inhibitors: synthesis and structure–activity relationship, *Bioorg Med Chem* 12, 5571–5578, 2004.
18. Opipari AW, Jr, Tan L, Boitano AE, Sorenson DR, Aurora A, and Liu JR, Resveratrol-induced autophagocytosis in ovarian cancer cells, *Cancer Res* 64, 696–703, 2004.
19. Wu SL, Sun ZJ, Yu L, Meng KW, Qin XL, and Pan CE, Effect of resveratrol and in combination with 5-FU on murine liver cancer, *World J Gastroenterol* 10, 3048–3052, 2004.
20. Olas B, Nowak P, and Wachowicz B, Resveratrol protects against peroxynitrite-induced thiol oxidation in blood platelets, *Cell Mol Biol Lett* 9, 577–587, 2004.
21. Baatout S, Derradji H, Jacquet P, Ooms D, Michaux A, and Mergeay M, Enhanced radiation-induced apoptosis of cancer cell lines after treatment with resveratrol, *Int J Mol Med* 13, 895–902, 2004.
22. Zoberi I, Bradbury CM, Curry HA, Bisht KS, Goswami PC, Roti Roti JL, and Gius D, Radiosensitizing and anti-proliferative effects of resveratrol in two human cervical tumor cell lines, *Cancer Lett* 175, 165–173, 2002.
23. Docherty JJ, Fu MM, Stiffler BS, Limperos RJ, Pokabla CM, and DeLucia AL, Resveratrol inhibition of herpes simplex virus replication, *Antiviral Res* 43, 145–155, 1999.
24. Docherty JJ, Smith JS, Fu MM, Stoner T, and Booth T, Effect of topically applied resveratrol on cutaneous herpes simplex virus infections in hairless mice, *Antiviral Res* 61, 19–26, 2004.
25. Walle T, Hsieh F, DeLegge MH, Oatis JE, Jr, and Walle UK, High absorption but very low bioavailability of oral resveratrol in humans, *Drug Metab Dispos* 32, 1377–1382, 2004.
26. Pisters KM, Newman RA, Coldman B, Shin DM, Khuri FR, Hong WK, Glisson BS, and Lee JS, Phase I trial of oral green tea extract in adult patients with solid tumors, *J Clin Oncol* 19, 1830–1838, 2001.
27. Omenn GS, Chemoprevention of lung cancer: the rise and demise of beta-carotene, *Annu Rev Public Health* 19, 73–99, 1998.
28. Cohn W, Thurmann P, Tenter U, Aebischer C, Schierle J, and Schalch W, Comparative multiple dose plasma kinetics of lycopene administered in tomato juice, tomato soup or lycopene tablets, *Eur J Nutr* 43, 304–312, 2004.
29. Szewczuk LM, Forti L, Stivala LA, and Penning TM, Resveratrol is a peroxidase-mediated inactivator of COX-1 but not COX-2: a mechanistic approach to the design of COX-1 selective agents, *J Biol Chem* 279, 22727–22737, 2004.

30. Diwadkar-Navsariwala V, Novotny JA, Gustin DM, Sosman JA, Rodvold KA, Crowell JA, Stacewicz-Sapuntzakis M, and Bowen PE, A physiological pharmacokinetic model describing the disposition of lycopene in healthy men, *J Lipid Res* 44, 1927–1939, 2003.
31. Aggarwal BB, Bhardwaj A, Aggarwal RS, Seeram NP, Shishodia S, and Takada Y, Role of resveratrol in prevention and therapy of cancer: preclinical and clinical studies, *Anticancer Res* 24, 2783–2840, 2004.

2 Sources and Chemistry of Resveratrol

Navindra P. Seeram, Vishal V. Kulkarni, and Subhash Padhye

CONTENTS

INTRODUCTION

Stilbenoids are phenol-based plant metabolites widely represented in nature and implicated with human health benefits against problems such as cancer, inflammation, neurodegenerative disease, and heart disease. Among stilbenes, the phytoalexin resveratrol (3,4′,5-trihydroxystilbene; Figure 2.1) has attracted immense attention from biologists and chemists due to its numerous biological properties. Resveratrol is a pivotal molecule in plant biology and plays an important role as the parent molecule of oligomers known as the viniferins [1]. It is also found in nature as closely related analogs, derivatives, and conjugates (Table 2.1) [1–80]. In addition, the inherent structural simplicity of the resveratrol molecule allows for the rational design of new chemotherapeutic agents, and hence a number of its synthetic adducts, analogs, derivatives, and conjugates have been reported (Table 2.1) [1–80].

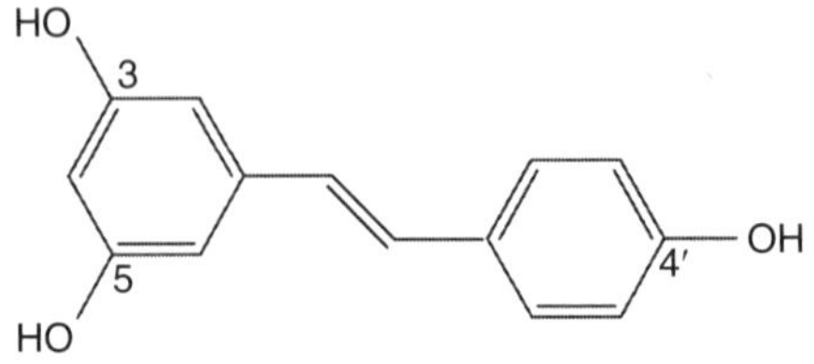

Trans-resveratrol (*trans*-3,4′,5-trihydroxystilbene)

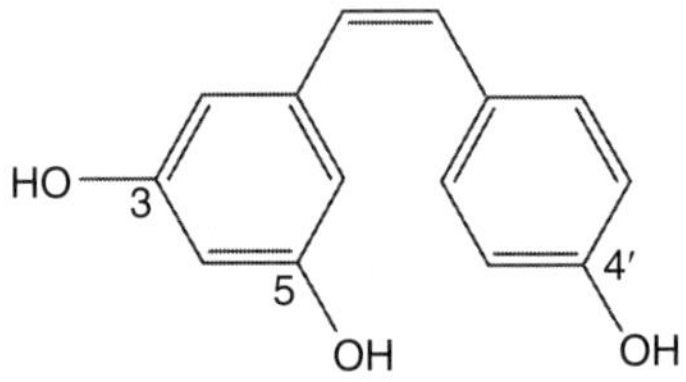

Cis-resveratrol (*cis*-3,4′,5-trihydroxystilbene)

FIGURE 2.1 Chemical structures of *trans*- and *cis*-resveratrol (3,4′, 5-trihydroxystilbene).

Numerous efforts have been directed to studies of structure–activity relationships (SARs) of resveratrol and its analogs with the goal of increasing and enhancing their *in vivo* biological potency and bioavailability. The pharmacological activity of resveratrol has also stimulated the development of numerous chemical analytical methods for its measurement in different matrices such as plant extracts, wines and other beverages, and food-derived products, as well as in biological fluids and tissues.

Because of the numerous biological properties and implications in health and disease associated with resveratrol, the focus of this chapter is on its occurrence, chemical analyses, synthesis, and studies of its chemistry.

SOURCES OF RESVERATROL

The sources of resveratrol and its related natural and synthetic derivatives, conjugates, and analogs are shown in Table 2.1 [1–80]. Resveratrol was first identified in 1940 from the white hellebore lily *Veratrum grandiflorum* O. Loes [81], although its richest known natural source is the Asian medicinal plant *Polygonum cuspidatum* (Japanese "Ko-jo-kon"). The occurrence of resveratrol was popularized in 1992 when it was discovered as a constituent of red wine, and implicated in the "French paradox," an epidemiological finding of an inverse relationship between red wine consumption and the incidence of heart disease. Resveratrol has also been implicated with

TABLE 2.1
Natural and Synthetic Sources of Resveratrol and its Analogs

Compound and sources	Ref.
Resveratrol (3,4′,5-trihydroxystilbene)[a]	1–13
Red grape, grapevine, grape leaf, and berry skin, muscadine grape, red wine, blueberry, cranberry, bilberry, lingonberry, sparkleberry, deerberry, partridgeberry; *Polygonum cuspidatum* (Japanese knotweed); *Morus* spp. (including mulberry); lily (*Veratrum* spp.); legumes (*Cassia* spp., *Pterolobium hexapetallum*); peanuts (*Arachis hypogaea*); *Rheum* spp. (including rhubarb); eucalyptus; spruce (*Picea* spp); pine (*Pinus* spp.); Poaceae (grasses, including *Festuca, Hordeum, Poa, Stipa,* and *Lolium* spp.); *Trifolium* spp.; *Nothofagus* spp.; *Artocarpus* spp.; *Gnetum* spp.; *Pleuropterus ciliinervis*; *Bauhinia racemosa*; *Paeonia lactiflora*; *Scilla nervosa*; *Tetrastigma hypoglaucum*; *Rumex bucephalophorus*; *Yucca* spp.; *Smilax* spp.	
Dihydroresveratrol (*trans*-3,5,4′-trihydroxybibenzylstilbene)[a]	14, 15
Dioscorea spp.; *Bulbophyllum triste*	
Piceatannol or astringinin (*trans*-3,4,3′,5′-tetrahydroxystilbene)	3, 15–20
White tea tree (*Melaleuca leucadendron*); Asian legume (*Cassia garrettiana*), *C. marginata*; rhubarb (*Rheum* spp.); *Euphorbia lagascae*; *Polygonum cuspidatum*; *Vitis vinifera*	
Dihydropiceatannol (*trans*-3,4,3′,5′-tetrahydroxybibenzylstilbene)[a]	17
Cassia garrettiana	
Gnetol (*trans*-2,6,3′,5′,-tetrahydroxystilbene)	7, 21, 22
Gnetum spp. (including *G. monatum*, *G. africanum*, *G. gnemon*, *G. ula*)	
Oxyresveratrol (*trans*-2,3′,4,5′-tetrahydroxystilbene)	9, 23–25
Morus spp.; *Maclura pomifera*; *Artocarpus gomezianus*; *Schoenocaulon officinale*	
Hydroxyresveratrol (*trans*-2,3,5,4′-tetrahydroxystilbene)	3
Polygonum cuspidatum	
Trans-3,4,5,4′-tetrahydroxystilbene[a]	26
Trans-3,3′,4′,5,5′-pentahydroxystilbene[a]	27, 28
Eucalyptus wandoo; *Vouacapoua americana*, *V. macropetala*	
Pinosylvin (*trans*-3,5-dihydroxystilbene)[a]	1, 26, 29–34
Gnetum cleistostachyum; *Alpinia katsumadai*; *Polyalthia longifolia*; *Polygonum nodosum*; *Pinus* spp. (including Scottish pine, *P. sylvestris*)	
Dihydropinosylvin (*trans*-3,5-dihydroxybibenzylstilbene)[a]	35–37
Dioscorea batatas	
Trans-2,4,4′-trihydroxystilbene[a]	36, 37
Trans-3,5,3′-trihydroxystilbene[a]	38, 39
Trans-3,4,5-trihydroxystilbene[a]	40
Trans-3,4,4′-trihydroxystilbene[a]	40, 41
Trans-3,4-dihydroxystilbene[a]	36, 37, 41
Trans-3,4′-dihydroxystilbene[a]	38, 39
Trans-3,3′-dihydroxystilbene[a]	38, 39

(continued)

TABLE 2.1
Continued

Compound and sources	Ref.
Trans-2,4-dihydroxystilbene[a]	36, 37
Trans-4,4′-dihydroxystilbene[a]	36, 37, 40, 41
Trans-3-hydroxystilbene[a]	38, 39
Trans-4-hydroxystilbene (*p*-hydroxystilbene)[a]	36, 37, 40
Trans-halogenated-3,5,4′-trihydroxystilbenes (fluoro-, chloro-, and iodoresveratrols)[a]	42, 43
Dimethoxypinosylvin (*trans*-3,5-dimethoxystilbene)[a]	26
Rhapontigenin or 3-methoxyresveratrol (*trans*-3,5,3′-trihydroxy-4′-methoxystilbene)[a]	6, 44, 45
Rheum spp. (*including R. rhaponticum, R. undulatum*); *Scilla nervosa*	
Isorhapontigenin (*trans*-3,5,4′-trihydroxy-3′-methoxystilbene)[a]	7, 46, 47
Gnetum spp.; *Belamcanda chinensis*	
Desoxyrhapontigenin or 4-methoxyresveratrol (*trans*-3,5-dihydroxy-4′-methoxystilbene)	29, 48–50
Gnetum cleistostachyum; *Rheum undulatum*; *Knema austrosiamensis*; *Rumex bucephalophorus*	
Pinostilbene or 3-methoxyresveratrol (*trans*-5,4′-dihydroxy-3-methoxystilbene)	50
Rumex bucephalophorus	
Trans-3,4′-dimethoxy-5-hydroxystilbene[a]	48, 49
Knema austrosiamensis	
Cis-3,5,3′-trihydroxy-4′-methoxystilbene[a]	51
Trimethylresveratrol (*trans*-3,5,4′-trimethoxystilbene)[a]	8, 26, 52
Pterolobium hexapetallum	
Gnetucleistol D or 2-methoxyoxyresveratrol (*trans*-2-methoxy-3′,4,5-trihydroxystilbene)	29
Gnetum cleistostachyum	
Gnetucleistol E or 3-methoxyisorhapontigenin (*trans*-3,3′-dimethoxy-5,4′-dihydroxystilbene)	29
Gnetum cleistostachyum	
Trans- and *cis*-3,5,4′-trimethoxy-3′-hydroxystilbene[a]	51
Trans- and *cis*-3,5,3′-trimethoxy-4′-hydroxystilbene[a]	51
Trans- and *cis*-3,5-dimethoxy-3′,4′-dihydroxystilbene[a]	51
Trans- and *cis*-3,5-dihydroxy-3′-amino-4′-methoxystilbene[a]	51
Trans- and *cis*-3,5-dimethoxy-4′-aminostilbene[a]	51
Trans- and *cis*-3,4′,5-trimethoxy-3′-aminostilbene[a]	51
Trans- and *cis*-3,5-dimethoxy-4′-nitrostilbene[a]	51
Trans- and *cis*-3,4′,5-trimethoxy-3′-nitrostilbene[a]	51
Trans-5,4′-dihydroxy-3-methoxystilbene	51
Rumex bucephalophorus	
Pterostilbene (*trans*-3,5-dimethoxy-4′-hydroxystilbene)[a]	8, 51, 53, 54
Dracena cochinchinensis; *Pterocarpus* spp. (including *P. santalinus*, *P marsupium*); *Vitis vinifera*; *Pterolobium hexapetallum*	
Cis-3,5-dimethoxy-4′-hydroxystilbene[a]	51

TABLE 2.1
Continued

Compound and sources	Ref.
3,4,5,4′-tetramethoxystilbene[a]	26
3,4,5,3′-tetramethoxystilbene[a]	26
3,4,5,3′,4′-pentamethoxystilbene[a]	26
Trans-3,4,3′,5′-tetramethoxystilbene	55
Crotalaria madurensis	
Trans- and *cis*-3,3′,5,5′-tetrahydroxy-4-methoxystilbene	56–60
Yucca periculosa, Y. schidigera; *Cassia pudibunda*	
Trans-4,4′-dihydroxystilbene	56
Yucca periculosa	
Trans-*3-hydroxy-5-methoxystilbene*	59
Cryptocarya idenburgensis	
Trans-4,3′-dihydroxy-5′-methoxystilbene	60
Dracaena loureiri	
Trans-4-hydroxy-3′,5′-dimethoxystilbene	60, 61
Dracaena loureiri, D. cochinchinensis	
Piceid or polydatin or resveratrol-3-glucoside (*trans*-3,5,4′-trihydroxystilbene-3-*O*-β-D-glucopyranoside)	2, 6, 62, 63
Polygonum cuspidatum; *Rheum rhaponticum*; *Picea* spp.; lentils (*Lens culinaris*)	
Rhapontin or rhaponticin (*trans*-3,3′,5-trihydroxy-4′-methoxystilbene-3-*O*-β-D-glucopyranoside)	2, 6
Rheum spp.; eucalyptus	
Deoxyrhapontin (*trans*-3,5-dihydroxy-4′-methoxystilbene-3-*O*-β-D-glucopyranoside)	6
Rheum rhaponticum	
Isorhapontin (*trans*-3,4′,5-trihydroxy-3′-methoxystilbene-3-*O*-β-D-glucopyranoside)	6, 62
Pinus sibirica; *Picea* spp.	
Piceatannol glucoside (3,5,3′,4′-tetrahydroxystilbene-4′-*O*-β-D-glucopyranoside)	2, 6
Rheum rhaponticum; *Polygonum cuspidatum*; spruce	
Pinostilbenoside (*trans*-3-methoxy-5-hydroxystilbene-4′-*O*-β-D-glucopyranoside)	64
Pinus koraiensis	
Resveratroloside or resveratrol-4′-glucopyranoside (*trans*-3,5,4′-trihydroxystilbene-4′-*O*-β-D-glucopyranoside)	2, 6, 3, 65
Polygonum cuspidatum; *Pinus* spp.; *Vitis vinifera*	
Astringin (*trans*-3,4,3′,5′-tetrahydroxystilbene-3′-*O*-β-D-glucopyranoside)	3, 62, 65
Picea spp.; *Vitis vinifera*	
Piceid-2″-*O*-gallate and -2″-*O*-coumarate	66
Pleuropterus ciliinervis	
Rhaponticin-2″-*O*-gallate and -6″-*O*-gallate	67
Rhubarb (*Rheum undulatum*)	
Piceatannol-6″-*O*-gallate	68
Chinese rhubarb (*Rhei rhizoma*)	

(*continued*)

TABLE 2.1
Continued

Compound and sources	Ref.
Cis-resveratrol-3,4′-*O*-β-diglucoside	69
Vitis vinifera (cell suspension culture)	
Combretastatinsa and their glycosides	70
(e.g., combretastain A = *trans*-2′,3′-dihydroxy-3,4,4′,5-tetramethoxystilbene)	
5-Methoxy-*trans*-resveratrol-3-*O*-rutinoside	71
Elephantorrhiza goetzei	
Oxyresveratrol-2-*O*-β-glucopyranoside	25
Schoenocaulon officinale	
Resveratrol-3,4′-*O*,*O*′-di-β-D-glucopyranoside	25
Schoenocaulon officinale	
Mulberrosides (e.g., *cis*-oxyresveratrol diglucoside)	72, 73
Morus alba (cell cultures), *M. lhou*	
Gnetupendins (isorhapontigenin dimer glucosides); gnemonosides (resveratrol oligomer glucosides)	73, 74
Gnetum pendulum, *G. gnemon*	
Gaylussacin (5-(β-D-glucosyloxy)-3-hydroxy-*trans*-stilbene-2-carboxylic acid)	75
Gaylussacia baccata, *G. frondosa*	
Resveratrol oligomers and oligostilbenes (including viniferins)	76–79
Dipterocarpaceae, Gnetaceae, Vitaceae, Cyperaceae, and Leguminosae plants (including *Vatica pauciflora*, *V. rassak*, *V. oblongifolia*; *Vateria indica*; *Shorea laeviforia*, *S. hemsleyana*; *Paeonia lactiflora*; *Sophora moorcroftiana*, *S. leachiana*; *Gnetum venosum*; *Cyperus longus*; *Upuna borneensis*; *Iris clarkei*)	
Bibenzyl derivatives (methoxy-hydroxy-dihydrostilbenes including alfoliol I, gigantol)[a]	80
Nidema boothi	

[a]Compounds obtained synthetically.

benefits against diseases such as cancer (reviewed by Aggarwal et al. [82]). Resveratrol has been identified from a number of dietary sources including red grapes, muscadine grapes, cranberries, bilberries, blueberries, lingonberries, sparkleberry, deerberry, partridgeberry, and peanuts. However, resveratrol is also consumed in the forms of botanical dietary supplements and herbal formulations used in traditional Chinese medicine (TCM) and Indian Ayurvedic medicine [83], where it is commonly used as an active ingredient. Other plant sources of resveratrol include *Vitis* spp. (including grapes, grapevines, leaves, and berry skins); *Yucca* spp.; *Smilax* spp.; *Morus* spp. (including mulberry); lily (*Veratrum* spp.); legumes (*Cassia* spp., *Pterolobium hexapetallum*); *Rheum* spp. (including rhubarb); eucalyptus; spruce (*Picea* spp); pine (*Pinus* spp.); Poaceae (grasses, including *Festuca*, *Hordeum*, *Poa*, *Stipa*, and *Lolium* spp.); *Trifolium* spp.; *Nothofagus* spp.;

Artocarpus spp; *Gnetum* spp.; *Pleuropterus ciliinervis*; *Bauhinia racemosa*; *Paeonia lactiflora*; *Scilla nervosa*; *Tetrastigma hypoglaucum*; and *Rumex bucephalophorus* (Table 2.1). In addition numerous synthetic analogs of resveratrol have been reported (Table 2.1).

STRUCTURE OF RESVERATROL

Resveratrol is an off-white powder (from MeOH) with a melting point of 253 to 255°C, molecular formula of $C_{14}H_{12}O_3$, and molecular weight of 228.25 g/mol. The essential structural skeleton of the molecule comprises two aromatic rings joined by a styrene double bond (Figure 2.1). The presence of the double bond facilitates *trans* and *cis* isomeric forms of resveratrol, which correspond to *E* and *Z* diasteromers, respectively (Figure 2.1). However, because *trans*-resveratrol is the preferred steric form and is relatively stable if protected from high pH and light [84], it is the commonly studied form of resveratrol as reported by most laboratories [82].

The ultraviolet (UV) absorption maxima (λ_{max}) for the *trans* and *cis* isomers are 308 and 288 nm, respectively, which allows for their detection and separation by high-performance liquid chromatography (HPLC) [85]. Besides these differences in spectrophotometric properties, the two isomers can also be distinguished by the chemical shifts in their nuclear magnetic resonance (NMR) spectra [84–86]. *Trans*-resveratrol is commercially available and on exposure to UV irradiation rapidly converts to the *cis* form [1–3,84–89]. *Trans*-resveratrol, studied under different conditions, has been shown to be stable for months, except in high pH buffers, when protected from light [84]. *Cis*-resveratrol, although extremely light sensitive, can also remain stable in the dark at ambient temperature in 50% EtOH for at least 35 days over the range 5.3 to 52.8 μmol/l [84]. Apart from photochemical conversion, low pH also causes *cis*-resveratrol to isomerize to *trans*-resveratrol. The free enthalpy difference between synthetic *trans*-resveratrol and photochemically prepared *cis*-resveratrol was estimated to be similar to common stilbenes, with the *trans* isomer being more stable by about 11 to 14 kJ/mol [85]. In addition, pK_a values of *trans*-resveratrol corresponding to the mono-, di-, and triprotonation of the system were 9.3, 10.0, and 10.6, respectively [85].

Hence, resveratrol occurs predominantly as the *trans* isomer and reports of the presence of the *cis* isomer, e.g., in certain wines, is attributed to photoisomeric conversion, enzyme action during fermentation, or release from viniferins [1–3,86–89]. *Trans*-resveratrol has been shown to be the more biologically active form of resveratrol. However, as regards the structure of the molecule, apart from the stereochemistry of the styrene bond, the positions of the phenolic substituents on the aromatic rings also play an important role in determining its biological activity. The molecular structure of resveratrol has been examined in detail and theoretical energy calculations for several excited states of *trans*-resveratrol and *trans*-σ-viniferin have shown the importance of the *p*-4′-OH group for biological activity [87,90]. Hence, natural and synthetic derivatives of resveratrol (Table 2.1) have been well examined from a SAR perspective in an effort to study resveratrol's impact on human health and disease.

CHEMICAL ANALYSES OF RESVERATROL

Over the past decade several methods have been developed to detect the presence and measure levels of resveratrol and its analogs based on the use of HPLC and gas chromatography (GC) [1–3,84–89]. Much attention has been focused on method development since studying the biological properties of resveratrol requires the analyses of complex mixtures containing very small amounts of stilbenes, and complete and quick extractions are required to minimize the loss from isomerization or denaturation. Generally, HPLC methods using reverse phase C18 columns coupled with UV detection (photodiode array [PDA] or diode array detector [DAD]) can adequately distinguish between resveratrol isomers and their analogs based on their different absorbance maxima. However, the use of mass spectrometry (MS), fluorimetry, and electrochemical detectors (ECDs), which are more specific than UV detection, has considerably improved sensitivity and decreased sample size [85,86]. GC methods with or without MS detection, although not as popular as HPLC, have also been frequently employed but require trimethylsilyl derivatization of resveratrol and its analogs [85,86].

SYNTHESIS OF RESVERATROL

Although resveratrol is a naturally occurring polyphenol that has been isolated from more than 70 plant species, it is not feasible to isolate this compound in sufficient quantities required for *in vitro* and *in vivo* biological tests. For example, it has been reported that 1 kg of dried grape skin can provide only 92 mg of resveratrol [1–3]. Hence, in the past decade great interest has arisen concerning resveratrol synthesis because of the numerous biological properties associated with this compound. There have also been reports on the production of stilbenes from cell culture and biotransformation studies [19,65] and from grapevine leaves that are stressed to increase the production of phytoalexins [89].

A survey of synthetic schemes reported for the production of resveratrol, although not exhaustive due to the large number of patented methods, follows. Many of the synthetic schemes described for resveratrol and its analogs rely on the Wittig or Wittig–Horner reaction. In the Wittig reaction, coupling of a benzyl anion with benzaldehdye forms a styrene double bond in 7 to 8 steps, and several attempts have been made to reduce the number of steps and increase the yield. The first reports of the synthesis of resveratrol were in 1940 by Takaoka [91], and in 1941 by Späth and Kromp [92], after Takaoka isolated resveratrol from the roots of *Veratrum grandiflorum* [80]. In 1940 Takaoka described the synthesis of resveratrol dimethyl and trimethyl ethers, which was carried out using Perkin condensation of *p*-anisyl acetic acid sodium salt with 1,3-dimethoxybenzaldehyde in acetic anhydride [91]. The product formed never crystallized so could not be compared with the natural product. Späth and Kromp reinvestigated the method by purifying a small sample of trimethoxystilbene carboxylic acid by sublimation and decarboxylation and isolating the trimethyl ether of resveratrol [92].

In 1997 Alonso et al. described the synthesis of resveratrol and its analogs pinosilvine and piceatannol [93]. A 3,5-dimethoxybenzyl trimethylsilyl ether was coupled with aldehydes in the presence of lithium powder and a catalytic amount of naphthalene. The expected alcohol was dehydrated and demethylated to yield the hydroxylated stilbene derivatives. Orsini et al. synthesized combreastatin and resveratrol and their corresponding glycosides via the Wittig reaction [70]. In 2001 Eddarir et al. described the organometallic synthesis of resveratrol, in which resveratrol was fluorinated on the styrene double bond [42]. Guiso et al., in 2002, employed the Heck reaction affording only the natural *E* isomer, i.e., the *trans* isomer of resveratrol in 70% yield [94]. A one-pot synthesis of 4-methoxyiodobenzene with vinyltrimethylsilane under arylation–desilylation conditions has recently been described by Jeffery and Ferber, which by removal of excess vinyltrimethylsilane and arylation of the 4-methoxystyrene by 4-methoxyiodobenzene in a one-flask reaction yields (*E*)-3,4′,5-trimethoxystilbene [95]. When demethylated this leads to resveratrol in 85% yield [95].

In the last few years synthetic chemistry has also branched out from classic chemistry to combinatorial chemistry. Hence, although resveratrol has been synthesized using conventional organic chemistry, recently researchers have carried out syntheses based on combinatorial methods. For example, resveratrol has been prepared by a method that involves a solid-phase cross metathesis reaction wherein a 4-vinylphenol was attached to a Merrifield resin affording a supported styreneyl ether [96,97]. This can then be coupled by a ruthenium carbene to various styrenes to yield selective (*E*)-stilbenoids [96,97].

THEORETICAL AND SAR STUDIES OF RESVERATROL

It has been well established that the interaction of biological molecules strongly depends upon the electrostatic fields generated in the process of charge transfer and is mainly determined by geometrical factors. A large number of theoretical or modeling studies have been carried out on resveratrol [87,98,99].

Del Nero and De Melo have reported a semiempirical calculation of the electronic and structural properties of *trans*-resveratrol, *trans*-stilbene, and diethylstilbestrol [98]. The analyses of the calculated bond lengths and chain rearrangements gave an insight of how chemical modifications of these molecules could affect the possible physiological properties of resveratrol. Semiempirical self-consistent field molecular orbital (SCFMO) calculations were used to calculate the structural and electronic properties of resveratrol and its analogs wherein the geometry of the systems was optimized and the electronic properties were calculated at the level of the AM1 method [99]. Stivala et al. have used the thermodynamic parameters and the formation enthalpies (ΔH_f) calculated by semiempirical methods to discuss the antioxidant activity of *cis*- and *trans*-resveratrol [100]. In addition, density functional theory (DFT) has also been proposed to be reliable in the study of energetics and geometrical properties of proton transfer and other ion–molecule reactions. Hence, Cao et al. have employed DFT calculations to obtain the geometry, the spin density, the highest occupied molecular orbital (HOMO), the lowest unoccupied molecular orbital (LUMO), and the single electron distribution of the 4′- and 5-radical of resveratrol [87]. It was found that resveratrol was a potent antioxidant with the 4′-OH group being more reactive than the 3- and 5-OH groups because of resonance effects. The dominant feature of the resveratrol radical was a semiquinone structure, which determined its stability. Delocalization of the unpaired electron density was mainly on the oxygen atom and its *ortho* and *para* positions. Hence, the antioxidant activity of resveratrol was found to be related to its spin density and unpaired electron distribution of the oxygen atom [87].

CONCLUSION

Resveratrol is a dietary polyphenol that is reported to have numerous biological properties and implications for human health and disease. However, given its low levels in food sources including red wine, it is unlikely that desired biological endpoints will be achieved from normal dietary consumption. In addition, its bioavailability and concentration in blood and tissues may fall well below levels required for most biological activities. Hence, continued research is necessary to evaluate the synergistic and/or additive effects of resveratrol with other food and food-related constituents. In addition, future studies should focus on the uptake and urinary excretion of its conjugated forms and metabolites formed *in vivo* by physiological changes and by enzymatic action of gut microflora. A thorough understanding of the chemistry of this molecule and its related conjugates and derivatives is important for correlation of its observed *in vitro* and *in vivo* biological properties and eventually for translation into practical benefits for human health.

REFERENCES

1. Soleas GJ, Diamandis EP, and Goldberg DM, The world of resveratrol, *Nutr Cancer Prev* 13, 159–82, 2000.
2. Sovak M, Grape extract, resveratrol and its analogs: a review, *J Med Food* 4, 93–105, 2001.
3. Pervaiz S, Resveratrol: from grapevines to mammalian biology, *FASEB J* 17, 1975–1985, 2003.
4. Powell RG, TePaske MR, Plattner RD, White JF, and Clement SL, Isolation of resveratrol from *Festuca versuta* and evidence for the widespread occurrence of this stilbene in the Poaceae, *Phytochemistry*, 35, 335–338, 1994.
5. Rimando AM, Kalt W, Magee JB, Dewey J, and Ballington JR, Resveratrol, pterostilbene, nd piceatannol in *Vaccinium* berries, *J Agric Food Chem* 52, 4713–4719, 2004.
6. Aaviskar A, Haga M, Kuzina K, Puessa T, Raal A, and Tsoupras G, Hydroxystilbenes in the roots of *Rheum rhaponticum*, *Proc Estonian Acad Sci* 52, 99–107, 2003.
7. Zulfiqar A, Toshiyuki T, Ibrahim I, Munekazu I, Furusawa M, Ito T, Nakaya K, Murata J, and Darnaedi D, Phenolic constituents of *Gnetum klossii*, *J Nat Prod* 66, 558–560, 2003.
8. Kumar RG, Jyostna D, Krupadanam GL, and Srimannarayana G, Phenanthrene and stilbenes from *Pterobolium hexapetallum*, *Phytochemistry* 27, 3625–3626, 2004.
9. Deshpande VH, Srinivasan R, and Rao AV, Wood phenolics of Morus species. IV. Phenolics of the heartwood of five Morus species, *Indian J Chem* 13, 453–457, 1975.
10. Montoro P, Piacente S, Oleszek W, and Pizza C, Liquid chromatography/tandem mass spectrometry of unusual phenols from *Yucca schidigera* bark: comparison with other analytical techniques, *J Mass Spectrosc* 39, 1131–1138, 2004.
11. Olas B, Wachowicz B, Stochmal A, and Oleszek W, Inhibition of oxidative stress in blood platelets by different phenolics from *Yucca schidigera* Roezl. bark, *Nutrition* 19, 633–640, 2003.
12. Feng F, Liu W, Chen Y, Liu J, and Zhao S, Flavonoids and stilbenes from *Smilax china*, *Zhonguo Yaoke Daxue Xuebao* 34, 119–121, 2003.
13. Cheng Y, Zhang D, Yu S, and Ding Y, Study on chemical constituents in rhizome of *Smilax perfoliate*, *Zhongguo Zhongyao Zazhi* 28, 233–235, 2003.
14. Adesanya SA, Ogundana SK, and Roberts MF, Dihydrostilbene phytoalexins from *Dioscorea bulbifera* and *D. dumentorum*, *Phytochemistry* 28, 773–774, 1989.
15. Ferrigni NR, Mclaughlin JL, Powell RG, and Smith CR, Use of potato disc and brine shrimp bioassays to detect activity and isolate piceatannol as the

antileukemic principle from the seeds of *Euphorbia lagascae*, *J Nat Prod* 47, 347–349, 1984.
16. Tsuruga T, Chun YT, Ebizuka Y, and Sankawa U, Biologically active constituents of *Melaleuca leucadendron*: inhibitors of induced histamine release from rat mast cells, *Chem Pharm Bull* 39, 3276–3278, 1991.
17. Inamori Y, Kato Y, Kubo M, Yasuda M, Baba K, and Kozawa M, Physiological activities of 3,3′,4,5′-tetrahydroxystilbene isolated from the heartwood of *Cassia garrettiana* Craib., *Chem Pharm Bull* 32, 213–218, 1984.
18. Ko S, Lee SM, and Whang WK, Anti-platelet aggregation activity of stilbene derivatives from *Rheum undulatum*, *Arch Pharm Res* 22, 401–403, 1999.
19. Teguo PW, Decendit S, Krisa S, Deffieux G, Vercauteren J, and Merillon JM, The accumulation of stilbene glycosides in *Vitis vinifera* cell suspension cultures, *J Nat Prod* 59, 1189–1191, 2001.
20. Rao VSS and Rajaduri S, Isolation of 3,4,3,5-tetrahydroxystilbene (piceatannol) from *Cassia marginata* heartwooood, *Aust J Chem* 21, 1921–1922, 1968.
21. Zaman A, Prakash S, Wizarat K, Joshi BS, Gawad DH, and Likhate MA, Isolation and structure of gnetol, a novel stilbene from *Gnetum ula*, *Indian J Chem* 22B, 101–104, 1983.
22. Ohguchi K, Tanaka T, Iliya I, Ito T, Iinuma M, Matsumoto K, Akao Y, and Nozawa Y, Gnetol as a potent tyrosinase inhibitor from genus *Gnetum*, *Biosci Biotech Biochem* 67, 663–665, 2003.
23. Djapic N, Djarmati Z, Filip S, and Jankov RM, A stilbene from the heartwood of *Maclura pomifera*, *J Serb Chem Soc* 68, 235–237, 2003.
24. Hakim EH, Ulinnuha UZ, Syah YM, and Ghisalberti EL, Artoindonesianins N and O, new prenylated stilbene and prenylated arylbenzofuran derivatives from *Atocarpus gomezianus*, *Fitoterapia* 73, 597–603, 2002.
25. Kanchanapoom T, Suga K, Kasai R, Yamasaki K, Kamel MS, and Mohamed MH, Stilbene and 2-arylbenzofuran glucosides from the rhizome of *Schoenocaulon officinale*, *Chem Pharm Bull* 50, 863–865, 2002.
26. Lu J, Ho CT, Ghai G, and Chen KY, Resveratrol analog, 3,4,5,4′-tetrahydroxystilbene, differentially induces pro-apoptotic p53/Bax gene expression and inhibits the growth of transformed cells but not their normal counterparts, *Carcinogenesis* 22, 321–328, 2001.
27. Hathway DE and Brit L, Hydroxystilbene compounds as taxonomic tracers in the genus eucalyptus, *Biochem J* 83, 80–84, 1962.
28. King FE, King TJ, Godson DH, and Manning LC, Chemistry of extractives from hardwoods. XXVII. The occurrence of 3,3′,4,5′-tetrahydroxy and 3,3′,4,5,5′-pentahydroxystilbene in *Vouacapoua* species, *J Chem Soc* 4477–4480, 1956.
29. Yao CS, Lin M, Liu X, and Wang YH, Stilbenes from *Gnetum cleistostachyum*, *Huaxue Xuebao* 61, 1331–1334, 2003.
30. Sofronova VE, Petrov KA, Sofronova V, Kriolitozony E, Petrov KA, and Yalutsk R, New phenolic growth inhibitor from buds of *Duschekia fruticosa* (Rupr), *Pouzar Rastitel'nye Resursy* 38, 92–97, 2002.
31. Ali MA and Debnath DC, Isolation and characterization of antibacterial constituent from devdaru (lignum of *Polyalthia longifolia* L.), *Bang J Sci Ind Res* 32, 20–24, 1997.
32. Ngo KS and Brown GD, Stilbenes, monoterpenes, diarylheptanoids, labdanes and chalcones from *Alpinia katsumadai*, *Phytochemistry* 47, 1117–1123, 1998.

33. Kuroyanagi M, Yamamoto Y, Fukushima S, Ueno A, Noro T, and Miyase T, Chemical studies on the constituents of *Polygonum nodosum*, *Chem Pharm Bull* 30, 1602–1628, 1982.
34. Rudloff E and Jorgensen E, Biosynthesis of pinosylvin in the sapwood *of Pinus resinosa*, *Phytochemistry* 2, 297–304, 1963.
35. Takasugi M, Kawashima S, Monde K, Katsui N, Masamune T, and Shirata A, Antifungal compounds from *Dioscorea batatas* inoculated with *Pseudomonas cichorii*, *Phytochemistry* 26, 371–375, 1987.
36. Lu M, Cai YJ, Fang JG, Zhou YL, Liu ZL, and Wu LM, Efficiency and structure–activity relationship of the antioxidant action of resveratrol and its analogs, *Pharmazie* 57, 474–478, 2002.
37. Cai YJ, Fang JG, Ma LP, Yang L, and Liu ZL, Inhibition of free radical-induced peroxidation of rat liver microsomes by resveratrol and its analogues, *Biochim Biophys Acta* 1637, 31–38, 2003.
38. Matsuoka A, Takeshita K, Furuta A, Ozaki M, Fukuhara K, and Miyata N, The 4′-hydroxy group is responsible for the *in vitro* cytogenetic activity of resveratrol, *Mutation Res* 521, 29–35, 2002.
39. Thakkar K, Geahlen RL, and Cushman M, Synthesis and protein-tyrosine kinase inhibitory activity of polyhydroxylated stilbene analogues of piceatannol, *J Med Chem* 36, 2950–2955, 1993.
40. Fang JG, Lu M, Chen ZH, Zhu HH, Li Y, Yang L, Wu LM, and Liu ZL, Antioxidant effects of resveratrol and its analogues against the free-radical-induced peroxidation of linoleic acid in micelles, *Chem Eur J* 8, 4191–4198, 2002.
41. Lu M, Fang JG, Liu ZL, and Wu LM, Effects of resveratrol and its analogs on scavenging hydroxyl radicals: evaluation of EPR spin trapping method, *Appl Magn Res* 22, 475–481, 2002.
42. Eddarir S, Zouanante A, and Rolando C, Fluorinated resveratrol and pterostilbene, *Tetrahedron Lett* 42, 9127–9130, 2001.
43. Lee HJ, Seo JW, Lee BH, Chung KH, and Chi DY, Synthesis and radical scavenging activities of resveratrol derivatives, *Bioorg Med Chem Lett* 14, 463–466, 2004.
44. Matsuda H, Tomohiro N, Hiraba K, Harima S, Ko S, Matsuo K, Yoshikawa M, and Kubo M, Study on anti-oketsu activity of rhubarb: II. Anti-allergic effects of stilbene components from *Rhei undulati* Rhizoma (dried rhizome of *Rheum undulatum* cultivated in Korea). *Bio Pharm Bull* 24, 264–267, 2001.
45. Bangani V, Crouch NR, and Mulholland DA, Homoisoflavanones and stilbenoids from *Scilla nervosa*, *Phytochemistry* 51, 947–951, 1999.
46. Wang QL, Lin M, and Liu CT, Antioxidative activity of natural isorhapontigenin, *Jpn J Pharm* 81, 61–66, 2001.
47. Feng Y, Bing W, Lin Z, and Zhi ZZ, Synthesis of the natural products resveratrol and isorhapotogenin (isorhapontigenin), *Chin Chem Lett* 9, 1003–1004, 1998.
48. Chun YJ, Ryu Sy, Jeong TC, and Mie YK, Mechanism based inhibition of human cytochrome P450 1A1 by rhapontigenin, *Drug Metab Disposition* 29, 389–393, 2001.
49. Gonzalez MJTG, Pinto MMM, Kijjoa A, Anantachoke C, and Herz W, Stilbenes and other constituents of *Knema austrosiamensis*, *Phytochemistry* 32, 433–438, 1993.

50. Kerem Z, Regev-Shoshani G, Flaishman MA, and Sivan L, Resveratrol and two monomethylated stilbenes from Israeli *Rumex bucephalophorus* and their antioxidant potential, *J Nat Prod* 66, 1270–1272, 2003.
51. Roberti M, Pizzirani D, Simoni D, Rondanin R, Baruchello R, Bonora C, Buscemi F, Grimaudo S, and Tolomeo M, Synthesis and biological evaluation of resveratrol and analogues as apoptosis-inducing agents, *J Med Chem* 46, 3546–3554, 2003.
52. Kim YM, Yun J, Lee CK, Lee H, Min KR, and Kim Y, Oxyresveratrol and hydroxystilbene compounds: inhibitory effect on tyrosinase and mechanism of action, *J Bio Chem* 277, 16340–16344, 2002.
53. Hu Y, Ning Z, and Liu D, Determination of pterostilbene in Dragon's Blood by RP-HPLC, *Yaowu Fenxi Zazhi* 22, 428–430, 2002.
54. Rimando AM, Cuendet M, Desmarchelier C, Mehta RG, Pezzuto JM, and Duke SO, Cancer chemopreventive and antioxidant activities of pterostilbene, a naturally occurring analogue of resveratrol, *J Agric Food Chem* 50, 3453–3457, 2002.
55. Bhakuni DS and Chaturvedi R, Chemical constituents of *Crotalaria madurensis*, *J Nat Prod* 47, 585–591, 1984.
56. Torres P, Avila JG, Romo De Vivar A, Garcia AM, Marin JC, Aranda E, and Cespedes CL, Antioxidant and insect growth regulatory activities of stilbenes and extracts from *Yucca periculosa*, *Phytochemistry* 64, 463–473, 2003.
57. Olas B, Wachowicz B, Stochmal A, and Oleszek W, Inhibition of oxidative stress in blood platelets by different phenolics from *Yucca schidigera* Roezl. bark, *Pol Nutr* 19, 633–640, 2003.
58. Messana I, Ferrari F, Cavalcanti MS, and Morace G, An anthraquinone and three naphthopyrone derivatives from *Cassia pudibunda*, *Phytochemistry* 30, 708–710, 1991.
59. Juliawaty LD, Kitajima M, Takayama H, Achmad SA, and Aimi N, A new type of stilbene-related secondary metabolite, idenburgene, from *Cryptocarya idenburgensis*, *Chem Pharm Bull* 48, 1726–1728, 2000.
60. Likhitwitayawuid K, Sawasdee K, and Kirtikara K, Flavonoids and stilbenoids with COX-1 and COX-2 inhibitory activity from *Dracaena loureiri*, *Planta Med* 68, 841–843, 2002.
61. Hu Y, Tu P, Li R, Wan Z, and Wang D, Studies on stilbene derivatives from *Dracaena cochinchinensis* and their antifungal activities, *Zhongcaoyao* 32, 104–106, 2001.
62. Aritomi M and Donnelly DMA, Stilbene glucosides in the bark of *Picea sitchensis*, *Phytochemistry* 15, 2006–2008, 1976.
63. Duenas M, Hernandez T, and Estrella I, Phenolic composition of the cotyledon and the seed coat of lentils (*Lens culinaris* L.), *Eur Food Res Technol* 51, 478–483, 2002.
64. Song HK, Jung J, Park KH, and Lim Y, Leukotriene D4 antagonistic activity of a stilbene derivative isolated from the bark of *Pinus koraiensis*, *Agric Chem Biotech* 44, 199–201, 2001.
65. Teguo PW, Fauconneau B, Deffieux G, Huguet F, Vercauteren J, and Merillon JM, Isolation, identification, and antioxidant activity of three stilbene glucosides newly extracted from *Vitis vinifera* cell cultures, *J Nat Prod* 61, 655–657, 1998.

66. Lee JP, Min BS, An RB, Na MK, Lee SM, Lee HK, Kim JG, Bae KH, and Kang SS, Stilbenes from the roots of *Pleuropterus ciliinervis* and their antioxidant activities, *Phytochemistry* 64, 759–763, 2003.
67. Kageura T, Matsuda H, Morikawa T, Toguchida I, Harima S, Oda M, and Yoshikawa M, Inhibitors from rhubarb on lipopolysaccharide-induced nitric oxide production in macrophages: structural requirements of stilbenes for the activity, *Bio Med Chem* 9, 1887–1893, 2001.
68. Kashiwada Y, Nonaka G, Nishioka I, Nishizawa M, and Yamagishi T, Studies on rhubarb (*Rhei rhizoma*): XIV. Isolation and characterization of stilbene glucosides from Chinese rhubarb. *Chem Pharm Bull* 36, 1545–1549, 1988.
69. Decendit A, Waffo-Teguo P, Richard T, Krisa S, Vercauteren J, Monti JP, Deffieux G, and Merillon JM, Galloylated catechins and stilbene diglucosides in *Vitis vinifera* cell suspension cultures, *Phytochemistry* 60, 795–798, 2002.
70. Orsini F, Pelizzoni F, Bellini B, and Miglierini G, Synthesis of biologically active polyphenolic glycosides (combretastatin and resveratrol series), *Carbohydr Res* 301, 95–109, 1997.
71. Wanjala CC and Majinda RR, A new stilbene glycoside from *Elephantorrhiza goetzei*, *Fitoterapia* 72, 649–655, 2001.
72. Hano Y, Goi K, Nomura T, and Ueda S, Sequential glucosylation determined by NMR in the biosynthesis of mulberroside D, a cis-oxyresveratrol diglucoside, in *Morus alba* cell cultures, *Life Sci* 53, 237–241, 1997.
73. Hirakura K, Fujimoto Y, Fukai T, and Nomura T, Constituents of the cultivated mulberry tree. Two phenolic glycosides from the root bark of the cultivated mulberry tree (*Morus lhou*), *J Nat Prod* 49, 218–224, 1986.
74. Iliya I, Tanaka T, Iinuma M, Furusawa M, Ali Z, Nakaya K, Murata J, and Darnaedi D, Five stilbene glucosides from *Gnetum gnemonoides* and *Gnetum africanum*, *Helv Chim Acta* 85, 2394–2402, 2002.
75. Iliya I, Ali Z, Tanaka T, Iinuma M, Furusawa M, Nakaya K, Murata J, Darnaedi D, Matsuura N, and Ubukata M. Stilbene derivatives from *Gnetum gnemon* Linn, *Phytochemistry* 62, 601–606, 2003.
76. Askari A, Worthen LR, and Shimizu Y, Gaylussacin, a new stilbene derivative from species of *Gaylussacia*, *Lloydia* 35, 49–54, 1972.
77. Cichewicz RH and Kouzi SA, Resveratrol oligomers: structure, chemistry, and biological activity, *Stud Nat Prod Chem* 26, 507–579, 2002.
78. Ito T, Ibrahim I, Tanaka T, Nakaya K, Iinuma M, Takahashi Y, Naganawa H, Akao Y, Nozawa Y, Ohyama M, Nakanishi Y, Bastow KF, and Lee KH, Chemical constituents of dipterocarpaceaeous and gnetaceaeous plants and their biological activities, *Tennen Yuki Kago Toron Koen Yosh* 43, 449–454, 2001.
79. Sotheeswaran S and Pasupathy V, Distribution of resveratrol oligomers in plants, *Phytochemistry* 32, 1083–1092, 1993.
80. Hernandez-Romero Y, Rojas JI, Castillo R, Rojas A, and Mata R, Spasmolytic effects, mode of action, and structure–activity relationships of stilbenoids from *Nidema boothii*, *J Nat Prod* 67, 160–167, 2004.
81. Takaoka MJ, *J Faculty Sci Hokkaido Imperial Univ* 3, 1–16, 1940.
82. Aggarwal BB, Bhardwaj A, Aggarwal RS, Seeram NP, Shishodia S, and Takada Y, Role of resveratrol in prevention and therapy of cancer: preclinical and clinical studies, *Anticancer Res* 24, 2783–2840, 2004.

83. Paul B, Masih I, Deopujari J, and Charpentier C, Occurrence of resveratrol and pterostilbene in age-old Darakchasava, an Ayurvedic medicine from India, *J. Ethnopharmocol* 68, 71–76, 1999.
84. Trela B and Waterhouse A, Resveratrol: isomeric molar absorptivities and stability, *J Agric Food Chem* 44, 1253–1257, 1996.
85. Deak M and Falk H, On the chemistry of resveratrol diasteromers, *Monat fur Chem* 134, 883–888, 2003.
86. Fremont L, Biological effects of resveratrol, *Life Sci* 66, 663–673, 2000.
87. Cao H, Pan X, Li Cong, Zhou C, Deng F, and Li T, Density functional theory calculations for resveratrol, *Bio Med Chem Lett* 13, 1869–1871, 2003.
88. Soleas GJ, Diamandis EP, and Goldberg DM, Resveratrol: a molecule whose time has come? And gone?, *Clin Biochem* 30, 91–113, 1997.
89. Langcake P and Pryce RJ, The production of resveratrol by *Vitis vinifera* and other members of the Vitaceae as a response to infection or injury, *Physiol Plant Pathol* 9, 77–86, 1976.
90. Caruso F, Tanski J, Villegas-Estrada A, and Rossi M, Structural basis for antioxidant activity of *trans*-resveratrol: ab initio calculations and crystal and molecular structure, *J Agric Food Chem* 52, 7304–7310, 2004.
91. Takaoka M, Phenolic substances of white hellebore (*Veratrum grandiflorum* Loes. fil.). II. Synthesis of resveratrol and its derivatives, *Proc Imperial Acad* (*Tokyo*) 16, 405–407, 1940.
92. Späth E and Kromp K, Natural stilbenes. III. Synthesis of resveratrol, *Ber Dtsch Chem Ges* 74B, 867–869, 1941.
93. Alonso E, Ramon DJ, and Yus M, Simple synthesis of 5-substituted resorcinols: a revisited family of interesting bioactive molecules, *J Org Chem* 62, 417–421, 1997.
94. Guiso M, Marra C, and Farina A, A new efficient resveratrol synthesis, *Tetrahedron Lett* 43, 597–598, 2002.
95. Jeffery T and Ferber B, One-pot palladium-catalyzed highly chemo-, regio-, and stereoselective synthesis of trans-stilbene derivatives. A concise and convenient synthesis of resveratrol, *Tetrahedron Lett* 44, 193–197, 2003.
96. Andrus MB, Nartey ED, and Meredith EL, Synthesis of Resveratrol, a Potent New Disease-Preventative Agent, book of abstracts, 219th ACS National Meeting, San Francisco, March 26–30, 2000.
97. Sako M, Hosokawa H, Ito T, and Iinuma M, Regioselective oxidative coupling of 4-hydroxystilbenes: synthesis of resveratrol and epsilon-viniferin (E)-dehydrodimers, *J Org Chem* 69, 2598–2600, 2004.
98. Del Nero J and De Melo CP, Investigation of the excited states of resveratrol and related molecules, *Int J Quantum Chem* 95, 213–218, 2003.
99. Erkoc S, Keskin N, and Erkoc F, Resveratrol and its analogues resveratrol-dihydroxyl isomers: semi-empirical SCF-MO calculations, *Theor Chem* 631, 67–73, 2003.
100. Stivala LA, Savio M, Carafoli F, Perucca P, Bianchi L, Maga G, Forti L, Pagnoni UM, Albini A, Prosperi E, and Vannini V, Specific structural determinants are responsible for the antioxidant activity and the cell cycle effects of resveratrol, *J Biol Chem* 276, 22586–22594, 2001.

3 Resveratrol as an Antioxidant

Catalina Alarcón de la Lastra, Isabel Villegas, and Antonio Ramón Martín

CONTENTS

INTRODUCTION

Trans-resveratrol, or *trans*-3,4′,5-trihydroxystilbene, is a phytoalexin present in a wide variety of plant species, including mulberries, peanuts, and grapes, and thus is a constituent of the human diet. Its stilbene structure is related to the synthetic estrogen diethylstilbestrol. The synthesis of *trans*-resveratrol in plants can be induced by microbial infections, ultraviolet radiation (UVB), and exposure to ozone [1].

Resveratrol has been the focus of a number of studies investigating its biological attributes, which include antioxidant activity, antiplatelet aggregation effect, antiatherogenic property, estrogen-like growth promoting effect, growth inhibiting activity, immunomodulation, and chemoprevention. A primary impetus for research on resveratrol has come from the paradoxical observation that a low incidence of cardiovascular diseases may coexist with intake of a high-fat diet, a phenomenon known as the "French paradox" [1]. More recently, since the first report on the apoptosis-inducing activity of resveratrol in human cancer cells, the interest in this molecule as a potential chemotherapy agent has significantly increased [2].

Over the last few years a number of studies have provided evidence of an important role of reactive oxygen species (ROS) or reactive oxygen metabolites (ROM) in mediating the development of oxidative stress. The term ROS refers to molecules or parts of molecules that possess unpaired electrons in valence orbitals. This electron-deficient state makes these agents highly reactive and they can damage adjacent molecules by abstracting an electron from or donating an electron to them. They are invariably produced in aerobic environments through a variety of mechanisms, which include electron "leakage" during biologic oxidations, action of flavin dehydrogenases, and specific membrane-associated secretion, as well as by physical activation of oxygen by irradiation, e.g., UVB in sunlight [3].

The more important ROS are superoxide anion ($O_2^{\bullet-}$), hydroxyl ($OH^{\bullet}$), peroxyl ($RO_2^{\bullet}$), and alkoxyl ($RO^{\bullet}$) radicals. There are also highly detrimental agents that are nitrogen-based, e.g., nitric oxide ($NO^{\bullet}$) and its metabolite peroxynitrite, this term referring to both the anion oxoperoxynitrate ($ONOO^-$) and its conjugate acid hydrogen oxoperoxynitrate (ONOOH), the product of the fast radical–radical reaction between $NO^{\bullet}$ and $O_2^{\bullet-}$ radicals [4,5]. In addition, certain nonradicals such as hydrogen peroxide (H_2O_2), singlet oxygen ($^1O_2$), ozone (O_3), or hypochlorous acid (HOCl) are oxidizing agents that are easily converted into radicals (Figure 3.1).

Physiological levels of ROS participate in cellular functions and in transcriptional and posttranscriptional control. Indeed, ROS have been recognized as important signal transduction intermediates that control gene expression, cell differentiation, immune activation, and apoptosis [6]. Likewise, recent studies have revealed that crosstalk of ROS regulates the circulation, energy metabolism, reproduction, embryonic development, and remodeling of tissues through apoptotic mechanisms, and functions as a major defense system against pathogens [7]. However, excessive ROS accumulation may induce the oxidative modification of cellular macromolecules (lipids, proteins, and nucleic acids) with deleterious potential, e.g., damaging DNA and acting as tumor promoters [8].

The potential sources of ROS within cells include mitochondria, xanthine oxidase (XO) activated under ischemic conditions, myeloperoxidase (MPO), cyclooxygenase (COX), nitric oxide synthase (NOS), and reduced nicotinamide adenine dinucleotide phosphate (NADPH) oxidases. Quantitatively, the principal free radical in tissues is $O_2^{\bullet-}$, which is converted to the secondary oxidant H_2O_2 by the enzyme superoxide dismutase (SOD). $O_2^{\bullet-}$ radicals can be produced by both endothelial cells through XO and activated neutrophils through NADPH oxidase, which reduces molecular oxygen to the $O_2^{\bullet-}$ radical, and through the enzyme MPO. This enzyme

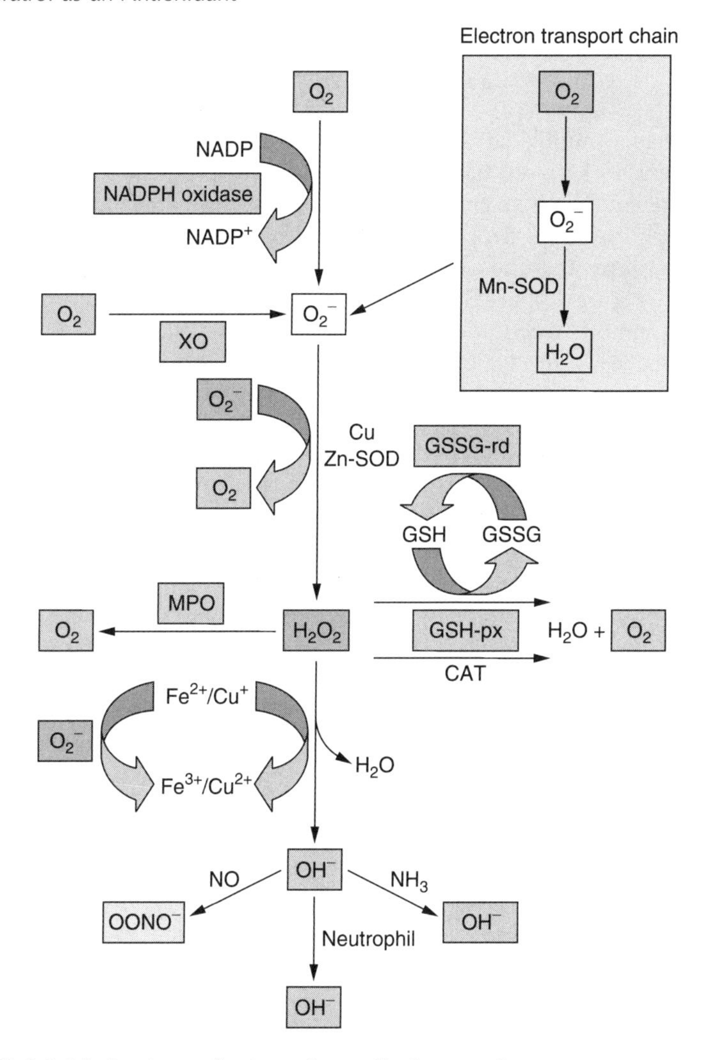

FIGURE 3.1 Mechanisms of oxygen free radical generation.

catalyzes the formation of such potent cytotoxic oxidants as HOCl from H_2O_2 and chloride ions and *N*-chloramines.

XO is found in tissues but not in neutrophils and is known to play a crucial role in ischemia-reperfusion injury. During ischemia, ATP is degraded to hypoxanthine and xanthine dehydrogenase is converted to XO. In the reperfusion state, XO catalyzes the reaction of hypoxanthine or xanthine and molecular oxygen to $O_2^{\bullet -}$ radicals. These radicals rapidly react with the free radical NO and peroxynitrite anion and other reactive species.

RESVERATROL AS FREE RADICAL SCAVENGER AND ANTIOXIDANT

Evidence has accumulated that resveratrol is both a free radical scavenger and a potent antioxidant because of its ability to promote the activities of a variety of antioxidative enzymes. The ability of polyphenolic compounds to act as antioxidants depends on the redox properties of their phenolic hydroxyl groups and the potential for electron delocalization across the chemical structure (Figure 3.2).

The common recognition of resveratrol as a natural antioxidant was clarified by Zini et al. [9] who suggested three different antioxidant mechanisms: (a) competition with coenzyme Q and to decrease the oxidative chain complex, the site of ROS generation, (b) scavenging $O_2^{\bullet-}$ radicals formed in the mitochondria, and (c) inhibition of lipid peroxidation (LP) induced by Fenton reaction products.

Recently the rate constant was reported for reaction of resveratrol with the $OH^{\bullet}$ radical in RAW 264.7 mouse peritoneal monocytes. This parameter is important in comparing resveratrol's antioxidant properties to those of other well-established antioxidants. Using the Fenton reaction, which involves Fe^{2+} and H_2O_2 for $OH^{\bullet}$ generation, resveratrol was found to be an effective scavenger of $OH^{\bullet}$ with a reaction rate constant of $9.45 \times 10^8\,M^{-1}\,sec^{-1}$, which was significant but somewhat less than recognized radical scavengers such as ascorbic acid and reduced glutathione (GSH) [8]. In addition, to demonstrate the ability of resveratrol to scavenge ROS, essentially $OH^{\bullet}$ generated by metal and enzymatic systems, Leonard et al. [8] utilized the Cr(VI), NADPH/GSSG-R system. Resveratrol affected this system, scavenging radicals produced by the above mentioned system. Similarly, scavenging of $O_2^{\bullet-}$ radicals by resveratrol was investigated using the reaction between xanthine and XO as a source of $O_2^{\bullet-}$ radicals. The results confirmed the ability of resveratrol to scavenge $O_2^{\bullet-}$ radicals. By contrast, in a study by Orallo et al. [10], using the enzymatic hypoxanthine oxidase–XO system, resveratrol did not modify the production of uric acid from XO, i.e., did not affect the XO activity and did not scavenge $O_2^{\bullet-}$ radicals in rat macrophage extracts.

In a study by Losa [11], using peripheral blood mononuclear cells isolated *ex vivo* from healthy humans incubated with the oxidant

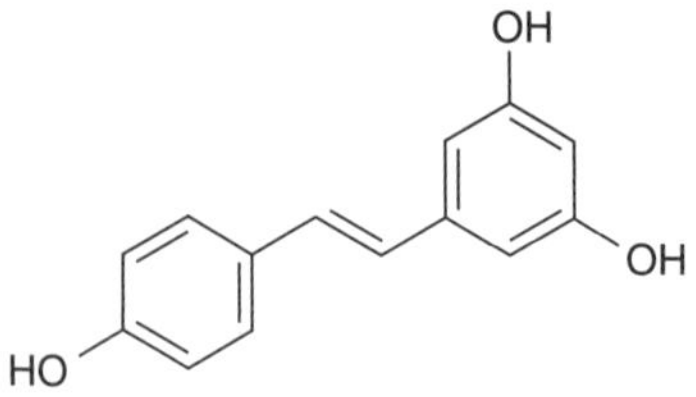

FIGURE 3.2 Chemical structure of resveratrol.

2-deoxy-D-ribose (dR), resveratrol appeared to act as a scavenger for $O_2^{\bullet-}$ generated by dR. The scavenger properties of resveratrol were also confirmed by Martínez and Moreno [12] who demonstrated a strong inhibitory effect on $O_2^{\bullet-}$ and H_2O_2 production in macrophages stimulated by lipopolysaccharide (LPS).

Other investigators have indicated that resveratrol can be an effective antioxidant. For example, peripheral blood mononuclear cells isolated *ex vivo* from healthy humans exposed to dR oxidative damage were remarkably protected in the presence of this compound [11]. Similarly, using unopsonized zymosan-stimulated murine macrophage RAW 264.7 cells, human monocytes, and neutrophils, it was demonstrated that resveratrol inhibited, in a dose-dependent manner, ROS production [13].

In order to protect tissues against the deleterious effects of oxygen radicals, all cells possess numerous antioxidant enzymes and free radical scavengers. Primary defenses include the enzymes SOD, catalase (CAT), and GSH-peroxidase. Three types of SOD isozymes, Cu,Zn-SOD, Mn-SOD, and extracellular SOD (EC-SOD), are known to accelerate the dismutation of $O_2^{\bullet-}$. The decrease in SOD activity may enhance LP as well as aggravate cellular injury. The product formed as a consequence of the enzymatic dismutation of $O_2^{\bullet-}$ is H_2O_2. Its direct toxicity to cells is limited unless it accumulates in high intracellular concentrations. It does, however, have a long half-life (>4 sec) and it can penetrate cell membranes. H_2O_2 levels are maintained at steady-state concentrations by several means, best known of which are the antioxidative enzymes CAT and GSH-peroxidase. CAT degrades H_2O_2 to H_2O while GSH-peroxidase utilizes H_2O_2 as a substrate during the metabolism of GSH to its oxidized form (glutathione disulfide or GSSG). At the same time, the antioxidant activity of GSH-peroxidase is coupled with the oxidation of GSH to GSSG, which can subsequently be reduced by GSSG-reductase with NADPH as the reducing agent (Figure 3.3). Not all H_2O_2 that is produced in a cell is detoxified. Some is converted to $OH^\bullet$, which, as noted above, is highly destructive. Fenton chemistry is the major means for generating $OH^\bullet$. H_2O_2 is readily converted to $OH^\bullet$ when it is in the presence of a transition metal, e.g., Fe^{2+}, Cu^+, etc. [14].

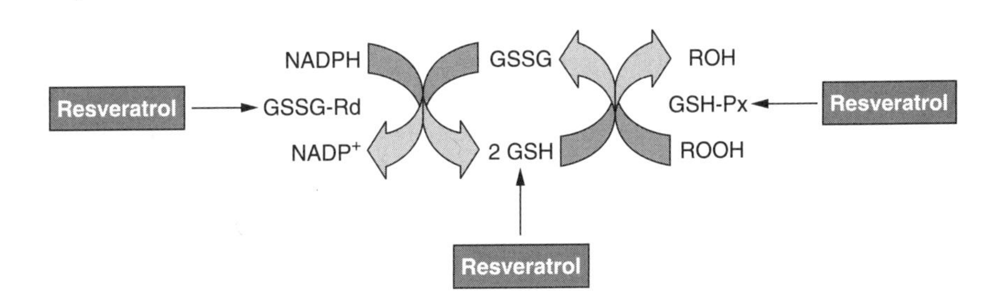

FIGURE 3.3 Influence of resveratrol on the glutathione redox cycle.

GSH is an important constituent of intracellular protective mechanisms against a number of noxious stimuli, including oxidative stress. GSH scavenges $O_2^{\bullet-}$ and protects protein thiol groups from oxidation. The GSH redox cycle catalyzed by the endogenous antioxidative enzyme GSH-peroxidase, as mentioned above, reduces H_2O_2, thus breaking the chain reaction leading from $O_2^{\bullet-}$ to the highly reactive $OH^{\bullet}$.

In a study by Losa [11] resveratrol appeared to maintain the GSH content in peripheral blood mononuclear cells isolated *ex vivo* from healthy humans against oxidative damage caused by dR, modulating the activities of two enzymes. It reduced the activity of *y*-glutamyltransferase, which contributes to the extracellular transpeptidation of aminoacid intermediates used in the *de novo* "salvage" synthesis of GSH, and GSH-S-transferase, which catalyzes the transfer of sulfhydryl groups to form GSH-conjugated metabolites that are less toxic and easily excreted [11].

Recently Olas et al. [15] investigated the influence of resveratrol in human blood platelets on changes of different thiols (GSH and free thiols of cysteine and cysteinylglycine) and protein thiol content induced by platinum compounds. The results showed that resveratrol reduced the platinum compound-induced decrease of platelet thiols, particularly thiols in acid-soluble fractions. Similarly, resveratrol induced an increase in GSH levels in a concentration-dependent manner in human lymphocytes activated with H_2O_2.

Jang and Pezzuto [16] have described the normalization of GSSG-reductase upon resveratrol treatment in a 12-*O*-tetradecanoylphorbol-13-acetate-induced oxidative event. Furthermore, the stilbene was able to modulate antioxidant enzymes (GSH-peroxidase, GSH-S-transferase, and GSSG-reductase) in H_2O_2-induced oxidative DNA damage in human lymphocytes [17].

EFFECTS OF RESVERATROL ON LIPID PEROXIDATION

Another aspect of free radical damage involves injury to cellular membranes. Measurement of LP is an indicator of possible free radical damage to cells. $OH^{\bullet}$ radicals, formed in the Fenton reaction ($Fe^{2+} + H_2O_2 \rightarrow Fe^{3+} + OH^{\bullet} + OH^{\bullet-}$), are able to initiate LP and cause cell membrane damage. Fe^{2+} ions can decompose LP intermediate hydroperoxides (ROOH) to $RO^{\bullet}$ (alkoxyl) radicals and Fe^{3+} to less potently oxidizing $ROO^{\bullet}$ (peroxyl) radicals [18]. ROOH are less hydrophilic than lipids themselves and their presence in membranes leaves the membranes increasingly permeable to protons and other ions. The products of LP, malondialdehyde (MDA) and other groups of aldehyde products, such as hexanol, 4-hydroxylnonenal, and related aldehydes, may also cause DNA damage [8] (Figure 3.4).

The ability of resveratrol to inhibit LP has well been documented in interesting and recently published studies. For example, Leonard et al. [8]

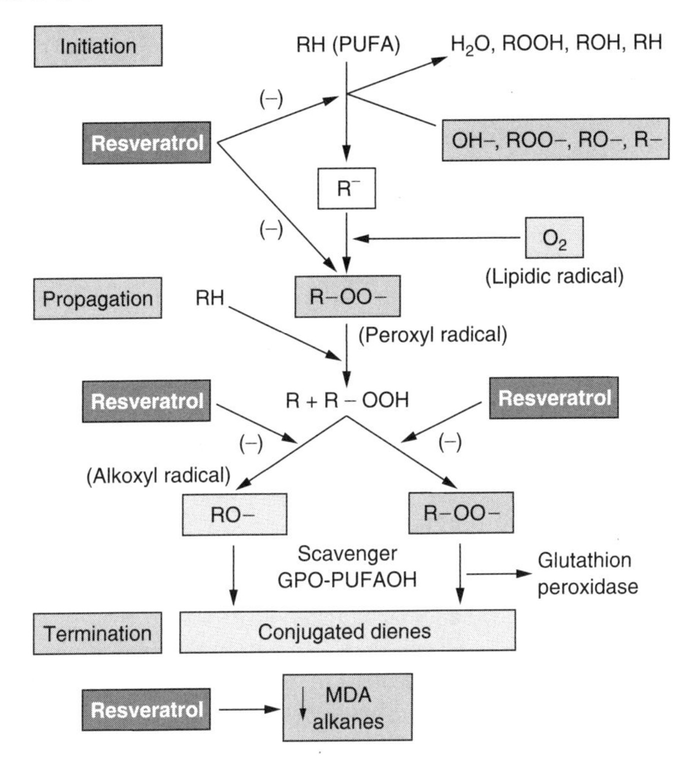

FIGURE 3.4 Inhibition of lipid peroxidation by resveratrol.

demonstrated in RAW 264.7 cells by exposure to $OH^{\bullet}$ radicals generated from the Fenton reaction the inhibitory effect of resveratrol against membrane LP. Likewise, resveratrol and its *trans*-3,4-dihydroxystilbene (3,4-DHS) and *trans*-4,4′-dihydroxystilbene (4,4′-DHS) analogs were effective antioxidants against both 2,2′-azobis(2-amidinopropane hydrochloride) or iron-induced peroxidation of rat liver microsomes with an activity sequence of 3,4-DHS > 4,4′-DHS > resveratrol. These data demonstrated that the *ortho*-diphenoxyl functionality imparts much higher antioxidative activity against ROS-induced peroxidation in membrane mimetic systems [19]. In similar experiments, Blond et al. [20] showed that in nonenzymatic or in NADPH-dependent peroxidation the concentration required to produce 50% inhibition was about three times lower with resveratrol than with the flavonoid quercetin. Additionally, resveratrol was found to prevent metal-induced LP in microsomes and low-density lipoproteins (LDLs). The authors compared the response of resveratrol to that of the polyphenols astringin and astringinin, and found that the presence of the *para*-hydroxy moiety in ring B and the *meta*-hydroxy moiety in ring A (Figure 3.2) is essential for the antioxidant activity of stilbenes [21]. Similarly, Sun et al. [22] demonstrated a significant protective effect of resveratrol from oxidative stress in rat adrenal pheochromocytoma cells exposed to lipid oxidation induced by iron and ethanol. Resveratrol has also been shown to scavenge peroxyl and hydroxyl radicals in reperfused postischemic isolated rat hearts, to reduce infarct size and to reduce the formation of MDA [23,24].

Further evidence for the anti-LP activity of resveratrol was reported in peripheral blood mononuclear cells isolated *ex vivo* from healthy humans. Addition of resveratrol to the culture medium reduced the intracellular LP caused by dR [11]. More recent data provide an interesting insight into the effect of this compound on rat liver LP in NADPH, Fe-ascorbate, and Fe-microsomal systems. Inhibition of LP by resveratrol was apparently due to the hydrogen donating properties rather than chelation of iron [18].

The inhibition of LP by resveratrol has also been shown in *in vivo* assays. For example, it has been reported that after spinal cord injury there is augmented ROS generation. In a recent study, the administration of resveratrol in spinal-cord-injured rats significantly reduced the MDA content as an index of LP, indicating that resveratrol might have a significant effect on the pathway [25].

To define the molecular mechanism(s) of resveratrol inhibition of LP, Tadolini et al. [26] utilized the Fe^{2+}-catalyzed lipid hydroperoxide-dependent peroxidation of sonicated phosphatidylcholine liposomes and peroxidation initiated by lipid-soluble AMVN peroxyl radicals. Their results indicated that resveratrol inhibits LP mainly by scavenging lipid peroxyl radicals within the membrane, like α-tocopherol.

EFFECTS OF RESVERATROL ON INTRACELLULAR REDOX STATE

Recent data have provided interesting insight into the effect of resveratrol on intracellular redox state. These data seem to support both anti- and pro-oxidant activity of this compound, depending on the concentration of resveratrol and the cell type. For example, Ahmad et al. [27] observed that exposure of human leukemia cells to low concentrations of resveratrol (4 to 8 μM) inhibited caspase activation and DNA fragmentation induced by incubation with H_2O_2. At these concentrations, resveratrol elicited prooxidant properties as evidenced by an increase in intracellular $O_2^{\bullet-}$ concentration. This prooxidant effect was further supported by the observation that the drop in intracellular $O_2^{\bullet-}$ and cytosolic acidification induced by H_2O_2 was completely blocked in cells preincubated with resveratrol. Likewise, in rat hepatocytes exposed to ferrylmyoglobin-induced oxidative stress, physiological concentrations (100 pM to 100 nM) of resveratrol exerted prooxidant activities [28]. In tumor cell line cultures, it has also shown a prooxidative effect on DNA damage during interaction with ADP-Fe^{3+} in the presence of H_2O_2 [29].

Similarly, the prooxidant effects of resveratrol were shown on rat liver microsomal LP and $OH^{\bullet}$ production in NADPH, Fe-ascorbate, and Fe-microsomal systems. Resveratrol inhibited LP; however, it increased $OH^{\bullet}$ generation, indicating that $OH^{\bullet}$ played a minor role in LP [18]. In addition, it is well known that heme (iron–protoporphyrin IX) is a prooxidant and its rapid degradation by heme oxygenase is believed to be neuroprotective. Using primary neuronal cultures, resveratrol was able to induce significantly heme oxygenase 1. This study indicated that the increase of heme oxygenase activity by resveratrol is a unique pathway by which this compound can exert its neuroprotective actions [30].

Further corroborating the prooxidant activity of resveratrol are data demonstrating its ineffectiveness in protecting proteins (bovine serum albumin) from oxidative damage induced by metal-catalyzed reactions or alkylperoxyl radicals [31].

ANTIOXIDANT ACTIVITY OF RESVERATROL AND CARDIOPROTECTION

Resveratrol is considered to be one of the major antioxidant constituents in red wine [21], which contains 6.5 mg/l of resveratrol [32]. A number of epidemiological studies have suggested that the mortality rate from coronary heart disease can be decreased by moderate consumption of 150 to 300 ml/day of red wines [33,34]. Antioxidant polyphenolic compounds, relatively abundant in red wines, have been considered to be mainly responsible for such epidemiological observations [35].

Atherosclerosis is a complex process between monocytes/macrophages, endothelial cells, and smooth muscle cells. This disease is characterized by lipid accumulation in the arterial wall, and starts with the attraction, recruitment, and activation of different cell types which provokes a local inflammatory response. Several significant targets contribute to the development of atherosclerotic lesions. The major ones are LDLs which are generated in the circulation by remodeling of very low-density lipoproteins (VLDLs) through the mechanisms of lipolysis and the exchange of lipids and proteins with high-density lipoproteins (HDLs). During transit in the circulation, oxidative changes may occur affecting the lipids. Although the exact mechanism whereby LDLs are oxidized *in vivo* is unclear, several studies indicate that LDL oxidative modification is a result of the interaction between the products of polyunsaturated fatty acid (PUFA) oxidation and apolipoprotein B-100 molecules (Apo B-100). The process is probably initiated by $OH^{\bullet}$ which shows high affinity for the PUFA of an LDL molecule. Besides, the oxidation may also affect the free lysine groups of apolipoprotein B (APO B) of LDL particles impairing their catabolism by the regulated APO B/E receptor system [36]. Oxidized LDL (ox-LDL) stimulates its own uptake by a class of receptors named "scavenger receptors" which are expressed in macrophages as well as in "nonprofessional" phagocytes. Unlike the physiologic LDL receptor, the scavenger receptor is not downregulated when the cell cholesterol content increases and the process leads to abnormal accumulation of cholesterol, cell activation, and transformation [37]. Uptake of ox-LDL (foam cells) by subendothelial macrophages is the initial process of the formation of atherosclerotic plaque [38].

In recent studies carried out with human hepatocarcinoma cell-line HepG2, which retain most of the functions of normal liver parenchymal cells, the addition of resveratrol to the culture medium resulted in a significant decrease in the intracellular concentration of APO B and its secretion which may be responsible for impaired LDL and part VLDL synthesis [39].

There is evidence to support the suggestion that resveratrol interferes with a number of antioxidant mechanisms leading to the inhibition of the development of atherosclerosis. As a matter of fact, resveratrol may protect LDL molecules against peroxidation through antioxidative activity and metal chelation [40]. Frankel et al. [41] were the first to demonstrate that *trans*-resveratrol added to human LDL reduced copper-catalyzed oxidation. The high capacity of the stilbene to chelate copper is potentially useful *in vivo* since LDLs are known to have a high ability to bind copper, preventing oxidative modification of LDL [36,41]. Fremont [36] found that the chelating capacity of the *cis* isomer was about half that of the *trans* isomer, suggesting that the special position of hydroxyl groups is of prime importance for chelation of copper.

Resveratrol was also shown to be a potent inhibitor of the oxidation of PUFA found in LDLs [42]. Further evidence for the protective effect this compound on atherogenic changes was reported in a study in which resveratrol was found to normalize dose-dependently the cellular uptake of oxy-LDL in the vascular wall via the B/E receptor system [36]. Ferroni et al. [37] showed that resveratrol can protect LDL aromatic residues against peroxynitrite-mediated oxidation, even in the presence of physiological concentrations of CO_2. However, some *in vivo* studies using hypercholesterolemic rabbits have not provided evidence for antilipogenic effects of resveratrol [43,44]. In addition, Ray et al. [23] found that resveratrol possessed cardioprotective effects through its peroxyl radical scavenging activity and inhibiting LP to reduce MDA content [23,45].

Blood platelets are found at the sites of early atherosclerotic lesions. When activated, blood platelets may produce ROS and secrete potent mitogenic factors such as platelet-derived growth factor, which lead to smooth muscle proliferation and progression of atherosclerotic lesions. Blood platelets are the target cells of ROS action in the local environment; ROS function as signaling molecules in the stimulation of platelet activation [46]. Preliminary results obtained by Olas et al. [47,48] showed that resveratrol had a protective effect against ROS production in resting and activated blood platelets. Olas et al. [46] have investigated the effects of resveratrol on oxidative stress in pig blood platelets. Resveratrol at doses close to those potentially achievable in blood after red wine consumption (0.05 to 2 μM) caused the inhibition of $O_2^{\bullet -}$ generation and reduced LP in a dose-dependent manner. Resveratrol also reduced oxidative stress induced by vitamin C at a pro-oxidative dose (100 μM).

EFFECTS OF RESVERATROL ON NITROGEN REACTIVE SPECIES GENERATION

It is now widely accepted that NO appears to play a cardioprotective role in the prevention of initiation of atherosclerosis. Under physiological conditions a moderate upregulation of eNOS is associated with beneficial effects, e.g., neuroprotective and cardiovascular effects. NO maintains coronary vasodilatory tone and inhibits adhesions of neutrophils and platelets to vascular endothelium. Several reports have shown a role for resveratrol in the regulation of the production of NO from vascular endothelium [49,50].

Abnormally high concentrations of NO and its derivatives peroxynitrite or nitrogen dioxide are devastating agents causing another form of cellular stress based on the generation of reactive nitrogen species (RNS) which causes cellular degeneration in various tissues, contributing to the development of neurodegenerative damage. In addition, increased NOS induction and activity have been associated with tumor growth and vascular invasion. Lorenz et al. [5] have investigated the effects of resveratrol and another natural hydroxystilbene, oxyresveratrol, on nitrosative and oxidative stress derived from microglial cells. Incubation with resveratrol considerably diminished NO production upon expression of the inducible isoform of NOS (iNOS). It also induced an inhibitory effect on the iNOS enzyme activity.

Bacterial endotoxic LPS is one of the most important stimuli for iNOS induction, resulting in NO production that has bactericidal effects. For example, in LPS-activated RAW 264.7 macrophages, preincubation of cells with resveratrol reduced inflammation by downregulation of iNOS and mRNA [51,52]. Similar data were found by Matsuda et al. [53] for mouse peritoneal exudate macrophages activated by LPS. In this model, resveratrol also inhibited NO production. Similarly, in another study using rat macrophages stimulated with thioglycollate, resveratrol at concentrations of 10 to 100 μM significantly and dose-dependently inhibited reactive nitrogen intermediate production. The results obtained demonstrate that resveratrol is a potent inhibitor of the antipathogen responses of rat macrophages and, thus, suggest that this agent may have applications in the treatment of diseases involving macrophage hyperresponsiveness [54].

ANTIOXIDANT ACTIVITY OF RESVERATROL AND HEPATOPROTECTION

Liver microsomes, especially smooth-surfaced endoplasmic reticulum, contain large amounts of PUFA. For this reason, they are particularly susceptible to oxidative stress and have been widely used as a model for oxidative stress and antioxidant studies [55,56]. Cai et al. [19] have shown the inhibition of free radical-induced peroxidation of rat liver microsomes by resveratrol and its analogs. The antioxidant mechanism involves trapping the initiating peroxyl ($RO_2^{\bullet}$) radicals and/or hydroxyl ($OH^{\bullet}$) radicals and reducing the α-tocopheroxyl radical ($TO^{\bullet}$) to generate endogenous TOH. More recently, Ozgová et al. [18] showed that resveratrol was able to prevent oxidative damage induced by LP in NADPH, Fe-ascorbate, and Fe-microsomal systems of rat liver, possibly through their hydrogen donating properties.

Nevertheless, in spite of little information available related to the hepatoprotective effects of resveratrol, there is evidence that this polyphenol has been used as a positive free radical scavenger control against hepatotoxic effects induced by tacrine [57] through a mechanism that involves oxidative stress [58].

Stellate cells are now known to play central roles in hepatic fibrogenesis induced by viral infection, alcohol, and various drugs. These cells are the major source of extracellular matrix components (ECM) in normal and pathological conditions [59]. There is evidence that oxidative stress enhances the proliferation of cultured stellate cells and their collagen synthesis [60,61]. Furthermore, an independent stimulation of ECM deposition seems to occur at a prenecrotic stage during oxidative stress-associated liver injury [62,63]. Kawada et al. [64] have shown that resveratrol might inhibit stellate cell activation by perturbing signal transduction pathways and cell cycle protein expression. This study also showed that resveratrol inhibited the production of NO and tumor necrosis factor (TNF)α by LPS-stimulated Kupffer cells. However, mRNA expression for iNOS and TNFα was not affected by resveratrol. This fact suggests that resveratrol might affect the posttranscriptional process of generating these proteins.

ANTIOXIDANT ACTIVITY OF RESVERATROL AND NEUROPROTECTION

Although there are not many studies of the neuroprotective effects of resveratrol, its antioxidant properties suggest a wide variety of possible protective actions.

Arterial thrombosis induced by atherosclerosis is a common cause of cerebral infarct. Several studies have suggested a relation between cerebral recirculation after an ischemic process and ROS [65,66]. The delayed neuronal death that appears as a consequence of a transient ischemia in cerebral ischemic damage, in rodent models, is the result of augmented NO production and the attenuated redox regulatory system. Either of them leads to the development of apoptosis and an increase of oxidative stress, respectively. Some investigators have indicated a potential neuroprotective activity for resveratrol based on its beneficial effects in cerebral ischemia models [67]. For example, it has been evaluated as a neuroprotective agent by virtue of its antioxidant properties, being effective in focal cerebral ischemia caused by middle cerebral artery occlusion [68], preventing motor impairment, increasing levels of MDA and GSH, and decreasing the volume of infarct as compared to control [69].

Other investigators have shown that resveratrol inhibits apoptotic neuron cell death induced by ox-LDLs [70,71] through inhibition of nuclear factor kappa B (NF-κB) and activator protein (AP)-1 pathways [70,72]. Its antiapoptotic effects also involve activation of caspase and degradation of poly(ADP-ribose)-polymerase route through inhibition of the mitochondrial cell death pathway [73]. Resveratrol is also able to protect against excitotoxic brain damage induced by kainic acid administration [74].

Chronic ethanol ingestion is known to cause oxidative damage in the brain, among other organs, as a consequence of the ability of ethanol to enhance ROS production and LP [75–77]. Ethanol may also enhance ROS production in the brain through other mechanisms including: induction of cytochrome P450 2E1 (CYP2E1), alteration of the cytokine signaling pathways for induction of iNOS and secretory phospholipase A_2 ($sPLA_2$), and production of prostanoids through the PLA_2/COX pathways and increasing generation of hydroxyethyl radicals [77]. These radicals are related to XO and 6-hydroxydopamine [78,79] and can cause more cellular damage than $OH^•$ radicals because they are more reactive and have a longer half-life.

The consumption of red wine containing high levels of polyphenolic compounds together with a high-fat diet in France are related to low incidences of coronary heart diseases. This fact, called the "French paradox" [33], has stimulated new investigations into the ability of grape polyphenols to ameliorate neuronal injuries due to chronic ethanol ingestion. Some authors have demonstrated that resveratrol protects neurons against oxidative stress [70,71], and also that it reduces neuron cell death induced by ethanol [22]. Sun and Sun [77] also reported that resveratrol protected the brain from neuronal damage due to chronic ethanol administration and suggested that it might be used as a therapeutic agent to ameliorate neurodegenerative processes.

There is evidence that chronic ethanol intake may contribute to the progression of Alzheimer's disease pathology [80]. However, moderate wine consumption correlates with a lower risk for this disease, suggesting an additional neuroprotective effect for resveratrol [81]. Alzheimer's disease is characterized by the progressive deposition of β-amyloid plaques enriched in amyloid β peptide. These plaques can mediate oxidative stress and induce neuronal toxicity through their ability to disrupt ion homeostasis [82,83] and so contribute to cell death. The protein oxidation and LP resulting from the increase in ROS generated from amyloid plaques affect the functionality of cell membranes [84]. There is evidence that resveratrol is able to restore intracellular GSH levels following oxidative stress induced by β-amyloid plaques [81].

Although the idiopathic cause of Parkinson's disease is unknown, post-mortem studies have shown an important role for oxidative damage and mitochondrial impairment of dopaminergic neurons as a mediator of nerve cell death in Parkinson's disease [85,86]. In this sense, Gélinas and Martinoli [87] have suggested that resveratrol may restore dopamine transporter protein expression to control levels, reversing 1-methyl-4-phenylpyridium (MPP^+) cytotoxicity and preventing neuronal degeneration caused by increased oxidative burden in neuronal PC12 cells.

Other effects of resveratrol in protecting brain cells from injury are related to the antiinflammatory activity that it exerts in astrocytes and their ability to inhibit NO production induced by cytokines [5]. Neuronal damage may be produced by an increase in oxidative and inflammatory mechanisms which are associated with glial cell activation. Astrocytes, the most abundant cell type in the central nervous system (CNS), are capable of responding to proinflammatory cytokines and, under pathological conditions, can generate large amounts of NO through induction of iNOS. The interaction between $O_2^{\bullet -}$ anions and the generated NO induces the production of a potent oxidant compound with cytotoxic effects, $ONOO^-$. This free radical may cause oxidation of biomolecules resulting in alteration of important enzymes and proteins such as glutamine synthetase and synaptophysin [77]. As mentioned above, the antioxidant properties of resveratrol ameliorate intracellular ROS and attenuate hippocampal cell death [81]. Furthermore, resveratrol is able to downregulate the expression of iNOS protein without altering iNOS activity derived from microglial cell activation [5].

In the CNS the heme oxygenase (HO) system has been reported to be active and to operate as a fundamental defensive mechanism for neurons exposed to an oxidant challenge. In this sense, exposure of astrocytes to resveratrol resulted in an increase of HO-1 mRNA, but it was not able to induce HO-1 protein expression and activity [88].

In another study, Czapski et al. [89] showed the efficacy of resveratrol in reducing oxidative stress induced in the brain cortex homogenate by $FeCl_2$ or $CuSO_4$ for determination of LP and protein tyrosine oxidation, respectively. They demonstrated that resveratrol was the most potent among the investigated antioxidants.

Kiziltepe et al. [90] have demonstrated that preischemic infusion of resveratrol protects the spinal cord from ischemia-reperfusion injury in rabbits, probably mediated by its antioxidant and NO promoting properties. For this reason, they suggest the use of resveratrol in humans as a new neuroprotective agent to avoid paraplegia that results from spinal cord ischemia caused by thoracic and thoracoabdominal aorta surgical procedures. Previously, Yang and Piao [25] showed that resveratrol has neuroprotective effects, including protecting axon, neuron, myelin, and subcellular organelles, and reducing local spinal tissue edema as secondary damage of spinal cord injury through improving the energy metabolism system and suppressing LP, preventing nerve functions being exacerbated progressively, preventing mitigating nerve damage, and maybe promoting nerve regeneration.

ANTIOXIDANT ACTIVITY OF RESVERATROL AND CARCINOGENESIS

The fact that antioxidants may prevent or delay the onset of some types of cancer is the basis for proposing that ROS and RNS participate in human cancer development. These agents could contribute to diverse cellular events associated with tumorigenesis causing structural alterations in DNA, affecting cytoplasmic and nuclear signal transduction pathways, and modulating the activity of proteins and genes that regulate cell proliferation-, differentiation-, and apoptosis-related genes (Figure 3.5) [91].

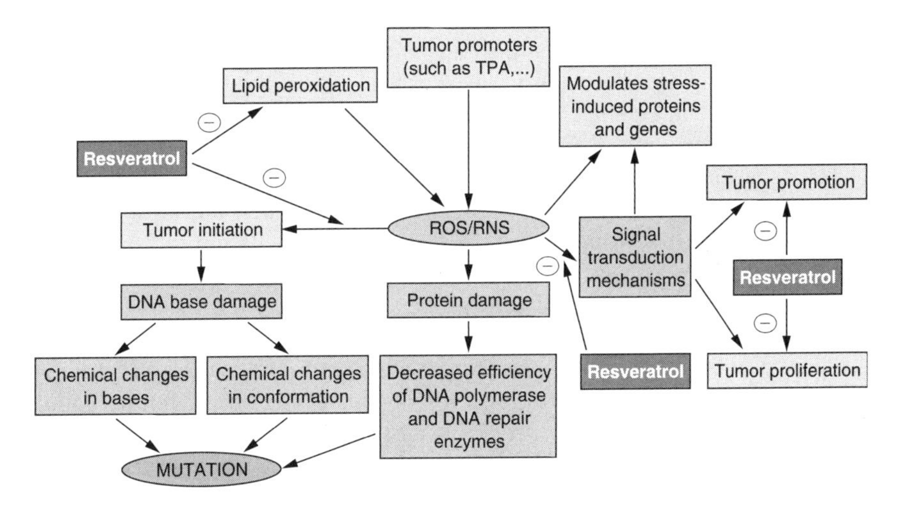

FIGURE 3.5 Antioxidant mechanisms of resveratrol in carcinogenesis.

DNA can undergo attack by different ROS in different ways. Thus, while $OH^{\bullet}$, which is especially damaging, attacks all four DNA bases, singlet oxygen, $^{1}O_2$, selectively attacks guanine. However, $O_2^{\bullet -}$ and H_2O_2 do not react with DNA bases at all, but both of them can release iron from ferritin or heme proteins, respectively, and this metal ion could then bind to DNA damaging it as a consequence of a chelating ability of DNA. Other radicals such as $RO_2^{\bullet}$, $RO^{\bullet}$, HNO_2, O_3, and $ONOO^-$ and its decomposition products are also effective in damaging DNA. For example, $ONOO^-$ itself is probably directly damaging to DNA bases or can induce the release of intracellular copper ions which may then bind to DNA. $ONOO^-$ can also inhibit phosphorylation through nitration of tyroxine residues [91].

Thus, DNA damage caused by ROS/RNS can induce multiple lesions, including strand breaks, AP sites, and modified purine and pyrimidines. The chemical modification of DNA bases induces a change in their hydrogen bonding specificity with subsequent mutation. The most frequent mutations found after the attack of DNA by ROS oxidation are C to T transitions and C to G transversions. Local DNA structure can change due to a nonplanar disposition of the oxidized bases. Thereby, not only changes in bases could be produced, but also changes in DNA conformation [91].

Since gene transcription can be regulated by oxidants, antioxidants, and other determinants of the intracellular redox state [92], ROS/RNS can also produce protein damage inducing other types of mutations. Thus, several RNS such as $ONOO^-$ and $NO^{\bullet}_2$ can attack proteins, nitrating aromatic amino acid residues and possibly affecting their ability to participate in signal transduction mechanisms. For example, H_2O_2 is able to induce the activation of the transcription factor NF-κB through the displacement of the inhibitory subunit, or stimulate the transcription or the activation of stress-induced proteins and genes. Furthermore, oxidation by this reactive agent not only could affect the activity of DNA repair enzymes as a consequence of protein damage, but also produce a range of mutagenic carbonyl products through ROS attack on lipids [91].

The suggestion that resveratrol affects tumor cells *in vitro* and *in vivo* is based on a large number of publications in recent years. Nevertheless, the precise mechanism through which resveratrol exerts its antitumor effects and the molecule targets remain unclear and seem to depend on cell type and tumor model. Thus, resveratrol has been suggested as a potential chemopreventive agent based on the inhibitory effects that it exerts on cellular events involved in the three stages of carcinogenesis: tumor initiation, promotion, and progression [16]. Resveratrol prevents the initial DNA damage by two different pathways: (i) acting as an antimutagen through the induction of phase II enzymes, such as quinone reductase, capable of metabolically detoxifying carcinogens by inhibiting the COX and cytochrome P450 1A1 enzymes known to be able to convert substances into mutagens; and (ii) acting as an antioxidant through inhibition of DNA damage by free radicals [93].

It has been proposed that free radicals derived from LP may function as tumor initiators. Thus, the products of LP, such as MDA and other aldehydes, may also cause DNA damage [8]. Recently, Leonard et al. [8] have shown that resveratrol exhibits a protective effect against LP in cell membranes and DNA damage caused by ROS. However, resveratrol has also demonstrated a prooxidative effect on DNA damage during its interaction with adenosine 5′-diphosphate (ADP)-Fe^{3+} in the presence of H_2O_2 because under these conditions strand breaks of DNA are stimulated [94].

The antipromotional and anticarcinogenic properties of resveratrol can be partly attributed to its ability to enhance gap-junctional intercellular communications in cells exposed to tumor promoters such as TPA [29]. The tumor promoting activity mediated by TPA has also been associated with oxidative stress by increased production of $O_2^{\bullet-}$ anion and H_2O_2, reduction of SOD activity, and interference with glutathione metabolism. In a model of TPA application to mouse skin, Jang and Pezzuto [16] showed that pretreatment with resveratrol induced the restoration of H_2O_2 and glutathione levels to control levels, as with myeloperoxidase, GSSG-reductase, and SOD activities, which were altered by TPA induction.

The development of skin cancer is related to cumulative exposure to solar UVB as well as the nuclear transcription factor NF-κB, which plays a critical role in skin biology. Resveratrol is able to block the damage caused by UVB exposure via its antioxidant properties inhibiting ultraviolet light-induced LP or blocking UVB-mediated activation of NF-κB. This important factor is involved in the inflammatory and carcinogenic signaling cascades, and resveratrol is able to prevent the activation of NF-κB caused by free radicals: resveratrol has a significant inhibitory effect on the NF-κB signaling pathway after cellular exposure to metal-induced radicals [8]. Resveratrol may also protect cells against endotoxin-induced inflammation by preventing NF-κB activation through the blockage of IκB kinase activity. As tissue inflammation is provoked by tumor promoters and serves as a driving force in tumor promotion, the use of antiinflammatory agents as chemopreventives at this stage of carcinogenesis has also been evaluated [95]. Resveratrol has been shown to exert antiinflammatory activity, reducing arachidonic acid release and induction of COX-2 by an antioxidant action [12].

Finally, resveratrol also possesses chemotherapeutic potential because it is able to inhibit tumor progression. Indeed, it suppresses growth of various cancer cell lines, partly by an inhibition of DNA polymerase and inhibition of deoxyribonucleotide synthesis through its ability to scavenge the essential tyrosyl radical of the ribonucleotide reductase [96] and partly by inducing cell cycle arrest [29].

REFERENCES

1. Ignatowicz E and Baer-Dubowska W, Resveratrol, a natural chemopreventive agent against degenerative diseases, *Pol J Pharmacol* 53(6), 557–569, 2001.
2. Pervaiz S, Resveratrol: from the bottle to the bedside?, *Leuk Lymphoma* 40, 491–498, 2001.
3. Bergamini CM, Gambetti S, Dondi A, and Cervellati C, Oxygen, reactive oxygen species and tissue damage, *Curr Pharm Design* 10, 1611–1626, 2004.
4. Beckman JS, Beckman TW, Chen J, Marshall PA, and Freeman BA, Apparent hydroxyl radical production by peroxynitrite: implications for endothelial injury from nitric oxide and superoxide, *Proc Natl Acad Sci USA* 87, 1620–1624, 1990.
5. Lorenz P, Roychowdhury S, Engelmann M, Wolf G, and Horn TF, Oxyresveratrol and resveratrol are potent antioxidants and free radical scavengers: effect on nitrosative and oxidative stress derived from microglial cells, *Nitric Oxide* 9, 64–76, 2003.
6. Nose K, Role of reactive oxygen species in the regulation of physiological functions, *Biol Pharm Bull* 23, 897–903, 2000.
7. Inoue M, Sato EF, Nishikawa M, Park AM, Kira Y, Imada I, and Utsumi K, Mitochondrial generation of reactive oxygen species and its role in aerobic life, *Curr Med Chem* 10, 2495–2505, 2003.
8. Leonard S, Xia C, Jiang BH, Stinefelt B, Klandorf H, Harris GK, and Shi X, Resveratrol scavenges reactive oxygen species and effects radical-induced cellular responses, *Biochem Biophys Res Commun* 309, 1017–1026, 2003.
9. Zini R, Morin C, Bertelli A, Bertelli AA, and Tillement JP, Effects of resveratrol on the rat brain respiratory chain, *Drugs Exp Clin Res* 25, 87–97, 1999.
10. Orallo F, Alvarez E, Camina M, Leiro JM, Gomez E, and Fernandez P, The possible implication of trans-resveratrol in the cardioprotective effects of long-term moderate wine consumption, *Mol Pharmacol* 61, 294–302, 2002.
11. Losa GA, Resveratrol modulates apoptosis and oxidation in human blood mononuclear cells, *Eur J Clin Invest* 33, 818–823, 2003.
12. Martínez J and Moreno JJ, Effect of resveratrol, a natural polyphenolic compound, on reactive oxygen species and prostaglandin production, *Biochem Pharmacol* 59, 865–870, 2000.
13. Jang DS, Kang BS, Ryu SY, Chang IM, Min KR, and Kim Y, Inhibitory effects of resveratrol analogs on unopsonized zymosan-induced oxygen radical production, *Biochem Pharmacol* 57, 705–712, 1999.
14. Reiter RJ, Tan DX, Gitto E, Sainz RM, Mayo JC, Leon J, Manchester LC, Vijaylaxmi, Kilic E, and Kilic U, Pharmacological utility of melatonin in reducing oxidative cellular and molecular damage, *Pol J Pharmacol* 56, 159–170, 2004.
15. Olas B, Wachowicz B, Bald E, and Glowacki R, The protective effects of resveratrol against changes in blood platelet thiols induced by platinum compounds, *J Physiol Pharmacol* 55, 467–476, 2004.
16. Jang M and Pezzuto JM, Cancer chemopreventive activity of resveratrol, *Drug Exp Clin Res* 25, 65–77, 1999.

17. Yen GC, Duh PD, and Lin CW, Effects of resveratrol and 4-hexylresorcinol on hydrogen peroxide-induced oxidative DNA damage in human lymphocytes, *Free Radical Res* 37, 509–514, 2003.
18. Ozgová S, Hermanek J, and Gut I, Different antioxidant effects of polyphenols on lipid peroxidation and hydroxyl radicals in the NADPH-, Fe-ascorbate- and Fe-microsomal systems, *Biochem Pharmacol* 66, 1127–1137, 2003.
19. Cai YJ, Fanf JG, Ma LP, Yang L, and Liu ZL, Inhibition of free radical-induced peroxidation of rat liver microsomes by resveratrol and its analogues, *Biochem Biophys Acta* 1637, 31–38, 2003.
20. Blond JP, Denis MP, and Bezard J, Antioxidant action of resveratrol in lipid peroxidation, *Sci Aliments* 15, 347–358, 1995.
21. Fauconneau B, Waffo-Teguo P, Huquet F, Barrier L, Decendit A, and Merillon JM, Comparative study of radical scavenger and antioxidant properties of radical scavenger and antioxidant properties of phenolic compounds from Vitis vinifera cell cultures using *in vitro* tests, *Life Sci* 61, 2103–2110, 1997.
22. Sun AY, Chen YM, James-Kracke M, Wixom P, and Cheng Y, Ethanol-induced cell death by lipid peroxidation in PC12 cells, *Neurochem Res* 22, 1187–1192, 1997.
23. Ray PS, Maulik G, Cordis GA, Bertelli AA, Bertelli A, and Das DK, The red wine antioxidant resveratrol protects isolated rat hearts from ischemia reperfusion injury, *Free Radical Biol Med* 27, 160–169, 1999.
24. Sato M, Maulik G, Bagchi D, and Das DK, Myocardial protection by protykin, a novel extract of trans-resveratrol and emodin, *Free Radical Res* 32, 135–144, 2000.
25. Yang YB and Piao YJ, Effect of resveratrol on secondary damages after acute spinal cord injury in rats, *Acta Pharmacol Sin* 24, 703–710, 2003.
26. Tadolini B, Juliano C, Piu L, Franconi F, and Cabrini L, Resveratrol inhibition of lipid peroxidation, *Free Radical Res* 33, 105–114, 2000.
27. Ahmad KA, Clement MV, and Pervaiz S, Pro-oxidant activity of low doses of resveratrol inhibits hydrogen peroxide-induced apoptosis, *Ann NY Acad Sci* 1010, 365–373, 2003.
28. Martínez R, Quintana K, Navarro R, Martin C, Hernandez ML, Aurrekoetxea I, Ruiz-Sanz JI, Lacort M, and Ruiz-Larrea MB, Pro-oxidant and antioxidant potential of catecholestrogens against ferrylmyoglobin-induced oxidative stress, *Biochim Biophys Acta* 1583, 167–175, 2002.
29. Gusman J, Malonne H, and Atassi G, A reappraisal of the potential chemopreventive and chemotherapeutic properties of resveratrol, *Carcinogenesis* 22, 1111–1117, 2001.
30. Zhuang H, Kim YS, Koehler RC, and Dore S, Potential mechanism by which resveratrol, a red wine constituent, protects neurons, *Ann NY Acad Sci* 993, 276–286; discussion 287–288, 2003.
31. Mayo JC, Tan DX, Sainz RM, Natarajan M, Lopez-Burillo S, and Reiter RJ, Protection against oxidative protein damage induced by metal-catalyzed reaction or alkylperoxyl radicals: comparative effects of melatonin and other antioxidants, *Biochim Biophys Acta* 1620, 139–150, 2003.
32. Privat C, Telo JP, Bernardes-Genisson V, Vieira A, Souchard JP, and Nepveu F, Antioxidant properties of trans-epsilon-vinifern as compared to

stilbene derivatives in aqueous and nonaqueous media, *J Agric Food Chem* 50, 1213–1217, 2002.
33. Renaud S and De Lorgeril M, Wine, alcohol, platelets and the French paradox for coronary heart disease, *Lancet* 339, 1523–1526, 1992.
34. Renaud SC, Gueguen R, Schenker J, and d'Houtaud A, Alcohol and mortality in middle-aged men from Eastern France, *Epidemiology* 9, 184–188, 1998.
35. Constant J, Alcohol, ischemic heart disease, and the French paradox, *Clin Cardiol* 20; 420–424, 1997.
36. Fremont L, Biological effects of resveratrol, *Life Sci* 66, 663–673, 2000.
37. Ferroni F, Maccaglia A, Pietraforte D, Turco L, and Minetti M, Phenolic antioxidants and the protection of low density lipoprotein from peroxynitrite-mediated oxidations at physiologic CO_2, *J Agric Food Chem* 52, 2866–2874, 2004.
38. Steinberg D, Parthasarathy S, Carew TE, Khoo JC, and Witztum JL, Beyond cholesterol. Modifications of low-density lipoprotein that increase its atherogenicity, *N Engl J Med* 320, 915–924, 1989.
39. Soleas GJ, Diamandis EP, and Goldberg DM, Resveratrol: a molecule whose time has come? And gone?, *Clin Biochem* 30, 91–113, 1997.
40. Belguendouz L, Fremont L, and Linard A, Resveratrol inhibits metal ion-dependent and independent peroxidation of porcine low-density lipoproteins, *Biochem Pharmacol* 53, 1347–1355, 1997.
41. Frankel EN, Waterhouse AL, and Kinsella JE, Inhibition of human LDL oxidation by resveratrol, *Lancet* 341, 1103–1104, 1993.
42. Miller NJ and Rice-Evans CA, Antioxidant activity of resveratrol in red wine, *Clin Chem* 41, 1789, 1995.
43. Wilson T, Knight TJ, Beitz DC, Lewis DS, and Engen RL, Resveratrol promotes atherosclerosis in hypercholesterolemic rabbits, *Life Sci* 59, PL15–21; 1996.
44. Turrens JF, Lariccia J, and Nair MG, Resveratrol has no effect on lipoprotein prolife and does not prevent peroxidation of serum lipids in normal rats, *Free Radical Res* 27, 557–562, 1997.
45. Mizutani K, Ikeda K, Kawai Y, and Yamori Y, Protective effect of resveratrol on oxidative damage in male and female stroke-prone spontaneously hypertensive rats, *Clin Exp Pharmacol Physiol* 28, 55–59, 2001.
46. Olas B, Wachowicz B, Stochmal A, and Oleszek W, Inhibition of oxidative stress in blood platelets by different phenolics from Yucca schidigera Roezl. bark, *Nutrition* 19, 633–640, 2003.
47. Olas B, Zbikowska HM, Wachowicz B, Krajewski A, and Magnuszewska A, Inhibitory effect of resveratrol on free radical generation in blood platelets, *Acta Biochem Pol* 46, 961–966, 1999.
48. Olas B, Wachowicz B, Szewczuk J, Saluk-Juszczak J, and Kaca W, The effect of resveratrol on the platelet secretory process induced by endotoxin and thrombin, *Microbios* 105, 7–13, 2001.
49. Hung LM, Chen JK, Huang SS, Lee RS, and Su MJ, Cardioprotective effect of resveratrol, a natural antioxidant derived from grapes, *Cardiovasc Res* 47, 549–555, 2000.
50. Hattori R, Otani H, Maulik N, and Das DK, Pharmacological preconditioning with resveratrol: role of nitric oxide, *Am J Physiol Heart Circ Physiol* 282, H1988–H1995, 2002.

51. Tsai SH, Lin-Shiau SY, and Lin JK, Suppression of nitric oxide synthetase and the down-regulation of the activation of NF Kappa B in macrophages by resveratrol, *Br J Pharmacol* 126, 673–680, 1999.
52. Wadsworth TL and Koop DR, Effects of the wine polyphenolics quercetin and resveratrol on pro-inflammatory cytokine expression in RAW 264.7 macrophages, *Biochem Pharmacol* 57, 941–949, 1999.
53. Matsuda H, Kageura T, Morikawa T, Toguchida I, Harima S, and Yoshikawa M, Effects of stilbene constituents from rhubarb on nitric oxide production in lipopolysaccharide-activated macrophages, *Bioorg Med Chem Lett* 10, 323–327, 2000.
54. Leiro J, Alvarez E, Garcia D, and Orallo F, Resveratrol modulates rat macrophage functions, *Int Immunopharmacol* 2, 767–774, 2002.
55. Mehlhorn RJ, Sumida S, and Packer L, Tocopheroxyl radical persistence and tocopherol consumption in liposomes and in vitamin E-enriched rat liver mitochondria and microsomes, *J Biol Chem* 264, 13448–13452, 1989.
56. Liebler DC, Stratton SP, and Kaysen KL, Antioxidant actions of beta-carotene in liposomal and microsomal membranes: role of carotenoid-membrane incorporation and alpha-tocopherol, *Arch Biochem Biophys* 338, 244–250, 1997.
57. Song EK, Cho H, Kim JS, Kim NY, An NH, Kim JA, Lee SH, and Kim YC, Diarylheptanoids with free radical scavenging and hepatoprotective activity *in vitro* from Curcuma longa, *Planta Med* 67, 876–877, 2001.
58. Osseni RA, Debbasch C, and Christen MO, Tacrine-induced reactive oxygen species in human liver cell line: the role of anethole dithiolethione as a scavenger, *Toxicol In Vitro* 13, 683–688, 1999.
59. Friedman SL, The cellular basis of hepatic fibrosis, *N Engl J Med* 328, 1828–1835, 1993.
60. Casini A, Ceni E, Salzano R, Milani S, Schuppan D, and Surrenti C, Acetaldehyde regulates the gene expression of matrix metalloproteinase-1 and -2 in human fat-storing cells, *Life Sci* 55, 1311–1316, 1994.
61. Lee KS, Buck M, Houglum K, and Chojkier M, Activation of hepatic stellate cells by TGFα and collagen type I is mediated by oxidative stress through c-*myb* expression, *J Clin Invest* 96, 2461–2468, 1995.
62. Pietrangelo A, Metals, oxidative stress and hepatic fibrogenesis, *Semin Liver Dis* 16, 13–30, 1996.
63. Reeves HL, Burt AD, Wood S, Day CI, Hepatic stellate cell activation occurs in the absence of hepatitis in alcoholic liver disease and correlates with the severity of steatosis, *J Hepatol* 25, 677–683, 1996.
64. Kawada N, Seki S, Inoue M, and Kuroki T, Effect of antioxidants, resveratrol, quercetin, and N-acetylcysteine, on the functions of cultured rat hepatic stellate cells and Kupffer cells, *Hepatology* 27, 1265–1274, 1998.
65. Spranger M, Krempien S, Schwab S, Donneberg S, and Hacke W, Superoxide dismutase activity in serum of patients with acute cerebral ischemic injury. Correlation with clinical course and infarct size, *Stroke* 28, 2425–2428, 1997.
66. Chang CY, Lai YC, Cheng TJ, Lau MT, and Hu ML, Plasma levels of antioxidant vitamins, selenium, total sulfhydryl groups and oxidative products in ischemic-stroke patients as compared to matched controls in Taiwan, *Free Radical Res* 28, 15–24, 1998.

67. Wang Q, Xu J, Rottinghaus GE, Simonyi A, Lubahn D, Sun GY, and Sun AY, Resveratrol protects against global cerebral ischemic injury in gerbils, *Brain Res* 958, 439–447, 2002.
68. Huang SS, Tsai MC, Chih CL, Hung LM, and Tsai SK, Resveratrol reduction of infarct size in Long-Evans rats subjected to focal cerebral ischemia, *Life Sci* 69, 1057–1065, 2001.
69. Sinha K, Chaudhary G, and Gupta YK, Protective effect of resveratrol against oxidative stress in middle cerebral artery occlusion model of stroke in rats, *Life Sci* 71, 655–665, 2002.
70. Draczynska-Lusiak B, Chen YM, and Sun AY, Oxidized lipoproteins activate NF-kappaB binding activity and apoptosis in PC12 cells, *Neuroreport* 9, 527–532, 1998.
71. Draczynska-Lusiak B, Doung A, and Sun AY, Oxidized lipoproteins may play a role in neuronal cell death in Alzheimer disease, *Mol Chem Neuropathol* 33, 139–148, 1998.
72. Yu R, Hebbar V, Kim DW, Mandlekar S, Pezzuto JM, and Kong AN, Resveratrol inhibits phorbol ester and UV-induced activator protein 1 activation by interfering with mitogen-activated protein kinase pathways, *Mol Pharmacol* 60, 217–224, 2001.
73. Nicolini G, Rigolio R, Miloso M, Bertelli AA, and Tredici G, Anti-apoptotic effect of trans-resveratrol on paclitaxel-induced apoptosis in the human neuroblastoma SH-SY5Y cell line, *Neurosci Lett* 302, 41–44, 2001.
74. Virgili M and Contestabile A, Partial neuroprotection of *in vivo* excitotoxic brain damage by chronic administration of the red wine antioxidant agent, trans-resveratrol in rats, *Neurosci Lett* 281, 123–126, 2000.
75. Montoliu C, Sancho-Tello M, Azorin I, Burgal M, Valles S, Renau-Piqueras J, and Guerri C, Ethanol increases cytochrome P4502E1 and induces oxidative stress in astrocytes, *J Neurochem* 65, 2561–2570, 1995.
76. Xia J, Simonyi A, and Sun GY, Chronic ethanol and iron administration on iron content, neuronal nitric oxide synthase, and superoxide dismutase in rat cerebellum, *Alcohol Clin Exp Res* 23, 702–707, 1999.
77. Sun AY and Sun GY, Ethanol and oxidative mechanisms in the brain, *J Biomed Sci* 8, 37–43, 2001.
78. Ahmad FF, Cowan DL, and Sun AY, Spin trapping studies of the influence of alcohol on lipid peroxidation, in *Biochemical Mechanisms of Alcoholism*, Sun GY et al., Eds, Humana Press, Clifton, NJ, 1989, pp. 215–226.
79. Oldfield FF, Cowan DL, and Sun AY, The involvement of ethanol in the free radical reaction of 6-hydroxydopamine, *Neurochem Res* 16, 83–87, 1991.
80. Freund G, and Ballinger WE Jr, Alzheimer's disease and alcoholism: possible interactions, *Alcohol* 9, 233–240, 1992.
81. Savaskan E, Olivieri G, Meier F, Seifritz E, Wirz-Justice A, and Muller-Spahn F, Red wine ingredient resveratrol protects from beta-amyloid neurotoxicity, *Gerontology* 49, 380–383, 2003.
82. Yankner BA, Mechanisms of neuronal degeneration in Alzheimer's disease, *Neuron* 16, 921–932, 1996.
83. Lovell MA, Robertson JD, Teesdale WJ, Campbell JL, and Markesbery WR, Copper, iron and zinc in Alzheimer's disease senile plaques, *J Neurol Sci* 158, 47–52, 1998.

84. Yatin SM, Aksenova M, Aksenov M, Markesbery WR, Aulick T, and Butterfield DA, Temporal relations among amyloid beta-peptide-induced free-radical oxidative stress, neuronal toxicity, and neuronal defensive responses, *J Mol Neurosci* 11, 183–197, 1998.
85. Jenner P, Oxidative mechanisms in nigral cell death in Parkinson's disease, *Movement Disorders* 13 (Suppl 1), 24–34, 1998.
86. Zhang Y, Dawson VL, and Dawson TM, Oxidative stress and genetics in the pathogenesis of Parkinson's disease, *Neurobiol Dis* 7, 240–250, 2000.
87. Gélinas S and Martinoli MG, Neuroprotective effect of estradiol and phytoestrogens on MPP^+-induced cytotoxicity in neuronal PC12 cells, *J Neurosci Res* 70, 90–96, 2002.
88. Scapagnini G, Butterfield DA, Colombrita C, Sultana R, Pascale A, and Calabrese V, Ethyl ferulate, a lipophilic polyphenol, induces HO-1 and protects rat neurons against oxidative stress, *Antioxid Redox Signal* 6, 811–818, 2004.
89. Czapski G, Cakala M, Kopczuk D, and Strosznajder JB, Effect of poly(ADP-ribose) polymerase inhibitors on oxidative stress evoked hydroxyl radical level and macromolecules oxidation in cell free system of rat brain cortex, *Neurosci Lett* 356, 45–48, 2004.
90. Kiziltepe U, Turan NH, Han U, Ulus AT, and Akar F, Resveratrol, a red wine polyphenol, protects spinal cord from ischemia-reperfusion injury, *J Vasc Surg* 40, 138–145, 2004.
91. Wiseman H and Halliwell B, Damage to DNA by reactive oxygen and nitrogen species: role in inflammatory disease and progression to cancer, *Biochem J* 313, 17–29, 1996.
92. Sen CK and Packer L, Antioxidant and redox regulation of gene transcription, *FASEB J* 10, 709–720, 1996.
93. Roemer K and Mahyar-Roemer M, The basis for the chemopreventive action of resveratrol, *Drugs Today* 38, 571–580, 2002.
94. Miura T, Muraoka S, Ikeda N, Watanabe M, and Fujimoto Y, Antioxidative and prooxidative action of stilbene derivatives, *Pharmacol Toxicol* 86, 203–208, 2000.
95. Hursting SD, Slaga TJ, Fischer SM, DiGiovanni J, and Phang JM, Mechanism-based cancer prevention approaches: targets, examples and the use of transgenic mice, *J Natl Cancer Inst* 91, 215–225, 1999.
96. Fontecave M, Lepoivre M, Elleingand E, Gerez C, and Guittet O, Resveratrol, a remarkable inhibitor of ribonucleotide reductase, *FEBS Lett* 421, 277–279, 1998.

4 Resveratrol as an Antiproliferative Agent for Cancer

Paola Signorelli and Riccardo Ghidoni

CONTENTS

INTRODUCTION

Tumors of diverse histological origin have specific as well as common molecular markers involved in their initiation and progression. Resveratrol has been shown to affect several intracellular targets that are key regulators of cell cycle and cell survival or death. Resveratrol may therefore counteract the altered functionality of proliferative versus apoptotic pathways, exerting an antitumor activity either as a cytostatic or as a cytotoxic agent that depends on cellular specific pattern of expression of proteins related to tumorigenesis processes. In this chapter the action of resveratrol on intracellular targets is reviewed in a variety of human cancer cell lines as well as in *in vivo* studies on tumor-bearing animals.

ANTIPROLIFERATIVE EFFECTS IN VARIOUS TYPES OF CANCER

BREAST CANCER

Breast carcinoma is a heterogeneous disease, related to a variety of molecular alterations affecting hormone responses and cellular growth. It is extensively studied and potentially curable by hormonal therapy or by drugs specifically directed against tumor antigens. Estrogens are prosurvival hormones and their withdrawal may result in development of enhanced survival pathways that increase cellular proliferation rate and resistance. Preliminary clinical studies indicate that estrogens are able to act as apoptotic agents and that high doses can cause tumor regression in postmenopausal women [1,2]. This applies to approximately 80% of breast cancers that are estrogen receptor positive (ER+: mainly ERα but also ERβ). The remaining hormone-independent tumors are associated with a higher rate of proliferation and metastatization, are less differentiated, and are not responsive to common therapy [3]. A therapeutic approach is offered by selective estrogen receptor modulators (SERMs) that may provide agonistic and antagonistic effects [2]. Retinoic acid receptor and peptide growth factor receptors are implicated (i.e. overexpressed) in breast carcinogenesis. The activity of the transcriptional factor NF-κB, implicated in the organogenesis of the mammary gland, is upregulated in many breast tumors, being a downstream target of hormone-activated receptors and an upstream modulator of breast cancer tumor promoters (such as cyclin D1, cyclooxygenase (COX)-2, matrix metalloproteinases (MMPs), etc.) [4,5]. Moreover gene mutations in the cell cycle controller p53 and in the oncosuppressors (DNA-repairing enzymes) BRCA1 and 2 [3,6] and prosurvival factors (survivin and Bcl-2 family members [7]) are frequently present and may serve for detection of the risk of cancer developing.

Resveratrol as a Phytoestrogen

Resveratrol binds ERα and ERβ with an affinity lower than 17β-estradiol [8], and is classified as a type I estrogen. Resveratrol was shown to exert, at low concentration ($<10\,\mu M$), agonistic or superagonistic activity in ER+ breast cancer cells, determining estrogen-regulated gene transcription (i.e., transforming growth factor alpha (TGFα) and insulin-like growth factor (IGF) [9]) and cellular proliferation [10]. Conversely, high concentrations of resveratrol induced cell cycle arrest and apoptosis. This latter effect may occur either dependently [11] or independently from receptor activation, being unaffected by antiestrogens and by protein synthesis inhibitors. This latter observation suggests that apoptosis may rely on parallel pathways not involving estrogen-related gene transcription events [12,13]. In this sense it is known that a number of hormone-induced effects depend on estrogen-receptor complexes but do not involve gene transcription [14]. Other authors reported a mixed action of resveratrol as an agonist in estrogen-depleted conditions and as an antagonist in estrogen-stimulated conditions [15]. In agreement with the hypothesis of hormone receptor-independent signaling triggered by resveratrol, a variety of proapoptotic effects have been described in ER− cells (see below). Levenson and co-workers [12] identified an agonistic and an antagonistic pool of genes that are activated by hormones or by SERMs. The agonistic set leads to proliferative pathways whereas the antagonistic set includes stress responsive, cell-cycle controlling, and apoptotic genes. Resveratrol, as well as other SERMs, activates a group of genes that overlap in both sets. The biological response depends on a network of events related to cellular basal gene expression and environmental conditions (e.g., presence or absence of estrogens). This response is based on transcriptional activation of a whole pool of genes triggering a pathway. Moreover there are several genes that are activated by resveratrol independently of ER.

In Vitro Studies

Resveratrol in the micromolar range of concentration was demonstrated to inhibit proliferation both in ER+ and ER− breast cancer cells [16,17].

Resveratrol induced cell cycle arrest in G1/S in a concentration-dependent manner, by inhibition of ribonucleotidase activity, downregulation of cyclin D1 and E, and upregulation of p21 [18] and p53 [19]. Other authors showed resveratrol-induced cycle arrest in late S, G2/M phase [20]. Resveratrol antiproliferative effects have been reported to lead frequently to apoptotic cell death [19,21–23]. Resveratrol was shown to increase the apoptotic and cytostatic effect of cyclic adenosine monophosphate (cAMP) acting via protein kinase A (PKA) [13], to inhibit phosphorylation and activity of PKB/Akt, a kinase involved in proliferation and tumor progression [24], to inhibit NF-κB [20], to activate mitogen-activated protein kinases (MAPKs, ERKs), and to induce p53 activation via phosphorylation and acetylation [21]. Inhibitor of p53 rescued from cell death whereas estradiol co-treatment did not, although this latter resulted in blocking ERK activation. Thus resveratrol-induced apoptosis via p53 seems to be independent of ERKs and of ER-mediated mechanisms [21]. Indeed mutation of p53 in several breast cancer cell lines impaired the apoptotic effects of resveratrol, whereas the wild type expressing cells were sensitive [22]. Resveratrol was reported to trigger sphingolipid-mediated signaling by activating *de novo* synthesis of ceramide, a well-known inducer of apoptosis. The block of ceramide accumulation impaired resveratrol-induced apoptosis [23]. Resveratrol-induced apoptosis was reported to be due to upregulation of proapoptotic and downregulation of prosurvival Bcl-2 family members [25], increase of cathepsin D expression [26], and upregulation of CD95L [27]. Resveratrol increased the expression of breast cancer oncosuppressors BRCA1 and 2 as well as of its regulator acetylase pCBP/p300. This modulation of gene expression, although enhanced by ER expression, occurs even in ER− cells [28]. Resveratrol was shown to inhibit efflux activity of the breast cancer resistance protein (BCRP), belonging to the multidrug transporters of the ATP-binding cassette superfamily (ABC) [29]. Moreover the compound was reported to increase vitamin D3 receptor gene expression in an ER-dependent fashion, thus sensitizing cells to vitamin D3 and its analogs [30]. These may represent a mechanism of resveratrol synergy with chemotherapeutics. Resveratrol may also be considered as a chemopreventive agent because of its ability to block induction of phase I enzymes that transform procarcinogen to carcinogen (CYP1 family of cytochrome P450) [31–33].

In Vivo Studies

In female rats, resveratrol (100 mg/kg/day) was shown to inhibit mammary tumorigenesis induced by *N*-methyl-*N*-nitrosourea [15] and 7,12-dimethylbenz(a)anthracene (DMBA) [20]. Decreased DMBA-induced tumor progression in breast cancer was associated with suppression of COX-2 and MMP-9 [20].

Prostate Cancer

Prostate cancer is frequent among elderly males. A lower incidence of prostate cancer among Asian men is attributed to a diet rich in phytoestrogens that may act as SERMs [34]. Prostate tumor cells at early stages require androgen for growth and hormone withdrawal represents the common therapy for androgen-dependent tumors. This approach does not offer a cure, since tumor cells often develop the ability to trigger androgen receptor (AR)-dependent signals and transcriptional activity even in unstimulated conditions [35]. Reports indicate that molecular changes follow androgen withdrawal as a balancing mechanism aimed at survival, such as increased activity of PKB/Akt [36] or downregulation of death receptors (e.g., TRAIL receptor [37]). Moreover tumor progression is often associated with the loss of receptor expression becoming hormone independent [34,37]. Prostate tumors express androgen and estrogen receptors. Modulation of different receptor expression regulates therapy responsiveness [37]. Many intracellular targets are known to be involved in prostate tumorigenesis and cancer progression. Mutation or altered functional activity (via acetylation/phosphorylation) of p53 has a central role in prostate cancer, especially in metastasis.

Resveratrol as a Phytoestrogen

Resveratrol was reported to induce S-phase entry and an increase in DNA synthesis at low concentration (10 μM) in prostate cancer cells. This proliferative effect was reversed at higher concentration, was occurring in AR+ cells, and was associated with downregulation of p21 and p27 [38]. Resveratrol (100 μM) was also shown to inhibit androgen-regulated gene expression [39], expression of steroid receptor coactivators [39,40] and downregulate AR expression [39].

In Vitro Studies

Prostate cancer cells were arrested at the transition from G0/G1 to S phase when treated with resveratrol at doses higher than 15 μM [38,41] and this resulted in apoptosis induction [42]. Resveratrol was shown to induce p53 expression in prostate cancer cells [43,44] as well as to induce its activation by serine phosphorylation via MAPK [45]. Apoptosis induced by p53 is the result of multiple activated pathways related to specific cellular settings [46]. Narayanan and co-workers identified a pattern of p53 primary or secondary targeted genes, involved in apoptosis induction, expressed in prostate cancer cells treated with resveratrol. In these conditions cells were arrested in G1 and underwent apoptosis. Interestingly, the authors evidenced the upregulation of the acetylase p300, involved in p53 activation, and the downregulation of NF-κB, androgen receptor coactivators, PSA. This scenario suggests a multibranched action of resveratrol, reprogramming genome transcriptional setting and harboring the apoptotic cascade [43]. Resveratrol was reported to impair tumor promoter (phorbol esters) effects by selectively blocking cPKC/MAPK (ERKs) activation [47]. Moreover resveratrol-induced growth inhibition and apoptosis in prostate cancer cells has been traced to the accumulation of the sphingolipid apoptotic mediator ceramide [48].

Colon and Digestive Tract Cancers

Cancers of the digestive tract — pancreatic, esophageal, gastric, and colorectal carcinomas — have high incidence and poor prognosis, due to the lack of an effective chemotherapeutic approach: combinations of 5-fluorouracil (5-FU) with other drugs (gemcitabine, irinotecam, etc.) have been only partially successful. Besides inherited mutations, gastrointestinal cancers often progress from normal tissue to adenoma and carcinoma through accumulation of genetic alterations and many molecular markers have been associated with tumor staging. These tumors show microsatellite instability (MSI) and this may be associated with mutation or methylation-silencing of mismatch repair gene (MMR), downregulation of E-cadherin, altered expression of Bcl-2 family members, cyclins and cyclin inhibitors, alteration of p53 expression and adenomatous polyposis coli (APC), oncogenic activation of β-cathenin, increased expression of COX-2, growth factor receptors (erbB family, EGFR), and MMPs, and resistance to death receptor-induced apoptosis (e.g., Fas, TRAIL, etc.) [49–51].

In Vitro Studies

Resveratrol treatment has been shown to arrest proliferation and to induce apoptosis in esophageal [52], pancreatic [53], gastric [54], and colon [55] cancer cells.

In esophageal cancer cells, resveratrol-induced growth inhibition and apoptosis have been associated with modulation of pro- and antiapoptotic members of the Bcl-2 family (Bcl-2 and Bax) [56].

In pancreatic cancer cells, resveratrol-induced apoptosis occurs via mitochondria alteration (membrane depolarization, cyt c release) and caspase activation [57].

Gastric adenocarcinoma cells treated with resveratrol arrested their cell cycle at the G0/G1 phase and underwent apoptosis, in association with inhibition of PKC and MAPK (ERKs) activity [54]. Antiproliferative effects at high concentration of resveratrol treatment have also been traced to increased basal levels of nitric oxide and reduced production of intracellular oxygen radicals [58]. Gastric cancer is a slow process initiated by primary exposure to carcinogens and/or *Helicobacter pylori* infection and consequent inflammatory response [59]. Resveratrol, a well-known antiinflammatory agent, has been reported to inhibit *in vitro* the growth of *Helicobacter pylori* [60] thus potentially offering a cure against infection-caused tumorigenesis. Resveratrol was reported to inhibit several enzyme activities involved in polyamine metabolism [61,62]. In gastric adenocarcinoma resveratrol reversed carcinogenesis induced by nitrosamines via PKC activity [54] and reversed induction of phase I enzyme [63].

Colon cancer cells arrested their cell cycle in S phase when treated with resveratrol at high concentration (100 to 200 μM), showing altered levels of cyclins and of the dependent kinases [64]. Block of proliferation was followed by a caspase-mediated apoptosis [64]. Proapoptotic action was also associated to modulation of Bcl-2 family members and their translocation to mitochondria as well as to relocation, clustering, and activation of DISC-forming elements (Fas, FADD, caspase-8) inside cholesterol-sphingomyelin rich domain of plasmamembrane, thus promoting apoptotic cascade even independently of Fas ligand [65]. Mahyar-Roemer and co-workers showed that resveratrol induced apoptosis in colon carcinoma cells independently of p53 expression by altering mitochondria functions [66]. In combination with butyrate (a known differentiating agent produced by intestinal bacteria from complex carbohydrates in dietary fibers) resveratrol enhanced differentiation-induced effects, although a single treatment with resveratrol did not inhibit colon cancer cell proliferation [55]. Moreover resveratrol caused a significant decrease of ornitine decarboxylase activity, a key enzyme of polyamine biosynthesis associated with cancer cell growth [67].

In Vivo Studies

Oral or intraperitoneal (2 mg/kg/day for prolonged time) administration of resveratrol to rats reduced significantly the number and the size of *N*-nitrosomethylbenzylamine-induced esophageal tumors. Moreover resveratrol downregulated the basal expression of COX-1 and impaired the increase of COX-2 expression as well as of prostaglandin production in tumor tissues [68]. Oral administration of resveratrol prevented the formation of colon tumors and reduced the formation of small intestinal tumors in mice genetically predisposed to develop intestinal cancers, by modulation of cell cycle- and tumor growth-related gene expression [69].

Lung Cancer

Lung cancer presents four main histological types of cells: adenocarcinoma, squamous or large-cell carcinoma, named nonsmall-cell lung cancer (NSCLC), and small-cell lung cancer (SCLC). NSCLC growth rate is slower than that of SCLC and they are chemo- and radioresistant, whereas the highly proliferating SCLC, normally present at advanced tumor stage, is sensitive to these treatments. Mutation or epigenetic modulation of expression of specific genes is frequently associated with lung cancer. Altered function of the oncosuppressor p53 and of the oncogenes Rb and myc, of MYO18B (myosine family), and loss of heterozygosity on chromosome 3p (carrying potential tumor suppressors) are common in NSCLC and SCLC, whereas alteration in k-ras, β-catenin, and p16 mainly occurs in NSCLC. Some of the above mentioned mutations are known to be early (p16) or late (p53) events in tumor staging. Among carcinogens involved in lung cancer induction polycyclic aromatic hydrocarbons (pollution, cigarette smoke) are particularly potent, bind AhR, and upregulate phase I enzymes, responsible for procarcinogen oxidation [70,71].

In Vitro Studies

Resveratrol has been proposed to block lung cancer initiation since it was shown to be a competitive inhibitor of AhR [72] and to inhibit receptor transcriptional activity, therefore preventing induction of phase I enzymes in human bronchial epithelial cells [73]. In lung carcinoma resveratrol was reported to block proliferation by arresting the cell cycle at the S phase, in association with hypophosphorylation of Rb and increase in p53 and p21 [74]. Growth inhibition resulted in apoptosis due to caspase activation and altered ratio between pro- and antiapoptotic Bcl-2 family members [75]. In several lines of NSCLC cells resveratrol was reported to inhibit NF-κB activation in response to tumor necrosis factor α (TNFα) by enhancing the activity of the deacetylase sirtuin1. This enzyme is responsible for deacetylating histone and nonhistone proteins thus regulating transcriptional activity. Resveratrol inhibition of NF-κB impairs its transcription of prosurvival genes (IAPs) thus sensitizing cells to apoptosis [76].

In Vivo Studies

Diverse administration of resveratrol (3 to 10 mg/kg/day) in lung tumor-bearing animals has been reported. Coinjection in mice lungs reduced significantly benzopyrene-induced formation of DNA adducts and increased CYP1 enzymes [77]. Resveratrol administration in mice bearing highly metastatic Lewis lung carcinoma (LLC) led to significant reduction of tumor size and lung metastases [78]. On the contrary, dietary administration of resveratrol did not show any chemopreventive activity [79,80].

Melanoma and Skin Cancer

Skin cancer can originate as a multistep process mainly due to repeated ultraviolet (UV) exposure, from epithelial cells (squamous and basal cell carcinoma), and from pigment-producing melanocytes (melanoma). Keratinocytes develop in moderately or highly aggressive carcinomas and melanocytes may form displastic nevi and melanoma, a highly metastasizing tumor. Skin cancer is usually resistant to radiotherapy and chemotherapy. Basal cell carcinoma is often associated with a specific mutation in the proliferation pathway triggered by membrane receptors Sonic Hedgehog (e.g., PTCH1 gene). The apoptotic response is linked to multiple alterations of oncogenes/oncosuppressors, aimed at impairing cell death and favoring cell survival. Although mutants or altered p53 can be associated with melanoma, dysfunction of this oncosuppressor occurs with a low incidence. On the contrary, mutations leading to downregulation of p14 (cyclin inhibitor) have been reported as well as upregulation in ras and raf activity that signal via ERKs to gene transcriptional factors. PTEN, a phosphatase implicated in reverting the proliferative cascade of PI3K/PKB, is frequently downregulated in melanoma. Bcl-2 downregulation and Bax overexpression have been reported in melanoma, and this evidence, contrasting with tumor resistance to apoptosis, may find explanation in altered expression of other members of the family. Moreover melanoma cells and nonmelanocytes express survivin and another inhibitor of apoptosis protein ML-IAP. Melanoma cells are resistant to TNFα, FasL, and TRAIL-induced apoptosis although expressing the receptors. TRAIL-R in melanoma triggers NF-κB activation: such prosurvival transcriptional factor activation may counterbalance the apoptotic stimulus [81,82].

In Vitro Studies

Epidermioid carcinoma cells treated with resveratrol have been shown to arrest their cell cycle. The activity of cyclins, related kinases, and inhibitors controlling the exit from G1 phase and entry in S phase were regulated by resveratrol (induction of p21, inhibition of cyclin D1, 2, and E, and cdk2, 4, and 6) [83] as well as the level of active-hyperphosphorylated Rb protein and E2F associated transcriptional factor being significantly reduced [84], causing cell cycle arrest at G0/G1 phase. Resveratrol has been shown to inhibit melanoma proliferation by blocking the cell cycle in S phase [85,86]. Growth arrest was concomitant with an increase in cyclin A, E, and B1, and it was an irreversible event, since removal of the compound did not restore normal cycling, followed by apoptosis [85,87]. Niles and co-workers reported that a melanotic cell line was more sensitive to resveratrol-induced apoptosis than an amelanotic cell line and this was associated with an enhanced phosphorylation of ERKs in the sensitive cells only. Moreover, neither cell lines increased p53 expression or activation (phosphorylation) [87].

In Vivo Studies

Pezzuto and co-workers published in 1997 a pioneer work on resveratrol antitumor activity. Topical application of resveratrol was able to impair tumor promotion of phorbol esters in animals treated with tumor initiator DMBA, decreasing significantly the amount of developing tumors per mouse [88]. Since then, other authors have shown similar results in two-stage tumorigenesis models [89,90]. Intraperitoneal injection of resveratrol in mice challenged with syngenic melanoma cells resulted in a dose-dependent delay of tumor growth whereas oral administration, although ineffective in blocking tumor growth, diminished significantly its metastatic potential [92]. Finally topical application of resveratrol on hairless mice resulted in a significant decrease in UVB exposure-hyperplastic response of epidermal cells by preventing activation of cyclines and cyclin-dependent kinases and enhancing expression of p21 and p53 [93].

Hepatoma

Hepatocellular carcinoma is a deadly tumor. The etiology is often associated with chronic viral infections (HVB, HVC) and alcohol abuse but multiple concurring agents as well as genetic alteration are often present. Among relevant genes there are p53, Rb, β-catenin, loss of heterozygosis on several loci carrying potential tumor suppressors, overexpression of c-myc, cyclin D1, growth factors (e.g., TGF), and matrix metalloproteases [94].

In Vitro Studies

Resveratrol has been reported to inhibit hepatocellular carcinoma cell line proliferation [65,95–97] and to induce apoptosis [65,96,97]. It has been proposed that resveratrol cycle arrest in G1 phase and cytoxicity is dependent on p53 and mediated by p21 and Bax increase [96]. Whereas cytostatic effects on hepatoma require a high dose (100 μM), resveratrol was shown to inhibit cancer cell migration even at low concentration (25 μM) [98,99].

In Vivo Studies

A number of studies indicated that resveratrol (approximately 10 mg/kg/day) administration to animals impairs growth of transplanted hepatoma [100–103]. Hepatoma cells arrested in S phase and showed increased levels of cell cycle-related proteins (cyclin B1 and p34CDK2) [102]. Moreover resveratrol enhanced immune response by promoting macrophage cytotoxicity against the tumor [103].

Cancer of the Brain and Other Nervous Tissues

Most of these tumors arise in the central nervous system (brain and spinal cord) and develop from glial cells and more rarely from neuronal cells. Childhood tumors, known as primitive neuroectodermal tumors, are most frequently medulloblastoma and neuroblastoma. Neuroblastoma develops at the embryonic stage and in more than 10% of cases regresses spontaneously in patients under one year of age, by cellular differentiation or apoptosis. In the remaining cases the tumor grows aggressively and there is no cure. Nervous system carcinogenesis is related to somatic genetic changes due to environmental exposure and to inherited mutations. Alteration patterns are extensively characterized, and are tumor and tumor-stage specific. Genes involved in astrocytomas and glioblastoma are p53, p16, Rb, growth factors and their receptor (e.g., PDGF and EGF), allelic loss of different loci (i.e., chromosome 22q), and PTEN. The common target genetic alterations in medulloblastoma are the Sonic Hedgehog receptor PTCH, APC (for DNA mismatch repair), Wnt and β-catenin, allelic loss in p17 and in addiction increased expression of erbB family members, and c-myc, ras, TRKC, multidrug resistance [104]. Neuroblastoma has inherited predisposition with a great heterogeneity of genetic aberrations. A number of allelic losses involving different loci, including putative neuroblastoma suppressor gene, and translocation involving 17q leading to segmental chromosome gain have been observed [105].

In Vitro Studies

Conversely from a few reports assessing that resveratrol impairs apoptosis in stress-stimulated neuronal cells [106,107], the polyphenol has been reported to have cytotoxic effects in human medulloblastoma [108] and neuroblastoma [109]. Resveratrol was demonstrated to arrest the proliferation of several medulloblastoma cell lines and to induce apoptosis, concomitantly with an increased expression in caspase-3 [108] and of phase I enzyme CYP1A1, possibly involved in resveratrol metabolic activation [110]. Neuroblastoma cells treated with resveratrol arrested cell cycle and died by apoptosis [109,111] showing upregulation of p53, p21, and Bax [111,112]. In neuroblastoma cells resveratrol treatment was shown to induce the expression of Egr1, a transcriptional factor regulating cell proliferation [113].

In Vivo Studies

Rats injected subcutaneously with glioma cells and treated intraperitoneally with resveratrol (40 mg/kg/day) showed a significant increase in survival time. A higher dose (100 mg/kg/day) was necessary to prolong survival rate of animals bearing intracerebrally inoculated tumors [114]. Similarly intraperitoneal resveratrol treatment (40 mg/kg/day) increased survival rate in mice subcutaneously injected with neuroblastoma [109].

Cancers of the Female Genital Apparatus

This complex of tissues is under the control of steroid hormones that regulate proliferation and differentiation. The majority of ovarian cancers are epithelial and prevalently ER+, with an increased ratio of ERα/ERβ versus normal tissue. Antiestrogenic compounds may regulate receptor expression and counteract estrogen-induced proliferation. Several ER target genes favor the carcinogenesis process (cyclin D1, c-myc, cathepsin D, kallicreins, pS2, Her2/Neu) [115]. Moreover genetic alterations are associated with tumor staging: BRAF or k-ras (mutually exclusive), PTEN, β-catenin are early events associated with a low-stage carcinoma; p53 mutation and HER-2/Neu overexpression are associated with high-stage carcinoma [116]. Endometrial carcinoma may arise as multistep process involving genetic changes. The low-stage of this tumor appears with high incidence, it is estrogen dependent (hyperestrogenic state), confined in the uterus, moderately differentiated, and associated with microsatellites instability, silencing (promoter hypermethylation) of DNA mismatch repair enzyme (MLH), PTEN, k-ras, and β-catenin mutation. A high-stage tumor originates from atrophic endometrium and develops in serous carcinoma, showing aggressiveness (peritoneal metastases), hormone independency, diverse genotype pattern with early p53 mutation, and Her2/Neu and c-myc overexpression. Cervical cancers are mainly related to human papillomavirus infection and associated with changes in p53, its homologue p63, and Rb, suppression of proapoptotic Bak, and modulation of cyclin inhibitors (p16, p27) [117].

Resveratrol as a Phytoestrogen

In endometrial adenocarcinoma cells resveratrol exhibited antagonistic effects by inhibiting estrogen-induced alkaline phosphatase activity and progesterone receptor expression. Moreover resveratrol treatment alone increased at low dose (1 μM) and down modulated at higher dose (15 μM) ERα expression [118].

In Vitro Studies

In endometrial adenocarcinoma cells resveratrol exerted cytostatic potential, modulating cyclin levels and arresting cells in S phase [118] and downregulating growth factors (EGF) [119]. Opipari and co-workers demonstrated in ovarian carcinoma that cell cycle arrest was followed by cell death via a nonapoptotic mechanism. Although resveratrol induced cyt c release, formation of apoptosome, and caspase activation, these latter were not the unique effectors of cellular death and cells died by autophagocytosis [120]. Analogous treatment revealed that growth arrest and cell death in this model affected the expression of more than 100 genes and particularly upregulated NAD(P)H quinone oxidoreductase, an enzyme that controls p53 [121]. Resveratrol was also shown to inhibit expression of HIF-1α and VEGF in ovarian cancer cells (by blocking the akt-MAPK pathway and translation proteins) and thus it may impair the angiogenesis process within tumors [122].

Resveratrol was shown to inhibit the activity of transcriptional factors involved in proliferation and survival. Radiation and phorbol ester induction of transcriptional factor AP-1 activity in cervical cancer cells was inhibited by resveratrol [123], as well as TNFα-induced NF-κB activity [124], via upstream inhibition of MAPKs. Although estradiol moderately activated AP-1, there was no competition between the hormone and the phytoestrogen, leading to the hypothesis that resveratrol inhibition was not receptor-dependent [123]. c-Jun proteasome degradation was shown to be favored by resveratrol via inhibition of COP9 signalosome (CK2, PKD) kinases that phosphorylate and stabilize AP-1 [125]. Resveratrol impairment of c-Jun and AP-1 activation prevented MMP-9 thus blocking cervical cancer invasiveness [126].

Hematological Cancers

Hematological malignancies develop as complications in genetic disorders. Chromosome instability, impairment of DNA repair, and bone marrow failure may facilitate proliferation of aberrant clones. The abnormal growth of bone marrow precursors occurs in leukemias and it may suppress proliferation of other clones of blood cells. Lymphomas are tumors of the lymphatic systems involving lymphocytes from lymphonodes, spleen, and thymus. Myeloma originates from plasmacells in bone marrow and it is associated with excessive production of a specific type of immunoglobulin.

Genetic alterations such as mutations or amplifications of oncogenes and oncosuppressors and rearrangement leading to gene fusion (reciprocal chromosomal translocation, chromosomal inversion, gene insertion, and deletion) are frequent and well characterized, so that genetic profiles are used for diagnosis, staging, and therapy selection [127].

In Vitro Studies

A common paradigm, assessed in a variety of lymphocytic and nonlymphocytic leukemias, is that resveratrol induces growth arrest [27,88,128–132]. Some authors demonstrated that a low dose (10 to 30 μM) is sufficient to arrest proliferation [113,131–133] and a higher dose (50 to 100 μM) is cytoxic [113,131–133]. Promyelocytic leukemia and histiocytic lymphoma treated with resveratrol arrested proliferation cycle in S phase [132–134] and increased cyclin A and E expression [132,133]. It has been also reported that resveratrol induced cycle arrest in G2/M phase in lymphoid and myeloid leukemia [135]. Removal of the treatment allowed cells to re-enter the cell cycle, meaning that the arrest did not involve irreversible processes. Conversely higher doses (starting from 50 μM) induced irreversible cell death by apoptosis [132,133]. Moreover resveratrol treatment induced growth arrest in non-Hodgkin's lymphoma and multiple myeloma [136,137]. Subtoxic doses of resveratrol (10 μM) were able to induce differentiation in promyelocytic leukemia [88] towards myelomonocytic phenotype and in myeloid leukemia by inducing the expression of adhesion molecules [138]. Apoptosis induced by resveratrol in promyelocytic leukemia was demonstrated to depend on expression of FasL [27], mediated by activation of the small GTP binding protein cdc42, mitogen-activated kinase ASK1, and JNK [139]. In contrast, other authors reported that resveratrol-induced apoptosis was independent of FasL-mediated events [130,134]. An important target involved in growth arrest and death induction is NF-κB, whose activation in acute myeloid leukemia and histiocytic lymphoma in response of diverse stimuli (e.g., IL1β, TNFα) is inhibited by resveratrol [124,131] via inhibition of IKK activity [140]. Similarly, resveratrol inhibited AP-1 and the downstream IL8 production in response to phorbol esters in histiocytic lymphoma [141]. Resveratrol-induced apoptosis was associated with modulation of pro- and antisurvival factors: in acute myeloid leukemia it significantly reduced intracellular levels of antiapoptotic proteins Bcl-2 and Bcl-xL [131]; in histiocytic lymphoma it increased caspase-3 activity and decreased levels of IAP1 and 2 [133]; in T-cell leukemia it downregulated the expression of survivin [142]. Down-regulation of prosurvival Bcl-xL and upregulation of the proapoptotic Apaf-1 was reported in resveratrol-induced apoptosis of non-Hodgkin's lymphoma [136]. In promyelocytic and erythroid leukemias, resveratrol was shown to induce the expression of Egr1, a transcriptional factor involved in cell cycle regulation, by upstream activation of ERKs. The induction is biphasic, with a first and direct induction within minutes of treatment and a delayed one that required protein synthesis [113]. In different leukemias, resveratrol induced a very early depolarization of mitochondrial membrane associated with caspase activation and apoptosis induction [143,144]. In B-lymphocyte leukemia resveratrol induced apoptosis by increasing caspase-3 activity, reducing expression of Bcl-2 and iNOS, and causing a drop in mitochondria transmembrane potential [145,146]. A few authors compared the effect of resveratrol on tumor cells versus normal peripheral blood lymphocytes reporting a stronger and more significant cytotoxicity on leukemia cells [27,135,144,145,147].

In Vivo Studies

A weak antileukemic affect of resveratrol at a high dose (80 mg/kg/day) was shown in mice [148].

Cancers from Other Sites

Kidney Cancer

The types of epithelial renal tumors are clear cell, types I and II papillary, chromophobe, and oncocytoma. The most frequent genetic alterations are loss of heterozygosity (LOH) in chromosome 3 (affecting expression of NRC, *RASSF1A*, *DRR1*, TGF-β type II receptor), inactivation of the tumor suppressor gene von Hippel-Lindau (*VHL*), gain of chromosomes 7 and 17 and loss of the Y chromosome, and mutations in the protooncogene c-Met and the fusion gene product PRCCTFE3 controlling proliferation and differentiation [149].

Resveratrol was reported to induce growth arrest and cell death in renal carcinoma cells. Resveratrol modulated the gene expression profile in this tumor in a concentration-dependent fashion, significantly affecting 29 genes that are directly or indirectly involved in proliferation inhibition or cell death (among others VDR, TRAF, CYP, and GADD45) [150].

Thyroid Cancer

Thyroid epithelial cancers include carcinomas of follicular cell origin and a more rare medullary thyroid cancer of the parafollicular C cells. Frequently occurring alteration concerns the oncogenic tyrosine kinase RET and NTRK1, PPARγ rearrangement, k-ras, and p53 [151].

In thyroid cancer cells resveratrol was shown to induce apoptosis by k-ras- and ERK-induced accumulation of p21, c-jun, and c-Fos and accumulation and phosphorylation of p53 [152].

Oral Carcinoma

Squamous cell carcinoma of the oral cavity still has a poor prognosis. Oral carcinoma is characterized by alteration of lately markers of epithelial differentiation and genomic markers such as mutations of k-ras and p53, deletions (e.g., 9p21–p22, including p16 locus), cyclin A, B, and D overexpression, EGFR/c-erb 1, members of the ras gene family, c-myc, int-2, hst-1, PRAD-1, and Bcl-1, Rb, downregulation of E-cadherin, and increased expression of MMPs [153–155].

Cell growth inhibition by resveratrol was shown to occur in an oral squamous carcinoma cell line [156].

CONCLUSIONS AND PERSPECTIVES

We have attempted here to survey the literature showing that resveratrol has antiproliferative effects against human tumors. If the *in vitro* evidences are solid and convincing, (Table 4.1) the studies in animal models are less frequent and occasionally controversial. However, most recently some epidemiological studies have demonstrated a reduced relative risk for prostate cancer associated with an increased level of red wine consumption [157] and an inverse relation between resveratrol and breast cancer risk [158]. All this makes resveratrol a very attractive and promising molecule, but further research is needed to evaluate the potential negative association between intake of resveratrol and cancer risk in humans.

TABLE 4.1
***In Vitro* Antiproliferative Effects of Resveratrol in Human Cancer Cells**

Cell line	Cellular effect	Mechanism	Ref.
Breast Cancer			
MCF-7	Growth inhibition	↑ ERα-associated PI3K activity ↑ Proteasome-dependent ERα degradation	11
MCF-7, MDA-MB-468	Growth inhibition	↓ E2 growth promoting effect ↓ mRNA expression of TGF-α and IGF-I R ↑ mRNA expression of TGF-β2	9
MCF-7, MDA-MB-231, MCF-10F	Growth inhibition	Unrelated to the ER status	12, 16, 17
MCF-7, MDA-MB-231	Growth inhibition/apoptosis	Alteration of cell cycle (cell specific)	18
MCF-7	Growth inhibition/apoptosis	↑ p53, p21, Bax ↓ Cyclin D, Cdk4, Bcl-2, Bcl-xL	19
MCF-7	Growth inhibition	↓ NF-κB activation	20
T47D, MDA-MB-231	Growth inhibition	Akt and FAK phosphorylation: ↓ in ER− cells, ↑ in ER+ cells	24
MCF-7	Growth inhibition/apoptosis	↑ MAPK-ERK1/2 ↑ p53 acetylation and phosphorylation	21
MDA-MB-231, MDA-MB-453, MDA-MB-468, BT20, SKRB3	Growth inhibition/apoptosis	Dependent on p53 cell status	22
MCF-7	Growth inhibition	↑ camp	13
KPL-1, MCF-7, MKL-F	Growth inhibition/apoptosis	↑ Bax, Bak ↓ Bcl-xL ↓ Linoleic acid cell stimulation	25
MCF-7, MDA-MB-435	Growth inhibition; effect related to the metastatic potential	↑ Cathepsin D	26

T47D	Growth inhibition/apoptosis	↑ CD95L	27
HBL-100, MCF-7, MDA-MB-231	Growth inhibition	↑ BRCA1, BRCA2, ERα, ERβ, p21, p53, pCBP/p300, RAD51, p52, ki67	28
MDA-MB-231	Growth inhibition/apoptosis	↑ Ceramide	23
MCF-7, T47D	Growth inhibition	↑ Effect of Vit D3 analog	30
MCF-7	Growth inhibition/apoptosis	↓ Cyclin A and B1, β-catenin	159
MDA-MB-231	Growth inhibition	↑ p21	12
MCF-7, T47D, MDA-MB-486	Growth inhibition	Unrelated to p53 expression	160
MCF-7, T47D, BT549	Growth inhibition/apoptosis	↓ Oxidative DNA damage	41
MCF-7, T47D, MDA-MB-231	Growth inhibition	–	161
4T1	Growth inhibition	Not studied	162
Prostate Cancer			
LNCaP	Growth inhibition	Opposite concentration- and time-dependent effects on DNA damage	38
LNCaP	Growth inhibition	↓ AR expression	39
LNCaP	Growth inhibition/apoptosis	↓ PSA, AR coactivator 24, NF-κB ↑ p53, PIG7, p21, pCBP/p300, Apaf-1	43
PC3, DU-145	Growth inhibition/apoptosis	↓ Oxidative DNA damage	41
LNCaP, DU-145, PC3, JCA-1	Growth inhibition/apoptosis	↓ PSA	42
DU-145	Growth inhibition/apoptosis	↓ D-type cyclins, Cdk4 expression ↑ p53, p21, Bax	44
DU-145	Growth inhibition/apoptosis	↑ MAPK ↑ p53 phosphorylation	45
PC3	Growth inhibition	↓ PKC-mediated Erk1/2 activation	47
PC3	Growth inhibition	↑ Ceramide	48
DU-145	Growth inhibition/apoptosis	↑ HSP70	163
LNCaP, DU-145, PC3	Growth inhibition	↓ NO secretion	164
DU-145, LNCaP	Growth inhibition/apoptosis	Unrelated to the ER status	165

(continued)

TABLE 4.1
Continued

Cell line	Cellular effect	Mechanism	Ref.
Esophageal Cancer			
EC-9706	Growth inhibition Apoptosis	↑ Bax ↓ Bcl-2	56
Pancreatic Cancer			
PANC-1, AsPC-1	Growth inhibition Apoptosis	↑ Caspases Membrane depolarization, cyt c release	57
Gastric Cancer			
KATO.III, RF-1	Growth inhibition Apoptosis	↓ PKCα, MAPK	54
SNU-1	Growth inhibition	↓ O_2– ↑ NOS	58
Colon Cancer			
CaCo2, HCT-116	Growth inhibition/apoptosis	↓ Cyclin D1, Cdk4 ↑ Cyclin A and E	64
HT-29	Growth inhibition	↓ Cdk7 activity	166
SW480	Growth inhibition/apoptosis	↑ Clustering of Fas in rafts and formation of a death-inducing signaling complex	65
HCT-116	Growth inhibition/apoptosis	Unrelated to p53 expression	66
CaCo-2	Growth inhibition/differentiation	Combined effect with butyrate	55
CaCo-2	Growth inhibition	↓ Ornithine decarboxylase activity	62
SW480	Growth inhibition	S-phase arrest	167
Lung Cancer			
A549	Growth inhibition/apoptosis	↑ p21 ↓ pRb phosphorylation, NF-κB	75
A549, EBC-1, Lu65	Growth inhibition/apoptosis	Combined effect with paclitaxel ↑ p21	74

NCI-H358, NCI-H460, NCI-H1299	Apoptosis	↓ NF-κB ↑ Sirtuin 1	76
Skin Cancer			
A431	Growth inhibition/apoptosis	↑ p21 ↓ pRb phosphorylation, cyclins D1, D2, and E	83, 84
Melanoma			
SK-Mel-28	Growth inhibition/apoptosis	↑ Cyclin A, E, and B1	85
SK-Mel-28, M14, PR-Mel	Growth inhibition/apoptosis	S-phase arrest	86
SK-Mel-28, A375	Growth inhibition/apoptosis	↑ ERK1/2 phosphorylation	87
Hepatoma			
HepG2	Growth inhibition	Non studied	95
HepG2, Hep3B	Growth inhibition/apoptosis	↑ p53, p21, Bax	96
H22	Growth inhibition/apoptosis	Not studied	97
HepG2	Growth inhibition	Block of cell cycle at the S–G2/M transition	168
AH109A	Growth inhibition/reversion of invasiveness	↓ ROS-induced invasiveness	98
HepG2	Growth inhibition/reversion of invasiveness	Not studied	99
H4	Growth inhibition	↓NF-κB and AP-1 activation ↓TNF-induced ROS and lipid peroxidation	124
Neuroblastoma			
SK-N-SH	Growth inhibition/differentiation	↓ Neutral endopeptidase ↓ Angiotensin-converting enzyme	169
N2a	Growth inhibition/apoptosis	↓ p21 ↑ Cyclin E	109
N2a	Growth inhibition/apoptosis	↑ p53, p21, Bax	111
SHEP	Growth inhibition/apoptosis	↑ Sensitization to TNF	112
Medulloblastoma			
Med-3, UW228-1, UW228-2, UW228-3	Growth inhibition/apoptosis/differentiation	↑ CYP1A1, caspase-3	108, 110

(*continued*)

TABLE 4.1
Continued

Cell line	Cellular effect	Mechanism	Ref.
Endometrial Cancer			
Ishikawa	Growth inhibition	↑ Cyclin A and E ↓ cdk2	15
Ishikawa	Growth inhibition	↓ EGF	119
Ovarian Cancer			
A2780, CaOV3, ES-2, TOV112D, A1947	Growth inhibition/autophagocytosis	Unrelated to Bcl-xL and Bcl-2 expression	120
A2780/CP70	Growth inhibition	↓ HIF-1α, VEGF	122
PA-1	Growth inhibition/apoptosis	↑ NQO-1	121
Cervical Cancer			
CaSki, HeLa	Growth inhibition	↓ PMA-induced MMP-9 expression, JNK, PKCδ activation	123, 126
HeLa	Growth inhibition	↓ NF-κB and AP-1 activation ↓ TNF-induced ROS and lipid peroxidation	124
HeLa	Growth inhibition	↑ c-Jun proteasome degradation	125
HeLa	Growth inhibition	↓ PK CKII	170
HeLa	Growth inhibition/apoptosis	↓ Oxidative DNA damage	41
HeLa	Growth inhibition/apoptosis	↑ Radiosensitivity	137
HeLa, SiHa	Growth inhibition	↑ Radiosensitivity	171
HeLa	Growth inhibition/apoptosis	Not studied	147
Hematological Cancers			
HL-60	Growth inhibition/apoptosis	↑ CD95L	27
U937, HL-60	Growth inhibition/apoptosis	↑ Caspase-3	129, 148
THP-1	Growth inhibition/apoptosis	Independent of Fas/FasL signaling pathway	130
OCIM2, OCI/AML3 AML[a]	Growth inhibition/apoptosis	↓ NF-κB activation induced by IL-1β ↑ S-phase arrest	131
HL-60	Growth inhibition/apoptosis	↑ Cyclin A and E	132

U937	Growth inhibition/apoptosis	↑ Caspase-3 ↓ Bcl-2 cyt c release	133
CEM-C7H2, Jurkat	Growth inhibition/apoptosis	↑ Caspase-6	134
T-cell leukemia cell lines	Growth inhibition/apoptosis	↓ Survivin	142
K562, KCL22, HL-U937, Jurkat, WSU-CLL	Growth inhibition/apoptosis	↑ Cyclin A and B	135
K562, IM-9	Growth inhibition/apoptosis	↑ Radiosensitivity	137
NHL[a], MM[a]	Sensitization of paclitaxel-induced apoptosis	↑ Bid, Apaf-1 ↓ Bcl-xL, Mcl-1	136
HL-60, NB4, U937, THP-1, ML-1, Kasumi-1 AML[a]	Growth inhibition/differentiation	↓ NF-κB binding activity ↑ Adhesion molecules	138
HL-60	Growth inhibition/apoptosis	↑ Cdc42	139
U937, Jurkat	Growth inhibition	↓ NF-κB and AP-1 activation ↓ TNF-induced ROS and lipid peroxidation	124
U937	Apoptosis	↓ NF-κB	140
CCRF-CEM	Apoptosis	↑ ROS, caspases, mitochondrial membrane depolarization	143
ALL[a]	Growth inhibition/apoptosis	↑ Caspase-9	144
CLL[a], HCL[a]	Growth inhibition/apoptosis	↑ Caspase-3 ↓ Bcl-2, iNOS	145
WSU-CLL, ESKOL CLL[a]	Growth inhibition/apoptosis	↓ iNOS, Bcl-2	146
HL-60	Growth inhibition/apoptosis	↓ CYP1B1	172
HL-60, K-562	Growth inhibition/apoptosis	Not studied	147
K562	Growth inhibition/aifferentiation	↑ p21	173
Thyroid Cancer			
BHP18-21, BHP2-7, FTC236, FTC238	Growth inhibition/apoptosis	↑ p53, c-Fos, c-jun, p21	152
Oral Cancer			
SCC-25	Growth inhibition	Not studied	156

[a]Cells freshly prepared from patients: NHL, non-Hodgkin's lymphoma; MM, multiple myeloma; AML, acute myeloid leukemia; ALL, acute lymphocytic leukemia; CLL, chronic lymphocytic leukemia; HCL, hairy cell leukemia.

REFERENCES

1. Lonning PE, *J Steroid Biochem Mol Biol* 79, 127–132, 2001.
2. Jordan VC, *Cancer Cell* 5, 207–213, 2004.
3. Keen JC and Davidson NE *Cancer* 97, 825–833, 2003.
4. Cao Y and Karin M, *J Mammary Gland Biol Neoplasia* 8, 215–223, 2003.
5. Surh YJ, Chun KS, Cha HH, Han SS, Keum YS, Park KK, and Lee SS, *Mutation Res* 480–481, 243–268, 2001.
6. Gasco M, Yulug IG, and Crook T, *Hum Mutation* 21, 301–306, 2003.
7. Coradini D, and Daidone MG, *Curr Opin Obstet Gynecol* 16, 49–55, 2004.
8. Bowers JL, Tyulmenkov VV, Jernigan SC, and Klinge CM, *Endocrinology* 141, 3657–3667, 2000.
9. Lu R and Serrero G, *J Cell Physiol* 179, 297–304, 1999.
10. Gehm BD, McAndrews JM, Chien PY, and Jameson JL, *Proc Natl Acad Sci USA* 94, 14138–14143, 1997.
11. Pozo-Guisado E, Lorenzo-Benayas MJ, and Fernandez-Salguero PM, *Int J Cancer* 109, 167–173, 2004.
12. Levenson AS, Gehm BD, Pearce ST, Horiguchi J, Simons LA, Ward JE, III, Jameson JL, and Jordan VC, *Int J Cancer* 104, 587–596, 2003.
13. El-Mowafy AM and Alkhalaf M, *Carcinogenesis* 24, 869–873, 2003.
14. Gray GA, Sharif I, Webb DJ, and Seckl JR, *Trends Pharmacol Sci* 22, 152–156, 2001.
15. Bhat KP, Lantvit D, Christov K, Mehta RG, Moon RC, and Pezzuto JM, *Cancer Res* 61, 7456–7463, 2001.
16. Mgbonyebi OP, Russo J, and Russo IH, *Int J Oncol* 12, 865–869, 1998.
17. Serrero G and Lu R, *Antioxid Redox Signal* 3, 969–979, 2001.
18. Pozo-Guisado E, Alvarez-Barrientos A, Mulero-Navarro S, Santiago-Josefat B, and Fernandez-Salguero PM, *Biochem Pharmacol* 64, 1375–1386, 2002.
19. Kim YA, Choi BT, Lee YT, Park DI, Rhee SH, Park KY, and Choi YH, *Oncol Rep* 11, 441–446, 2004.
20. Banerjee S, Bueso-Ramos C, and Aggarwal BB, *Cancer Res* 62, 4945–4954, 2002.
21. Zhang S, Cao HJ, Davis FB, Tang HY, Davis PJ, and Lin HY, *Br J Cancer* 91, 178–185, 2004.
22. Laux MT, Aregullin M, Berry JP, Flanders JA, and Rodriguez E, *J Alternative Complementary Med* 10, 235–239, 2004.
23. Scarlatti F, Sala G, Somenzi G, Signorelli P, Sacchi N, and Ghidoni R, *FASEB J* 17, 2339–2341, 2003.
24. Brownson DM, Azios NG, Fuqua BK, Dharmawardhane SF, and Mabry TJ, *J Nutr* 132, 3482S–3489S, 2002.
25. Nakagawa H, Kiyozuka Y, Uemura Y, Senzaki H, Shikata N, Hioki K, and Tsubura A, *J Cancer Res Clin Oncol* 127, 258–264, 2001.
26. Hsieh TC, Burfeind P, Laud K, Backer JM, Traganos F, Darzynkiewicz Z, and Wu JM, *Int J Oncol* 15, 245–252, 1999.
27. Clement MV, Hirpara JL, Chawdhury SH, and Pervaiz S, *Blood* 92, 996–1002, 1998.
28. Le Corre L, Fustier P, Chalabi N, Bignon YJ, and Bernard-Gallon D, *Clin Chim Acta* 344, 115–121, 2004.

29. Cooray HC, Janvilisri T, van Veen HW, Hladky SB, and Barrand MA, *Biochem Biophys Res Commun* 317, 269–275, 2004.
30. Wietzke JA and Welsh J, *J Steroid Biochem Mol Biol* 84, 149–157, 2003.
31. Ciolino HP and Yeh GC, *Mol Pharmacol* 56, 760–767, 1999.
32. Lee JE and Safe S, *Biochem Pharmacol* 62, 1113–1124, 2001.
33. Gerhauser C et al, *Mutation Res* 523–524, 163–172, 2003.
34. Ho SM, *J Cell Biochem* 91, 491–503, 2004.
35. Litvinov IV, De Marzo AM, and Isaacs JT, *J Clin Endocrinol Metab* 88, 2972–2982, 2003.
36. Murillo H, Huang H, Schmidt LJ, Smith DI, and Tindall DJ, *Endocrinology* 142, 4795–4805, 2001.
37. Guseva NV, Taghiyev AF, Rokhlin OW, and Cohen MB, *J Cell Biochem* 91, 70–99, 2004.
38. Kuwajerwala N, Cifuentes E, Gautam S, Menon M, Barrack ER, and Reddy GP, *Cancer Res* 62, 2488–2492, 2002.
39. Mitchell SH, Zhu W, and Young CY, *Cancer Res* 59, 5892–5895, 1999.
40. Narayanan BA, Narayanan NK, Stoner GD, and Bullock BP, *Life Sci* 70, 1821–1839, 2002.
41. Sgambato A, Ardito R, Faraglia B, Boninsegna A, Wolf FI, and Cittadini A, *Mutation Res* 496, 171–180, 2001.
42. Hsieh TC and Wu JM, *Exp Cell Res* 249, 109–115, 1999.
43. Narayanan BA, Narayanan NK, Re GG, and Nixon DW, *Int J Cancer* 104, 204–212, 2003.
44. Kim YA, Rhee SH, Park KY, and Choi YH, *J Med Food* 6, 273–280, 2003.
45. Lin HY, Shih A, Davis FB, Tang HY, Martino LJ, Bennett JA, and Davis PJ, *J Urol* 168, 748–755, 2002.
46. Slee EA, O'Connor DJ, and Lu X, *Oncogene* 23, 2809–2818, 2004.
47. Stewart JR and O'Brian CA, *Invest New Drugs* 22, 107–117, 2004.
48. Sala G, Minutolo F, Macchia M, Sacchi N, and Ghidoni R, *Drugs Exp Clin Res* 29, 263–269, 2003.
49. Scartozzi M, Galizia E, Freddari F, Berardi R, Cellerino R, and Cascinu S, *Cancer Treat Rev* 30, 451–459, 2004.
50. Ushijima T and Sasako M, *Cancer Cell* 5, 121–125, 2004.
51. Srivastava S, Verma M, and Henson DE, *Clin Cancer Res* 7, 1118–1126, 2001.
52. Zhou HB, Chen JJ, Wang WX, Cai JT, and Du Q, *World J Gastroenterol* 10, 1822–1825, 2004.
53. Ding XZ and Adrian TE, *Pancreas* 25, e71–e76, 2002.
54. Atten MJ, Attar BM, Milson T, and Holian O, *Biochem Pharmacol* 62, 1423–1432, 2001.
55. Wolter F and Stein J, *J Nutr* 132, 2082–2086, 2002.
56. Zhou HB, Yan Y, Sun YN, and Zhu JR, *World J Gastroenterol* 9, 408–411, 2003.
57. Mouria M, Gukovskaya AS, Jung Y, Buechler P, Hines OJ, Reber HA, and Pandol SJ, *Int J Cancer* 98, 761–769, 2002.
58. Holian O, Wahid S, Atten MJ, and Attar BM, *Am J Physiol Gastrointest Liver Physiol* 282, G809–G816, 2002.
59. Graham DY, *J Gastroenterol* 35 (Suppl 12), 90–97, 2000.
60. Mahady GB and Pendland SL, *Am J Gastroenterol* 95, 1849, 2000.

61. Wolter F, Turchanowa L, and Stein J, *Carcinogenesis* 24, 469–474, 2003.
62. Schneider Y, Vincent F, Duranton B, Badolo L, Gosse F, Bergmann C, Seiler N, and Raul F, *Cancer Lett* 158, 85–91, 2000.
63. Le Ferrec E et al, *J Biol Chem* 277, 24780–24787, 2002.
64. Wolter F, Akoglu B, Clausnitzer A, and Stein J, *J Nutr* 131, 2197–2203, 2001.
65. Delmas D et al, *J Biol Chem* 278, 41482–41490, 2003.
66. Mahyar-Roemer M, Katsen A, Mestres P, and Roemer K, *Int J Cancer* 94, 615–622, 2001.
67. Schneider Y, Fischer B, Coelho D, Roussi S, Gosse F, Bischoff P, and Raul F, *Cancer Lett* 211, 155–161, 2004.
68. Li ZG et al, *Carcinogenesis* 23, 1531–1536, 2002.
69. Schneider Y, Duranton B, Gosse F, Schleiffer R, Seiler N, and Raul F, *Nutr Cancer* 39, 102–107, 2001.
70. Moon C, Oh Y, and Roth JA, *Clin Cancer Res* 9, 5055–5067, 2003.
71. Mitsuuchi Y and Testa JR, *Am J Med Genet* 115, 183–188, 2002.
72. Casper RF, Quesne M, Rogers IM, Shirota T, Jolivet A, Milgrom E, and Savouret JF, *Mol Pharmacol* 56, 784–790, 1999.
73. Mollerup S, Ovrebo S, and Haugen A, *Int J Cancer* 92, 18–25, 2001.
74. Kubota T, Uemura Y, Kobayashi M, and Taguchi H, *Anticancer Res* 23, 4039–4046, 2003.
75. Kim YA, Lee WH, Choi TH, Rhee SH, Park KY, and Choi YH, *Int J Oncol* 23, 1143–1149, 2003.
76. Yeung F, Hoberg JE, Ramsey CS, Keller MD, Jones DR, Frye RA, and Mayo MW, *EMBO J* 23, 2369–2380, 2004.
77. Revel A, Raanani H, Younglai E, Xu J, Rogers I, Han R, Savouret JF, and Casper RF, *J Appl Toxicol* 23, 255–261, 2003.
78. Kimura Y and Okuda H, *J Nutr* 131, 1844–1849, 2001.
79. Berge GSOV, Eilertsen E, Haugen A, and Mollerup S, *Br J Cancer* 91, 1380–1383, 2004.
80. Hecht SS, Kenney PM, Wang M, Trushin N, Agarwal S, Rao AV, and Upadhyaya P, *Cancer Lett* 137, 123–130, 1999.
81. Hussein MR, Haemel AK, and Wood GS, *J Pathol* 199, 275–288, 2003.
82. Green CL and Khavari PA, *Semin Cancer Biol* 14, 63–69, 2004.
83. Ahmad N, Adhami VM, Afaq F, Feyes DK, and Mukhtar H, *Clin Cancer Res* 7, 1466–1473, 2001.
84. Adhami VM, Afaq F, and Ahmad N, *Biochem Biophys Res Commun* 288, 579–585, 2001.
85. Larrosa M, Tomas-Barberan FA, and Espin JC, *J Agric Food Chem* 51, 4576–4584, 2003.
86. Fuggetta MP, D'Atri S, Lanzilli G, Tricarico M, Cannavo E, Zambruno G, Falchetti R, and Ravagnan G, *Melanoma Res* 14, 189–196, 2004.
87. Niles RM, McFarland M, Weimer MB, Redkar A, Fu YM, and Meadows GG, *Cancer Lett* 190, 157–163, 2003.
88. Jang M et al, *Science* 275, 218–220, 1997.
89. Fu ZD, Cao Y, Wang KF, Xu SF, and Han R, *Ai Zheng* 23, 869–873, 2004.
90. Soleas GJ, Grass L, Josephy PD, Goldberg DM, and Diamandis EP, *Clin Biochem* 35, 119–124, 2002.

91. Caltagirone S, Rossi C, Poggi A, Ranelletti FO, Natali PG, Brunetti M, Aiello FB, and Piantelli M, *Int J Cancer* 87, 595–600, 2000.
92. Asensi M, Medina I, Ortega A, Carretero J, Bano MC, Obrador E, and Estrela JM, *Free Radical Biol Med* 33, 387–398, 2002.
93. Reagan-Shaw S, Afaq F, Aziz MH, and Ahmad N, *Oncogene* 23, 5151–5160, 2004.
94. Kim JW and Wang XW, *Carcinogenesis* 24, 363–369, 2003.
95. Kim HJ, Chang EJ, Bae SJ, Shim SM, Park HD, Rhee CH, Park JH, and Choi SW, *Arch Pharm Res* 25, 293–299, 2002.
96. Kuo PL, Chiang LC, and Lin CC, *Life Sci* 72, 23–34, 2002.
97. Sun ZJ, Pan CE, Liu HS, and Wang GJ, *World J Gastroenterol* 8, 79–81, 2002.
98. Kozuki Y, Miura Y, and Yagasaki K, *Cancer Lett* 167, 151–156, 2001.
99. De Ledinghen V et al, *Int J Oncol* 19, 83–88, 2001.
100. Miura D, Miura Y, and Yagasaki K, *Life Sci* 73, 1393–1400, 2003.
101. Wu SL, Sun ZJ, Yu L, Meng KW, Qin XL, and Pan CE, *World J Gastroenterol* 10, 3048–3052, 2004.
102. Yu L, Sun ZJ, Wu SL, and Pan CE, *World J Gastroenterol* 9, 2341–2343, 2003.
103. Liu HS, Pan CE, Yang W, and Liu XM, *World J Gastroenterol* 9, 1474–1476, 2003.
104. Louis DN, Pomeroy SL, and Cairncross JG, *Cancer Cell* 1, 125–128, 2002.
105. Schwab M, Westermann F, Hero B, and Berthold F, *Lancet Oncol* 4, 472–480, 2003.
106. Nicolini G, Rigolio R, Miloso M, Bertelli AA, and Tredici G, *Neurosci Lett* 302, 41–44, 2001.
107. Savaskan E, Olivieri G, Meier F, Seifritz E, Wirz-Justice A, and Muller-Spahn F, *Gerontology* 49, 380–383, 2003.
108. Wang Q, Li H, Wang XW, Wu DC, Chen XY, and Liu J, *Neurosci Lett* 351, 83–86, 2003.
109. Chen Y, Tseng SH, Lai HS, and Chen WJ, *Surgery* 136, 57–66, 2004.
110. Liu J, Wang Q, Wu DC, Wang XW, Sun Y, Chen XY, Zhang KL, and Li H, *Neurosci Lett* 363, 257–261, 2004.
111. Liontas A and Yeger H, *Anticancer Res* 24, 987–998, 2004.
112. Fulda S and Debatin KM, *Cancer Res* 64, 337–346, 2004.
113. Ragione FD, Cucciolla V, Criniti V, Indaco S, Borriello A, and Zappia V, *J Biol Chem* 278, 23360–23368, 2003.
114. Tseng SH, Lin SM, Chen JC, Su YH, Huang HY, Chen CK, Lin PY, and Chen Y, *Clin Cancer Res* 10, 2190–2202, 2004.
115. Cunat S, Hoffmann P, and Pujol P, *Gynecol Oncol* 94, 25–32, 2004.
116. Shih Ie M and Kurman RJ, *Am J Pathol* 164, 1511–1518, 2004.
117. Ellenson LH and Wu TC, *Cancer Cell* 5, 533–538, 2004.
118. Bhat KP and Pezzuto JM, *Cancer Res* 61, 6137–6144, 2001.
119. Kaneuchi M, Sasaki M, Tanaka Y, Yamamoto R, Sakuragi N, and Dahiya R, *Int J Oncol* 23, 1167–1172, 2003.
120. Opipari AW Jr, Tan L, Boitano AE, Sorenson DR, Aurora A, and Liu JR, *Cancer Res* 64, 696–703, 2004.
121. Yang SH et al, *Int J Oncol* 22, 741–750, 2003.
122. Cao Z, Fang J, Xia C, Shi X, and Jiang BH, *Clin Cancer Res* 10, 5253–5263, 2004.

123. Yu R, Hebbar V, Kim DW, Mandlekar S, Pezzuto JM, and Kong AN, *Mol Pharmacol* 60, 217–224, 2001.
124. Manna SK, Mukhopadhyay A, and Aggarwal BB, *J Immunol* 164, 6509–6519, 2000.
125. Uhle S et al, *EMBO J* 22, 1302–1312, 2003.
126. Woo JH et al, *Oncogene* 23, 1845–1853, 2004.
127. Kadin ME, *Hum Pathol* 34, 305–355, 2003.
128. Surh YJ, Hurh YJ, Kang JY, Lee E, Kong G, and Lee SJ, *Cancer Lett* 140, 1–10, 1999.
129. Gautam SC, Xu YX, Dumaguin M, Janakiraman N, and Chapman RA, *Bone Marrow Transplant* 25, 639–645, 2000.
130. Tsan MF, White JE, Maheshwari JG, Bremner TA, and Sacco J, *Br J Haematol* 109, 405–412, 2000.
131. Estrov Z, Shishodia S, Faderl S, Harris D, Van Q, Kantarjian HM, Talpaz M, and Aggarwal BB, *Blood* 102, 987–995, 2003.
132. Ragione FD et al, *Biochem Biophys Res Commun* 250, 53–58, 1998.
133. Park JW et al, *Cancer Lett* 163, 43–49, 2001.
134. Bernhard D, Tinhofer I, Tonko M, Hubl H, Ausserlechner MJ, Greil R, Kofler R, and Csordas A, *Cell Death Differentiation* 7, 834–842, 2000.
135. Ferry-Dumazet H et al, *Carcinogenesis* 23, 1327–1333, 2002.
136. Jazirehi AR and Bonavida B, *Mol Cancer Ther* 3, 71–84, 2004.
137. Baatout S, Derradji H, Jacquet P, Ooms D, Michaux A, and Mergeay M, *Int J Mol Med* 13, 895–902, 2004.
138. Asou H, Koshizuka K, Kyo T, Takata N, Kamada N, and Koeffier HP, *Int J Hematol* 75, 528–533, 2002.
139. Su JL, Lin MT, Hong CC, Chang CC, Shiah SG, Wu CW, Chen ST, Chau YP, and Kuo ML, *Carcinogenesis* 26, 1–10, 2005.
140. Holmes-McNary M and Baldwin AS Jr, *Cancer Res* 60, 3477–3483, 2000.
141. Shen F, Chen SJ, Dong XJ, Zhong H, Li YT, and Cheng GF, *J Asian Nat Prod Res* 5, 151–157, 2003.
142. Hayashibara T et al, *Nutr Cancer* 44, 193–201, 2002.
143. Tinhofer I et al, *FASEB J* 15, 1613–1615, 2001.
144. Dorrie J, Gerauer H, Wachter Y, and Zunino SJ, *Cancer Res* 61, 4731–4739, 2001.
145. Billard C, Izard JC, Roman V, Kern C, Mathiot C, Mentz F, and Kolb JP, *Leuk Lymphoma* 43, 1991–2002, 2002.
146. Roman V, Billard C, Kern C, Ferry-Dumazet H, Izard JC, Mohammad R, Mossalayi DM, and Kolb JP, *Br J Haematol* 117, 842–851, 2002.
147. Roy M, Chakraborty S, Siddiqi M, and Bhattacharya RK, *Asian Pac J Cancer Prev* 3, 61–67, 2002.
148. Gao X, Xu YX, Divine G, Janakiraman N, Chapman RA, and Gautam SC, *J Nutr* 132, 2076–2081, 2002.
149. Linehan WM, Walther MM, and Zbar B, *J Urol* 170, 2163–2172, 2003.
150. Shi T, Liou LS, Sadhukhan P, Duan ZH, Novick AC, Hissong JG, Almasan A, and DiDonato JA, *Cancer Biol Ther* 3, 882–888, 2004.
151. Bongarzone I and Pierotti MA, *Tumori* 89, 514–516, 2003.
152. Shih A, Davis FB, Lin HY, and Davis PJ, *J Clin Endocrinol Metab* 87, 1223–1232, 2002.
153. Nagler RM, *Anticancer Res* 22, 2977–2980, 2002.

154. Schwartz JL, *Crit Rev Oral Biol Med* 11, 92–122, 2000.
155. Sugerman PB, Joseph BK, and Savage NW, *Oral Dis* 1, 172–188, 1995.
156. Elattar TM and Virji AS, *Anticancer Res* 19, 5407–5414, 1999.
157. Schoonen WM, Salinas CA, Kiemeney LA, and Stanford JL, *Int J Cancer* 113, 133–140, 2005.
158. Levi F, Pasche C, Lucchini F, Ghidoni R, Terraroni M, and La Vecchia C, *Eur J Cancer Prev* 14, 139–142, 2005.
159. Joe AK, Liu H, Suzui M, Vural ME, Xiao D, and Weinstein IB, *Clin Cancer Res* 8, 893–903, 2002.
160. Soleas GJ, Goldberg DM, Grass L, Levesque M, and Diamandis EP, *Clin Biochem* 34, 415–420, 2001.
161. Damianaki A et al, *J Cell Biochem* 78, 429–441, 2000.
162. Bove K, Lincoln DW, and Tsan MF, *Biochem Biophys Res Commun* 291, 1001–1005, 2002.
163. Cardile V et al, *Anticancer Res* 23, 4921–4926, 2003.
164. Kampa M et al, *Nutr Cancer* 37, 223–233, 2000.
165. Morris GZ, Williams RL, Elliott MS, and Beebe SJ, *Prostate* 52, 319–329, 2002.
166. Liang YC, Tsai SH, Chen L, Lin-Shiau SY, and Lin JK, *Biochem Pharmacol* 65, 1053–1060, 2003.
167. Latruffe N, Delmas D, Jannin B, Malki MC, Passilly-Degrace P, and Berlot JP, *Int J Mol Med* 10, 755–760, 2002.
168. Delmas D, Jannin B, Malki MC, and Latruffe N, *Oncol Rep* 7, 847–852, 2000.
169. Melzig MF and Escher F, *Pharmazie* 57, 556–558, 2002.
170. Yoon SH, Kim YS, Ghim SY, Song BH, and Bae YS, *Life Sci* 71, 2145–2152, 2002.
171. Zoberi I, Bradbury CM, Curry HA, Bisht KS, Goswami PC, Roti Roti JL, and Gius D, *Cancer Lett* 175, 165–173, 2002.
172. Kang JH, Park YH, Choi SW, Yang EK, and Lee WJ, *Exp Mol Med* 35, 467–474, 2003.
173. Rodrigue CM, Arous N, Bachir D, Smith-Ravin J, Romeo PH, Galacteros F, and Garel MC, *Br J Haematol* 113, 500–507, 2001.

5 Mechanism of Apoptosis by Resveratrol

Shazib Pervaiz and Andrea Lisa Holme

CONTENTS

INTRODUCTION TO APOPTOSIS

Apoptosis is a term first used in 1972 by Kerr et al. to describe a tissue pathology known as "shrinking necrosis" [1]. Today, the term refers to a mechanism by which cells die in an orderly manner through a genetically governed event (programmed cell death) or via the triggering of a series of parallel, cross-talking pathways that induce orderly cell death. The hallmarks of apoptosis are DNA condensation and fragmentation, loss of cell volume, membrane blebbing, externalization of phosphatidylserine, protease activation, and the loss of mitochondria membrane potential; one or more of these hallmarks may be absent depending on the trigger and pathways adopted during apoptosis [2 and references therein]. As outlined in Figure 5.1, two central pathways lead to apoptosis: (1) positive induction by ligand binding to a plasma membrane death receptor, and (2) negative

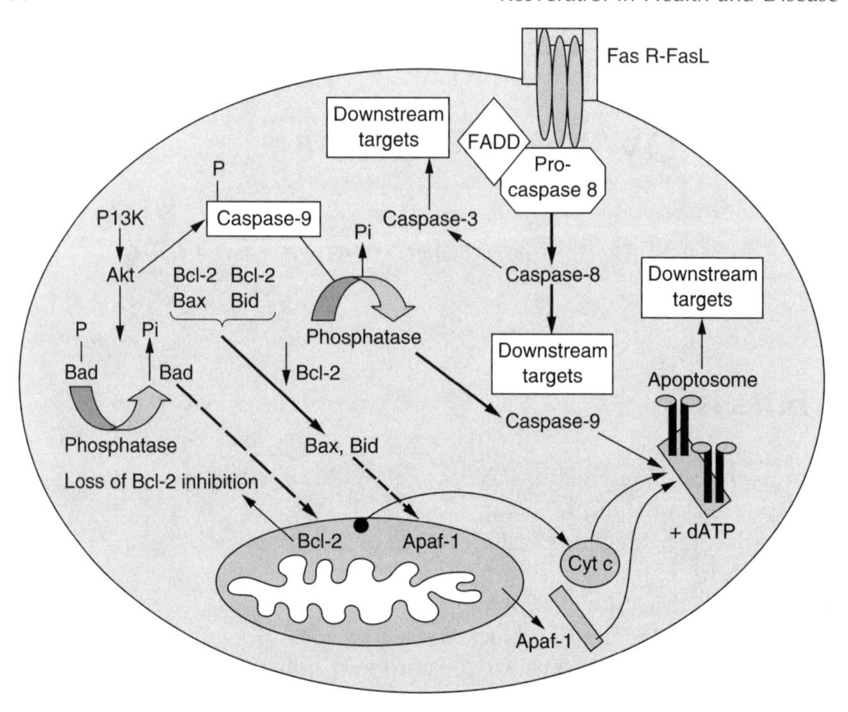

FIGURE 5.1 Overview of the intracellular apoptotic pathways. The extrinsic cell plasma membrane-bound death receptor triggers activation of the DISC, which feeds into downstream effector proteins. The intrinsic pathway can be indirectly activated by changes in the protein profile relating to a loss of mitochondria function or by the death receptors' cross talking with pro- and antiapoptotic proteins (see text for further details).

induction by loss of a suppressor activity involving the mitochondria [2,3]. Each pathway feeds into the activation of a pivotal group of thiol aspartate-specific proteases, i.e., caspases. Caspases are activated by a wide range of stimuli from death receptors, Granzyme B/perforin, loss of survival signals (hormones, cytokines, and serum withdrawal), redox stress, and ultraviolet light and irradiation, to the release of cytochrome c (cyt c) and other death promoting factors (such as AIF (apoptosis inducing factor) and Smac/DIABLO) from the mitochondria into the cytosol, all of which serve as triggers to activate caspases [4 and references therein]. In the classic death receptor model of Fas receptor (FasR(CD95/Apo1)) activation, the Fas ligand (FasL(CD95/Apo1L)) exists as a homotrimeric membrane molecule, with each FasL(CD95/Apo1L) trimer binding three FasR(CD95/Apo1) molecules on the surface of the target cell [5]. This results in the death inducing signaling complex (DISC) that is a complex of Fas (trimer), Fas-associated death domain (FADD), and procaspase-8. Upon recruitment by FADD, procaspase-8 oligomerization drives its self-activation through autoprocessing, and active caspase-8 then activates downstream caspases, committing the cell to apoptosis. The mitochondria play an active role in apoptosis, which is regulated by a group of proteins known as the Bcl-2 family of proteins (antiapoptotic, Bcl-2, Bcl-X_L; and proapoptotic, Bad, Bax, Bid, Bak) [6]. Whereas antiapoptotic members of the Bcl-2 family protect mitochondrial integrity by inhibiting membrane depolarization/ permeabilization and translocation of death amplifying protein, the proapoptotic counterparts neutralize these effects by dislodging antiapoptotic members from the mitochondria and forming oligomers that function as channels for the release of transmembrane proteins. The cytosolic cyt c triggers the recruitment of the protein Apaf-1 (apoptotic protease activating factor-1), procaspase-9, and dATP, to bring about the assembly of the apoptosome [7]. Apoptosome formation triggers activation of caspase-9, which accelerates apoptosis by activating other caspases.

Apoptosis in Cancer Therapy

In normal tissue homeostasis, it is well established that a very delicate balance exists between cell proliferation and cell death. An imbalance of this is seen in carcinogenesis that is a process that involves deregulated growth, with the outcome being enhanced proliferation and resistance to apoptotic triggers. Apoptosis is regarded as an active and progressive response to physiologic and pathologic stimuli, and cells dying by apoptosis cause minimal disturbance to the surrounding tissue. It has been evident that most of today's cancer cytotoxic agents act primarily by inducing apoptosis in cancer cells [8]. Therefore, understanding the discrete apoptotic pathways will not only contribute to our knowledge of the process of oncogenesis, but also pave the way for the development of new strategies toward the prevention and therapy of cancer.

In the quest for new cancer treatments, diet-derived polyphenols are attractive clinical candidates as the long-proven use of their dietary sources suggests low potential for unwanted side effects [9]. Resveratrol is a natural phytoestrogen, and reported to have many functions that strongly support beneficial effects against cancer, inflammatory, and cardiovascular systems [10 and references therein]. The doses of resveratrol able to induce cellular responses that result in growth inhibition, cell cycle arrest, and apoptosis can be divided into three different ranges: (1) in the 1 to 10 μM range specific biochemical effects in cell culture models are induced; (2) in the 10 to 30 μM range it decreases D-type cyclin expression, reduces [^{3}H] thymidine incorporation, inhibits phorbol ester-mediated cyclooxygenase (COX)-2 induction, decreases ornithin decarboxylase activity, and reduces indices of oxidative damage; and (3) in 25 to 100 and 100 to 200 μM concentrations, cytostatic and cytotoxic effects occur, respectively. There is an underlying observation that resveratrol induces a time- and dose-dependent induction of a variety of cellular machinery, with the result that a different pathway(s) dominates at lower dosages versus higher dosages. Thus, the results relating to resveratrol's apoptotic capability are confusing and remain controversial. In the following sections the mechanisms by which resveratrol appears to induce cell death, specifically apoptosis, will be presented with reference to its ability to act through (1) cell death receptors, (2) mitochondria pathway, (3) nuclear factors and cell cycle arrest, and (4) sensitization to drug-induced and free radical-associated apoptosis, in the context of its use as a cancer chemotherapeutic agent.

RESVERATROL AS A MEDIATOR OF CELL DEATH

DEATH RECEPTOR-MEDIATED APOPTOSIS

Initial investigations of resveratrol at relatively high concentrations in HL-60 cells showed a dose-dependent loss of membrane phospholipid asymmetry and DNA fragmentation [11]. This was shown to be due to resveratrol's ability to enhance FasL(CD95/Apo1L) expression in HL-60 cells (and T47D breast carcinoma cells) and resulted in a Fas-dependent cell death. In contrast, treatment of normal human peripheral blood lymphocytes (PBLs) with resveratrol had no effect on either Fas(CD95/Apo1) or FasL(CD95L/Apo1L) expression and did not result in cell death. This was the first report of resveratrol's ability to function via an autocrine loop to induce cell death by the positive induction of gene expression of extrinsic death triggers. Further independent data confirm resveratol's ability to work through a Fas-dependent pathway to induce death, and recent evidence has expanded this in a study using human SW480 colon cancer cells treated with 10 to 100 μM of resveratrol. In this model, various caspases (caspase-9, -3, and -8) are activated and trigger apoptosis via Fas(CD95/Apo1), tumor necrosis factor (TNF) and TRAIL (TNF-related apoptosis-inducing ligand/Apo2L) receptor activation [12,13]. Caspase activation was associated with the accumulation of the proapoptotic proteins Bax and Bak that relocalize to the mitochondria prior to membrane depolarization. Importantly, resveratrol first induced the clustering of Fas(CD95/Apo1) and its redistribution in cholesterol- and sphingolipid-rich fractions, together with FADD and procaspase-8, i.e., DISC formation [12]. However, FasL(CD95L/Apo1L) expression and localization remained unchanged leading to the suggestion that the accumulation of Fas(CD95/Apo1) in the membrane was sufficient to induce DISC activation. Investigations into the need for DISC recruitment was done by transient transfection of either a dominant-negative mutant of FADD, E8, or MC159 viral protein that interferes with the DISC function, and all were able to decrease the apoptotic response and partially blocked resveratrol-induced Bax and Bak relocalization. The cholesterol sequestering agent nystatin also prevents resveratrol-induced death receptor redistribution and cell sensitization to death receptor stimulation. Thus, in some cells resveratrol does not appear to increase the number of death receptors at the surface of tumor cells, but rather it induces their externalization via their redistribution into lipid rafts and facilitates the caspase cascade activation in response to death receptor stimulation.

Another recent report showed that in HL-60 cells exposed to 20 μM of resveratrol the activation of the MAPKKK apoptosis signal-regulating kinase 1 (Ask-1) was induced, which, in turn, activated the downstream kinases c-Jun N-terminal kinase (JNK) and p38 MAPK, but not the extracellular signal-regulated kinase (ERK) [14]. Only inhibition of JNK reduced FasL(CD95L/Apo1L) expression and DNA fragmentation induced by resveratrol. Resveratrol also activated the small GTP-binding protein Cdc42, rather than other members such as RhoA or Rac1, and expression of a mutant Cdc42 (N17Cdc42) reduced resveratrol-induced JNK activity, FasL(CD95L/Apo1L) expression, and apoptotic cell death. These results further substantiate a mechanism involving the Fas(CD95/Apo1) cell death pathway, and shows that resveratrol can induce apoptosis through the Cdc42/Ask-1/JNK/FasL(CD95L/Apo1L) signaling cascade in HL-60 cells.

TRAIL belongs to the tumor necrosis factor family, and is known to induce rapid apoptosis in tumor cells [15,16]. Resveratrol has recently been demonstrated to be a novel potentiator for TRAIL-induced apoptosis. This is discussed in more detail in the section on sensitization to drug-induced apoptosis.

Mitochondria-Mediated Apoptosis

Recruitment of the mitochondria is facilitated by the proapoptotic Bcl-2 family proteins, such as Bax and Bid, that translocate to the mitochondria and trigger mitochondrial permeability transition by changing the conformation of the mitochondrial inner membrane pore (PT pore) or through PT pore-independent localization to the outer membrane. Exposure of HL-60 cells to resveratrol resulted in changes in the mitochondria transmembrane potential and the release of cyt c, suggesting involvement of the mitochondria in resveratrol-induced apoptosis [17]. In another study, it has been reported that resveratrol-treated cells underwent S-phase arrest prior to Fas(CD95/Apo1)-independent apoptosis in CEM-C7H2 acute leukemia cells, and that the resultant apoptosis was mediated via conversion of procaspase-6 to its active form, while caspase-3 and caspase-2 were proteolytically activated to a much lesser extent [18].

Since these observations, a number of other reports have corroborated the mitochondria-dependent apoptotic activity of resveratrol in a variety of tumor cell lines, and the role of the mitochondria apoptotic pathway in the absence of the Fas(CD95/Apo1)-dependent pathway to induce cell death. A recent study using K562 and HSB-2 cells showed that, depending on the cell line, the caspase activation profile parallels the mitochondria activation or a lag period exists. This was demonstrated by exposing K562 cells to 25 μM of resveratrol, which resulted in cells arresting in the S phase of the cell cycle, and activation of caspase-3, -8, and -9 by 72 hours with a corresponding increase in Bax located at the mitochondria [19]. In contrast, HSB-2 cells were arrested in the G1 phase and caspase-8 and -3 increased by 48 hours. Caspase-9 increased by 24 hours and correlated with Bax translocated to the mitochondria and mitochondria activation; however, by 48 hours caspase-9 activity had decreased. In both cell lines, DNA fragmentation correlated with cell cycle inhibition and caspase-3 activation. K562 has a higher basal level of glutathione (GSH) than HSB-2, and the level of GSH, glutathione reductase, and peroxidase increased in both cells. This is likely the reason that resveratrol had the ability to act as an antioxidant when cells were exposed to *t*-butylhydroperoxide (a more stable analog of hydrogen peroxide) and then exposed to 25 μM of resveratrol for 10 to 20 minutes or 11 to 72 hours in this study.

There have been a number of other reports pointing to the upregulation of the proapoptotic protein Bax and downregulation of Bcl-2 upon exposure to resveratrol at a concentration range of 10 to 200 μM [20–24]. However, a recent report and our data (unpublished results) seem to suggest that resveratrol-mediated apoptosis is independent of Bax, as gene knockout of Bax did not alter tumor cell sensitivity to resveratrol [25]. Interestingly, resvertrol has been shown to bind to respiratory chain proteins, as well as a reductase involved in the detoxification of quinones, suggesting that resveratrol could act directly at the site of the mitochondria to induce mitochondrial stress [26,27].

Apoptosis Mediated by Nuclear Factors

Most of the effects seen with the use of resveratrol relate to its role as a modulator of transcription by either activating or repressing a sizable number of primary and secondary target genes via influencing transcription factors. Resveratrol has been shown to affect the expression of genes involved in cell cycle regulation and apoptosis, including cyclins, cyclin-dependent kinases (cdks), p53, and cdk inhibitors. One such target is the transcription factor nuclear factor kappa beta (NFκB), of which resveratrol blocks the activation without necessarily affecting its basal activity [28]. Resveratrol can also suppress TNF-induced phosphorylation and nuclear translocation of the p65 subunit of NFκB, and NFκB-dependent reporter gene transcription [29]. Suppression of TNF-induced NFκB activation by resveratrol is observed in myeloid cells (U-937), lymphoid (Jurkat), and epithelial (HeLa and H4) cells, and NFκB activation induced by PMA, LPS, H_2O_2, okadaic acid, and ceramide is also blocked by resveratrol.

Most agents that work through NFκB also affect the transcription factor activator protein-1 (AP-1). In HT29 cells stably transfected with AP-1, resveratrol causes an increase in AP-1 activity at low doses, while higher doses cause AP-1's activity to decrease and this in turn induces cell death. This decrease correlates with an increase in cyclin D1 and antiproliferative effects [30]. It is established that the activation of AP-1 is mediated by JNK and the upstream kinase MEK, and it has been reported that the TNF-induced activities of JNK and MEK are inhibited by resveratrol, thus providing a possible mechanism for AP-1 inhibition [31]. Interestingly, iNOS (inducible nitric oxide synthase) and COX-2 are among the proteins induced by NFκB and AP-1, both of which decrease with resveratrol treatment. The mechanism(s) by which this occurs remains unclear. However, several proposals have been put forward, including inhibition of Bcl-2 expression, accumulation of arachidonic acid, stimulation of ceramide production, dephosphorylation of Akt and ERK proteins, inhibition of PPARδ (peroxisome proliferator-activated receptor), and interference with the NFκB signaling pathway, all of which result in an antiproliferative effect and a decrease in cell survival [10,32 and references therein].

In a study using prostate androgen-sensitive and -insensitive cell lines, it was observed that an increase in DNA synthesis only occurred in androgen-sensitive LNCaP cells, and not in the androgen-insensitive prostate cells DU145 or in NIH3T3 fibroblast cells [33]. In LNCaP cells, the effect of resveratrol on DNA synthesis varies dramatically depending on the concentration and the duration of treatment. In cells treated for 1 hour, resveratrol showed only an inhibitory effect on DNA synthesis, which increased with increasing concentration. A dual effect of resveratrol was observed on DNA synthesis when treated for a 24-hour period: at 5 to 10 μM, a 2- to 3-fold increase in DNA synthesis was reported, and at 15 μM or more DNA synthesis was inhibited. The resveratrol-induced increase in DNA synthesis was associated with S-phase arrest, a concurrent decrease in nuclear p21Cipl and p27Kip1 levels, and an increase in nuclear Cdk2 activity associated with both cyclin A and cyclin E. Another study using DU145 cells showed that the loss of DNA synthesis was associated with the inhibition of the protein activity of D-type cyclins and Cdk4 expression, and the induction of p53 and p21 [34]. Resveratrol treatment also upregulated the Bax protein and mRNA expression in a dose-dependent manner that correlated with the activation of caspase-3 and -9. In another study, on the same cell line, resveratrol induced apoptosis via activated MAPK and increased serine-15 phosphorylation of p53 [35]. The PD 98059 MAPK inhibitor blocked the phosphorylation and apoptosis, implicating MAPK activation in the phosphorylation of p53 and a requirement for serine-15 phosphorylation of p53 for apoptosis.

Resveratrol also affects the cell cycle in other hormone-dependent and -independent cancers, as demonstrated in the human breast cancer cell lines estrogen receptor (ER)-positive MCF-7 and ER-negative MDA-MB-231, where it inhibits cell proliferation and viability by inducing death in both cell lines in a concentration- and cell-specific manner [36]. In MDA-MB-231, resveratrol (up to 200 μM) lowered the expression and kinase activities of positive G1/S and G2/M cell cycle regulators, and inhibited ribonucleotide reductase activity without a significant effect on the low expression of the tumor suppressors p21, p27, and p53. These cells reportedly died by a nonapoptotic process in the absence of a significant change in the cell cycle distribution. In contrast, resveratrol (less than 50 μM) added to MCF-7 produced a significant and transient increase in the expression and kinase activities of positive G1/S and G2/M regulators and the cell cycle inhibitor p21 expression was markedly induced in the presence of high levels of p27 and p53. These opposing effects resulted in cell cycle blockade at the S phase and apoptotic induction in MCF-7 cells. Thus, the antiproliferative activity of resveratrol rests in the concentration used and the characteristics of the target cell to allow differential regulation of the cell cycle to induce apoptosis or necrosis.

Resveratrol Affects Nuclear Receptors to Induce Apoptosis

The cytochrome P450 monooxygenases (CYP1) are important drug-metabolizing enzymes that play a major role in the detoxification and elimination of hydrophobic xenobiotics. Paradoxically, these enzymes also generate reactive metabolites which can form DNA adducts and lead to mutations. Studies using human medulloblastoma cell lines (Med-3, UW228-1, -2, and -3) showed that CYP1A1 upregulation is paralleled with resveratrol-induced differentiation and apoptosis [37,38]. Similar effects on other CYP450 isozymes, such as CYP1A2 and CYP3A4, have also been documented; thus their increased expression can be correlated to the detrimental accumulation of toxins [39]. One of the ways that resveratrol is known to regulate their activity is through their expression via the arylhydrocarbon receptor (AhR), which, once activated by ligands such as TCDD or planar aromatic hydrocarbons, translocates from the cytosol into the nucleus and heterodimerizes with Arnt (AhR nuclear translocator). There it binds to a class of promoter DNA sequences called the xenobiotic-responsive elements (XRE) and has been shown to be important in the induction of CYP1A1, CYP1A2, and CYP1B1, and other drug-metabolizing enzymes including glutathione *S*-transferase, UDP-glucuronyltransferase, aldehyde dehydrogenase, and NADP(H):oxidoreductase. Resveratrol is known to be able to inhibit the binding of the nuclear AhR to the xenobiotic response element of the CYP1A1 promoter without direct binding to the AhR, and others have showed that resveratrol inhibits AhR-induced CYP1A1 expression at the mRNA and protein levels [40–43]. Thus, their loss leads to toxic mediated death due to the accumulation of compounds that generate free radicals.

Resveratrol also has the ability to act as either an agonist or antagonist to nuclear steroid hormone receptors. One such receptor that resveratrol affects in an antagonistic manner is the androgen receptor (AR) which controls the transcription of androgen-inducible genes such as prostate specific antigen (PSA) and is implicated in the development of prostate cancer [44]. Treatment of LNCaP cells with resveratrol at 10 μM resulted in the inhibition of the androgen responsive growth via pathways that involve a decrease in the transcriptional activity of PSA, AR coactivator ARA 24, and NFκB p65 after 48 hours. Altered expression of these genes is associated with an activation of p53-responsive genes such as p53, PIG 7, p21(Waf1-Cip1), p300/CBP, and Apaf-1 [45].

RESVERATROL: A PLURIPOTENT PHYTOALEXIN

Resveratrol is also classified as a phytoestrogen, as it can bind the estrogen receptor (ER) [46–48]. ER is a nuclear steroid receptor that binds estrogens and regulates the transcription of estrogen-responsive genes by either binding directly to DNA, at particular sequences called estrogen response elements (EREs), or by interacting with other transcription factors, e.g., Sp1, bound to their cognate sites on DNA. There are two known ER subtypes, ERα and the more recently identified ERβ, which exhibit different patterns of tissue distribution and have select differences in biochemical properties. When activated by an agonist ligand, ERα interacts with coactivators, e.g., SRC-1 and CBP, that either acetylate lysine residues on histones to alter chromatin conformation and/or interact with components of the RNA polymerase II initiation complex to enhance target gene transcription. When MCF-7 cells are treated with resveratrol, transcription of estrogen responsive genes transfected into these cells occurs [47]. This activity was dependent upon the presence of the ERE sequence and the type of ER; higher transcriptional activity was seen with ERβ than ERα. However, the reported super agonist activity of resveratrol in the presence of estradiol was contradicted by reports demonstrating antiestrogenic activity of resveratrol; resveratrol suppressed estradiol-induced progesterone receptor expression. In another breast cancer cell line T47D (estrogen dependent), low concentrations of resveratrol (10 μM) resulted in an increase in cell proliferation and growth, whereas at relatively higher concentrations the cells underwent apoptotic death [11,48]. The estrogenic activity of resveratrol has also been demonstrated in ER + pituitary cells that undergo a significant increase in prolactin secretion, and in MC3T3E1 osteoblastic cells that respond to resveratrol by increasing alkaline phosphatase and hydroxylase activity, indicating estrogenic and bone loss preventive effects [49,50].

An interesting development is the report that resveratrol-induced apoptotic death through the ERα occurs by a novel physiologic, nonnuclear function of ERα, which can modulate phosphoinositide 3-kinase (PI3K) activity in MCF-7 cells [51]. Phosphoprotein kinase B (pPKB/AKT) followed the pattern of PI3K activity, and the AKT downstream target, glycogen synthase kinase 3 (GSK3), also showed a phosphorylation pattern that followed PI3K activity. Resveratrol's modulation of this pathway can be inhibited by the antiestrogen ICI 182780, and immunoprecipitation and kinase activity assays showed that resveratrol has a biphasic mechanism effect: at concentrations close to 10 μM PI3K activity was reported to be at the maximum, while above 50 μM there was a decrease in PI3K activity. These observations have led to some controversy with respect to the use of resveratrol or similar compounds as therapeutic agents against ER + breast cancer cells [52].

Resveratrol Sensitization to Drug-Induced Apoptosis

Resveratrol has received attention for its potential chemopreventive and antitumor effects in experimental systems. Recent evidence suggests that paclitaxel, alone or in combination with other drugs such as resveratrol, can be effectively used in the treatment of non-Hodgkin's lymphoma (NHL) and multiple myeloma (MM) [53]. Both resveratrol and paclitaxel arrested the cells at the G2/M phase of the cell cycle. Paclitaxel downregulated the expression of Bcl-X_L, and Mcl-1, and upregulated Bid and Apaf-1. The addition of low concentrations of resveratrol exerted a sensitizing effect on drug-refractory NHL and MM cells to apoptosis induced by paclitaxel, via selectively downregulating the expression of Bcl-X_L and myeloid cell differentiation factor-1 (Mcl-1) while upregulating the expression of Bax and Apaf-1. This inhibition of Bcl-X_L expression was due to the inhibition of the ERK1/2 pathway and diminished AP-1-dependent Bcl-X_L expression. This combination treatment resulted in apoptosis through the formation of tBid, mitochondrial membrane depolarization, release of cyt c and Smac/DIABLO, activation of the caspase cascade, and cleavage of poly(adenosine diphosphate-ribose) polymerase. Promisingly, resveratrol with paclitaxel had minimal cytotoxicity against quiescent and mitogenically stimulated human peripheral blood mononuclear cells.

Another recent report showed that resveratrol is a potent sensitizer of tumor cells for TRAIL-induced apoptosis through p53-independent induction of p21, and p21-mediated cell cycle arrest associated with survivin depletion [54]. Importantly, resveratrol sensitized various tumor cell lines but not normal human fibroblasts for apoptosis induced by death receptor ligation or anticancer drugs. Thus, this combined sensitizer (resveratrol)/inducer (e.g., TRAIL) strategy may be a novel approach to enhance the efficacy of TRAIL-based therapies in a variety of human cancers.

In addition, treatment of human ovarian carcinoma cells with 50 to 100 μM of resveratrol was reported to induce the biochemical hallmarks of apoptosis; however, interestingly, electron microscopic analysis of cells showed features consistent with autophagocytosis, thereby demonstrating a novel mechanism of death induction by this polyphenolic compound [55].

Clinical Relevance: A Conundrum

In 1996 Jang et al. demonstrated the ability of resveratrol to inhibit diverse cellular events associated with the three major stages of carcinogenesis, namely initiation, promotion, and progression [56]. This paper showed that resveratrol had the ability to inhibit the development of preneoplastic lesions in carcinogen-treated mouse mammary glands in culture and inhibit tumorigenesis in a mouse skin cancer model. Data from preclinical efficacy papers show that very low daily doses of resveratrol (between 200 μg/kg and 2 mg/kg), which are likely to give peak plasma concentrations of unmetabolized resveratrol of between 20 nM and 2 μM, are sufficient to exert potent cancer chemopreventive efficacy and pharmacodynamic activity in three chemical-induced rat carcinogenesis models (Table 5.1 and references therein) [57–59]. These data have turned resveratrol into one of the most effective diet-derived chemopreventive polyphenols ever investigated.

In contrast, most mechanistic studies *in vitro* suggest that carcinogenesis-modulating effects of resveratrol require the sustained presence of 5 to 100 μM. A study by Banerjee et al. demonstrated resveratrol's impressive breast cancer chemopreventive efficacy *in vivo* at a dose of 1 mg/kg, but in contrast in MCF-7 cells *in vitro* in the same study, to obtain similar growth inhibition and NFκB inactivation required resveratrol concentrations of 25 to 50 μM [60]. Indeed, the investigations of resveratrol's biochemical and cellular mechanisms of activity *in vitro* and *in vivo* (summarized in Table 5.1) suggest changes relevant to cancer chemoprevention at the cell or subcellular target system level, at concentrations between 10 and 100 μM.

Regulation of growth and proliferation in untransformed cells is maintained via regulation of the cell cycle by the cell cycle checkpoint proteins, in particular p53, Rb, p27, and the cdk inhibitor $p21^{Waf1/Cip1}$ [61]. Any alteration in the normal functioning of these proteins allows the cells to undergo unabated cycling resulting in accumulation of DNA mutations, and thereby carcinogenesis. A number of studies have now established that resveratrol inhibits cellular proliferation by inducing cell cycle arrest in the G1/S phase [9–20,28,54]. The phytoestrogen activity of resveratrol has also been inferred to account for its effects on hormone-regulated events contributing to cell proliferation and viability; in breast cancer cells, resveratrol has an antiproliferative effect by antagonizing the stimulatory effects of estrogen.

In theory antioxidants are effective anticancer drugs, as they can return the intracellular environment to that which is permissible to induce cell death. Resveratrol is a potent antioxidant, and in line with this observation, evidence from this laboratory showed that exposure of human leukemia cells to low concentrations of resveratrol (4 to 8 μM) inhibits caspase activation, DNA fragmentation, and translocation of cyt c induced by hydrogen peroxide (H_2O_2) or the anticancer drugs vincristine

TABLE 5.1
Mechanisms of Resveratrol in Cells *In Vitro* and *In Vivo* Related to Cancer

Mechanism	*In vitro* model	Effective dose (μM)	Ref.	
Growth inhibition	Multiple cell lines	5–10	73	
Apoptosis induction	Leukemia cells	32–100	11, 74	
p53-independent apoptosis	Colon tumor cells	100	75	
Estrogen agonism	Mammary cells	10–25	76	
Antiestrogenicity	Mammary cells	0.1–1	76	
Inhibition of oxygen radical formation	Macrophages	30	77	
Inhibition of cytochrome P450	Liver cells, microsomes, recombinant enzyme	1–20	40, 78, 88	
Activation of p53	Mouse epidermal cells	20	79	
Activation of c-jun kinase	Mouse epidermal cells	10–40	80	
Decrease in COX-2 expression	Mammary epithelial cells	5	81	
Increase in p21/Cip1, cyclins D1, D2, E; decrease in cdks 2, 4, 6	Epidermoid carcinoma cells	10	82	
Increase in cyclins A, B1, and cdks 1 and 2	Colon tumor cells	30	83	
Inhibition of protein kinase D activity	Fibroblasts	>100	84	
Inhibition of NFκB activation	Monocytes, macrophages	30	85	
Inhibition of NFκB and AP-1 activation	Myeloid, lymphoid, epithelial cells	5	86	
***In vivo* model**	**Daily dose[a]**	**Route**	**Efficacy**	**Ref.**
N-nitroso-*N*-methylurea-induced breast cancer in rat	100 mg/kg	Immunoglobin	+	70
Azoxymethane-induced colon cancer in rat	200 μg/kg	Drinking water	+	71
7,12-Dimethylbenz(a)anthracene-induced breast cancer in rat	1 mg/kg	Diet	+	57
N-nitrosomethylbenzylamine-induced esophageal in rat	1 or 2 mg/kg	i.g. or i.p.	+	72
$Apc^{Min+/+}$ mouse (intestinal neoplasia)	15 mg/kg	Drinking water	+	87

[a] Estimated doses of resveratrol admixed to the diet or drinking water.

and daunorubicin. However, in contrast to its apoptosis-inducing activity, resveratrol has also been shown to inhibit apoptosis in some systems. In this regard, an earlier report indicated that resveratrol interfered with H_2O_2-induced apoptotic signal [62]. We have previously shown that H_2O_2 triggers apoptosis by decreasing intracellular superoxide (O_2^-) and cytosolic pH, thus creating a permissive intracellular milieu for death execution [63,64]. A slight increase in intracellular O_2^- can inhibit receptor- or drug-induced apoptosis via direct or indirect effect on caspase activation pathways [65,66]. Our recent *in vitro* data seem to suggest a biphasic concentration–response relationship. Contrary to the apoptosis-inducing activity at relatively high concentrations (30 to 100 μM), we showed that prior exposure of cells to low concentrations of resveratrol (4 to 8 μM) could create an intracellular milieu resistant to apoptosis induced by either H_2O_2 or anticancer drugs [67,68]. These data sound a cautionary note for the use of resveratrol in combination chemotherapy regimens as the slight prooxidant effect of resveratrol could provide cancer cells with a survival advantage by impeding death execution signals. Obviously, these data are in contradiction to earlier reports describing the ability of resveratrol to enhance sensitivity of tumor cells to drug-induced apoptosis. However, in the reports demonstrating sensitizing effect of resveratrol, cells were preexposed for 24 hours with resveratrol followed by subsequent treatment with the chemotherapeutic agent [54]. Whereas preexposure with 30 to 100 μM resveratrol for 24 hours had some effect on cell proliferation, there was no significant effect induced at 10 μM. More importantly, this *in vitro* setting may not be predictive of the *in vivo* efficacy of this regimen, given the observations that the peak plasma levels of unmetabolized resveratrol in the rat are well below 10 μM (even after a high oral dose of 50 mg/kg body weight), and the relatively low bioavailability due to its biotransformation and rapid elimination [57]. In addition, bioavailability data from human subjects administered resveratrol at a dose of 360 μg/kg (20 times that associated with moderate wine intake) revealed plasma peak levels within 30 minutes of 20 nM of unmetabolized resveratrol (and 2 μM total), and a rapid plasma elimination half-life of 12 to 15 minutes [69 and references therein]. The relatively low bioavailability of resveratrol, its rapid clearance, and relatively poor accumulation in specific tissues in rats as detailed earlier, suggest that a long preincubation of cells with resveratrol may not represent the true clinical scenario. Indeed, the inhibitory effect of resveratrol at low concentrations (< 10 μM) on drug-induced apoptosis becomes significant in view of the stimulatory effect of 10 μM resveratrol on PI3K/Akt activity reported in MCF-7 breast cancer cells [51]. Thus, considering the perplexing discrepancy between the biologically relevant concentrations *in vitro* and the relatively low bioavailability *in vivo*, additional efficacy studies in animal models and bioavailability data in humans are required to derive any meaningful clinical relevance from *in vitro* studies.

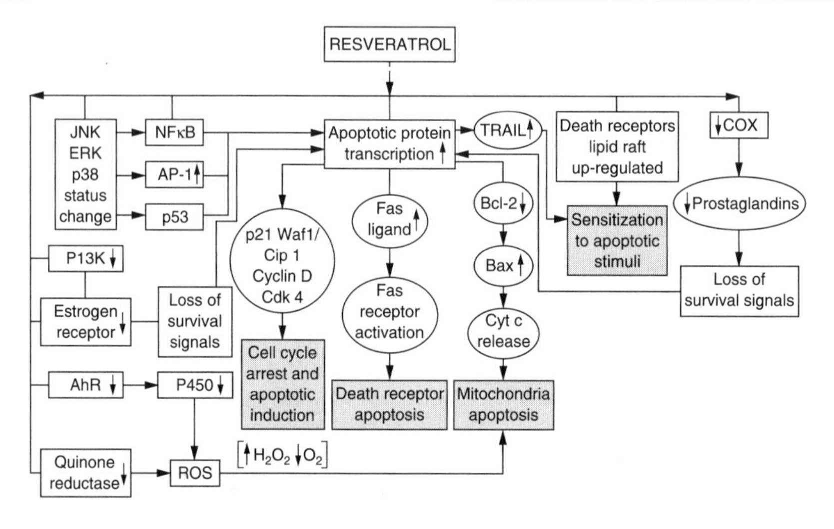

FIGURE 5.2 Overview of resveratrol's known targets (see text for further details).

SUMMARY

As outlined in this chapter and summarized in Figure 5.2, resveratrol can induce the transcription of a multitude of apoptotic proteins allowing the establishment of an autocrine loop to result in the activation of Fas(CD95/Apo1)-dependent apoptosis, or to activate/inactivate the intracellular survival pathways, transcription factors, and hormone receptors to induce mitochondria apoptosis. This ability to act as a pleiotropic agent of death is what renders resveratrol such an effective anticancer drug, but which also demands that its usage as a cytotoxic agent be done so with reference to the individual cancer phenotype.

REFERENCES

1. Kerr JF, Wyllie AH, and Currie AR, Apoptosis: a basic biological phenomenon with wide-ranging implications in tissue kinetics, *Br J Cancer* 26, 239–257, 1972.
2. Ashkenazi A and Dixit VM, Death receptors: signaling and modulation, *Science* 281, 1305–1308, 1998.
3. Green DR and Reed JC, Mitochondria and apoptosis, *Science* 281, 1309–1312, 1998.
4. Thornberry NA and Lazebnik Y, Caspases: enemies within, *Science* 281, 1312–1316, 1998.
5. Peter ME and Krammer PH, The CD95(APO-1/Fas) DISC and beyond, *Cell Death Differentiation* 10, 26–35, 2003.
6. Perfettini J-L, Romano T, Kroemer RT, and Guido Kroemer G, Fatal liaisons of p53 with Bax and Bak, *Nature Cell Biol* 6, 386–388, 2004.

7. Alnemri ES, Hidden powers of the mitochondria, *Nature Cell Biol* 1, E40–E42, 1999.
8. Denicourt C and Dowdy SF, Targeting apoptotic pathways in cancer cells, *Science* 305, 1411–1413, 2004.
9. Kelloff GJ, Crowell JA, Steele VE, Lubet RA, Malone WA, Boone CW, Kopelovich L, Hawk ET, Lieberman R, Lawrence JA, Ali I, Viner JL, and Sigman CC, Progress in cancer chemoprevention: development of diet-derived chemopreventive agents, *J Nutr* 130 (2S Suppl), 467S–471S, 2000.
10. Pervaiz S, Resveratrol: from grapevines to mammalian biology, *FASEB J* 14, 1975–1985, 2003.
11. Clement MV, Hirpara JL, Chawdhury SH, and Pervaiz S, Chemopreventive agent resveratrol, a natural product derived from grapes, triggers CD95 signaling-dependent apoptosis in human tumor cells, *Blood* 92, 996–1002, 1998.
12. Delmas D, Rebe C, Lacour S, Filomenko R, Athias A, Gambert P, Cherkaoui-Malki M, Jannin B, Dubrez-Daloz L, Latruffe N, and Solary E, Resveratrol-induced apoptosis is associated with Fas redistribution in the rafts and the formation of a death-inducing signaling complex in colon cancer cells, *J Biol Chem* 278, 41482–41490, 2003.
13. Delmas D, Rebe C, Micheau O, Athias A, Gambert P, Grazide S, Laurent G, Latruffe N, and Solary E, Redistribution of CD95, DR4 and DR5 in rafts accounts for the synergistic toxicity of resveratrol and death receptor ligands in colon carcinoma cells, *Oncogene* 23, 8979–8986, 2004.
14. Su JL, Lin MT, Hong CC, Chang CC, Shiah SG, Wu CW, Chen ST, Chau YP, and Kuo ML, Resveratrol induces FasL-related apoptosis through Cdc42 activation of ASK1/JNK-dependent signaling pathway in human leukemia HL-60 cells, *Carcinogenesis* 25, 2005–2013, 2004.
15. Wiley SR, Schooley K, Smolak PJ, Din WS, Huang CP, Nicholl JK, Sutherland GR, Smith TD, Rauch C, Smith CA, and Goodwin RG, Identification and characterization of a new member of the TNF family that induces apoptosis, *Immunity* 3, 673–682, 1995.
16. Walczak H, Miller RE, Ariail K, Gliniak B, Griffith TS, Kubin M, Chin W, Jones J, Woodward A, Le T, Smith C, Smolak P, Goodwin RG, Rauch CT, Schuh JC, and Lynch DH, Tumoricidal activity of tumor necrosis factor-related apoptosis-inducing ligand *in vivo*, *Nat Med* 5, 157–163, 1999.
17. Dorrie J, Gerauer H, Wachter Y, and Zunino SJ, Resveratrol induces extensive apoptosis by depolarizing mitochondrial membranes and activating caspase-9 in acute lymphoblastic leukemia cells, *Cancer Res* 61, 4731–4739, 2001.
18. Bernhard D, Tinhofer I, Tonko M, Hubl H, Ausserlechner MJ, Greil R, Kofler R, and Csordas A, Resveratrol causes arrest in the S-phase prior to Fas-independent apoptosis in CEM-C7H2 acute leukemia cells, *Cell Death Differentiation* 7, 834–842, 2000.
19. Luzi C, Brisdelli F, Cinque B, Cifone G, and Bozzi A, Differential sensitivity to resveratrol-induced apoptosis of human chronic myeloid (K562) and acute lymphoblastic (HSB-2) leukemia cells, *Biochem Pharmacol* 68, 2019–2030, 2004.
20. Kim YA, Lee WH, Choi TH, Rhee SH, Park KY, and Choi YH, Involvement of p21WAF1/CIP1, pRB, Bax and NF-kappaB in induction of growth arrest

and apoptosis by resveratrol in human lung carcinoma A549 cells, *Int J Oncol* 23, 1143–1149, 2003.
21. Kuo PL, Chiang LC, and Lin CC, Resveratrol-induced apoptosis is mediated by p53-dependent pathway in Hep G2 cells, *Life Sci* 72, 23–34, 2002.
22. Lu J, Ho CH, Ghai G, and Chen KY, Resveratrol analog, 3,4,5,4′-tetrahydroxystilbene, differentially induces pro-apoptotic p53/Bax gene expression and inhibits the growth of transformed cells but not their normal counterparts, *Carcinogenesis* 22, 321–328, 2001.
23. Park JW, Choi YJ, Suh SI, Baek WK, Suh MH, Jin IN, Min DS, Woo JH, Chang JS, Passaniti A, Lee YH, and Kwon TK, Bcl-2 overexpression attenuates resveratrol-induced apoptosis in U937 cells by inhibition of caspase-3 activity, *Carcinogenesis* 22, 1633–1639, 2001.
24. Ahmad KA, Iskandar KB, Hirpara JL, Clement MV, Pervaiz S, Hydrogen peroxide-mediated cytosolic acidification is a signal for mitochondrial translocation of Bax during drug-induced apoptosis of tumor cells, *Cancer Res* 64, 7867–7878, 2004.
25. Mahyar-Roemer M, Kohler H, and Roemer K, Role of Bax in resveratrol-induced apoptosis of colorectal carcinoma cells, *BMC Cancer* 2, 27–36, 2002.
26. Wang Z, Hsieh TC, Zhang Z, Ma Y, and Wu JM, Identification and purification of resveratrol targeting proteins using immobilized resveratrol affinity chromatography, *Biochem Biophys Res Commun* 323, 743–749, 2004.
27. Buryanovskyy L, Fu Y, Boyd M, Ma Y, Hsieh TC, Wu JM, and Zhang Z, Crystal structure of quinone reductase 2 in complex with resveratrol, *Biochemistry* 43, 11417–11426, 2004.
28. Gao X, Xu YX, Janakiraman N, Chapman RA, and Gautam SC, Immunomodulatory activity of resveratrol: suppression of lymphocyte proliferation, development of cell-mediated cytotoxicity, *Biochem Pharmacol* 62, 1299–1308, 2001.
29. Bhat KP and Pezzuto JM, Cancer chemopreventive activity of resveratrol, *Ann NY Acad Sci* 957, 210–229, 2002.
30. Jeong WS, Kim IW, Hu R, and Kong AN, Modulation of AP-1 by natural chemopreventive compounds in human colon HT-29 cancer cell line, *Pharm Res* 21, 649–660, 2004.
31. Yu R, Hebbar V, Kim DW, Mandlekar S, Pezzuto JM, and Kong AN, Resveratrol inhibits phorbol ester and UV-induced activator protein 1 activation by interfering with mitogen-activated protein kinase pathways, *Mol Pharmacol* 60, 217–224, 2001.
32. Aggarwal BB, Bhardwaj A, Aggarwal RS, Seeram NP, Shishodia S, and Takada Y, Role of resveratrol in prevention and therapy of cancer: preclinical and clinical studies, *Anticancer Res* 24, 2783–2840, 2004.
33. Kuwajerwala N, Cifuentes E, Gautam S, Menon M, Barrack ER, and Reddy GP, Resveratrol induces prostate cancer cell entry into S phase and inhibits DNA synthesis, *Cancer Res* 62, 2488–2492, 2002.
34. Kim YA, Rhee SH, Park KY, and Choi YH, Antiproliferative effect of resveratrol in human prostate carcinoma cells, *J Med Food* 6, 273–280, 2003.
35. Lin HY, Shih A, Davis FB, Tang HY, Martino LJ, Bennett JA, and Davis PJ, Resveratrol induced serine phosphorylation of p53 causes apoptosis in a mutant p53 prostate cancer cell line, *J Urol* 168, 748–755, 2002.

36. Pozo-Guisado E, Alvarez-Barrientos A, Mulero-Navarro S, Santiago-Josefat B, and Fernandez-Salguero PM, The antiproliferative activity of resveratrol results in apoptosis in MCF-7 but not in MDA-MB-231 human breast cancer cells: cell-specific alteration of the cell cycle, *Biochem Pharmacol* 64, 1375–1386, 2002.
37. Wang Q, Li H, Wang XW, Wu DC, Chen XY, and Liu J, Resveratrol promotes differentiation and induces Fas-independent apoptosis of human medulloblastoma cells, *Neurosci Lett* 351, 83–86, 2003.
38. Liu J, Wang Q, Wu DC, Wang XW, Sun Y, Chen XY, Zhang KL, and Li H, Differential regulation of CYP1A1 and CYP1B1 expression in resveratrol-treated human medulloblastoma cells, *Neurosci Lett* 363, 257–261, 2004.
39. Chan WK and Delucchi AB, Resveratrol, a red wine constituent, is a mechanism-based inactivator of cytochrome P4503A4, *Life Sci* 67, 3103–3112, 2000.
40. Chang TK, Chen J, and Lee WB, Differential inhibition and inactivation of human CYP1 enzymes by trans-resveratrol: evidence for mechanism-based inactivation of CYP1A2, *J Pharmacol Exp Ther* 299, 874–882, 2001.
41. Ciolino HP, Daschner PJ, and Yeh GC, Resveratrol inhibits transcription of CYP1A1 *in vitro* by preventing activation of the aryl hydrocarbon receptor, *Cancer Res* 58, 5707–5712, 1998.
42. Ciolino HP and Yeh GC, Inhibition of aryl hydrocarbon-induced cytochrome P-450 1A1 enzyme activity and CYP1A1 expression by resveratrol, *Mol Pharmacol* 56, 760–767, 1999.
43. Casper RF, Quesne M, Rogers IM, Shirota T, Jolivet A, Milgrom E, and Savouret JF, Resveratrol has antagonist activity on the aryl hydrocarbon receptor: implications for prevention of dioxin toxicity, *Mol Pharmacol* 56, 784–790, 1999.
44. Narayanan BA, Narayanan NK, Stoner GD, and Bullock BP, Interactive gene expression pattern in prostate cancer cells exposed to phenolic antioxidants, *Life Sci* 70, 1821–1839, 2002.
45. Gao S, Liu GZ, and Wang Z, Modulation of androgen receptor-dependent transcription by resveratrol and genistein in prostate cancer cells, *Prostate* 59, 214–225, 2004.
46. Abou-Zeid LA and El-Mowafy AM, Differential recognition of resveratrol isomers by the human estrogen receptor-alpha: molecular dynamics evidence for stereoselective ligand binding, *Chirality* 16, 190–195, 2004.
47. Bhat KP, Lantvit D, Christov K, Mehta RG, Moon RC, and Pezzuto JM, Estrogenic and antiestrogenic properties of resveratrol in mammary tumor models, *Cancer Res* 61, 7456–7463, 2001.
48. Gehm BD, McAndrews JM, Chien PY, and Jameson JL, Resveratrol, a polyphenolic compound found in grapes and wine, is an agonist for the estrogen receptor, *Proc Natl Acad Sci USA* 94, 14138–14143, 1997.
49. Stahl S, Chun TY, and Gray WG, Phytoestrogens act as estrogen agonists in an estrogen-responsive pituitary cell line, *Toxicol Appl Pharmacol* 152, 41–48, 1998.
50. Mizutani K, Ikeda K, Kawai Y, and Yamori Y, Resveratrol stimulates the proliferation and differentiation of osteoblastic MC3T3-E1 cells, *Biochem Biophys Res Commun* 253, 859–863, 1998.

51. Pozo-Guisado E, Lorenzo-Benayas MJ, and Fernandez-Salguero PM, Resveratrol modulates the phosphoinositide 3-kinase pathway through an estrogen receptor alpha-dependent mechanism: relevance in cell proliferation, *Int J Cancer* 109, 167–173, 2004.
52. Bowers JL, Tyulmenkov VV, Jernigan SC, and Klinge CM, Resveratrol acts as a mixed agonist/antagonist for estrogen receptors alpha and beta, *Endocrinology* 141, 3657–3667, 2000.
53. Jazirehi AR and Bonavida B, Resveratrol modifies the expression of apoptotic regulatory proteins and sensitizes non-Hodgkin's lymphoma and multiple myeloma cell lines to paclitaxel-induced apoptosis, *Mol Cancer Ther* 3, 71–84, 2004.
54. Fulda S and Debatin KM, Sensitization for tumor necrosis factor-related apoptosis-inducing ligand-induced apoptosis by the chemopreventive agent resveratrol, *Cancer Res* 64, 337–346, 2004.
55. Opipari AW Jr, Tan L, Boitano AE, Sorenson R, Aurora A, and Liu JR, Resveratrol-induced autophagocytosis in ovarian cancer cells, *Cancer Res* 64, 696–703, 2004.
56. Jang MS, Cai EN, Udeani GO, Slowing KV, Thomas CF, Beecher CWW, Fong HHS, Farnsworth NR, Kinghorn AD, Mehta RG, Moon RC, and Pezzuto JM, Cancer chemopreventive activity of resveratrol, a natural product derived from grapes, *Science* 275, 218–220, 1997.
57. Juan ME, Buenafuente J, Casals I, and Planas JM, Plasmatic levels of trans-resveratrol in rats, *Food Res Int* 35, 195–199, 2002.
58. Asensi M, Medina I, Ortega A, Carretero J, Bano MC, Obrador E, and Estrela JM, Inhibition of cancer growth by resveratrol is related to its low bioavailability, *Free Radical Biol Med* 33, 387–398, 2002.
59. Marier JF, Vachon P, Gritsas A, Zhang J, Moreau J-P, and Ducharme MP, Metabolism and disposition of resveratrol in rats: extent of absorption, glucuronidation, and enterohepatic recirculation evidenced by a linked-rat model, *J Pharmacol Exp Ther* 302, 369–373, 2002.
60. Banerjee S, Bueso-Ramos C, and Aggarwal BB, Suppression of 7,12-dimethylbenz(a)anthracene-induced mammary carcinogenesis in rats by resveratrol: role of nuclear factor-kappaB, cyclooxygenase 2, and matrix metalloprotease 9, *Cancer Res* 62, 4945–4954, 2002.
61. Liu S, Bishop WR, and Liu M, Differential effects of cell cycle regulatory protein p21waf/cip1 on apoptosis and sensitivity to cancer chemotherapy, *Drug Resist Update* 6, 183–195, 2003.
62. MacCarrone M, Lorenzon T, Guerrieri P, and Agro AF, Resveratrol prevents apoptosis in K562 cells by inhibiting lipoxygenase and cyclooxygenase activity, *Eur J Biochem* 265, 27–34, 1999.
63. Pervaiz S, Ramalingam JK, Hirpara JL, and Clement MV, Superoxide anion inhibits drug-induced tumor cell death, *FEBS Lett* 459, 343–348, 1999.
64. Pervaiz S and Clement MV, A permissive apoptotic environment: function of a decrease in intracellular superoxide anion and cytosolic acidification, *Biochem Biophys Res Commun* 290, 1145–1150, 2002.
65. Pervaiz S, Hirpara JL, and Clement MV, Caspase proteases mediate apoptosis induced by anticancer agent preactivated MC540 in human tumor cell lines, *Cancer Lett* 128, 11–22, 1998.

66. Clement MV, Hirpara JL, and Pervaiz S, Decrease in intracellular superoxide sensitizes Bcl-2-overexpressing tumor cells to receptor and drug-induced apoptosis independent of the mitochondria, *Cell Death Differentiation* 10, 1273–1285, 2003.
67. Ahmad KA, Clement MV, and Pervaiz S, Pro-oxidant activity of low doses of resveratrol inhibits hydrogen peroxide-induced apoptosis, *Ann NY Acad Sci* 1010, 365–373, 2003.
68. Ahmad KA, Clement MV, Hanif IM, and Pervaiz S, Resveratrol inhibits drug-induced apoptosis in human leukemia cells by creating an intracellular milieu nonpermissive for death execution, *Cancer Res* 64, 1452–1459, 2004.
69. Goldberg DM, Yan J, and Soleas GJ, Absorption of three wine-related polyphenols in three different matrices by healthy subjects, *Clin Biochem* 36, 79–87, 2003.
70. Bhat KPL, Lantvit D, Christov K, Mehta RG, Moon RC, and Pezzuto JM, Estrogenic and antiestrogenic properties of resveratrol in mammary tumor models, *Cancer Res* 61, 7456–7463, 2001.
71. Tessitore L, Davit A, Sarotto I, and Caderni G, Resveratrol depresses the growth of colorectal aberrant crypt foci by affecting bax and p21(CIP) expression, *Carcinogenesis* 21, 1619–1622, 2000.
72. Li ZG, Hong T, Shimada Y, Komoto I, Kawabe A, Ding Y, Kaganoi J, Hashimoto Y, and Imamura M, Suppression of N-nitrosomethylbenzylamine (NMBA)-induced esophageal tumorigenesis in F344 rats by resveratrol, *Carcinogenesis* 23, 1531–1536, 2002.
73. Roemer K and Mahyar-Roemer M, The basis for the chemopreventive action of resveratrol, *Drugs Today* 38, 571–580, 2002.
74. Surh YJ, Hurh YJ, Kang JY, Lee E, Kong G, and Lee SJ, Resveratrol, an antioxidant present in red wine, induces apoptosis in human promyelocytic leukemia (HL-60) cells, *Cancer Lett* 140, 1–10, 1999.
75. Mahyar-Roemer M, Katsen A, Mestres P, and Roemer K, Resveratrol induces colon tumor cell apoptosis independently of p53 and preceded by epithelial differentiation, mitochondrial proliferation and membrane potential collapse, *Int J Cancer* 94, 615–622, 2001.
76. Basly JP, Marre-Fournier F, LeBail JC, Habrioux G, and Chulia AJ, Estrogenic/antiestrogenic and scavanging properties of (E)- and (Z)-resveratrol, *Life Sci* 66, 769–777, 2000.
77. Martinez J and Moreno JJ, Effect of resveratrol, a natural polyphenolic compound, on reactive oxygen species and prostaglandin production, *Biochem Pharmacol* 59, 865–870, 2000.
78. Chan WK and Delucchi AB, Resveratrol, a red wine constituent, is a mechanism-based inactivator of cytochrome P450 3A4, *Life Sci* 67, 3103–3112, 2000.
79. She QB, Bode AM, Ma WY, Chen NY, and Dong ZG, Resveratrol-induced activation of p53 and apoptosis is mediated by extracellular-signal-regulated protein kinases and p38 kinase, *Cancer Res* 61, 1604–1610, 2001.
80. She QB, Huang CS, Zhang YG, and Dong ZG, Involvement of c-jun NH2-terminal kinases in resveratrol-induced activation of p53 and apoptosis, *Mol Carcinog* 33, 244–250, 2002.

81. Subbaramaiah K, Chung WJ, Michaluart P, Telang N, Tanabe T, Inoue H, Jang MS, Pezzuto JM, and Dannenberg AJ, Resveratrol inhibits cyclooxygenase-2 transcription and activity in phorbol ester-treated human mammary epithelial cells, *J Biol Chem* 273, 21875–21882, 1998.
82. Ahmad N, Adhami VM, Afaq F, Feyes DK, and Mukhtar H, Resveratrol causes WAF-l/p21-mediated G(1)-phase arrest of cell cycle and induction of apoptosis in human epidermoid carcinoma A431 cells, *Clin Cancer Res* 7, 1466–1473, 2001.
83. Delmas D, Passilly-Degrace P, Jannin B, Malki MC, and Latruffe N, Resveratrol, a chemopreventive agent, disrupts the cell cycle control of human SW480 colorectal tumor cells, *Int J Mol Med* 10, 193–199, 2002.
84. Haworth RS and Avkiran M, Inhibition of protein kinase D by resveratrol, *Biochem Pharmacol* 62, 1647–1651, 2001.
85. Holmes-McNary M and Baldwin A, Chemopreventive properties of trans-resveratrol are associated with inhibition of activation of the I κB kinase. *Cancer Res* 60, 3477–3483, 2000.
86. Manna SK, Mukhopadhyay A, and Aggarwal BB, Resveratrol suppresses TNF-induced activation of nuclear transcription factors NF-kappa B, activator protein-1, and apoptosis: potential role of reactive oxygen intermediates and lipid peroxidation, *J Immunol* 164, 6509–6519, 2000.
87. Schneider Y, Duranton B, Gosse F, Schleiffer R, Seiler N, and Raul F, Resveratrol inhibits intestinal tumorigenesis and modulates host-defense-related gene expression in an animal model of human familial adenomatous polyposis, *Nutr Cancer* 39, 102–107, 2001.
88. Chang TK, Lee WB, and Ko HH, Trans-resveratrol modulates the catalytic activity and mRNA expression of the procarcinogen-activating human cytochrome P4501B1, *Can J Physiol Pharmacol* 78, 874–881, 2000.

6 Resveratrol as Inhibitor of Cell Survival Signal Transduction

Simone Fulda and Klaus-Michael Debatin

CONTENTS

Tissue homeostasis is maintained by tight control of signaling events regulating cell survival and cell death. Apoptosis or programmed cell death plays a key role in maintaining tissue homeostasis under various conditions during normal development and in adult organisms. Thus, uncontrolled proliferation and/or failure to undergo cell death is involved in the pathogenesis of many human diseases, for example in cancer formation or in cardiovascular disorders. Moreover, current cancer therapies primarily act by triggering apoptosis in cancer cells. Natural products such as resveratrol have gained considerable attention as cancer chemopreventive or cardioprotective agents and also because of their antitumor properties. Among its wide range of biological activities, resveratrol has been reported to inhibit intracellular pathways that promote cell survival and/or block apoptosis. Further insights into the molecular mechanisms of how resveratrol may interfere with cell survival signaling may provide novel opportunities for the use of resveratrol in the prevention and treatment of human diseases.

INTRODUCTION

Physiological growth control and tissue homeostasis is regulated by a subtle balance between proliferation and cell death [1,2]. Accordingly, tipping this balance may result in uncontrolled proliferation or excessive cell loss, the molecular basis of many human diseases. For example, tissue damage in ischemic diseases or tissue loss in neurodegenerative disorders have been linked to excessive apoptosis, the cell's intrinsic death program [3]. Conversely, hyperproliferation and/or failure to undergo apoptosis are involved in cardiovascular or autoimmune diseases and also play a crucial role in tumor formation [3]. Moreover, killing of tumor cells by diverse cytotoxic approaches such as anticancer drugs, γ-irradiation, suicide genes, or immunotherapy is predominantly mediated through induction of apoptosis in tumor cells [4,5]. Thus, insufficient activation of apoptosis because of defects in apoptosis programs or because of the dominance of survival signals may result in cancer resistance. Despite aggressive therapies, resistance of many tumors to current treatment protocols still constitutes a major problem in cancer therapy. Thus, current attempts to improve cancer survival will have to include strategies that specifically target tumor cell resistance. The identification of naturally occurring antioxidant compounds such as resveratrol as signal transduction inhibitors of cell survival pathways may open new perspectives for their use in the treatment of a variety of human disorders.

RESVERATROL

Resveratrol (3,4′,5-trihydroxystilbene) is a polyphenolic natural product, synthesized by a wide variety of plant species including aliments such as grapes, and is present in red wine [6]. Its stilbene structure is related to the synthetic estrogen diethylstilbestrol [6]. Resveratrol has gained considerable attention because of its potential cancer chemopreventive or anticancer properties [6]. In addition, resveratrol may be beneficial in the control of atherosclerosis, heart disease, arthritis, or autoimmune disorders. Numerous biological activities have been ascribed to resveratrol, which may explain its antiinflammatory, anticarcinogenic, or anticancer properties [7]. Among its various actions, resveratrol has been demonstrated to inhibit cellular survival signaling. For example, resveratrol may interfere with apoptosis pathways both by directly triggering apoptosis-promoting signaling cascades and by blocking antiapoptotic mechanisms. By blocking survival and antiapoptotic pathways, resveratrol can sensitize cancer cells, which may result in synergistic antitumor activities when resveratrol is combined with conventional chemotherapeutic agents or cytotoxic compounds [8].

SIGNAL TRANSDUCTION IN CELL DEATH AND SURVIVAL

Apoptosis

Apoptosis or programmed cell death is the cell's intrinsic death program that occurs in various physiological and pathological situations [2,4,5,9,10]. In most cases, triggering of apoptosis eventually leads to activation of caspases, which act as death effector molecules in many forms of cell death [11]. Upon activation, caspases cleave a variety of substrates in the cytoplasm or nucleus leading to many of the morphological hallmarks of apoptotic cell death [11]. For example, polynucleosomal DNA fragmentation is the result of caspase-mediated activation of an endonuclease that cleaves DNA into the characteristic oligomeric fragments [12]. Likewise, degradation of lamin results in nuclear shrinking, while loss of overall cell shape is caused by proteolytic degradation of cytoskeletal proteins including actin or fodrin [13].

Caspases can be activated by triggering the extrinsic pathway (receptor pathway) or the intrinsic pathway of apoptosis (mitochondrial pathway) [13]. Stimulation of death receptors of the tumor necrosis factor (TNF) receptor superfamily such as CD95 (APO-1/Fas) or TNF-related apoptosis-inducing ligand (TRAIL) receptors at the plasma membrane results in activation of the initiator caspase-8 which can propagate the apoptosis signal by direct cleavage of downstream effector caspases such as caspase-3 [14]. The mitochondrial pathway is initiated by the release of apoptogenic factors such as cytochrome c, apoptosis-inducing factor (AIF), second mitochondria-derived activator of caspase (Smac)/DIABLO, or Omi/HtrA2 from the mitochondrial intermembrane space into the cytosol [15]. Upon release of cytochrome c into the cytosol, caspase-3 is activated through formation of the cytochrome c/Apaf-1/caspase-9-containing apoptosome complex. Smac/DIABLO or Omi/HtrA2 enhance caspase activation through antagonizing the inhibitory effects of inhibitor of apoptosis proteins (IAPs) [15]. AIF translocates into the nucleus upon apoptosis induction, where it induces large-scale DNA fragmentation in a caspase-independent manner [12].

Nonapoptotic Forms of Cell Death

Although caspases are crucial for cell death execution in many systems, caspase-independent apoptosis as well as nonapoptotic modes of cell death have also to be considered. For example, necrosis, autophagy, paraptosis, or some forms of cell death that cannot be easily classified at present have been described [16–18]. Although the signaling pathways and molecules involved in these alternative forms of cell death have not yet exactly been defined, noncaspase proteases such as calpains or cathepsins may be involved [19]. The relative contribution of these diverse cell death mechanisms under various conditions both *in vitro* and *in vivo* has to be addressed in future studies.

Inhibitor of Apoptosis Proteins (IAPs)

The family of endogenous caspase inhibitors known as IAPs comprises proteins such as XIAP, cIAP1, cIAP2, survivin, or livin (ML-IAP) [20]. In gene profiling studies, survivin was found to be expressed at high levels in the majority of human cancers representing the fourth most common transcriptome of the human genome [21,22]. In contrast, survivin was not detected in normal adult tissues, indicating that survivin may contribute to the malignant phenotype of cancer cells [21,22].

IAPs block apoptosis by directly inhibiting active caspase-3 and -7 and also by preventing activation of caspase-9 [20]. The role of survivin in regulation of apoptosis and proliferation is more complex compared to other IAP family proteins, since survivin is also involved in regulation of mitosis in addition to its role in apoptosis [21,22]. IAPs are negatively regulated, for example, by the mitochondrial proteins Smac/DIABLO or Omi/HtrA2, which translocate into the cytosol upon induction of apoptosis, and neutralize the antiapoptotic function of IAPs by binding to IAPs [23–25].

PI3K/AKT Pathway

The phosphoinositide 3-kinase (PI3K)/AKT pathway is a potent mediator of cell survival signals such as those delivered by growth factors or interactions with neighboring cells or with the extracellular matrix [26]. Upon growth factor binding, transmembrane receptor tyrosine kinases undergo auto- and transphosphorylation thereby recruiting PI3K to the plasma membrane where PI3K in turn recruits AKT via generation of phospholipids [27]. Once activated, AKT regulates multiple signaling pathways involved in cell proliferation, apoptosis, glucose metabolism, or angiogenesis [27]. The prosurvival function of AKT is mediated by phosphorylation of apoptosis signaling molecules such as Bad and caspase-9, or by inhibiting cytochrome c release from mitochondria [27]. A role of the PI3K/AKT pathway in treatment resistance has been suggested, since deregulated activation of AKT conferred resistance to apoptosis upon death receptor ligation or cytotoxic drug treatment [27]. Thus, targeting the

PI3K/AKT pathway, e.g., by small molecule inhibitors, may be useful to restore the sensitivity of tumor cells to cytotoxic therapies [28].

Nuclear Factor-kappaB

The transcription factor nuclear factor-kappaB (NF-κB) plays an important role in tumor formation and progression [29]. NF-κB is composed of hetero- or homodimers of NF-κB/Rel family of proteins, which mediate protein dimerization, nuclear import, and specific DNA binding. In most cell types, NF-κB is sequestered in the cytoplasm by its interaction with IκB proteins and therefore remains inactive. Upon activation of the IKK complex following stimulation, IκB becomes phosphorylated and is degraded via the proteasome thereby releasing NF-κB to translocate into the nucleus for transcription of target genes. NF-κB target genes include several antiapoptotic proteins, e.g., cIAP1, cIAP2, TRAF1, TRAF2, Bfl-1/A1, Bcl-X_L, FLIP, or TRAIL-R3 [29–31]. Interestingly, promoter activation of certain proapoptotic molecules, e.g., CD95 ligand, TRAIL-R1, TRAIL-R2, or TRAIL, is also controlled by NF-κB in line with the notion that NF-κB can promote apoptosis under certain circumstances [32–34]. Since certain types of anticancer treatments result in induction of NF-κB transcriptional activity, inhibition of NF-κB in parallel with anticancer drug treatment strongly enhanced the cytotoxic effect of chemotherapy [35]. Thus, NF-κB may play an important role in inducible chemoresistance, and inhibition of NF-κB may serve as a potential new adjuvant approach to chemotherapy [35].

RESVERATROL AS INHIBITOR OF CELL SURVIVAL SIGNALING

Inhibition of Antiapoptotic Pathways

Because of the potential detrimental effects on cell survival in the case of inappropriate triggering of apoptosis, apoptosis programs have to be tightly controlled at different levels. These antiapoptotic mechanisms regulating cell death have also been implicated to promote tumorigenesis and cancer resistance by allowing cancer cells to evade the cell's intrinsic death program [36]. Thus, signaling to cell death is often impaired in cancer cells, especially in resistant forms of cancer. Principally, tumor cells may acquire resistance through upregulation of antiapoptotic mechanisms or by downregulation of proapoptotic signaling molecules. IAPs such as survivin are expressed at high levels in many tumors and have been associated with refractory disease and poor prognosis [20,37]. Survivin is a member of the IAPs, which may contribute to resistance of tumors by facilitating both evasion from apoptosis and aberrant mitotic progression [37]. Since IAPs block apoptosis at the core of the apoptotic machinery by inhibiting caspases [20], therapeutic modulation of IAPs could target a key control point in deciding cell fate and therapy resistance.

Recently, resveratrol has been shown to trigger p53-independent induction of p21 and p21-mediated cell cycle arrest associated with survivin depletion [38]. Treatment with resveratrol resulted in downregulation of survivin expression through transcriptional and posttranscriptional mechanisms by inhibiting promoter activity, decreasing survivin protein stability, and also by enhancing proteasomal degradation of survivin [38]. Notably, resveratrol-mediated cell cycle arrest and survivin depletion as well as sensitization for TRAIL occurred independently of the p53 status, since these effects were also found in p53-deficient or p53 null cells [38]. Importantly, resveratrol sensitized various tumor cell lines for apoptosis induced by death receptor ligation or anticancer drugs [38]. Of note, resveratrol did not reverse the lack of toxicity of TRAIL on primary nonmalignant human cells, e.g., human fibroblasts, indicating some tumor specificity [38]. Thus, the dietary compound resveratrol may serve as a novel therapeutic agent to target survivin expression in cancers. Since the sensitization effect for death receptor- or anticancer drug-induced apoptosis provided by resveratrol was found at relatively low concentrations of resveratrol, which on their own only induced minimal apoptosis, the potential of resveratrol for anticancer therapy may largely reside in its ability to sensitize tumor cells for cell death. Thus, this combined sensitizer (resveratrol)/inducer (e.g., TRAIL or anticancer drugs) strategy may be a novel approach to render tumor cells more susceptible for death induction.

Downregulation of survivin expression by resveratrol has also been reported in adult T cell leukemia [39]. To this end, treatment with resveratrol resulted in a decrease in survivin expression and promoted apoptosis in adult T cell leukemia cells infected with the human T cell lymphotrophic virus-1 [39]. Moreover, recent evidence suggests that the chemopreventive activity of resveratrol may, at least in part, be mediated through blocking survivin expression. In a mouse model of ultraviolet B (UVB) radiation-induced skin carcinogenesis in SKH-1 hairless mice, resveratrol was shown to impart protection from acute UVB radiation-mediated cutaneous damages through inhibiting survivin [40]. In this model, topical pretreatment with resveratrol resulted in significant inhibition of UVB radiation-mediated increases in protein and mRNA levels of survivin and also in blockade of survivin phosphorylation in the skin of the SKH-1 hairless mice [40].

In addition to survivin, resveratrol has recently been reported to suppress expression of other antiapoptotic proteins. In non-Hodgkin's lymphoma and multiple myeloma cell lines, treatment with resveratrol selectively downregulated expression of the antiapoptotic proteins Bcl-x(L) and Mcl-1 [41]. Inhibition of Bcl-x(L) expression by resveratrol was mediated by inhibition of the extracellular signal-regulated kinase 1/2

(ERK1/2) pathway and diminished activator protein-1-dependent Bcl-x(L) expression [41]. Importantly, low concentrations of resveratrol exerted a sensitizing effect on drug-refractory non-Hodgkin's lymphoma and multiple myeloma cells to apoptosis induced by paclitaxel [41]. These findings indicate that resveratrol-based regimens may have potential clinical applications in drug-refractory B-cell malignancies.

Inhibition of PI3K/AKT or MAPK Pathway

Resveratrol has also been shown to interfere with the PI3K/AKT and mitogen-activated protein kinase (MAPK) pathway, two key survival signaling systems. For example, in a cell culture model of epidermal carcinogenesis based on mouse JB6 epidermal cells, resveratrol derivatives were found to inhibit cell transformation by blocking EGF-induced activation of phosphatidylinositol-3 kinase (PI3K) and AKT [42]. Moreover, in estrogen-responsive human breast cancer cells, resveratrol blocked activation of PI3K and thus cell survival and proliferation by interfering with an estrogen receptor alpha-associated PI3K pathway independent of the nuclear functions of the estrogen receptor alpha [43]. Also, resveratrol has been suggested to interfere with the growth factor receptor tyrosine kinase pathway. To this end, downregulation of EGF and suppression of cell growth was observed in endometrial cancer cells upon treatment with resveratrol [44]. Furthermore, control of tumor growth by resveratrol-mediated inhibition of MAPK signaling was reported in cervical carcinoma cells exposed to ultraviolet irradiation [45].

Notably, in addition to resveratrol-mediated blockade of MAPK survival signaling, the antitumor activity of resveratrol has also been shown to require MAPK-induced p53 activation and subsequent induction of apoptosis [42,46,47]. To this end, ERKs, c-Jun NH2-terminal kinases (JNKs), and p38 kinase were found to mediate resveratrol-induced activation of p53 and apoptosis through phosphorylation of p53 in JB6 mouse epidermal cells [42,47]. Similarly, activation of the MAPK cascade has been implicated to transmit the death signal initiated upon treatment with resveratrol through phosphorylation of p53 in papillary and follicular thyroid carcinoma cell lines [48]. The functional impact of resveratrol on MAPK signaling may, at least to some extent, depend on the concentrations of resveratrol used in different settings, since low concentrations of resveratrol (from 1 pM to 10 μM) were reported to induce phosphorylation of ERK1 and ERK2 in human neuroblastoma cells, while higher concentrations (50 to 100 μM) inhibited phosphorylation of MAP kinases [49].

Inhibition of PI3K/AKT and MAPK signaling pathways by resveratrol has also been implicated in mediating the beneficial effects of resveratrol in cardiovascular disease. To this end, resveratrol was reported to suppress angiotensin II-induced activation of PI3K/AKT, ERK1/2, and p70 S6 kinase and subsequent hypertrophy in rat aortic smooth muscle cells [50].

In addition, inhibition of proliferation induced by oxLDL was suppressed by resveratrol through blocking oxLDL-induced activation of ERK1/2 in bovine aortic smooth muscle cells [51]. Similarly, resveratrol remarkably attenuated endothelin-1-evoked protein tyrosine phosphorylation of ERK1/2, JNK-1, and p38 in coronary artery smooth muscle [52].

Inhibition of Protein Kinase C (PKC) Superfamily

The beneficial effects provided by resveratrol on endothelial functions as well as on tumor growth control have also been linked to its inhibitory effect on PKC activity [53]. PKC-controlled signaling has been suggested to play a key role in the regulation of endothelial functions [54]. The inhibitory effect of resveratrol was observed for membrane-associated PKCα and also for conventional PKCβ, whereas the activities of novel PKCε and atypical PKCζ were each unaffected [53]. In human androgen-independent prostate cancer cells, resveratrol was found to antagonize EGFR-dependent ERK1/2 activation by inhibiting PKCα in an isozyme-selective manner [55]. Moreover, resveratrol was reported to inhibit phorbol myristate acetate-induced matrix metalloproteinase-9 expression by inhibiting JNK and PKCδ signal transduction, thereby controlling growth and invasiveness of tumors [56]. In addition to PKC, resveratrol was found to inhibit PKD, a member of the PKC superfamily with distinctive structural, enzymological, and regulatory properties [57,58]. Since very high concentrations of resveratrol were required to achieve inhibition of PKD autophosphorylation in intact cells (IC(50) approximately 800 μM), its value as a tool to block cellular functions of PKD may, however, be questionable [58].

Inhibition of NF-κB Pathway

The antiinflammatory, antiproliferative, and anticarcinogenic effects provided by resveratrol have, at least in part, been ascribed to its inhibitory effect on activation of NF-κB, a nuclear transcription factor that regulates expression of various genes involved in inflammation, cytoprotection, and carcinogenesis [29,59–62]. Resveratrol was found to block TNFα-triggered NF-κB activation by suppressing TNFα-induced IκB kinase activity, phosphorylation and nuclear translocation of the p65 subunit of NF-κB, and NF-κB-dependent reporter gene transcription [59,60]. Resveratrol also blocked NF-κB activation induced by PMA, LPS, H_2O_2, okadaic acid, and ceramide [60]. Furthermore, there is recent evidence that resveratrol plays a role in controlling NF-κB activation by regulating chromatin remodeling through modulation of histone deacetylase activity [63]. To this end, the inhibitory effect of resveratrol on NF-κB-dependent signaling has been ascribed to its Sirtuin activity [64]. SIRT1, the mammalian ortholog of the yeast SIR2 (silencing information regulator) and a member of the Sirtuin family, has been implicated in modulating transcriptional silencing and cell survival, thereby promoting mammalian longevity [65]. Interestingly, SIRT1 was found to inhibit the transactivation potential of the RelA/p65 protein by deacetylating RelA/p65 at lysine 310 activity [63]. Thus, resveratrol, by its virtue as small-molecule agonist of Sirtuin activity, potentiated chromatin-associated SIRT1 protein on the cIAP-2 promoter region, an effect that correlated with a loss of NF-κB-regulated gene expression and sensitization of cells to TNFα-induced apoptosis activity [63].

Moreover, resveratrol has been reported to exhibit potent antileukemic activities against acute myeloid leukemia (AML) cell lines and also against fresh AML cells by blocking both the production of interleukin (IL)-1β and its effect on activation of NF-κB [66]. In leukemia cells, activation of NF-B appears to be an important step in the molecular events leading to IL-1β production and, as a result, also appears to stimulate proliferation [67]. Thus, by suppressing NF-κB activation resveratrol inhibited both IL-1β production and the IL-1β-mediated activation of NF-B, resulting in additional reduction in production of IL-1 [66]. These findings indicate that resveratrol may eliminate leukemia cells and become a potential agent in the treatment of AML.

The chemopreventive activities of resveratrol have also been attributed to its negative impact on NF-κB activation. For example, suppression of 7,12-dimethylbenz(*a*)anthracene-induced mammary carcinogenesis in rats by resveratrol have been linked to resveratrol-triggered inhibition of NF-κB and cyclooxygenase 2 [68].

CONCLUSIONS

Many studies over the last few years have provided evidence that the beneficial effects of resveratrol in chemoprevention, cardiovascular disorders, or cancer are mediated, at least in part, by its ability to interfere with cell survival signaling. To this end, resveratrol has been shown to promote apoptosis by blocking expression of antiapoptotic proteins or by inhibiting signal transduction through the PI3K/AKT, MAPK, or NF-κB pathway. Despite the enormous progress achieved, the relative impact of the various biological activities of resveratrol in specific diseases or individual types of cancer still needs to be explored. Nevertheless, these insights into the molecular mechanisms of action of resveratrol will facilitate the development of resveratrol or its derivatives as therapeutics in prevention and treatment of human diseases.

REFERENCES

1. Evan GI and Vousden KH, Proliferation, cell cycle and apoptosis in cancer, *Nature* 411, 342–348, 2001.

2. Hengartner MO, The biochemistry of apoptosis, *Nature* 407, 770–776, 2000.
3. Rudin CM and Thompson CB, Apoptosis and disease: regulation and clinical relevance of programmed cell death, *Annual Review of Medicine* 48, 267–281, 1997.
4. Herr I and Debatin KM, Cellular stress response and apoptosis in cancer therapy, *Blood* 98, 2603–2614, 2001.
5. Fulda S and Debatin KM, Death receptor signaling in cancer therapy, *Current Medicinal Chemistry – Anti-Cancer Agents* 3, 253–262, 2003.
6. Pervaiz S, Resveratrol: from grapevines to mammalian biology, *FASEB Journal* 17, 1975–1985, 2003.
7. Gusman J, Malonne H, and Atassi G, A reappraisal of the potential chemopreventive and chemotherapeutic properties of resveratrol, *Carcinogenesis* 22, 1111–1117, 2001.
8. Cal C, Garban H, Jazirehi A, Yeh C, Mizutani Y, and Bonavida B, Resveratrol and cancer: chemoprevention, apoptosis, and chemo-immunosensitizing activities, *Current Medicinal Chemistry – Anti-Cancer Agents* 3, 77–93, 2003.
9. Johnstone RW, Ruefli AA, and Lowe SW, Apoptosis: a link between cancer genetics and chemotherapy, *Cell* 108, 153–164, 2002.
10. Lowe SW and Lin AW, Apoptosis in cancer, *Carcinogenesis* 21, 485–495, 2000.
11. Degterev A, Boyce M, and Yuan J, A decade of caspases, *Oncogene* 22, 8543–8567, 2003.
12. Nagata S, Apoptotic DNA fragmentation, *Experimental Cell Research* 256, 12–18, 2000.
13. Hengartner MO, The biochemistry of apoptosis, *Nature* 407, 770–776, 2000.
14. Ashkenazi A and Dixit VM, Apoptosis control by death and decoy receptors, *Current Opinion in Cell Biology* 11, 255–260, 1999.
15. van Loo G, Saelens X, van Gurp M, MacFarlane M, Martin SJ, and Vandenabeele P, The role of mitochondrial factors in apoptosis: a Russian roulette with more than one bullet, *Cell Death and Differentiation* 9, 1031–1042, 2002.
16. Kanzawa T, Kondo Y, Ito H, Kondo S, and Germano I, Induction of autophagic cell death in malignant glioma cells by arsenic trioxide, *Cancer Research* 63, 2103–2108, 2003.
17. Sperandio S, Poksay K, de Belle I, Lafuente MJ, Liu B, Nasir J, and Bredesen DE, Paraptosis: mediation by MAP kinases and inhibition by AIP-1/Alix, *Cell Death and Differentiation* 11, 1066–1075, 2004.
18. Sperandio S, de Belle I, and Bredesen DE, An alternative, nonapoptotic form of programmed cell death, *Proceedings of the National Academy of Sciences of the USA* 97, 14376–14381, 2000.
19. Leist M and Jaattela M, Four deaths and a funeral: from caspases to alternative mechanisms, *Nature Reviews. Molecular Cell Biology* 2, 589–598, 2001.
20. Salvesen GS and Duckett CS, IAP proteins: blocking the road to death's door. *Nature Reviews. Molecular Cell Biology* 3, 401–410, 2002.
21. Altieri DC, Validating survivin as a cancer therapeutic target, *Nature Reviews. Cancer* 3, 46–54, 2003.
22. Velculescu VE, Madden SL, Zhang L. et al, Analysis of human transcriptomes, *Nature Genetics* 23, 387–388, 1999.
23. van Loo G, Saelens X, van Gurp M, MacFarlane M, Martin SJ, and Vandenabeele P, The role of mitochondrial factors in apoptosis: a Russian

roulette with more than one bullet, *Cell Death and Differentiation* 9, 1031–1042, 2002.

24. Du C, Fang M, Li Y, Li L, and Wang X, Smac, a mitochondrial protein that promotes cytochrome c-dependent caspase activation by eliminating IAP inhibition, *Cell* 102, 33–42, 2000.
25. Liston P, Fong WG, Kelly NL, Toji S, Miyazaki T, Conte D, Tamai K, Craig CG, McBurney MW, and Korneluk RG, Identification of XAF1 as an antagonist of XIAP anti-Caspase activity, *Nature Cell Biology* 3, 128–133, 2001.
26. Blume-Jensen P and Hunter T, Oncogenic kinase signalling, *Nature* 411, 355–365, 2001.
27. Vivanco I and Sawyers CL, The phosphatidylinositol 3-Kinase AKT pathway in human cancer, *Nature Reviews. Cancer* 2, 489–501, 2002.
28. West KA, Castillo SS, and Dennis PA, Activation of the PI3K/AKT pathway and chemotherapeutic resistance, *Drug Resistance Updates* 5, 234–248, 2002.
29. Karin M, Cao Y, Greten FR, and Li ZW, NF-kappaB in cancer: from innocent bystander to major culprit, *Nature Reviews. Cancer* 2, 301–310, 2002.
30. Bernard D, Quatannens B, Vandenbunder B, and Abbadie C, Rel/NF-kappaB transcription factors protect against tumor necrosis factor (TNF)-related apoptosis-inducing ligand (TRAIL)-induced apoptosis by up-regulating the TRAIL decoy receptor DcR1, *Journal of Biological Chemistry* 276, 27322–27328, 2001.
31. Kreuz S, Siegmund D, Scheurich P, and Wajant H, NF-kappaB inducers upregulate cFLIP, a cycloheximide-sensitive inhibitor of death receptor signaling, *Molecular and Cellular Biology* 21, 3964–3673, 2001.
32. Ravi R, Bedi GC, Engstrom LW, Zeng Q, Mookerjee B, Gelinas C, Fuchs EJ, and Bedi A, Regulation of death receptor expression and TRAIL/Apo2L-induced apoptosis by NF-kappaB, *Nature Cell Biology* 3, 409–416, 2001.
33. Baetu TM, Kwon H, Sharma S, Grandvaux N, and Hiscott J, Disruption of NF-kappaB signaling reveals a novel role for NF-kappaB in the regulation of TNF-related apoptosis-inducing ligand expression, *Journal of Immunology* 167, 3164–3173, 2001.
34. Siegmund D, Hausser A, Peters N, Scheurich P, and Wajant H, Tumor necrosis factor (TNF) and phorbol ester induce TNF-related apoptosis-inducing ligand (TRAIL) under critical involvement of NF-kappaB essential modulator (NEMO)/IKKgamma, *Journal of Biological Chemistry* 276, 43708–43712, 2001.
35. Shishodia S and Aggarwal BB, Nuclear factor-kappaB: a friend or a foe in cancer?, *Biochemical Pharmacology* 68, 1071–1080, 2004.
36. Igney FH and Krammer PH, Death and anti-death: tumour resistance to apoptosis, *Nature Reviews. Cancer* 2, 277–288, 2002.
37. Altieri DC, Validating survivin as a cancer therapeutic target, *Nature Reviews. Cancer* 3, 46–54, 2003.
38. Fulda S and Debatin K-M, Sensitization for tumor necrosis factor-related apoptosis-inducing ligand-induced apoptosis by the chemopreventive agent resveratrol, *Cancer Research* 64, 337–346, 2004.
39. Hayashibara T, Yamada Y, Nakayama S, Harasawa H, Tsuruda K, Sugahara K, Miyanishi T, Kamihira S, Tomonaga M, and Maita T, Resveratrol induces downregulation in survivin expression and apoptosis in HTLV-1-infected cell lines: a prospective agent for adult T cell leukemia chemotherapy, *Nutrition and Cancer* 44, 193–201, 2002.

40. Aziz M, Afaq F, and Ahmad N, Prevention of ultraviolet B radiation damage by resveratrol in mouse skin is mediated via modulation in survivin, *Photochemistry and Photobiology* 81, 25–31, 2005.
41. Jazirehi AR and Bonavida B, Resveratrol modifies the expression of apoptotic regulatory proteins and sensitizes non-Hodgkin's lymphoma and multiple myeloma cell lines to paclitaxel-induced apoptosis, *Molecular Cancer Therapeutics* 3, 71–84, 2004.
42. She QB, Huang C, Zhang Y, and Dong Z, Involvement of c-jun NH(2)-terminal kinases in resveratrol-induced activation of p53 and apoptosis, *Molecular Carcinogenesis* 33, 244–250, 2002.
43. Pozo-Guisado E, Lorenzo-Benayas MJ, and Fernandez-Salguero PM, Resveratrol modulates the phosphoinositide 3-kinase pathway through an estrogen receptor alpha-dependent mechanism: relevance in cell proliferation, *International Journal of Cancer* 109, 167–173, 2004.
44. Kaneuchi M, Sasaki M, Tanaka Y, Yamamoto R, Sakuragi N, and Dahiya R, Resveratrol suppresses growth of Ishikawa cells through down-regulation of EGF, *International Journal of Oncology* 23, 1167–1172, 2003.
45. Yu R, Hebbar, V, Kim DW, Mandlekar S, Pezzuto, JM, and Kong, A-NT, Resveratrol inhibits phorbol ester and UV-induced activator protein 1 activation by interfering with mitogen-activated protein kinase pathways, *Molecular Pharmacology* 60, 217–224, 2001.
46. Huang C, Ma W, Goranson A, and Dong Z, Resveratrol suppresses cell transformation and induces apoptosis through a p53-dependent pathway, *Carcinogenesis* 20, 237–242, 1999.
47. She Q-B, Bode AM, Ma W-Y, Chen N-Y, and Dong Z, Resveratrol-induced activation of p53 and apoptosis is mediated by extracellular signal-regulated protein kinases and p38 kinase, *Cancer Research* 61, 1604–1610, 2001.
48. Shih A, Davis FB, Lin H-Y, and Davis PJ, Resveratrol induces apoptosis in thyroid cancer cell lines via a MAPK- and p53-dependent mechanism, *Journal of Clinical Endocrinology and Metabolism* 87, 1223–1232, 2002.
49. Miloso M, Bertelli AA, Nicolini G, and Tredici G, Resveratrol-induced activation of the mitogen-activated protein kinases, ERK1 and ERK2, in human neuroblastoma SH-SY5Y cells, *Neuroscience Letters* 264, 141–144, 1999.
50. Haider UGB, Sorescu D, Griendling KK, Vollmar AM, and Dirsch VM, Resveratrol suppresses angiotensin II-induced AKT/protein kinase B and p70 S6 kinase phosphorylation and subsequent hypertrophy in rat aortic smooth muscle cells, *Molecular Pharmacology* 62, 772–777, 2002.
51. Liu Y and Liu G, Isorhapontigenin and resveratrol suppress oxLDL-induced proliferation and activation of ERK1/2 mitogen-activated protein kinases of bovine aortic smooth muscle cells, *Biochemical Pharmacology* 67, 777–785, 2004.
52. El-Mowafy AM and White RE, Resveratrol inhibits MAPK activity and nuclear translocation in coronary artery smooth muscle: reversal of endothelin-1 stimulatory effects, *FEBS Letters* 451, 63–67, 1999.
53. Slater SJ, Seiz JL, Cook AC, Stagliano BA, and Buzas CJ, Inhibition of protein kinase C by resveratrol, *Biochimica et Biophysica Acta* 1637, 59–69, 2003.
54. Yuan SY, Protein kinase signaling in the modulation of microvascular permeability, *Vascular Pharmacology* 39, 213–223, 2002.

55. Stewart JR and O'Brian CA, Resveratrol antagonizes EGFR-dependent Erk1/2 activation in human androgen-independent prostate cancer cells with associated isozyme-selective PKC alpha inhibition, *Investigational New Drugs* 22, 107–117, 2004.
56. Woo JH, Lim JH, Kim YH, Suh SI, Min do S, Chang JS, Lee YH, Park JW, and Kwon TK, Resveratrol inhibits phorbol myristate acetate-induced matrix metalloproteinase-9 expression by inhibiting JNK and PKC delta signal transduction, *Oncogene* 23, 1845–1853, 2004.
57. Van Lint J, Rykx A, Maeda Y, Vantus T, Sturany S, Malhotra V, Vandenheede JR, and Seufferlein T, Protein kinase D: an intracellular traffic regulator on the move, *Trends in Cell Biology* 12, 193–200, 2002.
58. Haworth RS and Avkiran M, Inhibition of protein kinase D by resveratrol, *Biochemical Pharmacology* 62, 1647–1651, 2001.
59. Holmes-McNary M and Baldwin AS Jr, Chemopreventive properties of trans-resveratrol are associated with inhibition of activation of the IkappaB kinase, *Cancer Research* 60, 3477–3483, 2000.
60. Manna SK, Mukhopadhyay A, and Aggarwal BB, Resveratrol suppresses TNF-induced activation of nuclear transcription factors NF-kappaB, activator protein-1, and apoptosis: potential role of reactive oxygen intermediates and lipid peroxidation, *Journal of Immunology* 164, 6509–6519, 2000.
61. Adhami VM, Afaq F, and Ahmad N, Suppression of ultraviolet B exposure-mediated activation of NF-kappaB in normal human keratinocytes by resveratrol, *Neoplasia* 5, 74–82, 2003.
62. Tsai S-H, Lin-Shiau S-Y, and Lin J-K, Suppression of nitric oxide synthase and the down-regulation of the activation of NFkappaB in macrophages by resveratrol, *British Journal of Pharmacology* 126, 673–680, 1999.
63. Yeung F, Hoberg JE, Ramsey CS, Keller MD, Jones DR, Frye RA, and Mayo MW, Modulation of NF-kappaB-dependent transcription and cell survival by the SIRT1 deacetylase, *EMBO J* 23, 2369–2380, 2004.
64. Howitz KT, Bitterman KJ, Cohen HY, Lamming DW, Lavu S, Wood JG, Zipkin RE, Chung P, Kisielewski A, Zhang LL, Scherer B, and Sinclair DA, Small molecule activators of sirtuins extend Saccharomyces cerevisiae lifespan, *Nature* 425, 191–196, 2003.
65. Blander G and Guarente L, The Sir2 family of protein deacetylases, *Annual Review of Biochemistry* 73, 417–435, 2004.
66. Estrov Z, Shishodia S, Faderl S, Harris D, Van Q, Kantarjian HM, Talpaz M, and Aggarwal BB, Resveratrol blocks interleukin-1{beta}-induced activation of the nuclear transcription factor NF-kappaB, inhibits proliferation, causes S-phase arrest, and induces apoptosis of acute myeloid leukemia cells, *Blood* 102, 987–995, 2003.
67. Rambaldi A, Torcia M, Bettoni S, Vannier E, Barbui T, Shaw AR, Dinarello CA, and Cozzolino F, Modulation of cell proliferation and cytokine production in acute myeloblastic leukemia by interleukin-1 receptor antagonist and lack of its expression by leukemic cells, *Blood* 78, 3248–3253, 1991.
68. Banerjee S, Bueso-Ramos C, and Aggarwal BB, Suppression of 7,12-dimethylbenz(a)anthracene-induced mammary carcinogenesis in rats by resveratrol: role of nuclear factor-kappaB, cyclooxygenase 2, and matrix metalloprotease 9, *Cancer Research* 62, 4945–4954, 2002.

7 Resveratrol as Inhibitor of Cell Cycle Progression

Jen-Kun Lin

CONTENTS

Resveratrol (RES; *trans*-3,4′,5-trihydroxystilbene) occurs naturally in the plant kingdom. RES has been found in dietary plants such as grapes, peanuts, mulberries, and various herbs. RES has been demonstrated to possess antioxidant and antiinflammatory activities, to induce phase II drug-metabolizing enzymes, and to inhibit cyclooxygenase (COX) and matrix metalloproteinase activities. RES has been shown to suppress *in vitro* proliferation of a variety of tumor cells. Its antiproliferative activity is associated with the blocking of the cell cycle at different phases, namely G0/G1, S, S/G2, and G2/M phases. It appeared that RES might execute its cell cycle arrest at different phases in different cancer cell lines.

RES has been suggested as a potential cancer chemopreventive agent based on its striking inhibitory effects on diverse cellular events associated with tumor initiation, promotion, and progression. The biochemical and molecular mechanisms of cancer chemoprevention by RES might occur through blocking or perturbing cell cycle progression and inducing cellular apoptosis in the programmed and sensitized cancer cells. It is of interest to note that several derivatives of RES have been synthesized and some of these derivatives were found to be more effective than RES itself for cancer chemopreventive action in cellular systems. Further clinical trials on testing these promising derivatives for cancer chemoprevention are urgently needed.

INTRODUCTION

Resveratrol (RES; *trans*-3,4′,5-trihydroxystilbene; Figure 7.1) is biosynthesized by several plants in response to adverse conditions such as environmental stress or pathogenic attack including fungal infection and disease resistance [1]. RES has been found in a multitude of dietary plants

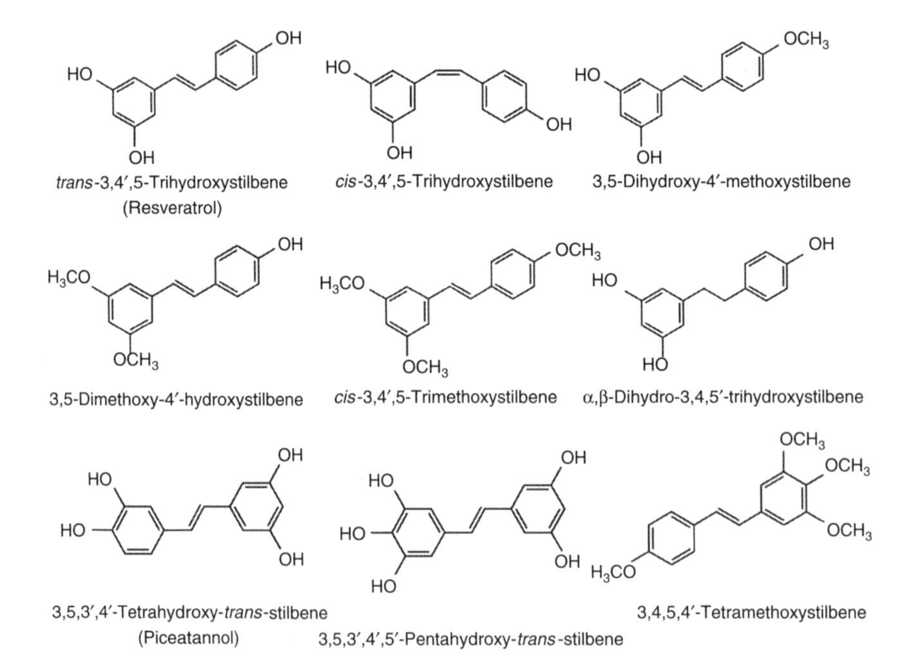

FIGURE 7.1 Chemical structures of resveratrol and its derivatives.

such as grape skins, peanuts, mulberries, and various herbs [2]. Red wine (1.5 to 3 mg/l) and grapes (50 to 100 μg/g grape skins) are probably its main sources in Western diets. One of its richest sources is the herb *Polygonum cuspidatum* (Chinese herb name is Fu-Chung), which has been used in Asian folk medicine for years to cure diseases including inflammation, allergy, and hyperlipemia [3]. RES has been demonstrated to possess antioxidant [4] and antiinflammatory [5] activities, to induce phase II drug-metabolizing enzymes, and to inhibit COX activity and transcription. Thus it has activity in regulating multiple cellular events associated with carcinogenesis [2,6]. A primary impetus for research on RES has come from the paradoxical observation that a low incidence of cardiovascular disease may coexist with the intake of a high-fat diet and RES, a phenomenon known as the "French paradox" [7]. The inhibitory effects of RES on high-fat-induced coronary heart disease have been attributed to its antioxidant and anticoagulative properties [8].

Recently, RES has been shown to act as a pleiotropic biological effector which regulates the multistep carcinogenesis process [2,6,9]. The studies add a new dimension to the expanding role of RES as a potential chemopreventive agent exhibiting antiinflammatory and anticarcinogenic effects. RES inhibited transformation in several animal models of carcinogenesis, including mouse 7,12-dimethylbenz(*a*)anthracene (DMBA)–12-*O*-tetradecanoyl phorbol-13-acetate (TPA)-induced skin cancers [2], azoxymethane-induced colon cancers [10], and transplanted Yoshida rat ascites hepatoma [9]. It is worth noting that in the DMBA–TPA-induced skin carcinogenesis model, RES inhibited the three major steps of carcinogenesis: initiation, promotion, and progression [2]. However, the exact mechanisms by which RES exerts these anticarcinogenic effects are not completely elucidated.

RES has been shown to suppress *in vitro* proliferation of variety of tumor cells [11]. Its antiproliferative activity is associated with the perturbed cell transit through G0/G1 phase of the cell cycle and with induction of apoptosis. For most cell types, both effects were observed at micromolar (20 to 100 μM) concentrations of RES. In some types of leukemia, the pro-apoptotic properties of RES were more pronounced for leukemic blasts than for normal lymphocytes [12].

CHEMICAL STRUCTURES OF RESVERATROL AND ITS DERIVATIVES

RES is a trihydroxystilbene phytopolyphenolic compound. Stilbene-based compounds are widely distributed in nature and have become of particular interest for chemists and biologists because of their wide range of biological activities [1,2,13]. Stilbene itself does not occur in nature, but hydroxylated stilbenes have been found in a multitude of medicinal plants.

RES, a phytoalexin present in grape and other food products, has been reported to play a role in the prevention of heart disease associated with red wine consumption because it inhibits platelet aggregation, alters eicosanoid synthesis, and modulates lipid and lipoprotein metabolism [14].

The simplicity of RES, associated with its interesting anticancer activity, offers promises for the rational design of new chemotherapeutic agents. In this context efforts have recently been devoted to the detailed study of structure–activity relationships (SARs) of these types of substituted stilbene derivatives as illustrated in Figure 7.1 [15,16]. RES has displayed *in vitro* growth inhibition in a number of human cancer cell lines [3,4]. In this program, a series of *cis*- and *trans*-stilbene-based RES derivatives were synthesized with the aim of discovering new lead compounds with clinical potential. All the synthesized compounds (32 compounds) were tested *in vitro* for cell growth inhibition and the ability to induce apoptosis in HL-60 promyelocytic leukemia cells [16]. The tested *trans*-stilbene derivatives were less potent than their corresponding *cis*-stilbene isomers, except for *trans*-resveratrol, whose *cis* isomer (Figure 7.1) was less active. The best results were obtained with *cis*-3,4′,5-trimethoxy-3′-hydroxystilbene and *cis*-3,4′,5-trimethoxy-3′-aminostilbene, which showed apoptotic activity at nanomolar concentrations. The corresponding *trans* isomers were less active both as antiproliferative and as apoptosis-inducing agents [16]. Of special interest, the above two *cis* derivatives were active toward resistant HL-60R cells and their activity was higher than that of several classic chemotherapeutic agents such as daunorubicin and etoposide [16]. Flow cytometry assay showed that at 50 nM *cis*-3,4′,5-trimethoxy-3′-hydroxystilbene and *cis*-3,4′,5-trimethoxy-3′-aminostilbene were able to recruit almost all cells in the apoptotic sub-G0/G1 peak; thus suggesting that the main mechanism of cytotoxicity of these compounds could be the activation of apoptosis. These results indicate that structural alteration of the stilbene motif of RES can be extremely effective in producing potent apoptosis-inducing agents [15,16].

CANCER CHEMOPREVENTIVE EFFECTS OF RESVERATROL

RES has been suggested as a potential cancer chemopreventive agent based on its striking inhibitory effects on diverse cellular events associated with tumor initiation, promotion, and progression. The cancer chemopreventive potential was demonstrated in 1997 when it was found that this compound inhibited DMBA-induced preneoplastic lesion formation in mouse mammary organ culture and reduced the incidence and multiplicity of DMBA–TPA-induced papilloma in the two-stage mouse skin model [2]. RES suppresses DMBA-induced rat mammary carcinogenesis which correlates with downregulation of NFκB, COX-2, and matrix metalloproteinase-9 expression [17].

The antimutagenic activity of RES was demonstrated against foodborne heterocyclic amine Trp-P-1 and 2-aminofluorene in salmonella bacterial tester strains [18]. Studies *in vitro* showed that RES affects the expression of cytochrome P450 and is one of the most selective inhibitors of human p450 1A1 [19]. Inhibition of CYP1A1 transcription may occur by preventing the activation of the aromatic hydrocarbon (Ah) receptor. RES promotes Ah receptor translocation to the nucleus and binding to DNA at dioxin-responsive elements, but transactivation does not take place [20]. RES also induces the phase II enzyme NAD(P)H– quinone oxidoreductase, which detoxifies many quinones. This enzyme is part of the gene battery activated through the Ah receptor.

RES has been shown to inhibit isolated and cellular protein kinase C (PKC) in model systems. Phosphorylation of substrates that resemble protamine sulfate was particularly affected [21]. RES has been incorporated into model phospholipid membranes, altering the phospholipids-phase polymorphism, and has inhibited PKCα enzyme activity *in vitro*.

RES inhibits the catalytic activity of the COX-2 in cultured human mammary epithelial cells with and without TPA treatment [21]. Likewise, human recombinant COX-2 expressed in baculovirus was inhibited by RES. Moreover, RES effectively suppressed the COX-2 promoter-dependent transcriptional activity in human colon cancer cells. RES also blocked TPA-mediated induction of COX-2 mRNA in cultured human mammary epithelial cells through repression of transcription factor-AP-1-dependent transactivation [22].

The tumor-promoting activity mediated by TPA has been associated with oxidative stress, as exemplified by increased production of superoxide anion radicals and hydrogen peroxide. Beside scavenging reactive oxygen species (ROS), RES was found to block the production of carbon- or nitrogen-centered free radicals, such as the phenylbutazone peroxyl radical and the benzidine-derived radical. RES, along with other antioxidants, like vitamin E and melatonin, prevent the oxidative DNA damage induced in rat kidney by potassium bromite ($KBrO_3$) [23].

In summary, RES promotes homeostasis and affects the earliest and later stages of carcinogenesis. Thus RES may be considered not only a potential chemopreventive, but also a chemotherapeutic agent to control tumor development.

APOPTOTIC EFFECTS OF RESVERATROL

RES has been shown to have growth-inhibiting activity in several human cancer cell lines and in animal models of carcinogenesis [3]. In HL-60 promyelocytic leukemia cells, treatment with RES led to growth inhibition, induction of apoptosis, S/G2 phase cell cycle arrest, and myelomonocytic differentiation [24]. RES also displayed antiproliferative activity in JB6 mouse epidermal, Caco-2 colorectal, and A431 epidermoid carcinoma cell lines [3].

The nuclear transcription factor NFκB modulates the effects of various transcription factors responsible for proliferation of normal myeloid and leukemic cells and its activation induces expression of various cytokines, including interleukin 1β (IL-1β). IL-1β itself plays major role in stimulating proliferation of acute myeloid leukemia (AML) cells. Meanwhile, IL-1β activates NFκB in AML cells [25]. RES suppressed both the production of IL-1β and its effect on activation of NFκB, inhibited AML cell proliferation, and activated caspase-3, thus inducing apoptotic cell death in AML cells. Whereas the ability of RES as a single agent to induce apoptosis was relatively low compared with its profound antiproliferative effect, the combination of RES as "sensitizer" together with the death ligand TRAIL (tumor necrosis factor-related apoptosis-inducing ligand) as "inducer" strongly cooperated to trigger apoptosis [26]. The potential of RES for anticancer therapy may therefore largely reside in its ability to sensitize tumor cells for cell death. The sensitizing effect of RES for apoptosis was not restricted to TRAIL, but was also found in response to CD95 ligation or anticancer drug treatment. Thus this combined sensitizer (RES)/inducer (TRAIL) strategy may be a novel approach to render tumor cells more susceptible to apoptotic induction [26].

INHIBITION OF CELL CYCLE PROGRESSION

RES has been shown to suppress *in vitro* proliferation of a variety of tumor cells [11]. Its antiproliferative activity is associated with the perturbed cell transit through G0/G1 phase of the cell cycle and with induction of apoptosis. For most cell types both effects were observed at micromolar concentrations of RES. Furthermore, it seems that the effect of RES on the cell cycle progression may be to arrest at different phases in the different cell types involved.

EFFECTS OF RESVERATROL ON THE G0/G1 TRANSITION IN HUMAN LYMPHOCYTES

RES (up to 50 μM) had no detectable effects on resting lymphocytes. With the mitogen phytohemagglutin (PHA), however, RES elicited concentration- and time-dependent responses in lymphocytes. RES (>50 μM) prevented cell entry into the cell cycle, resulting in 99% suppression at 100 μM. The arrested lymphocytes following 24 hour treatment with 50 μM RES had minimal RNA content, the feature characteristic of G0 cells, and were blocked at the stage past the induction of cyclin D2 and D3 and prior to induction of cyclin E. Prolonged treatment (72 hours) of PHA-stimulated lymphocytes with 100 μM RES showed a pronounced decrease in the expression of pRb, cyclin E and B1, and reduction in $p34^{cdc2}$ and PCNA [27].

RES treatment results in an induction of the cyclin kinase inhibitor $p21^{WAF/CIP1}$ which, by inhibiting cyclin (E, D1, and D2) and cyclin-dependent kinases (cdk2, cdk4, and cdk6), results in a G0/G1 phase arrest followed by apoptosis of A431 human epidermoid carcinoma cells [28]. The involvement of the pRb-E2F/DP pathway as an important contributor of RES-mediated cell cycle arrest and apoptosis has been demonstrated [29]. RES causes downregulation of hyperphosphorylated pRb protein with a relative increase in hypophosphorylated pRb that, in turn, compromises with the availability of free E2F. This series of events results in a stoppage of the cell cycle progression at the G0 to S phase transition, thereby leading to a G0/G1 arrest and subsequent apoptotic cell death [29].

Effects of Resveratrol on S-Phase Arrest in Prostate Carcinoma LNCaP Cells

The RES-induced increase in DNA synthesis was associated with enrichment of prostate carcinoma LNCaP cells in S phase and a concurrent decrease in nuclear $p21^{cip1}$ and $p27^{kip1}$ levels. Furthermore, consistent with entry of LNCaP cells into S phase, there was a dramatic increase in nuclear cdk 2 activity associated with both cyclin A and cyclin E. These findings indicate that LNCaP cells treated with RES are induced to enter into S phase, but subsequent progression through S phase is limited by the inhibitory effect of RES on DNA synthesis, particularly at concentrations above 15 μM [30]. Therefore, this unique ability of RES to exert opposite effects on two important processes in cell cycle progression, induction of S phase and inhibition of DNA synthesis, may be responsible for its apoptotic and antiproliferative effects.

Effects of Resveratrol on S/G2 Phase Transition in Human Colonic Adenocarcinoma Caco-2 Cells

The effect of RES on the cell cycle progression in the human colonic adenocarcinoma cell line Caco-2 was investigated [31]. The compound inhibited the growth and proliferation of Caco-2 cells in a dose-dependent manner (12.5 to 200 μM). Perturbed cell cycle progression from S to G2 phase was observed for concentrations up to 50 μM, whereas higher concentrations led to reversal of the S-phase arrest. These effects were specific for RES; they were not observed after incubation with the stilbene analogs stilbene methanol and rahpontin. The phosphorylation state of the Rb protein in Caco-2 cells was shifted from hyperphosphorylated to hypophosphorylated at 200 μM which may account for the reversal of the S-phase block at concentrations exceeding 50 μM. These findings suggest that RES exerts chemopreventive effects on colonic cancer cells by inhibition of the cell cycle [31].

Western blot analysis of positive cell cycle regulators (cdc2, cdk2, cdk4, cdk6, cyclin A, cyclin D1, and cyclin E) showed a dose-dependent increase in cyclin E levels and an increase in cyclin A levels only at concentrations up to 100 μM, suggesting the presence of an S to G2 block. RES treatment significantly reduced cyclin D1 levels and its related serine/threonine kinase cdk4. As a positive regulator of cdk4 and cdk6, cyclin D1 has been implicated in controlling the G1 phase of the cell cycle and is frequently overexpressed in human colon adenocarcinoma [32].

RES inhibits the proliferation of pulmonary artery endothelial cells, which, based on flow cytometric analysis, correlates with the suppression of cell progression through S and G2 phases of the cell cycle [33]. The perturbed progression through S and G2 phases is accompanied by an increase in the expression of tumor suppressor gene protein p53 and elevation of the level of cyclin-dependent kinase inhibitor $p21^{cip\text{-}1/WAF\text{-}1}$. All of the observed effects of RES, including induction of apoptosis at its higher concentration, are also compatible with its putative chemopreventive or antitumor activity.

Effects of Resveratrol on G2/M Arrest in Colon Carcinoma HT29 Cells

Recently, we have demonstrated RES-induced G2 arrest through the inhibition of CDK7 and $p34^{cdc2}$ kinases in colon carcinoma HT29 cells [34]. Based on flow cytometric analysis, RES inhibited the proliferation of HT29 colon cancer cells and resulted in their accumulation in the G2 phase of the cell cycle. Western blot analysis and kinase assays demonstrated that the perturbation of G2 phase progression by RES was accompanied by the inactivation of $p34^{cdc2}$ protein kinase, and increase in the tyrosine phosphorylated (inactive) form of $p34^{cdc2}$. Kinase assays revealed that the suppression of $p34^{cdc2}$ activity by RES was mediated through the inhibition of CDK7 kinase activity, while CDC25A phosphatase activity was not affected. In addition, RES-treated cells were shown to have a low level of CDK7 kinase-Thr^{161}-phosphorylated $p34^{cdc2}$. These results demonstrated that RES induced cell cycle arrest at the G2 phase through the inhibition of CDK7 kinase activity, suggesting that its antitumor activity might occur through the disruption of cell division at the G2/M phase [34].

The $p34^{cdc2}$ protein kinase is generally acknowledged to be the key mediator of G2/M phase transition in all eukaryotic cells [35,36]. The active mitotic kinase (MPF, or mitosis-promoting factor) is a dimer comprised of a catalytic subunit, $p34^{cdc2}$, and a regulatory subunit, a B-type cyclin [37]. The cyclins are a class of proteins that are synthesized during the interphase of each cell cycle and rapidly degraded at the end of mitosis [38]. The activity of the $p34^{cdc2}$ kinase not only depends on its association with cyclin B, but also on its phosphorylation state. Phosphorylation of either Thr^{14} or Tyr^{15} inhibits $p34^{cdc2}$ kinase activity, while phosphorylation of Thr^{161} by CDK7 kinase is required for kinase activity [39]. In addition, the dephosphorylation of Thr^{14} and Tyr^{15} by CDC25A phosphatase is a final step for performing $p34^{cdc2}$ kinase activity [40,41].

As revealed by flow cytometry, HT29 cells treated with RES accumulated in G2/M phase of the cell cycle. Because there was no evidence of an increased percentage of mitotic cells in RES-treated cultures, upon microscopic examination, the observed accumulation in G2/M indicates cell arrest in G2 rather than in mitosis (M phase). This finding is comparable (but not identical) with previous work, which showed that RES could induce HL60 cell [24] and bovine endothelial cell [33] arrest in S/G2 phase. It is obvious that the target sites for RES arresting in cell cycle progression is varied with different origins of cancer cell lines studied. Different exogenous and endogenous factors involving in the testing cultures might also affect the onset of RES arresting.

SOME RESVERATROL DERIVATIVES WITH UNIQUE BIOLOGICAL ACTIVITY

Piceatannol (*Trans*-3,3′,4′,5-tetrahydroxystilbene)

Piceatannol is a cancer-preventive agent that is found in red wine. Piceatannol differs from RES by having an additional aromatic hydroxyl group (Figure 7.1). RES undergoes metabolism by the cytochrome p450 CYP1B1 to give a metabolite which has been identified as the known antileukemic agent piceatannol [42]. This observation provides a novel explanation for the cancer chemopreventive properties of RES. It demonstrates that a natural dietary cancer-preventive agent can be converted to a compound with known anticancer activity by an enzyme that is found in human tumors [42]. This result gives an insight into the function role of xenobiotic metabolizing enzyme CYP1B1 and provides evidence for the concept that CYP1B1 in tumors may be functioning as a growth and tumor suppressor enzyme.

Trans-3,3′,4′,5,5′-pentahydroxystilbene

The effects of RES and its structurally related derivatives on epidermal growth factor (EGF)-induced cell transformation were investigated [43]. One of the RES derivatives, *trans*-3,3′,4′,5,5′-pentahydroxystilbene (Figure 7.1) exerted a more potent inhibitory effect than RES on EGF-induced cell transformation, but had less cytotoxic effects on normal nontransformed cells. Compared to RES, this compound also caused cell cycle arrest in the G1 phase, but did not induce p53 activation and apoptosis. Furthermore, this compound, but not RES, markedly inhibited EGF-induced phosphatidylinositol-3-kinase (PI3K) and Akt activation [43]. These data suggested that the greater antitumor effect of the compound compared to RES may act through a different mechanism by mainly targeting PI3K/Akt signaling pathways.

TRANS-3,4,4′,5-TETRAMETHOXYSTILBENE (DMU-212)

Recently, a variety of analogs of RES has been synthesized and tested in *in vitro* assays using human-derived colon cancer cells HCA-7 and HT29 [44]. One analog, *trans*-3,4,4′,5-tetramethoxystilbene (DMU-212, Figure 7.1) showed preferential growth-inhibiting and proapoptotic properties in transformed cells when compared with their untransformed counterparts. Based on pharmacokinetic studies, the RES afforded significantly higher levels than DMU-212 in the plasma and liver while DMU-212 exhibited superior availability compared to RES in the small intestine and colon. RES was metabolized to its sulfate or glucuronide conjugates, while DMU-212 underwent metabolic hydroxylation or single or double O-demethylation. DMU-212 and RES inhibited the growth of tumor-derived colon cancer cells HCA-7 and HT29 *in vitro* with IC_{50} values of between 6 and 26 μM. In the light of the superior levels achieved in the gastrointestinal tract after the administration of DMU-212 when compared to RES, the results provide a good rationale to evaluate DMU-212 as a colorectal cancer chemopreventive agent [44]. The increased drug levels in the liver, kidney, lung, and heart obtained after ingestion of RES in comparison to those after DMU-212 reflect the difference in availability observed in the plasma. In contrast, the levels of DMU-212 in the brain, small intestine, and colonic mucosa after DMU-212 administration exceeded levels of RES measured after RES intake. The higher availability of DMU-212 in the brain suggests that it is capable of crossing the blood–brain barrier more easily than RES, which is probably a consequence of the higher lipophilicity of DMU-212 [44]. In conclusion, the work described here provides an initial pharmacokinetic groundwork, which contributes to rational decision-making as to the choice of RES analogs that should be selected for comparative testing for cancer chemopreventive potency in preclinical models.

VATICANOL C (A RESVERATROL OLIGOMER)

In a recent study, 20 RES derivatives, which were isolated from stem bark of *Vatica rassak* (Dipterocarpaceae), were evaluated for *in vivo* cytotoxicity against a panel of human cancer cell lines [45]. Among them, seven compounds displayed marked cytotoxicity. Vaticanol C (Vat C, an oligomer derived from four molecules of RES) as a major component induced a considerable cytotoxicity in all cancer cell lines tested and exhibited growth suppression in colon cancer cell lines at low dose. Vat C caused two cell lines

(SW480 and HL60) to induce apoptosis at four to seven times lower concentrations compared with RES. The apoptosis in SW480 colon cancer cells was executed by the activation of caspase-3, which was shown by Western blot and apoptosis inhibition by caspase inhibitor assay. The mitochondrial membrane potential of apoptotic SW480 cells after 12 hours' treatment with Vat C was significantly lost, and concurrently the cytochrome c release and activation of caspase-9 were also detected by Western blot analysis. Overexpression of Bcl-2 protein in SW480 cells significantly prevented the cell death induced by Vat C. In summary, Vat C induced marked apoptosis in malignant cells mainly by affecting mitochondrial membrane potential [45].

A series of RES derivatives, namely Vat A, Vat B, Vat C, Vat D, Vat H, Vat I, Vat J, and others, were isolated from the *Vatica*, *Vateria*, and *Shorea* spp. of Dipterocarpaceae and their structures were determined, as previously described [46,47]. Further investigations on the cytotoxicity and cancer chemopreventive activities of these derivatives are highly recommended.

ACKNOWLEDGMENTS

This study was supported by the National Science Council NSC 93-2311-B-002-001, NSC 93-2320-B-002-111, and NSC 93-2320-B-002-127. The author would like to thank his research associates including Prof. S. Y. Lin-Shiau, Prof. Y. S. Ho, Prof. Y. C. Liang, Prof. S. H. Tsai and others for their excellent collaboration which made the completion of this chapter possible.

REFERENCES

1. Soleas GJ, Diamondis EP, and Goldberg DM, Resveratrol: a molecule whose time has come? And gone?, *Clin Biochem* 30, 91–113, 1997.
2. Jang M, Cai L, Udeani GO, Slowing KV, Thomas CF, Beecher CW, Fong HH, Famsworth NR, Kinghorn AD, Mehta RG, Moon RC, and Pezzuto JM, Cancer chemopreventive activity of resveratrol, a natural product derived from grapes, *Science* 275, 218–220, 1997.
3. Joe AK, Liu H, Suzui M, Vural ME, Xiao D, and Weinstein IB, Resveratrol induces growth inhibition, S-phase arrest, apoptosis and changes in biomarker expression in several human cancer cell lines, *Clin Cancer Res* 8, 893–903, 2002.
4. Stivala L, Savio M, Carfoli F, Perucca P, Bianchi L, Maga G, Forti L, Pagnonti UM, Albini A, Prosperi E, and Vannini V, Specific structural determinants are responsible for the anti-oxidant activity and the cell cycle effects of resveratrol, *J Biol Chem* 276, 22586–22594, 2001.
5. Tsai SH, Lin-Shiau SY, and Lin JK, Suppression of nitric oxide synthase and down-regulation of the activation of NFκB in macrophage by resveratrol, *Br J Pharmacol* 126, 673–680, 1999.

6. Lin JK and Tsai SH, Chemoprevention of cancer and cardiovascular disease by resveratrol, *Proc Natl Sci Counc ROC B* 23, 99–106, 1999.
7. Contant J, Alcohol, ischemic heart disease and the French paradox, *Coronary Artery Dis* 8, 643–649, 1997.
8. Sato M, Maulik G, Bagchi D, and Das DK, Myocardial protection by protykin, a novel extract of trans-resveratrol and emodin, *Free Radical Res* 32, 135–144, 2000.
9. Carbo N, Costelli P, Baccino FM, Lopez-Soriano F, and Argiles JM, Resveratrol, a natural product present in wine, decreases tumor growth in a rat mouse model, *Biochem Biophys Res Commun* 254, 739–743, 1999.
10. Tessitore L, Davito A, Saretto I, and Caderni G, Resveratrol depresses $p21^{cip-1}$ expression, *Carcinogenesis* 21, 1619–1622, 2000.
11. Gusmann J, Malonne H, and Atassi G, A reappraisal of the potential chemopreventive and chemotherapeutic properties of resveratrol, *Carcinogenesis* 22, 1111–1117, 2001.
12. Clement MV, Hirpara JL, Clawdbury SH, and Pervaiz S, Chemopreventive agent resveratrol, a natural product derived from grapes triggers CD95 signaling-dependent apoptosis in human tumor cells, *Blood* 92, 996–1002, 1998.
13. Burns J, Yokota T, Ashihara H, Lean MEJ, and Crozler A, Plant foods and herbal sources of resveratrol, *J Agric Food Chem* 50, 3337–3340, 2002.
14. Belguendouz L, Fremont L, and Gozzellino MT, Interaction of trans resveratrol with plasma lipoproteins, *Biochem Pharmacol* 55, 811–816, 1998.
15. Pettit GR, Grealish MP, Jung MK, Hamel E, Pettit RK, Chapuis JC, and Schmidt JM, Anti-neoplastic agents 465. Structural modification of resveratrol: sodium resvertrin phosphate, *J Med Chem* 45, 2534–2542, 2002.
16. Roberti M, Pizzirani D, Simoni D, Rondanin R, Baruchello R, Bonora C, Buscemi F, Grimaudo S, and Tolomeo M, Synthesis and biological evaluation of resveratrol and analogues as apoptotic-inducing agents, *J Med Chem* 46, 3546–3554, 2003.
17. Banerjee S, Bueso-Ramos C, and Aggarwal BB, Suppression of 7,12-dimethylbenz(a)anthracene-induced mammary carcinogenesis in rats by resveratrol: role of 19 nuclear factor-κB, cyclooxygenase-2 and matrix-metaloproteinase-9, *Cancer Res* 62, 4945–4954, 2002.
18. Uenobe F, Nakamura S, and Miyazawa W, Antimutagenic effect of resveratrol against TrpP-1, *Mutation Res* 373, 197–200, 1997.
19. Chun YJ, Kim MY, and Guengerich FP, Resveratrol is a selective human cytochrome p450 1A1 inhibitor, *Biochem Biophys Res Commun* 262, 20–24, 1999.
20. Ciolino HP, Daschner PJ, and Yeh GC, Resveratrol inhibits transcription of CYP1A1 *in vitro* by preventing activation of aryl hydrocarbon receptor, *Cancer Res* 58, 5707–5712, 1998.
21. Stewart JR, Ward NE, Ioannides CG, and O'Brian CA, Resveratrol preferentially inhibits protein kinase C-catalyzed phosphorylation of a co-factor-independent, arginine-rich protein substrate by a nobel mechanism, *Biochemistry* 38, 13244–13251, 1999.
22. Subbaramaiah K, Chung WJ, Michahuart P, Telang N, Tanabe T, Inoue H, Tang M, Pezzuto JM, and Dannenberg AJ, Resveratrol inhibits cyclooxygenase-2 transcription and activity in phorbol ester-treated human mammary epithelial cells, *J Biol Chem* 273, 21875–21882, 1998.

23. Cadenas S and Baija G, DNA damage induced by the kidney carcinogen $KBrO_3$, *Free Radical Biol Med* 26, 1531–1537, 1999.
24. Ragione FD, Cucciolla V, Borriello A, Pietra VD, Racioppi L, Soldati G, Manna C, Galletti P, and Zappia V, Resveratrol arrests the cell division cycle at S/G2 phase transition, *Biochem Biophys Res Commun* 250, 53–58, 1998.
25. Estrov Z, Shishodia S, Faderl S, Harris D, Van Q, Kantarjian HM, Talpaz M, and Aggarwal BB, Resveratrol blocks interleukin-1β-induced activation of the nuclear transcription factor NFκB, inhibits proliferation, causes G1-phase arrest, and induces apoptosis of acute myeloid leukemia cells, *Blood* 102, 987–995, 2003.
26. Fulda S and Debatin KM, Sensitization for tumor necrosis factor-related apoptosis-inducing ligand-induced apoptosis by the chemopreventive agent resveratrol, *Cancer Res* 64, 337–346, 2004.
27. Hsieh TC, Halicka D, Lu X, Kunicki J, Guo J, Darzynkiewicz Z, and Wu JM, Effects of resveratrol on the G0-G1 transition and cell cycle progression of mitogenically stimulated human lymphocyte, *Biochem Biophys Res Commun* 297, 1311–1317, 2002.
28. Ahmad N, Adhami VM, Afaq F, Feyes DR, and Mukhtar H, Resveratrol causes WAF-1/p21-mediated G1-phase arrest of cell cycle and induction of apoptosis in human apidermoid carcinoma A431 cells, *Clin Cancer Res* 7, 1466–1473, 2001.
29. Adhami VM, Afaq F, and Ahmad N, Involvement of the retinoblastoma (pRb)-E2F/DP pathway during anti-proliferative effects of resveratrol in human epidermoid carcinoma A431 cells, *Biochem Biophys Res Commun* 288, 579–585, 2001.
30. Kuwajerwala N, Cifuentes E, Gautam S, Menon M, Barrack ER, and Reddy GPV, Resveratrol induces prostate cancer cell entry into S phase and inhibits DNA synthesis, *Cancer Res* 62, 2488–2492, 2002.
31. Walter F, Akoglu B, Clausnitzer A, and Stein J, Downregulation of the cyclin D1/Cdk4 complex occurs during resveratrol-induced cell cycle arrest in colon cancer cell lines, *J Nutr* 131, 2197–2203, 2001.
32. Arber N, Hibshoosh H, Moss SF, Sutter T, Zhang Y, Begg M, Wang S, Weinstein IB, and Holt PR, Increased expression of cyclin D1 is an early event in multistage carcinogenesis, *Gastroenterology* 110, 669–674, 1996.
33. Hsieh TC, Juan G, Darzynkiewicz Z, and Wu JM, Resveratrol increases nitric oxide synthase, induces accumulation of p53 and $p21^{WAF\text{-}1/Cip\text{-}1}$ and suppresses cultured bovine pulmonary artery endothelial cell proliferation by perturbing progression through S and G2, *Cancer Res* 59, 2596–2601, 1999.
34. Liang YC, Tsai SH, Chen L, Lin-Shiau SY, and Lin JK, Resveratrol-induced G2 arrest through the inhibition of CDK7 and $p34^{cdc\text{-}2}$ kinases in colon carcinoma HT29 cells, *Biochem Pharmacol* 65, 1053–1060, 2003.
35. Nurse P, Universal control mechanism regulating onset of M-phase, *Nature* 344, 503–508, 1990.
36. Morgan DO, Principles of CDK regulation, *Nature* 374, 131–134, 1995.
37. Labbe JC, Capony JP, Caput D, Cavadore JC, Derancout J, Kagha M, Lelias JM, Picard A, and Doree M, MPF from starfish oocytes at first meiotic metaphase is a heterodimer containing one molecule of cdc 2 and one molecule of cyclin B, *EMBO J* 8, 3053–3058, 1989.

38. Hunt T, Maturation promoting factor, cyclin and control of M-phase, *Curr Opin Cell Biol* 1, 268–274, 1989.
39. Fesquet D, Labbe JC, Derancourt J, Capony JP, Galas S, Girard F, Lorea T, Shuttleworth J, Doree M, and Cavadore JC, The MO15 gene encodes the catalytic subunit of a protein kinase that activates cdc 2 and other cyclin-dependent kinase (CDKs) through phosphorylation of the Thr 161 and its homologues, *EMBO J* 12, 1311–1321, 1993.
40. Gautier J, Solomon MJ, Booher RN, Bazan JF, and Kirschner MW, cdc 25 is a specific tyrosine phosphatase that directly activates $p34^{cdc2}$, *Cell* 67, 197–211, 1991.
41. Strausfeld U, Labbe JC, Fesquet D, Cavadore JC, Picard A, Sadha K, Russel P, and Doree M, Dephosphorylation and activation of a $p34^{cdc2}$/cyclin B complex *in vitro* by human CDC25 protein, *Nature* 351, 242–244, 1991.
42. Potter GA, Patterson LH, Wanogho E, Perry PJ, Butler PC, Ijaz T, Ruparelia KC, Lamb JH, Farmer PB, Stanley LA, and Burke MD, The cancer preventive agent resveratrol is converted to the anti-cancer agent piceatannol by the cytochrome p450 enzyme CYP1B1, *Br J Cancer* 86, 774–778, 2002.
43. She QB, Ma WY, Wang M, Kaji A, Ho CT, and Dong Z, Inhibition of cell transformation by resveratrol and its derivatives: differential effects and mechanisms involved, *Oncogene* 22, 2143–2150, 2003.
44. Sale S, Verschoyle RD, Boocock D, Jones DJL, Wilsher N, Ruparelia KC, Pitter GA, Farmer PB, Steward WP, and Gescher AJ, Pharmacokinetics in mice and growth-inhibitory properties of the putative cancer chemo-preventive agent resveratrol and the synthetic analogues trans-3,4,5,4′-tetramethoxystilbene, *Br J Cancer* 90, 736–744, 2004.
45. Ito T, Akao Y, Yi H, Ohguchi K, Matsumoto K, Tanaka T, Iimura M, and Nozawa Y, Anti-tumor effect of resveratrol oligomers against human cancer cell lines and the molecular mechanism of apoptosis induced by vaticanol C, *Carcinogenesis* 24, 1489–1497, 2003.
46. Ito T, Tanaka T, Nakaya K, Iinuma M, Nakahashi Y, Naganawa H, Ohyama M, Nakanishi Y, Bastow KF, and Lee KH, A novel bridged stilbenoid trimer and four highly condensed stilbene oligmers in Vatica rassak, *Tetrahedron* 57, 7309–7321, 2001.
47. Ito T, Tanaka T, Nakaya K, Iimura M, Nakahashi Y, Naganawa H, Ohyama M, Nakanishi Y, Bastow KF, and Lee KH, A new resveratrol octomer, vateriaphenol A in Vateria indica, *Tetrahedron Lett* 42, 5909–5912, 2001.

8 Molecular Targets of Resveratrol: Implications to Health and Disease Prevention

Catherine A. O'Brian and Feng Chu

CONTENTS

MOLECULAR TARGETS OF RESVERATROL: FORESTALLING SENESCENCE AND PREVENTIVE INTERVENTION AGAINST CARDIOVASCULAR, METABOLIC, AND NEURODEGENERATIVE DISEASES

Calorie restriction (CR) extends lifespans across the spectrum of eukaryotic organisms, e.g., budding yeast, nematodes, flies, and mammals. While the conventional view has long attributed this to a decline in the metabolic rate and hence mitigation of oxidative damage, it turns out that a far more specific mechanism is at play in yeast and is also implicated in CR-triggered lifespan extension in higher organisms. Intriguingly, *trans*-3,4′, 5-trihydroxystilbene (*trans*-resveratrol, referred to hereafter as resveratrol) can substitute for CR to turn on this lifespan-extending mechanism.

In *Saccharomyces cerevisiae* longevity is calculated as the number of divisions that a mother cell undergoes prior to death, and glucose deprivation is a paradigm of CR-triggered lifespan extension in this organism. CR-triggered delayed senescence in *S. cerevisiae* requires expression of an intact nicotinamide adenine dinucleotide (NAD) synthesis pathway and the NAD-dependent deacetylase Sir2 (also called Sir2p) [1]. CR induction of nicotinamidase, which is encoded by the PNC1 gene, delays senescence by activating Sir2 through removal of the Sir2 inhibitor nicotinamide [2].

Resveratrol mimics the lifespan-extending effects of CR in *S. cerevisiae* by activating Sir2, but does so independently of nicotinamidase expression, as it is similarly effective in a PNC1-null mutant [3]. In contrast, CR-induced delayed senescence is abolished in PNC1-null *S. cerevisiae* [2]. Sir2 activation by resveratrol in yeast appears to be a consequence of direct interactions between Sir2 and resveratrol, based on analysis of purified recombinant Sir2.

S. cerevisiae Sir2 is a member of the sirtuin family of NAD-dependent deacetylases which are conserved from bacteria to humans and play important roles in gene silencing and DNA repair. Gene silencing by yeast, human, and other sirtuins involves their histone deacetylase (HDAC) activity [4,5], and the human sirtuin SIRT1 has also been shown to increase cell survival by deacetylating p53 [6]. Resveratrol stimulates the NAD-dependent deacetylase activity of purified recombinant human SIRT1 more than 5-fold and the activities of purified recombinant sirtuins from *Caenorhabditis elegans* (SIR-2.1) and *Drosophila melanogaster* (Sir2) 2.0- to 2.5-fold, primarily by enhancing NAD and substrate affinities [3,7].

Like CR, resveratrol enhancement of lifespan in *S. cerevisiae* was abolished in a Sir2-null mutant, and resveratrol failed to affect the yeast lifespan under CR conditions, indicating that CR and resveratrol both promote longevity by increasing Sir2 activity in yeast [3]. Interestingly, resveratrol produces analogous senescence delaying effects in multicellular eukaryotic organisms. Resveratrol extended the lifespan of the nematode *C. elegans* but was without effect in a SIR-2.1-null mutant [7]. Similarly, resveratrol extended the lifespan of *D. melanogaster*, but not under CR conditions and not in flies with defective or severely attenuated Sir2 [7,8]. Nature's downside of longevity enhancement by CR and other environmental cues is that it is generally accompanied by the tradeoff of reduced fertility. The effects of resveratrol diverge from CR in this respect, as resveratrol has the unusual advantage of delaying senescence in *C. elegans* and *D. melanogaster* without reducing fertility [7]. These studies raise the intriguing possibility that resveratrol or related sirtuin-activating phytochemicals might likewise delay aging in humans by a sirtuin-dependent mechanism.

In yeast, gene silencing by Sir2 is thought to delay senescence by repressing genomic instability [1]. Delayed senescence by CR in mammals is associated with a reduction in white adipocyte tissue (WAT), and mice engineered to have reduced WAT mass have been shown to live longer than *wt* mice [9]. A recent study in rodents suggests that a specific gene-silencing event, SIRT1 silencing of peroxisome proliferator-activated receptor-γ (PPAR-γ)-regulated genes, may contribute to CR protection against aging-related changes in mammals by inhibiting adipogenesis and promoting fat mobilization [10]. Enforced expression of SIRT1 attenuated adipogenesis and fat storage in cultured rodent adipocytes, whereas RNA silencing of SIRT1 expression enhanced these responses. In mice, fasting induced SIRT1 binding to PPAR-γ-regulated promoters *in vivo*, and knockdown of the SIRT1-interacting PPAR-γ cofactor NCoR in cultured adipocytes abolished the effects of SIRT1 on fat accumulation, identifying the mechanism of SIRT1 action as silencing of PPAR-γ-regulated genes. Resveratrol potently reduced fat accumulation in adipocytes, but was without effect when SIRT1 expression was knocked down in the cells [10]. Thus, resveratrol may to some extent recapitulate CR-mediated protection against the aging-related diseases diabetes and atherosclerosis in mammals by stimulating SIRT1 repression of PPAR-γ-regulated genes.

Neurodegenerative disorders constitute a devastating class of diseases related to aging. Neurons are subject to two types of self-destruction: apoptosis by canonical, caspase-mediated signaling programs, and axonal degeneration, which is a compartmentalized, orderly process that is triggered when an axon is mechanically cut from the cell body or injured by chemical or other means. While the outcome of apoptosis is suicide of the cell, axonal degeneration eliminates the axon without harming the cell body and thus produces what could be described as a viable, "amputated" neuron. Axonal degeneration contributes to the pathogenesis of Alzheimer's, Parkinson's, and Huntington's diseases as well as to the neurotoxicity of vincristine and other microtubule-targeting cancer chemotherapeutic drugs [11]. In an exciting development, resveratrol has been shown to protect against mammalian axonal degeneration, and the molecular target implicated is SIRT1 [12]. In *Wallerian degeneration slow* (*Wld^s*) mice, axonal degeneration is slowed and a fusion protein, which consists of an enzyme in the NAD synthesis pathway fused to a small fragment of a ubiquitination factor, is expressed as a result of a spontaneous dominant mutation. Expression of the fusion protein (*Wld^s* protein) is sufficient to impart the slowed axonal degeneration phenotype in *wt* mice [11].

Araki et al. established that expression of the nicotinamide mononucleotide adenylyl transferase 1 (Nmnat1) component of the *Wld^s* protein was as effective as the fusion protein itself in protecting against axonal degeneration triggered by axon transection or the addition of vincristine to primary cultures of mouse neurons. Incubation with NAD was similarly protective against axonal degeneration, but only when NAD was added to the primary neuron culture more than 8 hours before axonal transection. This indicated that the protective mechanism originated in the cell body, and suggested altered gene regulation by an NAD-dependent mechanism, e.g., SIRT1 activation, as the axon-sparing mechanism. Indeed, resveratrol proved as effective as NAD in forestalling axonal degeneration in the axon transection paradigm, and RNA silencing of SIRT1 eliminated protection against axonal degeneration by NAD [12]. Thus, SIRT1 activators such as resveratrol may simultaneously protect against neurological (axonal degeneration) [12], cardiovascular (atherosclerosis), and metabolic (diabetes) [10] aging-related pathological conditions in mammals.

In considering the potential relevance of resveratrol interactions with SIRT1 and other molecular targets to disease prevention, it is necessary to be aware that the oral bioavailability of resveratrol is very low in humans despite efficient absorption, due to rapid metabolism to sulfate and glucuronic conjugates [13]. It is therefore likely that accrual of health benefits of resveratrol, such as SIRT1 activation, will generally not be obtained by dietary means, e.g., red wine or grapes, and will instead require development of resveratrol analogs that are bioavailable in humans and retain the desired effects on specific molecular targets. One exception to this limitation is that dietary resveratrol may be available to engage molecular targets in the upper digestive tract, where its cancer-preventive effects could be beneficial.

Preliminary investigations suggest that resveratrol could forestall pathogenesis of the aging-related ocular disease macular degeneration through its effects on the molecular target singlet oxygen [14]. Pyridinium bisretinoid (A2E) and related autofluorescent pigments (lipofuscin) are formed as byproducts of the light-induced reaction of all-*trans*-retinal with phosphatidylethanolamine in the photoreceptor outer segment membrane [15]. Aging-related accumulation of the autofluorescent pigments in retinal pigment epithelial cells is implicated in macular degeneration pathogenesis [16]. A2E photoexcitation by blue-light illumination generates singlet oxygen, which in turn reacts with A2E to produce A2E epoxides; these reactive oxygen species (ROS) damage DNA in the retinal pigment epithelial cells, which leads to apoptosis. Resveratrol quenches singlet oxygen through its intrinsic antioxidant activity and inhibits irradiation-induced A2E epoxidation [14]. If further investigations reveal that resveratrol can, in fact, protect retinal pigment epithelial cells from aging-related, visible light-induced injury and death by quenching toxic A2E photoexcitation products, preventive intervention by local delivery of resveratrol, e.g., eye drops, may be possible if the polyphenol does not introduce any harmful side effects.

MOLECULAR TARGETS OF RESVERATROL: CANCER PREVENTION AND THERAPY

The risk of neoplastic disease in humans is strongly influenced by the extent to which DNA-damaged cells successfully persist in the body in a viable state, because DNA damage often leads to neoplastic transformation.

Thus, cancer risk is closely linked to the interplay of genetic predisposition and environmental factors (irradiation, chemical exposure, etc.) that produces DNA-damaged cells and the efficiency of innate physiological and molecular mechanisms that destroy and dispose of DNA-damaged cells. Resveratrol may potentiate the self-destruction of DNA-damaged cells by inhibiting the NF-κB survival pathway, and thus lower cancer risk.

NF-κB is a heterodimeric transcription factor composed of p50 and RelA/p65 subunits; RelA contains DNA-binding and transactivation domains and is responsible for promoter activation of NF-κB-regulated genes, while p50 contains a DNA-binding domain but lacks a transactivation domain and can repress expression of NF-κB target genes. NF-κB orchestrates a cell survival program by effecting transcriptional upregulation of functionally diverse survival genes, including Bcl-XL, cIAP, and cFLIP. In unstimulated cells, NF-κB is sequestered in the cytoplasm in a heterotrimeric complex with IκBα, where it is "on call" for activation by stimuli, such as TNFα and other proinflammatory cytokines. NF-κB activation entails a series of cytosolic and nuclear events. First, IκBα phosphorylation by IκBα kinase (IKK = IKKα, IKKβ, and IKKγ) enables IκBα ubiquitination and proteasomal degradation, releasing NF-κB for translocation to the nucleus. Full transcriptional activity requires post-translational modification of RelA by phosphorylation and acetylation. Several Ser/Thr protein kinases, some cytosolic and some nuclear, phosphorylate RelA at one or more of several phosphoregulatory sites, and histone acetyltransferases (HATs) catalyze RelA acetylation in the nucleus, primarily at three lysine residues (RelA K218, K221, K310). HAT-catalyzed histone-H3 and -H4 acetylation, in the region surrounding and within the NF-κB-regulated promoter target gene, is also important to NF-κB transcriptional activity [17,18].

Resveratrol inhibits the NF-κB-orchestrated survival program by impeding nuclear translocation of RelA in some cells [19] and by inducing RelA deacetylation in the nucleus in others [20]. Resveratrol may thus inhibit NF-κB transcriptional activation by redundant mechanisms in some cell types. In an analysis of human nonsmall-cell lung cancer (NSCLC) cells, resveratrol potentiated tumor necrosis factor α (TNFα)-induced apoptosis and inhibited TNFα-induced NF-κB transcriptional activation, without affecting either NF-κB nuclear translocation or DNA binding. Hypothesizing SIRT1 as the molecular target mediating the resveratrol effect, Yeung et al. established that ectopic SIRT1 expression was sufficient to suppress TNFα-induced NF-κB transcriptional activation in the NSCLC cells, and that endogenous SIRT1 and RelA had direct binding interactions in the cells [20]. Deacetylation at K310 markedly impairs RelA transcriptional activity without affecting its DNA-binding activity [21], and may thus produce a dominant negative effect. Analysis of HEK293 cells cotransfected with various RelA mutants, the HAT p300, and/or SIRT1 revealed that SIRT1 induced RelA deacetylation and mapped the deacetylation site to K310. Similar results obtained with recombinant SIRT1 and p300-acetylated RelA *in vitro* identified SIRT1 as the K310 deacetylase. Resveratrol likewise induced RelA deacetylation in cells cotransfected with p300 and RelA [20].

Resveratrol discriminated among NF-κB target genes in NSCLC cells, inhibiting the transcriptional upregulation of some (Bcl-XL, cIAP) but not others (IκBα). ChIP assays established that resveratrol suppressed cIAP transcriptional upregulation in association with stabilization of SIRT1 binding at the cIAP promoter. Furthermore, SIRT1 knockdown abrogated suppression of cIAP promoter activation as well as potentiation of TNFα-induced apoptosis by resveratrol in the NSCLC cells, clearly demonstrating that SIRT1 activation by resveratrol contributes significantly to suppression of the NF-κB survival pathway by the polyphenol [20]. This further persuades one that SIRT1 activators such as resveratrol may be beneficial in preventing disease, although it is too early to say whether indiscriminately augmenting deacetylation of SIRT1 effectors will produce harmful effects.

An in-depth analysis of resveratrol effects on NF-κB activation in human monocytic U937 leukemia cells established that resveratrol suppresses NF-κB activation by diverse stimuli, including the phorbol-ester diacylglycerol mimetic TPA, the Toll-receptor ligand LPS, the protein-phosphatase inhibitor okadaic acid, and TNFα. The relevance of these findings to cancer prevention was underscored by similar results obtained in an analysis of freshly isolated, normal human lymphocytes [19]. Resveratrol did not affect TNFα-induced IκBα degradation or the subsequent recovery of IκBα expression in U937 cells, nor did it affect IκBα phosphorylation. Resveratrol suppressed NF-κB activation in the cells by abrogating RelA nuclear translocation in association with inhibition of RelA phosphorylation, consistent with inhibition of a cytosolic RelA kinase impeding nuclear entry by NF-κB [19].

A third point of entry into the NF-κB activation pathway by resveratrol was revealed in an analysis of human monocytic THP-1 leukemia cells. Resveratrol suppressed TNFα-induced NF-κB activation in these cells by potent inhibition of an early event in the pathway, TNFα-induced IKK activation. Resveratrol inhibited IKK activation in THP-1 cells through effects on an unidentified molecular target hypothesized to be upstream of IKK in the TNFα-induced NF-κB activation pathway [22]. Exploiting the suppressive effects of resveratrol on NF-κB activation (deacetylation of RelA at K310, inhibition of a RelA cytosolic kinase, etc.) through development of bioavailable analogs for chemopreventive intervention may help thwart neoplastic disease development. In fact, efforts to develop analogs of resveratrol for use as antineoplastic drugs in humans are underway [23].

TNF-related apoptosis-inducing ligand (TRAIL) is a member of the TNF superfamily of cytokines and has attracted a great deal of attention in the field of experimental cancer therapeutics by virtue of its ability to induce apoptosis in a cancer cell-selective manner. The use of TRAIL preparations in the treatment of cancer shows promise, based on the marked antitumor efficacy of this approach against various human tumor xenografts in athymic or SCID mice without systemic toxicity [24], although some toxicity issues, e.g., TRAIL effects on human hepatocytes and esophageal epithelial cells [25], and the problem of tumor cell resistance [26], indicate a need for refinement of the treatment strategy.

The discovery that resveratrol potentiates TRAIL-induced apoptosis of a broad spectrum of cancer cell types [27,28] has raised two interesting questions. The first is whether bioavailable resveratrol analogs could serve as adjuvant therapeutics to improve the response to cancer therapy with TRAIL by mitigating cancer cell resistance to TRAIL-induced apoptosis [27,28]. A critical issue here will be whether the resveratrol analogs have the adverse effect of potentiating TRAIL toxicity in TRAIL-sensitive normal human tissues. Second, could chemopreventive intervention with a bioavailable resveratrol analog improve the efficiency of endogenous TRAIL in the apoptotic destruction of incipient neoplastic tumors? Here, efficacy rather than toxicity is the pressing issue. It is not clear whether endogenous TRAIL is adequate to support substantial improvement of the body's defenses against incipient neoplastic disease by chemopreventive intervention with a TRAIL-potentiating agent. However, immune cell expression or shedding of TRAIL does produce tumoricidal effects in some experimental models [29] and gives cause for optimism.

TRAIL engages four receptors in the TNFR superfamily with high affinity: the death-receptors, DR4 and DR5, and the decoy-receptors, DcR1 and DcR2 [24]. The observation that DR5 expression was generally much higher in diverse human tumor tissues versus a broadly representative set of human normal tissues, based on Western analysis with a DR5 mAb, suggests a basis for cancer cell-selective, TRAIL-induced apoptosis [30]. TRAIL engagement of DR4 and DR5 induces the extrinsic (death receptor-initiated) pathway of apoptosis, with amplification by the intrinsic (mitochondria-initiated) apoptosis pathway. Depending on factors that include the cell type [25] and the TRAIL dosage [31], the amplification loop may be required for TRAIL-induced apoptosis or dispensable. Upon receptor trimerization and activation, the cytoplasmic death domain in DR4/DR5 provides a scaffold for assembly of the death-inducing signaling complex (DISC) by recruiting the death-adaptor FADD (Fas-associated death domain), which in turn recruits procaspase-8 and activates the initiator caspase by a homodimerization mechanism [32,33]. Caspase-8 proteolytically activates its effectors caspase-3 and caspase-7 and also amplifies the apoptosis response through Bid cleavage, which inaugurates the canonical intrinsic pathway. This leads to initiator caspase-9 activation by dimerization and cleavage of effector caspase zymogens to active forms [24,33]. DcR1 and DcR2 do not recruit DISCs for lack of functional cytoplasmic death domains and produce resistance to DR4/DR5-induced apoptosis by competing for TRAIL at the cell surface [24]. Analogously, cFLIP is a catalytically inactive caspase-8 homolog that competes with procaspase-8 for occupancy of the DR4/DR5 DISC. cFLIP produces TRAIL resistance by this mechanism when expressed at relatively high levels (as in some tumor cells), but it activates procaspase-8 by heterodimerization at physiological concentrations [33].

In an analysis of over a dozen human cancer cell lines that encompassed neuroblastoma and melanoma cells, and breast, prostate, colon, and pancreatic cancer cells, resveratrol increased apoptosis induction by 30 ng/ml TRAIL in each cell line from <10% by TRAIL or resveratrol alone to 40–60% of the cells, suggesting a TRAIL-sensitization mechanism of broad relevance to the therapy of TRAIL-resistant neoplastic disease [27]. Abrogation of resveratrol-potentiated, TRAIL-induced apoptosis of SHEP neuroblastoma cells by the nonselective caspase inhibitor zVADfmk verified a caspase-dependent death mechanism, and strong synergy of resveratrol and TRAIL was observed in the activation of caspase-8 in the SHEP cells (inferred from cleavage of procaspase-8). SHEP cell sensitization to TRAIL was accompanied by resveratrol-induced G_1 arrest [27], compatible with previous observations that cancer cell sensitivity to TRAIL-induced apoptosis varies with the phase of the cell cycle and is optimal in G_0/G_1 [34].

The mechanism whereby resveratrol potentiated TRAIL-induced apoptosis involved downregulation of survivin, a member of the caspase-inhibitory family of inhibitor of apoptosis proteins (IAPs), by posttranscriptionally upregulated p21. This was demonstrated by observations of (1) an intact TRAIL-sensitization response to resveratrol in p53-deficient cancer cells that was associated with the characteristic G_1 arrest, p21 upregulation, and survivin downregulation events, (2) marked defects in the TRAIL-sensitization response to resveratrol in p21-deficient cancer cells, and (3) a comparable TRAIL-sensitization response accompanied by G_1 arrest and survivin downregulation in SHEP cells by enforcing p21 overexpression in lieu of resveratrol treatment. Furthermore, diminishing survivin expression with antisense oligos also sensitized SHEP cells to TRAIL [27]. The results provide evidence that enhancement of caspase-8 activation and p21-mediated survivin downregulation contribute to cancer cell sensitization to TRAIL by resveratrol; the direct molecular targets of resveratrol mediating these effects await identification.

Resveratrol-induced redistribution of DR4 and DR5 into lipid rafts is also implicated in cancer cell sensitization to TRAIL-induced apoptosis by resveratrol [28]. Lipid rafts are sphingolipid- and cholesterol-enriched, liquid-ordered microdomains in cell membranes (the name raft refers to flotation properties of isolated lipid-raft fractions) [35]. Signaling by CD95 (Fas) [36] and other TNFR family members, including DR4 and DR5 [28,36], is influenced by receptor partitioning into lipid rafts.

Analysis of human colon cancer HT29 cells established that resveratrol sensitized the cells to TRAIL-induced apoptosis [28], conforming with effects observed in other human cancer cell lines [27]. The HT29 analysis revealed that TRAIL was sufficient to induce DR4/DR5 DISC formation but apoptosis induction was marginal. Resveratrol increased TRAIL-induced HT29 apoptosis from <10% to 50–60% of the cells, but did not affect expression of TRAIL receptors and only modestly enhanced DISC formation. Instead, the marked apoptosis response of HT29 cells to resveratrol/TRAIL correlated with resveratrol induction of DR4/DR5 partitioning, together with the DISC components FADD and procaspase-8, into lipid rafts; lipid-raft partitioning was demonstrated by cosedimentation of the proteins on a sucrose density gradient with subcellular fractions enriched in cholesterol, sphingomyelin, and caveolin-2, a marker for lipid raft-associated caveolae. The significance of this correlative observation was indicated by the ability of the cholesterol-sequestering agent nystatin (which is a cholesterol-binding antibiotic) to abrogate redistribution of DISC-laden DR4/DR5 to lipid rafts in concert with inhibition of apoptosis induction by resveratrol/TRAIL. Thus, resveratrol analogs may help to overcome tumor resistance to TRAIL by promoting DR4/DR5 localization in lipid rafts, caspase-8 activation, and p21-mediated survivin down-regulation in cancer cells. Identification of the molecular targets of resveratrol involved in cancer cell sensitization to TRAIL by these mechanisms may help to cross the hurdle of developing bioavailable resveratrol analogs that retain TRAIL-sensitizing properties and can be used to test this therapeutic strategy *in vivo*.

The two-stage mouse skin carcinogenesis model is a paradigm of sequential initiation and promotion stages of neoplastic development. In the model, the initiation stage entails topical application of a subcarcinogenic dose of 7,12-dimethylbenz[*a*]anthracene (DMBA) to the mouse epidermis and is followed by the promotion stage, which involves repeated topical applications of 12-*O*-tetradecanoyl phorbol-13-acetate (TPA) for an extended period. The endpoint of the initiation–promotion sequence is benign papilloma formation in the mouse epidermis; neither DMBA nor TPA alone produces tumors in the model. Further TPA exposure is designated as the tumor progression stage and elicits squamous cell carcinomas. In the first demonstration that resveratrol has cancer chemopreventive activity *in vivo*, Pezzuto and colleagues found that topical application of resveratrol with TPA in the two-stage mouse skin model potently suppressed papilloma formation [37]; two subsequent reports confirmed the cancer-preventive activity of resveratrol in the DMBA/TPA mouse skin model [38,39].

TPA and other phorbol ester tumor promoters are mimetics of the second messenger sn-1,2-diacylglycerol (DAG), which is produced by receptor-linked phospholipase C activation; phorbol esters profoundly alter signaling by DAG effector proteins through nanomolar affinity at the

DAG binding-site (compared to micromolar for DAG) and much slower metabolic turnover than DAG. Resveratrol and phorbol esters have opposing effects on one of the principal DAG effector protein families involved in cell growth and survival regulation, the protein kinase C (PKC) isozyme family [40], suggesting a potential connection to the tumor-suppressing effects of resveratrol in the mouse epidermis.

PKC is a family of ten isozymes; of these, eight are activated by phosphatidylserine-dependent mechanisms that involve binding of the stimulatory cofactors DAG and Ca^{2+} or DAG alone to the regulatory domain [41]. Activation of DAG-responsive PKC isozymes by TPA at nanomolar concentrations [42] initially led to consideration of PKC as a target for cancer prevention or therapy. Subsequent studies revealed that the PKC family includes oncogenic isozymes, e.g., PKCε [43,44], and tumor-suppressive isozymes, e.g., PKCδ [45], indicating the need for isozyme-selective PKC targeting in designing strategies of cancer prevention and therapy. In addition to PKC, the protein kinase D (PKD) isozyme family (PKD-1, -2, and -3), has a DAG/TPA binding site in the regulatory domain. However, exposure to DAG/TPA is not sufficient to activate PKD; PKD activation requires phosphorylation at activation loop residues serine 744/748 by PKC. Thus, PKD is a PKC effector, and it is involved in the transmission of PKC-dependent cell proliferative and survival signals [46].

A study of resveratrol effects on purified PKC isozymes established that resveratrol inhibits PKC, but the inhibitory potency of the polyphenol was similar against the seven isozymes surveyed, which were the Ca^{2+}, DAG-activated isozymes PKCα, $PKC\beta_1$, $PKC\beta_2$, and PKCγ, the DAG-activated isozymes PKCδ and PKCε, and the DAG-independent isozyme PKCζ. PKC inhibition by resveratrol entailed competition with ATP substrate [40]. Analysis of purified PKD-1 (formerly called PKCμ) versus the above listed PKC isozymes determined that the polyphenol was somewhat more potent against PKD-1 than the PKC isozymes, and markedly so with respect to the kinase autophosphorylation reaction [47]. This suggested that resveratrol inhibition of PKC might be reinforced in cells by coordinate inhibition of the PKC effector PKD.

A recent analysis of growth-suppressive effects of resveratrol in the androgen-independent human prostate cancer cell line PC-3 revealed that, although resveratrol indiscriminately inhibits purified PKC/PKD species, the polyphenol selectively inhibits a subset of these targets in cells. Resveratrol blocked TPA-induced PKCα activation in the PC-3 cells without any effect on the other phorbol ester-regulated protein kinases in the cells, PKD and PKCε [48]. The lack of effect of resveratrol on both PKCε and PKD in the cells was consistent with observations identifying PKCε as the endogenous activator of PKD [49]. The strong suppression of tumor formation achieved by resveratrol in the two-stage mouse skin carcinogenesis model offers evidence that, in at least some tissues, the sum total of resveratrol inhibitory effects against TPA-responsive PKC isozymes and PKD may be cancer-preventive.

OTHER MOLECULAR TARGETS OF RESVERATROL

The multiplicity of resveratrol targets is an indication of the challenge posed in developing bioavailable resveratrol analogs that conserve the bioactivity of the polyphenolic phytoalexin in mammalian systems. However, dispensing with some features of resveratrol bioactivity may be beneficial. A case in point is the estrogen receptor (ER). ER agonists bind to the receptor and induce ER binding to estrogen response elements (EREs) in the promoters of target genes with the effect of promoter activation. Phytoestrogens are dietary nonsteroidal agents with ER agonist or mixed ER agonist/antagonist activity. Studies in human breast and uterine cancer cells have revealed that resveratrol is a mixed ER agonist/antagonist [50–52]. While phytoestrogens are widely viewed as healthful and cardioprotective, the ER agonist activity of resveratrol and other phytoestrogens suggests potential for increased risk of cancer and adverse cardiovascular effects over long-term usage by women.

Another target class warranting caution is the cyclooxygenase (COX) family. COXs catalyze the formation of prostaglandins from arachidonic acid. COX-1 is expressed constitutively as a housekeeping gene in most normal tissues, while COX-2 expression is induced by proinflammatory and other mitogenic signals. Resveratrol is a mechanism-based inactivator of COX-1 [53] and a potent suppressor of COX-2 induction [54]. Classic nonsteroidal antiinflammatory drugs (NSAIDs) inhibit both COX-1 and COX-2, and their use in the management of chronic inflammatory conditions such as osteoarthritis can be problematic due to serious gastrointestinal side effects. Development of selective COX-2 inhibitors was anticipated to improve the safety of antiinflammatory intervention in the treatment of chronic inflammatory disorders, but the selective COX-2 inhibitor rofecoxib (Vioxx) unfortunately increased risk of adverse cardiovascular events, for reasons still not understood but taken as a warning of the potential dangers of selective COX-2 inhibition [55]. Furthermore, long-term use of antiinflammatory drugs raises issues of susceptibility to infection. Thus, the potential hazards of COX targeting by resveratrol analogs warrant serious consideration.

ACKNOWLEDGMENTS

Supported by the Robert A. Welch Foundation and the Elsa U. Pardee Foundation.

REFERENCES

1. Lin SJ, Defossez PA, and Guarente L, Requirement of NAD and SIR2 for life-span extension by calorie restriction in *Saccharomyces cerevisiae*, *Science* 289, 2126–2128, 2000.
2. Anderson RM, Bitterman KJ, Wood JG, Medvedik O, and Sinclair DA, Nicotinamide and PNC1 govern lifespan extension by calorie restriction in *Saccharomyces cerevisiae*, *Nature* 423, 181–185, 2003.
3. Howitz KT, Bitterman KJ, Cohen HY, Lamming DW, Lavu S, Wood JG, Zipkin RE, Chung P, Kisielewski A, Zhang LL, Scherer B, and Sinclair DA, Small molecule activators of sirtuins extend *Saccharomyces cerevisiae* lifespan, *Nature* 425, 191–196, 2003.
4. Imai S, Armstrong CM, Kaeberlein M, and Guarente L, Transcriptional silencing and longevity protein Sir2 is an NAD-dependent histone deacetylase, *Nature* 403, 795–800, 2000.
5. Smith JS, Brachmann CB, Celic I, Kenna MA, Muhammad S, Starai VJ, Avalos JL, Escalante-Semerena JC, Grubmeyer C, Wolberger C, and Boeke JD, A phylogenetically conserved NAD+-dependent protein deacetylase activity in the Sir2 protein family, *Proc Natl Acad Sci* 97, 6658–6663, 2000.
6. Langley E, Pearson M, Faretta M, Bauer UM, Frye RA, Minucci S, Pelicci PG, and Kouzarides T, Human SIR2 deacetylates p53 and antagonizes PML/p53-induced cellular senescence, *EMBO J* 21, 2383–2396, 2002.
7. Wood JG, Rogina B, Lavu S, Howitz K, Helfand SL, Tatar M, and Sinclair D, Sirtuin activators mimic caloric restriction and delay ageing in metazoans, *Nature* 430, 686–689, 2004.
8. Bauer JH, Goupil S, Garber GB, and Helfand SL, An accelerated assay for the identification of lifespan-extending interventions in *Drosophila melanogaster*, *Proc Natl Acad Sci* 101, 12980–12985, 2004.
9. Bluher M, Kahn BB, and Kahn CR, Extended longevity in mice lacking the insulin receptor in adipose tissue, *Science* 299, 572–574, 2003.
10. Picard F, Kurtev M, Chung N, Topark-Ngarm A, Senawong T, Machado De Oliveira R, Leid M, McBurney MW, and Guarente L, Sirt1 promotes fat mobilization in white adipocytes by repressing PPAR-gamma, *Nature* 429, 771–776, 2004.
11. Raff MC, Whitmore AV, and Finn JT, Axonal self-destruction and neurodegeneration, *Science* 296, 868–871, 2002.
12. Araki T, Sasaki Y, and Milbrandt J, Increased nuclear NAD biosynthesis and SIRT1 activation prevent axonal degeneration, *Science* 305, 1010–1013, 2004.
13. Walle T, Hsieh F, DeLegge MH, Oatis JE, and Walle UK, High absorption but very low bioavailability of oral resveratrol in humans, *Drug Metab Disposition* 32, 1377–1382, 2004.
14. Sparrow JR, Vollmer-Snarr HR, Zhou J, Jang YP, Jockusch S, Itagaki Y, and Nakanishi K, A2E-epoxides damage DNA in retinal pigment epithelial cells. Vitamin E and other antioxidants inhibit A2E-epoxide formation, *J Biol Chem* 278, 18207–18213, 2003.
15. Mata NL, Weng J, and Travis GH, Biosynthesis of a major lipofuscin fluorophore in mice and humans with ABCR-mediated retinal and macular degeneration, *Proc Natl Acad Sci* 97, 7154–7159, 2000.

16. Parish CA, Hashimoto M, Nakanishi K, Dillon J, and Sparrow J, Isolation and one-step preparation of A2E and iso-A2E, fluorophores from human retinal pigment epithelium, *Proc Natl Acad Sci* 95, 14609–14613, 1998.
17. Aggarwal BB, NF-κB: the enemy within, *Cancer Cell* 6, 203–208, 2004.
18. Chen LF and Greene WC, Shaping the nuclear action of NF-κB, *Nat Rev Mol Cell Biol* 5, 392–401, 2004.
19. Manna SK, Mukhopadhyay A, and Aggarwal BB, Resveratrol suppresses TNF-induced activation of nuclear transcription factors NF-κB, activator protein-1, and apoptosis: potential role of reactive oxygen intermediates and lipid peroxidation, *J Immunol* 164, 6509–6519, 2000.
20. Yeung F, Hoberg JE, Ramsey CS, Keller MD, Jones DR, Frye RA, and Mayo MW, Modulation of NF-κB-dependent transcription and cell survival by the SIRT1 deacetylase, *EMBO J* 23, 2369–2380, 2004.
21. Chen LF, Mu Y, and Greene WC, Acetylation of RelA at discrete sites regulates distinct nuclear functions of NF-κB, *EMBO J* 21, 6539–6548, 2002.
22. Holmes-McNary M and Baldwin AS, Chemopreventive properties of trans-resveratrol are associated with inhibition of activation of the IκB kinase, *Cancer Res* 60, 3477–3483, 2000.
23. Pettit GR, Grealish MP, Jung MK, Hamel E, Pettit RK, Chapuis JC, and Schmidt JM, Antineoplastic agents: 465. Structural modification of resveratrol: sodium resverastatin phosphate, *J Med Chem* 45, 2534–2542, 2002.
24. Ashkenazi A, Targeting death and decoy receptors of the tumour-necrosis factor superfamily, *Nat Rev Cancer* 2, 420–430, 2002.
25. Kim SH, Kim K, Kwagh JG, Dicker DT, Herlyn M, Rustgi AK, Chen Y, and El-Deiry WS, Death induction by recombinant native TRAIL and its prevention by a caspase 9 inhibitor in primary human esophageal epithelial cells, *J Biol Chem* 279, 40044–40052, 2004.
26. Izeradjene K, Douglas L, Delaney AB, and Houghton JA, Casein kinase I attenuates tumor necrosis factor-related apoptosis-inducing ligand-induced apoptosis by regulating the recruitment of fas-associated death domain and procaspase-8 to the death-inducing signaling complex, *Cancer Res* 64, 8036–8044, 2004.
27. Fulda S and Debatin KM, Sensitization for tumor necrosis factor-related apoptosis-inducing ligand-induced apoptosis by the chemopreventive agent resveratrol, *Cancer Res* 64, 337–346, 2004.
28. Delmas D, Rebe C, Micheau O, Athias A, Gambert P, Grazide S, Laurent G, Latruffe N, and Solary E, Redistribution of CD95, DR4 and DR5 in rafts accounts for the synergistic toxicity of resveratrol and death receptor ligands in colon carcinoma cells, *Oncogene* 23, 8979–8986, 2004.
29. Takeda K, Hayakawa Y, Smyth MJ, Kayagaki N, Yamaguchi N, Kakuta S, Iwakura Y, Yagita H, and Okumura K, Involvement of tumor necrosis factor-related apoptosis-inducing ligand in surveillance of tumor metastasis by liver natural killer cells, *Nat Med* 7, 94–100, 2001.
30. Ichikawa K, Liu W, Zhao L, Wang Z, Liu D, Ohtsuka T, Zhang H, Mountz JD, Koopman WJ, Kimberly RP, and Zhou T, Tumoricidal activity of a novel anti-human DR5 monoclonal antibody without hepatocyte cytotoxicity, *Nat Med* 7, 954–960, 2001.

31. Rudner J, Jendrossek V, Lauber K, Daniel PT, Wesselborg S, and Belka C, Type I and type II reactions in TRAIL-induced apoptosis: results from dose-response studies, *Oncogene* 24, 130–140, 2005.
32. Bodmer JL, Holler N, Reynard S, Vinciguerra P, Schneider P, Juo P, Blenis J, and Tschopp J, TRAIL receptor-2 signals apoptosis through FADD and caspase-8, *Nat Cell Biol* 2, 241–243, 2000.
33. Boatright KM and Salvesen GS, Mechanisms of caspase activation, *Curr Opin Cell Biol* 15, 725–731, 2003.
34. Jin Z, Dicker DT, and El-Deiry WS, Enhanced sensitivity of G1 arrested human cancer cells suggests a novel therapeutic strategy using a combination of simvastatin and TRAIL, *Cell Cycle* 1, 82–89, 2002.
35. Munro S, Lipid rafts: elusive or illusive?, *Cell* 115, 377–388, 2003.
36. Muppidi JR and Siegel RM, Ligand-independent redistribution of Fas (CD95) into lipid rafts mediates clonotypic T cell death, *Nat Immunol* 5, 182–189, 2004.
37. Jang M, Cai L, Udeani GO, Slowing KV, Thomas CF, Beecher CW, Fong HH, Farnsworth NR, Kinghorn AD, Mehta RG, Moon RC, and Pezzuto JM, Cancer chemopreventive activity of resveratrol, a natural product derived from grapes, *Science* 275, 218–220, 1997.
38. Kapadia GJ, Azuine MA, Tokuda H, Takasaki M, Mukainaka T, Konoshima T, and Nishino H, Chemopreventive effect of resveratrol, sesamol, sesame oil and sunflower oil in the Epstein-Barr virus early antigen activation assay and the mouse skin two-stage carcinogenesis, *Pharmacol Res* 45, 499–505, 2002.
39. Soleas GJ, Grass L, Josephy PD, Goldberg DM, and Diamandis EP, A comparison of the anticarcinogenic properties of four red wine polyphenols, *Clin Biochem* 35, 119–124, 2002.
40. Stewart JR, Ward NE, Ioannides CG, and O'Brian CA, Resveratrol preferentially inhibits protein kinase C-catalyzed phosphorylation of a cofactor-independent, arginine-rich protein substrate by a novel mechanism, *Biochemistry* 38, 13244–13251, 1999.
41. Ron D and Kazanietz MG, New insights into the regulation of protein kinase C and novel phorbol ester receptors, *FASEB J* 13, 1658–1676, 1999.
42. Castagna M, Takai Y, Kaibuchi K, Sano K, Kikkawa U, and Nishizuka Y, Direct activation of calcium-activated, phospholipid-dependent protein kinase by tumor-promoting phorbol esters, *J Biol Chem* 257, 7847–7851, 1982.
43. Cacace AM, Ueffing M, Philipp A, Han EK, Kolch W, and Weinstein IB, PKC epsilon functions as an oncogene by enhancing activation of the Raf kinase, *Oncogene* 13, 2517–2526, 1996.
44. Reddig PJ, Dreckschmidt NE, Zou J, Bourguignon SE, Oberley TD, and Verma AK, Transgenic mice overexpressing protein kinase Cε in their epidermis exhibit reduced papilloma burden but enhanced carcinoma formation after tumor promotion, *Cancer Res* 60, 595–602, 2000.
45. Reddig PJ, Dreckschimdt NE, Ahrens H, Simsiman R, Tseng C-P, Zou J, Oberley TD, and Verma AK, Transgenic mice overexpressing protein kinase Cδ in the epidermis are resistant to skin tumor promotion by 12-O-tetradecanoyl-phorbol-13-acetate, *Cancer Res* 59, 5710–5718, 1999.
46. Rykx A, De Kimpe L, Mikhalap S, Vantus T, Seufferlein T, Vandenheede JR, and Van Lint J, Protein kinase D: a family affair, *FEBS Lett* 546, 81–86, 2003.
47. Stewart JR, Christman KL, and O'Brian CA, Effects of resveratrol on the autophosphorylation of phorbol ester-responsive protein kinases: inhibition of

protein kinase D but not protein kinase C isozyme autophosphorylation, *Biochem Pharmacol* 60, 1355–1359, 2000.
48. Stewart JR and O'Brian CA, Resveratrol antagonizes EGFR-dependent Erk1/2 activation in human androgen-independent prostate cancer cells with associated isozyme-selective PKC alpha inhibition, *Invest New Drugs* 22, 107–117, 2004.
49. Rey O, Reeve JR, Zhukova E, Sinnett-Smith J, and Rozengurt E, G protein-coupled receptor-mediated phosphorylation of the activation loop of protein kinase D: dependence on plasma membrane translocation and protein kinase Cε, *J Biol Chem* 279, 34361–34372, 2004.
50. Gehm BD, McAndrews JM, Chien PY, and Jameson JL, Resveratrol, a polyphenolic compound found in grapes and wine, is an agonist for the estrogen receptor, *Proc Natl Acad Sci* 94, 14138–14143, 1997.
51. Bhat KP and Pezzuto JM, Resveratrol exhibits cytostatic and antiestrogenic properties with human endometrial adenocarcinoma (Ishikawa) cells, *Cancer Res* 61, 6137–6144, 2001.
52. Levenson AS, Gehm BD, Pearce ST, Horiguchi J, Simons LA, Ward JE, Jameson JL, and Jordan VC, Resveratrol acts as an estrogen receptor (ER) agonist in breast cancer cells stably transfected with ERα, *Int J Cancer* 104, 587–596, 2003.
53. Szewczuk LM, Forti L, Stivala LA, and Penning TM, Resveratrol is a peroxidase-mediated inactivator of COX-1 but not COX-2: a mechanistic approach to the design of COX-1 selective agents, *J Biol Chem* 279, 22727–22737, 2004.
54. Mutoh M, Takahashi M, Fukuda K, Matsushima-Hibiya Y, Mutoh H, Sugimura T, and Wakabayashi K, Suppression of cyclooxygenase-2 promoter-dependent transcriptional activity in colon cancer cells by chemopreventive agents with a resorcin-type structure, *Carcinogenesis* 21, 959–963, 2000.
55. Davies NM and Jamali F, COX-2 selective inhibitors' cardiac toxicity: getting to the heart of the matter, *J Pharm Pharm Sci.* 7, 332–336, 2004.

9 Resveratrol as an Angiogenesis Inhibitor

Ebba Bråkenhielm, Renhai Cao, and Yihai Cao

CONTENTS

INTRODUCTION

The concept that tumor growth is dependent on angiogenesis and that suppression of neovascularization may be used for cancer therapy was proposed more than three decades ago [1]. It has taken nearly 35 years to prove this concept clinically [2]. Most of today's antiangiogenic approaches for cancer therapy are focused on the development of angiogenic factor antagonists that aim to neutralize angiogenic factors, or to block their receptors or receptor signaling pathways. Because of the unstable nature of the tumor cell genome, tumor cells may switch on several angiogenic factors during tumor progression [3]. Therefore, it is vitally important to identify angiogenesis inhibitors that block common signaling pathways triggered by several angiogenesis inhibitors. Although several endogenous angiogenesis inhibitors such as angiostatin, endostatin, and tumstatin have been found to inhibit endothelial cell growth, these protein inhibitors are generally difficult to manufacture in their biologically active forms. In addition, they have to be administered by repeated injections perhaps for the rest of patients' lives. Therefore, it may become an impractical approach for both the pharmaceutical industry and patients due to the high costs of production of protein drugs.

In contrast to protein drugs, small molecules in plants may have great advantages. They can be easily manufactured at low costs, often delivered orally, and their biological activities are less labile. In fact, more than 60% of the world's population relies almost entirely on plants for medication [4]. It is believed that these plant-derived small molecules contain a large number of angiogenesis inhibitors. Among natural product-derived small molecules, polyphenols, especially flavonoids, that are abundant in various fruits, soybeans, vegetables, herbs, roots, and leaves, have consistently been reported to act as active components in prevention of cancer, heart diseases, and diabetes [5]. Recently, more than a dozen polyphenol structures have been found to act as angiogenesis inhibitors, a finding that in part addresses the underlying mechanisms of the health beneficial effects of polyphenol compounds [6]. Epigallo-catechin-3-gallate (EGCG) in green tea was the first polyphenol discovered to act as an angiogenesis inhibitor [7]. Recently, it has been shown that resveratrol administrated orally also inhibits endothelial cell growth *in vitro* and angiogenesis *in vivo* [8]. Therefore, resveratrol could serve as a leading structure in the discovery of more potent and synthetic angiogenesis inhibitors.

ANTIANGIOGENIC ACTIVITY OF RESVERATROL

Resveratrol has been described as preventing tumor initiation, as well as tumor promotion and progression, making it a putative chemopreventive and therapeutic agent for use against cancer [9]. All these malignant steps are dependent on angiogenesis. To elucidate the anticancer mechanisms of resveratrol, we have found that it inhibits angiogenesis [8,10]. We have further shown that resveratrol could act as a direct angiogenesis inhibitor because it seems to block a common pathway for vascular endothelial growth factor (VEGF)- and fibroblast growth factor 2 (FGF-2)-induced angiogenesis in the mouse cornea when administered orally [8]. Further, angiogenesis-dependent processes, such as wound healing and tumor growth, are inhibited in mice drinking resveratrol [8,10,11]. However, the underlying molecular mechanisms of the antiangiogenic activity of resveratrol remain largely unknown. Some studies have shown that resveratrol may inhibit the synthesis of VEGF in response to stimuli such as hypoxia or inflammatory agents [12], whereas other reports show inhibitory effects on angiogenic growth factor signaling pathways in endothelial cells [8,13]. It has also been reported that resveratrol decreases the binding of VEGF to endothelial cells in response to 10 to 100 μM resveratrol [11].

ANTIANGIOGENIC MECHANISMS

INHIBITION OF SIGNALING PATHWAYS OF ANGIOGENESIS

The antiendothelial effects of resveratrol have been investigated *in vitro* using various different assays and cell lines, including human umbilical vein endothelial (HUVE) cells [10,11,13,14], bovine adrenocortical capillary endothelial (BCE) cells [8], bovine pulmonary artery endothelial (BPAE) cells [15,16], bovine aorta endothelial (BAE) cells [17], and porcine aorta endothelial (PAE) cells [8]. Generally, endothelial cells of microcapillary origin, such as BCE cells, are more responsive to angiogenic stimuli (such as angiogenic factors) and they are most relevant to *in vivo* angiogenesis because new blood vessels usually sprout from capillaries or microvessels. Indeed, in microcapillary endothelial cell cultures, low concentrations (0.1 to 5 μM) of resveratrol are sufficient to suppress cell proliferation, whereas higher concentrations (10 to 30 μM) are required to inhibit the growth of large-vessel endothelial cells, such as HUVE [8,11]. In addition to studies of cell proliferation and cell cycle distribution [15], other *in vitro* endothelial cell assays investigated for resveratrol include endothelial migration, chemotaxis, tube formation in three-dimensional systems, and stimulation of apoptosis [8,11,13,14]. Because different endothelial cells and stimulators (VEGF, FGF-2, or tumor necrosis factor alpha (TNFα)) have been used in published studies, these collective data suggest that resveratrol blocks a common pathway for endothelial cell growth. One of the common signaling pathways that are involved in these cellular activities includes the activation of mitogen-activated protein kinases (MAPKs; Erk-1/2, Jnk, and p38) that act directly downstream of growth factor tyrosine kinase receptors, such as VEGF receptors. It has been shown that resveratrol, at low concentrations, inhibits growth factor-induced MAPK activation (Erk-1/2 phosphorylation) in endothelial cells [8]. Another recent finding is the inhibition by resveratrol of VEGF-induced src phosphorylation [13], a kinase involved in activation of MAPKs [18]. In contrast, higher concentrations ($\sim$100 μM) of resveratrol have been reported to stimulate, or at least stabilize, Erk-1/2 phosphorylations in endothelial cells [16]. Similarly, in several studies high concentrations (20 to 100 μM) of resveratrol were reported to stimulate the activity of MAPKs in various cancer cell lines [19,20]. However, the serum levels of resveratrol, after consumption of red wine, is reported to be only in nanomolar ranges, indicating that effects in the higher micromolar range may be irrelevant for the *in vivo* biological activity of resveratrol [21–23]. Furthermore, because activated endothelial cells have proven more responsive to low concentrations of resveratrol *in vitro* [8,13,15,24], it is probable that the vascular endothelial cells as well as some circulatory cells (platelets and possibly some immune cells) will be the most sensitive *in vivo* targets of resveratrol following consumption of natural, resveratrol-rich dietary products.

Several cardioprotective effects have been ascribed to resveratrol, including antiinflammatory, profibrinolytic, and antiadhesive effects on the endothelium. The molecular mechanisms involved in these different activities have been investigated in some detail, much more so than the mechanisms behind resveratrol's specific antiangiogenic activity, and some common pathways have emerged. In many cases, these same pathways are also implicated in the complex regulation of endothelial cell responses to angiogenic stimuli. Furthermore, the *in vivo* anticancer effects of resveratrol may involve inhibition of the production of angiogenic growth factors by nonendothelial cells, thus making the pathways affected in tumor cells and parenchymal cells equally interesting for considerations of resveratrol's molecular targets in the suppression of angiogenesis.

Effects on Fibrinolysis and Extracellular Proteases as a Potential Antiangiogenic Mechanism

The processes of both fibrinolysis and extracellular proteolytic degradation are critical for angiogenesis [25]. Resveratrol has been shown to stimulate the fibrinolytic activity of endothelial cells (in response to agents such as lipopolysaccharides (LPS), interleukin-β, or TNFα) by suppression of the expression of tissue factor (TF). This effect was found to be dependent on inhibition of the degradation of IκBα, by inhibition of the Iκ kinases (IKKs) [24,26,27]. The stabilization of IκBα results in the direct inhibition of nuclear factor kappa B (NF-κB) p65/Rel-unit, which is prevented from translocation to the nucleus and thus from activating its target genes, including TF [28]. TF, a transmembrane cell surface receptor for proteins of the coagulation cascade, has been shown to play a role in tumor angiogenesis [29]. One of the mechanisms of the stimulation of angiogenesis is the upregulation of VEGF expression by TF [30]. It is thus possible that resveratrol, by preventing TNFα-induced TF expression, could decrease local VEGF production, leading to suppression of angiogenesis.

Fibrinolysis involves activation of tissue-type plasminogen activator (t-PA) and urokinase-type PA (u-PA). Resveratrol increases the expression levels and surface assembly of both enzymes in endothelial cells [31]. These enzymes are not only essential for the prevention and resolution of blood coagulation, but they are also implicated in the angiogenic process. Indeed, angiogenic growth factors such as VEGF and FGF-2 stimulate the expression of both t-PA and u-PA in endothelial cells *in vitro* [32]. It is known that the plasminogen and plasmin system, such as u-PA, is activated during the process of physiological and pathological angiogenesis. These and other extracellular proteases, including matrix metalloproteinases (MMPs), contribute to angiogenesis by releasing matrix-bound angiogenic growth factors, such as FGF-2 and VEGF [33]. However, if these enzymes remain in hyperactive state, they may lead to abnormal vessel growth by destruction of the provisional matrix and basement membrane necessary for the guidance and survival of the nascent endothelium. Interestingly, the same extracellular proteases could play an important role in activation of endogenous inhibitors of angiogenesis, such as endostatin and angiostatin, both proteolytic fragments of larger and circulating precursor proteins [34,35]. Thus, it is possible that resveratrol, in addition to its direct antiangiogenic effects, by stimulating the endothelial-associated protease activity may increase locally the production of other, endogenous angiogenesis inhibitors. This potential indirect inhibitory effect of resveratrol remains to be investigated.

In contrast to stimulation of proteases of the plasminogen–plasmin system, resveratrol inhibits the expression of MMP-2 and MMP-9 [36,37], both enzymes involved in the degradation of the vascular basement membrane collagen IV. Expression of these two MMPs is associated with aggressive tumor growth and metastasis in human cancers [38]. The inhibitory effect of resveratrol on MMP-9 expression in tumor cells, induced by phorbol myristate acetate (PMA, a synthetic cellular activator), was reported to involve suppression of protein kinase C delta (PKCδ) as well as reduced Jnk activation [39]. This effect was associated with reduced activity of transcription factors AP-1 and NF-κB, which regulate MMP-9 expression. Thus, it is possible that resveratrol, by reducing the levels of MMP production, could prevent the release of angiogenic growth factors from the extracellular matrix [33], leading to indirect suppression of angiogenesis. However, activities of MMPs depend also on posttranslational modifications in the form of proteolytic cleavage. Resveratrol suppresses MMP-2 activity, but only at high concentrations [40], whereas green tea polyphenols are more potent in inhibiting specific MMPs. It should be noted that clinical trials with nonselective MMP inhibitors, e.g., marimastat, as single agents for anticancer therapy have yielded inconclusive results [41], pointing to the delicate balance that exists between stimulatory versus inhibitory effects of decreased matrix proteolysis on the angiogenic cascade and tumor invasion process.

Stimulation of eNOS and Its Role in Regulation of Angiogenesis

Nitric oxide (NO) plays an important role in the regulation of vasotonus but also in endothelial responses to growth factors, such as VEGF [42]. Endothelial nitric oxide synthase (eNOS) catalyzes the formation of NO from L-arginine. Although eNOS is constitutively expressed at low levels, the mRNA expression and stability is increased in response to stimuli such as growth factors, e.g., VEGF [43,44], TGFβ, and estrogen [45], whereas other cytokines, e.g., TNFα, decrease eNOS levels. In a recent report, $VEGF^{121}$ was shown to be a more potent inducer of eNOS than the $VEGF^{165}$ isoform [46]. The molecular mechanisms may involve Hsp90 and Akt kinase [47]. NO is involved in endothelial activation leading to cell migration, but the role of NO in proliferative responses to VEGF is less clear [48]. Conversely, NO also stimulates HIF1α activity and VEGF production in cells during conditions of normoxia [49]. *In vivo* experiments show that eNOS stimulates angiogenic responses to tissue ischemia in part via upregulation of VEGF [50,51]. Resveratrol stimulates eNOS synthesis. Resveratrol increases eNOS transcription as well as mRNA stability, resulting in increased eNOS protein levels and enhanced NO production in endothelial cells [52]. This effect was unrelated to resveratrol's weak estrogenic activity (phytoestrogen), nor did it involve any of the well-known eNOS-promoter-binding transcription factors, such as Sp1, GATA, or Elf-1 [52]. This eNOS stimulatory effect is well suited to explain the general cardioprotective effects of resveratrol in inducing vasorelaxation. However, given the stimulatory role of NO in the angiogenic process, it seems contradictory that resveratrol, having potent antiangiogenic activity, should stimulate NO production. There are several interpretations to be made of these opposite effects. For instance it has been suggested that whereas low to moderate intracellular levels of NO stimulate HIF1α activity and VEGF production, higher NO levels seem to result in opposing effects [47]. If stimulation by resveratrol of eNOS synthesis results in substantially higher levels of NO in endothelial cells, this could result in inhibition of HIF1α activity, leading to reduced VEGF-induced angiogenesis. Further, since NO chemistry is highly influenced by the redox status of the cell, VEGF and resveratrol may have opposing effects on NO activity by shifting levels of cellular oxidants. Indeed, resveratrol is a known antioxidant, whereas VEGF has been reported to increase reactive oxygen species (ROS) production in endothelial cells [53,54].

Inhibition of Cellular Adhesion Molecules

The antiadhesive endothelial effect of resveratrol has been investigated *in vitro* in response to potent stimulators. Similar to VEGF, TNFα or LPS induce the expression of cell adhesion molecules on endothelial cells, including E-selectin, intercellular adhesion molecule-1 (ICAM-1), and vascular adhesion molecule-1 (VCAM-1) [55,56]. At physiologically relevant concentrations, resveratrol prevents the TNFα-induced expression of VCAM-1 as well as ICAM-1 [57,58]. These cell adhesion molecules are important for endothelial cell migration during angiogenesis [59]. Thus, one of the mechanisms behind the antimigration effect of resveratrol on endothelial cells might be suppression of these molecules [8,13,17,60].

Vascular endothelial cadherin (VE-cadherin) is another cellular adhesion molecule involved in cell-to-cell interactions, and it plays an important role in endothelial cell migration and tubular formation [61]. Stimulation of endothelial cells by various growth factors, such as VEGF, results in activation of VE-cadherin, leading to loosened cell–cell adhesion and increased cell migratory activity [62]. Inhibition of VE-cadherin by blocking antibodies has been reported to prevent tumor angiogenesis [63]. Recently, resveratrol at low micromolar concentrations was found to inhibit VEGF-induced VE-cadherin activation [13]. The molecular mechanism includes reduced src-mediated VE-cadherin tyrosine phosphorylation. Although resveratrol had a slight direct inhibitory effect on the src kinase activity, the fact that src previously was suggested to be activated by ROS argues for the possible involvement of the antioxidant properties of resveratrol as a mediator of the suppressive effect on VEGF-induced src signaling [13]. Indeed, VEGF is known to stimulate the production of ROS in endothelial cells, which further is linked to VE-cadherin activation *via* src-mediated phosphorylation [13,54]. Cell-to-cell contacts mediated by VE-cadherin are important for maintenance of vascular integrity, and activation of VE-cadherin, leading to internalization of the adhesion molecule, results in increased vascular permeability [63,64]. In an early phase of the angiogenic cascade, vascular permeability is usually increased, probably due to VEGF signaling. This vascular leakage aids in the formation of a provisional supportive matrix, which enables endothelial cell migration into avascular tissue. Recently, high levels of intracellular NO were found to reduce the cell surface levels of VE-cadherin, leading to stimulation of vascular permeability [65]. An additional pathway for VEGF-induced permeability may be mediated by stimulation of NO production, via upregulation of eNOS [43], leading to VE-cadherin activation and internalization. VEGF-induced NO-mediated permeability is found to depend on eNOS but not on iNOS [66]. Although resveratrol inhibits VEGF-induced VE-cadherin activation, which should lead to suppression of vascular permeability, it also, similar to VEGF, stimulates eNOS activity [52], indicating that resveratrol potentially could stimulate NO-induced VE-cadherin internalization resulting in increased permeability. However, resveratrol does not increase vascular permeability *in vivo*, but rather it prevents TNFα-stimulated vascular permeability in mice [58]. This finding supports the conclusion that resveratrol inhibits angiogenesis *in vivo*.

Inhibition of COX Enzymes and Inflammatory Pathways in Angiogenesis

Nonsteroidal antiinflammatory drugs (NSAIDs) have been demonstrated to suppress the occurrence and progression of several types of cancers [67]. NSAIDs block cellular prostanoid (e.g., prostaglandins and thromboxanes) synthesis during inflammation by inhibiting two cyclooxygenases (COXs): the constitutive COX-1 and the inducible COX-2. The latter enzyme is upregulated in most human tumors [68], where its expression is increased by inflammatory mediators but also by hypoxia, VEGF, and FGF-2 stimulation [69,70]. The involvement of the COX pathway in angiogenesis, as a result of inflammatory stimuli, has been well described for COX-2 [71]. Selective COX-2 inhibitors as well as nonselective COX NSAIDs, but not selective COX-1 inhibitors, in animal tumor models, downregulate the expression proangiogenic growth factors, such as VEGF, resulting in a suppression of tumor angiogenesis [72,73]. Indeed, the selective COX-2 inhibitor Celecoxib is currently in clinical trials for the treatment of cancer [67]. Resveratrol was first found to only inhibit COX-1 activity, being effective at 15 μM [9]. Although less is known about the role of COX-1 in tumor growth and angiogenesis, some studies indicate that certain selective COX-1 inhibitors also may inhibit angiogenesis [72]. Recently, resveratrol was found to inhibit the expression and activity of COX-2 [74]. Thus, resveratrol may prevent the angiogenic switch of inflammatory agents, such as TNFα, by inhibiting prostanoid-induced cellular activation via COX suppression. Interestingly, it has been suggested that COX-2 inhibitors, in addition to the direct suppressive effects on prostaglandin-mediated proangiogenic factor production, also may interfere *directly* with the $\alpha_v\beta_3$-integrin signaling in endothelial cells [75]. Thus, the molecular mechanisms for the protective effect of COX inhibitors in cancer and other pathologies could potentially include multiple pathways in suppression of angiogenesis.

Regulation of Transcription Factors and Nuclear Enzymes

Resveratrol influences the activity of several transcription factors, such as AP-1, NF-κB, HIF1α, and p53 [12,19]. Although these proteins are involved in many general cellular activation pathways, some of their target genes involve specific stimulators of angiogenesis. For instance, NF-κB upregulates VEGF, FGF-2, TNFα, COX-2, and other genes implicated in angiogenesis [69,70]. The molecular mechanisms of the effect of resveratrol on expression or activity of these transcription factors remain mostly elusive and indirect, possibly involving MAPK pathway disruption.

A novel interesting finding is the reported interaction of resveratrol with nuclear enzymes called sirtuins (Sir). These are histone deacetylases (HDACs) of the class III family, which can interact with histones as well as with some transcription factors. The yeast Sir-2 has been shown to prolong lifespan in yeast in response to caloric restriction [77]. Resveratrol, at 10 μM concentration, was found to stimulate the activity of Sir-2 in yeast cells, thereby significantly increasing the lifespan of this organism [78]. Resveratrol was thus able to mimic some of the cellular responses to calorie reduction. One molecular explanation for the observed effect might be that resveratrol altered the substrate-binding affinity of the sirtuin enzymes, thus promoting deacetylation of cellular targets. In mammalian cells, low concentrations of resveratrol stimulated SIRT-1 (the mammalian homologue of Sir-2), resulting in increased deacetylation of an important target protein, p53 [79]. This led to decreased stability of the tumor suppressor protein, and thus protection of the cells from p53-mediated apoptosis induced by various DNA damaging stimuli such as ionizing radiation. The effect was, however, reversed at higher resveratrol concentrations (50 μM), which is in agreement with previous data on the apoptosis-inducing effect of resveratrol in several different types of tumor cells. It remains to be investigated if resveratrol in endothelial cells also displays this indirect, dual p53-regulatory activity.

Recently, the effect of resveratrol on SIRT-1 activity was further investigated in carcinoma cells [80]. Resveratrol was found to prevent TNFα-induced NF-κB transactivation via stimulation of SIRT-1 activity leading to deacetylation, and thus inactivation, of the p65 unit of NF-κB. These data may help to explain the finding that in some cell types resveratrol treatment did not prevent TNFα-induced IκBα degradation and NF-κB nuclear translocation [28], although the transactivating potential of the transcription factor was still diminished.

The potential role of HDACs of type III, such as sirtuins, in endothelial cells and in regulation of angiogenesis is currently unknown. If the antiangiogenic effect of resveratrol *in vivo* involves stimulation of sirtuin activity it would be in line, at the organismal level, with the upregulation of sirtuins observed during times of starvation. However, some recent data suggest that nutrient deficiency, at the cellular level, might result in stimulation of angiogenesis [81], thus potentially involving sirtuins in positive regulation of angiogenesis. Interestingly, HDACs of type I and II have been implicated in stimulation of angiogenesis by inhibition of tumor suppressor genes such as p53 and vHL [82]. Furthermore, inhibitors of these HDACs were found to reduce angiogenesis via inhibition of HIF1α activity as well as by interference with VEGF signaling [83]. Inhibitors of HDACs are shown to reduce eNOS mRNA stability, a mechanism potentially contributing to the observed antiangiogenic effect of these agents [84].

Inhibition of Cell Cycle and Induction of Apoptosis

Resveratrol has antiproliferative effects in a number of different cell types, including endothelial cells [8,10,13–16], lymphocytes, fibroblasts, and many tumor cell lines [85–87]. The sensitivity of different cells to resveratrol is, however, very variable, with 2 to 5 μM being sufficient for the most responsive cells (BCE cells), whereas the most resistant cells require around 50 to 100 μM for inhibition of proliferation. The stimulating effects of resveratrol on the expression and activity of proapoptotic pathways, such as p53 or mitochondrial caspases, have been investigated in many different cell lines [11,14,19,86]. It is possible that resveratrol in addition to its direct effects on proapoptotic mediators also, by interfering with the signaling and/or synthesis of survival angiogenic stimulators, e.g., VEGF, decreases the cell survival signals (i.e., Akt and PI3K pathway), thus favoring cellular apoptosis. Furthermore, it was recently reported that resveratrol modulated cellular survival responses to TNFα [80]. TNFα signaling is complex in that it can stimulate either cell survival, by NF-κB-mediated induction of antiapoptotic genes such a Bcl-x_L and cIAP-2, or induction of apoptosis via Fas-associated pathway leading to caspase activation. Resveratrol, by inhibiting the NF-κB signaling, was found to decrease the TNFα-induced expression of antiapoptotic genes, thus potentially shifting the cellular response to TNFα to proapoptosis [80]. The molecular targets seem to involve SIRT-1, as enzyme levels were found to be specifically increased (together with NF-κB p65 unit) at the cIAP-2 promoter, indicating that SIRT-1-induced deacetylation of NF-κB may be one mechanism of resveratrol-stimulated antiapoptotic gene suppression [80]. In agreement, it was shown that resveratrol-stimulated sensitization of cells to TNFα-induced apoptosis was blocked in the absence of SIRT-1 enzyme.

Recently, resveratrol was also described to sensitize cells to another death receptor pathway leading to TNF-related apoptosis-inducing ligand (TRAIL)-induced apoptosis [88–90]. However, the concentration of resveratrol required for induction of apoptosis, e.g., in tumor cells, is generally very high (~100 μM), indicating that apoptosis of tumor cells in most cases would be unlikely to contribute to the *in vivo* suppressive effects of resveratrol on tumor growth. In contrast, the progression through the cell cycle has been found to be affected by slightly lower concentrations in many cells [85]. The changes induced by resveratrol treatment include decreased levels of several cyclins (cyclin D1, cyclin D2, and cyclin E), as well as cyclin-dependent kinases (cdks: cdk2, cdk4, and cdk6) [91]. Resveratrol has been found to induce cell cycle arrest of tumor cells during either the S/G_2 checkpoint or according to some reports during the G_1/S transition phase [11,85,91]. In endothelial cells, resveratrol was reported to increase p53 levels, which together with stimulation of cdk inhibitor $p21^{WAF1/CIP1}$, resulted in S/G_2 cell cycle arrest and apoptosis [14,15,19].

SUMMARY

Resveratrol seems to modify several cellular pathways that all may contribute to its antiangiogenic effect. Because angiogenesis is a multistep process that requires endothelial cells to proliferate, migrate, and form tubules, the observed inhibitory capacity of resveratrol on all these different activities should contribute to its potent antiangiogenic effect *in vivo*. Some of these putative molecular targets in endothelial cells are shared by resveratrol with other endogenous or synthetic angiogenesis inhibitors. Resveratrol is one of the simplest chemical structures in the polyphenol family that display antiangiogenic activity [6]. This finding may give us some clues to search for new and more potent angiogenesis inhibitors. Indeed, in addition to resveratrol, more than a couple of dozen other polyphenol compounds are found to inhibit angiogenesis and produce beneficial effects in the prevention and treatment of angiogenesis-dependent disorders such as cancer. These simple and orally available angiogenesis inhibitors have great advantages over endogenous angiogenesis inhibitors and have opened a new possibility for antiangiogenic drug development.

REFERENCES

1. Folkman J, Tumor angiogenesis: therapeutic implications, *N Engl J Med* 285, 1182–1186, 1971.
2. Yang JC, Haworth L, Sherry RM, Hwu P, Schwartzentruber DJ, Topalian SL, Steinberg SM, Chen HX, and Rosenberg SA, A randomized trial of bevacizumab, an anti-vascular endothelial growth factor antibody, for metastatic renal cancer, *N Engl J Med* 349, 427–434, 2003.
3. Folkman J, Role of angiogenesis in tumor growth and metastasis, *Semin Oncol* 29, 15–18, 2002.
4. Harvey A, Strategies for discovering drugs from previously unexplored natural products, *Drug Discov Today* 5, 294–300, 2000.
5. Dufresne CJ and Farnworth ER, A review of latest research findings on the health promotion properties of tea, *J Nutr Biochem* 12, 404–421, 2001.
6. Cao Y, Cao R, and Brakenhielm E, Antiangiogenic mechanisms of diet-derived polyphenols, *J Nutr Biochem* 13, 380–390, 2002.
7. Cao Y and Cao R, Angiogenesis inhibited by drinking tea, *Nature* 398, 381, 1999.
8. Brakenhielm E, Cao R, and Cao Y, Suppression of angiogenesis, tumor growth, and wound healing by resveratrol, a natural compound in red wine and grapes, *FASEB J* 15, 1798–1800, 2001.
9. Jang M, Cai L, Udeani GO, Slowing KV, Thomas CF, Beecher CW, Fong HH, Farnsworth NR, Kinghorn AD, Mehta RG, Moon RC, and Pezutto JM, Cancer chemopreventive activity of resveratrol, a natural product derived from grapes, *Science* 275, 218–220, 1997.

10. Tseng SH, Lin SM, Chen JC, Su YH, Huang HY, Chen CK, Lin PY, and Chen Y, Resveratrol suppresses the angiogenesis and tumor growth of gliomas in rats, *Clin Cancer Res* 10, 2190–2202, 2004.
11. Kimura Y and Okuda H, Resveratrol isolated from Polygonum cuspidatum root prevents tumor growth and metastasis to lung and tumor-induced neovascularization in Lewis lung carcinoma-bearing mice, *J Nutr* 131, 1844–1849, 2001.
12. Cao Z, Fang J, Xia C, Shi X, and Jiang BH, Trans-3,4,5′-trihydroxystilbene inhibits hypoxia-inducible factor 1alpha and vascular endothelial growth factor expression in human ovarian cancer cells, *Clin Cancer Res* 10, 5253–5263, 2004.
13. Lin MT, Yen ML, Lin CY, and Kuo ML, Inhibition of vascular endothelial growth factor-induced angiogenesis by resveratrol through interruption of Src-dependent vascular endothelial cadherin tyrosine phosphorylation, *Mol Pharmacol* 64, 1029–1036, 2003.
14. Szende B, Tyihak E, and Kiraly-Veghely Z, Dose-dependent effect of resveratrol on proliferation and apoptosis in endothelial and tumor cell cultures, *Exp Mol Med* 32, 88–92, 2000.
15. Hsieh TC, Juan G, Darzynkiewicz Z, and Wu JM, Resveratrol increases nitric oxide synthase, induces accumulation of p53 and p21(WAF1/CIP1), and suppresses cultured bovine pulmonary artery endothelial cell proliferation by perturbing progression through S and G2, *Cancer Res* 59, 2596–2601, 1999.
16. Bruder JL, Hsieh T, Lerea KM, Olson SC, and Wu JM, Induced cytoskeletal changes in bovine pulmonary artery endothelial cells by resveratrol and the accompanying modified responses to arterial shear stress, *BMC Cell Biol* 2, 1, 2001.
17. Igura K, Ohta T, Kuroda Y, and Kaji K, Resveratrol and quercetin inhibit angiogenesis *in vitro*, *Cancer Lett* 171, 11–16, 2001.
18. Hood JD, Frausto R, Kiosses WB, Schwartz MA, and Cheresh DA, Differential alphav integrin-mediated Ras-ERK signaling during two pathways of angiogenesis, *J Cell Biol* 162, 933–943, 2003.
19. She QB, Huang C, Zhang Y, and Dong Z, Involvement of c-jun NH(2)-terminal kinases in resveratrol-induced activation of p53 and apoptosis, *Mol Carcinog* 33, 244–250, 2002.
20. Miloso M, Bertelli AA, Nicolini G, and Tredici G, Resveratrol-induced activation of the mitogen-activated protein kinases, ERK1 and ERK2, in human neuroblastoma SH-SY5Y cells, *Neurosci Lett* 264, 141–144, 1999.
21. Turner RT, Evans GL, Zhang M, Maran A, and Sibonga JD, Is resveratrol an estrogen agonist in growing rats?, *Endocrinology* 140, 50–54, 1999.
22. Frankel EN, Waterhouse AL, and Kinsella JE, Inhibition of human LDL oxidation by resveratrol, *Lancet* 341, 1103–1104, 1993.
23. Bertelli A, Bertelli AA, Gozzini A, and Giovannini L, Plasma and tissue resveratrol concentrations and pharmacological activity, *Drugs Exp Clin Res* 24, 133–138, 1998.
24. Di Santo A, Mezzetti A, Napoleone E, Di Tommaso R, Donati MB, De Gaetano G, and Lorenzet R, Resveratrol and quercetin down-regulate tissue factor expression by human stimulated vascular cells, *J Thromb Haemost* 1, 1089–1095, 2003.

25. Hiraoka N, Allen E, Apel IJ, Gyetko MR, and Weiss SJ, Matrix metalloproteinases regulate neovascularization by acting as pericellular fibrinolysins, *Cell* 95, 365–377, 1998.
26. Holmes-McNary M and Baldwin AS Jr, Chemopreventive properties of trans-resveratrol are associated with inhibition of activation of the IkappaB kinase, *Cancer Res* 60, 3477–3483, 2000.
27. Pendurthi UR, Williams JT, and Rao LV, Resveratrol, a polyphenolic compound found in wine, inhibits tissue factor expression in vascular cells: a possible mechanism for the cardiovascular benefits associated with moderate consumption of wine, *Arterioscler Thromb Vasc Biol* 19, 419–426, 1999.
28. Pellegatta F, Bertelli AA, Staels B, Duhem C, Fulgenzi A, and Ferrero ME, Different short- and long-term effects of resveratrol on nuclear factor-kappaB phosphorylation and nuclear appearance in human endothelial cells, *Am J Clin Nutr* 77, 1220–1228, 2003.
29. Belting M, Dorrell MI, Sandgren S, Aguilar E, Ahamed J, Dorfleutner A, Carmeliet P, Mueller BM, Friedlander M, and Ruf W, Regulation of angiogenesis by tissue factor cytoplasmic domain signaling, *Nat Med* 10, 502–509, 2004.
30. Ollivier V, Bentolila S, Chabbat J, Hakim J, and de Prost D, Tissue factor-dependent vascular endothelial growth factor production by human fibroblasts in response to activated factor VII, *Blood* 91, 2698–2703, 1998.
31. Abou-Agag LH, Aikens ML, Tabengwa EM, Benza RL, Shows SR, Grenett HE, and Booyse FM, Polyphyenolics increase t-PA and u-PA gene transcription in cultured human endothelial cells, *Alcohol Clin Exp Res* 25, 155–162, 2001.
32. Pepper MS, Role of the matrix metalloproteinase and plasminogen activator-plasmin systems in angiogenesis. *Arterioscler Thromb Vasc Biol* 21, 1104–1117, 2001.
33. Bergers G, Brekken R, McMahon G, Vu TH, Itoh T, Tamaki K, Tanzawa K, Thorpe P, Itohara S, Werb Z et al, Matrix metalloproteinase-9 triggers the angiogenic switch during carcinogenesis, *Nat Cell Biol* 2, 737–744, 2000.
34. Pepper MS, Extracellular proteolysis and angiogenesis, *Thromb Haemost* 86, 346–355, 2001.
35. Cao Y, Endogenous angiogenesis inhibitors: angiostatin, endostatin, and other proteolytic fragments, *Prog Mol Subcell Biol* 20, 161–176, 1998.
36. Banerjee S, Bueso-Ramos C, and Aggarwal BB, Suppression of 7,12-dimethylbenz(a)anthracene-induced mammary carcinogenesis in rats by resveratrol: role of nuclear factor-kappaB, cyclooxygenase 2, and matrix metalloprotease 9, *Cancer Res* 62, 4945–4954, 2002.
37. Godichaud S, Krisa S, Couronne B, Dubuisson L, Merillon JM, Desmouliere A, and Rosenbaum J, Deactivation of cultured human liver myofibroblasts by trans-resveratrol, a grapevine-derived polyphenol, *Hepatology* 31, 922–931, 2000.
38. Sato H, Takino T, Okada Y, Cao J, Shinagawa A, Yamamoto E, and Seiki M, A matrix metalloproteinase expressed on the surface of invasive tumour cells, *Nature* 370, 61–65, 1994.
39. Woo JH, Lim JH, Kim YH, Suh SI, Min do S, Chang JS, Lee YH, Park JW, and Kwon TK, Resveratrol inhibits phorbol myristate acetate-induced matrix

metalloproteinase-9 expression by inhibiting JNK and PKC delta signal transduction, *Oncogene* 23, 1845–1853, 2004.
40. Demeule M, Brossard M, Page M, Gingras D, and Beliveau R, Matrix metalloproteinase inhibition by green tea catechins, *Biochim Biophys Acta* 1478, 51–60, 2000.
41. Pavlaki M and Zucker S, Matrix metalloproteinase inhibitors (MMPIs): the beginning of phase I or the termination of phase III clinical trials, *Cancer Metastasis Rev* 22, 177–203, 2003.
42. Babaei S and Stewart DJ, Overexpression of endothelial NO synthase induces angiogenesis in a co-culture model, *Cardiovasc Res* 55, 190–200, 2002.
43. Bouloumie A, Schini-Kerth VB, and Busse R, Vascular endothelial growth factor up-regulates nitric oxide synthase expression in endothelial cells, *Cardiovasc Res* 41, 773–780, 1999.
44. He H, Venema VJ, Gu X, Venema RC, Marrero MB, and Caldwell RB, Vascular endothelial growth factor signals endothelial cell production of nitric oxide and prostacyclin through flk-1/KDR activation of c-Src, *J Biol Chem* 274, 25130–25135, 1999.
45. Kleinert H, Wallerath T, Euchenhofer C, Ihrig-Biedert I, Li H, and Forstermann U, Estrogens increase transcription of the human endothelial NO synthase gene: analysis of the transcription factors involved, *Hypertension* 31, 582–588, 1998.
46. Jozkowicz A, Dulak J, Nigisch A, Funovics P, Weigel G, Polterauer P, Huk I, and Malinski T, Involvement of nitric oxide in angiogenic activities of vascular endothelial growth factor isoforms, *Growth Factors* 22, 19–28, 2004.
47. Kimura H and Esumi H, Reciprocal regulation between nitric oxide and vascular endothelial growth factor in angiogenesis, *Acta Biochim Pol* 50, 49–59, 2003.
48. Papapetropoulos A, Garcia-Cardena G, Madri JA, and Sessa WC, Nitric oxide production contributes to the angiogenic properties of vascular endothelial growth factor in human endothelial cells, *J Clin Invest* 100, 3131–3139, 1997.
49. Kimura H, Weisz A, Kurashima Y, Hashimoto K, Ogura T, D'Acquisto F, Addeo R, Makuuchi M, and Esumi H, Hypoxia response element of the human vascular endothelial growth factor gene mediates transcriptional regulation by nitric oxide: control of hypoxia-inducible factor-1 activity by nitric oxide, *Blood* 95, 189–197, 2000.
50. Murohara T, Asahara T, Silver M, Baufers C, Masuda H, Kalka C, Kearney M, Chen D, Symes JF, Fishman MC et al, Nitric oxide synthase modulates angiogenesis in response to tissue ischemia, *J Clin Invest* 101, 2567–2578, 1998.
51. Namba T, Koike H, Murakami K, Aoki M, Makino H, Hashiya N, Ogihara T, Kaneda Y, Kohno M, and Morishita R, Angiogenesis induced by endothelial nitric oxide synthase gene through vascular endothelial growth factor expression in a rat hindlimb ischemia model, *Circulation* 108, 2250–2257, 2003.
52. Wallerath T, Deckert G, Ternes T, Anderson H, Li H, Witte K, and Forstermann U, Resveratrol, a polyphenolic phytoalexin present in red wine, enhances expression and activity of endothelial nitric oxide synthase, *Circulation* 106, 1652–1658, 2002.
53. Maulik N and Das DK, Redox signaling in vascular angiogenesis, *Free Radical Biol Med* 33, 1047–1060, 2002.

54. Colavitti R, Pani G, Bedogni B, Anzevino R, Borrello S, Waltenberger J, and Galeotti T, Reactive oxygen species as downstream mediators of angiogenic signaling by vascular endothelial growth factor receptor-2/KDR, *J Biol Chem* 277, 3101–3108, 2002.
55. Jiang MZ, Tsukahara H, Ohshima Y, Todoroki Y, Hiraoka M, Maeda M, and Mayumi M, Effects of antioxidants and nitric oxide on TNF-alpha-induced adhesion molecule expression and NF-kappaB activation in human dermal microvascular endothelial cells, *Life Sci* 75, 1159–1170, 2004.
56. Kim I, Moon SO, Kim SH, Kim HJ, Koh YS, and Koh GY, Vascular endothelial growth factor expression of intercellular adhesion molecule 1 (ICAM-1), vascular cell adhesion molecule 1 (VCAM-1), and E-selectin through nuclear factor-kappa B activation in endothelial cells, *J Biol Chem* 276, 7614–7620, 2001.
57. Ferrero ME, Bertelli AE, Fulgenzi A, Pellegatta F, Corsi MM, Bonfrate M, Ferrara F, De Caterina R, Giovannini L, and Bertelli A, Activity *in vitro* of resveratrol on granulocyte and monocyte adhesion to endothelium, *Am J Clin Nutr* 68, 1208–1214, 1998.
58. Bertelli AA, Baccalini R, Battaglia E, Falchi M, and Ferrero ME, Resveratrol inhibits TNF alpha-induced endothelial cell activation, *Therapie* 56, 613–616, 2001.
59. Radisavljevic Z, Avraham H, and Avraham S, Vascular endothelial growth factor up-regulates ICAM-1 expression via the phosphatidylinositol 3 OH-kinase/AKT/Nitric oxide pathway and modulates migration of brain microvascular endothelial cells, *J Biol Chem* 275, 20770–20774, 2000.
60. Brakenhielm E, Veitonmaki N, Cao R, Kihara S, Matsuzawa Y, Zhivotovsky B, Funahashi T, and Cao Y, Adiponectin-induced antiangiogenesis and antitumor activity involve caspase-mediated endothelial cell apoptosis, *Proc Natl Acad Sci USA* 101, 2476–2481, 2004.
61. Nawroth R, Poell G, Ranft A, Kloep S, Samulowitz U, Fachinger G, Golding M, Shima DT, Deutsch U, and Vestweber D, VE-PTP and VE-cadherin ectodomains interact to facilitate regulation of phosphorylation and cell contacts, *EMBO J* 21, 4885–4895, 2002.
62. Chou MT, Wang J, and Fujita DJ, Src kinase becomes preferentially associated with the VEGFR, KDR/Flk-1, following VEGF stimulation of vascular endothelial cells, *BMC Biochem* 3, 32, 2002.
63. Corada M, Liao F, Lindgren M, Lampugnani MG, Breviario F, Frank R, Muller WA, Hicklin DJ, Bohlen P, and Dejana E, Monoclonal antibodies directed to different regions of vascular endothelial cadherin extracellular domain affect adhesion and clustering of the protein and modulate endothelial permeability, *Blood* 97, 1679–1684, 2001.
64. Weis S, Shintani S, Weber A, Kirchmair R, Wood M, Cravens A, McSharry H, Iwakura A, Yoon YS, Himes N et al, Src blockade stabilizes a Flk/cadherin complex, reducing edema and tissue injury following myocardial infarction, *J Clin Invest* 113, 885–894, 2004.
65. Gonzalez D, Herrera B, Beltran A, Otero K, Quintero G, and Rojas A, Nitric oxide disrupts VE-cadherin complex in murine microvascular endothelial cells, *Biochem Biophys Res Commun* 304, 113–118, 2003.
66. Fukumura D, Gohongi T, Kadambi A, Izumi Y, Ang J, Yun CO, Buerk DG, Huang PL, and Jain RK, Predominant role of endothelial nitric oxide synthase

in vascular endothelial growth factor-induced angiogenesis and vascular permeability, *Proc Natl Acad Sci USA* 98, 2604–2609, 2001.
67. Thun MJ, Henley SJ, and Patrono C, Nonsteroidal antiinflammatory drugs as anticancer agents: mechanistic, pharmacologic, and clinical issues, *J Natl Cancer Inst* 94, 252–266, 2002.
68. Soslow RA, Dannenberg AJ, Rush D, Woerner BM, Khan KN, Masferrer J, and Koki AT, COX-2 is expressed in human pulmonary, colonic, and mammary tumors, *Cancer* 89, 2637–2645, 2000.
69. Schmedtje JF Jr, Ji YS, Liu WL, DuBois RN, and Runge MS, Hypoxia induces cyclooxygenase-2 via the NF-kappaB p65 transcription factor in human vascular endothelial cells, *J Biol Chem* 272, 601–608, 1997.
70. Hernandez GL, Volpert OV, Iniguez MA, Lorenzo E, Martinez-Martinez S, Grau R, Fresno M, and Redondo JM, Selective inhibition of vascular endothelial growth factor-mediated angiogenesis by cyclosporin A: roles of the nuclear factor of activated T cells and cyclooxygenase 2, *J Exp Med* 193, 607–620, 2001.
71. Iniguez MA, Rodriguez A, Volpert OV, Fresno M, and Redondo JM, Cyclooxygenase-2: a therapeutic target in angiogenesis, *Trends Mol Med* 9, 73–78, 2003.
72. Tsujii M, Kawano S, Tsuji S, Sawaoka H, Hori M, and DuBois RN, Cyclooxygenase regulates angiogenesis induced by colon cancer cells, *Cell* 93, 705–716, 1998.
73. Yoshida S, Amano H, Hayashi I, Kitasato H, Kamata M, Inukai M, Yoshimura H, and Majima M, COX-2/VEGF-dependent facilitation of tumor-associated angiogenesis and tumor growth *in vivo*, *Lab Invest* 83, 1385–1394, 2003.
74. Subbaramaiah K, Chung WJ, Michaluart P, Telang N, Tanabe T, Inoue H, Jang M, Pezzuto JM, and Dannenberg AJ, Resveratrol inhibits cyclooxygenase-2 transcription and activity in phorbol ester-treated human mammary epithelial cells, *J Biol Chem* 273, 21875–21882, 1998.
75. Dormond O, Foletti A, Paroz C, and Ruegg C, NSAIDs inhibit alpha V beta 3 integrin-mediated and Cdc42/Rac-dependent endothelial-cell spreading, migration and angiogenesis, *Nat Med* 7, 1041–1047, 2001.
76. Shibata A, Nagaya T, Imai T, Funahashi H, Nakao A, and Seo H, Inhibition of NF-kappaB activity decreases the VEGF mRNA expression in MDA-MB-231 breast cancer cells, *Breast Cancer Res Treat* 73, 237–243, 2002.
77. Tissenbaum HA and Guarente L, Increased dosage of a sir-2 gene extends lifespan in Caenorhabditis elegans, *Nature* 410, 227–230, 2001.
78. Howitz KT, Bitterman KJ, Cohen HY, Lamming DW, Lavu S, Wood JG, Zipkin RE, Chung P, Kisielewski A, Zhang LL et al, Small molecule activators of sirtuins extend Saccharomyces cerevisiae lifespan, *Nature* 425, 191–196, 2003.
79. Brooks CL and Gu W, Ubiquitination, phosphorylation and acetylation: the molecular basis for p53 regulation, *Curr Opin Cell Biol* 15, 164–171, 2003.
80. Yeung F, Hoberg JE, Ramsey CS, Keller MD, Jones DR, Frye RA, and Mayo MW, Modulation of NF-kappaB-dependent transcription and cell survival by the SIRT1 deacetylase, *EMBO J* 23, 2369–2380, 2004.
81. Bobrovnikova-Marjon EV, Marjon PL, Barbash O, Vander Jagt DL, and Abcouwer SF, Expression of angiogenic factors vascular endothelial growth

factor and interleukin-8/CXCL8 is highly responsive to ambient glutamine availability: role of nuclear factor-kappaB and activating protein-1, *Cancer Res* 64, 4858–4869, 2004.
82. Kim MS, Kwon HJ, Lee YM, Baek JH, Jang JE, Lee SW, Moon EJ, Kim HS, Lee SK, Chung HY et al, Histone deacetylases induce angiogenesis by negative regulation of tumor suppressor genes, *Nat Med* 7, 437–443, 2001.
83. Deroanne CF, Bonjean K, Servotte S, Devy L, Colige A, Clausse N, Blacher S, Verdin E, Foidart JM, Nusgens BV et al, Histone deacetylases inhibitors as anti-angiogenic agents altering vascular endothelial growth factor signaling, *Oncogene* 21, 427–436, 2002.
84. Rossig L, Li H, Fisslthaler B, Urbich C, Fleming I, Forstermann U, Zeiher AM, and Dimmeler S, Inhibitors of histone deacetylation downregulate the expression of endothelial nitric oxide synthase and compromise endothelial cell function in vasorelaxation and angiogenesis, *Circulation Res* 91, 837–844, 2002.
85. Sgambato A, Ardito R, Faraglia B, Boninsegna A, Wolf FI, and Cittadini A, Resveratrol, a natural phenolic compound, inhibits cell proliferation and prevents oxidative DNA damage, *Mutation Res* 496, 171–180, 2001.
86. Park JW, Choi YJ, Suh SI, Baek WK, Suh MH, Jin IN, Min DS, Woo JH, Chang JS, Passaniti A, Lee YH, and Kwon TK, Bcl-2 overexpression attenuates resveratrol-induced apoptosis in U937 cells by inhibition of caspase-3 activity, *Carcinogenesis* 22, 1633–1639, 2001.
87. Hsieh T, Halicka D, Lu X, Kunicki J, Guo J, Darzynkiewicz Z, and Wu J, Effects of resveratrol on the G(0)–G(1) transition and cell cycle progression of mitogenically stimulated human lymphocytes, *Biochem Biophys Res Commun* 297, 1311–1317, 2002.
88. Fulda S and Debatin KM, Sensitization for tumor necrosis factor-related apoptosis-inducing ligand-induced apoptosis by the chemopreventive agent resveratrol, *Cancer Res* 64, 337–346, 2004.
89. Fulda S and Debatin KM, Sensitization for anticancer drug-induced apoptosis by the chemopreventive agent resveratrol, *Oncogene* 23, 6702–6711, 2004.
90. Clement MV, Hirpara JL, Chawdhury SH, and Pervaiz S, Chemopreventive agent resveratrol, a natural product derived from grapes, triggers CD95 signaling-dependent apoptosis in human tumor cells, *Blood* 92, 996–1002, 1998.
91. Ahmad N, Adhami VM, Afaq F, Feyes DK, and Mukhtar H, Resveratrol causes WAF-1/p21-mediated G(1)-phase arrest of cell cycle and induction of apoptosis in human epidermoid carcinoma A431 cells, *Clin Cancer Res* 7, 1466–1473, 2001.

10 Resveratrol Modulation of Gene Expression: The Role of Transcription Factors

Fulvio Della Ragione, Valeria Cucciolla, Adriana Borriello, and Vincenzo Zappia

CONTENTS

INTRODUCTION

Chemoprevention, defined as the pharmacological intervention with synthetic or naturally occurring agents to avoid, inhibit or, in some cases, reverse the various steps of carcinogenesis, has become increasingly recognized as an important strategy in the fight against cancer [1–3]. In recent years, many naturally occurring compounds, commonly present in the diet, have gained considerable attention as chemopreventive agents [1–3]. For example, the low incidence of prostate cancer in Asian men (from 30- to 120-fold less than that of the United States) has been attributed to the dietary consumption of polyphenols [4–6].

In this context, resveratrol (*trans*-3,4′,5 trihydroxystilbene), a phytoalexin found in a multitude of dietary plants, including particularly grapes and peanuts, has been shown to provide cancer chemopreventive effects in different systems based on its striking inhibition of diverse cellular events associated with tumor initiation, promotion, and progression [7]. Recently the molecule has also been definitely described as capable of slowing the aging process in a number of experimental models.

Although many resveratrol activities have been frequently attributed to the remarkable antioxidant properties of the compound [7,8], its precise mechanism of action is still poorly understood. Indeed, many pieces of evidence clearly indicate that the complex and frequently divergent resveratrol-dependent cellular phenotypes are linked to its pleiotropic molecular effects. These include the regulation of signal transduction pathways, the direct modulation of several enzymatic activities, and the interaction with ion channels [9 and references therein]. In addition, growing attention has been paid to the ability of resveratrol to modulate the activity of key transcription factors and, in turn, the cellular expression profiles.

Here, we briefly review the effects of resveratrol on transcription factors giving, when possible, major emphasis to the molecular mechanisms at the basis of resveratrol's activity.

RESVERATROL AND THE MODULATION OF INFLAMMATORY RESPONSE

A noteworthy biological feature of resveratrol is its remarkable antiinflammatory activity. Such an effect has been correlated to either the direct inhibition of enzymes involved in this process (particularly cyclooxygenase (COX)-2 [10,11]) or the ability to downregulate transduction pathways activated during the inflammatory response. Therefore, there is much interest in investigating the ability of resveratrol to modulate transcription factors that control the expression of genes encoding inflammatory mediators, particularly interleukins (IL-1 and IL-6), tumor necrosis factor

(TNFα), and inducible nitric oxide synthase (iNOS) [12,13]. These transcription factors mostly include nuclear factor kappa B (NF-κB, a member of the Rel family) and, in part, activator protein 1 (AP-1) and C/EBP.

NF-κB Pathway

NF-κB exists as a heterodimeric complex usually formed by two proteins of 50 and 65 kDa (RelA). In unstimulated cells, NF-κB is localized in the cytosol as an inactive non-DNA-binding complex associated with an inhibitor protein called inhibitory κB (IκB). IκB masks the nuclear translocation signal, thus preventing NF-κB from entering the nucleus [14,15]. Upon cell stimulation with various NF-κB inducers (including TNFα), IκB is rapidly phosphorylated on two serine residues (Ser32 and 36). This event targets the inhibitor protein for ubiquitination by the E3 ubiquitin-ligases (E3RSIκB) and subsequent degradation by the 26S proteasome. The released NF-κB dimer can then be translocated into the nucleus and might activate target genes by binding with high affinity to κB elements localized in their promoters [14,15].

Recently, further important findings have been obtained which must be merged into the scheme of the NF-κB pathway. In unstimulated cells, the *cis*-elements located in the promoter regions of NF-κB-regulated genes are occupied by p50/p50 homodimers [16]. These homodimers basally repress NF-κB-regulated genes by tethering corepressor complexes, such as NCoR and histone deacetylases (HDAC-1, -2, and -3) proteins [17–19]. SIRT1 (an HDAC of class III, see below) also contributes to basal repression by interacting with other proteins (CTIP2, HES1, and HEY2 [20]). The presence of HDAC activities causes a condensation of nucleosome structure which prevents transcription. The arrival of active NF-κB results in a tethering of histone acetyltransferase (HAT) proteins and the subsequent acetylation and activation of both p65 subunits (at lysine 310) and HATs themselves. Thus, in addition to a phosphorylation/dephosphorylation control, a complex acetylation/deacetylation regulation of NF-κB exists which appears mostly to occur after the nuclear entry of the transcription factor.

NF-κB controls the expression of genes that affect important cellular processes, such as adhesion, cell division cycle, angiogenesis, and apoptosis [14,22]. However, the phenotypical effects are strongly dependent on the experimental model employed (including its genetic background) and on the activating compounds. Thus, a simplistic generalization of data on NF-κB effects should be avoided.

Although resveratrol is unable to modulate NF-κB directly (Della Ragione et al., data not reported), the polyphenol affects the stimulation of NF-κB induced by different inflammatory agents in several cellular models, including HeLa, Jurkat, and U937 cell lines [12]. In particular, resveratrol addition prevented NF-κB activation in all these cells by lipopolysaccharides (LPS), okadaic acid, TNFα, H_2O_2, ceramide, and phorbol myristate acetate (PMA) [12].

The mechanism by which the molecule hampers NF-κB activation is different from that described in the case of other well-known inhibitors of the transcription factor (like herbimycin A and caffeic acid phenylethyl ester) that prevent NF-κB binding to DNA. Conversely, resveratrol appears to inhibit the phosphorylation and nuclear localization of the p65 subunit (Figure 10.1A). Alternatively, resveratrol has been proposed to hamper the phosphorylation and subsequent degradation of IκB [23], the protein that causes NF-κB localization in the cytoplasm (Figure 10.1A). Additional investigations have, however, challenged this last mechanism [24]. An additional intriguing hypothesis, which is analyzed in detail below, is correlated to the resveratrol ability of downregulating NF-κB acetylation (an activating mechanism) by stimulating a protein deacetylase activity (Figure 10.1B).

NF-κB is a pleiotropic transcription factor which plays important roles in diverse phenomena, including inflammation, cell proliferation, apoptosis, and oncogenesis. Thus, the capability of resveratrol to interfere with the activation of NF-κB is of particular interest to biologists and clinicians. However, the large number of target genes makes it difficult to propose a unique cellular response to NF-κB inhibition.

Finally, although a very large number of reports exist on the phytoalexin's capability of downregulating the activation of NF-κB (and thus the response to proinflammatory mediators), in some instances resveratrol does not affect the transcription factor upregulation. For example, in a macrophage model of LPS-induced inflammatory response resveratrol is unable to hamper NF-κB activation [25].

AP-1 Modulation

The activation of NF-κB is frequently associated with that of an additional transcription factor, namely AP-1. Accordingly, resveratrol has been shown to inhibit the ability of TNFα to activate AP-1 [12]. The effect on AP-1 is generally mediated by JNK (c-Jun N-terminal protein kinase) and/or the upstream kinase MEK (mitogen-activated protein kinase kinase or MAPKK) [26]. The TNFα-dependent JNK and MEK activation is prevented by resveratrol, thus allowing the hypothesis that this represents the mechanism of the polyphenol's AP-1 inhibition [12].

Among the genes activated by NF-κB and AP-1 are iNOS and COX-2 [27,28], two resveratrol potential effectors. Thus, it is likely that resveratrol inhibits iNOS and COX-2 via its effect on these two transcription factors. It is also conceivable that the expression of other genes modulated by NF-κB or AP-1, such as matrix metalloproteinase 9 (MMP-9) and cell surface adhesion molecules ICAM-1 (intercellular adhesion molecule 1) and VCAM-1 (vascular adhesion molecule 1) [29,30], frequently linked to the

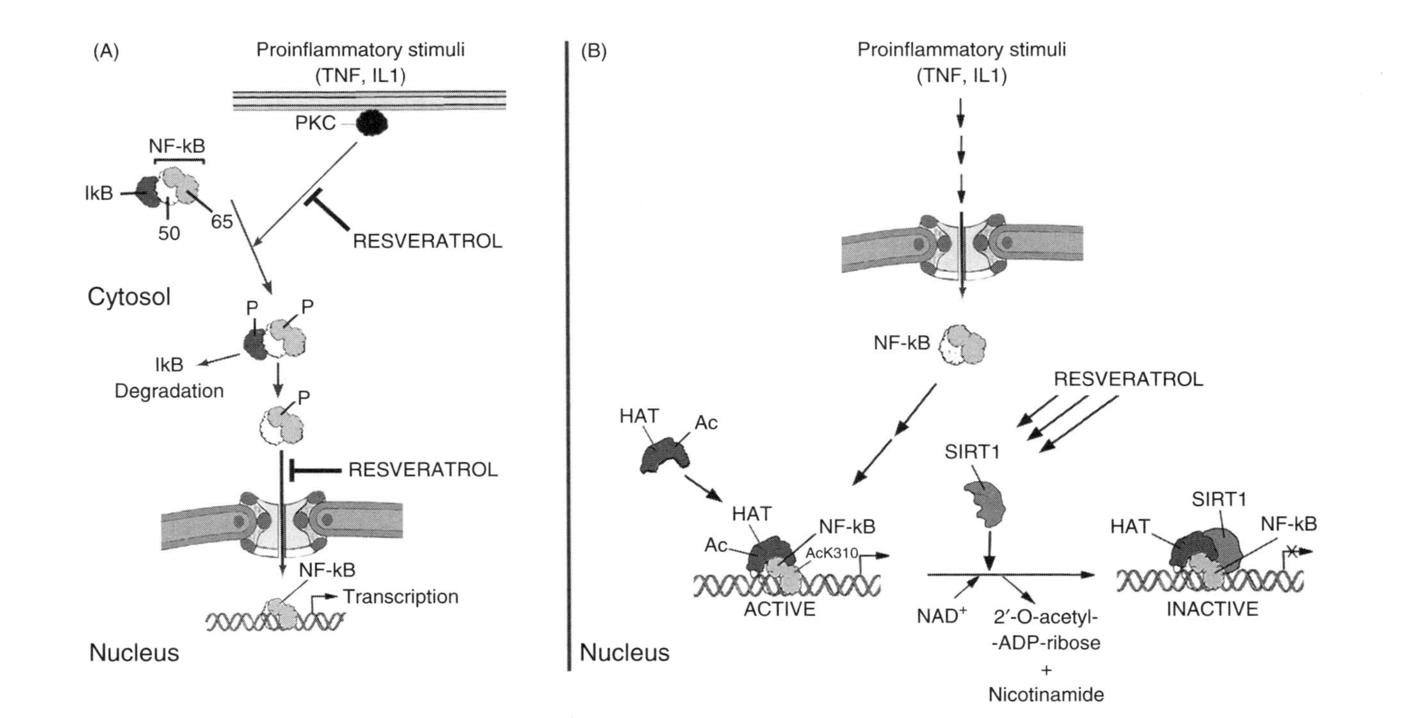

FIGURE 10.1 Effect of resveratrol on NF-κB pathway. (A) The activation of PKC (or other kinases), induced by inflammatory stimuli, causes the phosphorylation of IκB and of the 65 kDa NF-κB subunit. Phosphorylated IκB is then degraded, while active NF-κB translocates into the nucleus where it activates the expression of target genes. Resveratrol inhibits the kinase(s) activity as well as the nuclear import of active NF-κB. (B) Nuclear active NF-κB recognizes its consensus sequences and tethers histone acetylase activity (HAT). HAT acetylates and further activates NF-κB and HAT itself. Resveratrol upregulates SIRT1 (a protein deacetylase). SIRT1 interacts with the previous described complex and deacetylates and inactivates NF-κB and HAT.

carcinogenetic process, might be altered by resveratrol addition. Accordingly, it has recently been demonstrated that resveratrol suppresses malignant transformation in animal models by MMP-9 downregulation [31].

In conclusion, it is convincing that resveratrol modifies the cellular response to inflammatory mediators by interfering with the upregulation of pivotal mechanisms, mainly NF-κB transcription factor activity. However, it must be taken into consideration that since inflammatory cytokines, like TNFα, activate contemporaneously several cellular pathways, the inhibition of only part of them (as probably resveratrol does) might significantly modify the phenotypical cytokine effects, which might turn out to be unpredictable and dependent on the dose and the experimental model employed.

RESVERATROL AND INTRACELLULAR RECEPTORS

The description of resveratrol regulation of gene expression must take into consideration the well-known ability of the molecule to modulate the function of intracellular receptors (and functionally transcription factors) belonging to the superfamily of steroid receptors.

Resveratrol and the Androgen Receptor

Mitchell and colleagues reported that resveratrol inhibits, in a cell line established from prostate cancer, both the expression and the function of the androgen receptor (AR) [32]. AR is the mediator of androgen action, controlling the level of several androgen-regulated genes, including the prostate-specific antigen, the glandular kallikrein, and AR-specific coactivator ARA70. Therefore, in evaluating resveratrol's antitumoral activity it must be considered that prostate-specific antigen is supposed to be directly involved in the development of prostate cancer [33].

Recently it has been demonstrated that treatment of LNCaP (a prostate cancer cell line) with resveratrol significantly affects the growth rate. In particular, in cells treated for 1 hour, resveratrol showed only an inhibitory effect on DNA synthesis, which increased with increasing concentration (IC50 about 20 μM). However, when the treatment was extended to 24 hours, a dual effect was observed on DNA synthesis. At 5 to 10 μM the molecule caused a 2- to 3-fold increase in DNA synthesis, and at >15 μM it inhibited DNA synthesis. The increase in DNA synthesis was seen only in LNCaP cells, but not in androgen-independent DU145 prostate cancer.

The resveratrol-induced DNA synthesis upregulation was associated with the enrichment of LNCaP cells in S phase. Furthermore, consistent with the entry of LNCaP cells into S phase, there was a dramatic increase in nuclear cyclin-dependent kinase 2 activity associated with both cyclin A and cyclin E. Taken together, these observation indicate that LNCaP cells (an androgen-dependent cell line) treated with resveratrol are induced to enter into S phase, but subsequent progression through S phase is limited by the inhibitory effect of resveratrol on DNA synthesis, particularly at concentrations above 15 μM [34].

RESVERATROL AND THE ESTROGEN RECEPTOR

Resveratrol has been shown to have both estrogen agonist and antagonist activities [35,36]. As is well established, estrogens act via binding to the estrogen receptor (ER), another member of the nuclear receptor superfamily. The receptor engagement results in the nuclear translocation of the receptor and the subsequent transcriptional activation of estrogen responsive target genes [37]. However, more recently it has been demonstrated that estrogen receptor activation might also result in the activation of other classic transduction pathways (e.g., MAP kinases) with the resultant modulation of genes not showing ER consensus sequences at the promoter region.

In MCF-7 (an ER-positive human breast carcinoma cell line), resveratrol exposure resulted in the transcriptional activation of estrogen-responsive genes [38]. The effect was strictly dependent on the occurrence of ERE (estrogen responsive element) on the promoter of the target genes. Altogether, these findings suggest that the polyphenol might act as an efficacious phytoestrogen. Conversely, different findings indicate that the molecule, under distinct experimental conditions, shows antiestrogenic features. Indeed, resveratrol suppresses the estradiol-dependent progesterone receptor expression [35].

In T47D (a different estrogen-dependent breast cancer cell line), resveratrol at low concentration (around 10 μM) stimulates proliferation, while higher concentration causes activation of the apoptotic program [39]. Such a complex pattern has a negative effect on the possibility of using the molecule (or its analogs) in the treatment of ER-positive breast cancers.

The ability of resveratrol to activate the ER has also been demonstrated in nonbreast cell line. Indeed, the compound increases significantly prolactin secretion in pituitary cells [40] and induces the expression of alkaline phosphatase and active hydroxylase in an osteoblastic cell line (MC3T3-E1) [41]. Importantly, both the cellular models express high ER level.

Despite the overall reported evidence, which suggests that resveratrol might act as an estrogen-like compound, *in vivo* studies on rat models strongly argue against this mechanism of action [42–44]. Therefore, at present, the value of resveratrol as a phytoestrogen is questionable.

RESVERATROL AND ARYL HYDROCARBON RECEPTORS

One of the major definite *in vivo* chemopreventive activities of resveratrol is its ability to inhibit tumors caused by treatment with the aryl hydrocarbon DMBA (7,12-dimethylbenzanthracene) [7].

The molecular mechanisms of aryl hydrocarbon genotoxicity are quite complex and require an initial interaction with a specific receptor, a cytosolic aryl hydrocarbon receptor (AhR). Upon ligand binding, the activated AhR translocates into the nucleus where it interacts with the xenobiotic responsive element (XRE) [45,46], responsible for regulating the expression of several isoforms of cytochrome P450 (CYP) enzymes, including CYP1A1, CYP1A2, and CYP1B1, and some phase II detoxification enzymes [47]. CYP450 isoenzymes belong to a large family of either constitutive or inducible heme-containing enzymes which are involved in the metabolism of a wide variety of substances including the activation of a significant number of carcinogens [48,49]. Finally, the activated carcinogens might interact with DNA causing mutation or other cancer-promoting DNA alterations.

Taking into consideration the above reported mechanisms, molecules that inhibit CYP proteins or prevent their expression might be envisioned as potential anticarcinogenetic compounds and have been proposed for the development of novel cancer treatment.

The effect of resveratrol on AhR function has been the object of a large number of investigations. Several recent studies have demonstrated that resveratrol decreases CYP1A1 mRNA/protein expression or related activities in cultured cells, and that such inhibitory effects may be related to the AhR antagonist properties of the compound (Figure 10.2) [50–54]. Other reports describe that resveratrol can inhibit CYP activity by competing for the substrate binding site [55–58]. Recently it has been demonstrated that resveratrol inhibits AhR DNA binding activity and the resulting expression of CYP1A1 and CYP1B1 in MCF-10A cells. Moreover, additional assays revealed that resveratrol is also an effective inhibitor of CYP enzyme activities. Considering that CYP450s are overexpressed in a variety of human tumors, the strong inhibitory effect of resveratrol and similar compounds could have tremendous implications for the prevention and treatment of cancer.

These studies provide ample evidence that resveratrol inhibits aryl hydrocarbon-induced *CYP1A1* expression at the mRNA and protein levels [50,51,55]. In experiments done with DMBA, resveratrol inhibited the binding of the nuclear AhR to the XRE of the *CYP1A1* promoter without directly binding to the AhR [58]. Similar effects of resveratrol on other CYP450 isozymes, such as CYP1A2 and CYP3A4, have also been documented [59,60].

RESVERATROL AND p53

The transcription factor p53 has been described as the "guardian of the genome" since it arrests cell division cycle progression (mainly at level of G1/S phases) thus hampering damaged DNA from being replicated [61].

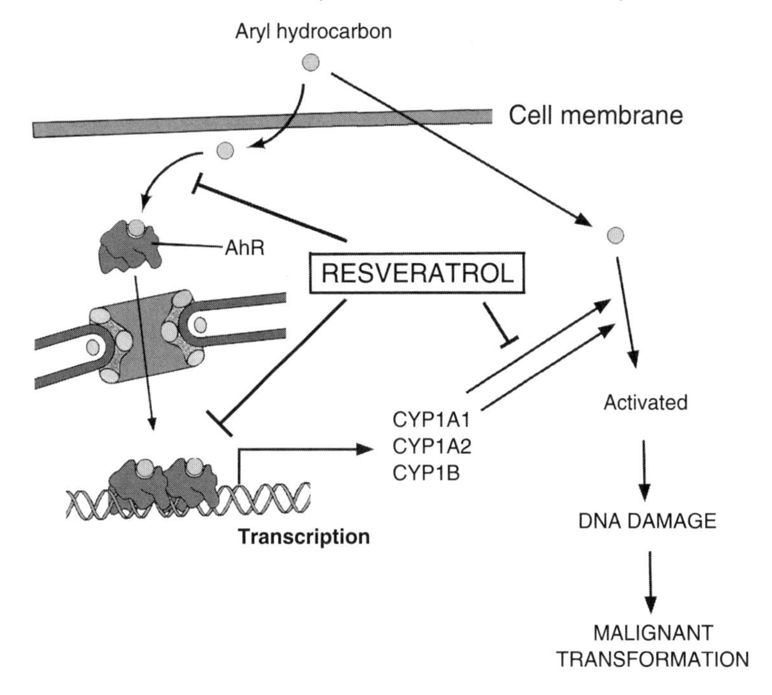

FIGURE 10.2 Effect of resveratrol on aryl hydrocarbon receptor function. Aryl hydrocarbon binds and activates its relative receptor, which, in turn, induces the expression of different CYP1 proteins. CYP1 proteins activate aryl hydrocarbons which damage DNA and might induce malignant transformation. Resveratrol hampers the upregulation of the receptor, its transcriptional activity, as well as the activation of aryl hydrocarbon.

In addition, p53 can trigger the apoptotic program, thus ensuring the removal of damaged or unwanted cells [62]. Loss-of-function mutations of p53 are very frequently associated with an increased incidence of tumor formation [63].

The transcription factor p53 is a short-lived protein whose activity is maintained at low levels in normal cells. Tight regulation of p53 is essential for its effect on tumorigenesis as well as for maintaining normal cell growth. Following DNA damage, p53 protein is protected from rapid degradation and acquires transcriptional-activating functions, largely as a result of posttranslational modifications [64–66]. Activation of the p53 protein as a transcription factor allows it, in turn, to upregulate the expression of genes whose products promote cell division cycle exit or induce apoptosis.

The p53 protein is phosphorylated in response to DNA damage [66]. For example, the ATM protein phosphorylates p53 at residue Ser15 [66] and Chk1/2 modify residue Ser20 [65,67]. Recent evidence, however, suggests that Ser15 phosphorylation does not lead directly to functional activation of p53 protein. Instead, it increases the affinity of the p300 protein acetylase for p53 [68]. This association leads, in turn, to the acetylation of p53. In particular, p53 is acetylated *in vitro* by p300 at Lys370–373, 381, and 382. Moreover, at least two of these sites, namely residues 370 and 382, are found to be acetylated *in vivo* in response to DNA damage [69].

The effect of resveratrol on p53 cellular level and transcriptional activity has been investigated in several instances, but the resulting picture is poorly defined. In mouse JB6 epidermal cell line, resveratrol is able to induce programmed cell death and this event is associated with the activation of p53 by inducing phosphorylation at serine 15 [70]. In addition, resveratrol-dependent p53 activation appears to be linked to the upregulation of several different transduction pathways including MAPK, JNK, and p38 kinase (Figure 10.3) [70]. Indeed, the inhibition of these pathways by means of different experimental approaches (specific drugs, antisense oligonucleotides, or dominant negative vectors) causes a diminished p53 activation and

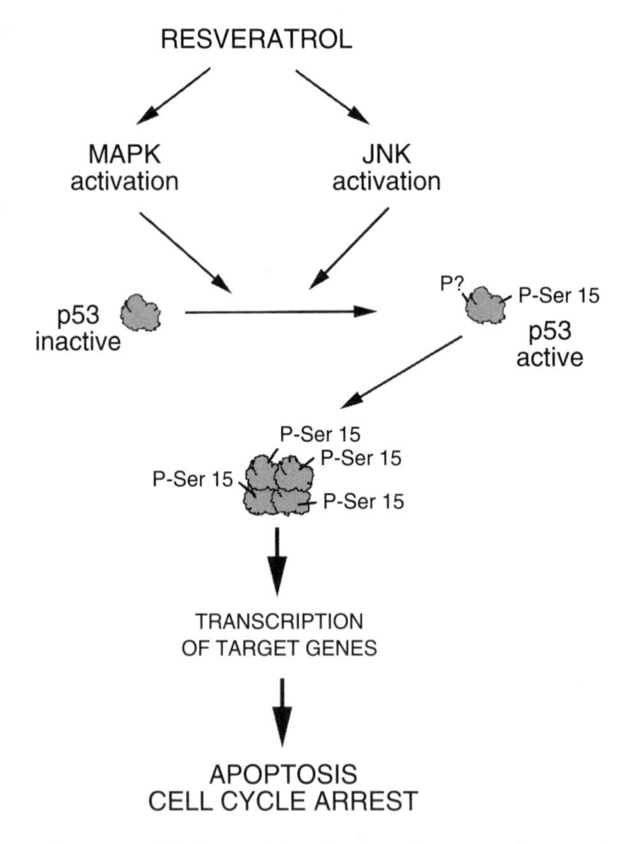

FIGURE 10.3 Resveratrol induces p53 activation. Resveratrol upregulates different signal transduction pathways (MAPK and JNK) which phosphorylate p53 (particularly on Ser15). Activated p53 tetramerizes and induces the expression of genes responsible for apoptosis or cell division cycle arrest.

downregulation of apoptosis [70]. Importantly, in thyroid cell lines, resveratrol-related p53 activation results in an enhanced expression of several genes (including *p21*^{Cip1}, *c-jun*, *c-fos*, and others), which might explain some of the phenotypical effects of resveratrol. Also, in this case the molecule induces apoptosis [71]. From these observations, it appears likely that the activation of p53 by resveratrol mostly results in the induction of the cell death program. A further proof in support of this hypothesis is the finding that mouse p53 null fibroblasts have a reduced apoptotic response to resveratrol compared to that of the normal counterpart [72].

However, the effects of resveratrol on p53 are certainly more complex than a simple activation of the transcription factor and, in turn, induction of apoptosis. Indeed, it must be considered that resveratrol stimulates enzymes that deacetylate p53 (see the section describing sirtuin proteins) and this results in the downregulation of the transcription factor activity. Thus, probably, the effect of resveratrol on p53 function is strictly context-specific.

Finally, in evaluating the importance of p53 in the resveratrol response, one must consider that the compound also induces remarkable phenotypical effects (cell growth arrest, differentiation, or apoptosis) on cells lacking a functional p53 protein. This is interesting not only for evaluating the molecular mechanisms of resveratrol but also since a large percentage of human cancers are p53-negative.

RESVERATROL AND EGR1 ACTIVATION

Egr1 (also known as *NGFI-A*, *TIS8*, *Krox-24*, and *Zif268*) is a member of the immediate early gene family and encodes a nuclear phosphoprotein involved in the regulation of cell growth and differentiation in response to a variety of growth factors and stress stimuli [73–77]. Moreover, *Egr1* is downregulated in several types of neoplasias, suggesting that it can act as a tumor suppressor gene in analogy with WT-1, another immediate early gene [78,79]. Indeed, Egr1 protein is decreased or undetectable in human breast and small-cell lung tumors [80,81] as well as in an array of tumor cell lines [78,82,83]. Gene deletions or mutations have also been reported in sporadic cancer cases [84]. It is worth mentioning that a recent study demonstrated that *Egr1* expression is strongly reduced in brain tumors compared with normal brain tissue, where the basal expression is high [85]. Particularly, in brain cancer, *Egr1* expression is suppressed in about 87% of cases, independently of other alterations of cell cycle genes. These results indicate that the loss of *Egr1* transcription might represent an important event of glial cell malignancy development and/or progression [86]. In addition, forced reexpression of *Egr1* suppresses the growth of transformed cells, both in soft agar and in athymic mice [87]. Accordingly, studies with antisense vectors indicated that the transformed phenotype is enhanced by the inhibition of *Egr1* expression [88].

It has been demonstrated, in HL60 cells, that several antioxidant molecules, including resveratrol, hydroxytyrosol, and pyrrolidine dithiocarbamate, are capable of inducing Egr1 [89]. A subsequent study, performed in K562 cells, has shown that resveratrol is capable of activating the Egr1 expression by a mechanism that involves the upregulation of MAPK pathway [90]. Moreover, target genes of Egr1, and particularly $p21^{Cip1}$, might then act as mediators of some of the effects of resveratrol, in particular for its antiproliferative effect. Indeed, $p21^{Cip1}$ is a pivotal inhibitor of active cyclin-dependent kinases and might then prevent the transition between the various cell division phases. Intriguingly, antisense molecules directed against Egr1 are able to downregulate the effect of resveratrol [90]. Since K562 cells are p53 negative, these findings identify a mechanism of action of resveratrol clearly independent of p53 pathway.

PROTEIN DEACETYLASE SIRTUINS: NEW EFFECTORS OF RESVERATROL ACTIVITIES

SIRTUIN FUNCTIONS AND INTERACTIONS

The *Saccharomyces cerevisiae* protein Sir2 (silent information regulator 2) acts as a NAD^+-dependent HDAC [91], and controls chromatin silencing [92,93] (Figure 10.4). Altered Sir2 levels cause, in yeast, aberrations in several functions, including DNA repair, senescence, and transcriptional silencing [92,93]. Although these Sir2 effects link the protein to the maintenance of genomic stability, its precise role still needs to be clarified. Preliminary investigations indicated that Sir2 plays a role in specific DNA repair processes [94–96]. Indeed, Sir2-deficient yeast strains showed alterations in the nonhomologous end-joining pathway of DNA [94]. Moreover, Sir2-containing complexes translocate to DNA double-strand breaks [95,96]. However, since homologs of Sir2 (forming the so-called sirtuin family) have been identified in all organisms examined including bacteria, which do not have histones, it is likely that Sir2 also targets nonhistone proteins for functional regulation.

Seven Sir2 homologs (SIRTs 1–7) have been so far identified in mammals. These genes might be orthologs of Sir2 with putative roles in regulating the silencing and stability of the genome, as well as the senescence of organisms [97,98]. Relatively few data are available concerning substrates and functions of SIRTs. Indeed, although several SIRTs are capable of deacetylating histones *in vitro* [91], none appears to function as a classic HDAC.

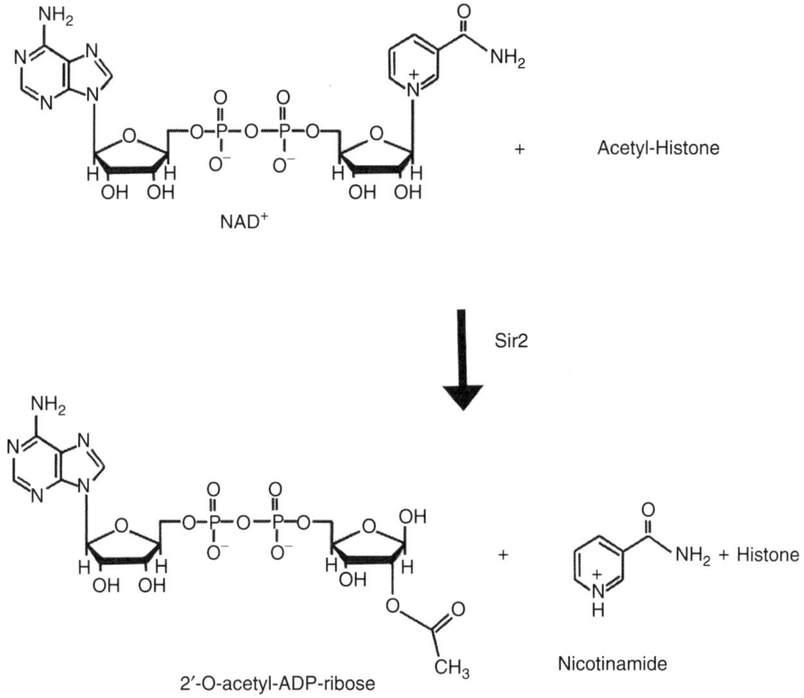

FIGURE 10.4 Reaction catalyzed by sirtuin proteins. Sirtuins are able to deacetylate acetylhistones (or other acetylated proteins) employing NAD^+ as acetyl acceptor. The products of the reaction are nicotinamide, deacetylated histones, and 2′-*O*-acetyl-ADP-ribose.

SIRTs 2 and 3 are localized at the cytoplasmic level. Particularly, SIRT2 is able to deacetylate tubulin, whereas in the case of SIRT3, which shows a mitochondrial localization, no substrates have been identified so far [99–101]. Among SIRTs, SIRT1 should represent the probable Sir2 ortholog on the basis of the extensive sequence similarity [97]. However, studies employing SIRT1-deficient mice do not suggest a role of the enzyme in genome silencing, as demonstrated in the case of the yeast enzyme [102–104]. Indeed the major phenotypical defects of SIRT1-ablated animals include smaller dimension, defects of the retina and heart, persistent eyelid closure, infertility, and, under a specific background, early postnatal lethality [103,104]. In addition, SIRT1-deficient cells exhibited p53 hyperacetylation after DNA damage and increased ionizing radiation-induced thymocyte apoptosis [104].

Recent studies have implicated SIRT proteins in the regulation of a number of transcription factors, including, in particular, p53, FOXOs, and NF-κB [105–110]. By means of strategies aimed at modulating SIRT1 content, it has been established that the protein can deacetylate p53 both

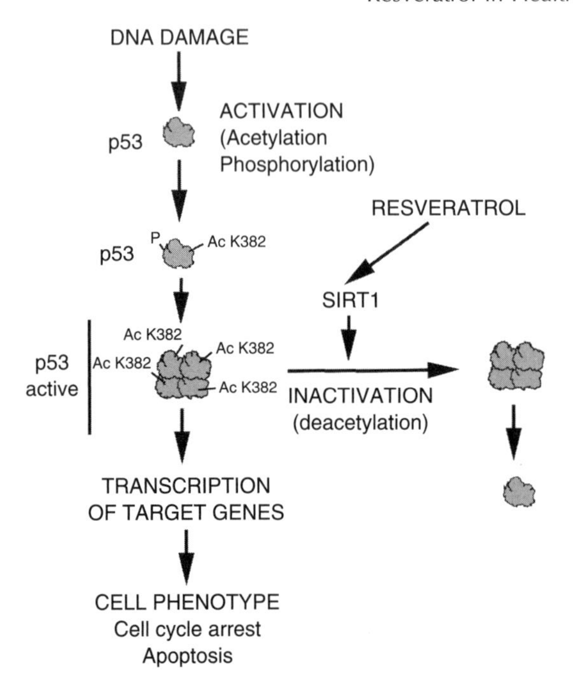

FIGURE 10.5 Resveratrol-dependent SIRT1 activation and p53 function. DNA damage induces different p53 modifications (acetylation and phosphorylation) which activate the protein. Activated p53 induces the expression of target genes, which, in turn, causes the activation of cell apoptosis or the arrest of proliferation. Resveratrol activates SIRT1 which deacetylates p53 and inactivates the protein.

in vitro and in cell culture [105–107]. As described before, p53 acetylation occurs at multiple lysine residues of the protein. Although the precise molecular mechanism has not been clarified, the modification is correlated with p53 stabilization and activation. Accordingly, SIRT1 upregulation inhibits p53 transcriptional activity and the apoptotic response to DNA damage and oxidative stresses (Figure 10.5). Conversely, the expression of a dominant negative form of SIRT1 protein potentiates the responses to cellular stresses [105,106]. SIRT1 is able to inhibit the activity of several FOXO family transcription factors. FOXOs control either cell cycle arrest or apoptosis by the transcriptional activation of several genes, including $p27^{Kip1}$, a pivotal cyclin-dependent kinase inhibitor [108–110]. Deacetylation of FOXOs results in the loss of their activity, particularly of the capacity to induce apoptosis [108–110]. The deacetylation of NF-κB by SIRT1 probably occurs at the level of DNA and includes both the deacetylation of p65 subunit and of HAT complexed with the transcription factor. Both these events result in a general loss of activity of NF-κB (Figure 10.1B) [111].

MECHANISTIC BASIS OF RESVERATROL ACTIVATORY FUNCTION

A few small molecules that stimulate sirtuin activity have been recently identified. Two of these activators, quercetin and piceatannol, are structurally similar and increase SIRT1 activity by more than several-fold [112]. A subsequent screening performed on the quercetin and piceatannol family identified additional powerful SIRT1 activators including particularly resveratrol [112].

The detailed mechanistic basis as to how resveratrol stimulates the activity of Sir2 proteins is not well understood, although kinetic data reveal that it stimulates Sir2 activity by lowering the Michaelis constant for both NAD^+ and the acetyllysine-bearing substrate by 35 and 5 times, respectively [113]. Since the β1–α2 loop of the Sir2 proteins undergoes significant conformational adjustments to facilitate NAD^+ binding and catalysis, it has been proposed that resveratrol (like other polyphenols) may somehow more optimally reconfigure the β1–α2 loop for NAD^+ binding. In addition, a comparison between the different Sir2 complexes reveals that the small zinc-binding domain of the catalytic core makes adjustments that appear to be important for binding to NAD^+ and acetyllysine, thus suggesting that the molecule may interact with this domain in a way that enhances binding to one or both of the substrates [113].

RESVERATROL AND SIRT1: EFFECTS ON TRANSCRIPTION FACTOR ACTIVITY

In several different species, the restriction of the diet (i.e., caloric restriction) results in a significant lifespan extension [114–118]. Initially, detailed genetic studies demonstrated that this event requires, in yeast, the presence of the Sir2 protein [119,120]. Thereafter, this observation was confirmed in *Caenorhabditis elegans* and *Drosophila melanogaster* [122], allowing the proposal of an important role of sirtuin proteins in life duration. The identification of resveratrol as a strong activator of Sir2 (and its orthologs, including SIRT1), prompted studies aimed at investigating the effect of resveratrol on life duration, in the hope of inducing a lifespan increase. Experimental observations confirmed this hypothesis, demonstrating that resveratrol is able to slow senescence in all the organisms investigated, including yeast, *C. elegans*, and *D. melanogaster* [112,123]. Moreover, the effect is abrogated when nutrients are limited, thus confirming that resveratrol modulates the aging process by mechanisms related to caloric restriction, i.e., sirtuin activation.

The agonistic activity of resveratrol on sirtuins might also explain the complexity and frequently the apparent peculiarity of its molecular action. Indeed, several of the phenotypical effects of the molecule could be, at least in part, explained on the basis of the SIRT1 effects on different transcription factors.

In the case of NF-κB, the ability of SIRT1 to downregulate the transcriptional factor matches with that of resveratrol, thus suggesting that the compound might act by stimulating SIRT1 activity which, in turn, hampers NF-κB upregulation (see also the section above on the NF-κB pathway) (Figure 10.1B). Conversely, as far as p53 modulation is concerned, the effects of resveratrol and SIRT1 on the transcription factor appear clearly distinct, being positive for the former (see the section on resveratrol and p53) and negative for the latter (see the section on sirtuin functions and interactions). However, very recently it has also been described that resveratrol is able to inactivate p53 via SIRT1 stimulation [112], thus suggesting a similarity between the two effectors. This confirms again that the activity of resveratrol on pivotal proteins might be strongly context dependent and strictly related to the cell phenotype and to the genetic background.

No data are so far available on the effect of resveratrol on FOXO proteins, but it is forecast that data will be soon available in this field of investigation.

CONCLUSIONS AND PERSPECTIVES

The large body of evidence reported here clearly allows the conclusion that the wine polyphenol resveratrol strongly affects the activity of a number of pivotal transcription factors, including NF-κB, AP-1, steroid receptors, p53, and Egr1. In turn, these activities might constitute, at least in part, the molecular bases of the complex phenotypical effects of the phytoalexin.

The simple chemical structure of resveratrol probably enables the molecule to interact with a large number of proteins resulting in a variety of biological effects, like modulation of cell division cycle, suppression of growth, induction of differentiation, upregulation of death-inducing factors, and inhibition of reactive oxygen species production. However, the complexity of the effects on cell cultures, albeit of clinical interest, is frequently unpredictable and strongly context specific. The importance of resveratrol is strengthened by *in vivo* studies which demonstrate its beneficial effects on cardiovascular, neurological, and hepatic systems, and, most excitingly, by data evidencing a cancer chemopreventive activity. In this context, the ability of resveratrol to modulate gene expression by affecting transcription factor activities is certainly of primary importance. However, further studies are necessary to identify *in vivo*, under well-defined genetic and environmental conditions, the transcriptional effect of the molecule on specific cell phenoypes.

An important finding is also the fact that resveratrol is able to activate the deacetylase activity of sirtuin proteins and, as a result, to extend life duration (at least in the investigated species). Interestingly, this effect occurs without an apparent cost of reproduction as, conversely, has been observed in many models of increased longevity.

So far, the precise functions of SIRT proteins (the human orthologs of yeast Sir2) have not been clearly defined. These proteins deacetylate a number of substrates (including particularly transcription factors) causing a complex modulation of gene expression. Thus, the ability of resveratrol to activate SIRT1 might be an important mechanism by which the phytoalexin controls genome transcription. However, so far, very little information is available on (1) the genes that are under the sirtuins' control, and, most generally, (2) the cellular processes that involve the different types of human sirtuins.

In addition, the molecular bases of lifespan extension due to resveratrol-dependent sirtuin activation remain quite elusive and it must be considered that possibly sirtuin activators may still cause tradeoffs that involve traits, not discovered, or that occur only under some environmental conditions. Intriguingly, the effect of resveratrol on sirtuin activity suggests that plants might synthesize the molecule in response to stress conditions and nutrient limitation possibly for activating their sirtuin pathways and for modulating the activity of their transcription factors. This hypothesis allows the development of new views on the physiological role of the molecule. Moreover, animals may use plant stress molecules as a cue to shift their genome transcription (and phenotype) in view of a decline in their environment or food supply.

In all the cases, understanding the importance of the endogenous function and evolutionary origin of sirtuin activators, including resveratrol, will lead to further insights into the underlying mechanisms of longevity regulation, and could aid in the development of intervention strategies that provide important health benefits.

ACKNOWLEDGMENTS

This work was supported in part by grants from Associazione Italiana per la Ricerca sul Cancro (AIRC), Progetti di Rilevante Interesse Nazionale (PRIN, MURST), and FIRB.

REFERENCES

1. Hong WK and Sporn MB, Recent advances in chemoprevention of cancer, *Science* 278, 1073–1077, 1997.
2. Sporn MB and Suh N, Chemoprevention of cancer, *Carcinogenesis* 21, 525–530, 2000.
3. Kelloff GJ, Crowell JA, Steele VE, Lubet RA, Malone WA, Boone CW, Kopelovich L, Hawk ET, Lieberman R, Lawrence JA, Ali I, Viner JL, and

Sigman CC, Progress in cancer chemoprevention: development of diet-derived chemopreventive agents, *J Nutr* 130, 467S–471S, 2000.
4. Greenlee RT, Murray T, Bolden S, and Wingo PA, Cancer statistics, 2000, *CA Cancer J Clin* 50, 7–33, 2000.
5. Giovannucci E, Ascherio A, Rimm EB, Stampfer MJ, Colditz GA, and Willett WC, Intake of carotenoids and retinol in relation to risk of prostate cancer, *J Natl Cancer Inst* 87, 1767–1776, 1995.
6. Makela S, Santti R, Salo L, and McLachlan JA, Phytoestrogens are partial estrogen agonists in the adult male mouse, *Environ Health Perspect* 103, 123–127, 1995.
7. Jang M, Cai L, Udeani GO, Slowing KV, Thomas CF, Beecher CW, Fong HH, Farnsworth NR, Kinghorn AD, Mehta RG, Moon RC, and Pezzuto JM, Cancer chemopreventive activity of resveratrol, a natural product derived from grapes, *Science* 275, 218–220, 1997.
8. Fremont L, Biological effects of resveratrol, *Life Sci* 66, 663–673, 2000.
9. Pervaiz S, Resveratrol: from grapevines to mammalian biology, *FASEB J* 17, 1975–1985, 2003.
10. MacCarrone M, Lorenzon T, Guerrieri P, and Agro AF, Resveratrol prevents apoptosis in K562 cells by inhibiting lipoxygenase and cyclooxygenase activity, *Eur J Biochem* 265, 27–34, 1999.
11. Subbaramaiah K, Chung WJ, Michaluart P, Telang N, Tanabe T, Inoue H, Jang M, Pezzuto JM, and Dannenberg AJ, Resveratrol inhibits cyclooxygenase-2 transcription and activity in phorbol ester-treated human mammary epithelial cells, *J Biol Chem* 273, 21875–21882, 1998.
12. Manna SK, Mukhopadhyay A, and Aggarwal BB, Resveratrol suppresses TNF-induced activation of nuclear transcription factors NF-kappa B, activator protein-1, and apoptosis: potential role of reactive oxygen intermediates and lipid peroxidation, *J Immunol* 164, 6509–6519, 2000.
13. Tsai SH, Lin-Shiau SY, and Lin JK, Suppression of nitric oxide synthase and the down-regulation of the activation of NFkappaB in macrophages by resveratrol, *Br J Pharmacol* 126, 673–680, 1999.
14. Baldwin AS, The NF-κB and IκB proteins: new discoveries and insights, *Annu Rev Immunol* 14, 649–681, 1996.
15. Ghosh S and Karin M, Missing pieces in the NF-κB puzzle, *Cell* 109, S81–S96, 2002.
16. Watanabe N, Iwamura T, Shinoda T, and Fujita T, Regulation of NFκB1 proteins by the candidate oncoprotein BCL-3: generation of NF-κB homodimers from the cytoplasmic pool of p50-p105 and nuclear translocation, *EMBO J* 16, 3609–3620, 1997.
17. Baek SH, Ohgi KA, Rose DW, Koo EH, Glass CK, and Rosenfeld MG, Exchange of N-CoR corepressor and Tip60 coactivator complexes links gene expression by NF-κB and beta-amyloid precursor protein, *Cell* 110, 55–67, 2002.
18. Ashburner BP, Westerheide SD, and Baldwin AS Jr, The p65 (RelA) subunit of NF-κB interacts with the histone deacetylase (HDAC) corepressors HDAC1 and HDAC2 to negatively regulate gene expression, *Mol Cell Biol* 21, 7065–7077, 2001.
19. Zhong HH, May MJ, Jimi E, and Ghosh S, The phosphorylation status of nuclear NF-κB determines its association with CBP/p300 or HDAC-1, *Mol Cell* 9, 625–636, 2002.

20. Senawong T, Peterson VJ, Avram D, Shepherd DM, Frye RA, Minucci S, and Leid M, Involvement of the histone deacetylase SIRT1 in chicken ovalbumin upstream promoter transcription factor (COUP-TF)-interacting protein 2-mediated transcriptional repression, *J Biol Chem* 278, 43041–43050. 2003.
21. Takehiko T and Fuyuki I, Human Sir2-related protein SIRT1 associates with the bHLH repressors HES1 and HEY2 and is involved in HES1- and HEY2-mediated transcriptional repression, *Biochem Biophys Res Commun* 301, 250–257, 2003.
22. Mayo MW and Baldwin AS, The transcription factor NF-κB: control of oncogenesis and cancer therapy resistance. *Biochim Biophys Acta Rev Cancer* 1470, M55–M62, 2000.
23. Holmes-McNary M and Baldwin AS Jr, Chemopreventive properties of trans-resveratrol are associated with inhibition of activation of the IkappaB kinase, *Cancer Res* 60, 3477–3483, 2000.
24. Ashikawa K, Majumdar S, Banerjee S, Bharti AC, Shishodia S, and Aggarwal BB, Piceatannol inhibits TNF induced NF-kappaB activation and NF-kappaB-mediated gene expression through suppression of IkappaBalpha kinase and p65 phosphorylation, *J Immunol* 169, 6490–6497, 2002.
25. Wadsworth TL and Koop DR, Effects of the wine polyphenolics quercetin and resveratrol on pro-inflammatory cytokine expression in RAW 264.7 macrophages, *Biochem Pharmacol* 57, 941–949, 1999.
26. Karin M and Delhase M, JNK or IKK, AP-1 or NF-kappaB, which are the targets for MEK kinase 1 action?, *Proc Natl Acad Sci USA* 95, 9067–9069, 1998.
27. von Knethen A, Callsen D, and Brune B, Superoxide attenuates macrophage apoptosis by NF-kappa B and AP-1 activation that promotes cyclooxygenase-2 expression, *J Immunol* 163, 2858–2866, 1999.
28. Hwang D, Jang BC, Yu G, and Boudreau M, Expression of mitogen-inducible cyclooxygenase induced by lipopolysaccharide, mediation through both mitogen-activated protein kinase and NF-kappaB signaling pathways in macrophages, *Biochem Pharmacol* 54, 87–96, 1997.
29. Sato H and Seiki M, Regulatory mechanism of 92 kDa type IV collagenase gene expression which is associated with invasiveness of tumor cells, *Oncogene* 8, 395–405, 1993.
30. Collins T, Read MA, Neish AS, Whitley MZ, Thanos D, and Maniatis T, Transcriptional regulation of endothelial cell adhesion molecules: NF-kappa B and cytokine inducible enhancers, *FASEB J* 9, 899–909, 1995.
31. Banerjee S, Bueso-Ramos C, and Aggarwal BB, Suppression of 7,12-dimethylbenz(a)anthracene-induced mammary carcinogenesis in rats by resveratrol: role of nuclear factor-kappaB, cyclooxygenase 2, and matrix metalloprotease 9, *Cancer Res* 62, 4945–4954, 2002.
32. Mitchell SH, Zhu W, and Young CY, Resveratrol inhibits the expression and function of the androgen receptor in LNCaP prostate cancer cells, *Cancer Res* 59, 5892–5895, 1999.
33. Culig Z, Klocker H, Bartsch G, and Hobisch A, Androgen receptors in prostate cancer, *Endocr Relat Cancer* 9, 155–170, 2002.
34. Kuwajerwala N, Cifuentes E, Gautam S, Menon M, Barrack ER, and Reddy GP, Resveratrol induces prostate cancer cell entry into S phase and inhibits DNA synthesis, *Cancer Res* 62, 2488–2492, 2002.

35. Lu R and Serrero G, Resveratrol, a natural product derived from grape, exhibits antiestrogenic activity and inhibits the growth of human breast cancer cells, *J Cell Physiol* 179, 297–304, 1999.
36. Basly JP, Marre-Fournier F, Le Bail JC, Habrioux G, and Chulia AJ, Estrogenic/antiestrogenic and scavenging properties of (E)- and (Z)-resveratrol, *Life Sci* 66, 769–777, 2000.
37. Gusman J, Malonne H, and Atassi G, A reappraisal of the potential chemopreventive and chemotherapeutic properties of resveratrol, *Carcinogenesis* 22, 1111–1117, 2001.
38. Gehm BD, McAndrews JM, Chien PY, and Jameson JL, Resveratrol, a polyphenolic compound found in grapes and wine, is an agonist for the estrogen receptor, *Proc Natl Acad Sci USA* 94, 14138–14143, 1997.
39. Clement MV, Hirpara JL, Chawdhury SH, and Pervaiz S, Chemopreventive agent resveratrol, a natural product derived from grapes, triggers CD95 signaling-dependent apoptosis in human tumor cells, *Blood* 92, 996–1002, 1998.
40. Stahl S, Chun TY, and Gray WG, Phytoestrogens act as estrogen agonists in an estrogen-responsive pituitary cell line, *Toxicol Appl Pharmacol* 152, 41–48, 1998.
41. Mizutani K, Ikeda K, Kawai Y, and Yamori Y, Resveratrol stimulates the proliferation and differentiation of osteoblastic MC3T3–E1 cells, *Biochem Biophys Res Commun* 253, 859–863, 1998.
42. Turner RT, Evans GL, Zhang M, Maran A, and Sibonga JD, Is resveratrol an estrogen agonist in growing rats?, *Endocrinology* 140, 50–54, 1999.
43. Ashby J, Tinwell H, Pennie W, Brooks AN, Lefevre PA, Beresford N, and Sumpter JP, Partial and weak oestrogenicity of the red wine constituent resveratrol: consideration of its superagonist activity in MCF-7 cells and its suggested cardiovascular protective effects, *J Appl Toxicol* 19, 39–45, 1999.
44. Freyberger A, Hartmann E, Hildebrand H, and Krotlinger F, Differential response of immature rat uterine tissue to ethinylestradiol and the red wine constituent resveratrol, *Arch Toxicol* 74, 709–715, 2001.
45. Nebert DW, Puga A, and Vasilious V, Role of the Ah receptor and the dioxin-inducible [Ah] gene battery in toxicity, cancer, and signal transduction, *Ann NY Acad Sci* 685, 624–640, 1993.
46. Okey AB, Riddick DS, and Harper PA, The Ah receptor: mediator the toxicity of 2,3,7,8-tetrachlorodibenzo-p-dioxin (TCDD) and related compound, *Toxicol Lett* 70, 1–22, 1994.
47. Rowlands JC and Gustafsson JA, Aryl hydrocarbon receptor-mediated signal transduction, *Crit Rev Toxicol* 27, 109–134, 1997.
48. Nelson DR, Koymans L, Kamataki T, Stegeman JJ, Feyereisen R, Waxman DJ, Waterman MR, Gotoh O, Coon MJ, Estabrook RW et al, P450 superfamily: update on new sequences, gene mapping, accession numbers and nomenclature, *Pharmacogenetics* 6, 1–42, 1996.
49. Murray GI, The role of cytochrome P450 in tumour development and progression and its potential in therapy, *J Pathol* 192, 419–426, 2000.
50. Ciolino HP, Daschner PJ, and Yeh GC, Resveratrol inhibits transcription of CYP1A1 *in vitro* by preventing activation of the aryl hydrocarbon receptor, *Cancer Res* 58, 5707–5712, 1998.
51. Casper RF, Quesne M, Rogers IM, Shirota T, Jolivet A, Milgrom E, and Savouret JF, Resveratrol inhibits transcription of CYP1A1 *in vitro* by

preventing activation of the aryl hydrocarbon receptor, *Mol Pharmacol* 56, 784–790, 1999.

52. Mollerup S, Ovrebo S, and Haugen A, Lung carcinogenesis: resveratrol modulates the expression of genes involved in the metabolism of PAH in human bronchial epithelial cells, *Int J Cancer* 92, 18–25, 2001.
53. Lee JE and Safe S, Involvement of a post-transcriptional mechanism in the inhibition of CYP1A1 expression by resveratrol in breast cancer cells, *Biochem Pharmacol* 62, 113–1124, 2001.
54. Singh SU, Casper RF, Fritz PC, Sukhu B, Ganss B, Girard BJr, Savouret JF, and Tenenbaum HC, Inhibition of dioxin effects on bone formation *in vitro* by a newly described aryl hydrocarbon receptor antagonist, resveratrol, *J Endocrinol* 167, 183–195, 2000.
55. Ciolino HP and Yeh GC, Inhibition of aryl hydrocarbon-induced cytochrome P-450 1A1 enzyme activity and CYP1A1 expression by resveratrol, *Mol Pharmacol* 56, 760–767, 1999.
56. Piver B, Berthou F, Dreano Y, and Lucas D, Inhibition of CYP3A, CYP1A and CYP2E1 activities by resveratrol and other non volatile red wine components, *Toxicol Lett* 125, 83–91, 2001.
57. Chun YJ, Kim MY, and Guengerich FP, Resveratrol is a selective human cytochrome P450 1A1 inhibitor, *Biochem Biophys Res Commun* 262, 20–24, 1999.
58. Chang TK, Chen J, and Lee WB, Differential inhibition and inactivation of human CYP1 enzymes by trans-resveratrol: evidence for mechanism-based inactivation of CYP1A2, *J Pharmacol Exp Ther* 299, 874–882, 2001.
59. Chan WK and Delucchi AB, Resveratrol, a red wine constituent, is a mechanism-based inactivator of cytochrome P450 3A4, *Life Sci* 67, 3103–3112, 2000.
60. Bhat KP and Pezzuto JM, Cancer chemopreventive activity of resveratrol, *Ann NY Acad Sci* 957, 210–229, 2002.
61. Steele RJ, Thompson AM, Hall PA, and Lane DP, The p53 tumour suppressor gene, *Br J Surg* 85, 1460–1467, 1998.
62. Flatt PM, Polyak K, Tang L J, Scatena CD, Westfall MD, Rubinstein LA, Yu J, Kinzler KW, Vogelstein B, Hill DE, and Pietenpol JA, p53-dependent expression of PIG3 during proliferation, genotoxic stress, and reversible growth arrest, *Cancer Lett* 156, 63–72, 2000.
63. Hussain SP, Hollstein MH, and Harris CC, p53 tumor suppressor gene: at the crossroads of molecular carcinogenesis, molecular epidemiology, and human risk assessment, *Ann NY Acad Sci* 919, 79–85, 2000.
64. Canman CE, Lim DS, Cimprich JA, and Taya Y, Activation of the ATM kinase by ionizing radiation and phosphorylation of p53, *Science* 281, 1677–1679, 1998.
65. Shieh SY, Ahn, J, and Tamai K, The human homologs of checkpoint kinases Chk1 and Cds1 (Chk2) phosphorylate p53 at multiple DNA damage-inducible sites, *Genes Dev* 14, 289–300, 2000.
66. Siciliano JD and Canman, CE, DNA damage induces phosphorylation of the amino terminus of p53, *Genes Dev* 11, 3471–3481, 1997.
67. Chehab NH, Malikzay A, and Stavridi ES, Phosphorylation of Ser-20 mediates stabilization of human p53 in response to DNA damage, *Proc Natl Acad Sci USA* 96, 13777–13782, 1999.

68. Lambert PF and Kashanchi F, Phosphorylation of p53 serine 15 increases interaction with CBP, *J Biol Chem* 273, 33048–33053, 1998.
69. Gu W and Roder RG, Activation of p53 sequence-specific DNA binding by acetylation of the p53 C-terminal domain, *Cell* 90, 595–606, 1997.
70. She QB, Bode AM, Ma WY, Chen NY, and Dong Z, Resveratrol-induced activation of p53 and apoptosis mediated by extracellular-signal-regulated protein kinases and p38 kinase, *Cancer Res* 61, 1604–1610, 2001.
71. Shih A, Davis FB, Lin H-Y, and Davis PJ, Resveratrol induces apoptosis in thyroid cancer cell lines via a MAPK- and p53-dependent mechanism, *J Clin Endocrinol Metab* 87, 1223–1232, 2002.
72. Huang C, Ma WY, Goranson A, and Dong Z, Resveratrol suppresses cell transformation and induces apoptosis through a p53-dependent pathway, *Carcinogenesis (Lond.)* 20, 237–242, 1999.
73. Bertelli AA, Giovannini L, Stradi R, Bertelli A, and Tillement PJ, Plasma, urine and tissue levels of trans- and cis-resveratrol (3,4′,5-trihydroxystilbene) after short-term or prolonged administration of red wine to rats, *Int J Tissue React* 18, 67–71, 1996.
74. Rodrigue CM, Arous N, Bachir D, Smith-Ravin J, Romeo PH, Galacteros F, and Garel MC, Resveratrol, a natural dietary phytoalexin, possesses similar properties to hydroxyurea towards erythroid differentiation, *Br J Haematol* 113, 500–507, 2001.
75. Sakamoto KM, Bardeleben C, Yates KE, Raines MA, Golde DW, and Gasson JC, 5′ upstream sequence and genomic structure of the human primary response gene, EGR-1/TIS8, *Oncogene* 6, 867–871, 1991.
76. Sukhatme VP, Cao X, Chang LC, Tsai-Morris CH, Stamenkovich D, Ferreira PC, Cohen DR, Edwards SA, Shows TB, and Curran T, A zinc finger-encoding gene coregulated with c-fos during growth and differentiation, and after cellular depolarization, *Cell* 53, 37–43, 1988.
77. Milbrandt J, A nerve growth factor-induced gene encodes a possible transcriptional regulatory factor, *Science* 238, 797–799, 1987.
78. Liu C, Rangnekar VM, Adamson E, and Mercola D, Suppression of growth and transformation and induction of apoptosis by EGR-1, *Cancer Gene Ther* 5, 3–28, 1998.
79. Huang RP, Wu JX, Fan Y, and Adamson ED, UV activates growth factor receptors via reactive oxygen intermediates, *J Cell Biol* 133, 211–220, 1996.
80. Huang RP, Liu CT, Fan Y, Mercola DA, and Adamson ED, Egr-1 negatively regulates human tumor cell growth via the DNA-binding domain, *Cancer Res* 55, 5054–5062, 1995.
81. Rauscher FJ III, The WT1 Wilms tumor gene product: a developmentally regulated transcription factor in the kidney that functions as a tumor suppressor, *FASEB J* 7, 896–903, 1993.
82. Huang RP, Fan Y, de Belle I, Niemeyer C, Gottardis MM, Mercola D, and Adamson ED, Decreased Egr-1 expression in human, mouse and rat mammary cells and tissues correlates with tumor formation, *Int J Cancer* 72, 102–109, 1997.
83. Levin WJ, Press MF, Gaynor RB, Sukhatme VP, Boone, TC, and Slamon DJ, Expression patterns of immediate early transcription factors in human

non-small cell lung cancer, The Lung Cancer Study Group, *Oncogene* 11, 1261–1269, 1995.
84. Calogero A, Cuomo L, D'Onofrio M, de Grazia U, Spinanti P, Mercola D, Faggioni A, Frati L, Adamson ED, and Ragona G, Expression of Egr-1 correlates with the transformed phenotype and the type of viral latency in EBV genome positive lymphoid cell lines, *Oncogene* 13, 2105–2112, 1996.
85. Liu CT, Yao J, Mercola D, and Adamson ED, The transcription factor EGR-1 directly transactivates the fibronectin gene and enhances attachment of human glioblastoma cell line U251, *J Biol Chem* 275, 20315–20323, 2000.
86. Le Beau MM, Espinosa R III, Neuman WL, Stock W, Roulston D, Larson RA, Keinanen M, and Westbrook CA, Cytogenetic and molecular delineation of the smallest commonly deleted region of chromosome 5 in malignant myeloid diseases, *Proc Natl Acad Sci USA* 90, 5484–5488, 1993.
87. Calogero A, Arcella A, De Gregorio G, Porcellini A, Mercola D, Liu C, Lombari V, Zani M, Giannini G, Gagliardi FM, Caruso R, Gulino A, Frati L, and Ragona G, The early growth response gene EGR-1 behaves as a suppressor gene that is down-regulated independent of ARF/Mdm2 but not p53 alterations in fresh human gliomas, *Clin Cancer Res* 7, 2788–2796, 2001.
88. Huang RP, Darland T, Okamura D, Mercola D, and Adamson ED, Suppression of v-sis-dependent transformation by the transcription factor, Egr-1, *Oncogene* 9, 1367–1377, 1994.
89. Della Ragione F, Cucciolla V, Borriello A, Della Pietra V, Racioppi L, Soldati G, Manna C, Galletti P, and Zappia V, Resveratrol arrests the cell division cycle at S/G2 phase transition, *Biochem Biophys Res Commun* 250, 53–58, 1998.
90. Della Ragione F, Cucciolla V, Criniti V, Indaco S, Borriello A, and Zappia V, p21Cip1 gene expression is modulated by Egr1: a novel regulatory mechanism involved in the resveratrol antiproliferative effect, *J Biol Chem* 278, 23360–23368, 2003.
91. Imai S, Armstrong CM, Kaeberlein M, and Guarente L, Transcriptional silencing and longevity protein Sir2 is an NAD-dependent histone deacetylase, *Nature* 403, 795–800, 2000.
92. Denu JM, Linking chromatin function with metabolic networks: Sir2 family of NAD(+)-dependent deacetylases, *Trends Biochem Sci* 28, 41–48, 2003.
93. Gasser SM and Cockell MM, The molecular biology of the SIR proteins, *Gene* 279, 1–16, 2001.
94. Tsukamoto Y, Kato J, and Ikeda H, Silencing factors participate in DNA repair and recombination in Saccharomyces cerevisiae, *Nature* 388, 900–903, 1997.
95. Mills KD, Sinclair DA, and Guarente L, MEC1-dependent redistribution of the Sir3 silencing protein from telomeres to DNA double-strand breaks, *Cell* 97, 609–620, 1999.
96. McAinsh AD, Scott-Drew S, Murray JA, and Jackson SP, DNA damage triggers disruption of telomeric silencing and Mec1p-dependent relocation of Sir3p, *Curr Biol* 9, 963–966, 1999.
97. Frye RA, Phylogenetic classification of prokaryotic and eukaryotic Sir2-like proteins, *Biochem Biophys Res Commun* 273, 793–798, 2000.
98. Frye RA, Characterization of five human cDNAs with homology to the yeast SIR2 gene: Sir2-like proteins (sirtuins) metabolize NAD and may have protein ADP-ribosyltransferase activity, *Biochem Biophys Res Commun* 260, 273–279, 1999.

99. North BJ, Marshall BL, Borra MT, Denu JM, and Verdin E, The human Sir2 ortholog, SIRT2, is an NAD+-dependent tubulin deacetylase, *Mol Cell* 11, 437–444, 2003.
100. Onyango P, Celic I, McCaffery JM, Boeke JD, and Feinberg AP, SIRT3, a human SIR2 homologue, is an NAD-dependent deacetylase localized to mitochondria, *Proc Natl Acad Sci USA* 99, 13653–13658, 2002.
101. Schwer B, North BJ, Frye RA, Ott M, and Verdin E, The human silent information regulator (Sir)2 homologue hSIRT3 is a mitochondrial nicotinamide adenine dinucleotide-dependent deacetylase, *J Cell Biol* 158, 647–657, 2002.
102. McBurney MW, Yang X, Jardine K, Hixon M, Boekelheide K, Webb JR, Lansdorp PM, and Lemieux M, The mammalian SIR2alpha protein has a role in embryogenesis and gametogenesis, *Mol Cell Biol* 23, 38–54, 2003.
103. McBurney MW, Yang X, Jardine K, Bieman M, Th'ng J, and Lemieux M, The absence of SIR2alpha protein has no effect on global gene silencing in mouse embryonic stem cells, *Mol Cancer Res* 1, 402–409, 2003.
104. Cheng HL, Mostoslavsky R, Saito S, Manis JP, Gu Y, Patel P, Bronson R, Appella E, Alt FW, and Chua KF, Developmental defects and p53 hyperacetylation in Sir2 homolog (SIRT1)-deficient mice, *Proc Natl Acad Sci USA* 100, 10794–10799, 2003.
105. Vaziri H, Dessain SK, Ng Eaton, E, Imai SI, Frye RA, Pandita TK, Guarente L, and Weinberg RA, hSIR2(SIRT1) functions as an NAD-dependent p53 deacetylase, *Cell* 107, 149–159, 2001.
106. Luo J, Nikolaev AY, Imai S, Chen D, Su F, Shiloh A, Guarente L, and Gu W, Negative control of p53 by Sir2alpha promotes cell survival under stress, *Cell* 107, 137–148, 2001.
107. Langley E, Pearson M, Faretta M, Bauer UM, Frye RA, Minucci S, Pelicci PG, and Kouzarides T, Human SIR2 deacetylates p53 and antagonizes PML/p53-induced cellular senescence, *EMBO J* 21, 2383–2396, 2002.
108. Motta MC, Divecha N, Lemieux M, Kamel C, Chen D, Gu W, Bultsma Y, McBurney M, and Guarente L, Mammalian SIRT1 represses forkhead transcription factors, *Cell* 116, 551–563, 2004.
109. Brunet A, Sweeney LB, Sturgill JF, Chua KF, Greer PL, Lin Y, Tran H, Ross SE, Mostoslavsky R, Cohen HY, Hu LS, Cheng HL, Jedrychowski MP, Gygi SP, Sinclair DA, Alt FW, and Greenberg ME, Stress-dependent regulation of FOXO transcription factors by the SIRT1 deacetylase, *Science* 303, 2011–2015, 2004.
110. van der Horst A, Tertoolen LG, de Vries-Smits LM, Frye RA, Medema RH, and Burgering BM, FOXO4 is acetylated upon peroxide stress and deacetylated by the longevity protein hSir2(SIRT1), *J Biol Chem* 279, 28873–28879, 2004.
111. Yeung F, Hoberg JE, Ramsey CS, Keller MD, Jones DR, Frye RA, and Mayo MW, Modulation of NF-kappaB-dependent transcription and cell survival by the SIRT1 deacetylase, *EMBO J* 23, 2369–2380, 2004.
112. Howitz KT, Bitterman KJ, Cohen HY, Lamming DW, Lavu S, Wood JG, Zipkin RE, Chung P, Kisielewski A, Zhang LL, Scherer B, and Sinclair DA, Small molecule activators of sirtuins extend Saccharomyces cerevisiae lifespan, *Nature* 425, 191–196, 2003.

113. Zhao K, Harshaw R, Chai X, and Marmorstein R, Structural basis for nicotinamide cleavage and ADP-ribose transfer by NAD-dependent Sir2 histone-protein deacetylases, *PNAS* 101, 8563–8568, 2004.
114. Jiang JC, Jaruga E, Repnevskaya MV, and Jazwinski SM, An intervention resembling caloric restriction prolongs life span and retards aging in yeast, *FASEB J* 14, 2135–2137, 2000.
115. Kenyon CA, Conserved regulatory mechanism for aging, *Cell* 105, 165–168, 2001.
116. Masoro EJ, Caloric restriction and aging: an update, *Exp Gerontol* 35, 299–305, 2000.
117. Koubova J and Guarente L, How does calorie restriction work?, *Genes Dev* 17, 313–321, 2003.
118. Sinclair DA, Paradigms and pitfalls of yeast longevity research, *Mech Ageing Dev* 123, 857–867, 2002.
119. Lin SJ, Defossez PA, and Guarente L, Requirement of NAD and SIR2 for life-span extension by calorie restriction in Saccharomyces cerevisiae, *Science* 289, 2126–2128, 2000.
120. Kaeberlein M, McVey M, and Guarente L, The SIR2/3/4 complex and SIR2 alone promote longevity in *Saccharomyces cerevisiae* by two different mechanisms, *Genes Dev* 13, 2570–2580, 1999.
121. Tissenbaum HA and Guarente L, Increased dosage of a sir-2 gene extends lifespan in *Caenorhabditis elegans*, *Nature* 410, 227–230, 2001.
122. Rogina B, Helfand SL, and Frankel S, Longevity regulation by Drosophila Rpd3 deacetylase and caloric restriction, *Science* 298, 1745, 2002.
123. Wood JG, Rogina B, Lavu S, Howitz K, Helfand SL, Tatar M, and Sinclair D, Sirtuin activators mimic caloric restriction and delay ageing in metazoans, *Nature* 430, 686–689, 2004.

11 Modulation of Gene Expression by Resveratrol

Bhagavathi A. Narayanan and
Narayanan K. Narayanan

CONTENTS

INTRODUCTION

Resveratrol is the key ingredient found in grapes, berries, and peanuts, and more than 70 species of plants contain this phenolic antioxidant. Resveratrol is most prevalent in the skin of grapes, and its presence in wine is thought to account for the fact that wine drinkers appear to have a lower risk for heart disease. At this point it is hard to say what dosage is needed for effective cancer prevention in humans as it surely depends on many factors, such as each person's genetic and physical makeup, total diet

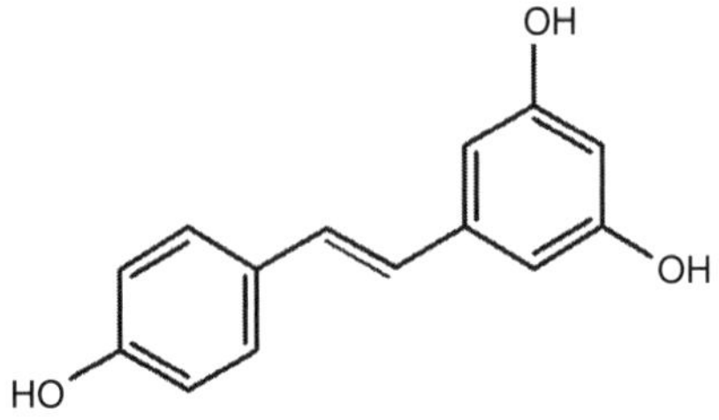

FIGURE 11.1 Structure of resveratrol.

composition, lifestyle, and environment. But it is safe to say that grapes and berries that contain resveratrol have several health benefits. Resveratrol and other phenolic antioxidants are attracting increasing attention for their potential as cancer chemopreventive agents [1–10]. Resveratrol is one of the most promising agents for human cancer prevention [11–18]. Resveratrol has antioxidant and antimutagenic activities that induce phase II drug-metabolizing enzymes; it mediates antiinflammatory effects and inhibits cyclooxygenase and hydroperoxidase functions [19–22].

The molecular structure of resveratrol is shown in Figure 11.1.

RESVERATROL AND MOLECULAR TARGETS

Cancer-preventing action by certain human nutrients derived from plants and fruits has been confirmed in various cell cultures and animal tumor models [11,23,24]. Our findings provide convincing evidence that resveratrol induces remarkable inhibitory effects against human cancer particularly against prostate cancer via modulation of cellular mechanisms associated with tumor initiation, promotion, and progression. Anticancer effects of phytochemicals such as resveratrol at the molecular level further suggest that these dietary components can modulate potential gene targets [5–7]. Hence the ideal chemoprevention strategy against human cancer with resveratrol would address genes related to both tumor initiating and promoting events, with less toxicity or side effects [1,3,4,25]. With regard to prostate cancer, the antiandrogenic effect induced by resveratrol is believed to regulate the expression of androgen receptor (AR) at the transcription level. In this connection, the AR ligand-binding domain (LBD) and N-terminal interacting proteins also play crucial roles [26,27]; however, not much is known about their regulation by resveratrol. Our earlier studies with resveratrol indicated an inhibition of a cofactor ARA 24 and PSA in LNCaP cells [15]. We have also reported that resveratrol represses different classes of androgen-regulated genes at the mRNA or protein level, including prostate-specific antigen, human glandular kallikrein-2, and the cyclin-dependent kinase inhibitor p27 in LNCaP cells [15]. This is likely attributable to a reduction in AR contents at the transcription level, inhibiting androgen-stimulated cell growth and gene expression. Recently, Mitchell et al. showed that resveratrol inhibits the expression and function of the androgen receptor in LNCaP cells [28].

In addition, resveratrol holds great promise for the prevention of several neoplastic disorders, including prostate cancer growth mechanisms, by targeting several cellular and molecular events involved in cancer growth [15,20,21,28–34]. These cited studies point to the importance of resveratrol as a potential chemopreventive agent that modulates gene targets and thus mediates anticancer effects. There are several key molecular targets of resveratrol that may participate in the modulation of early genes such as Nkx 3.1, AR, PSA, p63, p300, and other AR coactivators involved in prostate cancer development are yet to be investigated.

GENOMIC PROFILING OF RESVERATROL-MODULATED GENES AGAINST PROSTATE CANCER

Dietary phenolic compounds are known to elicit vital cellular responses such as cell cycle arrest, apoptosis, and differentiation by activating a cascade of molecular events. With the advent of microarray technology using high-density gene chips, we attempted to understand the impact of resveratrol on global genomic changes and particularly the changes associated with prostate cancer. We used total RNA extracted from LNCaP cells treated with resveratrol for 48 hours as described earlier [15]. Measurements on the levels of differentially expressed genes are presented as a scatter graph in Figure 11.2.

Upregulated genes amounted to 5.25% and downregulated genes accounted for 17.88% of all expressed genes out of a total of 2400 genes. To get reproducible results, the experiments were repeated with the same source of RNA from LNCaP cells treated with resveratrol and were compared with control cells. Genes expressed were confirmed with reverse transcription polymerase chain reaction (RT-PCR; Figure 11.3) using sequence-specific primers designed for selected genes.

A list of differentially expressed genes shows at least a moderate level of expression, varying more than twofold between the control and resveratrol-treated cells (Table 11.1). A twofold difference in the gene expression was preferred because (1) the signal intensities of nonhomologous plant genes as internal controls present in the array were used and (2) with respect to the intensity ratio of plant genes in our analysis, we decided to keep the signal-to-background ratio as 1.0. This ratio was used to normalize the entire array so that all the slides could be compared directly for a twofold increase. Although many of our microarray data analyses were performed with a cutoff for fivefold increase as preferred by other investigators [35], in our

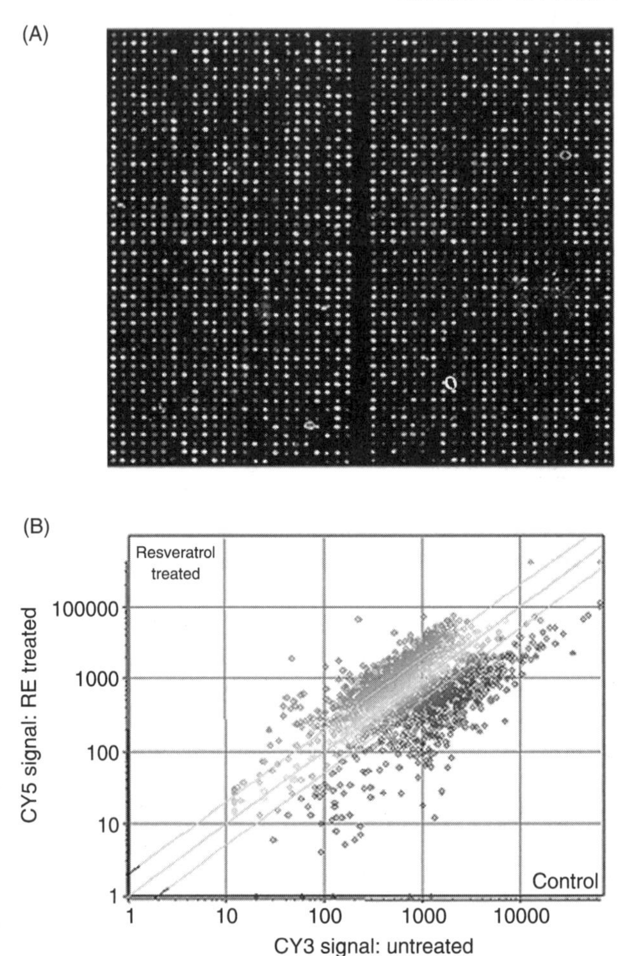

FIGURE 11.2 (A) Scanned image of hybridized human cDNA microarray containing 2400 genes. Total RNA from LNCaP cells treated with resveratrol for 48 hours was used for microarray analysis. (B) Scatter plot view showing the distribution of differentially expressed genes after 48 hours of resveratrol treatment (10^{-5} M) in LNCaP cells. Among the differentially expressed genes, upregulated (more than twofold) genes (5.25%) are shown above the median diagonal line and downregulated (less than twofold) genes (17.8%) are shown below the median diagonal line.

study we aimed to separate the genes of interest; for instance, with respect to p53-responsive genes a twofold increase permitted us to identify a large number of transcription factors related to their activation. Although a fivefold increase is a highly significant signal for the expressed genes, a twofold increase is an acceptable level of difference in expression as

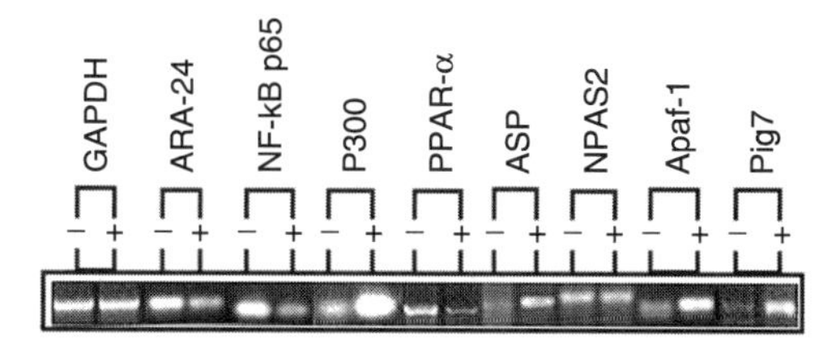

FIGURE 11.3 Validation of microarray data by RT-PCR. Agarose gel (2.5%) stained with ethidium bromide showing the amplified RT-PCR product with equal amount of RNA (2 μg). GAPDH was used as the internal control. Total RNA isolated from cells treated with resveratrol (10^{-5} M) for 48 hours and control cells were used for RT-PCR reactions using sequence-specific primers (+, treated; −, control). (From Narayanan BA, Re GG, and Narayanan NK, *Int J Cancer* 104, 204–212, 2003.)

TABLE 11.1
Differentially Expressed Genes in LNCaP Cells Exposed to Resveratrol Related to p53-Mediated Molecular Targets

Accession no.	Gene description	Mean Cy5/Cy3 ratio
U01877	p300 protein[a]	5.09 ± 1.21
AF013263	Apoptotic protease activating factor 1 (Apaf-1); cytochrome c-dependent activation of caspase-3[a]	4.40 ± 2.44
X63469	Transcription factor TFIIE beta	3.85 ± 0.15
L36645	Receptor protein-tyrosine kinase (HEK8)	3.82 ± 0.37
M35410	Insulin-like growth factor binding protein 2 (IGFBP2)	3.69 ± 0.20
X91257	Seryl-tRNA synthetase	3.60 ± 0.40
U81561	Protein tyrosine phosphatase receptor pi (PTPRP)	3.44 ± 0.52
Y11588	Apoptosis-specific protein[a]	3.41 ± 1.01
AF005654	Actin-binding double-zinc-finger protein (abLIM)	3.40 ± 0.46
U17714	Putative tumor suppressor ST13 (ST13)	3.38 ± 0.41
Z21943	Zinc finger protein	3.29 ± 0.30
U77970	Neuronal PAS2 (NPAS2)[a]	2.96 ± 0.66
X15722	Glutathione reductase	2.94 ± 0.49
U62433	Nicotinic acetylcholine receptor alpha4	2.87 ± 0.23
L13738	p21 (WAF1/CIP1)	2.70 ± 0.52
U13738	Cysteine protease CPP32 isoform beta; interleukin 1-beta converting enzyme; apoptotic protein	2.67 ± 0.90
AF012126	Zinc finger protein 198	2.64 ± 0.58
U09413	Zinc finger protein 135	2.46 ± 0.72
X62570	IFp53	2.40 ± 0.40
S78085	PDCD-2 factor[a]	2.38 ± 0.94
AF010312	p53 induced Pig 7[a]	2.23 ± 0.24
U22398	CdK-inhibitor p57KIP2[a]	2.16 ± 0.74

(*continued*)

TABLE 11.1
Continued

Accession no.	Gene description	Mean Cy5/Cy3 ratio
U23765	Bak protein; induction of apoptosis[a]	2.14 ± 0.56
U13737	Cysteine protease CPP32 isoform alpha; interleukin 1-beta converting enzyme; apoptotic protein[a]	2.13 ± 0.80
U91985	DNA fragmentation factor-45; triggers DNA fragmentation during apoptosis	2.13 ± 3.46
AF010313	p53 induced Pig 8	2.11 ± 1.09
AF010314	Pig10 (PIG10)	1.90 ± 1.25
AF016266	TRAIL receptor 2; binds cytotoxic ligand TRAIL; mediates apoptosis	1.87 ± 0.63
U15173	Bcl2 phosphorylated[a]	1.80 ± 0.98
U45879	Inhibitor of apoptosis protein 2; inhibition of apoptosis	1.57 ± 0.40
Z23116	Bcl-xS mRNA	1.11 ± 0.20
U83857	Aac11 (aac11); antiapoptosis	0.82 ± 0.49
U43399	14-3-3 protein epsilon isoform	0.78 ± 0.30
AF017987	Secreted apoptosis related protein 2 (SARP2)	0.68 ± 0.30
D86550	Serine/threonine protein kinase	0.50 ± 0.10
M34667	Phospholipase C-gamma	0.50 ± 0.17
U79269	Cyclin D protein kinase (CDPK)	0.49 ± 0.20
U68723	Checkpoint suppressor	0.48 ± 0.14
L19067	NF-kappa B transcription factor p65[a]	0.47 ± 0.13
D86970	TGFB1-antiapoptotic factor1	0.46 ± 0.20
M15798	ts11-G1 progression protein	0.46 ± 0.41
M54968	K-ras oncogene protein	0.46 ± 0.23
M93119	Zinc finger DNA binding motifs	0.46 ± 0.32
U17838	Zinc finger protein RIZ	0.45 ± 0.35
U28838	Transcription factor TFIIIB (hTFIIIB90)	0.45 ± 0.39
AB000468	Zinc finger protein RES4-26	0.44 ± 0.37
U67733	cGMP-stimulated phosphodiesterase PDE2A3	0.44 ± 0.14
AF030108	Regulator of G protein signaling (RGS5)	0.41 ± 0.16
M63488	Replication protein	0.41 ± 0.26
M64571	Microtubule associated protein 4	0.41 ± 0.17
S66427	RBP1, retinoblastoma binding protein 1	0.41 ± 0.28
U14577	Microtubule associated protein 1A	0.41 ± 0.20
U19251	Neuronal apoptosis inhibitory protein	0.39 ± 0.19
M86400	Phospholipase A2	0.38 ± 0.20
AF045581	BRCA1 protein	0.37 ± 0.60
AF062347	Zinc finger protein 216 splice variant	0.33 ± 0.24
M34668	PTP ase alpha mRNA	0.33 ± 0.14
J03778	Microtubule associated protein tau	0.31 ± 0.20
L07592	Peroxisome proliferator activated receptor (PPAR)	0.30 ± 0.17
U17040	Prostate specific antigen	0.10 ± 0.09
AF052578	Androgen receptor associated protein 24 (ARA24)[a]	0.01 ± 0.02

[a]RT-PCR confirmed.
From Narayanan BA, Re GG, and Narayanan NK, *Int J Cancer* 104, 204–212, 2003.

indicated by several earlier studies [36,37] within a 95% confidence interval as in the present study.

RESVERATROL-MODIFIED TRANSMEMBRANE PROTEINS, TRANSGLUTAMINASES, AND TRANSCRIPTION FACTORS

Androgen-responsive human prostate cancer cells LNCaP exposed to resveratrol modulated several set of genes related to transmembranes. Among a total of 246 expressed genes, 18.29% of the genes were transmembrane proteins comprising subsets of genes in the category of G protein-coupled receptors, endothelial cell protein receptors, integral membrane proteins, transmembrane secretary components, transmembrane glycoproteins, and cell surface glycoproteins. A large number of transmembrane proteins were downregulated by resveratrol, although their biological and functional significance is not well known. Among transcription factors a higher percentage of clones are differentially regulated. Nuclear factors such as NF-κB p50 and NF-κB p65 were downregulated by resveratrol (Figure 11.4).

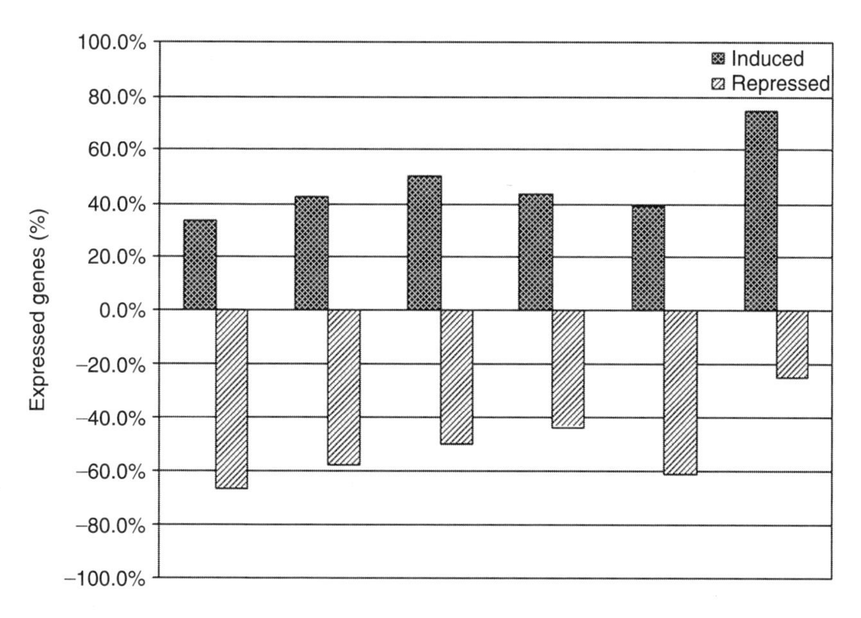

FIGURE 11.4 Histogram indicating the induction or repression of gene clusters after 48 hours of treatment with EA and resveratrol in LNCaP cells. (A) EA treatment; (B) resveratrol treatment (OG, oncogenes; PK, protein kinases; GP, glycoproteins; TMP, transmembrane proteins; TF, transcription factors; TG, transglutaminases). (From Narayanan BA, Narayanan KN, Stoner GD, and Bullock BP, *Life Sci* 70, 1–19, 2002.)

EFFECT OF RESVERATROL ON GROWTH FACTORS AND HORMONES

Androgen-responsive human prostate cancer cells LNCaP exposed to resveratrol indicated differential expression of transforming growth factor beta (TGFβ) and insulin-like growth factor (IGF), receptors, and growth factor binding proteins in conjunction with the coregulation of several related transcription factors, suggests their positive role in inducing cell cycle arrest and apoptosis in a p53-independent manner. Induction of TGFβ pathway by phenolic compounds is strengthened by the observed upregulation of cell cycle regulatory and apoptosis-inducing genes, along with an increase in the expression of TGFβ, IGF1 and 2, and binding proteins. Plant phenol-mediated cell cycle arrest in concert with the induction of cyclin-dependent kinase (cdk)-inhibitor p21 and downregulation of IGF2 [14] is likely related to orchestrated regulation which leads to growth arrest and apoptosis in human prostate cancer cells.

In vitro research and preclinical studies strongly support the anticancer effects of resveratrol by inducing cell cycle arrest, inhibiting DNA synthesis, and inducing apoptosis in human prostate cancer cells, and thus it holds great promise for development as a chemopreventive agent for prostate cancer [3–7,11,17,23,25]. Resveratrol is strongly linked with AR regulation. Recent studies by Mitchell et al. [28] and Hsieh and Wu [38] have shown that resveratrol represses several classes of androgen-upregulated genes at the mRNA level in LNCaP cells. Our earlier studies with resveratrol indicated an inhibition of the AR coactivator ARA-24 in LNCaP cells, consistent with previous reports [15].

RESVERATROL-INDUCED p53-MEDIATED MOLECULAR TARGETS

Numerous studies provide evidence for the anticarcinogenic activity of resveratrol [17,19,28,29,38–51], but the precise mechanisms involved in the modulation of oncogenic precursors of prostate carcinogenesis remain to be elucidated. Earlier studies indicated that the p53 protein is stabilized and activated after exposure of mammalian cells to DNA-damaging agents [52]. However, it is not known whether p53 activation by resveratrol follows the same pathway as that initiated by other agents that induce G1 arrest and apoptosis. There are indications that the nature of the p53 response depends on the levels of the p53 protein, the type of inducing agent, and the cell type employed. Significant cellular death was observed only when several of the genes controlled by p53 were expressed in concert, suggesting that p53 needs to activate parallel apoptotic pathways to induce programmed cell death [53,54].

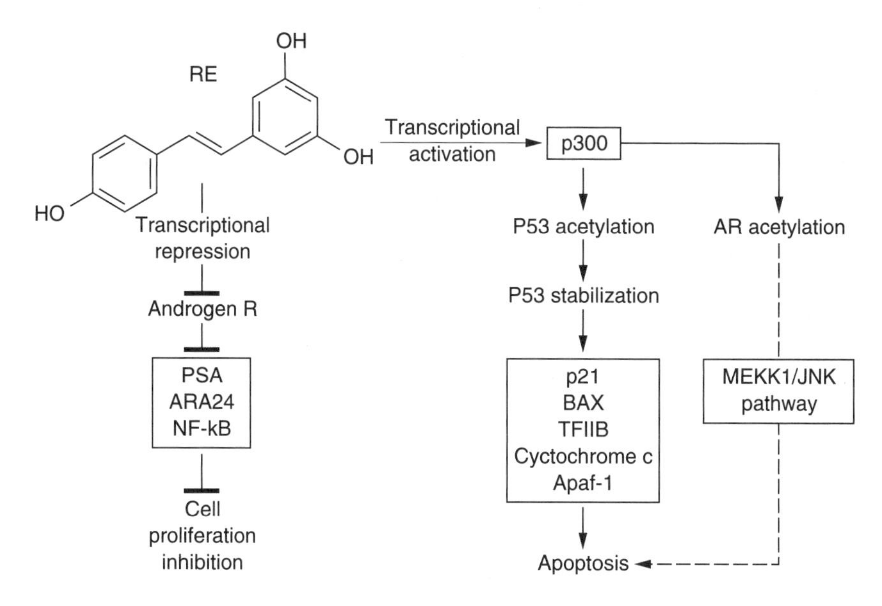

FIGURE 11.5 Mechanistic model illustrating the predicted cascade of molecular events elicited by resveratrol in LNCaP cells. AR conveys its transcription function by recruiting several coactivators shown in the left-hand box and induces cell proliferation. However, the functions of these coactivators are diverse and depend on the status of AR or p53 (wild-type or mutant). Resveratrol-mediated transcriptional repression and activation of AR and p53 target genes is predicted to be centered around p300. Consistent with earlier reports on AR acetylation and our observations, we predict that resveratrol-mediated p300 and p53 activation is associated with the expression of several proapoptotic genes. The dashed line indicates the pathway yet to be confirmed. (From Narayanan BA, Re GG, and Narayanan NK, *Int J Cancer* 104, 204–212, 2003.)

To delineate the complete cascade of molecular events in response to resveratrol treatment of prostate cancer requires a comprehensive study on gene expression at the transcription level. We employed human cDNA microarray analysis to obtain a genetic profiling of p53-targeted genes. We focused on determining whether resveratrol activates a cascade of p53-directed genes and transcription factors that are involved in apoptosis mechanism(s) in prostate cancer cells (Figure 11.5). In addition, we examined the p53-activated apoptotic pathway in conjunction with the modification of acetyltransferase p300 and caspase activator Apaf-1 that is induced by resveratrol in the LNCaP androgen-sensitive prostate cancer cell line.

RESVERATROL MODULATES DNA METHYLATION: INSIGHT FROM DNA MICROARRAY ANALYSIS

Dietary factors that influence DNA methylation and the risk for cancer development are evident from preclinical and clinical studies. Although a phytoestrogen, genistein, is believed to have an effect on DNA methylation in prostate carcinogenesis, the underlying mechanism is not known. There

TABLE 11.2
Genes Associated with Methylation Expressed in LNCaP Cells Exposed to Resveratrol

Genbank accession no.	Level of c expression (ratio)	Gene description
U01877	2.98	p300 protein
X68836	2.15	*S*-adenosylmethionine synthetase
X63692	0.34	DNA (cytosin-5) methyltransferase
AJ224442	0.50	Putative methyltransferase
Y10807	0.32	Arginine methyltransferase
X99209	0.69	Arginine methyltransferase
M11058	1.99	3-Hydroxy-3-methylglutaryl coenzyme A reductase
L25798	0.44	3-Hydroxy-3-methylglutaryl coenzyme A synthase
U23942	0.40	Lanosterol 14-demethylase cytochrome P450
D55653	1.54	Lanosterol 14-demethylase
D13892	3.03	Carboxyl methyltransferase
Z49878	0.27	Guanidinoacetate *N*-methyltransferase
J04031	0.21	Methylenetetrahydrofolate dehydrogenase
M58525	1.25	Catechol-*O*-methyltransferase (COMT)
U12387	0.49	Thiopurine methyltransferase (TPMT)
U12387	0.49	Thiopurine methyltransferase (TPMT)
X63563	1.61	RNA polymerase II 140 kDa subunit
J04965	0.25	RNA polymerase II 23 kD subunit (POLR2)
X64037	1.82	RNA polymerase II associated protein RAP74
Z47087	4.48	RNA polymerase II elongation factor-like protein
Z47727	0.96	RNA polymerase II subunit
L37127	8.20	RNA polymerase II
U75276	0.13	TFIIB related factor hBRF
M76766	2.62	Transcription factor (TFIIB)
U28838	0.29	Transcription factor TFIIIB 90 kDa
U02619	0.77	TFIIIC Box B-binding subunit
U18062	2.01	TFIID subunit TAFII55
X63469	3.85	Transcription factor TFIIE beta
Y07595	3.04	Transcription factor TFIIH 52 kD subunit
X57198	1.19	Transcription elongation factor (TFIIS)

are indications from our earlier studies that resveratrol may act via a newly envisioned mechanism to regulate methylation of DNA against prostate cancer in a dose-dependent manner. This involves the tumor suppressor gene p300, which plays a major role in acetylation and regulation of hormone receptors such as estrogen receptor (ER) and AR. An in-depth analysis of DNA microarray data indicated that several genes associated with DNA methylation have been modulated by resveratrol in prostate cancer LNCaP cells (Table 11.2). Results from differential expression of genes provide a precise pattern that defines the reprogramming of hundreds of genes in response to resveratrol. However, here we focus our attention on the influence of resveratrol on those genes that are related to methylation processes. Data presented in Table 11.2 demonstrate alterations in several enzymatic complexes including cytosine and arginin methyltransferases, along with p300 and several other transcription factors that could modify the chromatin architecture via methylation and acetylation processes.

EFFECT OF RESVERATROL ON METHYLATION OF p300 CpG SITES

To examine the methylation status of p300, we have treated three prostate cancer cell lines, namely LNCaP, DU145, and PC-3, with 10 μM of resveratrol for 48 hours; cells treated with DMSO served as a negative control as described earlier [15]. As in all of our earlier studies we have demonstrated the resveratrol-induced changes in global gene expression at 10 μM concentration, and also induced apoptosis in prostate cancer cells. Hence, we preferred to use the same nontoxic dose to determine the effect on the methylation pattern for p300.

All the three human prostate cancer cells types, LNCaP, DU145, and PC-3, examined for methylation changes of p300 CpG sites after treatment with resveratrol clearly indicated changes that were determined by PCR amplification of the CpG sites encompassing the promoter and exon 1 with DNA bisulfite modification with both methylated and unmethylated primers as depicted in Figure 11.6, suggesting that bisulfite modification was very clear and 100%, without any false positives.

RESVERATROL-MODULATED PROTEINS

In order to identify specific proteins that are altered or posttranslationally modified by resveratrol in cancer cells, we adopted the combination of two-dimensional sodium dodecylsulfate polyacrylamide gel electrophoresis (2D-SDS-PAGE) using immobilized pH gradients and mass spectrometry analysis and separated proteins solubilized from LNCaP cells treated with resveratrol and untreated. Figure 11.7A and Figure 11.7B show representa-

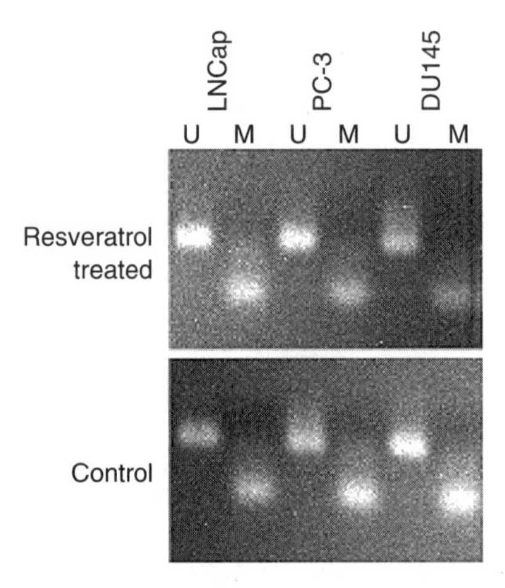

FIGURE 11.6 MSP analysis bands in the "U" lanes are products with unmethylated-specific primers and the bands in the "M" lanes are PCR products for methylation-specific primers. MSP analysis for the sites was performed in three prostate cancer cell lines, LNCaP, DU145, and PC-3, treated with 10 μM resveratrol and DMSO for control. The cells were treated for 48 hours. Trypsinized cells were washed with PBS and genomic DNA was isolated using a Dneasy DNA isolation kit (Qiagen). Bisulfite-modified DNA was used for MSP as a guide to differentiate ^{5m}C from C.

tive two-dimensional gels of untreated LNCaP cells and LNCaP cells treated with resveratrol, respectively. This approach revealed that an average of 225 polypeptide spots were matching in the control and resveratrol-treated cells. The overall differences in the expression level were calculated from spot percentages (individual spot density divided by total density of all measured spots). A total of 24 polypeptide spots showed a difference of $\geq$1.9 or $\leq$–1.9 were markedly increased or decreased in resveratrol-treated samples.

MODULATION OF PHOSPHOGLYCERATE MUTASE B BY RESVERATROL

Among the few selected differentially expressed candidate protein spots, we performed a mass spectrometric analysis for spot 178 that was expressed significantly in the control cells when compared to resveratrol-treated cancer cells. Although the pI and the MW range for spot 178 is closer to that of PSA, the mass spectrometric analysis identified spot 178 as a phosphoglycerate mutase (EC 5.4.2.1) that belongs to the binuclear metalloenzyme superfamily (Figure 11.8 and Table 11.3); this protein is acetylated at the N-terminus. Downregulation of phosphoglycerate mutase B by resveratrol is the significant finding of our study.

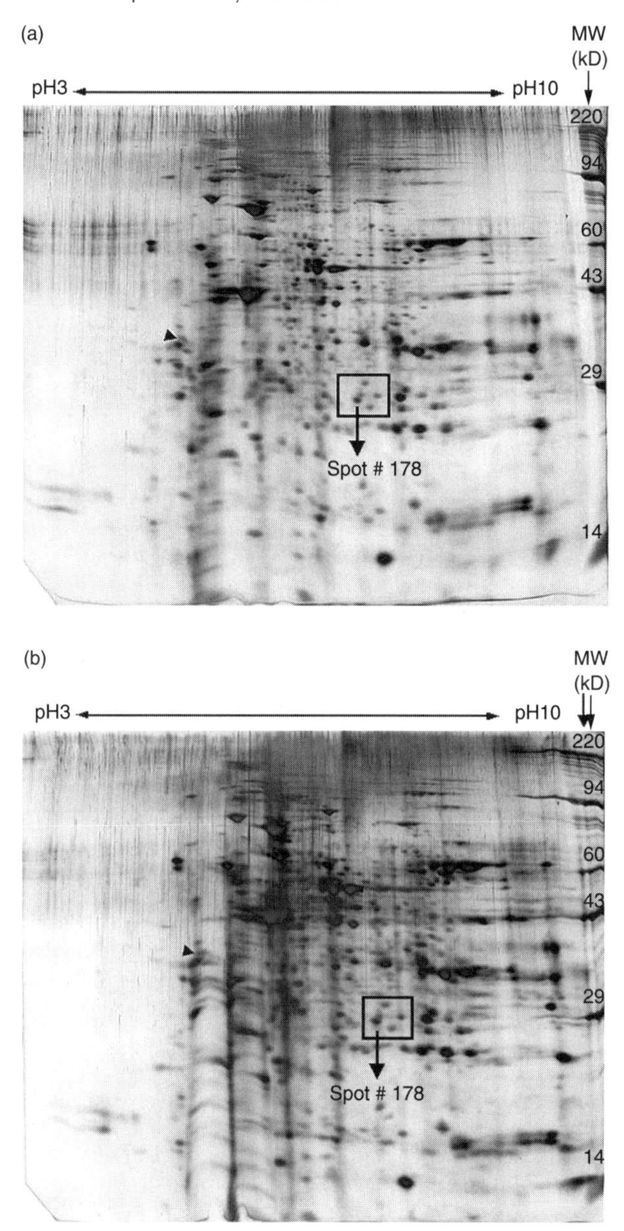

FIGURE 11.7 SDS-PAGE analysis. Two-dimensional gel electrophoresis was performed with total protein lysate extracted from LNCaP cells without and with resveratrol treatment (10^{-5} M) for 48 hours. Representative two-dimensional gels (silver stained) show the pattern of separated protein spots in (A) the control and (B) resveratrol-treated LNCaP cells at the 48-hour point. The square indicates spot 178 that was selected for mass spectrometric analysis. (From Narayanan NK, Narayanan BA, and Nixon DW, *Cancer Detection Prev* 28, 443–452, 2004.)

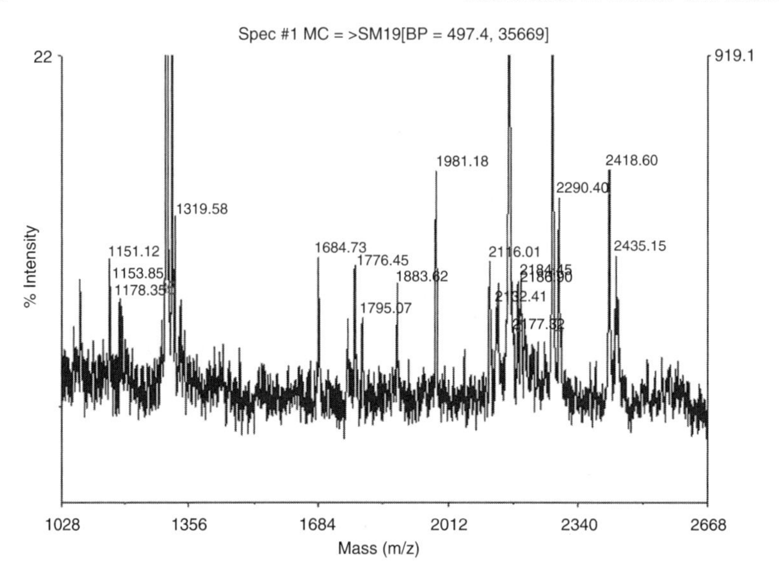

FIGURE 11.8 Mass spectrum of the major proteins of resveratrol-treated LNCaP cells. The proteins were identified on the basis of their molecular weights and peptide sequence. Peptide mass mapping of trypsin-digested proteins was analyzed by matrix-assisted laser desorption ionization (MALDI) mass spectrometry. (From Narayanan NK, Narayanan BA, and Nixon DW, *Cancer Detection Prev* 28, 443–452, 2004.)

TABLE 11.3
Summary of MALDI-MS Analysis Output of 2-D Spot 178

Protein identified	Mass observed (Da) $[M+H]^+$ obs.	Mass calculated (Da) $[M+H]^+$ calc.	Residues	Peptide sequence from database
Phosphoglycerate mutase B subunit	1151.12	1151.26	181–191	VLIAAHGNSLR
	1178.35	1178.53	1–10	MAAYKLVLIR
	1313.07	1313.38	11–21	HGESAWNLENR
	1684.73	1684.98	163–176	ALPFWNEEIVPQIK
	1795.07	1794.99	47–61	DAGYEFDICFTSVQK
	1883.62	1884.08	142–157	YADLTEDQLPSCESLK
	1981.18	1981.09	22–39	FSGWYDADLSPAGHEEAK
	2116.01	2116.5	223–240	NLKPIKPMQFLGDEETVR
	2418.6	2418.70	118–138	SYDVPPPPMEPDHPFYSNISK

Note: Values are averages. Sequence coverage = 53%.
From Narayanan NK, Narayanan BA, and Nixon DW, *Cancer Detection Prev* 28, 443–452, 2004.

We have previously identified the transcriptional profiles of genes involved in cell cycle regulation and apoptosis in prostate cancer cells exposed to resveratrol [55]. Knowing the potential role of several of these genes is crucial for cancer preventive effects; therefore, it is imperative to investigate the level of expression of their corresponding proteins. Towards this end, we used a two-pronged approach to examine the effect of resveratrol on metabolism-related proteins, particularly the enzymes involved in glycolysis. It is well known that in normal cells after glucose is broken down to pyruvic acid it is then carried into the mitochondria and totally combusted by some 12 enzymatic reactions into carbon dioxide and water. However, because cancer cells have a defective Krebs cycle, they must derive almost all of their energy needs from glycolysis. Crippling of the glycolytic pathway will literally starve the cancer cells to death [56–60]. cDNA microarray analysis revealed that 14 genes are specifically involved in the process of glycolysis as shown in Table 11.1. The application of 2D-SDS-PAGE followed by mass spectrometric analysis enabled us to characterize proteins that have been specifically altered by resveratrol [61]. The importance of such a comprehensive approach, along with cDNA microarray analysis, has been documented in several other studies that focus on the multistep process of carcinogenesis and on the identification of diagnostic parameters of prostate cancer [62–67]. We focus here on genes that are altered by resveratrol in LNCaP cells and we report for the first time a change in the expression of several genes involved in glycolysis.

The protein phosphoglycerate mutase (Figure 11.3) has conserved nucleophilic serine and metal-binding residues, yet the substrate-binding residues are not conserved [68]. This protein has been reported at a higher level in dermal fibroblast cells from healthy human subjects [69]. Phosphoglycerate mutase catalyzes the isomerization of 2- and 3-phosphoglycerates and is thus essential for glucose metabolism in most organisms [70]. The functional significance of this posttranslationally modified protein in resveratrol-treated LNCaP cells is yet to be investigated. Monoubiquitination of phosphoglycerate mutase B as well as formation of a noncovalent complex containing ubiquitin and phosphoglycerate mutase B are increased in colorectal cancer, which may suggest a potential pathophysiological event [1]. A decreased level of phosphoglycerate mutase isoenzymes was reported in breast carcinoma [71] indicating its differential expression. Modulation of phosphoglycerate mutase B by resveratrol in prostate cancer cells implies more than two different mechanisms, including the role in glycolysis, and a possible involvement of thiol groups and therefore mediation of protein–protein interactions [31]. Proteomic approaches to the investigations related to these modifications by resveratrol will reveal their clinical importance and unique functional mechanisms.

CONCLUSION

Gene expression analysis by DNA microarray presented here is part of our ongoing research on phenolic antioxidants against prostate cancer. Although a deeper biologic insight with respect to transcriptional regulation of specific genes is likely to develop from cDNA microarray analysis at multiple time points with multiple cell lines in a dose-dependent manner, some of the findings presented in this chapter are intended to give an overall view on the genomic response of cancer cells to resveratrol treatment which may vary with different doses and cell types. We preferred such a focused approach to identify those molecular targets with a specific dose in a given time that may lead to a different phenotype. It is very important to mention here that the extensive alteration in the expression of transglutaminases in prostate cancer cells by resveratrol is a first report. At least seven distinct transglutaminases have been characterized in mammals and are synthesized during terminal differentiation and cell death, suggesting that resveratrol-induced apoptosis is associated with differential expression of transglutaminases. Regulation of p300 by resveratrol indicates its role as an important component of p53 signaling, and thus providing new insight into the mechanism of cell growth inhibition. Exploring transcriptional and posttranslational modification of target genes and proteins modified by resveratrol may provide more insight into the differences in RNA versus protein expression. We are currently focusing on overcoming the limitations that are associated with detecting mRNA and the corresponding protein interactions via improved high-throughput genomic and proteomic approaches that relate the findings more accurately than confirming with several other methods.

ACKNOWLEDGMENTS

This study was supported in part by USPHS grant CA-17613 (NCI Cancer Center Grant) and AICR Grant No. 01A015. The authors thank Ilse Hoffmann for editing the manuscript.

REFERENCES

1. Wattenberg LW, Chemoprevention of cancer, *Cancer Res* 45, 1–8, 1985.
2. Wattenberg LW, An overview of chemoprevention: current status and future prospects, *Proc Soc Exp Biol Med* 216, 133–141, 1997.
3. Meyskens FL, Strategy for chemoprevention of human breast cancer, *Cancer Invest* 6, 609–613, 1988.
4. Meyskens FL, Cancer chemoprevention: progress and promise, *J Natl Cancer Inst* 91, 563–564, 1999.

5. Sporn MB, Carcinogenesis and cancer: different perspectives on the same disease, *Cancer Res* 51, 6215–6218, 1991.
6. Kelloff GJ, Boone CW, Crowell JA, Nayfield SG, Hawk E, Malone WF, Steele VE, Lubet RA, and Sigman CC, Risk biomarkers and current strategies for cancer chemoprevention, *J Cell Biochem Suppl* 25, 1–14, 1996.
7. Kelloff GJ, Hawk ET, Karp JE, Crowell JA, Boone CW, Steele VE, Lubet RA, and Sigman CC, Progress in clinical chemoprevention, *Semin Oncol* 24, 241–252, 1997.
8. Mukhtar H, Re: consumption of black tea and cancer risk: a prospective cohort study, *J Natl Cancer Inst* 88, 768, 1996.
9. Mukhtar H and Agarwal R, Skin cancer chemoprevention, *J Invest. Dermatol Symp Proc* 1, 209–214, 1996.
10. Stoner GD and Morse MA, Isothiocyanates and plant polyphenols as inhibitors of lung and esophageal cancer, *Cancer Lett* 114, 113–119, 1997.
11. Pezzuto JM, Plant-derived anticancer agents, *Biochem Pharmacol* 53, 121–133, 1997.
12. Dragsted LO, Natural antioxidants in chemoprevention, *Arch Toxicol Suppl* 20, 209–226, 1998.
13. Narayanan BA, Geoffroy O, Willingham CM, Re GG, and Nixon DW, p53/p21($^{WAF1/CIP1}$) expression and its possible role in G1 arrest and apoptosis in ellagic acid-treated cancer cells, *Cancer Lett* 136, 215–221, 1999.
14. Narayanan BA and Re GG, IGF II down regulation associated cell cycle arrest in colon cancer cells exposed to phenolic antioxidant ellagic acid, *Anticancer Res* 21, 359–364, 2001.
15. Narayanan BA, Narayanan KN, Stoner GD, and Bullock BP, Interactive gene expression pattern in prostate cancer cells exposed to phenolic antioxidants, *Life Sci* 70, 1–19, 2002.
16. Steele VE, Kelloff GJ, Balentine D, Boone CW, Mehta R, Bagheri D, Sigman CC, Zhu SY, and Sharma S, Comparative chemopreventive mechanisms of green tea, black tea, and selected polyphenol extracts measured by *in vitro* bioassays, *Carcinogenesis* 21, 63–67, 2000.
17. Agarwal C, Sharma Y, and Agarwal R, Anticarcinogenic effect of a polyphenolic fraction isolated from grape seeds in human prostate carcinoma DU145 cells: modulation of mitogenic signaling and cell cycle regulators and induction of G1-arrest and apoptosis, *Mol Carcinog* 28, 129–138, 2001.
18. Kong AN, Yu R. Hebbar V, Chen C, Owuor E, Hu R, Ee R, and Mandlekar S, Signal transduction events elicited by cancer prevention compounds, *Mutation Res* 480–481, 231–241, 2000.
19. Jang M, Cai L, Udeani GO, Slowing KV, Thomas CF, Beecher CW, Fong HH, Farnsworth NR, Kinghorn AD, Mehta RG, Moon RC, and Pezzuto JM, Cancer chemopreventive activity of resveratrol, a natural product derived from grapes, *Science* 275, 218–220, 1997.
20. Bhat KP, Lanvit D, Christov K, Mehta RG, Moon RC, and Pezzuto JM, Estrogenic and antiestrogenic properties of resveratrol in mammary tumor models, *Cancer Res* 61, 7456–7463, 2001.
21. Bhat KP and Pezzuto JM, Cancer chemopreventive activity of resveratrol, *Ann NY Acad Sci* 957, 210–229, 2002.
22. Mouria M, Gukovskaya AS, Jung Y, Buechler P, Hines OJ, Reber HA, and Pandol SJ, Food-derived polyphenols inhibit pancreatic cancer growth through

mitochondrial cytochrome C release and apoptosis, *Int J Cancer* 98, 761–769, 2002.
23. Dragsted LO, Strube M, and Larsen JC, Cancer-protective factors in fruits and vegetables: biochemical and biological background, *Pharmacol Toxicol* 72 (Suppl 1), 116–135, 1993.
24. Kong AN, Yu R, Hebbar V, Chen C, Owuor E, Hu R, Ee R, and Mandlekar S, Signal transduction events elicited by cancer prevention compounds, *Mutation Res* 480–481, 231–241, 2001.
25. Bertram JS, Kolonel LN, and Meyskens FL Jr, Rationale and strategies for chemoprevention of cancer in humans, *Cancer Res* 47, 3012–3031, 1987.
26. Brady ME, Ozanne DM, Gaughan L, Waite I, Cook S, Neal DE, and Robson CN, Tip60 is a nuclear hormone receptor coactivator, *J Biol Chem* 274, 17599–17604, 1999.
27. Sampson ER, Yeh SY, Chang HC, Tsai MY, Wang X, Ting HJ, and Chang C, Identification and characterization of androgen receptor associated coregulators in prostate cancer cells, *J Biol Regul Homeost Agents* 15, 123–139, 2001.
28. Mitchell SH, Zhu W, and Young CYF, Resveratrol inhibits the expression and function of the androgen receptor in LNCaP prostate cancer cells, *Cancer Res* 59, 5892–5895, 1999.
29. She QB, Bode AM, Ma WY, Chen NY, and Dong Z, Resveratrol-induced activation of p53 and apoptosis is mediated by extracellular-signal-regulated protein kinases and p38 kinase, *Cancer Res*; 61, 1604–1610, 2001.
30. She QB, Huang C, Zhang Y, and Dong Z, Involvement of c-jun NH(2)-terminal kinases in resveratrol-induced activation of p53 and apoptosis, *Mol Carcinog* 33, 244–250, 2002.
31. Su BN, Cuendet M, Hawthorne ME, Kardono LB, Riswan S, Fong HH, Mehta RG, Pezzuto JM, and Kinghorn AD, Constituents of the bark and twigs of Artocarpus dadah with cyclooxygenase inhibitory activity, *J Nat Prod* 65, 163–169, 2002.
32. Revel A, Raanani H, Younglari E, Xu J, Han R, Savouret JF, and Casper RF, Resveratrol, a natural aryl hydrocarbon receptor antagonist, protects sperm from DNA damage and apoptosis caused by benzo(a)pyrene, *Reprod Toxicol* 15, 479–486, 2001.
33. Jang M and Pezzuto JM, Effects of resveratrol on 12-O-tetradecanoylphorbol-13-acetate-induced oxidative events and gene expression in mouse skin, *Cancer Lett* 134, 81–89, 1998.
34. Jang M and Pezzuto JM, Cancer chemopreventive activity of resveratrol, *Drugs Exp Clin Res* 25, 65–77, 1999.
35. Waghray A, Feroze F, Schober MS, Yao F, Wood C, Puravs E, Krause M, Hanash S, and Chen YQ, Identification of androgen-regulated genes in the prostate cancer cell line LNCaP by serial analysis of gene expression and proteomic analysis, *Proteomics*, 1, 1327–1338, 2001.
36. Okabe H, Satoh S, Kato T, Kitahara O, Yanagawa R, Yamaoka Y, Tsunoda T, Furukawa Y, and Nakamura Y, Genome-wide analysis of gene expression in human hepatocellular carcinomas using cDNA microarray: identification of genes involved in viral carcinogenesis and tumor progression, *Cancer Res* 61, 2129–2137, 2001.

37. Belbin TJ, Singh B, Barber I, Socci N, Wenig B, Smith R, Prystowsky MB, and Childs G, Molecular classification of head and neck squamous cell carcinoma using cDNA microarrays, *Cancer Res* 62, 1184–1190, 2002.
38. Hsieh TC and Wu JM, Grape-derived chemopreventive agent resveratrol decreases prostate-specific antigen (PSA) expression in LNCaP cells by an androgen receptor (AR)-independent mechanism, *Anticancer Res* 20, 225–228, 2000.
39. Draczynska-Lusiak B, Chen YM, and Sun AY, Oxidized lipoproteins activate NF-kappaB binding activity and apoptosis in PC-12 cells, *Neuroreport* 9, 527–532, 1998.
40. Kawada N, Seki S, Inoue M, and Kuroki T, Effect of antioxidants, resveratrol, quercetin and N-acetyl-cysteine on the functions of cultured rat hepatic stellate cells and Kupffer cells, *Hepatology* 27, 1265–1274, 1998.
41. Rosenberg RS, Grass L, Jenkins DJ, Kendall CW, and Diamandis EP, Modulation of androgen and progesterone receptors by phytochemicals in breast cancer cell lines, *Biochem Biophys Res Commun* 248, 935–939, 1998.
42. Ciolino HP, Daschner PJ, and Yeh GC, Resveratrol inhibits transcription of CYP1A1 *in vitro* by preventing activation of the aryl hydrocarbon receptor, *Cancer Res* 58, 5707–5712, 1998.
43. Ciolino HP and Yeh GC, Inhibition of aryl hydrocarbon-induced cytochrome p-450 1A1 enzyme activity and CYP1A1 expression by resveratrol, *Mol Pharmacol* 56, 760–767, 1999.
44. Hsieh TC and Wu JM, Differential effects of growth, cell cycle arrest, and induction of apoptosis by resveratrol in human prostate cancer cell lines, *Exp Cell Res* 249, 109–115, 1999.
45. Tredici G, Miloso M, Nicolini G, Galbiati S, Cavaletti G, and Bertelli A, Resveratrol, map kinases and neuronal cells: might wine be a neuroprotectant?, *Drugs Exp Clin Res* 25, 99–103, 1999.
46. Ulsperger E, Hamilton G, Raderer M, Baumgartner G, Hejna M, Hoffmann O, and Mallinger R, Resveratrol pretreatment desensitizes AHTO-7 human osteoblasts to growth stimulation in response to carcinoma cell supernatants, *Int J Oncol* 15, 955–959, 1999.
47. Lu R and Serrero G, Resveratrol, a natural product derived from grape, exihibits antiestrogenic activity and inhibits the growth of human breast cancer cells, *J Cell Physiol* 179, 297–304, 1999.
48. Damianaki A, Bakogeorgou E, Kampa M, Notas G, Hatzoglou A, Panagiotou S, Gemetzi C, Kouroumalis E, Martin PM, and Castanas E, Potent inhibitory action of red wine polyphenols on human breast cancer cells, *J Cell Biochem* 78, 429–441, 2000.
49. Manna SK, Mukhopadhyay A, and Aggarwal BB, Resveratrol suppresses TNF-induced activation of nuclear transcription factors NF-kappa B, activator protein 1 and apoptosis: potential role of reactive oxygen intermediates and lipid peroxidation, *J Immunol* 164, 6509–6519, 2000.
50. Stewart JR, Christman KL, and and O'Brian CA, Effects of resveratrol on the autophosphorylation of phorbol ester-responsive protein kinases: inhibition of protein kinase D but not protein kinase C isozyme autophosphorylation, *Biochem Pharmacol* 60, 1355–1359, 2000.

51. Huang SS, Tsai MC, Chih CL, Hung LM, and Tsai SK, RE reduction of infarct size in Long-Evans rats subjected to focal cerebral ischemia, *Life Sci* 69, 1057–1065, 2001.
52. Stewart ZA and Pietenpol AP, p53 signaling and cell cycle checkpoints, *Chem Res Toxicol* 14, 243–263, 2000.
53. Burns TF and El-Diery WS, The p53 pathway and apoptosis, *J Cell Physiol* 181, 231–239, 1999.
54. Gu W, Shi XL, and Roeder RG, Synergistic activation of transcription by CBP and p53, *Nature* 387, 819–823, 1997.
55. Narayanan BA, Re GG, and Narayanan NK, Differential expression of genes induced by resveratrol in LNCaP cells: P53-mediated molecular targets, *Int J Cancer* 104, 204–212, 2003.
56. Koukourakis MI, Giatromanolaki A, and Sivridis E, Lactate dehydrogenase isoenzymes 1 and 5: differential expression by neoplastic and stromal cells in non-small cell lung cancer and other epithelial malignant tumors, *Tumor Biol* 24, 199–202, 2003.
57. Lee MG and Pedersen PL, Glucose metabolism in cancer: importance of transcription factor-DNA interactions within a short segment of the proximal region of the type II hexokinase promoter, *J Biol Chem* 278, 41047–41058, 2003.
58. Rivenzon-Segal D, Boldin-Adamsky S, Seger D, Seger R, and Degani H, Glycolysis and glucose transporter 1 as markers of response to hormonal therapy in breast cancer, *Int J Cancer* 107, 177–182, 2003.
59. Boros L G, Brackett DJ, and Harrigan GG, Metabolic biomarker and kinase drug target discovery in cancer using stable isotope-based dynamic metabolic profiling (SIDMAP), *Curr Cancer Drug Targets* 3, 445–453, 2003.
60. Gatenby RA and Gawlinski ET, The glycolytic phenotype in carcinogenesis and tumor invasion: insights through mathematical models, *Cancer Res* 63, 3847–3854, 2003.
61. Narayanan NK, Narayanan BA, and Nixon DW, Resveratrol induced cell growth inhibition and apoptosis is associated with modulation of phosphoglycerate mutase-B in human prostate cancer cells: 2D-SDS-PAGE and mass spectrometry evaluation, *Cancer Detection Prev*, 28, 443–452, 2004.
62. Meehan KL, Holland JW, and Dawkins HJS, Proteomic analysis of normal and malignant prostate tissue to identify novel proteins lost in cancer, *Prostate* 50, 54–63, 2002.
63. Ahram M, Best CJ, Flaig MJ, Gillespie JW, Leiva IM, Chuaqui RF, Zhou G, Shu H, Duray PH, Linehan WM, Raffeld M, Ornstein DK et al, Proteomic analysis of human prostate cancer, *Mol Carcinog* 33, 9–15, 2002.
64. Wulfkuhle JD, McLean KC, Paweletz CP, and Sgroi DC, New approaches to proteomic analysis of breast cancer, *Proteomics* 1, 1205–1215, 2001.
65. Verma M, Wright GL Jr, Hanash SM, Gopal-Srivastava R, and Srivastava S, Proteomic approaches within the NCI early detection research network for the discovery and identification of cancer biomarkers, *Ann NY Acad Sci* 945, 103–115, 2001.
66. Issaq HJ, Conrads TP, Janini GM, and Veenstra TD, Methods for fractionation, separation and profiling of proteins and peptides, *Electrophoresis* 23, 3048–3061, 2002.

67. Leman ES and Getzenberg RHJ, Nuclear matrix proteins as biomarkers in prostate cancer, *Cell Biochem* 86, 213–223, 2002.
68. Graham DE, Xu HM, and White RH, A divergent archaeal member of the alkaline phosphatase binuclear metalloenzyme superfamily has phosphoglycerate mutase activity, *FEBS Lett* 517, 190–194, 2002.
69. Boraldi F, Bini L, Liberatori S, Armini A, Pallini V, Tiozzo R, Pasquali-Ronchetti I, and Quaglino D, Proteome analysis of dermal fibroblasts cultured *in vitro* from human healthy subjects of different ages, *Proteomics* 3, 917–929, 2003.
70. Rigden DJ, Lamani E, Mello LV, Littlejohn JE, and Jedrzejas MJ, Insights into the catalytic mechanism of cofactor-independent phosphoglycerate mutase from X-ray crystallography, simulated dynamics and molecular modeling, *J Mol Biol* 328, 909–920, 2003.
71. Durany N, Joseph J, Jimenez OM, Climent F, Fernandez PL, Rivera F, and Carreras J, Phosphoglycerate mutase, 2,3-bisphosphoglycerate phosphatase, creatine kinase and enolase activity and isoenzymes in breast carcinoma, *Br J Cancer* 82, 20–27, 2000.

12 Resveratrol and Prostaglandin Biosynthesis

Lawrence M. Szewczuk and Trevor M. Penning

CONTENTS

INTRODUCTION

RESVERATROL

Red wine contains high levels of polyphenolic compounds, and daily consumption of this beverage has been correlated with a reduced risk of coronary heart disease [1]. This phenomenon is widely known as the "French paradox" and takes the form of an inverse association between the consumption of red wine and cardiac mortality in populations considered at high risk based on their diet [1]. Of the agents present in red wine, resveratrol (I; Figure 12.1) is of particular interest since it has cancer chemopreventive, antiinflammatory, and cardioprotective properties [2]. It is present at concentrations as high as 100 μM in red wine [3], and has been the focus of intense research aimed at elucidating its mechanisms of action. Resveratrol inhibits prostaglandin (PG) biosynthesis [2,4]. These lipid mediators are implicated in cancer, inflammation, and vascular homeostasis; therefore the ability of resveratrol to inhibit PG biosynthesis could account for its pharmacological properties. Our investigations identified resveratrol as a selective mechanism-based inactivator of the peroxidase activity of cyclooxygenase (COX)-1 only, and we propose that this mechanism may account for the cardioprotective effects of red wine [5]. In the sections that follow we will first review PG biosynthesis, COX

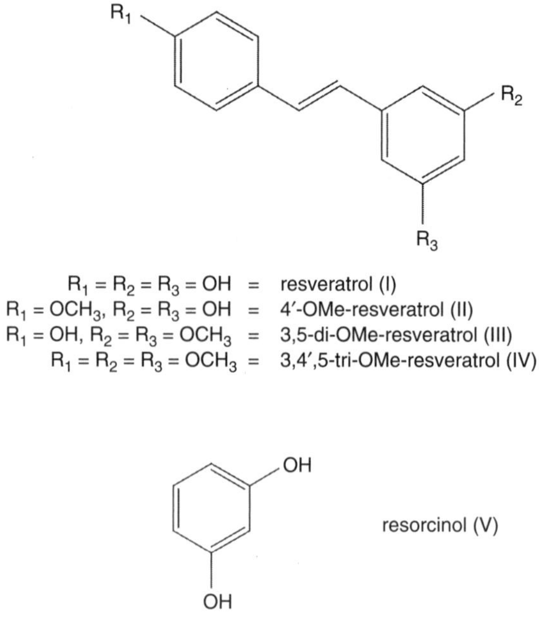

R_1 R_2 R_3

$R_1 = R_2 = R_3 = OH$ = resveratrol (I)
$R_1 = OCH_3$, $R_2 = R_3 = OH$ = 4′-OMe-resveratrol (II)
$R_1 = OH$, $R_2 = R_3 = OCH_3$ = 3,5-di-OMe-resveratrol (III)
$R_1 = R_2 = R_3 = OCH_3$ = 3,4′,5-tri-OMe-resveratrol (IV)

OH OH resorcinol (V)

FIGURE 12.1 Structure of resveratrol and its analogs.

enzymology and pharmacology, and then describe the mechanism of action of resveratrol.

Prostaglandin Biosynthesis

Prostaglandin H_2 synthases COX-1 and COX-2 catalyze the conversion of arachidonic acid (AA) to PGH_2 (PG biosynthesis is extensively reviewed by Funk [6] and Rouzer and Marnett [7]). This represents the first committed step in the biosynthesis of all PGs and thromboxanes (TXs). PGH_2 is subsequently converted to PGD_2, PGE_2, $PGF_{2\alpha}$, TXA_2, and prostacyclin (PGI_2) by the action of specific reductases and isomerases. These bioactive lipids mediate symptoms of inflammation like pyresis, edema, swelling, and pain (PGD_2, PGE_2, PGF_2); they can act as vasoconstrictors ($PGF_{2\alpha}$, TXA_2) and promote platelet aggregation (TXA_2) or they can act as vasodilators and inhibit platelet aggregation (PGI_2). All of the prostanoids are derived from the common precursor PGH_2 thus making COX an ideal drug target for modulating their production. There are two known COX isoforms, which differ mainly in expression patterns. COX-1 is constitutively expressed and involved in "housekeeping" functions, whereas COX-2 is induced in response to inflammatory stimuli and is hence the desired target for the nonsteroidal antiinflammatory drugs (NSAIDs) [8].

Catalytic Mechanism of COX

COX-1 and COX-2 convert AA to PGH_2 by two sequential reactions that occur at spatially distinct active sites on the enzymes. The first reaction involves the bis-dioxygenation of AA to yield the hydroperoxide PGG_2 (COX reaction; target of NSAID action), and the second reaction involves peroxidative cleavage of PGG_2 to yield PGH_2 (peroxidase reaction) (COX catalysis is extensively reviewed by Rouzer and Marnett [7]). The catalytic mechanism of these heme-containing enzymes is novel and requires that the peroxidase activity initiate the COX reaction by generating a tyrosyl radical in the COX active site. After initiation, the COX activity becomes autocatalytic with respect to the enzymes' ability to regenerate the tyrosyl radical for subsequent rounds of catalysis. In contrast, the peroxidase activity requires two electrons from a coreductant to return the heme Fe from the higher oxidation states generated during peroxidase catalysis [compound I (Fe^{5+}) and compound II (Fe^{4+})] back to its resting state (Fe^{3+}) before peroxide bond cleavage can occur again (Figure 12.2A). Both enzyme isoforms self-inactivate over time due to protein radical intermediates generated when there is insufficient coreductant to reduce the heme Fe back to its resting state. Coreductants protect against self-inactivation by providing a source of electrons to prevent the formation of damaging radical species. The normal phenomenon of enzyme self-inactivation is believed to limit PG synthesis *in vivo*.

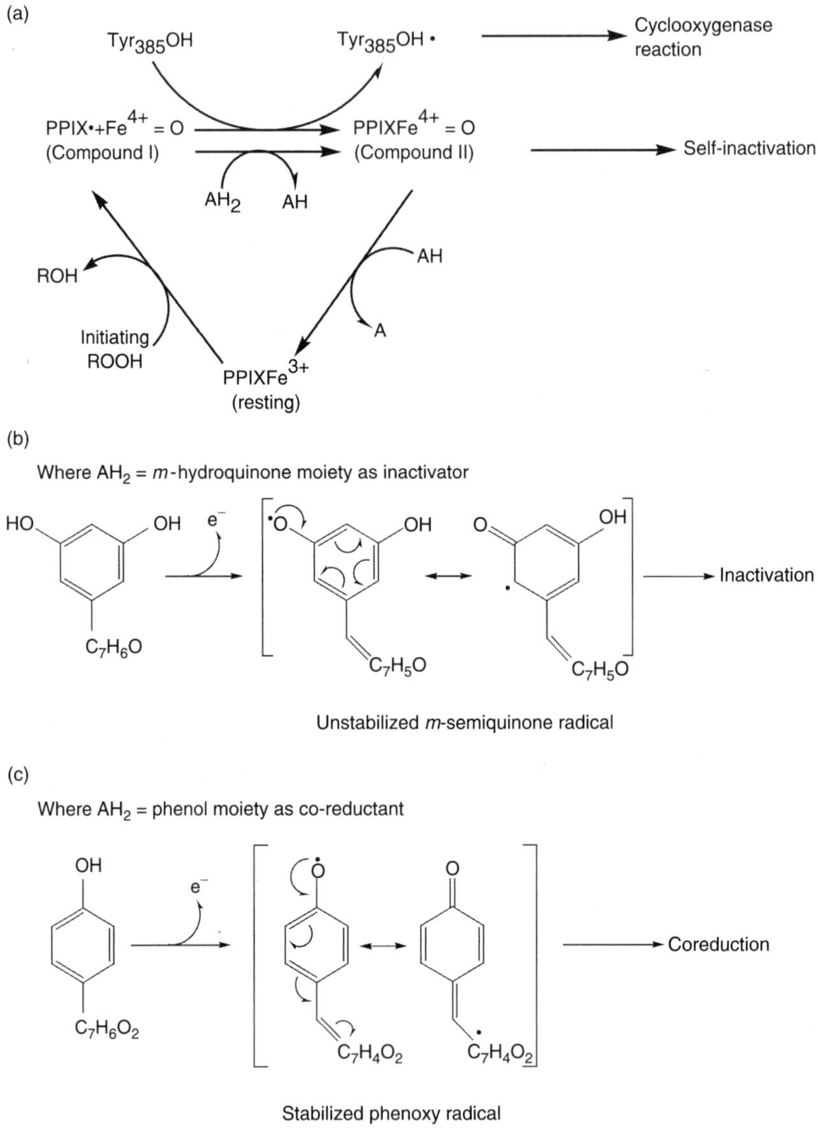

(a)

Tyr385OH → Tyr385OH • → Cyclooxygenase reaction

PPIX•+Fe4+ = O (Compound I) → PPIXFe4+ = O (Compound II) → Self-inactivation

AH2 AH

ROH AH

Initiating ROOH A

PPIXFe3+ (resting)

(b)

Where AH2 = *m*-hydroquinone moiety as inactivator

HO OH e− •O OH O OH → Inactivation

C7H6O C7H5O C7H5O

Unstabilized *m*-semiquinone radical

(c)

Where AH2 = phenol moiety as co-reductant

OH e− O → Coreduction

C7H6O2 C7H4O2 C7H4O2

Stabilized phenoxy radical

FIGURE 12.2 Mechanism of resveratrol action on COX-1.

COX ISOZYMES AS DRUG TARGETS

COX-1 AS A TARGET FOR CARDIOPROTECTIVE (ANTIPLATELET) AGENTS

Prostanoids are local mediators of vascular homeostasis; for example, TXA_2 is a potent vasoconstrictor and platelet aggregator synthesized in activated platelets, whereas prostacyclin (PGI_2) is an antiplatelet aggregator and potent vasodilator synthesized in the vascular endothelial cells (antiplatelet agents are extensively reviewed by Patrono et al. [9]). Vascular homeostasis results from a dynamic balance between TXA_2 and PGI_2 since they have opposing actions. Imbalances in this ratio can explain changes that occur in various pathological conditions including thrombosis. Both TXA_2 and PGI_2 are synthesized from the precursor PGH_2; however, different COX isoforms contribute to their formation. Platelets contain only COX-1, which is an obligate enzyme for TXA_2 formation. In contrast, COX-2 in the vascular endothelial cells is the primary source of systemic PGI_2 biosynthesis. Therefore, selective inhibition of COX-1 offers a viable mechanism for cardioprotective agents, which can act by tilting the TXA_2–PGI_2 balance in favor of PGI_2. Agents that target COX-1 for cardioprotection must eliminate platelet TXA_2 synthesis while sparing PGI_2 synthesis in the vascular endothelial cells.

Selective elimination of platelet TXA_2 synthesis is achieved by irreversible inhibition of COX. Aspirin is the prototypical antiplatelet (cardioprotective) agent that acetylates Ser530 in the COX active site resulting in loss of prostanoid synthesis by COX-1 and COX-2. Although aspirin is not selective for COX-1, extremely high efficacy as an antiplatelet agent results from its ability to irreversibly inhibit the enzymes. This inhibition can only be surmounted by new protein synthesis. Since platelets are unable to synthesize new protein the effect of aspirin is governed by the half-life of the platelet, which is 7 days. Thus a single low dose of aspirin can eliminate platelet TXA_2 synthesis for an extended period while PGI_2 synthesis in the vascular endothelial cells and prostanoid synthesis in the gastrointestinal (GI) track can recover quickly (hours). In this manner, low-dose aspirin shifts the TXA_2–PGI_2 balance to favor cardioprotection over thrombosis (Figure 12.3). Evidence will be provided that resveratrol is a selective inactivator of COX-1 but acts differently from aspirin in that it targets the peroxidase active site of the enzyme. Thus resveratrol may have the same pharmacological outcome as aspirin; however, the effect is achieved via an entirely different mechanism, namely mechanism-based inactivation of the peroxidase active site [5].

COX-2 as a Target for Antiinflammatory Agents

COX-2 is the desired target for antiinflammatory drugs since it is upregulated in response to inflammatory stimuli in macrophages, eisonophils, and neutrophils, and forms proinflammatory PGs [8]. In addition, selective COX-2 inhibitors should have decreased GI toxicity since COX-1 is the isoform responsible for synthesizing gastroprotective PGs and would not be inhibited. This has resulted in the development of multiple COX-2 selective NSAIDs, which are devoid of GI side effects (e.g., Celebrex and Vioxx), for the treatment of chronic inflammatory conditions [10].

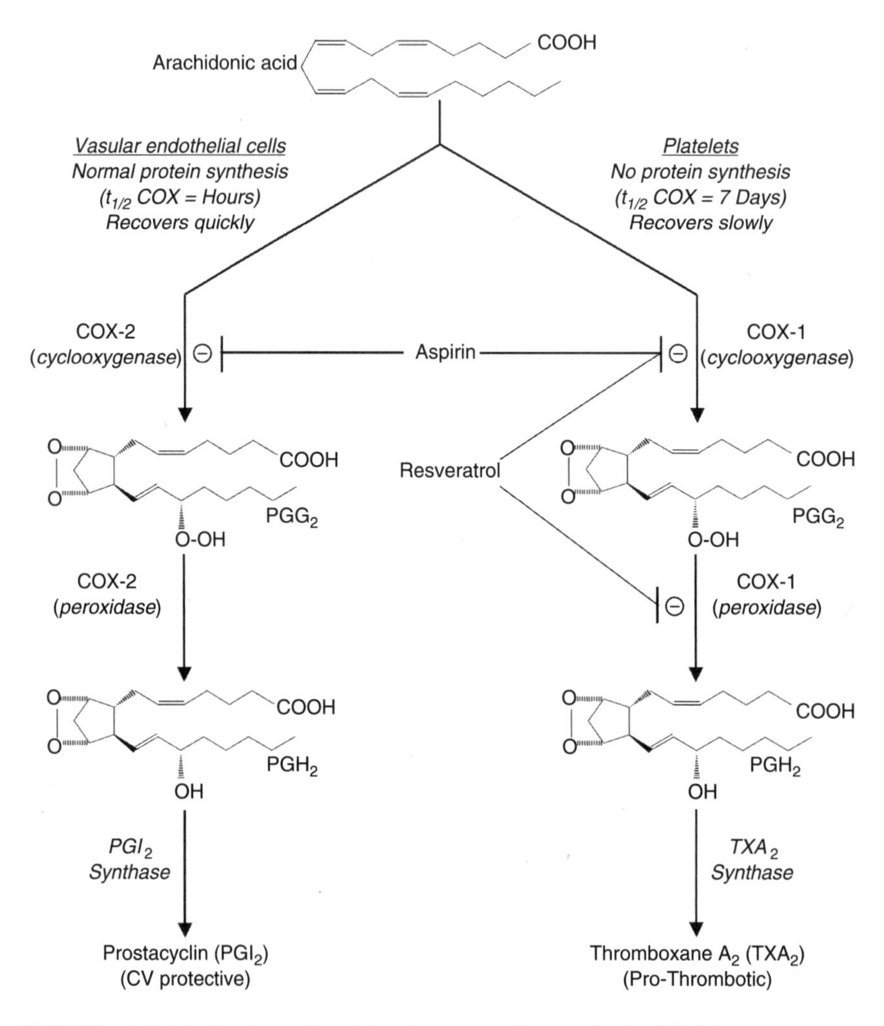

FIGURE 12.3 Mechanism of antiplatelet agents that work via COX-1.

COX-2 as a Target for Cancer Chemotherapeutic Agents

In addition to being the desired target of NSAIDs, COX-2 is a potential target for the chemoprevention of colon cancer. COX-2 expression is abnormally high in colon cancer cells and *in vivo* studies in a mouse model for colon cancer showed that both nonselective NSAIDs and COX-2-selective NSAIDs decreased precancerous polyps [11]. This suggests that COX-2 plays a role in the onset of this disease.

COX-2 has also been implicated as a potential target for the chemoprevention of breast cancer since its expression is elevated in breast carcinoma, and COX-2 conditional transgenic mice, in which gene expression was driven by the murine mammary tumor virus (MMTV) promoter, spontaneously developed breast carcinomas [12,13]. In addition, COX-2-derived PGE_2 induced angiogenesis at the onset of tumor development, and the nonspecific NSAID indomethacin inhibited angiogenesis and retarded tumor progression [14]. The role of COX-2 was confirmed by achieving the same effects with Celebrex. Furthermore, COX-2-derived PGE_2 has been linked to the induction of the aromatase gene in mammary epithelial cells leading to increased levels of estrogen [15]. The decrease in incidence of hormonal-dependent breast carcinomas following the frequent use of aspirin is consistent with inhibition of COX-2 resulting in downregulation of the aromatase gene and decreased estrogen biosynthesis [16].

DIRECT EFFECTS OF RESVERATROL ON COX CATALYSIS

EARLY FINDINGS

The observation that resveratrol and other polyphenolic compounds inhibited platelet aggregation and eicosanoid biosynthesis *in vitro* implicated inhibition of COX as a potential mechanism of action [17,18]. After the detection of resveratrol in red wine [19], its antiplatelet effects were confirmed and it was identified as a putative agent responsible for the "French paradox" [3,20,21]. Resveratrol was shown to be a potent inhibitor of both the cyclooxygenase and peroxidase reactions of COX-1 [2]. By contrast, classic NSAIDs target the cyclooxygenase reaction only and leave the peroxidase activity intact [22]. Furthermore, resveratrol was able to discriminate between the two COX isozymes since it weakly inhibited the peroxidase activity of COX-2 and left the cyclooxygenase activity of this isoform intact [2]. Johnson and Maddipati showed that resveratrol was a noncompetitive inhibitor of COX-1 using AA as a substrate indicating that drug binding occurred at a site other than the cyclooxygenase active site of this isozyme [4]. We built on these studies to further determine the discrete mechanism of resveratrol action on both COX-1 and COX-2 and to determine if its direct effects on these enzymes could account for its pharmacological activities [5].

RESVERATROL AS A MECHANISM-BASED INACTIVATOR OF THE PEROXIDASE ACTIVITY OF COX-1

Resveratrol was rapidly oxidized by COX-1 (1.26 μmol/min/mg) during the peroxidase cycle as evident by the disappearance of the absorbance spectrum of the drug in the presence of a peroxide cosubstrate (Figure 12.4A; time-resolved spectra with COX-2 were identical). Oxidation occurred at the peroxidase active site of COX-1 since blocking the cyclooxygenase active

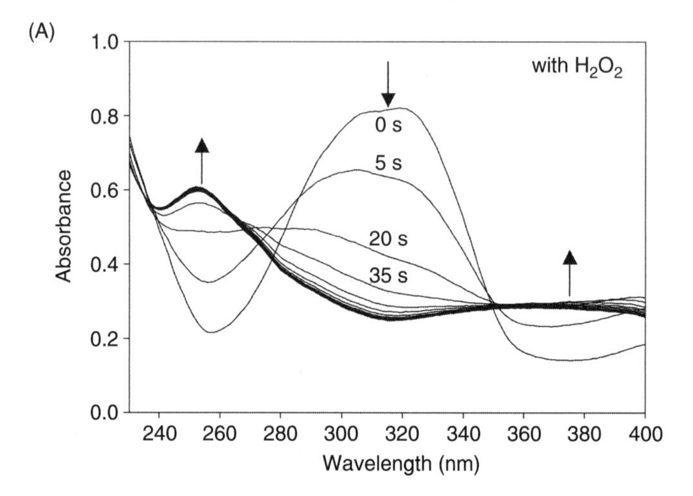

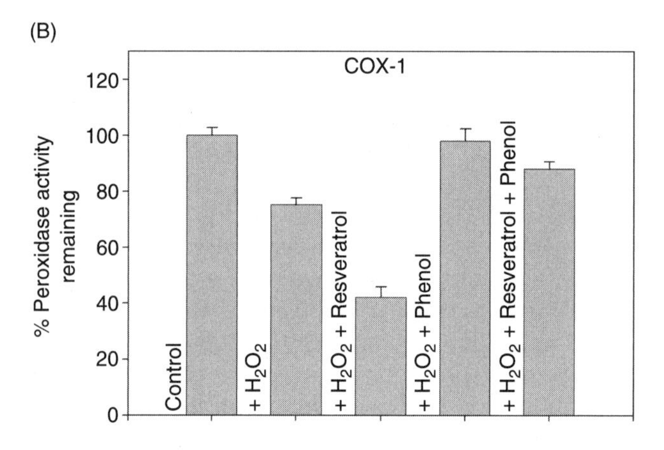

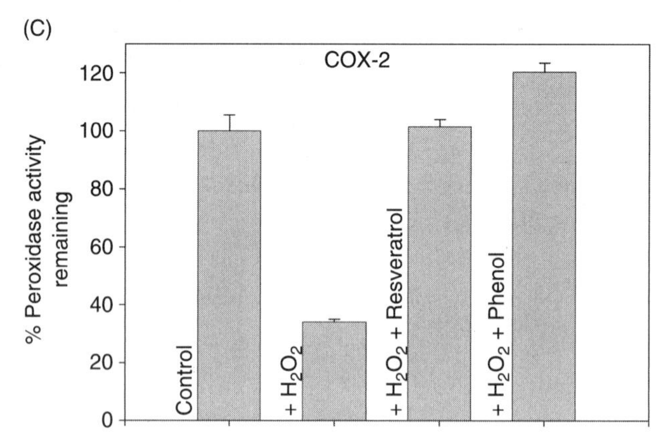

FIGURE 12.4 (A)–(C). See facing page for caption.

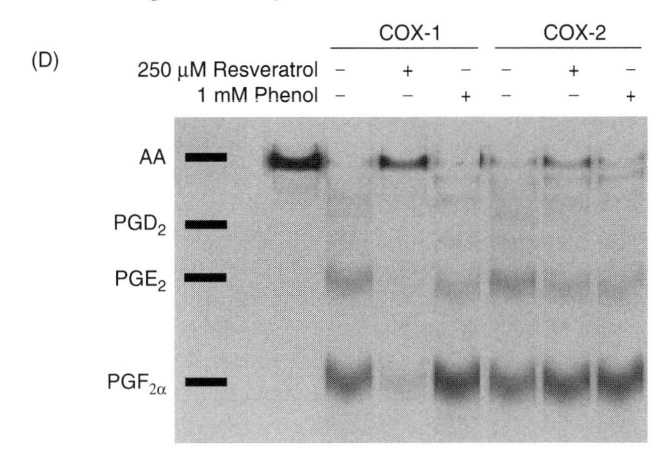

FIGURE 12.4 Mechanism-based inactivation of COX-1 by resveratrol. (A) Spectrophotometric analysis of resveratrol oxidation catalyzed by the COX peroxidase. Holo-COX-1 (0.3 μM) was incubated with 30 μM resveratrol in 100 mM Tris-HCl (pH 8.0) in the presence of H_2O_2 (300 μM) and time-resolved spectra were collected. Reactions were initiated with peroxide. (B, C) Mechanism-based inactivation of COX by resveratrol. Holo-COX-1 or holo-COX-2 (10 μM) was preincubated with mixtures of 100 μM resveratrol, 100 μM H_2O_2, and 1 mM phenol in 100 mM Tris-HCl (pH 8.0) for 5 min at 25°C. The samples were then diluted into the peroxidase activity assay (200-fold) and the percent activity remaining was determined. Values were corrected for resveratrol carryover based on IC_{50} curves. (D) Effect of resveratrol on COX-dependent prostanoid synthesis. Holo-COX-1 or holo-COX-2 (2 μM) was incubated with either 250 μM resveratrol or 1 mM phenol in 100 mM Tris-HCl (pH 8.0). Reactions were initiated by the addition of 150 μM [^{14}C]-AA (25 nCi/reaction) and quenched after 1 min with stannous chloride in HCl. PG products were separated by thin-layer chromatography (ethyl acetate: 2,2,4-trimethylpentane: acetic acid, 110:50:20 v/v/v saturated in water), visualized by autoradiography, and quantitated by scintillation counting. (From Szewczuk L, Forti L, Stivala L, and Penning T, *J Biol Chem* 279, 22727–22737, 2004.)

site with either indomethacin or aspirin had no effect on the specific activity for resveratrol turnover (1.56 and 1.12 μmol/min/mg, respectively). Both COX isoforms were capable of catalyzing the oxidation of resveratrol; however, COX-2 was the more robust catalyst with a specific activity for resveratrol turnover of 8.41 μmol/min/mg [5].

A series of preincubation/dilution studies were performed to determine whether the oxidation of resveratrol was coincident with irreversible inactivation of either COX-1 or COX-2 (see Figure 12.4B and Figure 12.4C). Several key findings were observed for COX-1 (Figure 12.4B). First, resveratrol alone had no effect on the enzyme during a five-minute preincubation period (control). Second, in the presence of H_2O_2 alone enzyme self-inactivation was observed. Third, under conditions in which resveratrol is rapidly oxidized (in the presence of H_2O_2) there was a significant increase in the amount of enzyme inactivation observed, which could not be accounted for by self-inactivation alone. Finally, the prototypical coreductant phenol was able to protect against both enzyme self-inactivation and resveratrol-mediated enzyme inactivation. This same pattern was not observed for COX-2 (Figure 12.4C), and instead resveratrol behaved identically to the coreductant phenol by protecting against enzyme self-inactivation. These findings identified resveratrol as a peroxidase-mediated mechanism-based inactivator of COX-1 and as a coreductant of COX-2. [^{14}C]-AA was used to show that as a consequence of inactivating the peroxidase activity, resveratrol was able to eliminate PG synthesis by COX-1 but had no effect on PG synthesis by COX-2 (Figure 12.4D) [5].

Progress curves of steady-state peroxidase assays performed in the presence of increasing amounts of resveratrol showed a time- and concentration-dependent inactivation event. The k_{obs} values for inactivation were corrected for the rate of self-inactivation ($0.014 \pm 0.001\,sec^{-1}$ in our assay system) and reploted in a Kitz–Wilson analysis, which showed saturation kinetics [5,23]. This analysis gave k_{inact} of $0.069 \pm 0.004\,sec^{-1}$, $K_{i\,inact}$ of $1.52 \pm 0.15\,\mu M$, and a calculated half-life for inactivation of 10.0 sec at saturating concentrations of resveratrol. Dividing k_{cat} for resveratrol oxidation by k_{inact} for resveratrol inactivation yielded a partition ratio for inactivation of 22. We also showed that the ratio between enzyme inactivated by resveratrol and enzyme inactivated by peroxide (self-inactivation) remained unchanged over a wide range of H_2O_2 concentrations (25 to 1000 μM). This implies that resveratrol can act as a mechanism-based inactivator of COX-1 over a dynamic range of peroxide concentrations, as would be expected *in vivo* [5].

STRUCTURE–ACTIVITY RELATIONSHIPS (SARs) WITH RESVERATROL ANALOGS

SAR data (Table 12.1) with methoxyresveratrol analogs (Figure 12.1) were revealing. First, the *m*-hydroquinone moiety (3,5-dihydroxy group) of resveratrol was required for mechanism-based inactivation of COX-1, since protecting this moiety as methyl ethers resulted in the elimination of inhibitory activity (see III and IV). This finding was confirmed by inhibition studies with resorcinol (V) and COX-1, which identified an unsubstituted *m*-hydroquinone as the minimal structure required for the inactivation of COX-1 (Table 12.1). Second, any of the three hydroxy groups on resveratrol could be oxidized by both COX-1 and COX-2, but the *trans*-alkene could not since the trimethoxy analog (IV) was not oxidized. The outcome of these oxidation reactions differed with both the position of the hydroxy group and with COX isozyme. With COX-2, all of the hydroxy groups on

TABLE 12.1
SAR Analysis of the Inhibition of COX by Resveratrol and Its Analogs

Analog	Enzyme	IC_{50} peroxidase (μM)	IC_{50} cyclooxygenase[a] (μM)	Oxidation of compound[b] (μmol/min/mg)	Mode of action	Peroxidase		
						k_{inact} (sec^{-1})	$K_{i\ inact}$ (μM)	$t_{1/2}$ (sec)
I	COX-1	2.8 ± 0.6	67 ± 19	1.26	Inactivator	0.069 ± 0.004	1.52 ± 0.15	10.0
	COX-2	ND	ND	8.41	Cosubstrate	ND	ND	ND
II	COX-1	5.1 ± 0.9	30 ± 5	0.06	Inactivator	0.046 ± 0.002	1.26 ± 0.16	15.1
	COX-2	ND	ND	2.98	Cosubstrate	ND	ND	ND
III	COX-1	ND	ND	9.75	Cosubstrate	ND	ND	ND
	COX-2	ND	ND	5.23	Cosubstrate	ND	ND	ND
IV	COX-1	ND	ND	0.00	No effect	ND	ND	ND
	COX-2	ND	ND	0.00	No effect	ND	ND	ND
V	COX-1	3.6 ± 0.6	30.8 ± 24.2	ND	Inactivator	0.018 ± 0.002	ND	38.5
	COX-2	ND	ND	ND	Cosubstrate	ND	ND	ND

Note: ND, not detectable.
[a]Values elevated because 1 mM phenol is present in the assay and no peroxide cosubstrate is present.
[b]Plots of velocity vs. enzyme amount (μg) gave linear lines with a correlation coefficient (r^2) of >0.975.
From Szewczuk L, Forti L, Stivala L, and Penning T, *J Biol Chem* 279, 22727–22737, 2004.

resveratrol serve as reducing cosubstrates to return the heme Fe to its resting state (Fe^{3+}) during peroxidase catalysis. However, with COX-1, oxidation of the *m*-hydroquinone moiety leads to inactivation while oxidation of the phenol moiety leads to reducing cosubstrate activity. Thus when the *m*-hydroquinone of resveratrol is protected as the dimethyl ether (III), coreductant activity is maintained and inactivation activity is lost. Therefore, with respect to COX-1, resveratrol contains moieties on opposing rings that make it both a mechanism-based inactivator and a reducing cosubstrate, namely a *m*-hydroquinone and a phenol moiety, respectively [5].

The SAR data with resveratrol analogs aided in identifying additional *m*-hydroquinones in red wine that were selective mechanism-based inactivators of COX-1. A follow-up SAR analysis of (±)-catechin and (±)-epicatechin (red wine *m*-hydroquinones) showed that they also acted as potent peroxidase-mediated mechanism-based inactivators of COX-1 but not of COX-2 (data not shown). These findings imply that resveratrol is not the sole agent responsible for modulating prostanoid synthesis in red wine, and suggest that many dietary *m*-hydroquinones may possess this property and thus have therapeutic value [24].

[^{3}H]-Resveratrol is Not Covalently Incorporated into COX-1

[^{3}H]-Resveratrol was used to determine if mechanism-based inactivation of COX-1 resulted in covalent modification of the enzyme. Sephadex G-25 gel filtration chromatography was used to separate bound and free [^{3}H]-resveratrol from enzyme inactivated by the drug in the presence of H_2O_2. Elution of tritium with the enzyme was observed as evident by a significant increase in the amount of radioactivity associated with the protein fractions. When COX-1 was inactivated by 60% an estimate of the stoichiometry indicated that 6.0 moles of [^{3}H]-compound were bound per mole of COX-1 monomer (i.e., 10.0 moles of [^{3}H]-compound bound per mole of inactivated COX-1 monomer). Nonspecific binding of [^{3}H]-resveratrol was observed in the absence of H_2O_2 yielding a stoichiometry of 1.4 moles of [^{3}H]-resveratrol bound per mole of COX-1 monomer; however, this nonspecific binding did not result in a loss of enzyme activity. These data indicated that inactivation of COX-1 resulted in coelution of the enzyme with radioactivity derived from [^{3}H]-resveratrol on a Sephadex G-25 gel filtration column [5].

Since gel filtration is facile, it was possible that the apparent tritium incorporation was due to hydrophobic interaction of an oxidation product with the enzyme. Therefore, a reverse-phased high-pressure liquid chromatography (RP-HPLC) analysis was performed on [^{3}H]-resveratrol-inactivated COX-1 isolated by the previously described Sephadex chromatography step. This analysis detected one peak of retained radioactivity that did not coelute with either the heme cofactor or COX-1 peaks. Furthermore, a liquid chromatography/mass spectrometry (LC/MS) analysis determined the mass of this product to be 454, which was consistent with the formation of a resveratrol dihydrodimer. These findings indicated that COX-1 was not covalently modified by resveratrol, and revealed that inactivation had occurred via a "hit-and-run" mechanism [5].

Trapping Resveratrol Radicals as Viniferins

Resveratrol functions as both a mechanism-based inactivator and a coreductant of the COX-1 peroxidase. These functions are mediated by the *m*-hydroquinone moiety (mechanism-based inactivator) and the phenol moiety (coreductant) [5]. Implicit in this bifunctionality is the notion that resveratrol is oxidized at the peroxidase active site of COX-1 resulting in the formation of two hypothetical radical species, namely a *m*-semiquinone radical and a phenoxy radical. The stability of these radicals may determine their effects on COX-1. For instance, the *m*-semiquinone radical cannot be stabilized through the ring structure to the *m*-hydroxy group or to the *trans*-stilbene scaffold and consequently formation of this radical leads to COX-1 inactivation (Figure 12.2B) [25]. In contrast, the phenoxy radical can be stabilized through the extended conjugation of the *trans*-stilbene scaffold (Figure 12.2C). These radicals have been trapped during mechanism-based inactivation of the COX-1 peroxidase as the resveratrol dihydrodimers *cis*-ε-viniferin (trapped *m*-semiquinone radical; Figure 12.5A) and *trans*-δ-viniferin (trapped phenoxy radical; Figure 12.5B). The identity of these products was established by LC/MS^n methods [26].

EFFECTS OF RESVERATROL ON COX-2 EXPRESSION

The cancer chemopreventive and antiinflammatory properties of resveratrol point to a mechanism whereby resveratrol regulates COX-2 expression [27]. This is consistent with our finding that resveratrol does not inactivate COX-2 [5]. Although resveratrol does not directly affect PG biosynthesis by COX-2, it impairs induction of enzyme expression in human colon adenocarcinoma cells in response to lipopolysaccharide, 12-*O*-tetradecanoylphorbol-13-acetate, and superoxide, and in human mammary epithelial cells [28–30]. In addition, resveratrol inhibited activator protein-1, protein kinase C_{α}, and extracellular signal-regulated kinase 1-mediated induction of COX-2 promoter activity [28]. Resveratrol also suppressed activation of the nuclear factor-kappa B, which has been implicated in regulating COX-2 expression [31].

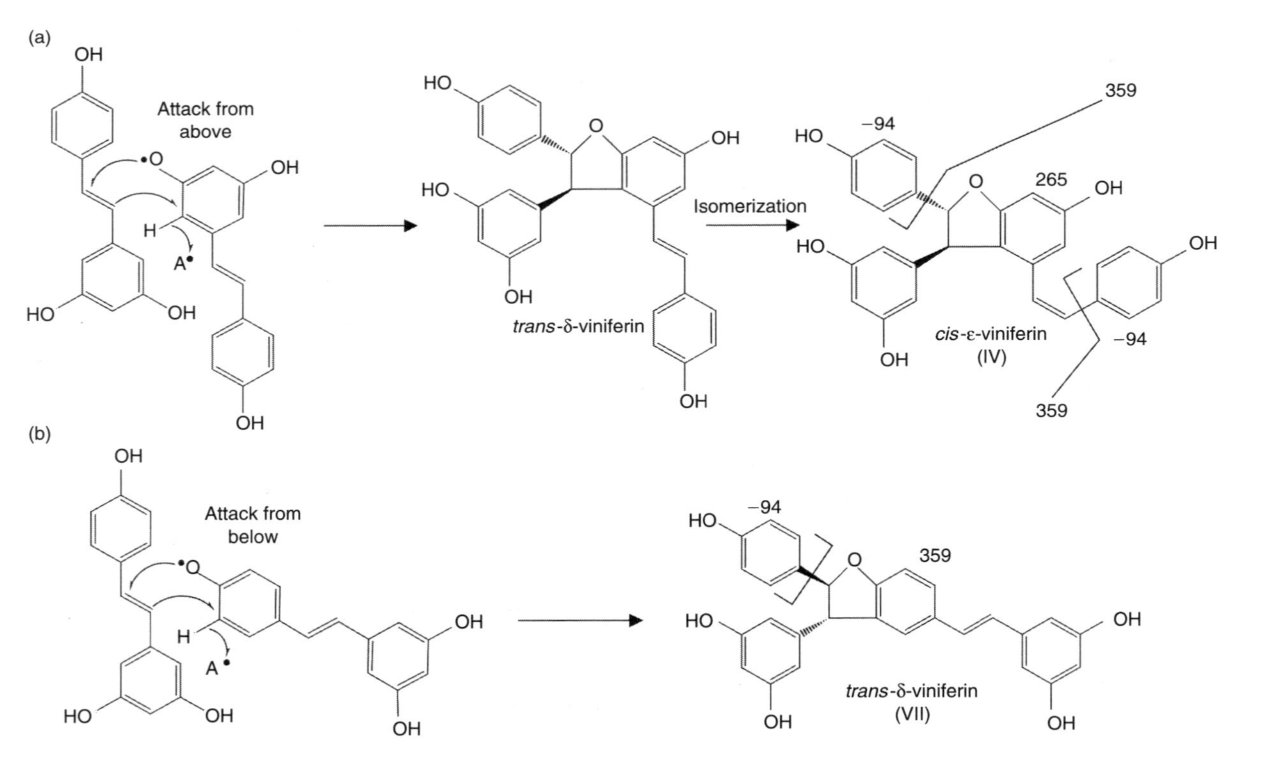

FIGURE 12.5 Mechanism of viniferin production by the COX-1 peroxidase. Fragment ions used in LC/MS identification of the viniferins are indicated. Note only the *cis*-ε-viniferin shows the loss of two phenol groups ($m/z = 359$ and $m/z = 265$).

PHARMACOKINETICS AND EFFICACY OF RESVERATROL (AND OTHER *m*-HYDROQUINONES)

The pharmacokinetics of resveratrol and its inhibitory properties of COX-1 satisfy the criteria of antiplatelet agents that target this isoform. The inactivation is irreversible, the half-life for inactivation is rapid (less than 25 sec), the $K_{i\ inact}$ value is well within the range of peak plasma concentrations ($K_{i\ inact}$ resveratrol = 1.26 μM), and the agents are highly selective for COX-1, the isoform present in the platelet [5,9,32]. Furthermore, higher peak plasma concentrations should be achievable for the mixture of *m*-hydroquinones that are present in red wine (e.g., catechins and epicatechins). While the cardioprotection afforded by red wine consumption is most likely the result of a combination of mechanisms (e.g., antioxidant activity), the irreversible inactivation of COX-1 by *m*-hydroquinones offers a route to long-lasting effects.

The efficacy of *m*-hydroquinones as COX-1 inactivators *in vivo* may be limited by the requirement of a peroxide cosubstrate and the coreductant tone of the platelet. Inactivation is favorable when the peroxide tone is high and the coreductant tone is low since under these conditions the *m*-hydroquinone is rapidly oxidized and inactivation of COX-1 proceeds unimpeded [5]. Due to this requirement, it is predicted that *m*-hydroquinones would be more effective in patients with prothrombotic conditions. These conditions are associated with increased levels of lipid hydroperoxides, specifically PGG_2 and 12-HpETE, which are both substrates for the COX-1 peroxidase [33,34]. In this manner, sufficient peroxide cosubstrate would be present to drive the mechanism-based inactivation of COX-1 by resveratrol. This mechanism is significantly different from low-dose aspirin therapy since aspirin can work in both the activated and resting platelet [9]. By doing so, aspirin is used to prevent the onset of prothrombotic conditions in high-risk patients and prevent primary and secondary myocardial infarct.

CONCLUSIONS

The data presented in this chapter offer a basis for the design of a new class of selective COX-1 inactivators, namely *m*-hydroquinones, which work via a mechanism-based event at the peroxidase active site. These compounds are unique for several reasons. First, they are natural products. Second, they prevent the formation of PGs by acting at a site different from where classic NSAIDs exert their effects. Third, they are highly selective for COX-1. Fourth, the inactivation is irreversible. Fifth, inactivation occurs via a radical mechanism and does not result in covalent modification of the enzyme. In this manner, resveratrol-mediated inactivation of COX-1 is reminiscent of enzyme self-inactivation. Furthermore, the irreversible inactivation of COX-1 in the platelet predicts the cardioprotective (antiplatelet) effects that have been reported for resveratrol and red wine. Although there are limits to the efficacy of resveratrol, it is likely that sufficient intake of dietary *m*-hydroquinones can yield therapeutic effects in a population at high risk for cardiovascular disease. By contrast, resveratrol may exert its cancer chemopreventive and antiinflammatory effects by downregulation of COX-2 gene expression.

REFERENCES

1. Renaud S and De Lorgeril M, Wine, alcohol, platelets, and the French paradox for coronary heart disease, *Lancet* 339, 1523—1526, 1992.
2. Jang M, Cai L, Udeani G, Slowing K, Thomas C, Beecher C, Fong H, Farnsworth N, Kinghorn A, Mehta R, Moon R, and Pezzuto J, Cancer chemopreventive activity of resveratrol, a natural product derived from grapes, *Science* 275, 218–220, 1997.
3. Pace-Asciak C, Hahn S, Diamandis E, Soleas G, and Goldberg D, The red wine phenolics trans-resveratrol and quercetin block human platelet aggregation and eicosanoid synthesis: implications for protection against coronary heart disease, *Clin Chim Acta* 235, 207–219, 1995.
4. Johnson J and Maddipati K, Paradoxical effects of resveratrol on the two prostaglandin H synthases, *Prostaglandins & Other Lipid Mediators* 56, 131–143, 1998.
5. Szewczuk L, Forti L, Stivala L, and Penning T, Resveratrol is a peroxidase-mediated inactivator of COX-1 but not COX-2: a mechanistic approach to the design of COX-1 selective agents, *J Biol Chem* 279, 22727–22737, 2004.
6. Funk C, Prostaglandins and leukotrienes: advances in eicosanoid biology, *Science* 294, 1871–1875, 2001.
7. Rouzer C and Marnett L, Mechanism of free radical oxygenation of polyunsaturated fatty acids by cyclooxygenases, *Chem Rev* 103, 2239–2304, 2003.
8. Hla T and Nielson K, Human cyclooxygenase-2 cDNA, *Proc Natl Acad Sci* 89, 7384–7388, 1992.
9. Patrono C, Coller B, Dalen J, FitzGerald G, Fuster V, Gent M, Hirsh J, and Roth G, Platelet-active drugs: the relationships among dose, effectiveness, and side effects, *Chest* 119 (1 Suppl), 39S–63S, 2001.
10. Bombardier C, Laine L, Reicin A, Shapiro D, Burgos-Vargas R, Davis B, Day R, Ferraz M, Hawkey C, Hochberg M, Kvien T, Schnitzer T, VIGOR Study Group, Comparison of upper gastrointestinal toxicity of rofecoxib and naproxen in patients with rheumatoid arthritis. VIGOR Study Group, *N Engl J Med* 343, 1520–1528, 2000.
11. Marx J, Cancer research. Antiinflammatories inhibit cancer growth: but how?, *Science* 291, 581–582, 2001.
12. DuBois R, Aspirin and breast cancer prevention: the estrogen connection, *JAMA* 291, 2488–2489, 2004.
13. Liu C, Chang S, Narko K, Trifan O, Wu M, Smith E, Haudenschild C, Lane T, and Hla T, Overexpression of cyclooxygenase-2 is sufficient to induce tumorigenesis in transgenic mice, *J Biol Chem* 276, 18563–18569, 2001.

14. Chang S, Liu C, Conway R, Han D, Nithipatikom K, Trifan O, Lane T, and Hla T, Role of prostaglandin E_2-dependent angiogenic switch in cyclooxygenase 2-induced breast cancer progression, *Proc Natl Acad Sci* 101, 591–596, 2004.
15. Zhao Y, Agarwal V, Mendelson C, and Simpson E, Estrogen biosynthesis proximal to a breast tumor is stimulated by PGE_2 via cyclic AMP, leading to activation of promoter II of the CYP19 (aromatase) gene, *Endocrinology* 137, 5739–5742, 1996.
16. Terry M, Gammon M, Zhang F, Tawfik H, Teitelbaum S, Britton J, Subbaramaiah K, Dannenberg A, and Neugut A, Association of frequency and duration of aspirin use and hormone receptor status with breast cancer risk, *JAMA* 291, 2433–2440, 2004.
17. Kimura Y, Okuda H, and Arichi S, Effects of stilbenes on arachidonate metabolism in leukocytes, *Biochim Biophys Acta* 834, 275–278, 1985.
18. Kitagawa S, Fujisawa H, Baba S, and Kametani F, Inhibitory effects of catechol derivatives on arachidonic acid-induced aggregation of rabbit platelets, *Chem Pharm Bull* 39, 1062–1064, 1991.
19. Siemann E and Creasy L, Concentration of the phytoalexin resveratrol in wine, *Am J Enol Vitic* 43, 49–52, 1992.
20. Goldberg D, Hahn S, and Parkes J, Beyond alcohol: beverage consumption and cardiovascular mortality, *Clin Chim Acta* 237, 155–187, 1995.
21. Pace-Asciak C, Rounova O, Hahn S, Diamandis E, and Goldberg D, Wines and grape juices as modulators of platelet aggregation in healthy human subjects, *Clin Chim Acta* 246, 163–182, 1996.
22. Flower R, Drugs which inhibit prostaglandin biosynthesis, *Pharmacol Rev* 26, 33–67, 1974.
23. Kitz R and Wilson I, Esters of methanesulfonic acid as irreversible inhibitors of acetylcholinesterase, *J Biol Chem* 237, 3245–3249, 1962.
24. Szewczuk L and Penning T, Mechanism-based inactivation of COX-1 by red wine *m*-hydroquinones: a structure–activity relationship study, *J Nat Prod* 67, 1777–1782, 2004.
25. Divi R and Doerge D, Mechanism-based inactivation of lactoperoxidase and thyroid peroxidase by resorcinol derivatives, *Biochemistry* 33, 9668–9674, 1994.
26. Szewczuk L, Lee SH, Blair IA, and Penning TM, Viniferin formation by COX-1: evidence for radical intermediates during cooxidation of resveratrol. *J Nat Prod* 68, 36–42, 2005.
27. Surh Y, Chun K, Cha H, Han S, Keum Y, Park K, and Lee S, Molecular mechanisms underlying chemopreventive activities of antiinflammatory phytochemicals: down-regulation of COX-2 and iNOS through suppression of NF-κB activation, *Mutation Res* 480–481, 243–268, 2001.
28. Subbaramaiah K, Chung W, Michaluart P, Telang N, Tanabe T, Inoue H, Jang M, Pezzuto J, and Dannenberg A, Resveratrol inhibits cyclooxygenase-2 transcription and activity in phorbol ester-treated human mammary epithelial cells, *J Biol Chem* 273, 21875–21882, 1998.
29. Martinez J and Moreno J, Effect of resveratrol, a natural pholyphenolic compound, on reactive oxygen species and prostaglandin production, *Biochem Pharmacol* 59, 865–870, 2000.
30. Mutoh M, Takahashi M, Fukuda K, Matsushima-Hibiya Y, Mutoh H, Sugimura T, and Wakabayashi K, Suppression of cyclooxygenase-2

promoter-dependent transcriptional activity in colon cancer cells by chemopreventive agents with a resorcin-type structure, *Carcinogenesis* 21, 959–963, 2000.

31. Tsai S, Lin-Shiau S, and Lin J, Suppression of nitric oxide synthase and the down-regulation of the activation of NF-κB in macrophages by resveratrol, *Br J Pharmacol* 126, 673–680, 1999.
32. Asensi M, Medina I, Ortega A, Carretero J, Bano M, Obrador E, and Estrela J, Inhibition of cancer growth by resveratrol is related to its low bioavailability, *Free Radical Biol Med* 33, 387–398, 2002.
33. Gorog P and Kovacs I, Lipid peroxidation by activated platelets: a possible link between thrombosis and atherogenesis, *Atherosclerosis* 115, 121–128, 1995.
34. Lagarde M, Lemaitre D, Calzada C, and Vericel E, Involvement of lipid peroxidation in platelet signaling, *Prostaglandins, Leukotrienes Essential Fatty Acids* 57, 489–491, 1997.

13 Resveratrol as an Inhibitor of Carcinogenesis

John M. Pezzuto

CONTENTS

Given the high probability of developing cancer over the period of a normal lifespan, cancer chemoprevention provides an attractive therapeutic strategy for the delay or reversal of this process. A variety of phytochemicals, such as sulfides, isothiocyanates, glucosinolates, flavonoids, carotenoids, phenols, and diarylhepanoids, are known to mediate chemopreventive responses. Resveratrol, a ubiquitous stilbene found in the diet of humans (e.g., as a component of grapes and wine), was uncovered by bioassay-guided fractionation and found to mediate cancer chemopreventive activity in a murine model with mechanisms involving various stages of the carcinogenic process. This work spurred a myriad of studies that are summarized in this chapter. As demonstrated with *in vitro* and cell culture models, resveratrol functions through a plethora of mechanisms, which can vary from model to model. Results from differential gene expression studies are daunting. Irrespective of the precise mechanism, however, efficacy has been demonstrated in some animal models, and a critical evaluation of resveratrol data relative to the characteristics of a promising cancer chemopreventive agent leads to favorable consideration. Animal studies have shown inhibitory activity in skin, breast, colon, and esophagus. Biomarkers are known and ample quantities of compound can be produced. Dietary administration is feasible. Following preclinical toxicological studies, it appears that human intervention trials are warranted. As learned by past experience, data from these trials are necessary prior to drawing any conclusions, but the current cancer chemopreventive profile of resveratrol provides promise for widespread use in the future.

INTRODUCTION

As summarized by the World Health Organization, cancer leads to about 12% of human deaths [1], claiming well over 6,000,000 lives each year. In the United States, cancer is the second leading cause of death, being responsible for approximately one in every four deaths. Interestingly, it is believed that at least one third of all cancers could be prevented [2,3]. As such, primary and secondary prevention strategies are reasonable approaches to reduce the occurrence of this disease [4–6] and subsequent deaths. Primary prevention strategies involve removing causative agents and other life-style modifications that decrease the risk of cancer, as exemplified by smoking cessation and screening tests to detect precancerous lesions. Unfortunately, not all causative agents are known and other suspected carcinogens are too widespread to prevent feasibly all exposure.

Secondary prevention, cancer chemoprevention, involves the use of nontoxic natural and/or synthetic agents to decrease the risk of malignant tumor development or spread [7,8]. Cancer chemoprevention is a multidisciplinary field of research that has evolved from numerous scientific observations [9]. For example, epidemiological studies have linked diets high in fresh fruits and vegetables to lower cancer rates. This dietary link is perhaps most strongly supported by studies reporting the cancer risk of migrants from areas of low incidence to high incidence. These studies demonstrated that the incidence of cancer among children of migrants is similar to that of the general population [10]. Another important breakthrough has been the prevention of experimentally induced cancer in laboratory animals. It was subsequently postulated that dietary components, particularly specific nutrients and/or phytochemicals found in fruits and vegetables, could be used to prevent cancer in human beings [8,11]. More recently, research in cancer biology has elucidated molecular mechanisms of cancer chemopreventive agents [5,9,12]. Much of the theoretical basis for cancer chemoprevention is the understanding that cancer develops over time through the process of carcinogenesis [13]. This process has been broken down into distinct yet overlapping stages, namely initiation, promotion, and progression. The evolution of these stages is believed to take 10 to 40 years, during which various genetic mutations must occur [9,14]. The field of cancer chemoprevention is focused on reversing, halting, or delaying these stages of carcinogenesis by means of secondary prevention [7–9].

Cancer chemopreventive agents have been classified according to the stage of carcinogenesis in which they have demonstrated activity, and broadly termed blocking and suppressing agents [7]. Blocking agents act by preventing the initiation stage through a variety of mechanisms such as directly detoxifying carcinogens, stimulating detoxifying enzymes, and inhibiting carcinogen formation. Suppressing agents act at the promotion and progression stages through mechanisms such as inhibition of arachidonic acid metabolism, induction of cell differentiation, and inhibition of ornithine decarboxylase activity [3,7,15]. In the case of hormone-dependent cancers, suppressing agents may act by preventing the hormone from binding to its receptor, as exemplified by the use of the selective estrogen receptor modulators tamoxifen and raloxifene for breast cancer prevention [3,14].

OVERVIEW OF CANCER CHEMOPREVENTION TRIALS INVOLVING PHYTOCHEMICALS

Many early cancer chemoprevention studies were focused on nutrients such as vitamin C, calcium, and retinoids [8,10]. In the last few decades, nonnutrient phytochemicals found in fruits and vegetables have been examined, and a number of promising natural product leads have resulted from this research effort [14,16,17]. For example, green tea extract and pure compounds such as caffeic acid phenethyl ester, capsaicin, curcumin, 6-gingerol, indole-3-carbinol, lycopene, and perillyl alcohol are undergoing clinical trials for their cancer chemopreventive activities [14,18,19]. The U.S. National Cancer Institute is supporting the evaluation of potential cancer chemopreventive agents at different levels of preclinical development and clinical trials [17]. Examples of natural products currently under preclinical or clinical development for cancer chemoprevention include curcumin and lycopene, which are in a phase I study for the prevention of colon cancer, while a soy protein supplement is in a phase II trial for the prevention of prostate cancer in patients with elevated prostate-specific antigens [20]. Moreover, soy isoflavones are also involved in a randomized study in preventing further development of cancer in patients with stage I or stage II prostate cancer [20]. Polyphenon E (green tea extract), in combination with low-dose aspirin, is in a phase II randomized study to prevent cancer in women at high risk for developing breast cancer [21,22]. Other natural products currently being investigated include *S*-allyl-L-cysteine, epigallocatechin gallate, genistein, folic acid, and quercetin [18,23].

DISCOVERY AND CHARACTERIZATION OF NATURAL PRODUCT INHIBITORS OF CARCINOGENESIS

With support provided by the National Cancer Institute, we have conducted a program project entitled "Natural Inhibitors of Carcinogenesis" since 1991. The major aim of this project has been the discovery of new cancer chemopreventive agents from plants, particularly those that are edible. We are now beginning to explore marine microorganisms for chemopreventive activity. The project involves botanical, biological, chemical, biostatistical, and administrative aspects [24–27]. Terrestrial plant materials selected for investigation are prioritized based on information obtained from the NAPRALERT database [28]. Edible plants or species with reported biological activity related to cancer chemoprevention, plants with no history of toxicity, and those poorly investigated phytochemically are selected for preliminary investigation, and a small amount of plant material is collected [24–26].

The panel of *in vitro* bioassays used for the discovery of potential cancer chemopreventive drugs includes screening tests that are typically enzyme- or cell-based assays [25,29]. These assays are adapted to high-throughput measurement techniques performed relatively rapidly in order to uncover the biological properties of a large number of candidate substances [25,29]. The initial bioassays afford a strategic framework for the evaluation of agents according to defined criteria, and to provide evidence of agent efficacy, and serve to generate valuable dose–response, toxicity, and pharmacokinetic data required prior to phase I clinical safety testing [25,29,30].

Thus, preliminary screening is performed with an ethyl acetate-soluble partition extract using a battery of short-term *in vitro* bioassays [25]. Bioactive extracts are further evaluated in a mouse mammary organ culture model as a secondary discriminator [31,32]. The battery of short-term *in vitro* assays was developed to monitor tumorigenesis at different stages. For example, antimutagenicity activity, antioxidant activity, and induction of NADPH:quinone reductase activity has been monitored to evaluate inhibition of carcinogenesis at the initiation stage [33–36]. Monitoring inhibition of carcinogenesis at the promotion stage has been performed by evaluating the inhibition of phorbol ester-induced ornithine decarboxylase activity, inhibition of cyclooxygenases-1 and -2 activity, inhibition of phorbol dibutyrate receptor binding, and inhibition of transformation of JB6 mouse epidermal cells [37–40]. Induction of HL-60 human promyelocytic leukemia cell differentiation, and inhibition of aromatase, antiestrogenic, estrogenic, and estrone sulfatase activities have been used to monitor inhibition of carcinogenesis at the progression stage [41–44].

Plant extracts showing potency and/or selectivity in preliminary biological screening procedures are selected for bioassay-guided fractionation to isolate the active principle or principles. Methanolic crude extracts are partitioned using solvents of varying polarities and then chromatographed by either gravity-, flash-, or low-pressure column over silica, alumina, ion-exchange resins, polyamide, reversed-phase silica gel, size-exclusion gels, or other solid-phase supporting material [26,45]. Analytical thin-layer and high-pressure liquid chromatography (HPLC) techniques are used to help determine optimal solvent systems for the maximal separation of active components of fractions [46]. Other separation techniques, such as droplet countercurrent chromatography (DCCC), high-speed countercurrent chromatography (HSCCC), and semipreparative HPLC are used occasionally for complex mixtures of active constituents [26,46,47].

After pure active isolates are evaluated in all of the available *in vitro* assays, selected compounds are evaluated in the *ex vivo* mouse mammary organ culture model [31,32]. Highly promising leads may be selected for testing in full-term, animal tumorigenesis models, such as the two-stage mouse skin model using 7,12-dimethylbenz(*a*)anthracene (DMBA) as initiator and 12-*O*-tetradecanoylphorbol 13-acetate (TPA) as promoter, and the rat and mouse mammary carcinogenesis models with DMBA or *N*-methyl-*N*-nitrosourea (MNU) as the carcinogens [24,25,39]. Other animal models may also be used.

POTENTIAL CANCER CHEMOPREVENTIVE AGENTS FROM PLANTS

As an example of the success of this program, over a recent period of approximately five years a total of 166 active compounds were isolated and biologically evaluated in our laboratories from 32 plant species [48]. The active metabolites were obtained using activity-guided fractionation with a preselected *in vitro* assay to monitor their purification process. These active compounds were found to represent 29 major secondary metabolite compound classes including alkaloids (of the β-carboline alkaloid, indoloquinoline alkaloids, and steroidal types), amides, benzenoids, benzofurans, cardiac glycosides, ceramides, a coumarin, diarylheptanoids, diterpenoids, fatty acids, flavonoids (of the aurone, bisaurone, chalcone, flavan, flavanone, flavone, flavonol, flavonone, and isoflavone types), glycerin esters, a β-ionone derivative, an iridoid, lignans, a monoterpenoid, a naphthopyran, norwithanolides, phenylphenalones, a porphyrin derivative, a rocaglamide derivative, rotenoids, sesquiterpene lactones, sesquiterpenoids, simaroubolides, a stilbenolignan, stilbenoids, triterpenoids, and withanolides. Active compounds based on three different types of novel carbon skeletons were obtained during this work, which included seven norwithanolides possessing a new C_{27} skeleton (as opposed to the 28 carbons of the more widespread withanolides) [46,49], a novel stilbenolignan containing a stilbene-phenylpropane unit with a dioxane moiety [50], and two triterpenes based on a 29-*nor*-3,4-*seco*-cycloartane skeleton [51]. Forty-nine new compounds from 19 species were found among the compound classes mentioned above and were classified into 16 major structural classes. A large number of known bioactive compounds were isolated from 32 species, and can be grouped into 23 major structural classes. Many of these known isolates were accompanied in their plant of origin by inactive substances with new structures.

Natural product lead isolates found active in the mouse mammary organ culture (MMOC) assay include an indoloquinoline alkaloid, a β-carboline alkaloid, an amide, six flavonoids, a porphyrin derivative, two rotenoids, a triterpene, and four withanolides The activity of chemopreventive agents with this *ex vivo* system is known to demonstrate a good correlation with *in vivo* animal models [31,32].

As summarized in Table 13.1, nine agents resulting from this project are considered promising leads for further development. Brassinin is an indole dithiocarbamate from Chinese cabbage [52]. We evaluated the effects on DMBA-induced mammary carcinogenesis in Sprague-Dawley rats. Results showed that a four-week intragastric treatment with brassinin beginning three weeks prior to DMBA to one-week post-DMBA treatment reduced the tumor incidence from 87% in the control to 52% in the treatment group, i.e., a 40% reduction in incidence. The number of tumors was reduced from 4 tumors per rat to 1 tumor per rat, respectively.

Another new chemopreventive agent identified in our program is deguelin. Deguelin is a rotenone found in an African plant, *Mundulea sericea* [37]. It was found to be a potent inhibitor of ornithine decarboxylase activity and inhibited carcinogen-induced mammary lesions in the MMOC assay. In the two-stage skin carcinogenesis model, deguelin remarkably suppressed induction of papillomas from 75% in control mice to 10% in the low-dose (33 μg) treatment group, and complete suppression (no tumors in any mice) was observed with 330 μg of topical deguelin [53]. Since deguelin is an analog of rotenone, which is toxic and used as a pesticide, it was essential to compare the mode of action of these two agents. Unlike rotenone, deguelin did not mediate its effects by affecting tubulin polymerization and therefore its action is microtubule independent. It was also observed that deguelin reduced the steady-state level of mRNA for c-fos protooncogene, which could contribute to the transcription of ornithine decarboxylase gene via the c-fos/jun (AP1) complex. We also showed that in c-*myc*ER cells expressing c-myc-estrogen receptor fusion protein the activation of TPA-independent ODC activity was suppressed [53]. The efficacy of deguelin was also determined in the MNU-induced rat mammary carcinogenesis model. There was no effect on tumor incidence. However, multiplicity was reduced

TABLE 13.1
Selective Chemopreventive Agents Identified from Natural Products

Plant/source	Compound	Initial targets	Carcinogenesis
Brucea javanica	Brusatol	Differentiation (HL-60)	HL-60
Casimiroa edulis	Zapotin	Differentiation (HL-60); apoptosis	Colon
Cassia quinquangulata	Resveratrol	Cyclooxygenase inhibition	Skin, mammary, colon, prostate
Mundulea sericea	Deguelin	Ornithine decarboxylase	Skin, mammary, colon, melanoma
Brassica spp.	Brassinin	Quinone reductase	Skin, mammary
Physalis philadelphica	Withanolide	Quinone reductase	To be determined
Broussonetia papyrifera	Abyssinone II (RAPID)	Aromatase	To be determined
Synthetic	4′-Bromoflavone (RAPID)	Quinone reductase	Mammary
Synthetic	Oxomate	Quinone reductase	Mammary

from 6.8 tumors per rat to 3.2 tumors per rat in the high deguelin dose group [54]. More recently, we have found deguelin to be effective against human melanoma growth in athymic mice and in experimental colon carcinogenesis (unpublished), and efficacy has been demonstrated in a model of human lung cancer [55].

Other agents of particular promise include zapotin [56] and brusatol [57,58], and studies are ongoing to facilitate clinical development. Further, various structural derivatives have been explored. For example, numerous analogs of brassinin and flavones have been synthesized as modulators of quinone reductase activity. Among these agents, including chloro-, bromo-, methyl-derivatives, as well as various chalcones, 4′-bromoflavone was found to exhibit extremely potent induction of quinone reductase activity. We further evaluated 4′-bromoflavone in the DMBA-induced mammary carcinogenesis model. The compound was given at 2 and 4 g/kg of diet for 2 weeks (−1 to +1 week in relation to DMBA treatment), and remarkable suppression of both tumor incidence and multiplicity was observed. Incidence was reduced from 94% in the control to 35 and 20% in the low- and high-dose groups, respectively. Tumors per rat were reduced from the control level of 2.6 to 0.65 and 0.20, respectively, at the two dose levels tested. There was no effect on body weight due to this treatment [35]. This compound is currently being developed through the RAPID program of the NCI, as is the aromatase inhibitor abyssinone II [59].

Using an approach similar to that described herein, sulforaphane was identified as a potent inducer of quinone reductase by Talalay and co-workers [60]. Since then, analogs that appear more efficacious have been synthesized. Hybrid molecules of brassinin and sulforaphane, such as oxomate and sulforamate, were synthesized and evaluated in our laboratories, with *in vitro* and *in vivo* models. Both sulforamate and oxomate are less toxic than sulforaphane at effective doses [61]. These agents inhibited development of preneoplastic lesions in MMOC and chemically induced mammary carcinogenesis. In murine Hepa1c1c7 cells, sulforamate induced glutathione by twofold, and enhanced quinone reductase activity at a transcriptional level [61]. Oxomate is active in the DMBA rat mammary carcinogenesis model [62].

PHENOMENON OF RESVERATROL

One of the most fascinating molecules we have "rediscovered" is resveratrol (Figure 13.1). Resveratrol is a natural phytoalexin that is expressed in plants as a defensive response against fungal infections and other environmental stressors [63]. The word "alexin" is from the Greek language, meaning to ward off or to protect. Resveratrol may also have alexin-like activity in humans, protecting against degenerative diseases. Synthesis of resveratrol in grapes is most likely associated with natural stress factors such as exposure to ultraviolet radiation [64], injury, or during fungal or mold invasion [65].

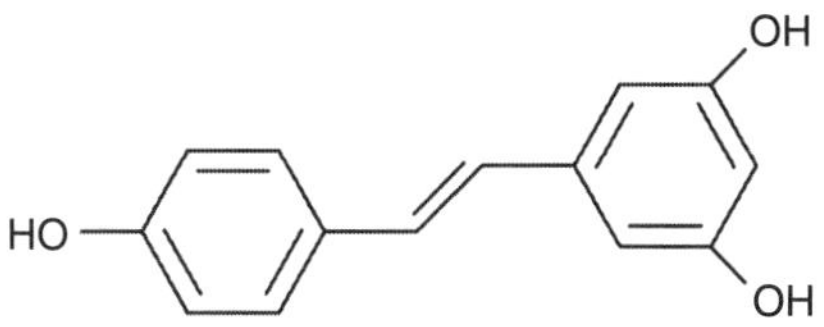

FIGURE 13.1 Structure of resveratrol.

Significant amounts of resveratrol were detected in healthy fruit clusters, prior to any detectable mold lesions. This suggested that the compound was synthesized soon after the recognition of the pathogen by the plant [66]. Montero et al. [67] investigated involvement of the plant hormone ethylene in resveratrol synthesis during fruit maturation. High resveratrol content correlated with low ethylene emission. Exogenous application of resveratrol on the fruit surface delayed the increase of ethylene emission, and doubled the normal shelf-life of grapes. This response is due to the antifungal activity of resveratrol, indicating the wide potential of such a compound for the control of the microbiota on fruits, and practical application as a natural chemical to prolong the shelf-life of fruits [68].

Resveratrol was first recognized as a biologically active compound by Siemann and Creasy [69]. The compound is found in several plants, chiefly in red grapes. The highest concentration (50 to 100 μg/g of grape wet weight) was determined in the grape skin. In wine, *cis* and *trans* isomers are present, in the free or glycosylated forms. *cis*-Resveratrol was not detected in grape skin and juices. Formation of the *cis* isomer by isomerization or breakdown of the *trans* form on exposure of wine to light and oxygen has been assumed [70]. In dietary supplements, the isomer is not always specified, but in most cases it is the *trans* form. In red wine, the concentration of the *trans* isomer ranges between 0.1 and 15 mg/l. The ratio of *cis*- and *trans*-resveratrol in wines varies by region. Climate, the type of grape, and the length of time the skin is kept with the grape during the winemaking process are some factors that influence the level of resveratrol and the ratio of isomers in wine [71]. Primarily, the compound is produced in the grape, grape shoots, and vines. Increasing irradiation of harvested grapes by ultraviolet B (UVB) or UVC light enhances yields of resveratrol [70]. Most resveratrol-containing supplements marketed in the U.S. contain extracts of the root of *Polygonium cuspidatum* Sieb. and Zucc., also known as the Japanese knotweed. The dried root and stem of this plant is used in traditional Japanese folk medicine (Ko-jo-kon) as a circulatory tonic, against fungal diseases, and various inflammatory and liver diseases [72]. Moreover, resveratrol synthase genes have been isolated and inserted into plants, creating transgenic varieties of alfalfa, tobacco, and other plant species with higher *trans*-resveratrol concentrations. Phytoalexins inserted into plants may provide defense against different pathogens [73]. Additionally, transgenic plants, e.g., alfalfa, transformed with resveratrol-synthesizing genes might become an economical source of the compound for scientific research or dietary supplements [74].

As part of our search for natural product cancer chemopreventive agents, acquisition number 46 (the current total is 4121), a nonedible legume identified as *Cassia quinquangulata* Rich. (Leguminosae), was extracted and found to demonstrate impressive inhibition with cyclooxygenase-1. Activity was also observed in the MMOC model, and the extract was selected for bioassay-guided fractionation. As a result, resveratrol was readily identified as the active principle. In addition to inhibiting cyclooxygenase activity, suggestive of antipromotional activity, the isolate was found to serve as an antioxidant and antimutagen. Further, it induced phase II drug-metabolizing enzymes involved chiefly in the detoxification of carcinogen metabolites (anti-initiation activity) and induced human promyelocytic leukemia cell differentiation (antiprogression activity). Finally, antitumor and anti-inflammatory effects were observed with mouse and rat models, respectively, providing support for the physiological significance of the *in vitro* and cell culture data [39].

When these data were published in 1997, a search of MEDLINE revealed a total of 21 manuscripts in the literature, largely relating to the natural occurrence of resveratrol, rather than biologic potential. There was a huge response by the media and public, perhaps because it was otherwise a slow news day, but clearly the public found comfort in the notion of food and beverages (such as grapes and wine) being of benefit to their health. Obviously, in addition to the general population, this notion attracted the attention of the scientific community. As indicated by a recent query of MEDLINE, from 1997 to the present a total of 1037 manuscripts investigating resveratrol have been published (Figure 13.2). Symposia have been conducted [75], funding streams have been created (California Table Grape Commission, http://www.tablegrape.com/), companies have been formed (Royalmount Pharma, http://www.royalmountpharma.com/), monographs and reviews have been written [76–81]. Of some importance, since this molecule is not complex, facile chemical syntheses have been devised, so abundant supplies of resveratrol are available [82–86].

In this review, a synopsis of the literature describing the cancer-related activity of resveratrol is presented. The results are presented in tabular form, roughly divided into reports studying resveratrol with *in vitro* models, cell culture systems, and *in vivo* systems. In studies where multiple models were employed, the paper is listed in the table representing the highest level of biological complexity.

IN VITRO STUDIES CONDUCTED WITH RESVERATROL

Relatively few reports have appeared wherein the primary tests were performed mainly with *in vitro* model systems. Some are presented in

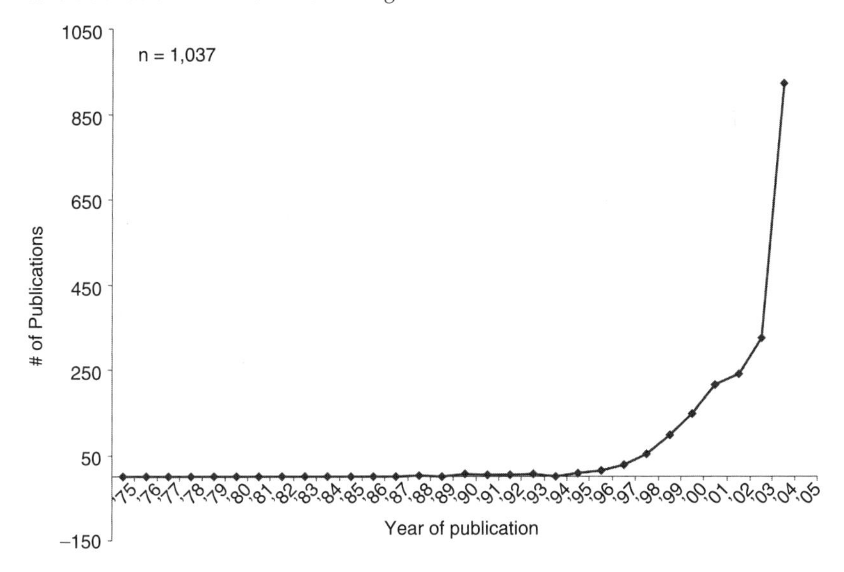

FIGURE 13.2 Line chart mapping the number of publications dealing with resveratrol from 1975 to 2004.

Table 13.2. Key observations have demonstrated antioxidant [34,87] activity, and ability to inhibit protein kinases [88–90], cyclooxygenases [91–93], cytochromes P450 [94], and tyrosinase [95]. Cr-induced damage to DNA can be prevented, probably through radical scavenging [96]. Metabolism has also been established, through glucuronidation and sulfonation, conversion to piceatannol, and metabolism can be modulated by flavonoids such as quercetin [97–99]. Mammalian proteins capable of binding resveratrol have been identified [100,101].

CELL CULTURE STUDIES CONDUCTED WITH RESVERATROL

Clearly, as summarized in Table 13.3, the majority of studies that have been performed to investigate the mode of resveratrol action involve cultured cells. In our original report [39], HL-60 and Hepa 1c1c7 cells were used, and these responses have been confirmed and expanded. As models of human cancers, prostate, colon, lung, breast, ovarian, renal, hepatoma, leukemic, bronchial, neuroblastoma, cervical, lymphoma, medulloblastoma, endometrial, esophageal, melanoma, pancreatic, gastric, epidermal, thyroid, fibroblast, retinoblastoma, and squamous cells, as well as macrophages, monocytes, myofibroblasts, transformed and transfected cells, and organ culture systems have been used. Activities may vary from system to system, but some generalizations apply. Certainly, apoptosis is a common mode of

TABLE 13.2
Evaluation of Resveratrol with *In Vitro* Model Systems

Model	What was measured	Effect	Ref.
Inhibition of COX-1	Some constituents found in red wine were tested for inhibition of COX-1	These agents were found to be as effective as resveratrol with respect to their ability of specifically inactivating COX-1	91
Resveratrol-targeting proteins (RTPs)	Resveratrol was immobilized on epoxy-activated agarose forming a resveratrol affinity column (RAC), which was used to detect and isolate RTPs	These results provide evidence for the existence of distinctive RTPs in mammalian cells and that RAC is a facile approach to identify and purify RTPs	100
Plasma protein binding	The bioavailability and bioabsorption of resveratrol in cells and tissues	Resveratrol was shown to interact with lipoproteins in plasma	101
The effect of resveratrol on several different systems involving the hydroxyl, superoxide, metal/enzymatic-induced, and cellular-generated radicals	The rate constant for reaction of resveratrol with the hydroxyl radical was determined, and resveratrol was found to be an effective scavenger of hydroxyl, superoxide, and metal-induced radicals, as well as showing antioxidant abilities in cells producing ROS. Resveratrol exhibits a protective effect against lipid peroxidation in cell membranes and DNA damage caused by ROS. Resveratrol was also found to have a significant inhibitory effect on the NF-κB signaling pathway after cellular exposure to metal-induced radicals	It was concluded that resveratrol in food plays an important antioxidant role	87
Metabolism of resveratrol by the cytochrome P450 enzyme CYP1B1 to produce piceatannol, a stilbene that has anti-leukemic activity and is also a tyrosine kinase inhibitor	The metabolite was identified by high-performance liquid chromatography analysis using fluorescence detection and the identity of the metabolite was further confirmed by derivatization followed by gas chromatography/mass spectrometry studies using authentic piceatannol for comparison	This observation provides a novel explanation for the cancer preventative properties of resveratrol. It demonstrates that a natural dietary cancer preventative agent can be converted to a compound with known anticancer activity by an enzyme that is found in human tumors. This result gives insight into the functional role of CYP1B1 and provides evidence for the concept that CYP1B1 in tumors may be functioning as a growth suppressor enzyme	97

Inhibitory effect on tyrosinase and mechanism of action for oxyresveratrol and hydroxystilbene compounds	To clarify the mechanism of the depigmenting property of hydroxystilbene compounds, inhibitory actions of oxyresveratrol and its analogs on tyrosinases from mushroom and murine melanoma B-16 have been elucidated in this study. Oxyresveratrol showed potent inhibitory effect with an IC_{50} value of 1.2 μM on mushroom tyrosinase activity, which was 32-fold stronger inhibition than kojic acid, a depigmenting agent used as the cosmetic material with skin-whitening effect and the medical agent for hyperpigmentation disorders. Hydroxystilbene compounds of resveratrol, 3,5-dihydroxy-4′-methoxystilbene, and rhapontigenin also showed more than 50% inhibition at 100 μM on mushroom tyrosinase activity, but other methylated or glycosylated hydroxystilbenes of 3,4′-dimethoxy-5-hydroxystilbene, trimethylresveratrol, piceid, and rhaponticin did not inhibit significantly. None of the hydroxystilbene compounds except oxyresveratrol exhibited more than 50% inhibition at 100 μM on L-tyrosine oxidation by murine tyrosinase activity; oxyresveratrol showed an IC_{50} value of 52.7 μM on the enzyme activity. The kinetics and mechanism for inhibition of mushroom tyrosinase exhibited the reversibility of oxyresveratrol as a noncompetitive inhibitor with L-tyrosine as the substrate. The interaction between oxyresveratrol and tyrosinase exhibited a high affinity reflected in a K_i value of $3.2–4.2 \times 10^7$M	Oxyresveratrol did not affect the promoter activity of the tyrosinase gene in murine melanoma B16 at 10 and 100 μM. Therefore, the depigmenting effect of oxyresveratrol works through reversible inhibition of tyrosinase activity rather than suppression of the expression and synthesis of the enzyme. The number and position of hydroxy substituents seem to play an important role in the inhibitory effects of hydroxystilbene compounds on tyrosinase activity	95

(*Continued*)

TABLE 13.2
Continued

Model	What was measured	Effect	Ref.
Reduction of DNA damage induced by Cr(III) using free radical scavengers [melatonin, N^1-acetyl-N^2-formyl-5-methoxykynuramine (AFMK), resveratrol, and uric acid]	The concentrations that reduced 8-hydroxydeoxyguanosine (8-OH-dG, an index for DNA damage) formation by 50% (IC_{50}) were 0.10 μM for both resveratrol and melatonin, and 0.27 μM for AFMK. The efficacy of uric acid, in terms of its inhibition of DNA damage in the same *in vitro* system, was about 60–150 times less effective than the other scavengers; the IC_{50} for uric acid was 15.24 μM	These findings suggest that three of the four antioxidants tested in these studies may have utility in protecting against the environmental pollutant Cr and that the protective effects of these free radical scavengers against Cr(III)-induced carcinogenesis may relate to their direct hydroxyl radical scavenging ability. The formation of 8-OH-dG was likely due to a Cr(III)-mediated Fenton-type reaction that generates hydroxyl radicals, which in turn damage DNA. Once formed, 8-OH-dG can mutate eventually leading to cancer; thus, the implication is that these antioxidants may reduce the incidence of Cr-related cancers	96
Resveratrol glucuronidation in human liver microsomes and to determine whether flavonoids inhibit resveratrol glucuronidation	The rate of resveratrol glucuronidation was measured in 10 liver samples. The mean ± SD and median of resveratrol glucuronidation rates were 0.69 ± 0.34 and 0.80 nmol/min/mg, respectively. Resveratrol glucuronosyl transferase followed Michaelis–Menten kinetics and the K_m and V_{max} (mean ± SD; $n = 5$) were 0.15 ± 0.09 mM and 1.3 ± 0.3 nmol/min/mg, respectively. The intrinsic clearance was $11 \pm 4 \times 10^{-3}$ ml/min/mg. The flavonoid quercetin inhibited resveratrol glucuronidation and its IC_{50} (mean ± SD; $n = 3$) was 10 ± 1 μM. Myricetin, catechin, kaempferol, fisetin, and apigenin (all at 20 μM) inhibited resveratrol glucuronidation and the percent of control ranged between 46% (catechin) and 72% (apigenin)	These results show that resveratrol is glucuronated in the human liver. Glucuronidation may reduce the bioavailability of this compound. However, flavonoids inhibit resveratrol glucuronidation and such an inhibition might improve the bioavailability of resveratrol	98

Investigation of the ability of resveratrol to inhibit protein kinase D (PKD)	The study compares the effects of resveratrol against the autophosphorylation reactions of PKC isozymes versus PKD. It was found that resveratrol inhibits PKD autophosphorylation in a concentration-dependent manner, but has only negligible effects against the autophosphorylation reactions of representative members of each PKC isozyme subfamily (cPKC-α, -β_1, and -γ, nPKC-Δ and -ε, and aPKC-ζ). Resveratrol was comparably effective against PKD autophosphorylation and PKD phosphorylation of the exogenous substrate syntide-2	The inhibitory potency of resveratrol against PKD may contribute to the cancer chemopreventive action of resveratrol	88
Investigation of the sulfation of resveratrol in the human liver and duodenum	A radiometric assay for resveratrol sulfation was developed. It employed 3′-phosphoadenosine-5′-phosphosulfate-[^{35}S] as the sulfate donor and the rates of resveratrol sulfation (mean ± SD, pmol/min/mg cytosolic protein) were 90 ± 21 (liver, $n = 10$) and 74 ± 60 (duodenum, $n = 10$, $p = 0.082$). Resveratrol sulfotransferase followed Michaelis–Menten kinetics and K_m (mean ± SD; çM) was 0.63 ± 0.03 (liver, $n = 5$) and 0.50 ± 0.26 (duodenum, $n = 5$, $p = 0.39$) and V_{max} (mean ± SD, pmol/min/mg cytosolic protein) were 125 ± 31 (liver, $n = 5$) and 129 ± 85 (duodenum, $n = 5$, $p = 0.62$). Resveratrol sulfation was inhibited by the flavonoid quercetin, mefenamic acid, and salicylic acid. IC_{50} of resveratrol sulfation for quercetin was 12 ± 2 pM (liver) and 15 ± 2 pM (duodenum), those for mefenamic acid were 24 ± 3 nM (liver) and 11 ± 0.6 nM (duodenum), and those for salicylic acid were 53 ± 9 μM (liver) and 66 ± 4 μM (duodenum)	The potent inhibition of resveratrol sulfation by quercetin suggests that compounds present in the diet may inhibit the sulfation of resveratrol, thus improving its bioavailability	99

(*Continued*)

TABLE 13.2
Continued

Model	What was measured	Effect	Ref.
Incorporation into model membranes and inhibition of protein kinase C α (PKCα) activity	Differential scanning calorimetry measured the effect of resveratrol on the gel to liquid-crystalline phase transition of multilamellar vesicles made of phosphatidylcholine/phosphatidylserine and the temperature at which the fluid lamellar to H(II) inverted hexagonal transition took place in multilamellar vesicles made of 1,2-dielaidoyl-*sn*-phosphatidylethanolamine. This effect on 1,2-dielaidoyl-*sn*-phosphatidylethanolamine polymorphism was confirmed through ^{31}P-NMR, which showed that an isotropic peak appeared at high temperature instead of the H(II) characteristic peak of 42 mM of resveratrol. The ability of resveratrol to inhibit PKCα when activated by phosphatidylcholine/phosphatidylserine vesicles was tested	The study reports that resveratrol is able to incorporate itself into model membranes in a location that is inaccessible to the fluorescence quencher acrylamide. These results indicate that the inhibition of PKCα by resveratrol can be mediated, at least partially, by membrane effects exerted near the lipid–water interface	89
Mechanism of protein kinase C (PKC) inhibition	Various systems were employed to determine inhibition by resveratrol: Ca^{2+}/phosphatidylserine-stimulated activity of a purified rat brain PKC isozyme mixture by competition with ATP; lipid-dependent activity of PKC isozymes with divergent regulatory domains; cofactor-independent catalytic domain fragment (CDF) of PKC generated by limited proteolysis. The effects of resveratrol were examined on PKC-catalyzed phosphorylation of the cofactor-independent substrate protamine sulfate, which is a polybasic protein that activates PKC by a novel mechanism	The results indicate that resveratrol has a broad range of inhibitory potencies against purified PKC that depend on the nature of the substrate and the cofactor dependence of the phosphotransferase reaction	90

Inhibition of cytochrome P450 1A1	To investigate the mechanism of anticarcinogenic activity of resveratrol, the effects on cytochrome P450 (P450) were determined in human liver microsomes and *Escherichia coli* membranes coexpressing human P450 1A1 or P450 1A2 with human NADPH-P450 reductase (bicistronic expression system)	Resveratrol slightly inhibited ethoxyresorufin *O*-deethylation (EROD) activity in human liver microsomes. Resveratrol exhibited potent inhibition of human P450 1A1 in a dose-dependent manner for EROD and for methoxyresorufin *O*-demethylation (MROD). The inhibition of human P450 1A2 by resveratrol was not very strong. Resveratrol showed over 50-fold selectivity for P450 1A1 over P450 1A2. The activities of human NADPH-P450 reductase were not significantly changed by resveratrol. In a human P450 1A1/reductase bicistronic expression system, resveratrol inhibited human P450 1A1 activity in a mixed-type inhibition (competitive/noncompetitive). These results suggest that resveratrol is a selective human P450 1A1 inhibitor	94
Identification of antioxidants in plant extracts	Test materials were assessed for potential to scavenge stable 1,2-diphenyl-2-picrylhydrazyl (DPPH) free radicals, reduce TPA-induced free radical formation in cultured HL-60 human leukemia cells, and inhibit responses observed with a xanthine/xanthine oxidase assay system. Based on secondary analyses performed to assess inhibition of 7,12-dimethylbenz(*a*)anthracene-induced preneoplastic lesion formation with a mouse mammary organ culture model, various plants were selected and subjected to bioassay-guided fractionation. Various compounds were identified	Approximately 700 plant extracts were evaluated, and 28 were found to be active in the DPPH free radical scavenging assay. The hydroxystilbenes piceatannol and resveratrol inhibited carcinogen-induced preneoplastic lesion formation in the mouse mammary gland organ culture model	34

Note: In addition to data obtained with cell culture or *in vivo* models, the following references contain data derived from *in vitro* systems: 78, 86, 103–107, 108–111, [113, 117–118, 121–124, 128, 130, 131, 133–139, 141, 144–146, 150, 153–156, 162, 163, 166, 167, 176, 177, 180–185, 187–189, 192–194, 196, 197, 199, 200–204, 206, 207, 209–231, 233–239, 241–257, 259, 260, 263, 267, 269–271, 273, 276, 277, 279, 280, 287–291.

TABLE 13.3
Evaluation of Resveratrol with Cell Culture Model Systems

Model	What was measured	Effect	Ref.
U937 cell growth	Inhibition of cell cycle progression	Cell growth was impaired due to reduced cell proliferation, without significant induction of apoptosis. There was an antiproliferative effect and duplication in the DNA of the resveratrol-treated U937 cells	168
Prostate cancer cells LNCaP, DU145, and PC-3	The role of phosphoglycerate mutase B using two-dimensional sodium dodecyl sulfate polyacrylamide gel electrophoresis (2D-SDS-PAGE) followed by mass spectrometric analysis of the prostate cells	Identified the role of phosphoglycerate mutase B on resveratrol-treated human prostate cells at the transcription level. There was an effect on metabolic enzymes in cancer cells, but did not affect normal cells	169
Polyamine metabolism in colorectal cancer cells	Inhibition of ornithine decarboxylase (ODC)	Due to resveratrol and some of its analogs interfering with signal transduction pathways, the activities of some protein kinases were inhibited, the expression of nuclear protooncogenes declines, and the activity of ODC is reduced	170
DU145 and LNCaP prostate cancer cells	Inhibition of epidermal growth factor (EGF) for up to 4 h resulted in brief activation of MAPK followed by inhibition of resveratrol-induced signal transduction, p53 phosphorylation, and apoptosis	Resveratrol-induced apoptosis in DU145 and LNCaP prostate cancer cells occurs through different PKC-mediated and MAPK-dependent pathways	171
DU145 prostate cancer cells	Using resveratrol and the ethanolic extract of propolis, a natural honeybee hive product, a comparison between the activity of these micronutrients and vinorelbine bitartrate (Navelbine), a semisynthetic drug normally used in the therapy of prostate cancer, was conducted. Several biochemical parameters were tested using the MTT assay, cell membrane integrity (lactate dehydrogenase release), cell redox status (nitric oxide formation, reactive oxygen species production, reduced glutathione levels), COMET assay with special attention on the presence of apoptotic DNA damage	Studies demonstrated the anticancer activity of resveratrol and propolis extract in human prostate cancer, exerting their cytotoxicity through two different types of cell death: necrosis and apoptosis, respectively. The data obtained suggest the possible use of these micronutrients both as an alternative to classic chemotherapy, and in combination with very low dosage of vinorelbine (5 μM)	172

	(TUNEL test), and possible mitochondrial transmembrane potential alteration ($\Delta \psi$)		
Colon carcinoma cells	In colon cancer cells that resist resveratrol-induced apoptosis, the polyphenol also induces a redistribution of death receptors into lipid rafts. This effect sensitizes these tumor cells to death receptor-mediated apoptosis. In resveratrol-treated cells, tumor necrosis factor (TNF), anti-CD95 antibodies, and TNF-related apoptosis-inducing ligand (TRAIL) activate a caspase-dependent death pathway that escapes Bcl-2-mediated inhibition	Resveratrol does not enhance the number of death receptors at the surface of tumor cells but induces their redistribution into lipid rafts and facilitates the caspase cascade activation in response to death receptor stimulation. The cholesterol sequestering agent nystatin prevents resveratrol-induced death receptor redistribution and cell sensitization to death receptor stimulation. Thus, whatever its ability to induce apoptosis in a tumor cell, resveratrol induces redistribution of death receptors into lipid rafts. This redistribution sensitizes the cells to death receptor stimulation	173
Inhibitors of NF-κB and AP-1	Due to the ability of resveratrol to interfere with the multistage carcinogenesis, numerous intracellular signaling cascades converge with the activation of NF-κB and AP-1, which act independently or coordinately to regulate expression of target genes. These eukaryotic transcription factors mediate pleiotropic effects on cellular transformation and tumor promotion	An update on the molecular mechanisms underlying chemoprevention by resveratrol with special focus on its effect on cellular signaling cascades mediated by NF-κB and AP-1	174
Multiple target signaling pathways	Efficacy of chemopreventive agents by preventing or reversing premalignant lesions and/or reducing second primary tumor incidence	This review focuses on recent work regarding three well-accepted cellular/molecular mechanisms that may at least partially explain the effectiveness of selected food factors as chemopreventive antipromotion agents. These food compounds may act by: (1) inducing apoptosis in cancer cells; (2) inhibiting neoplastic transformation through the inhibition of AP-1 and/or NF-κB activation; and/or (3) suppressing COX-2 overexpression in cancer cells	175

(Continued)

TABLE 13.3
Continued

Model	What was measured	Effect	Ref.
Cultured human lung cancer cells (A549)	Using resveratrol as the prototype stilbenoid, this group synthesized various analogs and evaluated their growth inhibitory effects in cultured human cancer cells. One of the stilbenoids, 3,4,5-trimethoxy-4′-bromo-*cis*-stilbene (BCS), was more effective than its corresponding *trans* isomer and resveratrol on the inhibition of cancer cell growth. Prompted by the strong growth inhibitory activity of BCS compared to its *trans* isomer and resveratrol in A549, the mechanism of action was investigated. BCS induced arrest at the G_2/M phase of the cell cycle at early times and subsequently increased in the sub-G_1 phase DNA contents in a time-dependent manner, indicating induction of apoptosis. Morphological observation with round-up shape and DNA fragmentation also revealed the apoptotic phenomena. BCS treatment elevated the expression levels of the proapoptotic protein p53, the cyclin-dependent kinase inhibitor p21, and the release of cytochrome c in the cytosol. The downregulation of checkpoint protein cyclin B1 by BCS was well correlated with the cell cycle arrest at G_2/M	These data suggest the potential of BCS to serve as a cancer chemotherapeutic or chemopreventive agent by virtue of arresting the cell cycle and induction of apoptosis of human lung cancer cells	115
ER-α and ER-β binding ability with quinone reductase (QR) expression in breast cancer cells	Used three phytoestrogens, biochanin A, genistein, and resveratrol, since they upregulate QR expression in breast cancer cells. It was reported that regulation can occur at the transcriptional level preferentially through ER-β transactivation at the electrophile response	The protective ability of resveratrol is partially dependent on the presence of ER-β and QR. Phytoestrogen-mediated induction of QR may represent an additional mechanism for breast cancer protection, although the effects may be specific for a given phytoestrogen	102

	element of the QR gene promoter. Chromatin immunoprecipitation analysis showed binding of ER-α and ER-β to the QR promoter, with increased ER-β binding in the presence of resveratrol. Antisense technology was used to determine whether such protection was dependent on ER-β or QR		
LNCaP and LAPC-4 prostate cancer cell lines	Quercetin and resveratrol caused an increase in expression of c-Jun as well as its phosphorylated form in a dose-dependent manner in prostatic cell lines using a transient transfection assay. Gel shift assays showed that induced c-Jun has specific DNA binding activity. Transient transfections demonstrated that c-Jun repressed prostate-specific antigen promoter activity and transcriptional activity of the androgen receptor (AR) promoter	These results support a mechanism in which overexpressed c-Jun mediates an inhibitory effect on the function of AR	176
Human ovarian cancer cells A2780/CP70 and OVCAR-3	Investigated the effect of resveratrol on hypoxia-inducible factor 1α (HIF-1α) and vascular endothelial growth factor (VEGF) expression	This study demonstrated that resveratrol inhibited HIF-1α and VEGF expression through multiple mechanisms. First, resveratrol inhibited AKT and mitogen-activated protein kinase activation, which played a partial role in the downregulation of HIF-1α expression. Second, resveratrol inhibited insulin-like growth factor 1-induced HIF-1α expression through the inhibition of protein translational regulators, including M_r 70,000 ribosomal protein S6 kinase 1, S6 ribosomal protein, eukaryotic initiation factor 4E-binding protein 1, and eukaryotic initiation factor 4E. Finally, it was found that resveratrol substantially induced HIF-1α protein degradation through the proteasome pathway. These data suggest that resveratrol may inhibit human ovarian cancer progression and angiogenesis by inhibiting HIF-1α and VEGF expression and thus provide a novel potential mechanism for the anticancer action of resveratrol	177

(*Continued*)

TABLE 13.3
Continued

Model	What was measured	Effect	Ref.
The expression of 2059 cancer-related genes in a renal cell carcinoma (RCC) cell line RCC54 treated with resveratrol	Biological functions of 633 genes were annotated based on biological process ontology and clustered into functional categories. Twenty-nine highly differentially expressed genes in resveratrol-treated RCC54 were identified and the potential implications of some gene expression alterations in RCC carcinogenesis were identified	The findings from this study support the hypothesis that resveratrol induces differential expression of genes that are directly or indirectly related to the inhibition of RCC cell growth and induction of RCC cell death. In addition, it is apparent that the gene expression alterations due to resveratrol treatment depend strongly on resveratrol concentration. This study provides a general understanding of the overall genetic response of RCC54 to resveratrol treatment and yields insights into the understanding of the cancer preventive mechanism of resveratrol in RCC	136
Used sensitizer for anticancer drug-induced apoptosis by inducing cell cycle arrest, which in turn resulted in survivin depletion	Analysis of cell cycle and apoptosis revealed that pretreatment with resveratrol resulted in cell cycle arrest in S phase and apoptosis induction preferentially out of S phase upon subsequent drug treatment. Likewise, cell cycle arrest in S phase by cell cycle inhibitors enhanced drug-induced apoptosis. Resveratrol-mediated cell cycle arrest sensitized for apoptosis by downregulating survivin expression through transcriptional and post-transcriptional mechanisms. Similarly, downregulation of survivin expression using survivin antisense oligonucleotides sensitized for drug-induced apoptosis. Downregulation of survivin and enhanced drug-induced apoptosis by resveratrol occurred in various human tumor cell lines irrespective of p53 status	The combined sensitizer (resveratrol)/inducer (cytotoxic drugs) concept may be a strategy to enhance the efficacy of anticancer therapy in a variety of human cancers	178

Hep3B hepatoma, Caki-1 renal carcinoma, SK-N-MC neuroblastoma, and HEK293 cell lines	Hypoxia-induced drug resistance is a major obstacle in the development of effective cancer chemotherapy. Examined whether drug resistance of various phenolic compounds (e.g., resveratrol) is acquired by hypoxia. The cell lines were cultured under normoxic or hypoxic conditions. Drug sensitivities to the phenolic compounds and expression of hypoxia-inducible factor-1α (HIF-1α) and the multidrug-resistance genes were examined in these cell lines	Drug resistance was acquired 24 h after hypoxia and subsided 8 h after reoxygenation. Protein synthesis inhibitors abolished this drug resistance. A transfection study demonstrated that HIF-1α enhanced this hypoxia-induced resistance and that its dominant-negative isoform suppressed resistance acquisition. However, MDR1 and MRP1, which provide multidrug resistance to conventional anticancer agents, were not induced by hypoxia. These results suggest that HIF-1α-dependent gene expression participates in the cellular process of the hypoxia-induced resistance to phenolic compounds	179
Human lymphoblastoid cell line TK6 and its p53-knockout counterpart (NH32)	The proapoptotic ability of (*Z*)-3,5,4′-tri-*O*-methyl-resveratrol (R3) was investigated using these cell lines. R3 induced the stimulation of caspase-3. Although R3 induced growth inhibition and apoptosis of both cell lines, two distinct mechanisms were observed	The p53-knockout NH32 cells were shown to override the G_2/M phase checkpoint with development of hyper-diploid cells, whereas TK6 cells accumulated at G_2/M. As p53 function is often altered in human cancer cells, these results show that the proapototic effects of R3 against tumor cells are independent of their p53 status	180
Leukemia B-cell lines and B-cell chronic lymphocytic leukemia (B-CLL) cells from patients	Resveratrol induces apoptosis of leukemic B-cells and simultaneously inhibits the production of endogenous nitric oxide (NO) through inducible NO synthase (iNOS) downregulation. In addition these results were observed with not only acetate derivatives of polyphenols, particularly the pentaacetate of viniferin (resveratrol dimer), but also with a synthetic flavone (a diaminomethoxyflavone) in both leukemia B-cell l and B-CLL cells from patients. Moreover, flavopiridol, another flavone already known for its proapoptotic properties in B-CLL cells, was found to downregulate both iNOS expression and NO production	Inhibition of the NO pathway during apoptosis of leukemia B-cells appears a common mechanism for several compounds belonging to two distinct families of phytoalexins, the flavones and grape-derived polyphenols	181

(Continued)

TABLE 13.3
Continued

Model	What was measured	Effect	Ref.
Human breast cancer MCF-7 cells	The mechanism of resveratrol to induce apoptosis in MCF-7 cells was dependent on mitogen-activated protein kinase (MAPK, ERK1/2) activation and was associated with serine phosphorylation and acetylation of p53. Treatment of MCF-7 cells with resveratrol in the presence of 17β-estradiol (E_2) further enhanced MAPK activation, but E_2 blocked resveratrol-induced apoptosis, as measured by nucleosome ELISA and DNA fragmentation assays. E_2 inhibited resveratrol-stimulated phosphorylation of serines 15, 20, and 392 of p53 and acetylation of p53 in a concentration- and time-dependent manner. These effects of E_2 on resveratrol action were blocked by ICI 182,780 (ICI), an inhibitor of the nuclear estrogen receptor-α (ER). ICI 182,780 did not block the actions of resveratrol, alone. Electrophoretic mobility studies of p53 binding to DNA and of p21 expression indicated that E_2 inhibited resveratrol-induced, p53-directed transcriptional activity	These results suggest that E_2 inhibits p53-dependent apoptosis in MCF-7 cells by interfering with posttranslational modifications of p53 which are essential for p53-dependent DNA binding and consequent stimulation of apoptotic pathways	182
CYP1A1 and 1B1 in UW228-3 medulloblastoma cells	The status of CYP1A1 and 1B1 in UW228-3 medulloblastoma cells without and with resveratrol treatments was elucidated in this study with the ethoxyresorufin *O*-deethylation assay, followed by RT-PCR, immunocytochemical staining, and Western blot hybridization. CYP1A1/1B1 enzymatic activity was low in UW228-3 cells but became several-fold higher upon resveratrol treatments. CYP1A1 was undetectable and CYP1B1 was	The results demonstrate that in the medulloblastoma cell system, CYP1A1 upregulation is paralleled with resveratrol-induced differentiation and apoptosis, while CYP1B1 may not be an essential element in metabolic activation of resveratrol in those cells. CYP1A1 and 1B1 are resveratrol response genes and potential chemo-sensitive markers of medulloblastoma cells	183

	cexpressed in normally cultured cells. Accompanied by the increased fraction of apoptosis, enhanced CYP1A1 and downregulated CYP1B1 were observed in resveratrol-treated cells in time- and dose-related fashions		
Various human breast cancer cells and a wild-type cell line, astrocytoma N 1321N1 was used as the control	Purpose of study was to identify the role of p53-dependent or p53-independent pathways in the induction of apoptosis in human breast cancer cells by resveratrol. A number of human breast cancer cell lines, as well as astrocytoma N 1321N1, were investigated for induction of apoptosis by resveratrol using both microscopic evaluation and DNA fragmentation assays. Concurrently, the p53 gene status (wild-type or mutant) of each cell line was established by Western blot using p53-specific antibody	Apoptosis induced by resveratrol was found to occur only in breast cancer cells expressing wild-type p53 but not in mutant p53-expressing cells. The study suggests that resveratrol induces apoptosis in breast cancer cells via p53-dependent pathways	184
Benzo(*a*)pyrene [B(*a*)P]-induced DNA adducts in human bronchial epithelial cells	The *in vitro* effect of resveratrol on B(*a*)P-induced DNA adducts in human bronchial epithelial cells. This was compared to the effect of resveratrol on the expression of the cytochrome P450 (CYP) genes CYP1A1 and CYP1B1 and the formation of B(*a*)P metabolites. Exposure of BEAS-2B and BEP2D cells to B(*a*)P and increasing concentrations of resveratrol resulted in a dose- and time-dependent inhibition of DNA adduct formation quantified by ^{32}P-postlabeling. Supporting this result, resveratrol was shown to inhibit CYP1A1 and CYP1B1 gene expression, as measured by RT-PCR. Also, a significant correlation was found between the number of DNA adducts and the mRNA levels of these genes. Using HPLC analysis, a concomitant decrease in the formation of B(*a*)P-derived metabolic products was detected	These data suggest that resveratrol has a chemopreventive role in polycyclic aromatic hydrocarbon-induced carcinogenesis	185

(*Continued*)

TABLE 13.3
Continued

Model	What was measured	Effect	Ref.
Stage 4 MYCN-amplified neuroblastoma (NB) cell lines	Stage 4 MYCN-amplified NB cell lines, with wild-type or mutant p53, were treated with curcumin and resveratrol and analyzed for effects on proliferation, cell cycle, induction of apoptosis, and p53 function	Observations suggest that the cytotoxicity, cell cycle arrest, and apoptosis induced by curcumin and resveratrol in NB cells may be mediated via functionally activated p53	186
Human breast tumor cell lines (HBL100, MCF7, and MBA-MB-231) and one breast cell line (MCF10a)	HBL100, MCF7, MBA-MB-231, and MCF10a cells were used to study the effect of resveratrol on the transcription of a group of genes whose proteins interact in different pathways with BRCA1. BRCA1, BRCA2, ERα, ERβ, p53, $p21^{waf1/cip1}$, CBP/P300, RAD51, pS2 and K*i*67 mRNA were quantified using real-time quantitative RT-PCR	Resveratrol modulated the expression of these genes in a pattern dependent on the status of α and β estrogen receptors. These results show that resveratrol regulates gene expression via the estrogen receptor pathway and also an undetermined pathway. Thus, resveratrol seems to have an effect on breast tumor cell lines by affecting several factors regulating the function of BRCA1	103
Human HT-29 colon cancer cells	Isothiocyanates, flavonoids, resveratrol, and curcumin were examined in this study. HT-29 cells were stably transfected with NF-κB luciferase construct, and stable clones were selected. HT-29 N9 cells were selected and treated with various concentrations of the natural chemopreventive agents and subsequently challenged with NF-κB stimulator lipopolysaccharide, and the luciferase activities were measured. Western blot analysis of phosphorylated IκBα was performed after treatments with the natural chemopreventive agents. The effects of these agents on cell viability and apoptosis were also evaluated by a nonradioactive cell proliferation MTS assay [3-(4,5-dimethylthiazol-2-yl)-5-(3-arboxymethoxyphenyl)-2-(4-sulfophenyl)-2H-tetrazolium, inner salt], Trypan blue staining, and caspase assay	Treatments resulted in different responses in the NF-κB-luciferase assay. ITCs such as phenethyl isothiocyanate, sulforaphane, allyl isothiocyanate, and curcumin strongly inhibited LPS-induced NF-κB-luciferase activation, whereas resveratrol increased activation at lower dose, but inhibited activation at higher dose, and tea flavonoids and procyanidin dimers had little or no effect. ITCs, curcumin, (−)-epigallocatechin-3-gallate, and resveratrol reduced LPS-induced IκBα phosphorylation. Furthermore, in the MTS assay, PEITC, SUL, and curcumin also potently inhibited cell growth. Caspase-3 activity was induced by chemopreventive compounds; however, the kinetics of caspase-3 activation varied between these compounds within the 48 h time period. These results suggest that natural chemopreventive agents have differential biological functions on the signal transduction pathways in the colon and/or colon cancer	187

HT-29 colon cancer cells	HT-29 cells were transfected with AP-1-luciferase reporter gene, and one of the stable clones (C-4) was used for subsequent experiments. The HT-29 C-4 cells were treated for 1 h with various natural chemopreventive agents and challenged with AP-1 stimulators such as 12-*O*-tetradecanoylphorbol-13-acetate (TPA) or hydrogen peroxide (H_2O_2) for 6 h. The c-Jun N-terminal kinase (JNK) was examined to understand the effect of these compounds on the upstream signaling activator of AP-1. The protein expression level of endogenous cyclin D1, a gene that is under the control of AP-1, was also analyzed after treatments with the agents. In addition, cell death induced by these compounds was evaluated by MTS assay [3-(4.5-dimethylthiazol-2-yl)-5-(3-arboxymethoxyphenyl)-2-(4-sulfophenyl)-2H-tetrazolium, inner salt]	TPA and H_2O_2 treatments strongly induced AP-1-luciferase activity as expected. Phenethyl isothiocyanate, sulforaphane, curcumin, and resveratrol increased AP-1-luciferase activity dose-dependently and then decreased at higher doses in the presence or absence of TPA. Allyl isothiocyanate and (−)-epigallocatechin-3-gallate increased AP-1-luciferase activity dose-dependently up to 50 and 100 μM. Other tea catechins and procyanidin dimers, however, had little or no effect on AP-1-luciferase activity. The JNK activity was induced by the isothiocyanates and EGCG. Most of the chemopreventive compounds induced cell death in a dose-dependent manner, with the exception of epicatechin and the procyanidins, which had little effect. The expression of endogenous cyclin D1 protein was well correlated with those of AP-1-luciferase assay. These results suggest that natural chemopreventive compounds may have differential biological functions on the signal transduction pathways such as AP-1 in the intervention of colon cancer progression and carcinogenesis	188
Human cancer cell lines, HL (cervix carcinoma), K-562 (chronic myeloid leukemia) and IM-9 (multiple myeloma)	The purpose of this study was to examine whether resveratrol can sensitize cancer cells to X-irradiation using human cancer cell lines (HeLa, K-562, and IM-9). The assays that were performed following X-irradiation (doses from 0 to 8 Gy) and/or incubation in the presence of resveratrol (concentrations ranging from 0 to 200 μM), were the following: Trypan blue exclusion test to determine cell viability, cell morphology after May-Grunwald Giemsa staining, DNA profile analysis by flow cytometry to assess cell cycle distribution and the presence of the sub-G_1 peak	The cell lines showed different radiation sensitivity (IM-9, high radiation sensitivity; K-562, intermediate radiation sensitivity; HeLa, low radiation sensitivity) as seen by the X-irradiation dose-related inhibition of cell growth and induction of apoptosis. The addition of resveratrol alone to the cell cultures induced apoptosis and inhibited cell growth from 50 (IM-9), 100 (EOL-1), or 200 μM (HeLa) resveratrol concentrations. Concomitant treatment of the cells with either resveratrol or X-irradiation induced a synergical effect at the highest dose of 200 μM. These results show that resveratrol can act as a potential radiation sensitizer at high concentrations	189

(*Continued*)

TABLE 13.3
Continued

Model	What was measured	Effect	Ref.
TNFα, IL-12, and IL-1β production from LPS activated phorbol myristate acetate (PMA) differentiated THP-1 human macrophages	The study demonstrates that resveratrol enhanced TNFα, IL-12, and IL-1β production from LPS activated phorbol myristate acetate (PMA) differentiated THP-1 human macrophages. Expression of CD86 on macrophages was enhanced by resveratrol alone and with LPS. When macrophages were primed with IFN-γ, resveratrol suppressed the expression of HLA-ABC, HLA-DR, CD80, CD86, and inhibited production of TNFα, IL-12, IL-6, and IL-1β induced by LPS	The differential impact of resveratrol on expression of CD14 might be correlated with differential response of macrophages to LPS with or without IFN-γ priming	190
The effects of resveratrol and estradiol (E_2) on expression of exogenous reporter genes and an endogenous estrogen-regulated gene (TGFα) in MDA-MB-231 cells stably transfected with wild-type (wt) ERα or mutants with deleted or mutated AF domains	To examine the role of the transcriptional activation function (AF) domains of ERα in resveratrol agonism, the effects of resveratrol and E_2 on expression of exogenous reporter genes and TGFα in MDA-MB-231 cells stably transfected with wtERα or mutants with deleted or mutated AF domains were compared. In reporter gene assays, cells expressing wtERα showed a superagonistic response to resveratrol. Deletion of AF-1 or mutation of AF-2 attenuated the effect of resveratrol disproportionately compared to that of E_2, while deletion of AF-2 abrogated the response to both ligands. In TGFα expression assays, resveratrol acted as a full agonist in cells expressing wtERα. Deletion of AF-1 attenuated stimulation by E_2 more severely than that by resveratrol, as did deletion of AF-2. In contrast, mutation of AF-2 left both ligands with a limited ability to induce TGFα expression	The effect of modifying or deleting AF domains depends strongly on the ligand and the target gene	104

Breast cancer resistance protein (BCRP/ ABCG2)	In two separate BCRP-overexpressing cell lines, accumulation of the established BCRP substrates mitoxantrone and bodipy-FL-prazosin was significantly increased by the flavonoids silymarin, hesperetin, quercetin, and daidzein, and resveratrol as measured by flow cytometry, although there was no corresponding increase in the respective wild-type cell lines. These compounds also stimulated the vanadate-inhibitable ATPase activity in membranes prepared from bacteria (*Lactococcus lactis*) expressing BCRP	Plant-derived polyphenols that interact with P-glycoprotein can also modulate the activity of the ABC transporter BCRP/ABCG2. Given the high dietary intake of polyphenols, such interactions with BCRP, particularly in the intestines, may have important consequences *in vivo* for the distribution of these compounds as well as other BCRP substrates	191
Transport in hepatic cells	A new technique was used to follow qualitatively resveratrol cell uptake and intracellular distribution, based on resveratrol fluorescent properties. A time-course study and the quantification of ^{3}H-labeled resveratrol uptake were performed using human hepatic-derived cells (HepG2 tumor cells) and hepatocytes	The temperature dependence of the kinetics of uptake as well as *cis*-inhibition experiments agree with the involvement of a carrier-mediated transport in addition to passive diffusion. The decrease of passive uptake resulted from resveratrol binding to serum proteins	125
Androgen receptor (AR)-mediated transcription in human prostate cancer cells (LNCaP and PC3)	Resveratrol and genistein activated AR-driven gene expression at low concentrations, whereas they repressed the AR-dependent reporter gene activity at high concentrations. Resveratrol and genistein induced AR-driven gene expression by activating the Raf-MEK-ERK kinase pathway. The ERK1 kinase phosphorylated the AR on multiple sites *in vitro*, but this phosphorylation event did not contribute to the resveratrol-induced AR transactivation	Due to evidence that AR pathways are involved in the development and progression of prostate cancer, these data show that the ability to modulate AR function would contribute to the observed chemopreventive activity of resveratrol and genistein	105

(*Continued*)

TABLE 13.3
Continued

Model	What was measured	Effect	Ref.
CYP1B1 gene expression in MCF-7 human breast carcinoma cells	MCF-7 cells were treated with β-naphthoflavone (BNF), emodin, resveratrol, or 0.1% dimethylsulfoxide (vehicle control). Total cellular RNA was isolated and reverse transcribed. cDNA samples were quantified by a fluorescence assay and a constant amount was amplified in a real-time DNA thermal cycler. Melting curve analysis and agarose gel electrophoresis of the amplicons resulted in a single peak and a single band, respectively. The identity of the amplicon was confirmed to be CYP1B1 by sequencing analysis. The standard curve for the real-time PCR amplification of CYP1B1 cDNA was log-linear for at least four orders of magnitude. The limit of quantitation (LOQ) of the assay was 100 copies	A method was described for the real-time PCR quantification of CYP1B1 gene expression in MCF-7 human breast carcinoma cells	192
Resveratrol inhibits drug-induced apoptosis in human leukemia cells	Studies show evidence that exposure of human leukemia cells to low concentrations of resveratrol inhibits caspase activation, DNA fragmentation, and translocation of cytochrome c induced by hydrogen peroxide or anticancer drugs C2, vincristine, and daunorubicin. At these concentrations, resveratrol induces an increase in intracellular superoxide and inhibits drug-induced acidification. Blocking the activation of NADPH oxidase	These data show that low concentrations of resveratrol inhibit death signaling in human leukemia cells via NADPH oxidase-dependent elevation of intracellular superoxide that blocks mitochondrial hydrogen peroxide production, thereby resulting in an intracellular environment nonconducive for death execution	130

	complex neutralized resveratrol-induced inhibition of apoptosis. The results implicate intracellular hydrogen peroxide as a common effector mechanism in drug-induced apoptosis that is inhibited by preincubation		
Analyzed the ability of resveratrol to modulate the ERα-dependent PI3K pathway using MCF-7 (human breast cancer cell)	Immunoprecipitation and kinase activity assays showed that resveratrol increased the ERα-associated PI3K activity with a maximum stimulatory effect at concentrations close to 10 μM; concentrations >50 μM decreased PI3K activity. Stimulation of PI3K activity by resveratrol was ERα-dependent since it could be blocked by the antiestrogen ICI 182,780. Resveratrol did not affect p85 protein expression but induced the proteasome-dependent degradation of the ERα. Nevertheless, the amount of PI3K immunoprecipitated by the ERα remained unchanged in presence of resveratrol, indicating that ERα availability was not limiting PI3K activity. Phosphoprotein kinase B (pPKB/AKT) followed the pattern of PI3K activity, whereas resveratrol did not affect total PKB/AKT expression. PKB/AKT downstream target glycogen synthase kinase 3 (GSK3) also showed a phosphorylation pattern that followed PI3K activity	The study suggests a mechanism through which resveratrol could inhibit survival and proliferation of estrogen-responsive cells by interfering with an ERα-associated PI3K pathway, following a process that could be independent of the nuclear functions of the ERα	106

(Continued)

TABLE 13.3
Continued

Model	What was measured	Effect	Ref.
Investigated whether resveratrol can sensitize non-Hodgkin's lymphoma (NHL) and multiple myeloma (MM) cell lines to paclitaxel-mediated apoptosis and to delineate the underlying molecular mechanism of sensitization	Both resveratrol and paclitaxel negatively modulated tumor cell growth by arresting the cells at the G_2-M phase of the cell cycle. Low concentrations of resveratrol exerted a sensitizing effect on drug-refractory NHL and MM cells to apoptosis induced by paclitaxel. Resveratrol selectively downregulated the expression of antiapoptotic proteins Bcl-x^L and myeloid cell differentiation factor-1 (Mcl-1) and upregulated the expression of proapoptotic proteins Bax and apoptosis protease activating factor-1 (Apaf-1). Paclitaxel downregulated the expression of Bcl-x^L, Mcl-1, and cellular inhibitor of apoptosis protein-1 antiapoptotic proteins and upregulated Bid and Apaf-1. Combination treatment resulted in apoptosis through the formation of tBid, mitochondrial membrane depolarization, cytosolic release of cytochrome c and Smac/DIABLO, activation of the caspase cascade, and cleavage of poly(adenosine diphosphate-ribose) polymerase. Combination of resveratrol with paclitaxel had minimal cytotoxicity against quiescent and mitogenically stimulated human peripheral blood mononuclear cells. Inhibition of Bcl-x^L expression by resveratrol was critical for chemosensitization and its functional impairment mimics resveratrol-mediated sensitization to paclitaxel-induced apoptosis. Inhibition of Bcl-x^L expression by resveratrol was due to the inhibition of the extracellular signal-regulated kinase 1/2 (ERK1/2) pathway and diminished activator protein-1-dependent Bcl-x^L expression. The findings by resveratrol were corroborated with inhibitors of the ERK1/2 pathway	This study demonstrates that in resistant NHL and MM cell lines resveratrol and paclitaxel selectively modify the expression of regulatory proteins in the apoptotic signaling pathway and the combination, via functional complementation, results in synergistic apoptotic activity	131

Response of ovarian cancer cells to resveratrol	Resveratrol inhibited growth and induced death in a panel of five human ovarian carcinoma cell lines. The response was associated with mitochondrial release of cytochrome c, formation of the apoptosome complex, and caspase activation. Surprisingly, even with these molecular features of apoptosis, analysis of resveratrol-treated cells by light and electron microscopy revealed morphology and ultrastructural changes indicative of autophagocytic rather than apoptotic death. This suggests that resveratrol can induce cell death through two distinct pathways. Consistent with the ability of resveratrol to kill cells via nonapoptotic processes, cells transfected to express high levels of the antiapoptotic proteins Bcl-x^L and Bcl-2 were equally sensitive as control cells to resveratrol	These findings show that resveratrol induces cell death in ovarian cancer cells through a mechanism distinct from apoptosis, therefore suggesting that it may provide leverage to treat ovarian cancer that is chemoresistant on the basis of ineffective apoptosis	193
Resveratrol suppresses EGFR-dependent Erk1/2 activation pathways stimulated by EGF and phorbol ester (12-*O*-tetradecanoyl phorbol 13-acetate, TPA) in human prostate cancer (AI PrCa PC-3) cells	Since protein kinase C (PKC) is the major cellular receptor for phorbol esters and taking into consideration that resveratrol is PKC-inhibitory, the effects of resveratrol on cellular PKC isozymes associated with the suppression of TPA-induced Erk1/2 activation were investigated. The PKC isozyme composition of PC-3 cells was defined by Western analysis of the cell lysate with a comprehensive set of isozyme-selective PKC antibiotics. PC-3 cells expressed PKCα, ε, ζ, ι, and PKD (PKCμ), as did another human AI PrCa cell line of distinct genetic origin, DU145. The effects of resveratrol on TPA-induced PKC isozyme activation were defined by monitoring PKC isozyme translocation and autophosphorylation. Under conditions where resveratrol suppressed TPA-induced Erk1/2 activation, the phytochemical produced isozyme-selective interference with TPA-induced translocation of cytosolic PKCα to the membrane/cytoskeleton and selectively diminished the amount of autophosphorylated PKCα in the membrane/cytoskeleton of the TPA-treated cells	These results demonstrate that resveratrol abrogation of a PKC-mediated Erk1/2 activation response in PC-3 cells correlates with isozyme-selective PKCα inhibition. The results provide evidence that resveratrol may have value as an adjuvant cancer therapeutic in advanced prostate cancer	194

(*Continued*)

TABLE 13.3
Continued

Model	What was measured	Effect	Ref.
Tumor necrosis factor-related apoptosis-inducing ligand (TRAIL)-induced apoptosis through p53-independent induction of p21 and p21-mediated cell cycle arrest associated with survivin depletion	Survivin expression and apoptosis revealed that resveratrol-induced G_1 arrest was associated with downregulation of survivin expression and sensitization for TRAIL-induced apoptosis. Accordingly, G_1 arrest using the cell cycle inhibitor mimosine or induced by p21 over-expression reduced survivin expression and sensitized cells for TRAIL treatment. Likewise, resveratrol-mediated cell cycle arrest followed by survivin depletion and sensitization for TRAIL was impaired in p21-deficient cells. Also, downregulation of survivin using survivin anti-sense oligonucleotides sensitized cells for TRAIL-induced apoptosis	Resveratrol is a potent sensitizer of tumor cells for TRAIL-induced apoptosis through p53-independent induction of p21 and p21-mediated cell cycle arrest associated with survivin depletion. Resveratrol sensitized various tumor cell lines, but not normal human fibroblasts, for apoptosis induced by death receptor ligation or anticancer drugs. Thus, this combined sensitizer (resveratrol)/inducer (e.g., TRAIL) strategy may be a novel approach to enhance the efficacy of TRAIL-based therapies in a variety of human cancers	195
Human breast cancer cell line MCF-7	Resveratrol treatment of MCF-7 cells resulted in a dose-dependent inhibition of the cell growth and the cells accumulated at the S phase transition of the cell cycle at low concentrations, but high concentrations do not induce S-phase accumulation. The antiproliferative effects of resveratrol were associated with a marked inhibition of cyclin D and cyclin-dependent kinase (Cdk) 4 proteins, and induction of p53 and Cdk inhibitor $p21^{WAF1/CIP}$. Growth suppression by resveratrol was also due to apoptosis, as seen by the appearance of a sub-G_1 fraction and chromatin condensation. In addition, the apoptotic process involves activation of caspase-9, a decrease of Bcl-2 as well as Bcl-x^L levels, and an increase of Bax levels	Using MCF-7, this study analyzed a possible mechanism by which resveratrol could interfere with cell cycle control and induce cell death	196

Determine the mechanism(s) by which 1,4-dihydropyridine Ca^{2+} channel blockers (DHPs) enhance the binding of neurotensin (NT) to prostate cancer PC3 cells and inhibit NT-induced inositol phosphate formation	Earlier work indicated that these effects, which involved the G protein-coupled NT receptor NTR1, were indirect and required cellular metabolism or architecture. At the micromolar concentrations used, DHPs can block voltage-sensitive and store-operated Ca^{2+} channels, K^+ channels, and Na^+ channels, and can inhibit lipid peroxidation. By varying $[Ca^{2+}]$ and testing the effects of stimulators and inhibitors of Ca^{2+} influx and internal Ca^{2+} release, we determined that although DHPs may have inhibited inositol phosphate formation partly by blocking Ca^{2+} influx, the effect on NT binding was Ca^{2+}-independent. By varying $[K^+]$ and $[Na^+]$, it was found that these ions did not contribute to either effect. For a series of DHPs, the activity order for effects on NTR1 function followed that for antioxidant ability. Antioxidant polyphenols (luteolin and resveratrol) mimicked the effects of DHPs and showed structural similarity to DHPs. Antioxidants with equal redox ability, but without structural similarity to DHPs (such as α-tocopherol, riboflavin, and *N*-acetyl-cysteine) were without effect. A flavoprotein oxidase inhibitor (diphenylene iodonium) and a hydroxy radical scavenger (butylated hydroxy anisole) also displayed the effects of DHPs	DHPs indirectly alter NTR1 function in live cells by a mechanism that depends on the ability of the drug to donate hydrogen but does not simply involve sulfhydryl reduction	197
Inhibition of phorbol myristate acetate-induced matrix metalloproteinase-9 expression by inhibiting JNK and PKC-Δ signal transduction	Resveratrol was found to significantly inhibit the PMA-induced increase in MMP-9 expression and activity. These effects of resveratrol are dose dependent and correlate with the suppression of MMP-9 mRNA expression levels. PMA caused about a 23-fold increase in MMP-9 promoter activity, which was suppressed by resveratrol. Transient transfection utilizing MMP-9 constructs, in which specific transcriptional factors were mutagenized, indicated that the effects of PMA and resveratrol were mediated via an activator protein-1 and NF-κB response element. Resveratrol inhibited PMA-mediated activation of c-Jun N-terminal kinase (JNK) and protein kinase C (PKC)-Δ activation	MMP-9 inhibitory activity of resveratrol and its inhibition of JNK and PKC-Δ may have therapeutic potential	198

(*Continued*)

TABLE 13.3
Continued

Model	What was measured	Effect	Ref.
Prostate cancer cell line (PC3)	This study compared the ability to induce both ceramide increase and growth inhibition in PC3 cells of resveratrol and three resveratrol analogs: piceatannol (*trans*-3,3′,4′,5-tetrahydroxystilbene), with an additional hydroxyl group in the 3′ position; *trans*-stilbene, the nonhydroxylated analog; and the semisynthetic *trans*-3,4′,5-trimethoxystilbene (TmS), with methoxyl groups in lieu of the hydroxyl groups. Of the three stilbenoids, only piceatannol (and not stilbene or TmS) produced ceramide-associated growth inhibition	This study demonstrates that resveratrol can exert antiproliferative/proapoptotic effects in association with the accumulation of endogenous ceramide in the androgen receptor (AR)-negative PC3. These data suggest the phenolic moiety of stilbenoids is a critical structural feature necessary to induce ceramide-associated growth inhibition	199
Human lymphoma B (DHL-4) cells	The effect of resveratrol on cell cycle and growth of DHL-4 cells was studied. MTT colorimetric test, Trypan blue dye exclusion assay, and cell cycle analysis showed that resveratrol has a dose-dependent antiproliferative and antiapoptotic action on DHL-4 cells	These results confirm the potential therapeutic role of resveratrol	200
HSP70 expression and cell death elicited by resveratrol in DU 145 human prostate cancer cells	DU 145 cells were treated with different concentrations of resveratrol, and cell viability and membrane breakdown were measured. The possible induction of oxidative stress was evidenced by performing a fluorescent analysis of intracellular reactive oxygen species (ROS) production, or evaluating the amount of nitrite/nitrate (NO) in culture medium. In addition, the expression of HSP70 level, evaluated by immunoblotting, was examined and compared with caspase-3 activity (fluorimetrically measured) and DNA damage, determined by single cell gel electrophoresis or COMET assay	These data indicate that the addition of resveratrol to DU 145 cells reduces cell viability and increases membrane breakdown, in a dose-dependent manner, without interfering with ROS production or NO synthesis, unless 200 μM resveratrol is added. Furthermore, at low concentration, resveratrol is able to raise HSP70 levels but, at high concentration, the measured levels of protective HSP70 were unmodified with respect to that of the control values. The results confirm the ability of resveratrol to suppress the proliferation of human prostate cancer cells with a typical apoptotic feature, interfering with the expression of HSPs70	201

Resveratrol inhibits the growth of human prostate carcinoma DU145 cells	Resveratrol treatment in DU145 cells resulted in a dose-dependent inhibition of cell growth and induced apoptotic cell death. The antiproliferative effect of resveratrol was associated with the inhibition of D-type cyclins and cyclin-dependent kinase (Cdk) 4 expression, and the induction of tumor suppressor p53 and Cdk inhibitor p21. Moreover, the kinase activities of cyclin E and Cdk2 were inhibited by resveratrol without alteration of their protein levels. Resveratrol treatment also upregulated the Bax protein and mRNA expression in a dose-dependent manner; however, Bcl-2 and Bcl-x^L levels were not significantly affected. These effects were found to correlate with an activation of caspase-3 and caspase-9	The study suggests that resveratrol has a strong potential for development as an agent for the prevention of human prostate cancer	202
Evaluated cellular effects of resveratrol derivatives, viniferin, gnetin H, and suffruticosol B, on proliferation and apoptosis with HL-60 cells	Resveratrol and its derivatives reduced viability of HL-60 cells in a dose-dependent manner with IC_{50} values of 20–90 μM. Ascending orders of IC_{50} values were suffruticosol B, gnetin H, viniferin and resveratrol, respectively. HL-60 cells treated with the four stilbenes exhibited distinct morphological changes characteristic of cell apoptosis such as chromatin condensation, apoptotic bodies, and DNA fragmentations. A time-dependent histogram of the cellular DNA analyzed by flow cytometry revealed a rapid increase in subdiploid cells and a concomitant decrease in diploid cells exposed to 100 μM resveratrol for 0–24 h. Cells treated with 25 μM of resveratrol, viniferin, gnetin H, and suffruticosol B for 24 h resulted in increment of sub-G_1 population by 51, 5, 11 and 59%, respectively. Treatment of cells with 0–20 μM resveratrol for 5 h produced a concentration-dependent decrease in cytochrome P450 (CYP) 1B1 mRNA levels. Suffruticosol B also suppressed CYP1B1 gene expression	These results demonstrated that resveratrol oligomers also strongly suppressed HL-60 cell proliferation, and induced DNA damage. In addition, CYP1B1 gene suppression may suggest an involvement in the resveratrol-induced apoptosis in HL-60 cells	203

(*Continued*)

TABLE 13.3
Continued

Model	What was measured	Effect	Ref.
Increasing the throughput of Caco-2 cell monolayer assays and expanding the scope of this assay to include modeling intestinal drug metabolism	A state-of-the-art Caco-2 cell monolayer permeability assay combines multiwell plates fitted with semipermeable inserts on which Caco-2 cells have been cultured with liquid chromatography–mass spectrometry (LC-MS) or LC–tandem mass spectrometry (LC-MS-MS) for the quantitative analysis of test compounds, and the identification of their intestinal metabolites. Application of LC-MS and LC-MS-MS for the measurement of resveratrol permeability and metabolism in the Caco-2 model was demonstrated. The apparent permeability coefficient for apical (AP) to basolateral (BL) movement of resveratrol was 2.0×10^{-5} cm/sec	Resveratrol was not a substrate for P-glycoprotein or multidrug resistance-associated proteins (MRP). No phase I metabolites were observed, but the phase II conjugates resveratrol-3-glucuronide and resveratrol-3-sulfate were identified based on LC-MS and LC-MS-MS analysis and comparison with synthetic standards. Although these data indicate that resveratrol diffuses rapidly across the intestinal epithelium, extensive phase II metabolism during absorption might reduce resveratrol bioavailability	126
Effect on proliferation and inducing apoptosis in lung cancer cell lines (A549, EBC-1, Lu65)	Resveratrol inhibited the growth of A549, EBC-1, and Lu65 lung cancer cells by 50% (ED_{50}) at concentrations of 5–10 μM. The combined effect of resveratrol and paclitaxel was examined in these cells. Although simultaneous exposure to resveratrol plus paclitaxel did not result in significant synergy, resveratrol significantly enhanced the subsequent antiproliferative effect of paclitaxel. In addition, resveratrol as well as paclitaxel induced apoptosis in EBC-1 and Lu65 cells, as measured by TUNEL and caspase assays, as well as flow cytometry. Resveratrol enhanced the subsequent apoptotic effects of paclitaxel. The effects of resveratrol and paclitaxel were examined on levels of $p21^{waf1}$, $p27^{kip1}$, E-cadherin, EGFR, and Bcl-2 in EBC-1 cells. Resveratrol prior to paclitaxel induced $p21^{waf1}$ expression approximately 4-fold	These results suggest that resveratrol may be a promising alternative therapy for lung cancer and that lung cancer cells exposed to resveratrol have a lowered threshold for killing by paclitaxel	132

Resveratrol was used to treat four human medulloblastoma cell lines (Med-3, UW228-1, -2 and -3) and its effects on cell growth, differentiation, and death were examined by multiple approaches	Expression of Fas, FasL, and caspase-3 in the cells without and with resveratrol treatments was examined by immunocytochemical staining and mRNA *in situ* hybridization, and the influence of anti-Fas antibody in cell growth and survival was determined as well. The results demonstrated that resveratrol could suppress growth, promote differentiation, and commit target cells to apoptosis in time- and dose-related fashions. Fas was constitutively expressed but FasL was undetectable in the four lines in spite of resveratrol treatment. Anti-Fas antibody neither inhibited growth nor induced apoptosis of those cell lines. Upregulated caspase-3 was found in resveratrol-treated populations and appearance of its cleaved form was closely associated with the apoptotic event	These findings suggest that resveratrol is an effective antimedulloblastoma agent that kills medulloblastoma cells through a Fas-independent pathway	204
MDA-MB-231, a highly invasive and metastatic breast cancer cell	Accumulation of ceramide derives from both *de novo* ceramide synthesis and sphingomyelin hydrolysis. It was demonstrated that ceramide accumulation induced by resveratrol can be associated with the activation of serine palmitoyltransferase (SPT), the key enzyme of *de novo* ceramide biosynthetic pathway, and neutral sphingomyelinase (nSMase), a main enzyme involved in the sphingomyelin/ceramide pathway. However, by using specific inhibitors of SPT, myriocin and L-cycloserine, and nSMase, gluthatione, and manumycin, only the SPT inhibitors could counteract the biological effects induced by resveratrol	The study shows that resveratrol can induce growth inhibition and apoptosis in MDA-MB-231 in concomitance with a dramatic endogenous increase of growth inhibitory/proapoptotic ceramide. Resveratrol seems to exert its growth inhibitory/apoptotic effect on MDA-MB-231 by activating the *de novo* ceramide synthesis pathway	205

(*Continued*)

TABLE 13.3
Continued

Model	What was measured	Effect	Ref.
Suppress the growth of endometrial cancer cells (Ishikawa cell line) through downregulation of EGF	Ishikawa cells were treated with resveratrol (1, 10, 50, and 100 μM) for 1, 3, 5, and 7 days, and analyzed for growth signal genes (EGF and VEGF), cell cycle regulatory genes (p53 and p21), and apoptosis-related genes (Bcl-2 and Bax). Results of these experiments demonstrate that after resveratrol treatment, the growth of Ishikawa cells was inhibited in a dose-dependent manner. The gene and protein expression data suggest that resveratrol treatment significantly decreased EGF, whereas VEGF was upregulated in Ishikawa cell lines. Interestingly, protein expressions of p21 and Bax were decreased, even though their mRNA expressions did not show significant changes	This study suggests that resveratrol can suppress proliferation of Ishikawa cells through downregulation of EGF	206
Involvement of $p21^{WAF1/CIP1}$, pRB, Bax, and NF-κB in induction of growth arrest and apoptosis by resveratrol in human lung carcinoma A549 cells	Resveratrol treatment of A549 cells resulted in a concentration-dependent induction of S-phase arrest in cell cycle progression. This antiproliferative effect of resveratrol was associated with a marked inhibition of the phosphorylation of the retinoblastoma protein (pRB) and concomitant induction of cyclin-dependent kinase (Cdk) inhibitor $p21^{WAF1/CIP}$, which appears to be transcriptionally upregulated and is p53-dependent. In addition, resveratrol treatment resulted in induction of apoptosis as determined by fluorescence microscopy and flow cytometric analysis. These effects were found to correlate with an activation of caspase-3 and a shift in Bax/ $Bcl\text{-}x^{L}$ ratio more towards apoptosis. Resveratrol treatment also inhibited the transcriptional activity of NF-κB	These findings suggest that resveratrol has potential for development as an agent for prevention of human lung cancer	207

Chemopreventive properties on intestinal carcinogenesis using a methylated derivative of resveratrol (*Z*-3,5,4′-trimethoxy-stilbene: R3)	R3 at 0.3 μM exerted an 80% growth inhibition of Caco-2 cells and arrested growth completely at 0.4 μM (R3 was 100-fold more active than resveratrol). The *cis* conformation of R3 was also 100-fold more potent than the *trans* isomer. R3 (0.3 μM) caused cell cycle arrest at the G_2/M phase transition. The drug inhibited tubulin polymerization in a dose-dependent manner ($IC_{50} = 4$ μM), and it reduced also by 2-fold ornithine decarboxylase and *S*-adenosylmethionine decarboxylase activities. This caused the depletion of the polyamines, putrescine and spermidine, which are growth factors for cancer cells. R3 inhibited partially colchicine binding to its binding site on tubulin, indicating that R3 either partially overlaps with colchicine binding or that R3 binds to a specific site of tubulin that is not identical with the colchicine binding site modifying colchicine binding by allosteric influences	R3 is an interesting antimitotic drug that exerts cytotoxic effects by depleting the intracellular pool of polyamines and by altering microtubule polymerization. Such a drug may be useful for the treatment of neoplastic diseases	116
Pleiotropic effects of resveratrol	Resveratrol has several pharmacological properties which include inhibition of arachidonate metabolism in leukocytes and platelets, modulation of lipid metabolism, inhibition of platelet aggregation and lipid peroxidation, and reduction of expression and activity of cyclooxygenase-2. Resveratrol induces cell cycle arrest at S/G_2 phase transition and a proapoptotic cell death in several types of cancer cells	These biological activities help to explain the vasorelaxing, anticancer, and antiinflammatory activity of resveratrol. This article summarizes the wide range of biological activities of resveratrol in three disease states: cancer, coronary heart disease, and pain. In addition, the mechanisms underlying these promising properties of resveratrol are reviewed	208

(*Continued*)

TABLE 13.3
Continued

Model	What was measured	Effect	Ref.
Resveratrol-induced apoptosis in human colon cancer cells (SW480), with special attention to the role of the death receptor Fas in this pathway	The study shows that resveratrol activates various caspases and triggers apoptosis in SW480 cells. Caspase activation is associated with accumulation of the proapoptotic proteins Bax and Bak that undergo conformational changes and relocalization to the mitochondria. Resveratrol does not modulate the expression of Fas and Fas-ligand (FasL) at the surface of cancer cells, and inhibition of the Fas/FasL interaction does not influence the apoptotic response to the molecule. Resveratrol induces the clustering of Fas and its redistribution in cholesterol and sphingolipid-rich fractions of SW480 cells, together with FADD and procaspase-8. This redistribution is associated with the formation of a death-inducing signaling complex (DISC). Transient transfection of either a dominant-negative mutant of FADD, E8, or MC159 viral proteins that interfere with the DISC function, decreases the apoptotic response of SW480 cells to resveratrol and partially prevents resveratrol-induced Bax and Bak conformational changes	These results indicate that the ability of resveratrol to induce the redistribution of Fas receptor in membrane rafts may contribute to the ability of the molecule to trigger apoptosis in colon cancer cells	209

Cell growth inhibition and ability to induce apoptosis in HL60 promyelocytic leukemia cells	A series of *cis*- and *trans*-stilbene-based resveratrols were prepared with the aim of discovering new lead compounds with clinical potential. All the synthesized compounds were tested *in vitro* for cell growth inhibition and the ability to induce apoptosis in HL60 promyelocytic leukemia cells. The tested *trans*-stilbene derivatives were less potent than their corresponding *cis* isomers, except for *trans*-resveratrol, whose *cis* isomer was less active. The best results were obtained with the *cis*-3,5-dimethoxy derivatives of rhapontigenin (*trans*-3,5,3′-trihydroxy-4′methoxystilbene) and its 3′-amino derivative, respectively, which showed apoptotic activity at nanomolar concentrations. The corresponding *trans* isomers were less active both as antiproliferative and as apoptosis-inducing agents. Some compounds were active toward resistant HL60R cells and their activity was higher than that of several classic chemotherapeutic agents	The flow cytometry assay showed that low concentrations (50 nM) of test compounds were able to recruit almost all cells in the apoptotic sub-G_0-G_1 peak, thus suggesting that the main mechanism of cytotoxicity of these compounds could be the activation of apoptosis. These data indicate that structural alteration of the stilbene motif of resveratrol can be extremely effective in producing potent apoptosis-inducing agents	86

(*Continued*)

TABLE 13.3
Continued

Model	What was measured	Effect	Ref.
Antitumor effect of resveratrol oligomers against human cancer cell lines and the molecular mechanism of apoptosis induced by vaticanol C	Vaticanol C, a resveratrol derivative that was isolated from the stem bark of *Vatica rassak* (Dipterocarpaceae), induced considerable cytotoxicity in all cell lines tested and exhibited growth suppression in colon cancer cell lines at low doses. Vaticanol C caused two cell lines (SW480 and HL60) to induce cell death at 4–7 times lower concentrations, compared with resveratrol. The growth suppression by vaticanol C was found to be due to apoptosis, which was assessed by morphological findings (nuclear condensation and fragmentation) and DNA ladder formation in the colon cancer cell lines. The apoptosis in SW480 colon cancer cells was executed by the activation of caspase-3, which was shown by Western blot and apoptosis inhibition assays. Furthermore, the mitochondrial membrane potential of apoptotic SW480 cells after 12 h treatment with vaticanol C was significantly lost, and concurrently the cytochrome c release and activation of caspase-9 were also detected by Western blot analysis. Overexpression of Bcl-2 protein in SW480 cells significantly prevented the cell death induced by vaticanol C	These findings indicate that vaticanol C induced marked apoptosis in malignant cells mainly by affecting mitochondrial membrane potential	117

Increase in BRCA1 and BRCA2 mRNA expression in human breast tumor cell lines (MCF7, HBL 100 and MDA-MB 231)	These studies investigated the effects of resveratrol on BRCA1 and BRCA2 expression in MCF7, HBL 100, and MDA-MB 231 cells using quantitative real-time RT-PCR, and by perfusion chromatography of the proteins. All cell lines were treated with 30 μM resveratrol	The expression of BRCA1 and BRCA2 mRNAs was increased although no change in the expression of the proteins was found. These data indicated that resveratrol at 30 μM can increase expression of genes involved in the aggressiveness of human breast tumor cell lines	210
Effects of resveratrol on the activity of adenylate- and guanylate-cyclase (AC, GC) enzymes, two known cytostatic cascades in MCF-7 breast cancer cells	Resveratrol increased cAMP levels in both time- and concentration-dependent manners. In contrast, it had no effect on cGMP levels. The stimulatory effects for resveratrol on AC were not altered either by a protein synthesis inhibitor or estrogen receptor (ER) blockers. Likewise, cAMP formation by resveratrol was insensitive to either the broad-spectrum phosphodiesterase (PDE) inhibitor or the cAMP-specific PDE inhibitor. Instead, these PDE inhibitors significantly augmented maximal cAMP formation by resveratrol. Parallel experiments showed that either resveratrol or rolipram inhibited the proliferation of these cells in a concentration-responsive manner. Further, concurrent treatment with resveratrol and rolipram significantly enhanced their individual cytotoxic responses. The antiproliferative effects were appreciably reversed by the kinase-A inhibitors Rp-cAMPS or KT-5720. Pretreatment with the $cPLA_2$ inhibitor arachidonyl trifluoromethyl ketone markedly antagonized the cytotoxic effects of resveratrol, but had no effect on that of rolipram	The study demonstrates that resveratrol is an agonist for the cAMP/kinase-A system, a documented proapoptotic and cell cycle suppressor in breast cancer cells	211

(Continued)

TABLE 13.3
Continued

Model	What was measured	Effect	Ref.
Inhibition of proliferation of CEM-C7H2 lymphocytic leukemia cells	Inhibition of proliferation by resveratrol of CEM-C7H2 lymphocytic leukemia cells was paradoxically associated with an enhanced cellular 3-(4,5-dimethylthiazol-2-yl)-2,5-diphenyltetrazolium bromide (MTT)-reducing activity. This phenomenon was most pronounced at the subapoptotic concentration range of resveratrol	The results of the study show that the MTT-reducing activity can be increased by resveratrol without a corresponding increase in the number of living cells and that this occurs at a concentration range of resveratrol which is not sufficient to induce apoptosis but suffices to slow down cell growth. This phenomenon appears to be restricted to proliferation inhibitors with antioxidant properties and is cell type-specific. Thus, in determining the effects of flavonoids and polyphenols on proliferation in certain cell types, this might represent a pitfall in the MTT proliferation assay	212
A series of 43 stilbene derivatives that showed cytotoxicity against human lung carcinoma (A549) was analyzed using comparative molecular field analysis (CoMFA) for defining the hypothetic pharmacophore model	The polyoxylated stilbenes were found to be active inhibitors of tubulin polymerization. Several *cis*-stilbenes are structurally similar to combretastatins. However, the *trans*-stilbenes are assumed to be close to resveratrol. With several synthesized compounds that were evaluated for antitumor cytotoxicity against human lung tumor cells (A549), the stilbene derivatives were subjected to CoMFA. To perform systematic molecular modeling of these compounds, a conformational search was carried out based on the precise dihedral angle analysis of the lead compound. The x-ray	The three dimensional (3D) quantitative structure–activity relationship study resulted in reasonable cross-validated, conventional r^2 values equal to 0.640 and 0.958, respectively	118

	crystallographic structure of combretastatin A-1 was also used for defining the active conformers of the compounds. After determining the energy-minimized conformers of the lead compound, CoMFA was performed using five different alignments		
Phytoestrogen regulation of a vitamin D_3 (VDR) receptor promoter and 1,25-dihydroxyvitamin D_3 actions in human breast cancer cells (T47D and MCF-7)	These studies examined regulation of VDR promoter region by two phytoestrogens, resveratrol and genistein. The group transiently transfected a VDR promoter luciferase construct into the estrogen receptor (ER) positive human breast cancer cell lines T47D and MCF-7, and treated with resveratrol or genistein. Both phytoestrogens upregulated the transcription of the VDR promoter, as measured by reporter gene activity, approximately twofold compared to vehicle treated cells. Cotreatment with the antiestrogen tamoxifen in T47D cells and transfection in an estrogen receptor negative breast cancer cell line demonstrated that the effects of phytoestrogens on the VDR promoter are dependent on ER. Resveratrol and genistein also increased VDR protein expression as detected by Western blotting	Treatment with resveratrol had no effect on cell number or cell cycle profile, while treatment with genistein increased cell number. Because resveratrol could upregulate VDR without increasing breast cancer cell growth, the group hypothesized that resveratrol-mediated increase in VDR expression would sensitize breast cancer cells to the effects of 1,25-dihydroxyvitamin D_3 and vitamin D_3 analogs. In support of this hypothesis, both T47D and MCF-7 cells pretreated with resveratrol exhibited increased VDR-mediated transactivation of a vitamin D_3 responsive promoter compared to cells pretreated with vehicle. In addition, cotreatment with resveratrol enhanced the growth inhibitory effects of 1,25-dihydroxyvitamin D_3 and the vitamin D_3 analog EB1089. These data support the concept that dietary factors, such as phytoestrogens, may impact on breast cancer cell sensitivity to vitamin D_3 analogs through regulation of the VDR promoter	107

(*Continued*)

TABLE 13.3
Continued

Model	What was measured	Effect	Ref.
Activity of resveratrol against fresh acute myeloid leukemia (AML) cells and its mechanism of action	Since interleukin 1β (IL-1β) plays a key role in proliferation of AML cells, the effect of resveratrol was first tested on the AML cell lines OCIM2 and OCI/AML3, both of which produce IL-1β and proliferate in response. Resveratrol inhibited proliferation of both cell lines in a dose-dependent fashion by arresting the cells at S phase, thus preventing their progression through the cell cycle; IL-1β partially reversed this inhibitory effect. Resveratrol significantly reduced production of IL-1β in OCIM2 cells. It also suppressed the IL-1β-induced activation of transcription factor NF-κB, which modulates an array of signals controlling cellular survival, proliferation, and cytokine production. Incubation of OCIM2 cells with resveratrol resulted in apoptotic cell death. Because caspase inhibitors Ac-DEVD-CHO or z-DEVD-FMK partially reversed the antiproliferative effect of resveratrol, the group tested its effect on the caspase pathway and found that resveratrol	Having shown resveratrol is an effective *in vitro* inhibitor of AML cells, the data suggest this compound may have a role in future therapies for AML	213

	induced the activation of the cysteine protease caspase-3 and subsequent cleavage of the DNA repair enzyme poly(adenosine diphosphate [ADP]-ribose) polymerase. Resveratrol suppressed colony-forming cell proliferation of fresh AML marrow cells from five patients with newly diagnosed AML in a dose-dependent fashion		
Inhibition of cell transformation by resveratrol and its derivatives	To develop more effective agents with fewer side effects for the chemoprevention of cancer, the group investigated the effect of resveratrol and its structurally related derivatives on epidermal growth factor (EGF)-induced cell transformation. The results provided the first evidence that one of the resveratrol derivatives exerted a more potent inhibitory effect than resveratrol on EGF-induced cell transformation, but had less cytotoxic effects on normal nontransformed cells. Compared to resveratrol, this compound also caused cell cycle arrest in the G_1 phase, but did not induce p53 activation and apoptosis. Furthermore, this compound markedly inhibited EGF-induced phosphatidylinositol-3 kinase (PI-3K) and Akt activation	These data suggested that the greater antitumor effect of the compound compared to resveratrol may act through a different mechanism by mainly targeting PI-3K/Akt signaling pathways	119

(*Continued*)

TABLE 13.3
Continued

Model	What was measured	Effect	Ref.
Resveratrol-induced modification of polyamine metabolism is accompanied by induction of c-Fos	The objective of the study was to investigate the effect of resveratrol on polyamine metabolism in the human cell line Caco-2. The group demonstrated that inhibition of ornithine decarboxylase (ODC) was due to attenuated ODC protein and mRNA levels. The resveratrol analog piceatannol also diminished ODC activity, protein and mRNA levels, whereas (–)-epigallocatechin gallate exerted only weak effects on ODC. The transcription factor c-Myc was attenuated by resveratrol treatment and to a lesser extent by piceatannol and EGCG. *S*-Adenosylmethionine decarboxylase was concomitantly inhibited by resveratrol and piceatannol treatment, whereas EGCG did not affect its activity. In addition, resveratrol, piceatannol, and EGCG enhanced spermidine/spermine N^1-acetyltransferase activity. Intracellular levels of spermine and spermidine were not affected, whereas putrescine and N^8-acetylspermidine concentrations increased after incubation with resveratrol. These events were paralleled by an increase of the activator protein-1 constituents c-Fos and c-Jun. Whereas DNA-binding activity of c-Jun remained unchanged, DNA-binding activity of c-Fos was significantly enhanced by resveratrol and piceatannol, but inhibited by EGCG	The data suggest that growth arrest by resveratrol is accompanied by inhibition of polyamine synthesis and increased polyamine catabolism. c-Fos seems to play a role in this context. Effects of piceatannol on polyamine synthesis were similar, but not as potent as those exerted by resveratrol	127

Resveratrol-induced G_2 arrest through the inhibition of CDK7 and $p34^{CDC2}$ kinases in colon carcinoma HT29 cells	These studies present an explanation for the antitumor effect of resveratrol. Based on flow cytometric analysis, resveratrol inhibited the proliferation of HT29 colon cancer cells and resulted in their accumulation in the G_2 phase of the cell cycle. Western blot analysis and kinase assays demonstrated that the perturbation of G_2 phase progression by resveratrol was accompanied by the inactivation of $p34^{CDC2}$ protein kinase, and an increase in the tyrosine phosphorylated (inactive) form of $p34^{CDC2}$. Kinase assays revealed that the reduction of $p34^{CDC2}$ activity by resveratrol was mediated through the inhibition of CDK7 kinase activity, while CDC25A phosphatase activity was not affected. In addition, resveratrol-treated cells were shown to have a low level of CDK7 kinase-Thr^{161}-phosphorylated $p34^{CDC2}$	These results demonstrated that resveratrol induced cell cycle arrest at the G_2 phase through the inhibition of CDK7 kinase activity, suggesting that its antitumor activity might occur through the disruption of cell division at the G_2/M phase	214
Investigation of apoptosis in esophageal cancer cells (EC-9706) induced by resveratrol, and the relation between this apoptosis and expression of Bcl-2 and Bax	With *in vitro* experiments, the MTT assay was used to determine the cell growth inhibitory rate. Transmission electron microscopy and the TUNEL staining method were used to detect quantitatively and qualitatively the apoptosis status of esophageal cancer cell line EC-9706 before and after the resveratrol treatment. Immunohistochemical staining was used to detect the expression of apoptosis-regulated gene Bcl-2 and Bax	Resveratrol inhibited growth of the esophageal cancer cell line EC-9706 in a dose- and time-dependent manner. Resveratrol induced EC-9706 cells to undergo apoptosis with typically apoptotic characteristics, including morphological changes of chromatin condensation, chromatin crescent formation, nucleus fragmentation, and apoptotic body formation. TUNEL assay showed that after the treatment of EC-9706 cells with resveratrol for 24 to 96 h, the AIs were apparently increased with treated time. Immunohistochemical staining showed that after the treatment of EC-9706 cells with resveratrol for 24 to 96 h, the PRs of Bcl-2 proteins were apparently reduced with treatment time and the PRs of Bax proteins were apparently increased with treatment time	215

(*Continued*)

TABLE 13.3
Continued

Model	What was measured	Effect	Ref.
Examined whether resveratrol has any effect on growth and gene expression in human ovarian cancer PA-1 cells	The studies show that resveratrol inhibits cell growth and induces apoptosis in PA-1 human ovarian cancer cells. The group also investigated the effect of resveratrol on changes of global gene expression during resveratrol-induced growth inhibition and apoptosis in PA-1 cells using a human cDNA microarray with 7448 sequence-verified clones. Out of the 7448 genes screened, 118 genes were founded to be affected in their expression levels by more than 2-fold after 24 h treatment with 50 μM resveratrol. Resveratrol treatment of PA-1 cells at the final concentration of 50 μM for 6, 12, 24, and 48 h and gene expression patterns were analyzed by microarray. Clustering of the genes modulated more than 2-fold at three of the above times points divided the genes into two groups. Within these groups, there were specific subgroups of genes whose expression were substantially changed at the specified time points	One of the most highly upregulated genes found in this study was NAD(P)H quinone oxidoreductase 1 (NQO-1), which has recently been shown to be involved in p53 regulation. Although the precise roles of genes whose expression levels were found to fluctuate after resveratrol treatment remain to be elucidated, gene expression in human ovarian cancer cells following resveratrol exposure, as offered by this study, may provide clues for the mechanism of resveratrol action	137
The effects of resveratrol at both the molecular (TGFα gene activation) and the cellular (cell growth) levels in breast cancer cells stably transfected with wild-type (wt) ER (D351) and mutant (mut) ER (D351Y)	TGFα mRNA induction was used as a specific marker of estradiol (E_2) responsiveness. Resveratrol caused a concentration-dependent (10^{-8}–10^{-4} M) stimulation of TGFα mRNA, indicating that it acts as an estrogen agonist in these cell lines. The pure antiestrogen ICI 182,780 (ICI) blocked resveratrol-induced activation of TGFα, consistent with action through an ER-mediated pathway. Further studies that combined treatments with E_2 and resveratrol showed that resveratrol does not act as an antagonist in the presence of various	Resveratrol acts as an ER agonist at low doses but also activates ER-independent pathways, some of which inhibit cell growth	108

(10^{-11}–10^{-8} M) concentrations of E_2. To determine whether resveratrol can be classified as a type I or type II estrogen, resveratrol was examined with the D351G ER in the TGFα assay and it was found that resveratrol belongs to the type I estrogens. Both resveratrol and E_2 had concentration-dependent growth inhibitory effects in cells expressing wtER and D351Y ER. Although the pure antiestrogen ICI blocked the growth inhibitory effects of E_2, it did not block the inhibitory effects of resveratrol, suggesting that the antiproliferative effects of resveratrol also involve ER-independent pathways. Resveratrol differentially affected the levels of ER protein in these two cell lines: Resveratrol downregulated wtER levels while significantly upregulating the amount of mutD351Y ER. Cotreatment with ICI resulted in strongly reduced ER levels in both cell lines. Gene array studies revealed resveratrol-induced upregulation of more than 80 genes, among them a profound activation of $p21^{CIP1/WAF1}$, a gene associated with growth arrest. The $p21^{CIP1/WAF1}$ protein levels measured by Western blotting confirmed resveratrol-induced significant upregulation of this protein in both cell lines

(*Continued*)

TABLE 13.3
Continued

Model	What was measured	Effect	Ref.
Activation of p53-directed genes that are involved in apoptosis mechanism(s) or whether resveratrol modifies the androgen receptor and its coactivators directly or indirectly and induces cell growth inhibition. These were conducted using androgen-sensitive prostate cancer cells (LNCaP)	The study demonstrates by DNA microarray, RT-PCR, Western blot, and immunofluorescence analyses that treatment of androgen-sensitive prostate cancer cells (LNCaP) with 10^{-5} M RE for 48 h downregulates prostate-specific antigen (PSA), AR coactivator ARA 24, and NF-κB p65. Altered expression of these genes is associated with an activation of p53-responsive genes such as p53, PIG 7, $p21^{Waf1\text{-}Cip1}$, p300/CBP, and Apaf-1. The effect of RE on p300/CBP plays a central role in its cancer preventive mechanisms in LNCaP cells	The results implicate activation of more than one set of functionally related molecular targets. Key molecular targets have been identified that are associated with AR and p53 target genes. These findings suggest the need for further extensive studies on AR coactivators, such as p300, its central role in posttranslational modifications such as acetylation of p53 and/or AR by resveratrol in a time- and dose-dependent manner at different stages of prostate cancer that will fully elucidate the role of resveratrol as a chemopreventive agent for prostate cancer in humans	138
Induction of apoptosis in human melanoma cells (A375 and SK-mel28)	This study examined the effect of resveratrol on growth of two human melanoma cell lines. It was found that resveratrol inhibited growth and induced apoptosis in both cell lines, with the amelanotic cell line A375 being more sensitive. The potential involvement of different MAP kinases in the action of resveratrol was also examined. Although resveratrol did not alter the phosphorylation of p38 or JNK MAP kinases in either cell line, it induced phosphorylation of ERK1/2 in A375, but not in SK-mel28 cells	These results suggest that *in vivo* studies of the effect of resveratrol on melanoma are warranted and that it might have effectiveness as either a therapeutic or chemopreventive agent against melanoma	216

Effect of resveratrol on signal transduction pathways involved in paclitaxel-induced apoptosis in human neuroblastoma SH-SY5Y cells	Resveratrol was able to reduce significantly paclitaxel-induced apoptosis in the human neuroblastoma (HN) SH-SY5Y cell line, acting on several cellular signaling pathways that are involved in paclitaxel-induced apoptosis. Resveratrol reverses phosphorylation of Bcl-2 induced by paclitaxel and concomitantly blocks Raf-1 phosphorylation, also observed after paclitaxel exposure, thus suggesting that Bcl-2 inactivation may be dependent on the activation of the Raf/Ras cascade. Resveratrol also reverses the sustained phosphorylation of JNK/SAPK, which specifically occurs after paclitaxel exposure	The study demonstrates that the toxic action of paclitaxel on neuronal-like cells is not only related to the effect of the drug on tubulin, but also to its capacity to activate several intracellular pathways leading to inactivation of Bcl-2, thus causing cells to die by apoptosis, and resveratrol significantly reduces paclitaxel-induced apoptosis by modulating the cellular signaling pathways which commit the cell to apoptosis	133
Induction of downregulation in survivin expression and apoptosis in HTLV-1-infected cell lines	These studies investigated the effect of resveratrol in adult T cell leukemia. Observations showed that resveratrol induced growth inhibition in all five human T cell lymphotrophic virus-1-infected cell lines examined, with 50% effective doses of 10.4–85.6 μM. In the resveratrol-treated cells, induction of apoptosis was confirmed by annexin V-based analyses and morphological changes. The most surprising observation was that resveratrol treatment resulted in a gradual decrease in the expression of survivin, an antiapoptotic protein, during cell apoptosis	These findings indicate that resveratrol inhibits the growth of human T cell lymphotrophic virus-1-infected cell lines, at least in part, by inducing apoptosis mediated by downregulation in survivin expression. In view of the accumulating evidence that survivin may be an important determinant of a clinical response in adult T cell leukemia, these data suggest that resveratrol merits further investigation as a potential therapeutic agent for this incurable disease	217

(*Continued*)

TABLE 13.3
Continued

Model	What was measured	Effect	Ref.
Action of some phenolic compounds (curcumin, yakuchinone B, resveratrol, and capsaicin) in four human tumor cell lines: acute myeloblastic leukemia (HL-60), chronic myelogenic leukemia (K-562), breast adenocarcinoma (MCF-7), and cervical epithelial carcinoma (HeLa)	The phenolics exhibited growth inhibition as assessed by Trypan blue dye exclusion. It was evident from the results of the MTT reduction assay and [^{3}H]thymidine incorporation into nuclear DNA that the phenolics were cytotoxic and inhibited cell proliferation. Dose-response studies indicated curcumin to be most cytotoxic towards HL-60, K-562, and MCF-7 but did not show much activity in HeLa cells. On the other hand, yakuchinone B, although less active than curcumin, displayed cytotoxicity towards all four cell lines. Resveratrol was cytotoxic only in leukemic cells, while capsaicin was marginally cytotoxic. All these phenolics did not elicit any cytotoxic activity as judged by the above parameters towards lymphocytes purified from normal human blood. When cells treated with phenolics were stained with propidium iodide and examined under a fluorescent microscope, characteristic apoptotic features such as chromatin condensation and nuclear fragmentation were observed. Scoring of cells with apoptotic and nonapoptotic features showed positive correlation of apoptotic index with dose of phenolic, and fragmented DNA extracted free of genomic DNA displayed on gel electrophoresis a typical ladder pattern	The actions of the phenolics as inducers of apoptosis in tumor cells suggest their potential use in a strategy for cancer control	218

Comparative antiproliferative and apoptotic effects of resveratrol, ε-viniferin, and vine-shoot-derived polyphenols (vineatrols) on chronic B cell malignancies (B-cell chronic lymphocytic leukemia, B-CLL or hairy cell leukemia, HCL) and normal peripheral blood-derived mononuclear cells (PBMC) as control	Resveratrol, its dimer ε-viniferin, and two preparations of vineatrol (a grape-derived polyphenol fraction isolated from vine-shoot extracts) were compared for their effects on the proliferation and survival of normal and leukemic human lymphocytes. Two different batches of vineatrol (vineatrol 10 and 25%) were obtained by HPLC fractionation and contained 10 and 25% *trans*-resveratrol, respectively. The different polyphenols were added to cultures of leukemic cells from chronic B cell malignancies or PBMC cells as a control. The different polyphenols displayed antiproliferative effect on the leukemic cells, as estimated by the observed inhibition of ^{3}H-thymidine uptake and the reduction of cell recovery. Vineatrol 10% was the most potent whereas vineatrol 25% and resveratrol displayed comparable activity, ε-viniferin only exhibited slight effects. The same order of potency was observed for their capacity to induce apoptosis in leukemic B cells. In contrast, the survival of PBMC cells was little affected in the presence of these polyphenolic compounds and higher concentrations were required in order to elicit cell death. Polyphenol-driven apoptosis in chronic leukemic B cells was shown to correlate with an activation of caspase 3, a drop in the mitochondrial transmembrane potential, a reduction in the expression of the antiapoptotic protein Bcl-2, as well as a reduction in the expression of the inducible nitric oxide synthase (iNOS)	These data indicate that vine-shoots may be a convenient and natural source of material for the purification of resveratrol and other polyphenolic compounds of putative therapeutic interest	219

(*Continued*)

TABLE 13.3
Continued

Model	What was measured	Effect	Ref.
Inhibition of cell proliferation in hepatoblastoma HepG2 and colorectal tumor SW480 cells. Analysis of the biochemical mechanism of resveratrol	In order to determine if the amount of resveratrol taken up during food or drink consumption is sufficient to ensure the whole body relevance of *in vitro* described beneficial effects, the ratio between plasma level of resveratrol and its cell bioabsorption was evaluated. The study reports a higher uptake of resveratrol in the human hepatic derived HepG2 cells than in colorectal derived SW480 cells. In contrast, resveratrol is conjugated in these cells and derivatives are released in large amounts in the cell medium. Based on present knowledge, resveratrol appears to be a promising bioactive natural molecule with potential applications in phytotherapy, pharmacology, or in nutriprotection (nutraceutic food) area	The results show that resveratrol strongly inhibits cell proliferation at the micromolar range in a time- and dose-dependent manner. Resveratrol appears to block the cell cycle at the transition S to G_2/M since there is no inhibition of ^{3}H-thymidine incorporation observed, while there is an increase of the cell number in S phase	128
Inhibition of cell proliferation using two *in vitro* assay systems, i.e., binding to human estrogen receptor α and stimulation of MCF-7 cell proliferation. L5178Y mouse lymphoma and Chinese hamster V79 cells were used	Resveratrol was analyzed for genotoxic potential. Resveratrol induced cellular toxicity, micronuclei, and metaphase chromosome displacement in L5178Y mouse lymphoma cells. The induction of micronuclei was observed in Chinese hamster V79 cells. Determination of kinetochore signals in micronuclei and cell cycle analysis suggested that resveratrol did not cause a direct disturbance of mitosis. In support of this notion, cell-free tubulin polymerization studies indicated no direct effect of resveratrol on microtubule assembly	According to an estimation of daily intake and bioavailability, concentrations that were found genotoxic *in vitro* might be reached in human exposure. On the other hand, the estrogenic acitivity might be beneficial	166

Elucidation of rapid phytoestrogen action on ER-positive and -negative breast cancer cells using various estrogenic compounds	As a preliminary step toward elucidating rapid phytoestrogen action on breast cancer cells, the effect of 17β-estradiol (E_2), genistein, daidzein, and resveratrol was investigated for their activation status of signaling proteins that regulate cell survival and invasion, the cell properties underlying breast cancer progression. The effect of these estrogenic compounds on the activation, via phosphorylation, of Akt/protein kinase B (Akt) and focal adhesion kinase (FAK) were analyzed in ER-positive and -negative human breast cancer cell lines. E_2, genistein, and daidzein increased whereas resveratrol decreased both Akt and FAK phosphorylation in non-metastatic ER-positive T47D cells. In metastatic ER-negative MDA-MB-231 cells, all estrogenic compounds tested increased Akt and FAK phosphorylation. The inhibitory action of resveratrol on cell survival and proliferation is ER-dependent	All estrogenic compounds tested, including resveratrol, may exert supplementary ER-independent non-genomic effects on cell survival and migration in breast cancer cells	109
Evaluation of the potential role of resveratrol on pancreatic cancer cell proliferation using two human pancreatic cancer cell lines, PANC-1 and AsPC-1	Resveratrol inhibited proliferation of both PANC-1 and AsPC-1 in a concentration- and time-dependent manner as measured by ^{3}H-thymidine incorporation. Cell number of both PANC-1 and AsPC-1 was also significantly decreased following 48 and 72 h of treatment with 100 μmol resveratrol. The growth inhibition induced by resveratrol was accompanied by apoptotic morphologic changes, characterized by cell rounding and cell membrane blebbing suggesting apoptosis. Propidium iodide staining of DNA, measured by flow cytometry, showed a dramatic increase in the fraction of sub-G_0/G_1 cells following resveratrol treatment in both PANC-1 and AsPC-1	The substantial apoptosis inducted by resveratrol on these two cell lines was confirmed by the terminal deoxynucleotidyl transferase-mediated deoxyuridine triphosphate nick-end labeling assay. These findings suggest that resveratrol may have a potent antiproliferative effect on human pancreatic cancer with induction of apoptosis. Resveratrol may be of value for the management and prevention of human pancreatic cancer	220

(Continued)

TABLE 13.3
Continued

Model	What was measured	Effect	Ref.
Antiproliferation effect of resveratrol in two human liver cancer cell lines, Hep G2 and Hep 3B	The results showed that resveratrol inhibited cell growth in p53-positive Hep G2 cells only. This anticancer effect was a result of cellular apoptotic death induced by resveratrol via the p53-dependent pathway. The group demonstrated that the resveratrol-treated cells were arrested in the G_1 phase and were associated with the increase of p21 expression. Resveratrol-treated cells had enhanced Bax expression but they were not involved in Fas/APO-1 apoptotic signal pathway. In contrast, the p53-negative Hep 3B cells treated with resveratrol did not show the antiproliferation effect nor did they show significant changes in p21 or Fas/APO-1 levels	The study demonstrated that resveratrol effectively inhibited cell growth and induced programmed cell death in hepatoma cells on a molecular basis. These results implied that resveratrol might also be a new potent chemopreventive drug candidate for liver cancer as it played an important role to trigger p53-mediated molecules involved in the mechanism of p53-dependent apoptotic signal pathway	221
Tested antiproliferative activity and induction of apoptosis using MCF-7 and MDA-MB-231 human breast cancer cells	Using human breast cancer cell lines MCF-7 and MDA-MB-231, a possible mechanism by which resveratrol could interfere with cell cycle control and induce cell death was analyzed. The results showed that although resveratrol inhibited cell proliferation and viability in both cell lines, apoptosis was induced in a concentration- and cell-specific manner. In MDA-MB-231, resveratrol reduced the expression and kinase activities of positive G_1/S and G_2/M cell cycle regulators and inhibited ribonucleotide reductase activity in a concentration dependent manner, without a significant effect on the low expression of tumor suppressors p21, p27, and p53.	The antiproliferative activity of resveratrol could take place through the differential regulation of the cell cycle leading to apoptosis or necrosis. This could be influenced, among other factors, by the concentration of this molecule and by the characteristics of the target cell	222

	These cells died by a nonapoptotic process in the absence of a significant change in cell cycle distribution. In MCF-7, resveratrol produced a significant and transient increase in the expression and kinase activities of positive G_1/S and G_2/M regulators. Simultaneously, p21 expression was markedly induced in presence of high levels of p27 and p53. These opposing effects resulted in cell cycle blockade at the S-phase and apoptosis induction in MCF-7 cells		
Induction of apoptosis in human HCT116 colon carcinoma cells	The expression, subcellular localization, and importance of Bax for resveratrol-provoked apoptosis were assessed in human HCT116 colon carcinoma cells and derivatives with both Bax alleles inactivated. Low to moderate concentrations of resveratrol induced colocalization of cellular Bax protein with mitochondria, collapse of the mitochondrial membrane potential, activation of caspases 3 and 9, and, finally, apoptosis. In the absence of Bax, membrane potential collapse was delayed, and apoptosis was reduced but not absent. Resveratrol inhibited the formation of colonies by both HCT116 and HCT116 Bax -/- cells	Resveratrol, at physiological doses, can induce a Bax-mediated and a Bax-independent mitochondrial apoptosis. Both can limit the ability of the cells to form colonies	223

(*Continued*)

TABLE 13.3
Continued

Model	What was measured	Effect	Ref.
Cytostatic effects on peripheral blood human lymphocytes	Resveratrol had no detectable effects on resting lymphocytes. With the mitogen phytohemagglutin (PHA), however, resveratrol elicited concentration- and time-dependent responses in lymphocytes. Resveratrol prevented cell entry into the cell cycle, resulting in 99% suppression at 100 μM. The arrested lymphocytes following 24 h treatment with 50 μM resveratrol had minimal RNA content, the feature characteristic of G_0 cells, and were blocked at the stage past the induction of cyclins D2 and D3 and prior to induction of cyclin E. Prolonged treatment (72 h) of PHA-stimulated lymphocytes with 100 μM resveratrol showed a pronounced decrease in the expression of pRb, cyclins E and B, and reduction in $p34^{cdc2}$ and PCNA. The activation-induced apoptosis was also reduced in the presence of $\geq$50 μM resveratrol	These data suggest that studies designed to test resveratrol's efficacy as a chemopreventive agent should include evaluation of its immunomodulatory effect revealed by suppression of lymphocyte stimulation as well as its effect on apoptosis of stimulated lymphocytes	167
Inhibition of neutral endopeptidase (NEP) and angiotensin-converting enzyme (ACE) activity of the neuroblastoma cell line SK-N-SH	The long-term incubation of the cells for 4 days with quercetin, resveratrol, and a combination of both substances in concentrations lower than necessary for inhibition of NEP and ACE activity induced the cellular enzyme activity of NEP and ACE associated with an inhibition of cellular proliferation. The long-term treatment of neuroblastoma cells with quercetin and resveratrol enhanced the differentiation state of the cells	Taking into account the significance of NEP and ACE for the degradation of amyloid β peptides, the effect of quercetin and resveratrol as constituents of red wine for neuroprotective activity is suggested	224

Induction of apoptosis using prostate cancer cells (LNCaP and DU 145) and the significance of the three hydroxyl groups on resveratrol to the measured effect	Hormone-sensitive LNCaP cells and hormone-insensitive DU145 cells were treated with resveratrol, trimethoxy-resveratrol, or diethylstilbestrol (the positive control for toxicity and apoptosis). Cell viability was determined by using an MTS assay. Apoptosis was determined by the appearance of apoptotic morphology, annexin V-FITC-positive intact cells, and caspase activation	Resveratrol and diethylstilbestrol decreased viability in LNCaP cells, but only resveratrol-treated cells expressed apoptotic morphology, annexin V-FITC-positive cells, and caspase activation. Trimethoxyresveratrol had no effect on DU145 cell viability and was less toxic to LNCaP cells than resveratrol. Resveratrol was toxic to cells regardless of whether the cells were hormone-responsive or -unresponsive. This finding suggests that the hormone-responsive status of the cell is not an important determinant of the response to resveratrol. Furthermore, the hydroxyl groups on resveratrol are required for cell toxicity. Finally, resveratrol but not diethylstilbestrol induced caspase-mediated apoptosis	110
Investigation of the effects of resveratrol on DNA binding via esterification reactions with 2-hydroxyamino-1-methyl-6-phenyl-imidazo[4,5-*b*]pyridine (N-OH-PhIP) — a metabolite of a mammary gland carcinogen present in cooked meats	Treatment of primary cultures of human mammary epithelial cells with 50 μM resveratrol led to a decrease in PhIP-DNA adducts ranging from 31 to 69%. Using substrate-specific assays and mammary gland tissue cytosols, resveratrol inhibited PhIP-DNA adduct formation by *O*-acetyltransferase and sulfotransferase catalysis. Cytosols from tumor tissue and breast reduction tissue were similarly affected. Resveratrol also suppressed *O*-acetyltransferase and sulfotransferase activities from the breast cancer cell lines MCF-7 and ZR-75-1. It was also observed that resveratrol stimulated ATP-dependent cytosolic activation of N-OH-PhIP in all human samples but not in mouse liver samples	The data suggest that *O*-acetyltransferases and sulfotransferases may represent antioncogenic targets for resveratrol	225

(*Continued*)

TABLE 13.3
Continued

Model	What was measured	Effect	Ref.
Inhibition of the growth and the induction of apoptosis of both normal and leukemic hematopoietic cells	Resveratrol inhibited the proliferation and induced the apoptosis of all tested lymphoid and myeloid leukemia cells. Prior to apoptosis, resveratrol induced caspase activity in a dose-dependent manner and cell cycle arrest in G_2/M phase, correlating with a significant accumulation of cyclins A and B. Leukemia cell death with resveratrol required both caspase-dependent and -independent proteases, as it was significantly inhibited by simultaneous addition of z-VAD-FMK and leupeptin to these cultures. While resveratrol did not affect nonactivated normal lymphocytes, it decreased the growth and induced the apoptosis of cycling normal human peripheral blood lymphocytes at lower concentrations than those required for most leukemia cells. Resveratrol also induced the apoptosis of early normal human $CD34^+$ cells and decreased the number of colonies generated by these precursor cells in a dose-dependent manner	These data suggest the complexity of resveratrol-mediated signaling pathways and revealed the high antiproliferative and proapoptotic activities of resveratrol in normal cycling hemopoietic cells	226
Cytotoxic and antimutagenic effects of resveratrol, ε-viniferin, viniferin, gnetin H, and suffruticosols A and B were determined against five different cancer cell lines, C6 (mouse glioma),	Resveratrol showed significant cytotoxic activity against HepG2 (liver hepatoma) and HT-29 (colon) human cancer cell lines. In contrast, *trans*-ε-viniferin and *cis*-viniferin, and gnetin H exhibited marked cytotoxic activity against HeLa (cervical) and MCF-7 (breast) human cancer cell lines. However, suffruticosol A and B had a reduced cytotoxic effect against all cancer cells except C6	The six stilbenes showed cytotoxic activity in a dose-dependent manner, and were especially potent against the C6 (mouse glioma) cancer cell. These stilbenes exerted antimutagenic activity in a dose-dependent fashion. Of them, resveratrol exhibited the strongest antimutagenic effect against MNNG, while the other five resveratrol oligomers also mediated moderate antimutagenic activity	120

HepG2 (liver hepatoma), HT-29 (colon), HeLa (cervicse), MCF-7 (breast), and mutagenicity of *N*-methyl-*N′*-nitro-*N*-nitrosoguanidine (MNNG) in *Salmonella typhimurium* TA100			
Induction of serine phosphorylation of p53 causes apoptosis in a mutant p53 prostate cancer cell line (DU 145)	The effect of resveratrol was determined in the androgen-insensitive DU145 prostate cancer cell line. Induction of apoptosis and activation of apoptosis related signal transduction pathways were measured. DU145 cells were treated with resveratrol and apoptosis was measured by determining nucleosome content. Activation of mitogen-activated protein kinase (MAPK) (extracellular signal-regulated kinase 1/2), p53 content, and serine-15 phosphorylation of p53 were measured by immunoblot. Electrophoretic mobility shift assay of p53 binding to DNA, and measurement of p21 and glyceraldehyde-3-phosphate dehydrogenase messenger RNA were also done	Resveratrol induced apoptosis in DU145 cells. It activated MAPK and caused increased abundance of p53 and serine-15 phosphorylated p53. Resveratrol-induced serine-15 phosphorylation of p53 was blocked by PD 98059, a MAPK kinase inhibitor, implicating MAPK activation in the phosphorylation of p53. PD 98059 also inhibited resveratrol-induced apoptosis. These results suggest that apoptosis induction by resveratrol in DU145 cells requires serine-15 phosphorylation of p53 by MAPK. Inhibition of MAPK-dependent serine-15 phosphorylation resulted in reduced p53 binding to a p53-specific oligonucleotide on electrophoretic mobility shift assay. Pifithrin-α, a p53 inhibitor, blocked resveratrol-induced serine-15 phosphorylation of p53 and p53 binding to DNA. Resveratrol caused a p53-stimulated increase in p21 messenger RNA. Transfection of additional wild-type p53 into DU145 cells induced apoptosis, which was further enhanced by resveratrol treatment. Resveratrol causes apoptosis in DU145 prostate cancer cells. This action depends on the activation of MAPK, increase in cellular p53 content, serine-15 phosphorylation of p53, and increased p53 binding to DNA	227

(*Continued*)

TABLE 13.3
Continued

Model	What was measured	Effect	Ref.
Inhibition of cell proliferation in human colorectal tumor SW480 cell line	The mechanism by which resveratrol inhibits cell proliferation was studied in human colorectal tumor SW480 cell line. Resveratrol strongly inhibits cell proliferation at the micromolar range in a time- and dose-dependent manner. Resveratrol appears to block the cell cycle at the transition → G_2/M since inhibition of [^{3}H]-thymidine incorporation is not observed, while there is an increase of the cell number in S phase. During this inhibition process, resveratrol increases the content of cyclins A and B1 as well as cyclin-dependent kinases Cdk1 and Cdk2. Moreover, resveratrol promotes Cdk1 phosphorylation	Resveratrol exerts a strong inhibition of SW480 human colorectal tumor cell proliferation and modulates cyclin and cyclin-dependent kinase activities	228
Differentiation induced by butyrate in Caco-2 colon cancer cells	The aim of this study was to determine whether resveratrol modulates the effects of butyrate on Caco-2, a colonic adenocarcinoma cell line. The growth inhibitory effect of resveratrol was more powerful than that of butyrate. Butyrate did not intensify the inhibition of proliferation exerted by resveratrol. Although the polyphenol enhanced the differentiation-inducing effect of butyrate, it did not elevate alkaline phosphatase activity or E-cadherin protein expression, markers of epithelial differentiation, when applied alone. Butyrate-induced transforming growth factor-β1 secretion was inhibited by resveratrol. Treatment with the combination of resveratrol and butyrate attenuated levels of $p27^{Kip1}$, whereas resveratrol enhanced the effect butyrate had on the induction of $p21^{Waf1/Cip1}$ expression	These data demonstrate a possible combined chemopreventive effect of two substances naturally occurring in the colonic lumen after ingestion of fibers and resveratrol-containing food	229

Inducer of differentiation in human myeloid leukemias (HL-60, NB4, U937, THP-1, ML-1, Kasumi-1) and fresh samples from 17 patients with acute myeloid leukemia	This group studied the *in vitro* biological activity of resveratrol by examining its effect on proliferation and differentiation in myeloid leukemia cell lines (HL-60, NB4, U937, THP-1, ML-1, Kasumi-1) and fresh samples from 17 patients with acute myeloid leukemia. Resveratrol alone inhibited the growth in liquid culture of each of the six cell lines. Resveratrol enhanced the expression of adhesion molecules (CD11a, CD11b, CD18, CD54) in each of the cell lines except for Kasumi-1. Moreover, resveratrol induced 37% of U937 cells to produce superoxide as measured by the ability to reduce nitroblue tetrazolium (NBT). The combination of resveratrol and all-*trans*-retinoic acid (ATRA) induced 95% of the NB4 cells to become NBT-positive, whereas $<1\%$ and 12% of the cells became positive for NBT after a similar exposure to either resveratrol or ATRA alone, respectively. In U937 cells exposed to resveratrol, the binding activity of NF-κB protein was suppressed. Eight of 19 samples of fresh acute leukemia cells reduced NBT after exposure to resveratrol	These findings show that resveratrol inhibits proliferation and induces differentiation of myeloid leukemia cells	230
Induction of apoptosis in human B-cell lines derived from chronic B-cell malignancies (WSU-CLL and ESKOL), and in leukemic lymphocytes from patients with B-cell chronic lymphocytic leukemia (B-CLL)	Resveratrol displayed antiproliferative activity on both B-cell lines, as estimated by the decrease in cell recovery and inhibition of thymidine uptake. Furthermore, resveratrol induced apoptosis in the two cell lines as well as in B-CLL patients' cells, as evidenced by the increase in annexin V binding, caspase activation, DNA fragmentation, and decrease of the mitochondrial transmembrane potential $\Delta \psi$. The group previously reported that nitric oxide (NO), endogenously released by an iNO synthase (iNOS) spontaneously expressed in these leukemic cells, contributed to their resistance towards apoptosis. The group shows that resveratrol inhibited both iNOS protein expression and *in situ* NO release in WSU-CLL, ESKOL, and B-CLL patients' cells. In addition, Bcl-2 expression was inhibited by resveratrol	Downregulation of the two antiapoptotic proteins iNOS and Bcl-2 can contribute to the apoptotic effects of resveratrol in leukemic B cells from chronic leukemia. These data suggest that this drug is of potential interest for the therapy of B-CLL	231

(*Continued*)

TABLE 13.3
Continued

Model	What was measured	Effect	Ref.
Evaluation of pterostilbene for antioxidative potential	The peroxyl-radical scavenging activity of pterostilbene was the same as that of resveratrol, having total reactive antioxidant potentials of 237 ± 58 and 253 ± 53 μM, respectively. Both compounds were found to be more effective than Trolox as free radical scavengers. Using a plant system, pterostilbene also was shown to be as effective as resveratrol in inhibiting electrolyte leakage caused by herbicide-induced oxidative damage, and both compounds had the same activity as α-tocopherol. Pterostilbene showed moderate inhibition ($IC_{50} = 19.8$ μM) of cyclooxygenase (COX)-1, and was weakly active ($IC_{50} = 83.9$ μM) against COX-2, whereas resveratrol strongly inhibited both isoforms of the enzyme with IC_{50} values of approximately 1 μM	Using a mouse mammary organ culture model, carcinogen-induced preneoplastic lesions were, similarly to resveratrol, significantly inhibited by pterostilbene ($ED_{50} = 4.8$ μM), suggesting antioxidant activity plays an important role in this process	232
Impact of phenolic compounds, ellagic acid (EA) and resveratrol (RE), on target genes in prostate cancer (LNCaP) cells	Human cDNA microarrays were used with 2400 clones consisting of 17 prosite motifs to characterize alterations in gene expression pattern in response to EA and RE. Over a 48 h exposure of androgen-sensitive LNCaP cells to EA and RE, a total of 593 and 555 genes, respectively, showed more than a two fold difference in expression. A distinct set of genes in both EA- and RE-treated cells may represent the signature profile of phenolic antioxidant-induced gene expression in LNCaP cells. Although extensive similarity was found between effects of EA- and RE-responsive genes in prostate cancer cells, out of 246	In-depth analysis of the data from this study provides insight into the alterations in the p53-responsive genes, p300, Apaf-1, NF-κBp50, and p65 and PPAR families of genes, suggesting the activation of multiple signaling pathways leads to growth inhibition of LNCaP cells.	139

	genes with overlapping responses, 25 genes showed an opposite effect. Quantitative RT-PCR was used to verify and validate the differential expression of selected genes identified from cDNA microarrays		
Induction of prostate cancer cells (LNCaP and DU145) entry into S phase and inhibition of DNA synthesis	The studies revealed that, in androgen-sensitive LNCaP cells, the effect of resveratrol on DNA synthesis varied dramatically depending on the concentration and the duration of treatment. In cells treated for 1 h, resveratrol showed only an inhibitory effect on DNA synthesis, which increased with increasing concentration ($IC_{50} = 20\ \mu M$). However, when treatment duration was extended to 24 h, a dual effect of resveratrol on DNA synthesis was observed. At 5 to 10 μM it caused a 2- to 3-fold increase in DNA synthesis, and at ≥15 μM, it inhibited DNA synthesis. The increase in DNA synthesis was seen only in LNCaP cells, but not in androgen-independent DU145 prostate cancer cells or in NIH3T3 fibroblast cells. The resveratrol-induced increase in DNA synthesis was associated with enrichment of LNCaP cells in S phase, and a concurrent decrease in nuclear $p21^{Cipl}$ and $p27^{Kip1}$ levels. Furthermore, consistent with the entry of LNCaP cells into S phase, there was a dramatic increase in nuclear Cdk2 activity associated with both cyclin A and cyclin E	The observations indicate that LNCaP cells, treated with resveratrol, are induced to enter into S phase, but subsequent progression through S phase is limited by the inhibitory effect of resveratrol on DNA synthesis, particularly at concentrations above 15 μM. Therefore, the ability of resveratrol to exert opposing effects on two important processes in cell cycle progression, induction of S phase and inhibition of DNA synthesis, may be responsible for its apoptotic and antiproliferative effects	233

(Continued)

TABLE 13.3
Continued

Model	What was measured	Effect	Ref.
Inhibition of gastric cancer cell (gastric adenocarcinoma SNU-1 cells) proliferation	Low levels of exogenous reactive oxygen (H_2O_2) stimulated [^{3}H]thymidine uptake in SNU-1 cells, whereas resveratrol suppressed both synthesis of DNA and generation of endogenous O^{2-} but stimulated nitric oxide (NO) synthase (NOS) activity. To address the role of NO in the antioxidant action of resveratrol, the effect of sodium nitroprusside (SNP) was measured, an NO donor, on O^{2-} generation, and on [^{3}H]thymidine incorporation. SNP inhibited DNA synthesis and suppressed ionomycin-stimulated O^{2-} generation in a concentration-dependent manner	The results revealed that the antioxidant action of resveratrol toward gastric adenocarcinoma SNU-1 cells may reside in its ability to stimulate NOS to produce low levels of NO, which, in turn, exert antioxidant action. Resveratrol-induced inhibition of SNU-1 proliferation may be partly dependent on NO formation, and the group hypothesizes that resveratrol exerts its antiproliferative action by interfering with the action of endogenously produced reactive oxygen. These data are supportive of the action of NO against reactive oxygen and suggest that a resveratrol-rich diet may be chemopreventive against gastric cancer	234
Induced activation of p53 and apoptosis using the JB6 mouse epidermal cell line	This study determined that c-jun NH_2-terminal kinases (JNKs) are involved in resveratrol-induced p53 activation and induction of apoptosis. In the JB6 mouse epidermal cell line, resveratrol activated JNKs dose-dependently within a dose range of 10–40 μM, the same dosage responsible for the inhibition of tumor promoter-induced cell transformation. Stable expression of a dominant negative mutant of JNK1 or disruption of the Jnk1 or Jnk2 gene markedly inhibited resveratrol-induced p53-dependent transcription activity and induction of apoptosis. Furthermore, resveratrol-activated JNKs were shown to phosphorylate p53 *in vitro*, but this	These data suggested that JNKs act as mediators of resveratrol-induced activation of p53 and apoptosis, which may occur partially through p53 phosphorylation	235

	activity was repressed in the cells expressing a dominant negative mutant of JNK1 or in Jnk1 or Jnk2 knockout (*Jnk1*$^{-/-}$ or *Jnk2*$^{-/-}$) cells		
Investigation of S-phase arrest, apoptosis, and changes in biomarker expression in six human cancer cell lines (MCF7, SW480, HCE7, Seg-1, Bic-1, and HL60)	Resveratrol induced marked growth inhibition in five of these cell lines, with IC_{50} values of approximately 70–150 μM. However, only partial growth inhibition was seen in Bic-1 cells. After treatment with 300 μM resveratrol for 24 h, most of the cell lines were arrested in the S phase of the cell cycle. In addition, induction of apoptosis was demonstrated by the appearance of a sub-G_1 peak and confirmed using an annexin V-based assay. Cyclin B1 expression levels were decreased in all cell lines after 48 h of treatment. In SW480 cells, cyclin A, cyclin B1, and β-catenin expression levels were decreased within 24 h. There was a decrease in cyclin D1 expression after only 2 h of treatment, and this persisted for up to 48 h. This decrease was partially blocked by concurrent treatment with the proteasome inhibitor calpain inhibitor I. Using a luciferase-based reporter assay, resveratrol did not inhibit cyclin D1 promoter activity in SW480 cells. Furthermore, using a reverse transcription PCR-based assay, only a higher dose of resveratrol (300 μM) appeared to decrease cyclin D1 mRNA. Seg-1 cells expressed basal levels of cyclooxygenase-2 (COX-2), which was further induced by resveratrol. Neither basal levels nor induction of COX-2 was detectable in the remaining cell lines. Thus, COX-2 does not appear to be a critical target of this compound	These studies provide support for the use of resveratrol in chemoprevention and cancer therapy trials. Cyclin D1, cyclin B1, β-catenin, and apoptotic index could be useful biomarkers to evaluate treatment efficacy	236

(*Continued*)

TABLE 13.3
Continued

Model	What was measured	Effect	Ref.
Induction of apoptosis in two papillary thyroid carcinoma (PTC) and two follicular thyroid carcinoma (FTC) cell lines via a MAPK- and p53-dependent mechanism	Two PTC and FTC cell lines were treated with resveratrol, which showed activation and nuclear translocation of MAPK (extracellular signal-regulated kinase 1/2). Cellular abundance of the oncogene suppressor protein p53, serine phosphorylation of p53, and abundance of c-fos, c-jun, and p21 mRNAs were also increased by resveratrol. Inhibition of the MAPK pathway by either H-ras antisense transfection or PD 98059, an MAPK kinase inhibitor, blocked these resveratrol-induced effects. Addition of pifithrin-α, a specific inhibitor of p53, or transfection of p53 antisense oligonucleotides caused decreased resveratrol-induced p53 and p21 expression in PTC and FTC cells. Studies of nucleosome levels estimated by ELISA and of DNA fragmentation showed that resveratrol induced apoptosis in both papillary and follicular thyroid cancer cell lines; these effects were inhibited by pifithrin-α and by p53 antisense oligonucleotide transfection. PD 98059 and H-ras antisense transfection also blocked induction of apoptosis by resveratrol	Resveratrol acts via a Ras-MAPK kinase-MAPK signal transduction pathway to increase p53 expression, serine phosphorylation of p53, and p53-dependent apoptosis in PTC and FTC cell lines	237
Induction of apoptosis by 3,4′-dimethoxy-5-hydroxystilbene in human promyeloid leukemic HL-60 cells	Treatment of HL-60 cells with DMHS suppressed the cell growth in a concentration-dependent manner with an IC_{50} value of 25 μM. DMHS increased internucleosomal DNA fragmentation in a time-dependent manner. The cell death by DMHS was partially prevented by the caspase inhibitor zVAD-fmk. DMHS caused activation of caspases such as caspase-3, -8, and -9. Immunoblot experiments revealed that DMHS-induced apoptosis was associated with the induction of Bax expression.	The results indicated that DMHS leads to apoptotic cell death in HL-60 cells through increased Bax expression and release of cytochrome c into cytosol and may be considered as a good candidate for a cancer chemopreventive agent in humans	238

	The release of cytochrome c from mitochondria into the cytosol was increased in response to DMHS		
Mechanism of resveratrol-mediated suppression of tissue factor gene expression	The mechanism was examined by which resveratrol inhibits the expression of TF in monocytes by using a monocytic cell line, THP-1, as a model cell. Northern blot analysis, gel mobility shift assays, and transfection studies with various TF promoter constructs, as well as other transcription regulatory constructs, were used to elucidate the inhibitory mechanism of resveratrol. The data show that resveratrol inhibited lipopolysaccharide (LPS)-induced expression of TF in human monocytes and monocytic cell line THP-1 in a dose dependent manner. Resveratrol did not significantly alter the binding of various transcription factors involved in TF gene expression to DNA. However, resveratrol suppressed the transcription of cloned human TF promoter. Further experiments revealed that resveratrol reduced κB- but not AP-1-driven transcriptional activity. Additional experiments showed that resveratrol suppressed the phosphorylation of p65 and its transactivation	The results indicate that resveratrol does not inhibit the activation or translocation of NF-κB/Rel proteins but inhibits NF-κB/Rel-dependent transcription by impairing the transactivation potential of p65	239
Induction of apoptosis using resveratrol with 5-fluorouracil (5-FU) on the growth of hepatoma cell line H22	The number of cells was measured by the MTT method and morphological changes of H22 cells were investigated under microscopic examination. Resveratrol inhibited the growth of H22 cells in a dose- and time-dependent manner. The synergistic antitumor effects of resveratrol with 5-FU increased to a greater extent than for H22 cells treated with 5-FU alone. Characteristics of apoptosis such as typical apoptotic bodies were commonly found in tumor cells in the drug-treated groups	Resveratrol can suppress the growth of H22 cells, its antitumor activity may occur through the induction of apoptosis	134

(*Continued*)

TABLE 13.3
Continued

Model	What was measured	Effect	Ref.
Induction of apoptosis using human colon cancer cell lines (SW480, DLD-1 and COLO201)	Vaticanol C (resveratrol tetramer) was characterized by nuclear changes and DNA ladder formation and tested in SW480, DLD-1, and COLO201 cells	Vaticanol C suppressed cell growth through induction of apoptosis	121
Inhibition of progression through the S phase of the cell cycle in colorectal cancer cell lines (Caco-2 and HCT-116) using the compound piceatannol	Growth of Caco-2 and HCT-116 cells was analyzed by crystal violet assay, which demonstrated dose- and time-dependent decreases in cell numbers. Treatment of Caco-2 cells with piceatannol reduced proliferation rate. No effect on differentiation was observed. Determination of cell cycle. Immunoblotting demonstrated that cyclin-dependent kinases (cdk) 2 and 6, as well as cdc2 were expressed at steady-state levels, whereas cyclin D1, cyclin B1, and cdk 4 were downregulated. The abundance of $p27^{Kip1}$ was also reduced, whereas the protein level of cyclin E was enhanced. Cyclin A levels were enhanced only at concentrations up to 100 μM. These changes also were observed in studies with HCT-116 cells	Piceatannol can be considered to be a promising chemopreventive or anticancer agent	240
Radiosensitizing and antiproliferative effects tested in human cervical tumor cell lines (HeLa, Me180, A2780, and SiHa, and the mouse normal	The group hypothesized that tumor cells may exhibit changes in the cellular response to ionizing radiation (IR) following exposure to resveratrol. Clonogenic cell survival assays were performed using irradiated HeLa and SiHa cells pretreated with resveratrol prior to IR exposure, and resulted in enhanced tumor cell killing by IR in a dose-dependent manner. Further analysis of COX-1 inhibition indicated that resveratrol	These results suggest that resveratrol alters both cell cycle progression and the cytotoxic response to IR in two cervical tumor cell lines	241

fibroblast cell line, NIH 3T3)	pretreatment: (1) inhibited cell division as assayed by growth curves; and (2) induced an early S-phase cell cycle checkpoint arrest, as demonstrated by fluorescence-activated cell sorting, as well as bromodeoxyuridine pulse-chase analysis		
Evaluation of 12 phenols (*trans*-astringin, *trans*-piceid, *trans*-resveratroloside, *trans*-resveratrol, *trans*-piceatannol, *cis*-resveratroloside, *cis*-piceid, *cis*-resveratrol, (+)-catechin, (–)-epicatechin, epicatechin 3-*O*-gallate, and procyanidin B2 3′-*O*-gallate) for potential to inhibit cyclooxygenases and preneoplastic lesion formation in carcinogen-treated mouse mammary glands in organ culture	At 10 μg/ml, *trans*-astringin and *trans*-piceatannol inhibited development of 7,12-dimethylbenz(*a*)anthracene-induced preneoplastic lesions in mouse mammary glands with 68.8% and 76.9% inhibition, respectively, compared with untreated glands. The latter compound was the most potent of the 12 compounds tested in this assay, with the exception of *trans*-resveratrol (87.5% inhibition). In the cyclooxygenase (COX)-1 assay, *trans* isomers of the stilbenoids appear to be more active than *cis* isomers: *trans*-resveratrol [50% inhibitory concentration (IC_{50}) = 14.9 μM, 96%] vs. *cis*-resveratrol (IC_{50} = 55.4 μM). In the COX-2 assay, among the compounds tested, only *trans*- and *cis*-resveratrol exhibited significant inhibitory activity (IC_{50} = 32.2 and 50.2 μM, respectively). This is the first report showing the potential cancer chemopreventive activity of *trans*-astringin. *Trans*-astringin and its aglycone *trans*-piceatannol were active in the mouse mammary gland organ culture assay but did not exhibit activity in COX-1 and COX-2 assays. *Trans*-resveratrol was active in all three of the bioassays used in this investigation	These findings suggest that *trans*-astringin and *trans*-piceatannol may function as potential cancer chemopreventive agents by a mechanism different from that of *trans*-resveratrol	92

(*Continued*)

TABLE 13.3
Continued

Model	What was measured	Effect	Ref.
Expression of autocrine growth modulators in human breast cancer cells (MCF-7 and MDA-MB-46)	Resveratrol maximally inhibited the growth stimulatory effect mediated by 10^{-9} M estradiol without affecting cell viability. At the molecular level, resveratrol in a dose-dependent fashion antagonized the stimulation by estradiol of an estrogen response element reporter gene construct and of progesterone receptor gene expression in MCF-7 cells. Resveratrol also inhibited the proliferation of the estrogen receptor negative human breast carcinoma cell line MDA-MB-468. These latter data suggest that resveratrol can also inhibit breast cancer cell proliferation by another mechanism besides estrogen receptor antagonism. The study shows resveratrol altered the expression of several autocrine growth modulators and their receptors in MCF-7 cells. Resveratrol at 10^{-5} M inhibited the expression of the autocrine growth stimulators transforming growth factor-α (TGF-α), PC cell-derived growth factor, and insulin-like growth factor I receptor mRNA. In addition, resveratrol significantly elevated the expression of the growth inhibitor TGF-β2 mRNA without changes in TGF-β1 and TGF-β3 expression	Resveratrol inhibited the growth of estrogen receptor-positive MCF-7 cells cultivated in the presence of estradiol in a dose-dependent fashion. These data suggest that resveratrol inhibits proliferation by altering autocrine growth modulator pathways in breast cancer cells	242

Thirteen stilbene-related compounds tested in the mouse hepatoma Hepa 1c1c7 cells	The compounds were tested for their ability to be inducers of phase II detoxifying metabolic enzyme quinone reductase (QR) in the mouse hepatoma Hepa 1c1c7 cells	Several of the compounds were found potentially to induce QR activity in this cell line. In addition, substitution with 3-thiofurane ring instead of phenyl ring in the stilbene skeleton also exhibited potential induction of QR activity. This study provides primary information to design the potential inducers of QR activity in the stilbene analogs	122
Induction of apoptosis using the HCT116 colon carcinoma cell line	Cell death is primarily mitochondria-mediated and not receptor-mediated. No cells survived in cultures continuously exposed to 100 μM resveratrol for 120 h. When compared with 5-fluorouracil, resveratrol stimulated p53 accumulation and activity only weakly and with delayed kinetics and neither the increased levels nor the activity affected apoptosis detectably. The apoptosis agonist Bax was overproduced in response to resveratrol regardless of p53 status, yet the kinetics of Bax expression were influenced by p53. Remarkably, apoptosis was preceded by mitochondrial proliferation and signs of epithelial differentiation	Using the human wild-type p53-expressing HCT116 colon carcinoma cell line and HCT116 cells with both p53 alleles inactivated by homologous recombination, the study shows that resveratrol at concentrations comparable to those found in some foods can induce apoptosis independently of p53. Resveratrol triggers a p53-independent apoptotic pathway in HCT116 cells that may be linked to differentiation	243

(*Continued*)

TABLE 13.3
Continued

Model	What was measured	Effect	Ref.
Determination of the chemopreventive potential of resveratrol against human gastric adenocarcinoma cells (KATO-III and RF-1)	The study shows the action of resveratrol on cellular function and cellular integrity by measuring DNA synthesis, cellular proliferation, cell cycle distribution, cytolysis, apoptosis, and phosphotransferase activities of two key signaling enzymes, protein kinase C (PKC) and mitogen-activated protein kinases (ERK1/ERK2), in KATO-III and RF-1 cells. Resveratrol inhibited [^{3}H]thymidine incorporation into cellular DNA of normally proliferating KATO-III cells and of RF-1 cells whose proliferation was stimulated with carcinogenic nitrosamines. Treatment with resveratrol arrested KATO-III cells in the G_0/G_1 phase of the cell cycle and eventually induced apoptotic cell death, but had a minimal effect on cell lysis. Resveratrol treatment had no effect on ERK1/ERK2 activity but significantly inhibited PKC activity of KATO-III cells and of human recombinant PKCα	Results indicate that resveratrol has potential as a chemopreventive agent against gastric cancer since it exerts an overall deactivating effect on human gastric adenocarcinoma cells. Resveratrol-induced inhibition of PKC activity and of PKCα, without any change in ERK1/ERK2 activity, suggests that resveratrol utilizes a PKC-mediated mechanism to deactivate gastric adenocarcinoma cells	244
Inhibition of cell growth and apoptosis in human cancer cells (Col2) using a resveratrol analog, *trans*-3,5,2′,4′-tetramethoxystilbene	Prompted by the strong growth inhibitory activity of the compound compared to resveratrol in cultured Col2, a mechanistic study was performed using the analog. It induced the accumulation of cellular DNA content in the sub-G_0 phase of the cell cycle in a time-dependent manner. The morphological changes were also consistent with an apoptotic process	Trans-3,5,2′,4′-tetramethoxystilbene potentiated the inhibition of cancer cell growth. This result indicated that the compound induced apoptosis of cancer cells, and may be a candidate for use as a cancer chemotherapeutic or cancer chemopreventive agent	123

Induction of apoptosis in the lymphoma cell line BJAB and in primary, leukemic lymphoblasts	Using BJAB cells overexpressing a dominant-negative mutant of the Fas-associated death domain (FADD) adaptor protein to block death receptor-mediated apoptosis, the study demonstrates that resveratrol- and piceatannol-induced cell death is independent of the CD95/Fas signaling pathway. To explore the antileukemic properties of both compounds in more detail, primary leukemic lymphoblasts were investigated. Piceatannol but not resveratrol is a very efficient inducer of apoptosis in this *ex vivo* assay with leukemic lymphoblasts of 21 patients suffering from childhood lymphoblastic leukemia (ALL)	Resveratrol and piceatannol are potent inducers of apoptotic cell death in BJAB Burkitt-like lymphoma cells with an ED_{50} of 25 μM. Experiments revealed that treatment of BJAB cells with both substances led to a concentration-dependent activation of caspase-3 and mitochondrial permeability transition	245
Antiproliferative effects using human epidermoid carcinoma (A431) cells and involvement of the retinoblastoma (pRb)-E2F/DP pathway	Immunoblot analysis demonstrated that resveratrol treatment of A431 cells results in a dose- as well as time-dependent decrease in the hyperphosphorylated form of pRb with a relative increase in hypophosphorylated pRb. This response was accompanied by downregulation of protein expression of all five E2F family members of transcription factors studied and their heterodimeric partners DP1 and DP2. This suggests that resveratrol causes a downregulation of hyperphosphorylated pRb protein with a relative increase in hypophosphorylated pRb that, in turn, compromises with the availability of free E2F. It is suggested that this series of events results in a stoppage of the cell cycle progression at the $G_1 \rightarrow S$ phase transition thereby leading to a G_0/G_1 arrest and subsequent apoptotic cell death	Evidence is provided for the involvement of the pRb-E2F/DP pathway as an important contributor of resveratrol-mediated cell cycle arrest and apoptosis. This study shows the involvement of the pRb-E2F/DP pathway as a mechanism of the cancer-chemopreventive effects of resveratrol	246

(*Continued*)

TABLE 13.3
Continued

Model	What was measured	Effect	Ref.
Inhibition of CYP1A1 expression in breast cancer cells (T47D and MCF-7)	Resveratrol inhibited TCDD-induced reporter gene activity in cells transfected with an Ah-responsive construct containing a human CYP1A1 gene promoter insert, whereas 3′-methoxy-4′-nitroflavone, a "pure" AhR antagonist, inhibited this response. Resveratrol induced transformation of the rat cytosolic AhR and, after treatment of T47D and MCF-7 cells with resveratrol, a transformed nuclear AhR complex was observed. In contrast to 3′-methoxy-4′-nitroflavone, resveratrol did not block TCDD-induced AhR transformation *in vitro* or nuclear uptake of the AhR complex in breast cancer cells. The action of resveratrol on the AhR was consistent with that of an AhR agonist; however, resveratrol did not exhibit functional AhR agonist or antagonist activities in breast cancer cells. Actinomycin D chase experiments in T47D cells showed that resveratrol and dehydroepiandrosterone both increased the rate of CYP1A1 mRNA degradation, whereas resveratrol did not affect CYP1A1-dependent activity in cells pretreated with TCDD for 18 h	These data suggest that resveratrol inhibits CYP1A1 via an AhR-independent posttranscriptional pathway	247

Bcl-2 overexpression reduction and induction of apoptosis in U937/vector and U937/Bcl-2 cells by inhibition of caspase-3 activity	U937/vector and U937/Bcl-2 cells were used, which were generated by transfection of the cDNA of the Bcl-2 gene. As compared with U937/vector, U937/Bcl-2 cells exhibited a 4-fold greater expression of Bcl-2. Treatment with 60 or 100 μM resveratrol for 24 h produced morphological features of apoptosis and DNA fragmentation in U937/vector cells, respectively. This was associated with caspase-3 activation and PLC-gamma1 degradation. In contrast, resveratrol-induced caspase-3 activation and PLC-gamma1 degradation and apoptosis were significantly inhibited in U937/Bcl-2 cells	The effect of high intracellular levels of the antiapoptosis protein Bcl-2 on caspase-3 activation, PLC-γ1 degradation, and cytochrome c release during resveratrol-induced apoptosis was determined. Bcl-2 overexpressing cells exhibited less cytochrome c release and sustained expression levels of the IAP proteins during resveratrol-induced apoptosis. In addition, these findings indicate that Bcl-2 inhibits resveratrol-induced apoptosis by a mechanism that interferes with cytochrome c release and activity of caspase-3 that is involved in the execution of apoptosis	248
Inhibition of cell proliferation and prevention of oxidative DNA damage using a panel of cell lines of various histogenetic origin, including normal rat fibroblasts and mouse mammary epithelial cells compared to human breast, colon, and prostate cancer cells	The concentration of resveratrol inhibiting cell growth by 50% (IC_{50}) ranged from about 20 to 100 μM. At such concentrations, a significant increase was observed in the apoptotic index in most of the cell lines analyzed. There was a reduction in the percentage of cells in the G_2/M phase which was most frequently associated with an increase of cells in the S phase of the cell cycle. Resveratrol was able to prevent the increase in reactive oxygen species (ROS) following exposure to oxidative agents. Resveratrol also reduced nuclear DNA fragmentation, as assessed by single cell gel electrophoresis (COMET test)	The results suggest that resveratrol can act as an antimutagenic/anticarcinogenic agent by preventing oxidative DNA damage which plays a pivotal role in the carcinogenic activity of many genotoxic agents	249

(Continued)

TABLE 13.3
Continued

Model	What was measured	Effect	Ref.
Cytostatic and antiestrogenic properties using human endometrial adenocarcinoma (Ishikawa) cells	Treatment of cultured Ishikawa cells with resveratrol did not significantly increase the levels of an estrogen-inducible marker enzyme, alkaline phosphatase. When alkaline phosphatase was induced by treatment of 17β-estradiol (E_2), resveratrol exhibited a dose-dependent decrease in activity. When Ishikawa cells were treated with resveratrol as a single agent, estrogen-inducible progesterone receptor (PR) was not enhanced, and PR expression induced by treatment with E_2 was inhibited by resveratrol in a dose-dependent fashion at both the mRNA and protein levels. Resveratrol mediated suppression of a functional activity of PR as demonstrated by downregulation of α_1-integrin expression induced by E_2 plus progesterone. With transient transfection experiments conducted with Ishikawa cells, antiestrogenic effects were confirmed by dose-dependent inhibition of E_2-induced estrogen response element-luciferase transcriptional activity. Because resveratrol antagonized estrogenic effects in Ishikawa cells, competitive binding analyses were performed to examine the potential of displacing $[^3H]E_2$ from human estrogen receptor (ER). Resveratrol showed no discernable activity with ER-α, but with ER-β, E_2 was displaced with an IC_{50} of 125 μM. However, mRNA and protein expression of ER-α but not ER-β were suppressed by resveratrol in Ishikawa cells, in the concentration range of 5–15 μM. In the presence or absence of E_2, resveratrol inhibited Ishikawa cell proliferation in a time-dependent manner with cells accumulating in the S phase of the cycle ≤48 h. This effect was reversible. Analysis of some critical cell cycle proteins revealed a specific increase in expression of cyclins A and E but a decrease in cyclin-dependent kinase 2	These data suggest resveratrol exerts an antiproliferative effect in Ishikawa cells, and the effect may be mediated by both estrogen-dependent and -independent mechanisms	111

Investigate the effect of resveratrol on the human colonic adenocarcinoma cell line Caco-2 and the colon carcinoma cell line HCT-116	The compound inhibited cell growth and proliferation of Caco-2 cells in a dose-dependent manner as assessed by crystal violet assay and [^{3}H]thymidine and [^{14}C]leucine incorporation. Apoptosis was determined by measuring caspase-3 activity, which increased significantly after treatment with resveratrol. Perturbed cell cycle progression from the S to G_2 phase was observed for concentrations up to 50 μmol/l, whereas higher concentrations led to reversal of the S phase arrest. These effects were specific for resveratrol; they were not observed after incubation with the stilbene analogs stilbenemethanol and rhapontin. Levels of cyclin D1 and cyclin-dependent kinase (cdk) 4 proteins were decreased, as revealed by immunoblotting. Resveratrol enhanced the expression of cyclin E and cyclin A. The protein levels of cdk2, cdk6, and proliferating cell nuclear antigen were unaffected. Similar results were obtained for HCT-116 cells, indicating that cell cycle inhibition by resveratrol is independent of cyclooxygenase inhibition. The phosphorylation state of the retinoblastoma protein in Caco-2 cells was shifted from hyperphosphorylated to hypophosphorylated at 200 μM, which may account for reversal of the S phase block at concentrations exceeding 50 μM	These findings suggest that resveratrol exerts chemopreventive effects on colonic cancer cells by inhibition of the cell cycle	250

(*Continued*)

TABLE 13.3
Continued

Model	What was measured	Effect	Ref.
Induction of apoptotic cell death investigation using CD95-sensitive leukemia lines, B-lineage leukemic cells that are resistant to CD95-signaling, and leukemia lines derived from patients with pro-B t(4;11), pre-B, and T-cell ALL	Multiple dose treatments of the leukemic cells with resveratrol resulted in cell death with no statistically significant cytotoxicity against normal peripheral blood mononuclear cells under identical conditions. Resveratrol treatment did not increase CD95 expression or trigger sensitivity to CD95-mediated apoptosis in the ALL lines. Inhibition of CD95 signaling with a CD95-specific antagonistic antibody indicated that CD95–CD95 ligand interactions were not involved in initiating resveratrol-induced apoptosis. However, in each ALL line, resveratrol induced progressive loss of mitochondrial membrane potential as measured by the dual emission pattern of the mitochondria-selective dye JC-1. The broad-spectrum caspase inhibitor benzyloxycarbonyl-Val-Ala-Asp-fluoromethylketone failed to block the depolarization of mitochondrial membranes induced by resveratrol, further indicating that resveratrol action was independent of upstream caspase-8 activation via receptor ligation. However, increases in caspase-9 activity ranged from 4- to 9-fold in the eight cell lines after treatment with resveratrol	This study shows that resveratrol induces extensive apoptotic cell death not only in CD95-sensitive leukemia lines, but also in B-lineage leukemic cells that are resistant to CD95 signaling. This was investigated using leukemia lines derived from patients with pro-B t(4;11), pre-B, and T-cell ALL. These results suggest a general mechanism of apoptosis induction by resveratrol in ALL cells that involves a mitochondria/caspase-9-specific pathway for the activation of the caspase cascade and is independent of CD95 signaling	251

Evaluation of the effects of resveratrol on invasion of the human hepatoma cell line HepG2	Cell invasion was assessed using a Boyden chamber assay. Activation of the HGF signal transduction pathways was evaluated by Western blot with phospho-specific antibodies. Urokinase expression was measured by RT-PCR and zymography. Resveratrol decreased hepatocyte growth factor-induced cell scattering and invasion. It also decreased cell proliferation without evidence for cytotoxicity or apoptosis. Resveratrol did not decrease the level of the hepatocyte growth factor receptor c-met and did not impede the hepatocyte growth factor-induced increase in c-met precursor synthesis. Resveratrol did not decrease hepatocyte growth factor-induced c-met autophosphorylation, or Akt-1 or extracellular-regulated kinases-1 and -2 activation. Resveratrol did not decrease urokinase expression and did not block the catalytic activity of urokinase	The results demonstrate that resveratrol decreases hepatocyte growth factor-induced HepG2 cell invasion by an unidentified postreceptor mechanism	252
Suppression of hepatoma cells (rat ascites hepatoma cell line of AH109A cells) invasion independently of antiproliferative action	Resveratrol (100 and 200 μM) inhibited the proliferation of hepatoma cells, although it exerted little influence up to 50 μM. Resveratrol suppressed the invasion of the hepatoma cells even at a concentration of 25 μM. Sera from rats orally given resveratrol restrained only the invasion of AH109A cells. Resveratrol and resveratrol-loaded rat serum suppressed reactive oxygen species-potentiated invasive capacity	These results suggest that the antiinvasive activity of resveratrol is independent of the antiproliferative activity, and that the antioxidative property of resveratrol may be involved in its antiinvasive action	253

(Continued)

TABLE 13.3
Continued

Model	What was measured	Effect	Ref.
Modulation in cyclin-dependent kinase (cdk) inhibitor-cyclin-cdk machinery, WAF-1/p21-mediated G_1-phase arrest of the cell cycle and induction of apoptosis of human epidermoid carcinoma (A431) cells	Resveratrol treatment of A431 cells resulted in a dose-dependent (1) inhibition of cell growth as shown by 3-(4,5-dimethylthiazol-2-yl)-2,5-diphenyltetrazolium bromide assay, (2) G_1-phase arrest of the cell cycle as shown by DNA cell cycle analysis, and (3) induction of apoptosis as assessed by ELISA. The immunoblot analysis revealed that resveratrol treatment causes a dose- and time-dependent (1) induction of WAF1/p21; (2) decrease in the protein expression of cyclin D1, cyclin D2, and cyclin E; and (3) decrease in the protein expression of cdk2, cdk4, and cdk6. Resveratrol treatment was also found to result in a dose- and time-dependent decrease in kinase activities associated with all of the cdks examined	Taken together, the study suggests that resveratrol treatment of the cells causes an induction of WAF1/p21 that inhibits cyclin D1/D2-cdk6, cyclin D1/D2-cdk4, and cyclin E-cdk2 complexes, thereby imposing an artificial checkpoint at the $G_1 \rightarrow S$ transition of the cell cycle. This series of events results in a G_1-phase arrest of the cell cycle, which is an irreversible process that ultimately results in the apoptotic death of cancer cells. This study shows the involvement of each component of cdk inhibitor-cyclin-cdk machinery during cell cycle arrest and apoptosis of cancer cells by resveratrol	254
Effect of synthetic resveratrol on the growth of estrogen receptor (ER)-positive (KPL-1 and MCF-7) and -negative (MKL-F) human breast cancer cell lines	Resveratrol, at low concentrations, caused cell proliferation in ER-positive cell lines. At high concentrations, it caused suppression of cell growth in all three cell lines. Growth suppression was due to apoptosis as seen by the appearance of a sub-G_1 fraction. The apoptosis cascade upregulated Bax and Bak protein, downregulated $Bcl\text{-}x^L$ protein, and activated caspase-3. Resveratrol antagonized the effect of linoleic acid and suppressed the growth of both ER-positive and -negative cell lines	The study shows that resveratrol could be a promising anticancer agent for both hormone-dependent and hormone-independent breast cancers, and may mitigate the growth stimulatory effect of linoleic acid in the Western-style diet	255

Expression of the cytochrome P450 1A1 (CYP1A1) and 1B1 (CYP1B1), microsomal epoxide hydrolase (mEH), and glutathione *S*-transferase P1 (GSTP1) genes were tested to determine involvement in the metabolism of polycyclic aromatic hydrocarbons in the human bronchial epithelial cell line BEP2D	Expression of the genes was measured by quantitative reverse transcriptase polymerase chain reaction. The cells were treated either with benzo(*a*)pyrene or 2,3,7,8-tetrachlorodibenzo-*p*-dioxin in the presence or absence of resveratrol. Resveratrol inhibited both the constitutive and the induced expression of CYP1A1 and CYP1B1 in a dose-dependent manner. The expression of the mEH gene was increased in response to resveratrol and no change in the expression of GSTP1 was found. The altered gene expression in response to resveratrol was reflected in a reduced overall level of benzo(*a*)pyrene metabolism	These data indicate that resveratrol may exert lung cancer chemopreventive activity through altering the expression of genes involved in the metabolism of polycyclic aromatic hydrocarbons, resulting in altered formation of carcinogenic benzo(*a*)pyrene metabolites in human bronchial epithelial cells	256
Antiapoptotic effect of *trans*-resveratrol on paclitaxel (an anticancer drug)-induced apoptosis in the human neuroblastoma SH-SY5Y cell line	Paclitaxel induces apoptosis in human neuroblastoma cell line SH-SY5Y. The addition of resveratrol to SH-SY5Y cultures exposed to paclitaxel significantly reduces cellular death. The neuroprotective action of resveratrol is due neither to its antioxidant capacity nor to interference with the polymerization of tubulin induced by paclitaxel. Resveratrol is able to inhibit the activation of caspase-7 and degradation of poly-(ADP-ribose)-polymerase which occur in SH-SY5Y exposed to paclitaxel	Resveratrol exerts its antiapoptotic effect by modulating the signal pathways that commit these neuronal-like cells to apoptosis	135

(*Continued*)

TABLE 13.3
Continued

Model	What was measured	Effect	Ref.
Using a mouse JB6 epidermal cell line, elucidation of the potential signaling components underlying resveratrol-induced p53 activation and induction of apoptosis are investigated	In a mouse JB6 epidermal cell line, resveratrol activated extracellular-signal-regulated protein kinases (ERKs), c-Jun NH_2-terminal kinases (JNKs), and p38 kinase, and induced serine 15 phosphorylation of p53. Stable expression of a dominant negative mutant of ERK2 or p38 kinase, or their respective inhibitors, PD98059 or SB202190, repressed the phosphorylation of p53 at serine 15. In contrast, overexpression of a dominant negative mutant of JNKI had no effect on the phosphorylation. Most importantly, ERKs and p38 kinase formed a complex with p53 after treatment with resveratrol. Strikingly, resveratrol-activated ERKs and p38 kinase, but not JNKs, phosphorylated p53 at serine 15 *in vitro*. Furthermore, pretreatment of the cells with PD98059 or SB202190 or stable expression of a dominant negative mutant of ERK2 or p38 kinase impaired resveratrol-induced p53-dependent transcriptional activity and apoptosis, whereas constitutively active MEK1 increased the transcriptional activity of p53	These data strongly suggest that both ERKs and p38 kinase mediate resveratrol-induced activation of p53 and apoptosis through phosphorylation of p53 at serine 15	257

Effect of cell transformation and gene expression using a resveratrol analog, 3,4,5,4′-tetrahydroxy-stilbene (R-4). WI38VA and WI38 cells were used	RNase protection assay showed that R-4 significantly induced the expression of p53, GADD45, and Bax genes and concomitantly suppressed the expression of Bcl-2 gene in WI38VA, but not in WI38 cells. A large increase in p53 DNA binding activity and the presence of p53 in the Bax promoter binding complex suggested that p53 was responsible for the Bax gene expression induced by R-4 in transformed cells. Within 4 h of treatment with R-4, the Bax to Bcl-2 protein ratio in WI38 and WI38VA cells was a difference of three orders of magnitude. While R-4 prominently induced the p53/Bax proapoptotic genes, it also concomitantly suppressed the expression of COX-2 in WI38VA cells	R-4 inhibited the growth of SV40 virally transformed WI38 cells (WI38VA), but had no effect on normal WI38 cells. R-4 induced apoptosis in WI38VA cells, but not in WI38 cells. The study suggests that the induction of p53 gene by R-4 in transformed cells may play a key role in the differential growth inhibition and apoptosis of transformed cells	124
Reverse inhibition and progression through S and G_2 phases of the cell cycle in human leukemia U937 cells	Resveratrol induces arrest in the S phase at low concentrations, but high concentrations do not induce S-phase accumulation in U937 cells. Removal of resveratrol from the culture medium stimulates U937 cells to reenter the cell cycle synchronously, as judged by the expression patterns of cyclin E and A and by fluorescent activated cell sorting analysis. These data demonstrate that resveratrol causes S-phase arrest and reversible cell cycle arrest	The report shows that resveratrol induces antiproliferation and arrests the S phase in human histiocytic lymphoma U937 cells. Resveratrol is an important cell cycle blocker	258

(Continued)

TABLE 13.3
Continued

Model	What was measured	Effect	Ref.
Investigation of the effect of catechin, epicatechin, quercetin, and resveratrol on the growth of prostate cancer cell lines (LNCaP, PC3, and DU145)	A dose- and time-dependent inhibition of cell growth by polyphenols was found at nanomolar concentrations. The proliferation of LNCaP and PC3 cells was preferentially inhibited by catechin, epicatechin, and quercetin, whereas resveratrol was the most potent inhibitor of DU145 cell growth. Possible mechanisms of action were investigated: (1) The competition of polyphenols for androgen binding in LNCaP cells revealed significant interaction only in the case of high concentrations of quercetin, at least at five orders of magnitude higher than the concentrations needed for cell growth inhibition. All other phenols showed low interactions. (2) Oxygen species production after mitogen stimulation and H_2O_2 sensitivity of these cell lines did not correlate with the observed antiproliferative effects, ruling out such a mode of action. (3) NO production revealed two different patterns: LNCaP and DU145 cells produced high concentrations of NO, whereas PC3 cells produced low concentrations. Phorbol ester stimulation of cells did not reveal any additional effect in LNCaP and DU145 cells, whereas it enhanced the secretion of NO in PC3 cells. Polyphenols decreased NO secretion. This effect correlates with their antiproliferative action and the inhibition of inducible NO synthase	It is proposed that the antiproliferative effect of polyphenols is mediated through the modulation of NO production. The data show a direct inhibitory effect of low concentrations of antioxidant wine phenols on the proliferation of human prostate cancer cell lines mediated by the production of NO, further suggesting potential beneficial effects of wine and other phenol-containing foods or drinks for the control of prostate cancer cell growth	259
Arrest in the S-phase prior to Fas-independent apoptosis in CEM-C7H2 lymphocytic leukemia cells	Resveratrol induced arrest in the S phase and apoptosis in the T cell-derived T-ALL lymphocytic leukemia cell line CEM-C7H2 which is deficient in functional p53 and p16. Expression of transgenic p16/INK4A reduced the percentage of apoptotic cells. Antagonist antibodies to Fas or FasL, or constitutive expression of crmA did not diminish the extent of resveratrol-induced apoptosis. Furthermore, a caspase-8-negative, Fas-resistant Jurkat cell line was sensitive to resveratrol-induced apoptosis which could be strongly inhibited in the	Resveratrol causes arrest in the S phase prior to Fas-independent apoptosis in CEM-C7H2 acute leukemia cells	260

	Jurkat as well as in the CEM cell line by z-VAD-fmk and z-IETD-fmk. The almost complete inhibition by z-IETD-fmk and the lack of inhibition by crmA suggested caspase-6 to be the essential initiator caspase. Western blots revealed the massive conversion of procaspase-6 to its active form, while caspase-3 and caspase-2 were proteolytically activated to a much lesser extent		
Effect of resveratrol on estrogen receptors α and β	Resveratrol was shown to bind ER in cytosolic extracts from MCF-7 and rat uteri. However, the contribution of ERα vs. ERβ in this binding is unknown. Thus, resveratrol differs from other phytoestrogens that bind ERβ with higher affinity than ERα. Resveratrol acts as an estrogen agonist and stimulates ERE-driven reporter gene activity in CHO-K1 cells expressing either ERα or ERβ. The estrogen agonist activity of resveratrol depends on the ERE sequence and the type of ER. Resveratrol-liganded ERβ has higher transcriptional activity than E_2-liganded ERβ at a single palindromic ERE. This indicates that those tissues that uniquely express ERβ or that express higher levels of ERβ than ERα may be more sensitive to the estrogen agonist activity of resveratrol. For the natural, imperfect EREs from the human c-fos, pS2, and progesterone receptor (PR) genes, resveratrol shows activity comparable to that induced by E_2. Resveratrol exhibits E_2 antagonist activity for ERα with select EREs. In contrast, resveratrol shows no E_2 antagonist activity with ERβ	These data indicate that resveratrol differentially affects the transcriptional activity of ERα and ERβ in an ERE sequence-dependent manner. The study reports that resveratrol binds ERβ and ERα with comparable affinity, but with 7000-fold lower affinity than estradiol (E_2)	112
Investigation of gap-junctional intercellular communication (GJIC) in WB-F344 rat liver epithelial cells	Seventeen to 50 μM resveratrol increased GJIC significantly by a factor of 1.3 compared with solvent vehicle controls, when the WB-F344 cells were exposed to resveratrol for 6 h. Most tumor promoters, including the phorbol ester TPA and the insecticide DDT, block GJIC. Resveratrol at 17–50 μM also significantly prevented downregulation of GJIC by TPA and DDT, by a factor of 2.7 and 1.8, respectively. This recovery of GJIC from TPA inhibition was partly correlated with hindered hyperphosphorylation of Cx43	Resveratrol was found to enhance GJIC and counteract the effects of tumor promoters on GJIC, and this is likely a mechanism that contributes to the antipromotional and anticarcinogenic properties of resveratrol	261

(*Continued*)

TABLE 13.3
Continued

Model	What was measured	Effect	Ref.
Investigation of pharmacological effects using Caco-2 human colon cancer cells	The study investigated the effects of resveratrol on the growth and polyamine metabolism of Caco-2 human colon cancer cells. Treatment of the Caco-2 cells with 25 μM resveratrol caused a 70% growth inhibition. The cells accumulated at the S/G_2 phase transition of the cell cycle. No signs of cytotoxicity or apoptosis were detected. Resveratrol caused a significant decrease of ornithine decarboxylase (ODC) activity	ODC inhibition resulted in the reduction of the intracellular putrescine content, indicating that polyamines might represent one of several targets involved in the antiproliferative effects of resveratrol	262
Investigation of inflammatory and antioncogenic properties of resveratrol	Because the transcription factor NF-κB is involved in inflammatory diseases and oncogenesis, resveratrol was tested to see if it could modulate NF-κB activity	Resveratrol was shown to be a potent inhibitor of both NF-κB activation and NF-κB-dependent gene expression through its ability to inhibit IκB kinase activity, the key regulator in NF-κB activation, likely by inhibiting an upstream signaling component. In addition, resveratrol blocked the expression of mRNA-encoding monocyte chemoattractant protein-1, a NF-κB-regulated gene. Relative to cancer chemopreventive properties, resveratrol induced apoptosis in fibroblasts after the induced expression of oncogenic H-Ras	263
Inhibition of cell proliferation in rat hepatoma Fao cell line and human hepatoblastoma HepG2 cell line	The ability of resveratrol to inhibit cell proliferation was studied in rat hepatoma Fao cell line and human hepatoblastoma HepG2 cell line. The results show that resveratrol strongly inhibits cell proliferation at the micromolar range in a time- and dose-dependent	Resveratrol shows a strong inhibition of hepatic cell proliferation where alcohol may act as an enhancing agent	264

	manner. Concentrations higher than 50 μM become toxic. Fao cells are more sensitive than HepG2 cells. The presence of ethanol lowers the threshold of resveratrol effect. Resveratrol appears to prevent or to delay the entry to mitosis since no inhibition of [^{3}H]thymidine incorporation is observed, while there is an increase of cell number in S and G_2/M phases		
Induction of Fas signaling-independent apoptosis in THP-1 human monocytic leukemia cells	Resveratrol inhibits the growth of THP-1 human monocytic leukemia cells in a dose-dependent manner with a median effective dose of 12 μM. It did not induce differentiation of THP-1 cells and had no toxic effect on THP-1 cells that had been induced to differentiate into monocytes/macrophages by phorbol myristate acetate. A significant fraction of resveratrol-treated cells underwent apoptosis as judged by flow cytometric analysis of DNA content, DNA fragmentation and caspase-specific cleavage of poly(ADP-ribosyl) polymerase. Resveratrol treatment had no effect on the expression of Fas receptor or Fas ligand (FasL) in THP-1 cells, nor did it induce clustering of Fas receptors. In addition, THP-1 cells were resistant to activating anti-Fas antibody, and neutralizing anti-Fas and/or anti-FasL antibodies had no protective effect against resveratrol-induced inhibition of THP-1 cell growth. The effect of resveratrol on THP-1 cells was reversible after its removal from the culture medium	These results suggest that (1) resveratrol inhibits the growth of THP-1 cells, at least in part, by inducing apoptosis; (2) resveratrol-induced apoptosis of THP-1 cells is independent of the Fas/FasL signaling pathway; and (3) resveratrol does not induce differentiation of THP-1 cells and has no toxic effect on differentiated THP-1 cells. Resveratrol may be a potential chemotherapeutic agent for the control of acute monocytic leukemia	265

(Continued)

TABLE 13.3
Continued

Model	What was measured	Effect	Ref.
Suppression of cyclooxygenase-2 (COX-2) promoter-dependent transcriptional activity in human colon cancer DLD-1 cells by 14 chemopreventive agents, which include resveratrol, with a resorcin-type structure	A β-galactosidase reporter gene system in human colon cancer DLD-1 cells was constructed, and measured COX-2 promoter-dependent transcriptional activity in the cells. Interferon gamma suppressed this COX-2 promoter activity, while 12-*O*-tetradecanoylphorbol-13-acetate and transforming growth factor α (TGFα) exerted enhancing effects. The influence of 14 compounds on COX-2 promoter activity was tested	The compounds, all having a common resorcin moiety, were found to effectively suppress the COX-2 promoter activity with and without TGFα-stimulation in DLD-1 cells. Since all these compounds have a resorcin moiety as a common structure, a resorcin-type structure may play an active role in the inhibition of COX-2 expression in colon cancer cells	93
Inhibition of cell proliferation is tested using prostate cancer cells (LNCaP) and the expression of a prostate specific gene, PSA	A 4-day treatment with resveratrol reduced the levels of intracellular and secreted PSA by approximately 80%, as compared to controls. To test whether this decrease was coordinated with changes in androgen receptor (AR) expression, levels of AR were assayed by Western blot analysis, using the cognate antibody, or by binding with the radioactive ligand methyltrienolone [^{3}H]R1881	With either assay, little or no change in AR expression could be detected between control and resveratrol-treated cells. Thus, it would appear that the prostate tumor marker PSA is downregulated by resveratrol by a mechanism independent of changes in AR	266
Effect of resveratrol and grapevine polyphenols on cultured human liver myofibroblasts	Resveratrol profoundly affects myofibroblast phenotype: it induced morphological modifications. Resveratrol markedly reduced proliferation of myofibroblasts in a dose-dependent manner. Resveratrol also decreased the expression of α smooth muscle actin (α-SMA) without affecting vimentin or β-cytoplasmic actin expression. It decreased myofibroblast migration in a monolayer wounding assay. In addition, resveratrol inhibited the messenger RNA (mRNA) expression of type I collagen.	Resveratrol can deactivate human liver myofibroblasts. Neither *trans*-piceid nor *trans*-piceatannol reproduces resveratrol effects on liver myofibroblasts. Although *trans*-resveratrol decreases the proliferation of skin fibroblast and vascular smooth muscle cells, it does not affect their expression of α-SMA, which indicates some cell specificity	267

	Resveratrol decreased the secretion of matrix metalloproteinase 2 (MMP-2)		
Growth and proliferation of human oral squamous carcinoma cell (SCC-25)	Resveratrol, quercetin, the combination of the two, and diluted red wine were tested for dose-dependent inhibition in human oral squamous carcinoma cell (SCC-25) growth and DNA synthesis	Resveratrol induced significant dose-dependent inhibition in human oral squamous carcinoma cell (SCC-25) growth and DNA synthesis. Quercetin exhibited a biphasic effect, stimulation and minimal inhibition in cell growth and DNA synthesis. Combining resveratrol with quercetin resulted in a gradual and significant increase in the inhibitory effect of the two compounds. Diluted red wine had a significantly more inhibitory effect on cell growth, DNA synthesis, and changes in cell morphology than each compound alone or in combination. Resveratrol by itself or a combination of resveratrol and quercetin are effective inhibitors of SCC-25 growth and DNA synthesis. The presence of other wine phenolic phytochemicals enhances significantly the effect of resveratrol and quercetin on inhibition of cancer cell growth and DNA synthesis	268
Suppression of phorbol ester (PMA)-mediated induction of COX-2 in human mammary and oral epithelial cells	Human mammary and oral epithelial cells were treated with PMA and tested for inhibition of COX-2 mRNA, COX-2 protein, and prostaglandin synthesis	Treatment of cells with PMA induced COX-2 mRNA, COX-2 protein, and prostaglandin synthesis. Nuclear runoffs revealed increased rates of COX-2 transcription after treatment with PMA. These effects were inhibited by resveratrol. Resveratrol inhibited PMA-mediated activation of protein kinase C and the induction of COX-2 promoter activity by c-Jun. Phorbol ester-mediated induction of AP-1 activity was blocked by resveratrol	269

(Continued)

TABLE 13.3
Continued

Model	What was measured	Effect	Ref.
Inhibition of the expression and function of the androgen receptor (AR) in prostate cancer cells (LNCaP)	Effects of resveratrol were tested on androgen-stimulated growth and gene expression in LNCaP cells. The group transfected a construct containing a 6-kb PSA promoter fragment in front of a luciferase reporter gene into LNCaP cells with or without Mib to test whether resveratrol can directly affect androgen-mediated transcriptional activity of the *PSA* gene. Northern or western analysis were performed to see whether different classes of the androgen-regulated genes are affected	Resveratrol represses different classes of androgen up-regulated genes at the protein or mRNA level including prostate-specific antigen, human glandular kallikrein-2, AR-specific coactivator ARA70, and the cyclin-dependent kinase inhibitor p21. This inhibition is likely attributable to a reduction in AR contents at the transcription level, inhibiting androgen-stimulated cell growth and gene expression	270
Effects of resveratrol on the increased proliferation of human AHTO-7 osteoblastic cell line induced by conditioned media (CM) from a panel of carcinoma cell lines [pancreas (BxPC3, Panc-1), breast (ZR75-1), renal (ACHN), colon	A tamoxifen-sensitive mechanism was used to test the ability of resveratrol to modulate AHTO-7 proliferation. Resveratrol was tested for its proliferative response of AHTO-7 cells to CM from carcinoma in the panel of carcinoma cell lines	Resveratrol was found to modulate AHTO-7 proliferation in a tamoxifen-sensitive mechanism at lower concentrations, but failed to induce the osteoblast differentiation marker alkaline phosphatase (ALP) in contrast to vitamin D_3. The proliferative response of AHTO-7 cells to CM from carcinoma cell lines was diminished upon pretreatment with resveratrol. Highest inhibition was demonstrated for BxPC3, Panc-1, ZR75-1, and ACHN carcinoma cell line supernatants whereas the effect on SW620, Colo320DM, PC3, DU145, LNCaP CM was less pronounced. Direct addition of resveratrol affected	271

(SW620, Colo320DM) and prostate (PC3, DU145 and LNCaP) cancer]		only supernatants of cell lines exhibiting growth stimulatory activity for normal WI38 lung fibroblasts. Resveratrol inhibited proliferation of DU145 and LNCaP cells, altered cell cycle distribution of all prostate cancer cell lines, but did not inhibit the production of osteoblastic factors by these lines. In sum, resveratrol failed to induce ALP activity as marker of osteoblast differentiation in human osteoblastic AHTO-7 cells; however, it inhibited their response to osteoblastic carcinoma-derived growth factors in concentrations significantly lower than those to reduce growth of cancer cells, thus effectively modulating tumor–osteoblast interaction	
Induction of apoptosis in human promyelocytic leukemia (HL-60) cells	Resveratrol was tested to determine if it reduces viability and DNA synthesis capability in cultured HL-60 cells	The growth inhibitory and antiproliferative properties of resveratrol appear to be attributable to its induction of apoptotic cell death as determined by morphological and ultrastructural changes, internucleosomal DNA fragmentation, and increased proportion of the subdiploid cell population. Resveratrol treatment resulted in a gradual decrease in the expression of antiapoptotic Bcl-2	272

(*Continued*)

TABLE 13.3
Continued

Model	What was measured	Effect	Ref.
Ability to control growth and cell cycle transition in MDA-MB-435 and MCF-7 breast carcinoma cells. Effect on a panel of MDA-MB-435 cells transfected with nm23-H1 and nm23-H2 genes	Resveratrol was tested in MDA-MB-435 and MCF-7 breast carcinoma cells lines to determine ability to control growth and cell cycle transition. In addition, resveratrol was tested to determine its effect on a panel of MDA-MB-435 cells transfected with nm23-H1 and nm23-H2 genes. The responses of these cells to resveratrol were assessed by measuring proliferation, cell cycle phase distribution, and changes in expression of several genes	The data revealed that resveratrol exerted a greater inhibitory effect on the MDA-MB-435 cells. A diminution of percentage of cells in G_1 phase and a corresponding accumulation of cells in S phase of the cell cycle was observed. These studies have shown that resveratrol reduced growth of all cell types. Overexpression of both wild-type and catalytically inactive nm23-H1 but not nm23-H2 reduced the proportion of cells in G_1 phase compared to the control cells. Little change in expression of PCNA, Rb, p53, and Bcl-2 was observed in the five cell types treated with resveratrol, compared to untreated cells. Noted exceptions included reduced expression of Rb protein and increased expression of p53 in two of the cells, and increased expression of Bcl-2 in one treated with resveratrol. Resveratrol upregulated expression of cathepsin D by 50–100% in all cell lines except one	273
Inhibition of carcinogen-induced preneoplastic lesion formation in mouse mammary organ culture and tumorigenesis in the	Antiinitiation activity was determined by inhibition of the hydroperoxidase function of cyclooxygenase (COX), and induction of phase II drug-metabolizing enzymes. Antipromotion activity was determined by inhibition of production of arachidonic acid metabolites catalyzed by either COX-1 or COX-2, and chemical carcinogen-induced neoplastic transformation of mouse embryo	Resveratrol reduced the generation of hydrogen peroxide, and normalized levels of myeloperoxidase and oxidized-glutathione reductase activities. It also restored glutathione levels and superoxide dismutase activity. Based on the reverse transcriptase polymerase chain reaction, resveratrol selectively inhibited TPA-induced expression of c-fos and transforming growth factor-β1	78

two-stage mouse skin model. Cancer chemopreventive potential on the three major stages of carcinogenesis	fibroblasts. Antiprogression activity was determined by induction of human promyelocytic leukemia (HL-60) cell differentiation. Treatment of mouse skin with resveratrol to determine the effect of 12-*O*-tetradecanoylphorbol-13-acetate (TPA)-induced oxidative stress	(TGF-β1), but did not affect other TPA-induced gene products including COX-1, COX-2, c-myc, c-jun, and tumor necrosis factor-α. These data indicate that resveratrol may interfere with reactive oxidant pathways and/or modulate the expression of c-fos and TGF-β1 to inhibit tumorigenesis in mouse skin. Resveratrol inhibited the *de novo* formation of inducible nitric oxide synthase (iNOS) in mouse macrophages stimulated with lipopolysaccharide	
Investigation of the effects of resveratrol on growth, induction of apoptosis, and modulation of prostate-specific gene expression using cultured prostate cancer cells that mimic the initial (hormone-sensitive) and advanced (hormone-refractory) stages of prostate carcinoma cell lines (LNCaP, DU145, PC-3, and JCA-1)	Androgen-responsive LNCaP and androgen-nonresponsive DU145, PC-3, and JCA-1 cells were cultured with different concentrations of resveratrol. Cell growth, cell cycle distribution, and apoptosis were determined	Addition of resveratrol led to a substantial decrease in growth of LNCaP and in PC-3 and DU145 cells, but only had a modest inhibitory effect on proliferation of JCA-1 cells. Flow cytometric analysis showed resveratrol to disrupt partially G_1/S transition in all three androgen-nonresponsive cell lines, but had no effect in the androgen-responsive LNCaP cells. In contrast to the androgen-nonresponsive prostate cancer cells, however, resveratrol causes a significant percentage of LNCaP cells to undergo apoptosis and significantly lowers both intracellular and secreted prostate-specific antigen (PSA) levels without affecting the expression of the androgen receptor (AR)	274

(Continued)

TABLE 13.3
Continued

Model	What was measured	Effect	Ref.
Resveratrol-induced activation of the mitogen-activated protein kinases ERK1 and ERK2 in human neuroblastoma SH-SY5Y cells	Phosphorylation of the mitogen-activated protein (MAP) kinases, extracellular signal-regulated kinase 1 (ERK1), and extracellular signal-regulated kinase 2 (ERK2) induced by resveratrol has been studied *in vitro* on undifferentiated and differentiated (induction by retinoic acid) SH-SY5Y human neuroblastoma cells	In undifferentiated cells, resveratrol induced phosphorylation of ERK1 and ERK2. A wide range of resveratrol concentrations were able to induce phosphorylation of ERK1 and ERK2, while higher concentrations inhibited MAP kinases phosphorylation. In retinoic acid (RA)-differentiated cells, resveratrol induced an evident increase in ERK1 and ERK2 phosphorylation. This study demonstrates that resveratrol may have a biological effect on neuron-like cells	275
Antiestrogenic activity and inhibition of the growth of human breast cancer cells (MCF-7)	The effect of resveratrol on the growth of human breast cancer cells was examined	Resveratrol inhibits the growth of estrogen receptor (ER)-positive MCF-7 cells in a dose-dependent fashion. Detailed studies with MCF-7 cells demonstrate that resveratrol antagonized the growth-promoting effect of 17β-estradiol (E_2) in a dose-dependent fashion at both the cellular (cell growth) and the molecular (gene activation) levels. Resveratrol abolished the growth-stimulatory effect mediated by concentrations of E_2. The antiestrogenic effect of resveratrol was demonstrated at the molecular level. Resveratrol, in a dose-dependent fashion, antagonized the stimulation by E_2 of progesterone receptor gene expression in MCF-7 cells. Expression of transforming growth factor-α and insulin-like growth factor I receptor mRNA was inhibited while the expression of transforming growth factor β2 mRNA	113

		was significantly elevated in MCF-7 cells cultivated in the presence of resveratrol. The results show that resveratrol, a partial ER agonist, acts as an ER antagonist in the presence of estrogen leading to inhibition of human breast cancer cells	
Effect of resveratrol and quercetin on human oral squamous carcinoma cells (SCC-25) growth and proliferation	Resveratrol and quercetin were incubated with human oral squamous carcinoma cells SCC-25. Cell growth was determined by counting the number of viable cells with a hemocytometer. Cell proliferation was measured by means of incorporation of [^{3}H]thymidine	Resveratrol induced significant dose-dependent inhibition in cell growth as well as in DNA synthesis. Quercetin exhibited a biphasic effect, stimulation and minimal inhibition in cell growth and DNA synthesis. Combining resveratrol and quercetin resulted in a gradual and significant increase in the inhibitory effect of quercetin on cell growth and DNA synthesis. Resveratrol or a combination of resveratrol and quercetin, in concentrations equivalent to that present in red wines, are effective inhibitors of oral squamous carcinoma cell (SCC-25) growth and proliferation	276
Suppression of cell transformation and induction of apoptosis through a p53-dependent pathway. JB6 P^+ mouse epidermal cell line C1 41 and its stable p53-luciferase reporter plasmid transfect C1 41 p53 cells were used	Inhibition of resveratrol on TPA- or EGF-induced cell transformation was investigated in JB6 C1 41 cells. The levels of p53 protein after resveratrol treatment were measured by Western blot for immunoprecipitation with specific antibodies against p53	Resveratrol suppresses tumor promoter-induced cell transformation and induces apoptosis, transactivation of p53 activity, and expression of p53 protein in the same cell line and at the same dosage. Also, resveratrol-induced apoptosis occurs only in cells expressing wild-type p53 (p53+/+), but not in p53-deficient (p53−/−) cells, while there is no difference in apoptosis induction between normal lymphoblasts and sphingomyelinase-deficient cell lines. These results demonstrate that resveratrol induces apoptosis through activation of p53 activity, suggesting that its antitumor activity may occur through the induction of apoptosis	277

(Continued)

TABLE 13.3
Continued

Model	What was measured	Effect	Ref.
Inhibition of phorbol ester (PMA)-mediated induction of COX-2 in human mammary and oral epithelial cells	Cells were treated with PMA to determine COX-2 induction and the rate of production of prostaglandin E_2. Nuclear runoffs were tested for COX-2 transcription after treatment with PMA	Resveratrol blocked PMA-dependent activation of AP-1-mediated gene expression. Rsveratrol directly inhibited the activity of COX-2	278
Induction of apoptotic cell death in HL60 human leukemia cell line	Resveratrol-treated tumor cells exhibit a dose-dependent increase in externalization of inner membrane phosphatidylserine and in cellular content of subdiploid DNA, indicating loss of membrane phospholipid asymmetry and DNA fragmentation. Resveratrol-induced cell death is mediated by intracellular caspases as observed by the dose-dependent increase in proteolytic cleavage of caspase substrate poly (ADP-ribose) polymerase (PARP) and the ability of caspase inhibitors to block resveratrol cytotoxicity. Resveratrol treatment enhances CD95L expression on HL60 cells, as well as T47D breast carcinoma cells, and resveratrol-mediated cell death is specifically CD95-signaling dependent. On the contrary, resveratrol treatment of normal human peripheral	These data show specific involvement of the CD95–CD95L system in the anticancer activity of resveratrol and highlight the chemotherapeutic and chemopreventive potential	279

	blood lymphocytes (PBLs) does not affect cell survival for up to 72 h, which correlates with the absence of a significant change in either CD95 or CD95L expression on treated PBLs		
Inhibitor of ribonucleotide reductase using L1210-R2 murine lymphoblastic leukemia cells	The ability of resveratrol to destroy the tyrosyl radical was correlated with its strong dose-dependent inhibitory effects on enzyme activity, as assayed in soluble extracts of murine leukemia cells containing high protein R2 expression and high cytidine diphosphate reductase activity suitable for sensitive and reproducible assays. The antiproliferative properties of resveratrol and its inhibitory effects on DNA synthesis were evaluated by [^{3}H]thymidine incorporation	Resveratrol is an inhibitor of ribonucleotide reductase and DNA synthesis in mammalian cells, which might have further applications as an antiproliferative or a cancer chemopreventive agent in humans	280
C3H10T1/2 CL8 mouse embryo fibroblasts were used to determine the ability of resveratrol to block eicosanoid production and chemically induced cellular transformation	Cyclooxygenase metabolites were investigated using HPLC analysis. COX-1 and -2 activities were determined by measuring arachidonic acid metabolite production. Two-stage transformation assays with C3H10T1/2 cells were performed	Resveratrol is capable of inhibiting eicosanoid production catalyzed by both COX-1 and -2, and transformation of chemically initiated 10T1/2 cells	77

(Continued)

TABLE 13.3
Continued

Model	What was measured	Effect	Ref.
Investigation of resveratrol to determine if it is a phytoestrogen using MCF-7 MDA-MB-231, and T47D cells	Diethylstilbestrol, a synthetic estrogen and structurally similar to resveratrol, was examined to determine whether resveratrol might be a phytoestrogen. At concentrations comparable to those required for its other biological effects, resveratrol inhibited the binding of labeled estradiol to the estrogen receptor and it activated transcription of estrogen-responsive reporter genes transfected into human breast cancer cells. This transcriptional activation was estrogen receptor-dependent, required an estrogen response element in the reporter gene, and was inhibited by specific estrogen antagonists. In some cell types (e.g., MCF-7 cells), resveratrol functioned as a superagonist (i.e., produced a greater maximal transcriptional response than estradiol) whereas in others it produced activation equal to or less than that of estradiol. Resveratrol also increased the expression of native estrogen-regulated genes, and it stimulated the proliferation of estrogen-dependent T47D breast cancer cells	Resveratrol is a phytoestrogen and that it exhibits variable degrees of estrogen receptor agonism in different test systems	114

Note: In addition to data obtained with *in vivo* models, the following references describe results obtained with cell culture: 39, 129, 147, 153, 154, 156, 157, 162, 289.

action, and the response is dependent on p53. A number of related factors can be modulated by resveratrol, such as activation of caspases, decreases in Bcl-2 and Bcl-x^L, increases in Bax, inhibition of D-type cyclins and cyclin-dependent kinases, activation of c-jun NH_2-terminal kinase, and interference with NF-κB- and AP-1-mediated cascades. Of course, nonapoptotic cell death pathways have also been observed, as well as induction of cell differentiation.

In hormone-responsive cell types, a variety of studies have been performed to assess the hormonal (estrogenic or androgenic) potential of resveratrol [102–114], largely due to the structural similarity with diethylstilbestrol (DES). Data range from superagonistic in transient transfection studies with reporter genes, to completely inactive. Most typically, weak hormonal activity has been observed in the absence of the native steroid, and antihormonal activity has been observed on the addition of native hormone.

As might be expected, further studies have been performed to investigate structural derivatives of resveratrol, either naturally occurring stilbenes or synthetic analogs, as well as *cis* and *trans* isomers [86,110,115–124]. These data are of interest, as is the generation and subsequent biologic potential of resveratrol metabolites and results obtained with cell culture models of transport [125–128]. In the area of cancer chemotherapy, the ability of resveratrol to modulate the toxic side effects of taxol, vincristine, and 5-fluorouracil have been investigated [129–135].

The overall mechanism that is facilitated by resveratrol is undoubtedly complex. As demonstrated by differential expression studies in various cell cultures [108,136–139], hundreds of genes are affected by treatment with resveratrol. These results are quite profound and are consistent with the raft of responses observed in numerous model systems. The overall physiological significance remains to be defined [140].

IN VIVO STUDIES CONDUCTED WITH RESVERATROL

On an intuitive level, data obtained with studies performed with *in vivo* models (Table 13.4) appear to be of greatest relevance to the human situation: "the proof is in the pudding." It was clear from the outset that resveratrol is capable of mediating physiological responses in animal models. In our original report [39], antiinflammatory activity was observed in rats, and inhibition of tumorigenesis was observed in the two-stage mouse skin model. Importantly, in the rat inflammation model, resveratrol was administered orally, so a preliminary indication of bioavailability and systemic activity was also provided.

Clearly, however, experimental outcomes are dependent on the particular model and protocol that is applied. Inhibition in the two-stage mouse skin system has been confirmed [141] and greatly expanded with activity being observed in UV-induced skin cancer models [142–145]. These data

TABLE 13.4
Evaluation of Resveratrol with *In Vivo* Model Systems

Model	What was measured	Effect	Ref.
Sprague-Dawley female rats	This study was carried out to determine whether grape-seed extract (GSE), using genistein as the control, added to rodent diets protected against carcinogen-induced mammary tumorigenesis in rats and whether this was affected by the composition of the whole diet	The results demonstrated that GSE is chemopreventive in an animal model of breast cancer, and suggests that chemopreventive activity in GSE and genistein may depend on the diet	281
Inhibitors of NF-κB activation. Mice	Curcumin and resveratrol attenuated total protein degradation in murine myotubes at all concentrations of proteolysis-inducing factor (PIF), and attenuated the PIF-induced increase in expression of the ubiquitin-proteasome proteolytic pathway, as determined by the "chymotrypsin-like" enzyme activity, proteasome subunits, and E2(14k)	Curcumin was ineffective in preventing weight loss and muscle protein degradation in mice bearing the MAC16 tumor, whereas resveratrol significantly attenuated weight loss and protein degradation in skeletal muscle, and produced a significant reduction in NF-κB DNA-binding activity. The inactivity of curcumin was probably due to a low bioavailability. These results suggest that agents that inhibit nuclear translocation of NF-κB may prove useful for the treatment of muscle wasting in cancer cachexia	282
SKH-1 hairless mice	Assess the involvement of IAP family protein survivin during resveratrol-mediated protection from multiple exposures of UVB radiations in the SKH-1 hairless mouse skin	It was demonstrated that topical pretreatment of resveratrol resulted in significant inhibition of UVB exposure-mediated increases in (1) cellular proliferation (Ki-67 immunostaining); (2) protein levels of epidermal cyclooxygenase (COX)-2 and ornithine decarboxylase (ODC), established markers of tumor promotion; (3) protein and mRNA levels of survivin; and	142

		(4) phosphorylation of survivin in the skin of SKH-1 hairless mouse. Resveratrol pretreatment also resulted in (1) reversal of UVB-mediated decrease of Smac/DIABLO, and (2) enhancement of UVB-mediated induction of apoptosis in mouse skin. The study suggested that resveratrol imparts chemopreventive effects against UVB exposure-mediated damage in SKH-1 hairless mouse skin by inhibiting survivin and the associated events	
Effect of resveratrol and in combination with 5-fluorouracil (5-FU) on murine hepatoma22 cell (liver cancer)	Transplantable murine hepatoma22 model was used to evaluate the antitumor activity of resveratrol alone or in combination with 5-FU *in vivo*. H22 cell cycles were analyzed with flow cytometry	Resveratrol could induce the S-phase arrest of H22 cells and enhance the antitumor effect of 5-FU on murine hepatoma22 and antagonize its toxicity markedly. These results suggest that resveratrol, as a biochemical modulator to enhance the therapeutic effects of 5-FU, may be potentially useful in cancer chemotherapy	129
Examined the absorption, bioavailability, and metabolism of ^{14}C-resveratrol after oral and i.v. doses in six human volunteers	Plasma and urine were collected and tested for absorption. Liquid chromatography/mass spectrometry analysis identified three metabolic pathways, i.e., sulfate and glucuronic acid conjugation of the phenolic groups and, interestingly, hydrogenation of the aliphatic double bond, the latter likely produced by intestinal microflora. Extremely rapid sulfate conjugation by the intestine/liver appears to be the rate-limiting step in the bioavailability of resveratrol	Accumulation of resveratrol (the systemic bioavailability is very low) in epithelial cells along the aerodigestive tract and potentially active resveratrol metabolites may still produce cancer-preventive and other effects	283

(*Continued*)

TABLE 13.4
Continued

Model	What was measured	Effect	Ref.
Male and female CD virus antibody free (VAF) rats	To evaluate the potential toxicity of resveratrol, rats were administered by gavage 0, 300, 1000, and 3000 mg resveratrol per kilogram body weight per day for 4 weeks	Most of the adverse events occurred in the rats treated with 3000 mg per kilogram body weight per day. These included increased clinical signs of toxicity; reduced final body weights and food consumption; elevated BUN, creatinine, alkaline phosphatase, alanine aminotransferase, total bilirubin, and albumin; reduced hemoglobin, hematocrit, and red cell counts; and increased white cell counts. Increases in kidney weights and clinically significant renal lesions, including an increased incidence and severity of nephropathy, were observed. Diffuse epithelial hyperplasia in the bladder was considered equivocal and of limited biological significance. No histological effects on the liver were observed, despite the clinical chemistry changes and increased liver weights in the females. Effects seen in the group administered 1000 mg resveratrol per kilogram body weight per day included reduced body weight gain (females only) and elevated white blood cell count (males only). Plasma resveratrol concentrations in blood collected 1 h after dose administration during week 4 were dose related but were relatively low given the high dosage levels; conjugates were not measured. Under the conditions of this study, no adverse effects were observed at a dose of 300 mg resveratrol per kilogram body weight per day	284

Benzo(*a*)pyrene-induced lung tumorigenesis in A/J mice	Inhibition of PAH bioactivation through reduced expression of the CYP1A1 and CYP1B1 genes in human bronchial epithelial cells	Resveratrol added to the diet showed no effect on benzo(*a*)pyrene-induced lung tumorigenesis. Resveratrol did not change CYP1A1 and CYP1B1 gene expression or benzo[a]pyrene protein adduct levels in the lung tissue	147
Mice	Ames assay and micronucleus formation assay were used to test the antimutagenic activities of resveratrol. Croton oil-induced enhancement of ODC activities of dorsal epidermis cells in mouse and mouse ear edema model were used to investigate the antipromotion effect of resveratrol. DMBA/croton oil-induced mouse skin tumor model was used to evaluate chemopreventive effect of resveratrol to cancer *in vivo*	The Ames test showed resveratrol exhibited 42.2% inhibition on the reversion of *Salmonella typhimurium* TA100 induced by methylmethansulfonate, and resveratrol exhibited 91.8% inhibition on the reversion induced by benzo(*a*)pyrene. Pretreatment of resveratrol prevented cyclophosphamide (CTX)-induced micronucleus formation of polychromatic erythrocytes of mouse bone marrow in dose-dependent manner. Mice treated with 30 mg/kg of resveratrol for 6 days before croton oil exposure have palliative ear edema. Treatment of 180 mg/kg resveratrol for 3 days caused a 69.3% decrease of ODC activities in croton oil-induced dorsal epidermis. It was shown that resveratrol could inhibit DMBA/croton oil-induced mouse skin papilloma, which includes prolonging the latent period of tumor occurrence, decreased incidence of papilloma, and reduced tumor number per mouse, in a dose-dependent manner	285

(*Continued*)

TABLE 13.4
Continued

Model	What was measured	Effect	Ref.
Neuroblastoma (Neuro-2a) cells and determined effects on neuroblastoma tumors in syngeneic A/J mice	Cytotoxic effects, cellular apoptosis, and alterations in the cell cycle were determined in Neuro-2a neuroblastoma cells exposed for varying lengths of time to a series of resveratrol concentrations. Expression of associated cell cycle regulatory proteins, cyclin E and p21, was detected by Western blot analysis, and the antitumor effects of resveratrol were investigated by treating subcutaneous neuroblastoma tumors with intraperitoneal injections of 40 mg/kg resveratrol daily for 28 days	Resveratrol caused significant cytotoxicity and increased apoptosis and S-phase accumulation of neuroblastoma cells. S-phase accumulation was related to the downregulation of p21 and upregulation of cyclin E. In addition, resveratrol exerted antitumor effects on neuroblastomas in mice. Thus, resveratrol shows promise for the treatment of neuroblastoma	156
Pretreatment of mice with resveratrol at a dose of 2.5 mg/kg body weight for two weeks blocked the *N*-nitrosodiethylamine (NDEA)-induced cytosolic ODC levels in the liver and lungs	The blockage was pronounced in hepatic tissue compared to pulmonary tissue. Resveratrol feeding caused a significant reduction in microsomal cyclooxygenase (COX) activities in the liver and lungs, while the dosage of NDEA (200 mg/kg body weight) induced COX activity 24 h after its administration	Resveratrol pretreatment effectively blocked the induction of COX activity in the lungs by NDEA	286

SKH-1 hairless mouse skin	This study evaluated the involvement of cell cycle regulatory molecules during resveratrol-mediated protection from multiple exposures of UVB radiation in the SKH-1 hairless mouse skin. Resveratrol was topically applied on the skin of SKH-1 hairless mice. Studies were performed at 24 h following the last UVB exposure	Topical application of resveratrol resulted in a significant decrease in UVB-induced bifold skin thickness, hyperplasia, and infiltration of leukocytes. The data from immunoblot and/or immunohistochemical analyses revealed that multiple exposures to UVB radiation causes significant upregulation in (1) proliferating cell nuclear antigen (PCNA), a marker of cellular proliferation, and (2) cyclin-dependent kinase (cdk)-2, -4, and -6, cyclin-D1, and cyclin-D2. Resveratrol treatment resulted in significant downregulation in UV-mediated increases in these critical cell cycle regulatory proteins. Resveratrol treatment resulted in a further stimulation of UVB-mediated increases in cyclin kinase inhibitor WAF1/p21 and tumor suppressor p53. Further, resveratrol was found to cause significant decreases in UVB-mediated upregulation of (1) the mitogen-activated protein kinase kinase, and (2) the 42 kDa isotype of mitogen-activated protein kinase (MAPK). These data suggested that the antiproliferative effects of resveratrol might be mediated via modulation in the expression and function of cell cycle regulatory proteins cyclin-D1 and -D2, cdk-2, -4, and -6, and WAF1/p21. Also, the modulation of cki–cyclin–cdk network by resveratrol may be associated with inhibition of the MAPK pathway. Resveratrol may be useful for the prevention of UVB-mediated cutaneous damage including skin cancer	143

(Continued)

TABLE 13.4
Continued

Model	What was measured	Effect	Ref.
The pharmacokinetic properties of the resveratrol analog 3,4,5,4′-tetramethoxy-stilbene (DMU 212) were compared with those of resveratrol in the plasma, liver, kidney, lung, heart, brain, and small intestinal and colonic mucosa of mice	DMU 212 or resveratrol were administered intragastrically, and drug concentrations were measured by HPLC. Metabolites were characterized by cochromatography with authentic reference compounds and were identified by mass spectrometry. The ratios of area of plasma or tissue concentration vs. time curves of resveratrol over DMU 212 (AUC_{res}/AUC_{DMU212}) for the plasma, liver, and small intestinal and colonic mucosa were 3.5, 5, 0.1, and 0.15, respectively. Thus, resveratrol afforded significantly higher levels than DMU 212 in the plasma and liver, while DMU 212 exhibited superior availability compared to resveratrol in the small intestine and colon. Resveratrol was metabolised to its sulfate or glucuronate conjugates, while DMU 212 underwent metabolic hydro-xylation or single and double *O*-demethylation. DMU 212 and resveratrol inhibited the growth of human-derived colon cancer cells HCA-7 and HT-29 *in vitro* with IC_{50} values of between 6 and 26 μM	Due to the superior levels achieved in the gastrointestinal tract after the administration of DMU 212, when compared to resveratrol, the results provide a good rationale to evaluate DMU 212 as a colorectal cancer chemopreventive agent	287
Determine the effect of resveratrol on intest-inal tumorigenesis and the protumorigenic COX pathway in $Apc^{Min/+}$ mice	Resveratrol was administered as a powdered admixture in the diet at 0, 4, 20, or 90 mg/kg body weight for 7 weeks. In two separate experiments, resveratrol did not affect intestinal tumor load. It was stable in the diet under experimental conditions, circulated in the plasma as the glucuronide-conjugated form, and reached the tumors as	This study demonstrates that resveratrol consumed *ad libitum* in the diet does not modify tumorigenesis in $Apc^{Min/+}$ mice	149

	evidenced by significant decreases in PGE_2 levels. However, immunohistochemical staining of intestinal tumors revealed no changes in COX-2 expression		
Antitumor effect of stilbenoids from *Vateria indica* against allografted sarcoma S-180 in DDY mice	Examined the antitumor activity of the ethanol extract from the stem bark of *Vateria indica,* which contains various resveratrol oligomers. High-performance liquid chromatography analysis showed that the extract contains bergenin, hopeaphenol, vaticanol B, vaticanol C, and ε-viniferin. The *in vitro* assay displayed the anticancer activity of the extract against mouse sarcoma 180 cells ($IC_{50} = 29.5\,\mu M$). In the animal study, the tumor growth of sarcoma S-180 cells subcutaneously allografted in DDY mice was significantly retarded by oral administration of the extract. The extract did not show significant toxicity to mice	These results demonstrated that the ethanol extract containing various stilbenoids from the stem bark of *V. indica* has potent antitumor activity	288
Effect of prepubertal resveratrol exposure on *N*-methyl-*N*-nitrosourea (MNU)-induced mammary carcinogenesis in female Sprague-Dawley rats	Prepubertal rats were treated daily with either 10 or 100 mg/kg resveratrol for 5 days, and were compared with resveratrol-untreated animals. Six rats in each group were autopsied at 49 days of age, and their growth was evaluated. All remaining rats were given 50 mg/kg MNU, followed by monitoring for occurrence of mammary carcinoma. A dose of 100 mg/kg resveratrol significantly increased incidence of rat with mammary carcinomas and multiplicity (all histologically detected mammary carcinomas per rat), but did not affect latency, compared with untreated controls. Resveratrol did not affect body weight increase, but 100 mg/kg resveratrol caused slightly earlier vaginal opening	Although all rats cycled, resveratrol-treated animals exhibited significantly increased irregularity of estrous cycle, spending more time in the estrus phase. Thus, short resveratrol treatment of prepubertal female rats affected endocrine function, and accelerated development of MNU-induced mammary carcinomas	152

(Continued)

TABLE 13.4
Continued

Model	What was measured	Effect	Ref.
Determine the antitumor activity of resveratrol and its effect on the expression of cell cycle proteins including cyclin D1, cyclin B1, and $p34^{cdc2}$ in transplanted liver cancer using the murine transplanted hepatoma H22 model	Murine transplanted hepatoma H22 model was used to evaluate the *in vivo* antitumor activity of resveratrol. Following i.p. administration of resveratrol, the change in tumor size was recorded and the protein expression of cyclin D1, cyclin B1, and $p34^{cdc2}$ in the tumor and adjacent noncancerous liver tissues were measured by immunohistochemistry. Following treatment of H22 tumor-bearing mice with resveratrol, the growth of murine transplantable liver cancer was inhibited by 36.3 or 49.3%, respectively. The inhibitory effect was significant compared to that in control group. The level of expression of cyclin B1 and $p34^{cdc2}$ protein was decreased in the transplantable murine hepatoma H22 treated with resveratrol whereas the expression of cyclin D1 protein did not change	Resveratrol exhibits antitumor activities on murine hepatoma H22. The underlying antitumor mechanism of resveratrol might involve the inhibition of the cell cycle progression by decreasing the expression of cyclin B1 and $p34^{cdc2}$ protein	157
Investigation in male F344 rats for the potential beneficial or adverse effect of prolonged dietary administration of moderate to high doses of lycopene, quercetin, and resveratrol, or a	Selected markers for toxicity and defense mechanisms were assayed in blood, liver, and colon and the impact of the antioxidant administrations on putative preneoplastic changes in liver and colon was assessed. The dietary carcinogen, 2-amino-3-methylimidazo[4,5-*f*]quinoline (IQ) served as a prooxidant, genotoxicity, and general toxicity control. IQ increased the levels of protein and DNA oxidation products in plasma, the area of glutathione S-transferase-placental form positive	These results indicate that moderate to high doses of common dietary antioxidants can damage lymphocyte DNA and induce low levels of preneoplastic liver lesions in experimental animals. Therefore, long-term exposure to moderate to high doses of antioxidants may modulate carcinogenesis via prooxidative mechanisms and nonoxidative mechanisms	289

mixture of lycopene and quercetin	(GST-P) foci in the liver as well as the number of colonic aberrant crypt foci (ACF). All antioxidants and the antioxidant combination significantly increased the level of lymphocytic DNA damage, to an extent comparable with the effect induced by IQ. In contrast to the control group where no GST-P foci were detected, GST-P foci were detected in animals exposed to quercetin, lycopene, and the combination of the two. However, the increase in the volume of GST-P foci did not reach statistical significance		
Prevention of the effects of benzo(*a*)pyrene (BaP) on the lung of Balb/c mice	Balb/c mice were injected for 5 weeks with corn oil, BaP, resveratrol, or BaP + resveratrol. Immunohistochemistry was performed on lung sections for the determination of CYP1A1 protein, BPDE-DNA adducts, and apoptosis. A semiquantitative immunohistochemistry score (H score) was used for data analysis. Mice exposed to BaP had a significant induction of lung BPDE-DNA adducts when compared with controls. The BPDE-DNA adduct induction by BaP was abrogated significantly by resveratrol. A similar pattern was found by immunohistochemistry for apoptosis and CYP1A1. Western blotting confirmed that resveratrol prevented BaP-induced CYP1A1 expression. This increase in CYP1A1 expression in response to BaP administration most likely causes BaP metabolism, BPDE-DNA adduct formation, and subsequent apoptosis	All BaP-induced effects could be prevented by resveratrol, suggesting a possible chemopreventive role against the development of lung cancer	146

(*Continued*)

TABLE 13.4
Continued

Model	What was measured	Effect	Ref.
The antitumor and immunomodulatory activity of resveratrol on experimentally implanted tumor of H22 in Balb/c mice	The cytotoxicity of peritoneal macrophages (Mφ) against H22 cells was measured by treating mice with H22 tumors with different concentrations of resveratrol, and the inhibitory rates were calculated and IgG contents were determined by a single immunodiffusion method. The plaque-forming cell (PFC) was measured by an improved Cunningham method, and the levels of serum tumor necrosis factor-α (TNFα) were measured by a cytotoxic assay against L929 cells. Resveratrol promoted the cytotoxicity of Mφ against H22 cells. Resveratrol (500, 1000, and 1500 mg/kg) could curb the growth of the implanted tumor of H22 in mice. The inhibitory rates were 31.5, 45.6, and 48.7%, respectively, which could raise the level of serum IgG and PFC response to sheep red blood cells. Resveratrol (1000 and 1500 mg/kg) and BCG (200 mg/kg) could increase the production of serum TNFα in mice H22 tumor. However, the effect of resveratrol was insignificant	Resveratrol could inhibit the growth of H22 tumor in Balb/c mice. The antitumor effect of resveratrol might be related to directly inhibiting the growth of H22 cells and indirectly inhibiting its potential effect on nonspecific host immunomodulatory activity	160

This study investigates whether dietary resveratrol could inhibit the proliferation and metastasis of tumors and hyperlipidemia in Donryu rats subcutaneously implanted with an ascites hepatoma cell line AH109A	Hepatoma-bearing rats were fed resveratrol in the diet for 20 days, solid tumor growth and metastasis tended to be suppressed dose-dependently. Resveratrol significantly suppressed the serum lipid peroxide level, indicating its antioxidative properties or those of its metabolite(s) *in vivo*. Resveratrol dose-dependently suppressed both the serum triglyceride and very-low-density lipoprotein + low-density lipoprotein (VLDL + LDL)-cholesterol levels	The hypocholesterolemic action of resveratrol is attributed, at least in part, to an increased excretion of neutral sterols and bile acids into feces. These results suggest that dietary resveratrol is hypolipidemic with a tendency for antitumor-growth and antimetastasis effects in hepatoma-bearing rats	158
Suppression of ultraviolet B exposure-mediated activation of NF-κB in normal human keratinocytes by resveratrol, by investigating the effects of cutaneous damage in SKH-1 hairless mice	These studies demonstrated that resveratrol imparts protection from UVB-mediated cutaneous damage in SKH-1 hairless mice. The mechanism of action of resveratrol is not clearly understood. The involvement of NF-κB was investigated as the mechanism of chemoprevention of UV damage by resveratrol. In the normal human epidermal keratinocytes, resveratrol blocked UVB-mediated activation of NF-κB in a dose- as well as time-dependent fashion. Resveratrol treatment of keratinocytes also inhibited UVB-mediated phosphorylation and degradation of IκBα and activation of IKKα	These studies suggest that the NF-κB pathway plays a critical role in the chemopreventive effects of resveratrol against the adverse effects of UV radiation including photocarcinogenesis	144

(Continued)

TABLE 13.4
Continued

Model	What was measured	Effect	Ref.
Investigation of the absorption and tissue distribution of ^{14}C-*trans*-resveratrol following oral administration to male Balb/c mice	Male Balb/c mice were given a single oral dose of ^{14}C-*trans*-resveratrol and were sacrificed at 1.5, 3, or 6 h postdose. The distribution of radioactivity in tissues was evaluated using whole-body autoradiography, quantitative organ-level determination, and microautoradiography. In addition, identification of radioactive compounds in kidney and liver was done with high-performance liquid chromatography	Autoradiographic survey of mouse sections as well as radioactivity quantification in various organs revealed a preferential fixation of ^{14}C-*trans*-resveratrol in the organs and biological liquids of absorption and elimination (stomach, liver, kidney, intestine, bile, urine). Moreover, the studies show that ^{14}C-*trans*-resveratrol-derived radioactivity is able to penetrate the tissues of liver and kidney, a finding supported by microautoradiography. The presence of intact ^{14}C-*trans*-resveratrol together with glucurono- and/or sulfoconjugates in these tissues was also shown. This study demonstrates that *trans*-resveratrol is bioavailable following oral administration and remains mostly in intact form. The results also suggest a wide range of target organs for cancer chemoprevention by wine polyphenols in humans	290
Investigation of the inhibitory effects of vitamin C (VC), vitamin E (VE), tea polyphenols (TP), garlic squeeże, curcumin, and grapeseed extract on NB-DNA and	These dietary constituents showed inhibitory effects on DNA or Hb adduction. VC, VE, TP, and grapestone extract could efficaciously inhibit the adductions by 33–50%, and all of these six agents could inhibit Hb adduction by 30–64%. Resveratrol, curcumin, VC, and VE were also investigated as inhibitors of NB-DNA adduction *in vitro* using a liquid scintillation counting technique	These agents in the presence of NADPH and S9 components also pronouncedly blocked DNA adduction in a dose-dependent fashion. The study suggests that these seven constituents may interrupt the process of NB-induced chemical carcinogenesis	291

NB-hemoglobin (Hb) adductions in mice using an ultrasensitive method of accelerator mass spectrometry (AMS) with ^{14}C-labeled nitrobenzene			
Prevention of short-term ultraviolet B radiation-mediated damage by resveratrol in SKH-1 hairless mice	This study was designed to examine whether resveratrol possesses the potential to ameliorate the damage caused by short-term UVB exposure to mouse skin. Single topical application of resveratrol to SKH-1 hairless mice was found to result in significant inhibition of UVB-mediated increase in bifold skin thickness and skin edema. Treatment of mouse skin with resveratrol was also found to result in significant inhibition of UVB-mediated induction of cyclooxygenase and ornithine decarboxylase (ODC) enzyme activities and protein expression of ODC, which are well-established markers for tumor promotion. Resveratrol inhibits UVB-mediated increased level of lipid peroxidation, a marker of oxidative stress	The results suggest that resveratrol may afford substantial protection against the damage caused by UVB exposure, and these protective effects may be mediated via its antioxidant properties	145

(*Continued*)

TABLE 13.4
Continued

Model	What was measured	Effect	Ref.
Mammary carcinogenesis induction by 7,12-dimethylbenz(*a*)-anthracene (DMBA) in female Sprague-Dawley rats. Analyzed effect on MCF-7 human breast cancer cell lines	Investigated the chemopreventive potential of resveratrol by testing it against mammary carcinogenesis induced by DMBA in female Sprague-Dawley rats. Dietary administration of resveratrol had no effect on body weight gain and tumor volume but produced striking reductions in the incidence, multiplicity, and extended latency period of tumor development relative to DMBA-treated animals. Histopathological analysis of the tumors revealed that DMBA induced ductal carcinomas and focal microinvasion *in situ*, whereas treatment with resveratrol suppressed DMBA-induced ductal carcinoma. Immunohistochemistry and Western blot analysis revealed that resveratrol suppressed the DMBA-induced cyclooxygenase-2 and matrix metalloprotease-9 expression in the breast tumor. Gel shift analysis showed suppression of DMBA-induced NF-κB activation by resveratrol. Treatment of human breast cancer MCF-7 cells with resveratrol also suppressed the NF-κB activation and inhibited proliferation at S-G_2-M phase	The results suggest that resveratrol suppresses DMBA-induced mammary carcinogenesis, which correlates with downregulation of NF-κB, cyclooxygenase-2, and matrix metalloprotease-9 expression	154
Suppression of *N*-nitrosomethylbenzylamine (NMBA)-induced rat esophageal tumorigenesis in F344 male rats	Resveratrol was administered orally or intraperitoneally (i.p.) to F344 male rats. In the groups in which resveratrol was administered orally, the number of NMBA-induced esophageal tumors per rat was significantly reduced, and the size of maximum tumors in each group with resveratrol treatment was also significantly smaller than that in NMBA alone group. Although the pathological examination did not indicate significantly	The higher expression of COX-1, the upregulated COX-2 expression, and the increased levels of PGE_2 synthesis were all significantly decreased by administering resveratrol. This study suggests that resveratrol suppressed NMBA-induced rat esophageal tumorigenesis by targeting COXs and PGE_2, and therefore may be a promising natural anticarcinogenesis agent for the prevention and treatment of human esophageal cancer	155

	decreased incidence of carcinomas by administering resveratrol, the tendency of carcinogenesis suppression was observed. Semiquantitative RT-PCR and ELISA analysis demonstrated that following NMBA treatment, the expression of COX-1 mRNA was strongly present in tumor tissues, while weakly present in nontumor tissues; the expression of COX-2 mRNA was induced in both tumor and nontumor tissues. The production of prostaglandin E_2 (PGE_2) increased approximately 6-fold, compared with the normal esophageal mucosa		
Assessment of the anticancer potential of resveratrol, sesamol, sesame oil, and sunflower oil in the promotion stage of cancer development, employing the *in vitro* Epstein-Barr virus early antigen activation assay induced by the tumor promoter 12-*O*-tetradecanoylphorbol 13-acetate (TPA)	The group studied the activities of these compounds in the brine shrimp cytotoxicity assay as well as on the stable 1,1-diphenyl-2-picrylhydrazyl (DPPH) free radical scavenging bioassay to compare some mechanisms of anticancer activity. The group compared the observed chemoprotective capabilities of the four products with the *in vivo* 7,12-dimethylbenz(*a*)anthracene-initiated and TPA-promoted mouse skin two-stage carcinogenesis protocols	All the products tested showed a profound inhibitory effect on the Epstein-Barr virus early antigen induction using Raji cells. Comparatively, sesame oil was the most potent followed by sesamol and then resveratrol. Only sesamol and resveratrol showed a remarkable cytotoxic activity in the brine shrimp lethality assays as well as profound free radical scavenging activity in the DPPH bioassay. In both test systems, sesamol exhibited a more remarkable activity than resveratrol while sesame oil and sunflower oil did not exhibit any appreciable activity even at the highest concentrations tested. In the *in vivo* assay at a 50-fold molar ratio to TPA, sesamol offered 50% reduction in mouse skin papillomas at 20 weeks after promotion with TPA. Under an identical molar ratio to TPA, resveratrol offered a 60% reduction in the papillomas in mouse at 20 weeks. Sesamol seems to be an almost equally potent chemopreventive agent. Sesame oil and sunflower oil offered 20 and 40% protection, respectively, in the mouse skin tumor model. The antioxidant capabilities of these compounds could not solely explain the observed anticancer characteristics	141

(*Continued*)

TABLE 13.4
Continued

Model	What was measured	Effect	Ref.
Bioavailability of resveratrol and its effect on tumor growth were investigated using rabbits, rats, and mice	Tissue levels of resveratrol were studied after i.v. and oral administration of resveratrol to rabbits, rats, and mice. The half-life of resveratrol in plasma, after i.v. administration of 20 mg resveratrol/kg body weight, was very short. The highest concentration of resveratrol in plasma, either after i.v. or oral administration, was reached within the first 5 min in all animals studied. Extravascular levels (brain, lung, liver, and kidney) of resveratrol, which paralleled those in plasma, were always < 1 nmol/g fresh tissue. Resveratrol measured in plasma or tissues was in the *trans* form (at least 99%). Hepatocytes metabolized resveratrol in a dose-dependent fashion, which means that the liver can remove circulating resveratrol very rapidly. *In vitro* B16 melanoma (B16M) cell proliferation and generation of reactive oxygen species (ROS) was inhibited by resveratrol in a concentration-dependent fashion (100% inhibition of tumor growth was found in the presence of 5 μM resveratrol). Addition of 10 μM H_2O_2 to B16M cells, cultured in the presence of 5 μM resveratrol, reactivated cell growth. Oral administration of resveratrol did not inhibit growth of B16M inoculated into the footpad of mice (solid growth)	Oral administration of resveratrol decreased hepatic metastatic invasion of B16M cells inoculated intrasplenically. The antimetastatic mechanism involves a resveratrol (1 μM)-induced inhibition of vascular adhesion molecule 1 (VCAM-1) expression in the hepatic sinusoidal endothelium (HSE), which consequently decreased *in vitro* B16M cell adhesion to the endothelium via very late activation antigen 4 (VLA-4)	292
Effect and mechanism of action of quercetin, rutin, resveratrol, and genistein on human pancreatic carcinoma cell line	The group measured effects of quercetin on pancreatic cancer in a nude mouse model. They investigated the effects of quercetin, rutin, resveratrol, and genistein on apoptosis and underlying signaling in pancreatic carcinoma cells *in vitro*. Quercetin decreased primary tumor growth, increased apoptosis, and prevented metastasis in a model of pancreatic cancer. *In vitro* quercetin and resveratrol, but	The results suggest that food-derived polyphenols inhibit pancreatic cancer growth and prevent metastasis by inducing mitochondrial dysfunction, resulting in cytochrome c release, caspase activation, and apoptosis	293

(Mia PACA-2) and rat pancreatic carcinoma (BSp73AS) using a nude mouse model to indicate effect	not rutin, markedly enhanced apoptosis, causing mitochondrial depolarization and cytochrome c release followed by caspase-3 activation. In addition, the effect of a combination of quercetin and resveratrol on mitochondrial cytochrome c release and caspase-3 activity was greater than the expected additive response. The inhibition of mitochondrial permeability transition prevented cytochrome c release, caspase-3 activation, and apoptosis caused by polyphenols. Nuclear factor-κB activity was inhibited by quercetin and resveratrol, but not genistein, indicating that this transcription factor is not the only mediator of the effects of polyphenols on apoptosis		
Antileukemic activity of resveratrol *in vitro* and *in vivo* was examined using a mouse myeloid leukemia cell line (32Dp210) and tested in C3H (H-2^k) mice	Treatment of 32Dp210 leukemia cells with resveratrol at micromolar concentrations significantly and irreversibly inhibited their clonal growth *in vitro*. The clonal growth inhibition by resveratrol was associated with extensive cell death and an increase in hypodiploid (sub-G_1) cells. Resveratol caused internucleosomal DNA fragmentation, suggesting apoptosis as the mode of cell death in 32Dp210 cells. DNA fragmentation was associated with activation of caspase-3, because cleavage of procaspase-3 was detected in resveratrol-treated cells. Although 32Dp210 cells treated with resveratrol *in vitro* did not produce leukemia *in vivo*, only a weak antileukemic effect of resveratrol was observed when administered orally. At doses of 8 mg or 40 mg/kg body daily, five times/week, resveratrol did not affect the survival of mice injected with leukemia cells. Weak potential antileukemic activity of resveratrol was suggested only at a dose of 80 mg/kg body (2 survivors of 14 mice treated)	Despite strong antiproliferative and proapoptotic activities of resveratrol against 32Dp210 cells *in vitro*, a potential antileukemia effect *in vivo*, if present, occurs only in a small fraction of mice	163

(*Continued*)

TABLE 13.4
Continued

Model	What was measured	Effect	Ref.
Effect of resveratrol on growth of 4T1 mammary carcinoma cells *in vitro* and *in vivo* (female Balb/c mice)	*In vitro*, resveratrol inhibited growth of 4T1 breast cancer cells in a dose- and time-dependent manner. *In vivo*, however, resveratrol had no effect on time to tumor take, tumor growth, or metastasis when administered intraperitoneally daily (1, 3, or 5 mg/kg) for 23 days starting at the time of tumor inoculation. Resveratrol had no effect on body weight, organ histology, or estrous cycling of the tumor-bearing mice	These studies indicate that resveratrol is a potent inhibitor of 4T1 breast cancer cells *in vitro*, it is nontoxic to mice at 1–5 mg/kg, and has no growth-inhibitory effect on 4T1 breast cancer *in vivo*	162
Characterization of the estrogen-modulatory effects of resveratrol in a variety of *in vitro* and *in vivo* mammary models using effect of resveratrol and in combination with 17β-estradiol (E2) in MCF-7, T47D, LY2, and S30 mammary cancer cell lines	With cells transfected with reporter gene systems, the activation of estrogen response element-luciferase was studied, and using Western blot analysis, the expression of E2-responsive progesterone receptor (PR) and presnelin 2 protein was monitored. Furthermore, the effect of resveratrol on formation of preneoplastic lesions (induced by 7,12-dimethylbenz(*a*)anthracene) and PR expression (with or without E2) was evaluated with mammary glands of Balb/c mice placed in organ culture. Finally, the effect of p.o. administered resveratrol on *N*-methyl-*N*-nitrosourea-induced mammary tumors was studied in female Sprague-Dawley rats. As a result, in transient transfection studies with MCF-7 cells, resveratrol showed a weak estrogenic response, but when resveratrol was combined with E2 (1 nM), a clear dose-dependent antagonism was observed. Similar mixed estrogenic/antiestrogenic effects were noted with S30 cells, whereas resveratrol functioned as a pure estrogen antagonist with T47D and LY2 cells. Furthermore, in MCF-7 cells, resveratrol induced PR protein expression, but when resveratrol was combined with E2, expression	In the absence of E2, resveratrol exerts mixed estrogen agonist/antagonist activities in some mammary cancer cell lines, but in the presence of E2, resveratrol functions as an antiestrogen. In rodent models, carcinogen-induced preneoplastic lesions and mammary tumors are inhibited. These data suggest that resveratrol may have beneficial effects if used as a chemopreventive agent for breast cancer	153

	of PR was suppressed. With T47D cells, resveratrol significantly downregulated steady-state and E2-induced protein levels of PR. With LY2 and S30 cells, resveratrol downregulated presnelin 2 protein expression. Using the mouse mammary organ culture model, resveratrol induced PR when administered alone, but expression was suppressed in the presence of E2 (1 nM). Furthermore, resveratrol inhibited the formation of estrogen-dependent preneoplastic ductal lesions induced by 7,12-dimethylbenz(*a*)anthracene in these mammary glands ($IC_{50} = 3.2\ \mu M$) and reduced *N*-methyl-*N*-nitrosourea-induced mammary tumorigenesis when administered to female Sprague-Dawley rats by gavage		
Inhibition of intestinal tumorigenesis and modulation of host-defense-related gene expression in Min mice using human familial adenomatous polyposis	Resveratrol was administered for seven weeks to Min mice starting at five weeks of age. The control group was fed the same diet and received water containing 0.4% ethanol. Resveratrol prevented the formation of colon tumors and reduced the formation of small intestinal tumors by 70%. Comparison of the expression of 588 genes in the small intestinal mucosa showed that resveratrol downregulated genes that are directly involved in cell cycle progression or cell proliferation (cyclins D1 and D2, DP-1 transcription factor, and Y-box binding protein). Resveratrol upregulated several genes that are involved in the recruitment and activation of immune cells (cytotoxic T lymphocyte Ag-4, leukemia inhibitory factor receptor, and monocyte chemotactic protein 3) and in the inhibition of the carcinogenic process and tumor expansion (tumor susceptibility protein TSG101, transforming growth factor-β, inhibin-β A subunit, and desmocollin 2)	The data show the complexity of the events associated with intestinal tumorigenesis and the multiplicity of the molecular targets of resveratrol. The high potency and efficacy of resveratrol support its use as a chemopreventive agent in the management of intestinal carcinogenesis	150

(*Continued*)

TABLE 13.4
Continued

Model	What was measured	Effect	Ref.
Prevention of tumor growth and metastasis to lung and tumor-induced neovascularization in Lewis lung carcinoma-bearing mice (LLC)	Resveratrol significantly reduced the tumor volume, tumor weight and metastasis to the lung in mice bearing highly metastatic LLC tumors. Resveratrol did not affect the number of $CD4^+$, $CD8^+$, and natural killer (NK)1.1.(+) T cells in the spleen. Therefore, the inhibitory effects of resveratrol on tumor growth and lung metastasis could not be explained by natural killer or cytotoxic T-lymphocyte activation. In addition, resveratrol inhibited DNA synthesis most strongly in LLC cells. Resveratrol increased apoptosis in LLC cells, and decreased the S-phase population. Resveratrol inhibited tumor-induced neovascularization in an *in vivo* model. Moreover, resveratrol significantly inhibited the formation of capillary-like tube formation from human umbilical vein endothelial cells (HUVEC). Resveratrol inhibited the binding of vascular endothelial growth factor (VEGF) to HUVEC	The study suggests that the antitumor and antimetastatic activities of resveratrol might be due to the inhibition of DNA synthesis in LLC cells and the inhibition of LLC-induced neovascularization and tube formation (angiogenesis) of HUVEC by resveratrol	161
Investigation of whether resveratrol affects azoxymethane (AOM)-induced colon carcinogenesis using male F344 rats	Resveratrol was administered to male F344 rats. Aberrant crypt foci (ACF) were isolated and proliferation, apoptosis, and expression of the cell cycle genes Bax and p21 were determined	Resveratrol significantly reduced the number of ACF/colon and their multiplicity, and also abolished large ACF. In resveratrol-treated rats, Bax expression was enhanced in ACF but not in the surrounding mucosa. In both controls and resveratrol-treated rats, proliferation was higher in ACF than in normal mucosa. p21 was	151

		expressed in ACF of controls and of resveratrol-treated rats and in normal mucosa of controls, but was lost in normal mucosa of resveratrol-treated animals. The results suggest a protective role of resveratrol in colon carcinogenesis with a mechanism involving changes in Bax and p21 expression	
Potential activities of six compounds — butylated hydroxyanisole (BHA), *myo*-inositol, curcumin, esculetin, resveratrol, and lycopene — as chemopreventive agents against lung tumor induction in A/J mice by benzo(*a*)pyrene (BaP) and 4-(methyl-nitrosamino)-17-(3-pyridyl)-1-butanone (NNK)	Groups of 20 A/J mice were treated weekly by gavage with a mixture of BaP and NNK (3 μmol each) for 8 weeks, then sacrificed 26 weeks after the first carcinogen treatment. Mice treated with BHA (20 or 40 μmol) by gavage 2 h before each dose of BaP and NNK had significantly reduced lung tumor multiplicity. Treatment with BHA (20 or 40 μmol) by gavage weekly or with dietary BHA (2000 ppm), curcumin (2000 ppm), or resveratrol (500 ppm) from 1 week after carcinogen treatment until termination had no effect on lung tumor multiplicity. Treatment with dietary *myo*-inositol (30,000 ppm) or esculetin (2000 ppm) from 1 week after carcinogen treatment until termination significantly reduced lung tumor multiplicity, with the effect of *myo*-inositol being significantly greater than that of esculetin. Treatment with dietary LTO (167, 1667 or 8333 ppm) from 1 week before carcinogen treatment until termination had no effect on lung tumor multiplicity	The results of this study demonstrate that BHA is an effective inhibitor of BaP plus NNK-induced lung tumorigenesis in A/J mice when administered during the period of carcinogen treatment and that, among the compounds tested, *myo*-inositol is most effective after carcinogen treatment	148

(*Continued*)

TABLE 13.4
Continued

Model	What was measured	Effect	Ref.
Inhibition of tumor growth in a rat tumor model	Resveratrol was administered to rats that were inoculated with a fast-growing tumor (the Yoshida AH-130 ascites hepatoma). Tumor cell content was measured. Flow cytometric analysis of the tumor cell population was employed	Resveratrol administration to rats caused a very significant decrease (25%) in tumor cell content. The effects of this compound were associated with an increase in the number of cells in the G_2/M cell cycle phase. Flow cytometric analysis of the tumor cell population revealed the existence of an aneuploid peak (representing 28% of total), which suggests that resveratrol causes apoptosis in the tumor cell population resulting in a decreased cell number	159
Investigation of cancer chemopreventive activity in the three major stages of carcinogenesis	Resveratrol was tested for antiinitiation activity, antipromotion activity, and antiprogression activity. It was tested in carcinogen-treated mouse mammary glands in culture, a mouse skin cancer model, and a rat antiinflammation model	Resveratrol was found to act as an antioxidant and antimutagen and to induce phase II drug-metabolizing enzymes; it mediated antiinflammatory effects in a rat model and inhibited cyclooxygenase and hydroperoxidase functions; and it induced human promyelocytic leukemia cell differentiation. It inhibited the development of preneoplastic lesions in carcinogen-treated mouse mammary glands in culture and inhibited tumorigenesis in the two-stage mouse skin cancer model	39

are very promising and suggest utility for the prevention of skin cancer. Other results are less clear.

Resveratrol reduced biomarkers of lung carcinogenesis produced in benzo(*a*)pyrene-treated mice [146] but not tumorigenesis [147]. We also found that resveratrol was not active in the benzo(*a*)pyrene mouse lung tumorigenesis model (unpublished data), nor was it active in a mixed-carcinogen lung cancer model [148]. A lack of activity was also observed in the *Min* mouse model [149], but a similar study reported a reduction of intestinal tumors when resveratrol was administered [150]. Aberrant crypts were also reduced in carcinogen-treated rats [151].

An increase in tumorigenesis was reported when resveratrol was administered to rats treated with *N*-methyl-*N*-nitrosourea [152], but this is contrary to our results wherein an inhibition was observed [153]. Activity was also reported in the DMBA rat mammary carcinogenesis model [154] as well as the NMBA esophageal model [155]. In addition, activity was observed against a variety of transplanted tumors, including neuroblastoma [156], hepatoma [129,157–159], H22 cells [160], and Lewis lung [161]. The compound was not active against 4T1 breast cancer [162] and leukemia [163].

CONCLUSIONS

As summarized above, a great deal of work has been performed over the past few years to characterize the cancer chemopreventive potential of resveratrol. The ultimate objective of this work is to answer one question: Is resveratrol of value to alleviate any type of cancer in humans? Since humans are already consuming resveratrol, either as a constituent of the diet or as a dietary supplement, data could already exist to suggest the potential of resveratrol to function in this capacity. Consumption of red wine, for example, implies the ingestion of resveratrol, and correlations can be examined between consumption and cancer incident. However, no clear answers can be derived from such epidemiological considerations, so the possible efficacy of resveratrol remains an open question. As learned by the failure of β-carotene to prevent lung cancer [164,165], human clinical trials are necessary to understand the true efficacy of experimental agents, irrespective of compelling laboratory data that may suggest effectiveness.

Spearheaded by Waun Ki Hong and Michael B. Sporn, a chemopreventive working group recently provided a report describing the prevention of cancer in the current millennium [3]. Included in this report was a list of seven desirable/acceptable characterizations of cancer chemopreventive agents. In brief, these will be considered in the context of resveratrol:

1. Efficacy in preventing cancer. Resveratrol has been shown to demonstrate efficacy in multiple animal models. Activity in humans is unknown.

2. Knowledge about mechanism of inhibition. As summarized in this chapter, a great deal of information is available concerning the mechanism of action of resveratrol. Although a straightforward sequence of critical events cannot be defined due to the overtly pleiotropic mode of action, some existing data are certainly valuable.
3. Information as to likely efficacy in the human. The most compelling data are derived from animal studies in which resveratrol is administered by the oral route. This has been accomplished. Therefore, although absorption and metabolism [125–128] requires additional investigation and remains moot to some extent, the potential of efficacy in humans does appear likely. Some indication of toxicity has been suggested [166,167] and needs to be further defined, but most studies suggest favorable therapeutic indices.
4. Demonstration of efficacy in experimental animals. In general, it is not possible to predict efficacy in only one animal model. Efficacy has been demonstrated in breast, skin, esophagus, and colon models, but further tests should be performed in additional models such as bladder, prostate, uterus, and kidney. Resveratrol does not appear active in some mouse lung cancer models.
5. Lack of toxicity and undesirable side effects. Certainly, long-term feeding studies have been performed with resveratrol without untoward toxic side effects [153], but some suggestions of potential toxicity and/or hormonal activity have been put forth. Overall, it seems highly likely that a therapeutic regimen could be devised with an acceptable risk/benefit ratio. Nonetheless, thorough preclinical assessment of resveratrol in acceptable models of toxicity will be required prior to advocating long-term human investigation trials.
6. Compounds already approved by the U.S. Food and Drug Administration (FDA) for human use or likely to be approved readily. To obtain FDA approval for clinical trails, it is likely that comprehensive preclinical toxicity trials are necessary. However, these studies are straightforward and a notable advantage is an ample supply of resveratrol through chemical synthesis. The work needs to be completed prior to drawing conclusions, but existing data suggest acceptable dose regimens could be devised, and it seems likely that FDA approval would follow.
7. Occurrence of the agent in foods or beverages. The occurrence of resveratrol in foods or beverages is an obvious advantage in terms of development. It can already be stated with a high degree of confidence that consumption of limited quantities of resveratrol is not harmful to humans, and great flexibility becomes available in terms of long-term dosing strategies.

In sum, a great deal of time, money, and intellectual capital has been invested in the exploration of resveratrol. From a purely academic point of view, considering the structural simplicity of resveratrol, the extent of this effort is incredible. Implicitly, however, the sheer magnitude of investigation supports the intrinsic value of this compound. Based on the criteria discussed above, and the overall favorable characteristics of resveratrol, it is reasonable to advocate further development as a cancer chemopreventive agent. Perhaps this review will have some value in facilitating the process.

ACKNOWLEDGMENTS

The author is grateful to faculty colleagues associated with this research project, namely Drs. N.R. Farnsworth, H.H.S. Fong, S. Hedayat, A.D. Kinghorn, J.W. Kosmeder II, R.G. Mehta, A.D. Mesecar, R.C. Moon, R.M. Moriarty, B.D. Santarsiero, D.D. Soejarto, and R.B. van Breemen, and to many postdoctoral associates, graduate students, and research assistants who worked in the laboratory in support to this research. Special thanks are extended to Elizabeth Ryan for help in organizing this manuscript and collating the descriptions given in the tables. The support of collaborators throughout the world who have participated in the selection, collection, and identification of plant materials used in the present work is also gratefully acknowledged. The present work is supported by program project grant P01 CA48112, funded by the National Cancer Institute, NIH, Bethesda, Maryland.

REFERENCES

1. Stewart BW and Kleihues P, *World Cancer Report*, IACR Press, Lyon, 2003, pp. 9–19.
2. Jemal A, Murray T, Samuels A, Ghafoor A, Ward E, and Thun MJ, Cancer statistics. 2003, *CA Cancer J Clin* 53, 5–26, 2003.
3. Alberts DS, Conney AH, Ernster VL, Garber JE, Greenwald P, Gudas LJ, Hong WK, Kelloff GJ, Kramer RA, Lerman CE, Mangelsdorf DJ, Matter A, Minna JD, Nelson WG, Pezzuto JM, Prendergast F, Rusch VW, Sporn MB, Wattenberg LW, and Weinstein B, Prevention of cancer in the next millennium. Report of the Chemoprevention Working Group to the American Association for Cancer Research, *Cancer Res* 59, 4743–4758, 1999.
4. Kelloff GJ, Perspectives on cancer chemoprevention research and drug development, in Van de Woude GF and Klein G, Eds, Academic Press, San Diego, 1999, pp. 199–334.
5. Sporn M, The war on cancer, *Lancet* 347, 1377–1381, 1996.
6. Greenwald P, Kelloff GJ, Burch-Whitman C, and Kramer BS, *CA Cancer J Clin* 45, 31–49, 1995.
7. Wattenberg L, Chemoprevention of cancer, *Cancer Res* 45, 1–8, 1985.

8. Sporn MB, Dunlop NM, Newton DL, and Smith JM, Prevention of chemical carcinogenesis by vitamin A and its synthetic analogs (retinoids), *Fed Proc* 35, 1332–1338, 1976.
9. Kelloff GJ, Hawk ET, Karp JE, Crowell JA, Boone CW, Steele VE, Lubet RA, and Sigman CC, Progress in clinical chemoprevention, *Semin Oncol* 24, 241–252, 1997.
10. Willett WC and MacMahon B, Diet and cancer: an overview [second of two parts], *N Engl J Med* 310, 697–703, 1984.
11. Harris CC, Chemical and physical carcinogenesis: advances and perspectives for the 1990s, *Cancer Res* 51 (Suppl), 5023–5044, 1991.
12. De Flora S, Mechanisms of inhibitors of mutagenesis and carcinogenesis, *Mutation Res* 402, 151–158, 1998.
13. Sporn MB, Carcinogenesis and cancer: different perspectives on the same disease, *Cancer Res* 51, 6215–6218, 1991.
14. Kelloff GJ, Crowell JA, Steele VE, Lubet RA, Boone CW, Malone WA, Hawk ET, Lieberman R, Lawrence JA, Kopelovich L, Ali I, Viner JL, and Sigman CC, Progress in cancer chemoprevention, *Ann NY Acad Sci* 889, 1–13, 1999.
15. Morse MA and Stoner GD, Cancer chemoprevention: principles and prospects, *Carcinogenesis* 14, 1737–1746, 1993.
16. Surh Y-J, Molecular mechanisms of chemopreventive effects of selected dietary and medicinal phenolic substances, *Mutation Res* 428, 305–327, 1999.
17. Reddy L, Odhav B, and Bhoola KD, Natural products for cancer prevention: a global perspective, *Pharmacol Ther* 99, 1–13, 2003.
18. Fujiki H, Suganuma M, Imai K, and Nakachi K, Green tea: cancer preventive beverage and/or drug, *Cancer Lett* 188, 9–13, 2002.
19. Surh Y-J, Cancer chemoprevention with dietary phytochemicals, *Nature Rev Cancer* 3, 768–780, 2003.
20. http://clinicaltrials.gov/ct/search?term=cancer+chemoprevention&submit.
21. Chow H-HS, Cai Y, Hakim IA, Crowell JA, Shahi F, Brooks CA, Dorr RT, Hara Y, and Alberts DS, Pharmacokinetics and safety of green tea polyphenols after multiple-dose administration of epigallocatechin gallate and polyphenon E in healthy individuals, *Clin Cancer Res* 9, 3312–3319, 2003.
22. Fujiki H, Two stages of cancer prevention with green tea, *J Cancer Res Clin*, 125, 589–597, 1999.
23. Ren W, Qiao Z, Wang H, Zhu L, and Zhang L, Flavonoids: promising anticancer agents, *Med Res Rev* 23, 519–534, 2003.
24. Pezzuto JM, Natural product cancer chemopreventive agents, in *Recent Advances in Phytochemistry*, Vol. 29, Arnason JT, Mata R, and Romeo JT, Eds, Plenum Press, New York, 1995, pp. 19–45.
25. Pezzuto JM, Song LL, Lee SK, Shamon LA, Mata-Greenwood E, Jang J, Jeong H-J, Pisha E, Mehta RG, and Kinghorn AD, Bioassay methods useful for activity-guided isolation of natural product cancer chemopreventive agents, in *Chemistry, Biological and Pharmacological Properties of Medicinal Plants from the Americas*, Hostettmann K, Gupta MP, Marston A, Eds, Harwood Academic, Chur, Switzerland, 1998, pp. 81–110.
26. Kinghorn AD, Su B-N, Lee D, Gu J-Q, and Pezzuto JM, Cancer chemopreventive agents discovered by activity-guided fractionation: an update, *Curr Org Chem* 7, 213–226, 2003.

27. Pezzuto JM, Kosmeder II JW, Park EJ, Lee SK, Cuendet M, Gills J, Bhat K, Grubjesic S, Park H-S, Mata-Greenwood E, Tan YM, Yu R, Lantvit DD, and Kinghorn AD, Characterization of chemopreventive agents in natural products, in *Strategies for Cancer Chemoprevention*, Kelloff GJ, Hawk ET, Sigman CC, Eds, Humana Press, Totowa, NJ, 2005, pp. 3–37.
28. Loub WD Farnsworth NR, Soejarto DD, and Quinn ML, NAPRALERT: computer handling of natural product research data, *J Chem Inf Computer Sci* 25, 99–103, 1985.
29. Kosmeder JW II and Pezzuto JM, Intermediate biomarkers, *Cancer Treat Res* 106, 31–61, 2001.
30. Crowell JA and Holmes CJ, Agent identification and preclinical testing, *Cancer Treat Res* 106, 1–30, 2001.
31. Mehta RG and Moon RC, Characterization of effective chemopreventive agents in mammary gland *in vitro* using an initiation-promotion protocol, *Anticancer Res* 11, 593–596, 1991.
32. Mehta RG, Bhat KP, Hawthorne ME, Kopelovich L, Mehta RR, Christov K, Kelloff GJ, Steele VE, and Pezzuto JM, Induction of atypical ductal hyperplasia in mouse mammary gland organ culture, *J Natl Cancer Inst* 93, 1103–1106, 2001.
33. Shamon L and Pezzuto J, Assessment of antimutagenic activity with *Salmonella typhimurium* strain TM677, *Methods Cell Sci* 19, 57–62, 1997.
34. Lee SK, Mbwambo ZH, Chung H, Luyengi L, Gamez EJC, Mehta RG, Kinghorn AD, and Pezzuto JM, Evaluation of the antioxidant potential of natural products, *Comb Chem High-Throughput Screening* 1, 35–46, 1998.
35. Song LL, Kosmeder II JW, Lee SK, Gerhäuser C, Lantvit D, Moon RC, Moriarty RM, and Pezzuto JM, Cancer chemopreventive activity mediated by 4′-bromoflavone, a potent inducer of phase II detoxification enzymes, *Cancer Res* 59, 578–585, 1999.
36. Kang YH and Pezzuto JM, Induction of quinone reductase as a primary screen for natural product anticarcinogens, *Meth Enzymol* 382, 380–414, 2004.
37. Gerhäuser C, Mar W, Lee SK, Suh N, Luo Y, Kosmeder II JW, Moriarty RM, Luyengi L, Kinghorn AD, Fong HHS, Mehta RG, Constantinou A, Moon RC, and Pezzuto JM, Rotenoids mediate potent cancer chemopreventive activity through transcriptional regulation of ornithine decarboxylase, *Nature Med* 1, 260–266, 1995.
38. Mbwambo ZH, Lee SK, Mshiu EN, Pezzuto JM, and Kinghorn AD, Constituents from the stem wood of *Euphorbia quinquecostata* with phorbol dibutyrate receptor-binding inhibitory activity, *J Nat Prod* 59, 1051–1055, 1996.
39. Jang M, Cai L, Udeani GO, Slowing KV, Thomas CF, Beecher CW, Fong HH, Farnsworth NR, Kinghorn AD, Mehta RG, Moon RC, and Pezzuto JM, Cancer chemopreventive activity of resveratrol, a natural product derived from grapes, *Science* 275, 218–220, 1997.
40. El-Sayed KA, Hamann MT, Waddling CA, Jensen C, Lee SK, Dunstan CA, and Pezzuto JM, Structurally novel bioconversion products of the marine natural product sarcophine effectively inhibit JB6 cell transformation, *J Org Chem* 63, 7449–7455, 1998.

41. Suh N, Luyengi L, Fong HHS, Kinghorn AD, and Pezzuto JM, Discovery of natural product chemopreventive agents utilizing HL-60 cell differentiation as a model, *Anticancer Res* 15, 233–240, 1995.
42. Pisha E and Pezzuto JM, Cell-based assay for the determination of estrogenic and anti-estrogenic activities, *Methods Cell Sci* 19, 37–43, 1997.
43. Jeong H-J, Shin YG, Kim I-H, and Pezzuto JM, Inhibition of aromatase activity by flavonoids, *Arch Pharm Res* 22, 309–312, 1999.
44. Chang LC, Gills JJ, Bhat KPL, Luyengi L, Farnsworth NR, Pezzuto JM, and Kinghorn AD, Activity-guided isolation of constituents of *Cerbera manghas* with antiproliferative and antiestrogenic activities, *Bioorg Med Chem Lett* 10, 2431–2434, 2000.
45. Gamez EJC, Luyengi L, Lee SK, Zhu L-F, Zhou B-N, Fong HHS, Pezzuto JM, and Kinghorn AD, Antioxidant flavonoid glycosides from *Daphniphyllum calycinum*, *J Nat Prod* 61, 706–708, 1998.
46. Su BN, Park EJ, Nikolic D, Santarsiero BD, Mesecar AD, Vigo JS, Graham JG, Cabieses F, van Breemen RB, Fong HH, Farnsworth NR, Pezzuto JM, and Kinghorn AD, Activity-guided isolation of novel norwithanolides from *Deprea subtriflora* with potential cancer chemopreventive activity, *J Org Chem* 68, 2350–2361, 2003.
47. Gu JQ, Park EJ, Luyengi L, Hawthorne ME, Mehta RG, Farnsworth NR, Pezzuto JM, and Kinghorn AD, Constituents of *Eugenia sandwicensis* with potential cancer chemopreventive activity, *Phytochemistry* 58, 121–127, 2001.
48. Kinghorn AD, Su B-N, Jang DS, Chang LC, Lee D, Gu J, Carcache-Blanco EJ, Pawlus AD, Lee SK, Park EJ, Cuendet M, Gills JJ, Bhat K, Park HS, Mata-Greenwood E, Song LL, Jang M, and Pezzuto JM, Natural inhibitors of carcinogenesis, *Planta Med* 70, 691–705, 2004.
49. Su B-N, Park EJ, Nikolic D, Schunke Vigo J, Graham JG, Cabieses F et al, Isolation and characterization of miscellaneous secondary metabolites of *Deprea subtriflora*, *J Nat Prod* 66, 1089–1093, 2003.
50. Lee D, Cuendet M, Schunke Vigo J, Graham JG, Cabieses F, Fong HHS et al, A novel cyclooxygenase-inhibitory stilbenolignan from the seeds of *Aiphanes aculeata*, *Org Lett* 3, 2169–2171, 2001.
51. Lee D, Park EJ, Cuendet M, Axelrod F, Chavez PI, Fong HHS et al, Cyclooxygenase-inhibitory and antioxidant constituents of the aerial parts of *Antirhea acutata*, *Bioorg Med Chem Lett* 11, 1565–1568, 2001.
52. Mehta RG, Liu J, Constantinou A, Thomas CF, Hawthorne M, You M, Gerhäuser C, Pezzuto JM, Moon RC, and Moriarty RM, Cancer chemopreventive activity of brassinin, a phytoalexin from cabbage, *Carcinogenesis* 16, 399–404, 1995.
53. Gerhäuser C, Lee SK, Kosmeder J, Moriarty RM, Hamel E, Mehta RG, Moon RC, and Pezzuto JM, Regulation of ornithine decarboxylase induction by deguelin, a natural product cancer chemopreventive agent, *Cancer Res* 57, 3429–3435, 1997.
54. Udeani GO, Gerhäuser C, Thomas CF, Moon RC, Kosmeder JW, Kinghorn AD, Moriarty RM, and Pezzuto JM, Cancer chemopreventive activity mediated by deguelin, a naturally occurring rotenoid, *Cancer Res* 57, 3424–3428, 1997.

55. Lee HY, Suh YA, Kosmeder JW, Pezzuto JM, Hong WK, and Kurie JM, Deguelin-induced inhibition of cyclooxygenase-2 expression in human bronchial epithelial cells, *Clin Cancer Res* 10, 1074–1079, 2004.
56. Murillo G, Hirschelman WH, Kinghorn AD, Moriarty RM, Pezzuto JM, and Mehta RG, Prevention of colon carcinogenesis by zapotin, *Cancer Chemoprev*, submitted.
57. Mata-Greenwood E, Cuendet M, Sher D, Gustin D, Stock W, and Pezzuto JM, Brusatol-mediated induction of leukemic cell differentiation and G_1 arrest is associated with down-regulation of c-*myc*, *Leukemia* 16, 2275–2284, 2002.
58. Cuendet M, Christov K, Lantvit DD, Deng Y, Hedayat S, Helson L, McChesney JD, and Pezzuto JM, Multiple myeloma regression mediated by bruceantin, *Clin Cancer Res* 10, 1170–1179, 2004.
59. Lee D, Bhat KP, Fong HH, Farnsworth NR, Pezzuto JM, and Kinghorn AD, Aromatase inhibitors from *Broussonetia papyrifera*, *J Nat Prod* 64, 1286–1293, 2001.
60. Zhang Y, Kensler TW, Cho CG, Posner GH, and Talalay P, Anticarcinogenic activities of sulforaphane and structurally related synthetic norbornyl isothiocyanates, *Proc Natl Acad Sci USA* 91, 3147–3150, 1994.
61. Gerhäuser C, You M, Liu J, Moriarty RM, Hawthorne M, Mehta RG, Moon RC, and Pezzuto JM, Cancer chemopreventive potential of sulforamate, a novel analogue of sulforaphane that induces phase 2 drug-metabolizing enzymes, *Cancer Res* 57, 272–278, 1997.
62. Kosmeder II JW, Hirschelman WH, Song LS, Park EJ, Tan Y, Yu R, Hawthorne M, Mehta RG, Grubbs CJ, Lubet RA, Moriarty RM, and Pezzuto JM, Cancer Chemopreventive Activity of Oxomate, a Monofunctional Inducer of Phase II Detoxification Enzymes [abstr], 224th American Chemical Society National Meeting, Boston, MA, August 18–22, 2002.
63. Stewart JR, Artime MC, and O'Brian CA, Resveratrol: a candidate nutritional substance for prostate cancer prevention, *J Nutr* 133, 2440S–2443S, 2003.
64. Creasy LL and Coffee M, Phytoalexin production potential of grape berries, *J Am Soc Hortic Sci* 113, 230, 1998.
65. Schwekendiek A, Pfeffer G, and Kindl H, Pine stilbene synthase cDNA, a tool for probing environmental stress, *FEBS Lett* 301, 41–44, 1992.
66. Jeandet P, Douillet-Breuil AC, Bessis R, Debord S, Sbaghi M, and Adrian M, Phytoalexins from the *Vitaceae*: biosynthesis, phytoalexin gene expression in transgenic plants, antifungal activity, and metabolism, *J Agric Food Chem* 50, 2731–2741, 2002.
67. Montero C, Cristescu SM, Jimenez JB, Orea JM, te Lintel Hekkert S, Harren FJ, and Gonzalez Urena A, *trans*-Resveratrol and grape disease resistance, a dynamic study by high-resolution laser-based techniques, *Plant Physiol* 131, 129–138, 2003.
68. Hawksworth DL, Micological research news, *Mycol Res* 107, 769–770, 2003.
69. Siemann GJ and Creasy LL, Concentration of the phytoalexin resveratrol in wine, *Am J Ecol Viticul* 43, 49, 1992.
70. Cantos E, Garcia-Viguera C, de Pascual-Teresa S, and Tomas-Barberan FA, Effect of postharvest ultraviolet irradiation on resveratrol and other phenolics of cv. Napoleon table grapes, *J Agric Food Chem* 48, 4604–4612, 2000.

71. Careri M, Corradini C, Elviri L, Nicoletti I, and Zagnoni I, Direct HPLC analysis of quercetin and *trans*-resveratrol in red wine, grape, and winemaking byproducts, *J Agric Food Chem* 51, 5226–5231, 2003.
72. Nonomura S, Kanagawa H, and Makimoto A, Chemical constituents of polygonaceous plants: I. Studies on the components of Ko-jo-kon (*Polygonum cuspidatum* Sieb. et Zucc.), *Yakugaku Zasshi* 83, 983, 1963.
73. Hain R, Reif HJ, Krause E, Langebartels R, and Kindl H, Disease resistance results from foreing phytoalexin expression in a novel plant, *Nature Rev Cancer* 361, 153, 1993.
74. Paiva NL, Engineering resveratrol accumulation into alfalfa and other food plants, International Molecular Farming Conference, London, Ontario, Canada, 1999, p. 134.
75. Pezzuto JM and Steele VE, Eds, Proceedings of a Conference Exploring the Power of Phytochemicals: Research Advances on Grape Compounds, Swets and Zeitlinger, Lisse, The Netherlands, 1998 (A supplement of *Pharm Biol*).
76. Bhat KPL, Kosmeder JW II, and Pezzuto JM, Biological effects of resveratrol, *Antioxid Redox Signal* 3, 1041–1064, 2001.
77. Jang M and Pezzuto JM, Resveratrol blocks eicosanoid production and chemically induced cellular transformation: implications for cancer chemoprevention, *Pharm Biol* 36, 28–34, 1998.
78. Jang M and Pezzuto JM, Cancer chemopreventive activity of resveratrol, *Drugs Exp Clin Res* 25, 65–77, 1999.
79. Bhat KP and Pezzuto JM, Cancer chemopreventive activity of resveratrol, *Ann NY Acad Sci* 957, 210–229, 2002.
80. Kondratyuk T, Shalaev E, and Pezzuto JM, Cancer chemoprevention by resveratrol, in *Carcinogenic and Anticarcinogenic Food Components*, Bartoszek AB, Ed, CRC Press, Boca Raton, FL, 2005.
81. Bagchi D, *Resveratrol and Human Health*, McGraw Hill, Columbus, OH, 2000.
82. Pettit GR, Grealish MP, Jung MK, Hamel E, Pettit RK, Chapuis JC, and Schmidt JM, Antineoplastic agents. 465. Structural modification of resveratrol: sodium resveratrin phosphate, *J Med Chem* 45, 2534–2542, 2002.
83. Kim S, Ko H, Park JE, Jung S, Lee SK, and Chun YJ, Design, synthesis, and discovery of novel trans-stilbene analogues as potent and selective human cytochrome P450 1B1 inhibitors, *J Med Chem* 45, 160–164, 2002.
84. Thakkar K, Geahlen RL, and Cushman M, Synthesis and protein-tyrosine kinase inhibitory activity of polyhydroxylated stilbene analogues of piceatannol, *J Med Chem* 36, 2950–2955, 1993.
85. Cushman M, Nagarathnam D, Gopal D, He HM, Lin CM, and Hamel E, Synthesis and evaluation of analogues of (*Z*)-1-(4-methoxyphenyl)-2-(3,4,5-trimethoxyphenyl)ethene as potential cytotoxic and antimitotic agents, *J Med Chem* 35, 2293–2306, 1992.
86. Roberti M, Pizzirani D, Simoni D, Rondanin R, Baruchello R, Bonora C, Buscemi F, Grimaudo S, and Tolomeo M, Synthesis and biological evaluation of resveratrol and analogues as apoptosis-inducing agents, *J Med Chem* 46, 3546–3554, 2003.
87. Leonard SS, Xia C, Jiang BH, Stinefelt B, Klandorf H, Harris GK, and Shi X, Resveratrol scavenges reactive oxygen species and effects radical-induced cellular responses, *Biochem Biophys Res Commun* 309, 1017–1026, 2003.

88. Stewart JR, Christman KL, and O'Brian CA, Effects of resveratrol on the autophosphorylation of phorbol ester-responsive protein kinases: inhibition of protein kinase D but not protein kinase C isozyme autophosphorylation, *Biochem Pharmacol* 60, 1355–1359, 2000.
89. Garcia-Garcia J, Micol V, de Godos A, and Gomez-Fernandez JC, The cancer chemopreventive agent resveratrol is incorporated into model membranes and inhibits protein kinase C alpha activity, *Arch Biochem Biophys* 372, 382–388, 1999.
90. Stewart JR, Ward NE, Ioannides CG, and O'Brian CA, Resveratrol preferentially inhibits protein kinase C-catalyzed phosphorylation of a cofactor-independent, arginine-rich protein substrate by a novel mechanism, *Biochemistry* 38, 13244–13251, 1999.
91. Szewczuk LM and Penning TM, Mechanism-based inactivation of COX-1 by red wine *m*-hydroquinones: a structure–activity relationship study, *J Nat Prod* 67, 1777–1782, 2004.
92. Waffo-Teguo P, Hawthorne ME, Cuendet M, Merillon JM, Kinghorn AD, Pezzuto JM, and Mehta RG, Potential cancer-chemopreventive activities of wine stilbenoids and flavans extracted from grape (*Vitis vinifera*) cell cultures, *Nutr Cancer* 40, 173–179, 2001.
93. Mutoh M, Takahashi M, Fukuda K, Matsushima-Hibiya Y, Mutoh H, Sugimura T, and Wakabayashi K, Suppression of cyclooxygenase-2 promoter-dependent transcriptional activity in colon cancer cells by chemopreventive agents with a resorcin-type structure, *Carcinogenesis* 21, 959–963, 2000.
94. Chun YJ, Kim MY, and Guengerich FP, Resveratrol is a selective human cytochrome P450 1A1 inhibitor, *Biochem Biophys Res Commun* 262, 20–24, 1999.
95. Kim YM, Yun J, Lee CK, Lee H, Min KR, and Kim Y, Oxyresveratrol and hydroxystilbene compounds. Inhibitory effect on tyrosinase and mechanism of action, *J Biol Chem* 277, 16340–16344, 2002.
96. Burkhardt S, Reiter RJ, Tan DX, Hardeland R, Cabrera J, and Karbownik M, DNA oxidatively damaged by chromium(III) and H_2O_2 is protected by the antioxidants melatonin, N^1-acetyl-N^2-formyl-5-methoxykynuramine, resveratrol and uric acid, *Int J Biochem Cell Biol* 33, 775–783, 2001.
97. Potter GA, Patterson LH, Wanogho E, Perry PJ, Butler PC, Ijaz T, Ruparelia KC, Lamb JH, Farmer PB, Stanley LA, and Burke MD, The cancer preventative agent resveratrol is converted to the anticancer agent piceatannol by the cytochrome P450 enzyme CYP1B1, *Br J Cancer* 86, 774–778, 2002.
98. de Santi C, Pietrabissa A, Mosca F, and Pacifici GM, Glucuronidation of resveratrol, a natural product present in grape and wine, in the human liver, *Xenobiotica* 30, 1047–1054, 2000.
99. de Santi C, Pietrabissa A, Spisni R, Mosca F, and Pacifici GM, Sulphation of resveratrol, a natural product present in grapes and wine, in the human liver and duodenum, *Xenobiotica* 30, 609–617, 2000.
100. Wang Z, Hsieh TC, Zhang Z, Ma Y, and Wu JM, Identification and purification of resveratrol targeting proteins using immobilized resveratrol affinity chromatography, *Biochem Biophys Res Commun* 323, 743–749, 2004.
101. Jannin B, Menzel M, Berlot JP, Delmas D, Lancon A, and Latruffe N, Transport of resveratrol, a cancer chemopreventive agent, to cellular

targets: plasmatic protein binding and cell uptake, *Biochem Pharmacol* 68, 1113–1118, 2004.

102. Bianco NR, Chaplin LJ, and Montano MM, Differential induction of quinone reductase by phytoestrogens and protection against estrogen-induced DNA damage, *Biochem J* 385, 279–287, 2005.
103. Le Corre L, Fustier P, Chalabi N, Bignon YJ, and Bernard-Gallon D, Effects of resveratrol on the expression of a panel of genes interacting with the BRCA1 oncosuppressor in human breast cell lines, *Clin Chim Acta* 344, 115–121, 2004.
104. Gehm BD, Levenson AS, Liu H, Lee EJ, Amundsen BM, Cushman M, Jordan VC, and Jameson JL, Estrogenic effects of resveratrol in breast cancer cells expressing mutant and wild-type estrogen receptors: role of AF-1 and AF-2, *J Steroid Biochem Mol Biol* 88, 223–234, 2004.
105. Gao S, Liu GZ, and Wang Z, Modulation of androgen receptor-dependent transcription by resveratrol and genistein in prostate cancer cells, *Prostate* 59, 214–225, 2004.
106. Pozo-Guisado E, Lorenzo-Benayas MJ, and Fernandez-Salguero PM, Resveratrol modulates the phosphoinositide 3-kinase pathway through an estrogen receptor α-dependent mechanism: relevance in cell proliferation, *Int J Cancer* 109, 167–173, 2004.
107. Wietzke JA and Welsh J, Phytoestrogen regulation of a vitamin D3 receptor promoter and 1,25-dihydroxyvitamin D3 actions in human breast cancer cells, *J Steroid Biochem Mol Biol* 84, 149–157, 2003.
108. Levenson AS, Gehm BD, Pearce ST, Horiguchi J, Simons LA, Ward JE 3rd, Jameson JL, and Jordan VC, Resveratrol acts as an estrogen receptor (ER) agonist in breast cancer cells stably transfected with ER alpha, *Int J Cancer* 104, 587–596, 2003.
109. Brownson DM, Azios NG, Fuqua BK, Dharmawardhane SF, and Mabry TJ, Flavonoid effects relevant to cancer, *J Nutr* 132, 3482S–3489S, 2002.
110. Morris GZ, Williams RL, Elliott MS, and Beebe SJ, Resveratrol induces apoptosis in LNCaP cells and requires hydroxyl groups to decrease viability in LNCaP and DU 145 cells, *Prostate* 52, 319–329, 2002.
111. Bhat KP and Pezzuto JM, Resveratrol exhibits cytostatic and antiestrogenic properties with human endometrial adenocarcinoma (Ishikawa) cells, *Cancer Res* 61, 6137–6144, 2001.
112. Bowers JL, Tyulmenkov VV, Jernigan SC, and Klinge CM, Resveratrol acts as a mixed agonist/antagonist for estrogen receptors α and β, *Endocrinology* 141, 3657–3667, 2000.
113. Lu R and Serrero G, Resveratrol, a natural product derived from grape, exhibits antiestrogenic activity and inhibits the growth of human breast cancer cells, *J Cell Physiol* 179, 297–304, 1999.
114. Gehm BD, McAndrews JM, Chien PY, and Jameson JL, Resveratrol, a polyphenolic compound found in grapes and wine, is an agonist for the estrogen receptor, *Proc Natl Acad Sci USA* 94, 14138–14143, 1997.
115. Lee EJ, Min HY, Joo Park H, Chung HJ, Kim S, Nam Han Y, and Lee SK, G_2/M cell cycle arrest and induction of apoptosis by a stilbenoid, 3,4,5-trimethoxy-4′-bromo-*cis*-stilbene, in human lung cancer cells, *Life Sci* 75, 2829–2839, 2004.

116. Schneider Y, Chabert P, Stutzmann J, Coelho D, Fougerousse A, Gosse F, Launay JF, Brouillard R, and Raul F, Resveratrol analog (*Z*)-3,5, 4′-trimethoxystilbene is a potent anti-mitotic drug inhibiting tubulin polymerization, *Int J Cancer* 107, 189–196, 2003.
117. Ito T, Akao Y, Yi H, Ohguchi K, Matsumoto K, Tanaka T, Iinuma M, and Nozawa Y, Antitumor effect of resveratrol oligomers against human cancer cell lines and the molecular mechanism of apoptosis induced by vaticanol C, *Carcinogenesis* 24, 1489–1497, 2003.
118. Kim S, Min SY, Lee SK, and Cho WJ, Comparative molecular field analysis study of stilbene derivatives active against A549 lung carcinoma, *Chem Pharm Bull* (Tokyo) 51, 516–521, 2003.
119. She QB, Ma WY, Wang M, Kaji A, Ho CT, and Dong Z, Inhibition of cell transformation by resveratrol and its derivatives: differential effects and mechanisms involved, *Oncogene* 22, 2143–2150, 2003.
120. Kim HJ, Chang EJ, Bae SJ, Shim SM, Park HD, Rhee CH, Park JH, Choi SW, Cytotoxic and antimutagenic stilbenes from seeds of *Paeonia lactiflora*, *Arch Pharm Res* 25, 293–299, 2002.
121. Ito T, Akao Y, Tanaka T, Iinuma M, and Nozawa Y, Vaticanol C, a novel resveratrol tetramer, inhibits cell growth through induction of apoptosis in colon cancer cell lines, *Biol Pharm Bull* 25, 147–148, 2002.
122. Heo YH, Kim S, Park JE, Jeong LS, and Lee SK, Induction of quinone reductase activity by stilbene analogs in mouse Hepa 1c1c7 cells, *Arch Pharm Res* 24, 597–600, 2001.
123. Nam KA, Kim S, Heo YH, and Lee SK, Resveratrol analog, 3,5,2′, 4′-tetramethoxy-*trans*-stilbene, potentiates the inhibition of cell growth and induces apoptosis in human cancer cells, *Arch Pharm Res* 24, 441–445, 2001.
124. Lu J, Ho CH, Ghai G, and Chen KY, Resveratrol analog, 3,4,5, 4′-tetrahydroxystilbene, differentially induces pro-apoptotic p53/Bax gene expression and inhibits the growth of transformed cells but not their normal counterparts, *Carcinogenesis* 22, 321–328, 2001.
125. Lancon A, Delma D, Osman H, Thenot JP, Jannin B, and Latruffe N, Human hepatic cell uptake of resveratrol: involvement of both passive diffusion and carrier-mediated process, *Biochem Biophys Res Commun* 316, 1132–1137, 2004.
126. Li Y, Shin YG, Yu C, Kosmeder JW, Hirschelman WH, Pezzuto JM, and van Breemen RB, Increasing the throughput and productivity of Caco-2 cell permeability assays using liquid chromatography-mass spectrometry: application to resveratrol absorption and metabolism, *Comb Chem High Throughput Screen* 6, 757–767, 2003.
127. Wolter F, Turchanowa L, and Stein J, Resveratrol-induced modification of polyamine metabolism is accompanied by induction of c-Fos, *Carcinogenesis* 24, 469–474, 2003.
128. Latruffe N, Delmas D, Jannin B, Malki MC, Passilly-Degrace P, and Berlot JP, Molecular analysis on the chemopreventive properties of resveratrol, a plant polyphenol microcomponent, *Int J Mol Med* 10, 755–760, 2002.
129. Wu SL, Sun ZJ, Yu L, Meng KW, Qin XL, and Pan CE, Effect of resveratrol and in combination with 5-FU on murine liver cancer, *World J Gastroenterol* 10, 3048–3052, 2004.

130. Ahmad KA, Clement MV, Hanif IM, and Pervaiz S, Resveratrol inhibits drug-induced apoptosis in human leukemia cells by creating an intracellular milieu nonpermissive for death execution, *Cancer Res* 64, 1452–1459, 2004.
131. Jazirehi AR and Bonavida B, Resveratrol modifies the expression of apoptotic regulatory proteins and sensitizes non-Hodgkin's lymphoma and multiple myeloma cell lines to paclitaxel-induced apoptosis, *Mol Cancer Ther* 3, 71–84, 2004.
132. Kubota T, Uemura Y, Kobayashi M, and Taguchi H, Combined effects of resveratrol and paclitaxel on lung cancer cells, *Anticancer Res* 23, 4039–4046, 2003.
133. Nicolini G, Rigolio R, Scuteri A, Miloso M, Saccomanno D, Cavaletti G, and Tredici G, Effect of *trans*-resveratrol on signal transduction pathways involved in paclitaxel-induced apoptosis in human neuroblastoma SH-SY5Y cells, *Neurochem Int* 42, 419–429, 2003.
134. Sun ZJ, Pan CE, Liu HS, and Wang GJ, Anti-hepatoma activity of resveratrol *in vitro*, *World J Gastroenterol* 8, 79–81, 2002.
135. Nicolini G, Rigolio R, Miloso M, Bertelli AA, and Tredici G, Anti-apoptotic effect of *trans*-resveratrol on paclitaxel-induced apoptosis in the human neuroblastoma SH-SY5Y cell line, *Neurosci Lett* 302, 41–44, 2001.
136. Shi T, Liou LS, Sadhukhan P, Duan ZH, Novick AC, Hissong JG, Almasan A, and Didonato JA, Effects of resveratrol on gene expression in renal cell carcinoma, *Cancer Biol Ther* 3, 882–888, 2004.
137. Yang SH, Kim JS, Oh TJ, Kim MS, Lee SW, Woo SK, Cho HS, Choi YH, Kim YH, Rha SY, Chung HC, and An SW, Genome-scale analysis of resveratrol-induced gene expression profile in human ovarian cancer cells using a cDNA microarray, *Int J Oncol* 22, 741–750, 2003.
138. Narayanan BA, Narayanan NK, Re GG, and Nixon DW, Differential expression of genes induced by resveratrol in LNCaP cells: P53-mediated molecular targets, *Int J Cancer* 104, 204–212, 2003.
139. Narayanan BA, Narayanan NK, Stoner GD, and Bullock BP, Interactive gene expression pattern in prostate cancer cells exposed to phenolic antioxidants, *Life Sci* 70, 1821–1839, 2002.
140. Pezzuto JM, Resveratrol: a whiff that induces a biologically specific tsunami, *Cancer Biol Ther* 3, 889–890, 2004.
141. Kapadia GJ, Azuine MA, Tokuda H, Takasaki M, Mukainaka T, Konoshima T, and Nishino H, Chemopreventive effect of resveratrol, sesamol, sesame oil and sunflower oil in the Epstein-Barr virus early antigen activation assay and the mouse skin two-stage carcinogenesis, *Pharmacol Res* 45, 499–505, 2002.
142. Aziz M, Afaq F, and Ahmad N, Prevention of ultraviolet B radiation damage by resveratrol in mouse skin is mediated via modulation in survivin, *Photochem Photobiol* 81, 25–31, 2005.
143. Reagan-Shaw S, Afaq F, Aziz MH, and Ahmad N, Modulations of critical cell cycle regulatory events during chemoprevention of ultraviolet B-mediated responses by resveratrol in SKH-1 hairless mouse skin, *Oncogene* 23, 5151–5160, 2004.
144. Adhami VM, Afaq F, and Ahmad N, Suppression of ultraviolet B exposure-mediated activation of NF-κB in normal human keratinocytes by resveratrol, *Neoplasia* 5, 74–82, 2003.

145. Afaq F, Adhami VM, and Ahmad N, Prevention of short-term ultraviolet B radiation-mediated damages by resveratrol in SKH-1 hairless mice, *Toxicol Appl Pharmacol* 186, 28–37, 2003.
146. Revel A, Raanani H, Younglai E, Xu J, Rogers I, Han R, Savouret JF, and Casper RF, Resveratrol, a natural aryl hydrocarbon receptor antagonist, protects lung from DNA damage and apoptosis caused by benzo[*a*]pyrene, *J Appl Toxicol* 23, 255–261, 2003.
147. Berge G, Ovrebo S, Eilertsen E, Haugen A, and Mollerup S, Analysis of resveratrol as a lung cancer chemopreventive agent in A/J mice exposed to benzo[*a*]pyrene, *Br J Cancer* 91, 1380–1383, 2004.
148. Hecht SS, Kenney PM, Wang M, Trushin N, Agarwal S, Rao AV, and Upadhyaya P, Evaluation of butylated hydroxyanisole, *myo*-inositol, curcumin, esculetin, resveratrol and lycopene as inhibitors of benzo[*a*]pyrene plus 4-(methylnitrosamino)-1-(3-pyridyl)-1-butanone-induced lung tumorigenesis in A/J mice, *Cancer Lett* 137, 123–130, 1999.
149. Ziegler CC, Rainwater L, Whelan J, and McEntee MF, Dietary resveratrol does not affect intestinal tumorigenesis in $Apc^{Min/+}$ mice, *J Nutr* 134, 5–10, 2004.
150. Schneider Y, Duranton B, Gosse F, Schleiffer R, Seiler N, and Raul F, Resveratrol inhibits intestinal tumorigenesis and modulates host-defense-related gene expression in an animal model of human familial adenomatous polyposis, *Nutr Cancer* 39, 102–107, 2001.
151. Tessitore L, Davit A, Sarotto I, and Caderni G, Resveratrol depresses the growth of colorectal aberrant crypt foci by affecting Bax and $p21^{CIP}$ expression, *Carcinogenesis* 21, 1619–1622, 2000.
152. Sato M, Pei RJ, Yuri T, Danbara N, Nakane Y, and Tsubura A, Prepubertal resveratrol exposure accelerates *N*-methyl-*N*-nitrosourea-induced mammary carcinoma in female Sprague-Dawley rats, *Cancer Lett* 202, 137–145, 2003.
153. Bhat KP, Lantvit D, Christov K, Mehta RG, Moon RC, and Pezzuto JM, Estrogenic and antiestrogenic properties of resveratrol in mammary tumor models, *Cancer Res* 61, 7456–7463, 2001.
154. Banerjee S, Bueso-Ramos C, and Aggarwal BB, Suppression of 7,12-dimethylbenz(*a*)anthracene-induced mammary carcinogenesis in rats by resveratrol: role of nuclear factor-κB, cyclooxygenase 2, and matrix metalloprotease 9, *Cancer Res* 62, 4945–4954, 2002.
155. Li ZG, Hong T, Shimada Y, Komoto I, Kawabe A, Ding Y, Kaganoi J, Hashimoto Y, and Imamura M, Suppression of *N*-nitrosomethylbenzylamine (NMBA)-induced esophageal tumorigenesis in F344 rats by resveratrol, *Carcinogenesis* 23, 1531–1536, 2002.
156. Chen Y, Tseng SH, Lai HS, and Chen WJ, Resveratrol-induced cellular apoptosis and cell cycle arrest in neuroblastoma cells and antitumor effects on neuroblastoma in mice, *Surgery* 136, 57–66, 2004.
157. Yu L, Sun ZJ, Wu SL, and Pan CE, Effect of resveratrol on cell cycle proteins in murine transplantable liver cancer, *World J Gastroenterol* 9, 2341–2343, 2003.
158. Miura D, Miura Y, and Yagasaki K, Hypolipidemic action of dietary resveratrol, a phytoalexin in grapes and red wine, in hepatoma-bearing rats, *Life Sci* 73, 1393–1400, 2003.

159. Carbo N, Costelli P, Baccino FM, Lopez-Soriano FJ, and Argiles JM, Resveratrol, a natural product present in wine, decreases tumour growth in a rat tumour model, *Biochem Biophys Res Commun* 254, 739–743, 1999.
160. Liu HS, Pan CE, Yang W, and Liu XM, Antitumor and immunomodulatory activity of resveratrol on experimentally implanted tumor of H22 in Balb/c mice, *World J Gastroenterol* 9, 1474–1476, 2003.
161. Kimura Y and Okuda H, Resveratrol isolated from *Polygonum cuspidatum* root prevents tumor growth and metastasis to lung and tumor-induced neovascularization in Lewis lung carcinoma-bearing mice, *J Nutr* 131, 1844–1849, 2001.
162. Bove K, Lincoln DW, and Tsan MF, Effect of resveratrol on growth of 4T1 breast cancer cells *in vitro* and *in vivo*, *Biochem Biophys Res Commun* 291, 1001–1005, 2002.
163. Gao X, Xu YX, Divine G, Janakiraman N, Chapman RA, and Gautam SC, Disparate *in vitro* and *in vivo* antileukemic effects of resveratrol, a natural polyphenolic compound found in grapes, *J Nutr* 132, 2076–2081, 2002.
164. Blumberg J and Block G, The α-tocopherol, beta-carotene cancer prevention study in Finland, *Nutr Rev* 52, 242–250, 1994.
165. Omenn GS, Goodman GE, Thornquist MD, Balmes J, Cullen MR, Glass A, Keogh JP, Meyskens FL Jr, Valanis B, Williams JH Jr, Barnhart S, Cherniack MG, Brodkin CA, and Hammar S, Risk factors for lung cancer and for intervention effects in CARET, the β-carotene and retinol efficacy trial, *J Natl Cancer Inst* 88, 1550–1559, 1996.
166. Schmitt E, Lehmann L, Metzler M, and Stopper H, Hormonal and genotoxic activity of resveratrol, *Toxicol Lett* 136, 133–142, 2002.
167. Hsieh T, Halicka D, Lu X, Kunicki J, Guo J, Darzynkiewicz Z, and Wu J, Effects of resveratrol on the G_0-G_1 transition and cell cycle progression of mitogenically stimulated human lymphocytes, *Biochem Biophys Res Commun* 297, 1311–1317, 2002.
168. Castello L and Tessitore L, Reservatol inhibits cell cycle progression in U937 cells, *Oncol Rep* 13, 133–137, 2005.
169. Narayanan NK, Narayanan BA, and Nixon DW, Resveratrol-induced cell growth inhibition and apoptosis is associated with modulation of phosphoglycerate mutase B in human prostate cancer cells: two-dimensional sodium dodecyl sulfate-polyacrylamide gel electrophoresis and mass spectrometry evaluation, *Cancer Detection Prev* 28, 443–452, 2004.
170. Wolter F, Ulrich S, and Stein J, Molecular mechanisms of the chemopreventive effects of resveratrol and its analogs in colorectal cancer: key role of polyamines?, *J Nutr* 134, 3219–3222, 2004.
171. Shih A, Zhang S, Cao HJ, Boswell S, Wu YH, Tang HY, Lennartz MR, Davis FB, Davis PJ, and Lin HY, Inhibitory effect of epidermal growth factor on resveratrol-induced apoptosis in prostate cancer cells is mediated by protein kinase C-α, *Mol Cancer Ther* 3, 1355–1364, 2004.
172. Scifo C, Cardile V, Russo A, Consoli R, Vancheri C, Capasso F, Vanella A, and Renis M, Resveratrol and propolis as necrosis or apoptosis inducers in human prostate carcinoma cells, *Oncol Res* 14, 415–426, 2004.
173. Delmas D, Rebe C, Micheau O, Athias A, Gambert P, Grazide S, Laurent G, Latruffe N, and Solary E, Redistribution of CD95, DR4 and DR5 in rafts

accounts for the synergistic toxicity of resveratrol and death receptor ligands in colon carcinoma cells, *Oncogene* 23, 8979–8986, 2004.
174. Kundu JK and Surh YJ, Molecular basis of chemoprevention by resveratrol: NF-κB and AP-1 as potential targets, *Mutation Res* 555, 65–80, 2004.
175. Bode AM and Dong Z, Targeting signal transduction pathways by chemopreventive agents, *Mutation Res* 555, 33–51, 2004.
176. Yuan H, Pan Y, and Young CY, Overexpression of c-Jun induced by quercetin and resveratrol inhibits the expression and function of the androgen receptor in human prostate cancer cells, *Cancer Lett* 213, 155–163, 2004.
177. Cao Z, Fang J, Xia C, Shi X, and Jiang BH, *trans*-3,4,5′-Trihydroxystilbene inhibits hypoxia-inducible factor 1α and vascular endothelial growth factor expression in human ovarian cancer cells, *Clin Cancer Res* 10, 5253–5263, 2004.
178. Fulda S and Debatin KM, Sensitization for anticancer drug-induced apoptosis by the chemopreventive agent resveratrol, *Oncogene* 23, 6702–6711, 2004.
179. Hyun JY, Chun YS, Kim TY, Kim HL, Kim MS, and Park JW, Hypoxia-inducible factor 1alpha-mediated resistance to phenolic anticancer, *Chemotherapy* 50, 119–126, 2004.
180. Schneider Y, Fischer B, Coelho D, Roussi S, Gosse F, Bischoff P, and Raul F, (*Z*)-3,5,4′-tri-*O*-Methyl-resveratrol, induces apoptosis in human lymphoblastoid cells independently of their p53 status, *Cancer Lett* 211, 155–161, 2004.
181. Quiney C, Dauzonne D, Kern C, Fourneron JD, Izard JC, Mohammad RM, Kolb JP, and Billard C, Flavones and polyphenols inhibit the NO pathway during apoptosis of leukemia B-cells, *Leukemia Res* 28, 851–861, 2004.
182. Zhang S, Cao HJ, Davis FB, Tang HY, Davis PJ, and Lin HY, Estrogen inhibits resveratrol-induced post-translational modification of p53 and apoptosis in breast cancer cells, *Br J Cancer* 91, 178–185, 2004.
183. Liu J, Wang Q, Wu DC, Wang XW, Sun Y, Chen XY, Zhang KL, and Li H, Differential regulation of CYP1A1 and CYP1B1 expression in resveratrol-treated human medulloblastoma cells, *Neurosci Lett* 363, 257–261, 2004.
184. Laux MT, Aregullin M, Berry JP, Flanders JA, and Rodriguez E, Identification of a p53-dependent pathway in the induction of apoptosis of human breast cancer cells by the natural product, resveratrol, *J Alternative Complementary Med* 10, 235–239, 2004.
185. Berge G, Ovrebo S, Botnen IV, Hewer A, Phillips DH, Haugen A, and Mollerup S, Resveratrol inhibits benzo[*a*]pyrene-DNA adduct formation in human bronchial epithelial cells, *Br J Cancer* 91, 333–338, 2004.
186. Liontas A and Yeger H, Curcumin and resveratrol induce apoptosis and nuclear translocation and activation of p53 in human neuroblastoma, *Anticancer Res* 24, 987–998, 2004.
187. Jeong WS, Kim IW, Hu R, and Kong AN, Modulatory properties of various natural chemopreventive agents on the activation of NF-κB signaling pathway, *Pharm Res* 21, 661–670, 2004.
188. Jeong WS, Kim IW, Hu R, and Kong AN, Modulation of AP-1 by natural chemopreventive compounds in human colon HT-29 cancer cell line, *Pharm Res* 21, 649–660, 2004.
189. Baatout S, Derradji H, Jacquet P, Ooms D, Michaux A, and Mergeay M, Enhanced radiation-induced apoptosis of cancer cell lines after treatment with resveratrol, *Int J Mol Med* 13, 895–902, 2004.

190. Feng YH, Zhu YN, Liu J, Ren YX, Xu JY, Yang YF, Li XY, and Zou JP, Differential regulation of resveratrol on lipopolysacchride-stimulated human macrophages with or without IFN-γ pre-priming, *Int Immunopharmacol* 4, 713–720, 2004.
191. Cooray HC, Janvilisri T, van Veen HW, Hladky SB, and Barrand MA, Interaction of the breast cancer resistance protein with plant polyphenols, *Biochem Biophys Res Commun* 317, 269–275, 2004.
192. Cheung CY, Chen J, and Chang TK, Evaluation of a real-time polymerase chain reaction method for the quantification of CYP1B1 gene expression in MCF-7 human breast carcinoma cells, *J Pharmacol Toxicol Methods* 49, 97–104, 2004.
193. Opipari AW Jr, Tan L, Boitano AE, Sorenson DR, Aurora A, and Liu JR, Resveratrol-induced autophagocytosis in ovarian cancer cells, *Cancer Res* 64, 696–703, 2004.
194. Stewart JR and O'Brian CA, Resveratrol antagonizes EGFR-dependent Erk1/2 activation in human androgen-independent prostate cancer cells with associated isozyme-selective PKC α inhibition, *Invest New Drugs* 22, 107–117, 2004.
195. Fulda S and Debatin KM, Sensitization for tumor necrosis factor-related apoptosis-inducing ligand-induced apoptosis by the chemopreventive agent resveratrol, *Cancer Res* 64, 337–346, 2004.
196. Kim YA, Choi BT, Lee YT, Park DI, Rhee SH, Park KY, and Choi YH, Resveratrol inhibits cell proliferation and induces apoptosis of human breast carcinoma MCF-7 cells, *Oncol Rep* 11, 441–416, 2004.
197. Carraway RE, Hassan S, and Cochrane DE, Polyphenolic antioxidants mimic the effects of 1,4-dihydropyridines on neurotensin receptor function in PC3 cells, *J Pharmacol Exp Ther* 309, 92–101, 2004.
198. Woo JH, Lim JH, Kim YH, Suh SI, Min do S, Chang JS, Lee YH, Park JW, and Kwon TK, Resveratrol inhibits phorbol myristate acetate-induced matrix metalloproteinase-9 expression by inhibiting JNK and PKC δ signal transduction, *Oncogene* 23, 1845–1853, 2004.
199. Sala G, Minutolo F, Macchia M, Sacchi N, and Ghidoni R, Resveratrol structure and ceramide-associated growth inhibition in prostate cancer cells, *Drugs Exp Clin Res* 29, 263–269, 2003.
200. Bruno R, Ghisolfi L, Priulla M, Nicolin A, and Bertelli A, Wine and tumors: study of resveratrol, *Drugs Exp Clin Res* 29, 257–261, 2003.
201. Cardile V, Scifo C, Russo A, Falsaperla M, Morgia G, Motta M, Renis M, Imbriani E, and Silvestre G, Involvement of HSP70 in resveratrol-induced apoptosis of human prostate cancer, *Anticancer Res* 23, 4921–4926, 2003.
202. Kim YA, Rhee SH, Park KY, and Choi YH, Antiproliferative effect of resveratrol in human prostate carcinoma cells, *J Med Food* 6, 273–280, 2003.
203. Kang JH, Park YH, Choi SW, Yang EK, and Lee WJ, Resveratrol derivatives potently induce apoptosis in human promyelocytic leukemia cells, *Exp Mol Med* 35, 467–474, 2003.
204. Wang Q, Li H, Wang XW, Wu DC, Chen XY, and Liu J, Resveratrol promotes differentiation and induces Fas-independent apoptosis of human medulloblastoma cells, *Neurosci Lett* 351, 83–86, 2003.
205. Scarlatti F, Sala G, Somenzi G, Signorelli P, Sacchi N, and Ghidoni R, Resveratrol induces growth inhibition and apoptosis in metastatic

breast cancer cells via *de novo* ceramide signaling, *FASEB J* 17, 2339–2341, 2003.

206. Kaneuchi M, Sasaki M, Tanaka Y, Yamamoto R, Sakuragi N, and Dahiya R, Resveratrol suppresses growth of Ishikawa cells through down-regulation of EGF, *Int J Oncol* 23, 1167–1172, 2003.
207. Kim YA, Lee WH, Choi TH, Rhee SH, Park KY, and Choi YH, Involvement of $p21^{WAF1/CIP1}$, pRB, Bax and NF-κB in induction of growth arrest and apoptosis by resveratrol in human lung carcinoma A549 cells, *Int J Oncol* 23, 1143–1149, 2003.
208. Granados-Soto V, Pleiotropic effects of resveratrol, *Drug News Perspect* 16, 299–307, 2003.
209. Delmas D, Rebe C, Lacour S, Filomenko R, Athias A, Gambert P, Cherkaoui-Malki M, Jannin B, Dubrez-Daloz L, Latruffe N, and Solary E, Resveratrol-induced apoptosis is associated with Fas redistribution in the rafts and the formation of a death-inducing signaling complex in colon cancer cells, *J Biol Chem* 278, 41482–41490, 2003.
210. Fustier P, Le Corre L, Chalabi N, Vissac-Sabatier C, Communal Y, Bignon YJ, and Bernard-Gallon DJ, Resveratrol increases BRCA1 and BRCA2 mRNA expression in breast tumour cell lines, *Br J Cancer* 89, 168–172, 2003.
211. El-Mowafy AM and Alkhalaf M, Resveratrol activates adenylyl-cyclase in human breast cancer cells: a novel, estrogen receptor-independent cytostatic mechanism, *Carcinogenesis* 24, 869–873, 2003.
212. Bernhard D, Schwaiger W, Crazzolara R, Tinhofer I, Kofler R, and Csordas A, Enhanced MTT-reducing activity under growth inhibition by resveratrol in CEM-C7H2 lymphocytic leukemia cells, *Cancer Lett* 195, 193–199, 2003.
213. Estrov Z, Shishodia S, Faderl S, Harris D, Van Q, Kantarjian HM, Talpaz M, and Aggarwal BB, Resveratrol blocks interleukin-1β-induced activation of the nuclear transcription factor NF-κB, inhibits proliferation, causes S-phase arrest, and induces apoptosis of acute myeloid leukemia cells, *Blood* 102, 987–995, 2003.
214. Liang YC, Tsai SH, Chen L, Lin-Shiau SY, and Lin JK, Resveratrol-induced G_2 arrest through the inhibition of CDK7 and $p34^{CDC2}$ kinases in colon carcinoma HT29 cells, *Biochem Pharmacol* 65, 1053–1060, 2003.
215. Zhou HB, Yan Y, Sun YN, and Zhu JR, Resveratrol induces apoptosis in human esophageal carcinoma cells, *World J Gastroenterol* 9, 408–411, 2003.
216. Niles RM, McFarland M, Weimer MB, Redkar A, Fu YM, and Meadows GG, Resveratrol is a potent inducer of apoptosis in human melanoma cells, *Cancer Lett* 190, 157–163, 2003.
217. Hayashibara T, Yamada Y, Nakayama S, Harasawa H, Tsuruda K, Sugahara K, Miyanishi T, Kamihira S, Tomonaga M, and Maita T, Resveratrol induces downregulation in survivin expression and apoptosis in HTLV-1-infected cell lines: a prospective agent for adult T cell leukemia chemotherapy, *Nutr Cancer* 44, 193–201, 2002.
218. Roy M, Chakraborty S, Siddiqi M, and Bhattacharya RK, Induction of apoptosis in tumor cells by natural phenolic compounds, *Asian Pac J Cancer Prev* 3, 61–67, 2002.

219. Billard C, Izard JC, Roman V, Kern C, Mathiot C, Mentz F, and Kolb JP, Comparative antiproliferative and apoptotic effects of resveratrol, ε-viniferin and vine-shots derived polyphenols (vineatrols) on chronic B lymphocytic leukemia cells and normal human lymphocytes, *Leuk Lymphoma* 43, 1991–2002, 2002.
220. Ding XZ and Adrian TE, Resveratrol inhibits proliferation and induces apoptosis in human pancreatic cancer cells, *Pancreas* 25, 71–76, 2002.
221. Kuo PL, Chiang LC, and Lin CC, Resveratrol-induced apoptosis is mediated by p53-dependent pathway in Hep G_2 cells, *Life Sci* 72, 23–34, 2002.
222. Pozo-Guisado E, Alvarez-Barrientos A, Mulero-Navarro S, Santiago-Josefat B, and Fernandez-Salguero PM, The antiproliferative activity of resveratrol results in apoptosis in MCF-7 but not in MDA-MB-231 human breast cancer cells: cell-specific alteration of the cell cycle, *Biochem Pharmacol* 64, 1375–1386, 2002.
223. Mahyar-Roemer M, Kohler H, and Roemer K, Role of Bax in resveratrol-induced apoptosis of colorectal carcinoma cells, *BMC Cancer* 2, 27, 2002.
224. Melzig MF and Escher F, Induction of neutral endopeptidase and angiotensin-converting enzyme activity of SK-N-SH cells *in vitro* by quercetin and resveratrol, *Pharmazie* 57, 556–558, 2002.
225. Dubuisson JG, Dyess DL, and Gaubatz JW, Resveratrol modulates human mammary epithelial cell *O*-acetyltransferase, sulfotransferase, and kinase activation of the heterocyclic amine carcinogen *N*-hydroxy-PhIP, *Cancer Lett* 182, 27–32, 2002.
226. Ferry-Dumazet H, Garnier O, Mamani-Matsuda M, Vercauteren J, Belloc F, Billiard C, Dupouy M, Thiolat D, Kolb JP, Marit G, Reiffers J, and Mossalayi MD, Resveratrol inhibits the growth and induces the apoptosis of both normal and leukemic hematopoietic cells, *Carcinogenesis* 23, 1327–1333, 2002.
227. Lin HY, Shih A, Davis FB, Tang HY, Martino LJ, Bennett JA, and Davis PJ, Resveratrol induced serine phosphorylation of p53 causes apoptosis in a mutant p53 prostate cancer cell line, *J Urol* 168, 748–755, 2002.
228. Delmas D, Passilly-Degrace P, Jannin B, Malki MC, and Latruffe N, Resveratrol, a chemopreventive agent, disrupts the cell cycle control of human SW480 colorectal tumor cells, *Int J Mol Med* 10, 193–199, 2002.
229. Wolter F and Stein J, Resveratrol enhances the differentiation induced by butyrate in caco-2 colon cancer cells, *J Nutr* 132, 2082–2086, 2002.
230. Asou H, Koshizuka K, Kyo T, Takata N, Kamada N, and Koeffier HP, Resveratrol, a natural product derived from grapes, is a new inducer of differentiation in human myeloid leukemias, *Int J Hematol* 75, 528–533, 2002.
231. Roman V, Billard C, Kern C, Ferry-Dumazet H, Izard JC, Mohammad R, Mossalayi DM, and Kolb JP, Analysis of resveratrol-induced apoptosis in human B-cell chronic leukaemia, *Br J Haematol* 117, 842–851, 2002.
232. Rimando AM, Cuendet M, Desmarchelier C, Mehta RG, Pezzuto JM, and Duke SO, Cancer chemopreventive and antioxidant activities of pterostilbene, a naturally occurring analogue of resveratrol, *J Agric Food Chem* 50, 3453–3457, 2002.

233. Kuwajerwala N, Cifuentes E, Gautam S, Menon M, Barrack ER, and Reddy GP, Resveratrol induces prostate cancer cell entry into S phase and inhibits DNA synthesis, *Cancer Res* 62, 2488–2492, 2002.
234. Holian O, Wahid S, Atten MJ, and Attar BM, Inhibition of gastric cancer cell proliferation by resveratrol: role of nitric oxide, *Am J Physiol Gastrointest Liver Physiol* 282, G809–816, 2002.
235. She QB, Huang C, Zhang Y, and Dong Z, Involvement of c-jun NH_2-terminal kinases in resveratrol-induced activation of p53 and apoptosis, *Mol Carcinog* 33, 244–250, 2002.
236. Joe AK, Liu H, Suzui M, Vural ME, Xiao D, and Weinstein IB, Resveratrol induces growth inhibition, S-phase arrest, apoptosis, and changes in biomarker expression in several human cancer cell lines, *Clin Cancer Res* 8, 893–903, 2002.
237. Shih A, Davis FB, Lin HY, and Davis PJ, Resveratrol induces apoptosis in thyroid cancer cell lines via a MAPK- and p53-dependent mechanism, *J Clin Endocrinol Metab* 87, 1223–1232, 2002.
238. Lee SH, Ryu SY, Kim HB, Kim MY, and Chun YJ, Induction of apoptosis by 3,4′-dimethoxy-5-hydroxystilbene in human promyeloid leukemic HL-60 cells, *Planta Med* 68, 123–127, 2002.
239. Pendurthi UR, Meng F, Mackman N, and Rao LV, Mechanism of resveratrol-mediated suppression of tissue factor gene expression, *Thromb Haemost* 87, 155–162, 2002.
240. Wolter F, Clausnitzer A, Akoglu B, and Stein J, Piceatannol, a natural analog of resveratrol, inhibits progression through the S phase of the cell cycle in colorectal cancer cell lines, *J Nutr* 132, 298–302, 2002.
241. Zoberi I, Bradbury CM, Curry HA, Bisht KS, Goswami PC, Roti Roti JL, and Gius D, Radiosensitizing and anti-proliferative effects of resveratrol in two human cervical tumor cell lines, *Cancer Lett* 175, 165–173, 2002.
242. Serrero G and Lu R, Effect of resveratrol on the expression of autocrine growth modulators in human breast cancer cells, *Antioxid Redox Signal* 3, 969–979, 2001.
243. Mahyar-Roemer M, Katsen A, Mestres P, and Roemer K, Resveratrol induces colon tumor cell apoptosis independently of p53 and precede by epithelial differentiation, mitochondrial proliferation and membrane potential collapse, *Int J Cancer* 94, 615–622, 2001.
244. Atten MJ, Attar BM, Milson T, and Holian O, Resveratrol-induced inactivation of human gastric adenocarcinoma cells through a protein kinase C-mediated mechanism, *Biochem Pharmacol* 62, 1423–1432, 2001.
245. Wieder T, Prokop A, Bagci B, Essmann F, Bernicke D, Schulze-Osthoff K, Dorken B, Schmalz HG, Daniel PT, and Henze G, Piceatannol, a hydroxylated analog of the chemopreventive agent resveratrol, is a potent inducer of apoptosis in the lymphoma cell line BJAB and in primary, leukemic lymphoblasts, *Leukemia* 15, 1735–1742, 2001.
246. Adhami VM, Afaq F, and Ahmad N, Involvement of the retinoblastoma (pRb)-E2F/DP pathway during antiproliferative effects of resveratrol in human epidermoid carcinoma (A431) cells, *Biochem Biophys Res Commun* 288, 579–585, 2001.
247. Lee JE and Safe S, Involvement of a post-transcriptional mechanism in the inhibition of CYP1A1 expression by resveratrol in breast cancer cells, *Biochem Pharmacol* 62, 1113–1124, 2001.

248. Park JW, Choi YJ, Suh SI, Baek WK, Suh MH, Jin IN, Min DS, Woo JH, Chang JS, Passaniti A, Lee YH, and Kwon TK, Bcl-2 overexpression attenuates resveratrol-induced apoptosis in U937 cells by inhibition of caspase-3 activity, *Carcinogenesis* 22, 1633–1639, 2001.
249. Sgambato A, Ardito R, Faraglia B, Boninsegna A, Wolf FI, and Cittadini A, Resveratrol, a natural phenolic compound, inhibits cell proliferation and prevents oxidative DNA damage, *Mutation Res* 496, 171–180, 2001.
250. Wolter F, Akoglu B, Clausnitzer A, and Stein J, Downregulation of the cyclin D1/Cdk4 complex occurs during resveratrol-induced cell cycle arrest in colon cancer cell lines, *J Nutr* 131, 2197–2203, 2001.
251. Dorrie J, Gerauer H, Wachter Y, and Zunino SJ, Resveratrol induces extensive apoptosis by depolarizing mitochondrial membranes and activating caspase-9 in acute lymphoblastic leukemia cells, *Cancer Res* 61, 4731–4739, 2001.
252. De Ledinghen V, Monvoisin A, Neaud V, Krisa S, Payrastre B, Bedin C, Desmouliere A, Bioulac-Sage P, and Rosenbaum J, *trans*-Resveratrol, a grapevine-derived polyphenol, blocks hepatocyte growth factor-induced invasion of hepatocellular carcinoma cells, *Int J Oncol* 19, 83–88, 2001.
253. Kozuki Y, Miura Y, and Yagasaki K, Resveratrol suppresses hepatoma cell invasion independently of its anti-proliferative action, *Cancer Lett* 167, 151–156, 2001.
254. Ahmad N, Adhami VM, Afaq F, Feyes DK, and Mukhtar H, Resveratrol causes WAF-1/p21-mediated G_1-phase arrest of cell cycle and induction of apoptosis in human epidermoid carcinoma A431 cells, *Clin Cancer Res* 7, 1466–1473, 2001.
255. Nakagawa H, Kiyozuka Y, Uemura Y, Senzaki H, Shikata N, Hioki K, and Tsubura A, Resveratrol inhibits human breast cancer cell growth and may mitigate the effect of linoleic acid, a potent breast cancer cell stimulator, *J Cancer Res Clin Oncol* 127, 258–264, 2001.
256. Mollerup S, Ovrebo S, and Haugen A, Lung carcinogenesis: resveratrol modulates the expression of genes involved in the metabolism of PAH in human bronchial epithelial cells, *Int J Cancer* 92, 18–25, 2001.
257. She QB, Bode AM, Ma WY, Chen NY, and Dong Z, Resveratrol-induced activation of p53 and apoptosis is mediated by extracellular-signal-regulated protein kinases and p38 kinase, *Cancer Res* 61, 1604–1610, 2001.
258. Park JW, Choi YJ, Jang MA, Lee YS, Jun DY, Suh SI, Baek WK, Suh MH, Jin IN, and Kwon TK, Chemopreventive agent resveratrol, a natural product derived from grapes, reversibly inhibits progression through S and G_2 phases of the cell cycle in U937 cells, *Cancer Lett* 163, 43–49, 2001.
259. Kampa M, Hatzoglou A, Notas G, Damianaki A, Bakogeorgou E, Gemetzi C, Kouroumalis E, Martin PM, and Castanas E, Wine antioxidant polyphenols inhibit the proliferation of human prostate cancer cell lines, *Nutr Cancer* 37, 223–233, 2000.
260. Bernhard D, Tinhofer I, Tonko M, Hubl H, Ausserlechner MJ, Greil R, Kofler R, and Csordas A, Resveratrol causes arrest in the S-phase prior to Fas-independent apoptosis in CEM-C7H2 acute leukemia cells, *Cell Death Differentiation* 7, 834–842, 2000.
261. Nielsen M, Ruch RJ, and Vang O, Resveratrol reverses tumor-promoter-induced inhibition of gap-junctional intercellular communication, *Biochem Biophys Res Commun* 275, 804–809, 2000.

262. Schneider Y, Vincent F, Duranton B, Badolo L, Gosse F, Bergmann C, Seiler N, and Raul F, Anti-proliferative effect of resveratrol, a natural component of grapes and wine, on human colonic cancer cells, *Cancer Lett* 158, 85–91, 2000.
263. Holmes-McNary M and Baldwin AS Jr, Chemopreventive properties of trans-resveratrol are associated with inhibition of activation of the IκB kinase, *Cancer Res* 60, 3477–3483, 2000.
264. Delmas D, Jannin B, Malki MC, and Latruffe N, Inhibitory effect of resveratrol on the proliferation of human and rat hepatic derived cell lines, *Oncol Rep* 7, 847–852, 2000.
265. Tsan MF, White JE, Maheshwari JG, Bremner TA, and Sacco J, Resveratrol induces Fas signalling-independent apoptosis in THP-1 human monocytic leukaemia cells, *Br J Haematol* 109, 405–412, 2000.
266. Hsieh TC and Wu JM, Grape-derived chemopreventive agent resveratrol decreases prostate-specific antigen (PSA) expression in LNCaP cells by an androgen receptor (AR)-independent mechanism, *Anticancer Res* 20, 225–228, 2000.
267. Godichaud S, Krisa S, Couronne B, Dubuisson L, Merillon JM, Desmouliere A, and Rosenbaum J, Deactivation of cultured human liver myofibroblasts by *trans*-resveratrol, a grapevine-derived polyphenol, *Hepatology* 31, 922–931, 2000.
268. Elattar TM and Virji AS, Thc effect of red wine and its components on growth and proliferation of human oral squamous carcinoma cells, *Anticancer Res* 19, 5407–5414, 1999.
269. Subbaramaiah K, Michaluart P, Chung WJ, Tanabe T, Telang N, and Dannenberg AJ, Resveratrol inhibits cyclooxygenase-2 transcription in human mammary epithelial cells, *Ann NY Acad Sci* 889, 214–223, 1999.
270. Mitchell SH, Zhu W, and Young CY, Resveratrol inhibits the expression and function of the androgen receptor in LNCaP prostate cancer cells, *Cancer Res* 59, 5892–5895, 1999.
271. Ulsperger E, Hamilton G, Raderer M, Baumgartner G, Hejna M, Hoffmann O, and Mallinger R, Resveratrol pretreatment desensitizes AHTO-7 human osteoblasts to growth stimulation in response to carcinoma cell supernatants, *Int J Oncol* 15, 955–959, 1999.
272. Surh YJ, Hurh YJ, Kang JY, Lee E, Kong G, and Lee SJ, Resveratrol, an antioxidant present in red wine, induces apoptosis in human promyelocytic leukemia (HL-60) cells, *Cancer Lett* 140, 1–10, 1999.
273. Hsieh TC, Burfeind P, Laud K, Backer JM, Traganos F, Darzynkiewicz Z, and Wu JM, Cell cycle effects and control of gene expression by resveratrol in human breast carcinoma cell lines with different metastatic potentials, *Int J Oncol* 15, 245–252, 1999.
274. Hsieh TC and Wu JM, Differential effects on growth, cell cycle arrest, and induction of apoptosis by resveratrol in human prostate cancer cell lines, *Exp Cell Res* 249, 109–115, 1999.
275. Miloso M, Bertelli AA, Nicolini G, and Tredici G, Resveratrol-induced activation of the mitogen-activated protein kinases, ERK1 and ERK2, in human neuroblastoma SH-SY5Y cells, *Neurosci Lett* 264, 141–144, 1999.
276. ElAttar TM and Virji AS, Modulating effect of resveratrol and quercetin on oral cancer cell growth and proliferation, *Anticancer Drugs* 10, 187–193, 1999.

277. Huang C, Ma WY, Goranson A, and Dong Z, Resveratrol suppresses cell transformation and induces apoptosis through a p53-dependent pathway, *Carcinogenesis* 20, 237–242, 1999.
278. Subbaramaiah K, Chung WJ, Michaluart P, Telang N, Tanabe T, Inoue H, Jang M, Pezzuto JM, and Dannenberg AJ, Resveratrol inhibits cyclooxygenase-2 transcription and activity in phorbol ester-treated human mammary epithelial cells, *J Biol Chem* 273, 21875–21882, 1998.
279. Clement MV, Hirpara JL, Chawdhury SH, and Pervaiz S, Chemopreventive agent resveratrol, a natural product derived from grapes, triggers CD95 signaling-dependent apoptosis in human tumor cells, *Blood* 92, 996–1002, 1998.
280. Fontecave M, Lepoivre M, Elleingand E, Gerez C, and Guittet O, Resveratrol, a remarkable inhibitor of ribonucleotide reductase, *FEBS Lett* 421, 277–279, 1998.
281. Kim H, Hall P, Smith M, Kirk M, Prasain JK, Barnes S, and Grubbs C, Chemoprevention by grape seed extract and genistein in carcinogen-induced mammary cancer in rats is diet dependent, *J Nutr* 134, 3445S–3452S, 2004.
282. Wyke SM, Russell ST, and Tisdale MJ, Induction of proteasome expression in skeletal muscle is attenuated by inhibitors of NF-κB activation, *Br J Cancer* 91, 1742–1750, 2004.
283. Walle T, Hsieh F, Delegge MH, Oatis JE Jr, and Walle UK, High absorption but very low bioavailability of oral resveratrol in humans, *Drug Metab Disposition* 32, 1377–1382, 2004.
284. Crowell JA, Korytko PJ, Morrissey RL, Booth TD, and Levine BS, Resveratrol-associated renal toxicity, *Toxicol Sci* 82, 614–619, 2004.
285. Fu ZD, Cao Y, Wang KF, Xu SF, and Han R, Chemopreventive effect of resveratrol to cancer, *Ai Zheng* 23, 869–873, 2004.
286. Khanduja KL, Bhardwaj A, and Kaushik G, Resveratrol inhibits *N*-nitrosodiethylamine-induced ornithine decarboxylase and cyclooxygenase in mice, *J Nutr Sci Vitaminol* (Tokyo) 50, 61–65, 2004.
287. Sale S, Verschoyle RD, Boocock D, Jones DJ, Wilsher N, Ruparelia KC, Potter GA, Farmer PB, Steward WP, and Gescher AJ, Pharmacokinetics in mice and growth-inhibitory properties of the putative cancer chemopreventive agent resveratrol and the synthetic analogue *trans*-3,4,5,4′-tetramethoxystilbene, *Br J Cancer* 90, 736–744, 2004.
288. Mishima S, Matsumoto K, Futamura Y, Araki Y, Ito T, Tanaka T, Iinuma M, Nozawa Y, and Akao Y, Antitumor effect of stilbenoids from *Vateria indica* against allografted sarcoma S-180 in animal model, *J Exp Ther Oncol* 3, 283–288, 2003.
289. Breinholt VM, Molck AM, Svendsen GW, Daneshvar B, Vinggaard AM, Poulsen M, and Dragsted LO, Effects of dietary antioxidants and 2-amino-3-methylimidazo[4,5-f]-quinoline (IQ) on preneoplastic lesions and on oxidative damage, hormonal status, and detoxification capacity in the rat, *Food Chem Toxicol* 41, 1315–1323, 2003.
290. Vitrac X, Desmouliere A, Brouillaud B, Krisa S, Deffieux G, Barthe N, Rosenbaum J, and Merillon JM, Distribution of [^{14}C]-*trans*-resveratrol, a cancer chemopreventive polyphenol, in mouse tissues after oral administration, *Life Sci* 72, 2219–2233, 2003.

291. Li H, Cheng Y, Wang H, Sun H, Liu Y, Liu K, and Peng S, Inhibition of nitrobenzene-induced DNA and hemoglobin adductions by dietary constituents, *Appl Radiat Isot* 58, 291–298, 2003.
292. Asensi M, Medina I, Ortega A, Carretero J, Bano MC, Obrador E, and Estrela JM, Inhibition of cancer growth by resveratrol is related to its low bioavailability, *Free Radical Biol Med* 33, 387–398, 2002.
293. Mouria M, Gukovskaya AS, Jung Y, Buechler P, Hines OJ, Reber HA, and Pandol SJ, Food-derived polyphenols inhibit pancreatic cancer growth through mitochondrial cytochrome C release and apoptosis, *Int J Cancer* 98, 761–769, 2002.

14 Resveratrol as an Antitumor Agent *In Vivo*

Francis Raul

CONTENTS

INTRODUCTION

The polyphenol resveratrol (3,4′,5-trihydroxystilbene) is a phytoalexin used by plants to defend themselves against fungal and other forms of aggression. Of the two isomers, *trans*-resveratrol is usually produced in higher amounts than *cis*-resveratrol, which is considered to be less biologically active. Major sources of resveratrol in the human diet are grapes, berries, peanuts, and red wine [1,2]. In grape skin resveratrol is expressed in response to fungal (*Botrytis cinerea*) infection [3,4]. There is considerable variation in the resveratrol content among red wines depending on grape cultivar, vintage, climate, fungal pressure, and duration of grape skin maceration. Low concentrations (~0.1 mg/l) are detected in white wines owing to the short period of grape skin maceration. Among red wines the highest concentrations (5 to 10 mg/l) are found in Burgundy and Oregon, where Pinot Noir is the predominant cultivar. Red wines produced in regions with a warm and dry climate have the lowest resveratrol concentrations [5].

Of biomedical interest was the identification of resveratrol in the roots of *Polygonum cuspidatum* [6]. Root extracts of this plant have been used in traditional Chinese and Japanese folk medicine against inflammations and diseases of the liver, heart, and blood vessels [6]. The anticancer properties of resveratrol were first recognized when Jang et al. [7] showed that this polyphenol exerted activity against all major stages of the carcinogenic process: initiation, promotion, and progression. Resveratrol was found to act as an antioxidant and antimutagen; it induces phase II drug-metabolizing enzymes (antiinitiation activity), mediates antiinflammatory effects, and inhibits cyclooxygenase (COX) and hyperperoxidase functions (antipromotion activity), and it induces human promyelocytic leukemia cell differentiation (antiprogression activity).

ANTITUMOR EFFECTS AND ANTITUMOR MECHANISMS *IN VIVO*

SKIN CANCER

Depending on the cellular origin, human skin cancers are classified as melanocytic or epithelial. Melanomas are less common but more lethal than epithelial skin cancers [7].

Several studies have shown that resveratrol inhibits the development of skin cancer. In their pioneering work on the antitumor activity of resveratrol, Jang et al. [7] studied tumorigenesis in the two-stage mouse skin cancer model, which uses 7,12-dimethyl-benz(*a*)anthracene (DMBA) as initiator and 12-*O*-tetradecanoylphorbol 13-acetate (TPA) as promoter. During an 18-week study, mice treated with DMBA and TPA developed an average of two tumors per mouse with 40% tumor incidence. Topical application of 1 to 25 μmol of resveratrol together with TPA twice a week reduced the number of skin tumors per mouse by 68 to 98%, and the tumor incidence was lowered by 50 to 88%, with no signs of resveratrol-induced cytotoxicity.

The antitumor activities of resveratrol, sesamol, sesame oil, and sunflower oil were compared in the *in vivo* DMBA-initiated and TPA-promoted mouse skin two-stage carcinogenesis protocols. Resveratrol was found to be the most effective among these compounds [8].

In another study, the effect of resveratrol was investigated in a murine model of carcinogenesis. Mice were treated by topical applications of DMBA, followed one week later by twice-weekly applications of Tween-60, until small papillomas appeared. Topical application of 10 μmol of resveratrol twice a week for 16 weeks caused a significant reduction (40%) of the tumor diameter [9].

Nonmelanoma skin cancer is the most common cancer among humans and solar ultraviolet B (UVB) radiation is its major cause. In a study designed to examine whether resveratrol possesses the potential to ameliorate damages caused by short-term UVB exposure of mouse skin, it was found that a single topical application of resveratrol (25 μmol) to SKH-1 hairless mice significantly inhibited the UVB-mediated increase in skin thickness and skin edema [10]. The resveratrol treatment was also found to result in significant inhibition of the UVB-triggered induction of COX and ornithine decarboxylase (ODC) activities in mouse skin. ODC is a key enzyme of polyamine biosynthesis and is a well-known marker for tumor promotion and progression.

Blood Cancer

Several studies evaluated the antiproliferative effects of resveratrol on leukemia, and they were recently reviewed [11]. Using leukemia cells derived from patients, it was demonstrated that resveratrol induced apoptotic cell death by depolarizing mitochondrial membranes and activating caspase-9 [12]. Although resveratrol inhibits the growth of both normal hematopoietic progenitor cells and leukemia cells in a dose-related manner, the antiproliferative effect of resveratrol on normal hematopoietic progenitor cells is less dramatic and reversible compared to leukemia cells [13]. Resveratrol induces apoptosis in leukemia cells, but not in normal hematopoietic cells. These cells maintain their capacity to hematologically reconstitute lethally irradiated mice.

Resveratrol undergoes metabolism by the cytochrome P450 enzyme CYPIBI, which is highly expressed in tumor cells, to give a metabolite which has been identified as the known antileukemic agent piceatannol [14]. The only difference between resveratrol and piceatannol is the presence of an additional hydroxyl group in one of the aromatic rings of piceatannol. This compound has well-known anticancer properties and was identified as the antileukemic agent obtained from a plant extract [15]. It was shown that piceatannol, but not resveratrol, acted as an efficient inducr of apoptosis in an *ex vivo* assay with leukemic lymphoblasts of 21 patients suffering from childhood lymphoblastic leukemia [16].

The strong antiproliferative and proapoptotic effects of resveratrol against leukemia cells suggested testing its antileukemic effects in animals. When mice were inoculated with 32Dp210 leukemia cells and treated orally with resveratrol at a daily dose of 8 mg/kg body weight, no antileukemic activity was observed [17]. Even when the size of the leukemia cell inoculum was reduced and the dose of resveratrol increased to 40 mg/kg, no diminution of the progression of leukemia was observed. At a dose of 80 mg/kg, resveratrol protected only a small fraction of mice from leukemia-induced death [17]. These findings suggest that resveratrol, despite its antileukemic effect *in vitro*, is not very effective in inhibiting the progression of leukemia *in vivo*. It remains to be investigated whether the antileukemic effect of resveratrol can be increased by administering even higher doses without toxicity, or by administering it via other routes.

Breast Cancer

Disregarding skin cancer, breast cancer is the most common cancer in women. It was first shown that resveratrol inhibits in a dose-dependent manner the development of DMBA-induced preneoplastic lesions in a mouse mammary gland culture model of carcinogenesis [7].

Resveratrol is structurally similar to estradiol and diethylstilbestrol [18]. Estrogens are known to be important mitogens in the breast, and are associated with increased breast cancer risk. Resveratrol was considered a phytoestrogen due to its potent estrogenic and even superestrogenic properties in MCF-7 mammary cancer cells, when combined with 17β-estradiol [18]. However, other studies conducted with these cells demonstrated antagonist activity in the presence of 17β-estradiol [19,20]. Hence, it is still unclear whether resveratrol is an estrogen receptor agonist or antagonist.

In order to determine the effect of resveratrol on mammary tumorigenesis, the *N*-methyl-*N*-nitrosourea (MNU) rat model was used. The etiology of this model is generally considered relevant to human breast cancer. Resveratrol was administered by gavage 5 days/week for the entire period of the study, starting 7 days before MNU administration. Doses of 10 and 100 mg/kg body weight were administered to female Sprague-Dawley rats [21]. Resveratrol (100 mg/kg) caused a significant suppression of tumor multiplicity and increased the latency period by 28 days. Tumor incidence was reduced by approximately 50% at 69 days of treatment. It increased gradually until the end of the experiment, when it reached control values. At a low dose of resveratrol (10 mg/kg), no significant differences were observed between control and treated groups. The histomorphology examination of the tumors revealed increased alveolar and adipocyte differentiation in response to resveratrol treatment with the occurrence of apoptotic cells within peripheral tumor areas, which were rare in controls. From these results it is difficult to speculate that the antitumor effect of resveratrol is due to its antiestrogenic effect. However, the observation that resveratrol delays the occurrence of mammary tumors and suppresses tumor formation at early stages suggests that the agent may be more efficient against premalignant lesions than against established tumors [21].

In another study [22], resveratrol was fed to female Sprague-Dawley rats. It inhibited tumor formation in comparison with vehicle treatment in DMBA-initiated mammary carcinogenesis. In this experimental design, resveratrol was dissolved in 70% ethanol and was then added into the diet, and ethanol in the diet was allowed to evaporate at room temperature for 1 day. Resveratrol (0.1 mg/rat) mixed to the diet was given to the rats, starting 7 days before exposure to DMBA and during the whole experimental period (120 days). Tumor incidence was calculated as the percentage of animals with one or more palpable tumors per treatment group. It was reduced by 45% due to resveratrol supplementation. The average number of tumors that developed per animal per week in each treatment group was reduced from 2.4 to 1.0 tumors per animal by resveratrol feeding at the end of the experiment. Relative to control animals, the latency of tumor development was increased by 3 weeks with resveratrol [22].

It has been shown that during the prepubertal period and childhood, estrogen reduces breast cancer risk through estrogen-induced activation of certain tumor suppressor genes including BRCA1 and p53 [23]. In this study, resveratrol treatment was initiated before the onset of puberty and it is therefore likely that resveratrol treatment accelerated mammary tissue differentiation, leading to the accumulation of refractory cell phenotypes during the period of carcinogen sensitivity. It was also found [22] that resveratrol suppressed the DMBA-induced COX-2 and matrix metalloprotease-9 (MMP-9) expression in breast tumors. In addition, a resveratrol-triggered downregulation of the transcription factor NF-κB was observed. Both COX-2 and MMP-9 genes are regulated by the activation of NF-κB, and both COX-2 and MMP-9 have been implicated in tumor development and metastasis. Because NF-κB activation was suppressed in resveratrol-treated animals, it is suggested that NF-κB may play an important role, but, as described in this chapter, it is also clear that NF-κB-independent mechanisms are involved in the antitumor effects of resveratrol.

In contrast to the previously described studies, it was reported that prepubertal exposure of female rats to resveratrol accelerates MNU-induced mammary carcinoma [24]. In this experimental setting, prepubertal female rats were treated by subcutaneous injections with daily 10 or 100 mg/kg resveratrol for 5 days. After 30 days, rats were given 50 mg/kg of MNU, after which the occurrence of mammary carcinoma was monitored. Several weeks after acute exposure to resveratrol, the dose of 100 mg/kg (but not of 10 mg/kg) significantly increased the incidence of mammary carcinoma, and cancer multiplicity was higher than in untreated controls, but the treatment did not affect latency. Resveratrol treatment increased significantly the number of rats with an irregular, prolonged estrous cycle.

No effects of resveratrol on mammary carcinogenesis were reported in another study. The daily intraperitoneal injection of resveratrol at doses of 1, 3, and 5 mg/kg (starting at the time of tumor inoculation) had no effect on the growth of 4T1 mammary tumors in female Balb/c mice, despite a potent antiproliferative effect of resveratrol (30 μM) on 4T1 breast cancer cells *in vitro* [25]. At the doses used, resveratrol was not toxic to the animals. In the 4T1 mammary tumor-bearing mice, resveratrol had no effect on the estrous cycle frequency or uterine histology, suggesting that it has little or no estrogen agonist activity [25].

Lung Cancer

Kimura and Okuda [26] studied the effect of stilbene glucosides, among them the resveratrol derivative resveratrol 3-O6D-glucoside (piceid) found in grapes and wines, on tumor growth and lung metastasis in Lewis lung carcinoma (LLC)-bearing mice. Piceid (300 mg/kg) or 2,3,4′,5-tetrahydroxystilbene-2-O-D-glucoside (150 mg/kg) were administered orally twice a day for 32 days, starting 12 hours after tumor implantation. Tumor growth in the right hind paw and lung metastasis was inhibited by oral administration of the stilbene glucosides. This effect cannot be explained by natural killer cell or cytotoxic T lymphocyte activation. Piceid inhibited DNA synthesis in LLC cells at a concentration of 1000 μM, but not at lower concentrations (10 to 100 μM). In addition both stilbene glucosides were found to inhibit the formation of capillary-like tube networks (angiogenesis) at concentration of 100 to 1000 μM.

In a similar work [27], the same authors showed that resveratrol (2.5 or 10 mg/kg) administered intraperitoneally once a day for 21 days, starting 12 hours after implantation of the tumor, significantly reduced the tumor volume (42%), tumor weight (44%), and metastasis to the lung (56%) of LLC-bearing mice. Resveratrol did not affect the number of $CD4^+$, $CD8^+$, and natural killer (NK)$1.1.^+$ T cells in the spleen. Therefore, the inhibitory effects of resveratrol on tumor growth and lung metastasis could not be explained by natural killer or cytotoxic T lymphocyte activation. In addition, resveratrol inhibited DNA synthesis most strongly in LLC cells; its 50% inhibitory concentration (IC_{50}) was 6.8 μmol/l. Resveratrol inhibited tumor-induced neovascularization at doses of 2.5 and 10 mg/kg in an *in vivo* model. Moreover, resveratrol significantly inhibited the formation of capillary-like tube formation from human umbilical vein endothelial cells (HUVEC) at concentrations of 10 to 100 μmol/l. The authors suggested that the antitumor and antimetastatic activities of resveratrol might be due to the inhibition of DNA synthesis in LLC cells and the inhibition of LLC-induced angiogenesis by resveratrol.

Liver Cancer

Only a few laboratories investigated the antitumor effects of resveratrol against liver cancer. Resveratrol administered intraperitoneally (1 mg/kg daily) to rats inoculated with a fast-growing tumor (Yoshida AH-130 ascites hepatoma) caused a 25% decrease in tumor cell number [28]. The effect was associated with an increase in the number of cells in the G2/M cell cycle phase. Interestingly, flow cytometric analysis of the tumor cell population revealed the existence of an aneuploid peak (representing 28% of total cell number), which suggests that resveratrol causes apoptosis in tumor cell population.

Resveratrol suspensions of 0, 3, 15, 30, and 300 mg/ml were given by gavage (0.5 ml/150 g body weight) to male Donryu rats (5 weeks old) and blood samples were taken 1 hour after oral intubation for *ex vivo* proliferation and invasion assays. The resveratrol-loaded rat serum restrained hepatoma cell invasion but not proliferation [29]. Resveratrol and resveratrol-loaded rat serum suppressed reactive oxygen species-mediated invasive capacity. The antiinvasive activity of resveratrol was found to be independent of its antiproliferative activity [29].

By feeding 10 or 50 ppm resveratrol in the diet of hepatoma-bearing rats for 20 days, solid tumor growth and metastasis formation diminished dose-dependently. Resveratrol (50 ppm) significantly decreased the serum lipid peroxide level, indicating its antioxidative properties or those of its metabolite *in vivo*. Resveratrol dose-dependently decreased both the serum triglyceride and cholesterol levels. The hypocholesteremic action of resveratrol is at least in part attributed to an increased excretion of neutral sterols and bile acids into the intestinal lumen. These results suggest that dietary resveratrol is hypolipidemic with a tendency for antitumor growth and antimetastatic effects in hepatoma-bearing rats [30].

Liu et al. [31] examined the antitumor and immunomodulatory activity of resveratrol in Balb/C mice with mouse carcinoma cells H22 xenografts. One day after tumor implantation, mice were injected intraperitoneally with resveratrol (500, 1000, or 1500 mg/kg). Resveratrol inhibited the growth of H22 tumors by 31.5, 45.6, and 48.7%, respectively. The antiproliferative effect was unrelated to changes in the immune response. However, the huge amount of resveratrol used in this study makes interpretation of these results difficult.

Colorectal and Intestinal Cancers

Two animal models were used in order to assess the antitumor effects of resveratrol: azoxymethane (AOM)-treated rats as a model reproducing the development of sporadic colorectal cancer in humans, and Min mice as a model for familial adenomatous polyposis.

Administration of AOM causes numerous morphological changes ranging from normal intestinal epithelium up to carcinomas that are biologically and histologically similar to those seen in human colon tumors [32,33]. AOM-induced tumors resembling human tumors are often mutated on K-ras and β-catenin genes, and show microsatellite instability, but unlike human tumors they are rarely mutated at the Apc gene (15%), and they have a low tendency to metastasize.

Tessitore et al. [34] investigated whether resveratrol affected AOM-induced colon carcinogenesis by administering resveratrol (200 μg/kg/day in drinking water) to male F344 rats for 100 days, beginning 10 days before AOM treatment. The number and multiplicity of aberrant crypt foci (ACF), which are preneoplastic alterations of the colonic mucosa, were significantly reduced in the colon of rats receiving resveratrol. In resveratrol-treated rats, bax expression was enhanced in ACF but not in the surrounding mucosa. The cyclin kinase inhibitor WAF1/p21 was expressed in ACF of controls and of resveratrol-treated rats and in normal mucosa of controls, but was lost in normal mucosa of resveratrol-treated animals. These results suggest a protective effect of resveratrol to colon carcinogenesis via mechanisms involving bax and p21 expression [34].

The mutant Min mice, a model of multiple intestinal neoplasia, were shown to have a mutated Apc gene similar to that of patients developing familial adenomatous polyposis (FAP) and to a majority of sporadic colon cancer. The congenic Min mice mimic the rapid development of adenomatous polyps that affect humans with germ-line inactivation of one Apc gene [32]. A major drawback of these mutants as model for FAP is that the tumors occur predominantly in the small intestine and only a few tumors appear in the colon, whereas in the human disease the tumors are located almost exclusively in the colon.

Our laboratory investigated the effects of oral administration of resveratrol on intestinal tumorigenesis in Min mice [35]. Resveratrol (0.01%) was administered in drinking water containing 0.4% ethanol for 7 weeks starting at an age of 5 weeks. Controls received drinking water with 0.4% ethanol. In this study, resveratrol prevented the formation of colonic tumors and reduced the formation of small intestinal tumors by 70%. Comparative gene expression studies showed that resveratrol downregulated those genes in the small intestinal mucosa that are directly involved in cell cycle progression or cell proliferation. In addition resveratrol upregulated the expression of several genes that are involved in the recruitment and activation of immune cells and in the inhibition of tumor expansion. These observations highlight the multiplicity of molecular targets involved in the antitumor effects of resveratrol.

In another study in which resveratrol was given to Min mice as a powdered mixture in the diet at 0, 4, 20, or 90 mg/kg body weight for 7 weeks, no changes in intestinal tumorigenesis were observed [36]. In this experimental design, resveratrol circulated in the plasma as glucuronide-conjugate and reached the tumors as was evidenced by a significant decrease in PGE_2 levels. However, surprisingly, immunohistochemical staining of intestinal tumors revealed no changes in COX-2 expression.

More recently, another study performed on Min mice confirmed the antitumor effects of resveratrol [37]. In this study, Min mice received resveratrol in the diet at 0.05, 0.2, and 0.5% from weaning (4 weeks) to 18 weeks of age. At this time adenoma numbers were scored. It was shown that resveratrol decreased significantly the number of adenomas at the two higher doses when compared to controls and suggests that resveratrol exerts adenoma-retarding activity in Min mice.

Prostate Cancer

In the Western world, prostate cancer is the third most frequently diagnosed cancer.

The effects of resveratrol as an antitumor agent have so far not been evaluated *in vivo*. However, some reports suggest the potential of resveratrol as a new therapeutic candidate for prostate cancer [38].

It has been hypothesized [39] that resveratrol may be especially suitable as an antitumor agent for prostate cancer owing to its ability to: (1) inhibit each stage of multistage carcinogenesis, (2) scavenge incipient populations of androgen-dependent prostate cancer cells through androgen receptor antagonism, and (3) scavenge incipient populations of androgen-independent prostate cancer cells by short-circuiting the epidermal growth factor receptor (EGFR)-dependent autocrine signaling to extracellular signal-regulated kinase 1/2 (Erk1/2) activation. Activated Erk1/2 play a major role in the hormone insensitivity and autonomous growth of the tumors *in vivo* [39].

PSA is a prostate specific antigen used as marker to monitor responsiveness of prostate cancer patients to various treatment modalities. PSA was found to be significantly downregulated by resveratrol via an androgen receptor-independent mechanism [40]. The central role of estrogens in prostate cancer development has been well supported by epidemiological findings and animal data [38]. Data from animal studies support the hypothesis that formation of genotoxic metabolites and reactive intermediates are a probable cause of prostate cancer initiation. Individuals expressing specific polymorphisms of estrogen-metabolizing enzymes, and local production of estrogens via aromatization presumably determine the lifetime exposure and influence prostate cancer risks among different human populations. A variety of new estrogenic/antiestrogenic selective estrogen receptor modulator (SERM)-like compounds, including resveratrol, are being evaluated for their antitumor potential in the next generation of prostate cancer therapies [38].

LIMITATIONS IN THE USE OF RESVERATROL AS AN ANTITUMOR AGENT *IN VIVO*

From the studies described above, it is clear that resveratrol shows potent *in vivo* antitumor activity against a limited number of tumors. Its efficacy depends on the localization of the tumor in the body. Thus, resveratrol seems efficient against skin cancers by direct application (a situation close to the *in vitro* situation), blood cancer (resveratrol can be metabolized into piceatannol, a well-known antileukemic agent), breast cancer (this seems to be related to its antiestrogenic activity), and also potentially against liver cancer (where resveratrol is highly metabolized). Resveratrol seems to exert more a preventive than a therapeutic effect against intestinal and lung cancers and its effects against prostate cancers are to be demonstrated.

The observed variations in the antitumor efficacy of resveratrol may be related to the rapid metabolic processing of the compound by various organs, which causes a low bioavailability. Asensi et al. [41] measured tissue levels of resveratrol after intravenous (i.v.) and oral administration to rabbits, rats, and mice. After i.v. administration of 20 mg/kg body weight, the half-life of resveratrol in plasma was very short (14 min in rabbits). The highest plasma concentration of resveratrol administered by either i.v. or oral route was reached within 5 min in all animals. The amount of resveratrol in brain, lung, liver, and kidney was less than 1 nmol/g fresh tissue. Hepatocytes metabolized resveratrol in a dose-dependent fashion, which means that the liver removes circulating resveratrol very rapidly. Another study [42] investigated the absorption and tissue distribution in mice of ^{14}C-*trans*-resveratrol. The animals received a single oral dose of the compound (5 mg/kg) and were sacrificed 1.5, 3, or 6 h later. The data revealed a preferential accumulation of ^{14}C-*trans*-resveratrol in stomach, liver, kidney, intestine, bile, and urine. The presence of intact ^{14}C-*trans*-resveratrol together with glucurono- and/or sulfoconjugates in these tissues was also shown. It seems that resveratrol is most likely to be absorbed in the form of a glucuronide at the level of the small intestine [43]. Marier et al. [44] showed that resveratrol is bioavailable at 38% when administered in a solution of hydroxypropyl β-cyclodextrin and undergoes extensive first-pass glucuronidation. Urinary excretion of resveratrol and its conjugates appears to be minimal compared with the biliary excretion.

Taken together, all metabolic and pharmacokinetic studies suggest that resveratrol is satisfactorily absorbed from the gastrointestinal tract and efficiently conjugated in intestine and liver. Plasma levels of unmetabolized resveratrol are well below 10 μM, even after an oral dose of 50 mg/kg. In contrast, resveratrol conjugates seem to reach much higher plasma levels than resveratrol [45].

CONCLUSIONS AND OUTLOOK

In order to gain a reliable view on the efficacy of resveratrol as an antitumor agent *in vivo*, precise preclinical investigations on the pharmacodynamics and pharmacokinetics of this agent should be carried out in addition to mechanistic studies. In this context, Gescher and Steward [45] proposed a rational approach:

- Efficacy studies of resveratrol in animals should include measurement of parent compound and metabolites in the target tissues.
- The bioavailability of resveratrol in humans needs to be determined.
- Metabolites of resveratrol should be characterized and quantified in humans.
- Mechanistic *in vitro* studies should focus on resveratrol metabolites, especially on its conjugates.

Two other strategic approaches are presently developed and might well represent the future of resveratrol as an antitumor drug. The first approach is related to the use of resveratrol as a chemosensitizing drug (see Chapter 15), since it has been reported that resveratrol acts as a potent sensitizer for anticancer drug-induced apoptosis. Thus, the combined sensitizer (resveratrol)/inducer (cytotoxic drugs) concept may become a novel strategy to enhance the efficacy of anticancer therapy in a variety of human cancers [46].

The second approach is based on the synthetic manipulations of the trihydroxystilbene motif with the aim to develop resveratrol analogs and derivatives of pharmacological interest. These manipulations generated new compounds with more potent anticancer properties than the parent compound. Chemical synthetic attempts have predominantly been concerned with the introduction of additional hydroxyl groups and/or with the introduction of methyl groups into the trihydroxystilbene framework [47,48]. Among these new compounds the *O*-methylated analog of resveratrol (*Z*)-3,4′,5-tri-*O*-methylresveratrol has been shown to exert antimitotic activity on human colon cancer cells in the nanomolar range by inhibiting tubulin polymerization [49], and inducing apoptosis in cancer cells independently of their p53 status [50]. More recently it was shown that another resveratrol analog, 3,4,4′,5-tetramethoxystilbene, exhibited preferential growth-inhibitory and proapoptotic properties in transformed cells when compared to resveratrol [51]. This analog showed more favorable pharmacokinetic properties than resveratrol, in that it yielded higher drug levels in the small intestinal and colonic mucosa and brain. The compound showed no toxicity in rats when administered orally at doses up to 400 mg/kg. The efficacy of these promising resveratrol analogs needs now to be demonstrated *in vivo*.

REFERENCES

1. Dercks W and Creasy LL, Influence of fosetyl-A1 on phytoalexin accumulation in the Plasmopara viticola-grapevine interaction, *Physiol Mol Plant Pathol* 34, 203–213, 1989.
2. Siemann EH and Creasy LL, Concentration of the phytoalexin resveratrol in wine, *Am J Enol Vitic* 43, 49–52, 1992.
3. Langcake P and Pryce RJ, The production of resveratrol by *Vitis vinifera* and other members of the Vitaceae as a response to infection or injury, *Physiol Plant Pathol* 9, 77–86, 1976.
4. Jeandet P, Bessis R, Sbaghi M, and Meunier P, Production of the phytoalexin resveratrol by grape berries as a response to Botrytis attack under natural conditions, *J Phytopathol* 143, 135–139, 1995.
5. Soleas GJ, Diamandis EP, and Goldberg DM, Resveratrol: a molecule whose time has come? And gone?, *Clin Biochem* 30, 91–113, 1997.
6. Kimura Y, Ohminami H, Okuda H, Baba K, Kozawa M, and Arichi S, Effects of stilbene components of roots of Polygonum cuspidatum on liver injury in peroxidized oil fed rats, *Planta Med* 49, 51–54, 1983.
7. Jang M, Cai L, Udeani GO, Slowing KV, Thomas CF, Beecher CW, Fong HH, Farnsworth NR, Kinghorn AD, Mehta RG, Moon RC, and Pezzuto JM,

Cancer chemopreventive activity of resveratrol, a natural product derived from grapes, *Science* 275, 218–220, 1997.
8. Kapadia GJ, Azuine MA, Tokuda H, Takasaki M, Mukainaka T, Konoshima T, and Nishino H, Chemopreventive effect of resveratrol, sesamol, sesame oil and sunflower oil in the Epstein-Barr virus early antigen activation assay and the mouse skin two-stage carcinogenesis, *Pharmacol Res* 45, 499–505, 2002.
9. Pervaiz S, Resveratrol: from the bottle to the bedside?, *Leuk Lymphoma* 40, 491–498, 2001.
10. Afaq F, Adhami VM, and Ahmad N, Prevention of short-term ultraviolet B radiation-mediated damages by resveratrol in SKH-1 hairless mice, *Toxicol Appl Pharmacol* 186, 28–37, 2003.
11. Aziz MH, Kumar R, and Ahmad N, Cancer chemoprevention by resveratrol: *in vitro* and *in vivo* studies and the underlying mechanisms [review], *Int J Oncol* 23, 17–28, 2003.
12. Dorrie J, Gerauer H, Wachter Y, and Zunino SJ, Resveratrol induces extensive apoptosis by depolarizing mitochondrial membranes and activating caspase-9 in acute lymphoblastic leukaemia cells, *Cancer Res* 61, 4731–4739, 2001.
13. Gautam SC, Xu YX, Dumaguin M, Janakiraman N, and Chapman RA, Resveratrol selectively inhibits leukaemia cells: a prospective agent for ex vivo bone marrow purging, *Bone Marrow Transplant* 25, 639–645, 2000.
14. Potter GA, Patterson LH, Wanogho E, Perry PJ, Butler PC, Ijaz T, Ruparelia KC, Lamb JH, Farmer PB, Stanley LA, and Burke MD, The cancer preventative agent resveratrol is converted to the anticancer agent piceatannol by the cytochrome P450 enzyme CYPIBI, *Br J Cancer* 86, 774–778, 2002.
15. Ferrigni NR, McLaughlin JL, Powell RG, and Smith CR, Isolation of piceatannol as the antileukemic principle from the seeds of euphorbia lagascae, *J Nat Prod* 47, 347–352, 1984.
16. Wieder T, Prokop A, Bagci B, Essmann F, Bernicke D, Schulze-Osthoff K, Dorken B, Schmalz HG, Daniel PT, and Henze G, Piceatannol, a hydroxylated analog of the chemopreventive agent resveratrol, is a potent inducer of apoptosis in the lymphoma cell line BJAB and in primary leukemic lymphoblasts, *Leukemia* 15, 1735–1742, 2001.
17. Gao X, Xu YX, Divine G, Janakiraman N, Chapman RA, and Gautam SC, Disparate *in vitro* and *in vivo* antileukemic effects of resveratrol, a natural polyphenolic compound found in grapes, *J Nutr* 132, 2076–2081, 2002.
18. Gehm BD, McAndrews JM, Chien PY, and Jameson JL, Resveratrol, a polyphenolic compound found in grapes and wine, is an agonist for the estrogen receptor, *Proc Natl Acad Sci USA* 94, 14138–14143, 1997.
19. Mghonyebi OP, Russo J, and Russo IH, Antiproliferative effect of synthetic resveratrol on human breast epithelial cells, *Int J Oncol* 12, 865–869, 1998.
20. Lu R and Serrero G, Resveratrol, a natural product derived from grape, exhibits antiestrogenic activity and inhibits the growth of human breast cancer cells, *J Cell Physiol* 179, 297–304, 1999.
21. Bhat KPL, Lantvit D, Christov K, Mehta RG, Moon RC, and Pezzuto JM, Estrogenic and antiestrogenic properties of resveratrol in mammary tumor models, *Cancer Res* 61, 7456–7463, 2001.
22. Banerjee S, Bueso-Ramos C, and Aggarwal BB, Suppression of 7,12-dimethylbenz(a)anthracene-induced mammary carcinogenesis in rats by

resveratrol: role of nuclear factor-kappaB, cyclooxygenase 2, and matrix metalloprotease 9, *Cancer Res* 62, 4945–4954, 2002.
23. Hilakivi-Clarke L, Estrogens, BRCA1, and breast cancer, *Cancer Res* 60, 4993–5001, 2000.
24. Sato M, Pei RJ, Yuri T, Danbara N, Nakane Y, and Tsubura A, Prepubertal resveratrol exposure accelerates N-methyl-N-nitrosourea-induced mammary carcinoma in female Sprague-Dawley rats, *Cancer Lett* 202, 137–145, 2003.
25. Bove K, Lincoln DW, and Tsan MF, Effect of resveratrol on growth of 4T1 breast cancer cells *in vitro* and *in vivo*, *Biochem Biophys Res Commun* 291, 1001–1005, 2002.
26. Kimura Y and Okuda H, Effects of naturally occurring stilbene glucosides from medicinal plants and wine, on tumour growth and lung metastasis in Lewis lung carcinoma-bearing mice, *J Pharm Pharmacol* 52, 1287–1295, 2000.
27. Kimura Y and Okuda H, Resveratrol isolated from Polygonum cuspidatum root prevents tumor growth and metastasis to lung and tumor-induced neovascularization in Lewis lung carcinoma-bearing mice, *J Nutr* 131, 1844–1849, 2001.
28. Carbo N, Costelli P, Baccino FM, Lopez-Soriano FJ, and Argiles JM, Resveratrol, a natural product present in wine, decreases tumour growth in a rat tumour model, *Biochem Biophys Res Commun* 254, 739–743, 1999.
29. Kozuki Y, Miura Y, and Yagasaki K, Resveratrol suppresses hepatoma cell invasion independently of its anti-proliferative action, *Cancer Lett* 167, 151–156, 2001.
30. Miura D, Miura Y, and Yagasaki K, Hypolipidemic action of dietary resveratrol, a phytoalexin in grapes and red wine, in hepatoma-bearing rats, *Life Sci* 73, 1393–1400, 2003.
31. Liu HS, Pan CE, Yang W, and Liu XM, Antitumor and immunomodulatory activity of resveratrol on experimentally implanted tumor of H22 in Balb/c mice, *World J Gastroenterol* 9, 1474–1476, 2003.
32. Corpet DE and Pierre F, Point: from animal models to prevention of colon cancer. Systematic review of chemoprevention in Min mice and choice of the model system, *Cancer Epid Biom Prev* 12, 391–400, 2003.
33. Reddy BS, Studies with azoxymethane-rat preclinical model for assessing tumor development and chemoprevention, *Environ Mol Mutagen* 44, 26–35, 2004.
34. Tessitore L, Davit A, Sarotto I, and Caderni G, Resveratrol depresses the growth of colorectal aberrant crypt foci by affecting bax and $p21^{CIP}$ expression, *Carcinogenesis* 21, 1619–1622, 2000.
35. Schneider Y, Duranton B, Gossé F, Schleiffer R, Seiler N, and Raul F, Resveratrol inhibits intestinal tumorigenesis and modulates host-defense-related gene expression in an animal model of human familial adenomatous polyposis, *Nutr Cancer* 39, 102–107, 2001.
36. Ziegler CC, Rainwater L, Whelan J, and McEntee MF, Dietary resveratrol does not affect intestinal tumorigenesis in $Apc^{Min/+}$mice, *J Nutr* 134, 5–10, 2004.
37. Sale SL, Tunstall RG, Potter GA, Steward WP, and Gescher AJ, Comparison of the Effects of Trans-3,4′,5-trihydrostilbene (Resveratrol) and 3,4,4′,5-Tetramethoxystilbene (DMU212) in the ApcMin/+ Mouse and on Cyclooxygenase-2 (COX-2) Expression in Cells *In Vitro*, 95th AACR Annual Meeting, Orlando, FL, March 27–31, 2004.

38. Ho SM, Estrogens and anti-estrogens: key mediators of prostate carcinogenesis and new therapeutic candidates, *J Cell Biochem* 491–503, 2004.
39. Stewart JR, Artime MC, and O'Brian CA, Resveratrol: a candidate nutritional substance for prostate cancer prevention, *J Nutr* 133, 2440S–2443S, 2003.
40. Hsieh TC and Wu JM, Grape-derived chemopreventive agent resveratrol decreases prostate-specific antigen (PSA) expression in LNCaP cells by an androgen receptor (AR)-independent mechanism, *Anticancer Res* 20, 225–228, 2000.
41. Asensi M, Medina I, Ortega A, Carretero J, Bano MC, Obrador E, and Estrela J, Inhibition of cancer growth by resveratrol is related to its low bioavailability, *Free Radical Biol Med* 33, 387–398, 2002.
42. Vitrav X, Desmoulière A, Brouillaud B, Krisa S, Deffieux G, Barthe N, Rosenbaum J, and Mérillon JM, Distribution of (^{14}C)-trans-resveratrol, a cancer chemopreventive polyphenol, in mouse tissues after oral administration, *Life Sci* 72, 2219–2233, 2003.
43. Andlauer W, Kolb J, Siebert K, and Furst P, Assessment of resveratrol bioavailability in the perfused small intestine of the rat, *Drugs Exp Clin Res* 26, 47–55, 2000.
44. Marier JF, Vachon P, Gritsas A, Zhang J, Moreau JP, and Ducharme MP, Metabolism and disposition of resveratrol in rats: extent of absorption, glucuronidation, and enterohepatic recirculation evidenced by a linked-rat model, *JPET* 302, 369–373, 2002.
45. Gescher AJ and Steward WP, Relationship between mechanisms, bioavailability, and preclinical chemopreventive efficacy of resveratrol: a conundrum, *Br J Cancer* 12, 953–957, 2003.
46. Fulda S and Debatin KM, Sensitization for tumor necrosis factor-related apoptosis-inducing ligand-induced apoptosis by the chemopreventive agent resveratrol, *Cancer Res* 64, 337–346, 2004.
47. Pettit GR, Grealish MP, Jung K, Hamel E, Pettit RK, Chapuis JC, and Schmidt JM, Antineoplastic agents. 465. Structural modification of resveratrol: sodium resverastatin phosphate, *J Med Chem* 45, 2534–2542, 2002.
48. Roberti M, Pizzirani D, Simoni D, Rondanin R, Baruchello R, Bonora C, Buscemi F, Grimaudo S, and Tolomeo M, Synthesis and biological evaluation of resveratrol and analogues as apoptosis-inducing agents, *J Med Chem* 46, 3546–3554, 2003.
49. Schneider Y, Chabert P, Stutzmann J, Coelho D, Fougerousse A, Gossé F, Launay JF, Brouillard R, and Raul F, Resveratrol analog (Z)-3,5,4′-trimethoxystilbene is a potent anti-mitotic drug inhibiting tubulin polymerization, *Int J Cancer* 107, 189–196, 2003.
50. Schneider Y, Fischer B, Coelho D, Roussi S, Gossé F, Bischoff P, and Raul F, (Z)-3,5,4′-Tri-*O*-methyl-resveratrol, induces apoptosis in human lymphoblastoid cells independently of their p53 status, *Cancer Lett* 211, 155–161, 2004.
51. Sale S, Verschoyle RD, Boocock D, Jones DJL, Wilsher N, Ruparelia KC, Potter GA, Farmer PB, Steward WP, and Gescher AJ, Pharmacokinetics in mice and growth-inhibitory properties of the putative cancer chemopreventive agent resveratrol and the synthetic analogue trans 3,4,5,4′-tetramethoxystilbene, *Br J Cancer* 90, 736–744, 2004.

15 Resveratrol as a Sensitizer to Apoptosis-Inducing Stimuli

Ali R. Jazirehi and Benjamin Bonavida

CONTENTS

INTRODUCTION

Several notable advances have been made in the treatment of cancer by chemotherapeutic and hormonal drugs. However, the development and/or acquisition of tumor resistance to both hormonal and chemotherapeutics presents a major problem [1]. Failure to cure resistant tumors with traditional therapeutic approaches has led to the introduction of immunotherapy. Immunotherapeutic strategies under investigation are based on the premise that chemotherapeutic- and hormonal-resistant tumors are invariably sensitive to immunotherapy. Unfortunately, immunotherapy still fails to deliver significant curative rates. The failure of immunotherapy to overcome drug resistance and the realization that both chemotherapy and immunotherapy mediate their effects by killing tumor cells by apoptosis suggest that both therapies share common signaling pathways for the induction of apoptosis. Apoptosis is the principal mechanism employed by the immune system and chemotherapeutic drugs in eradicating tumor cells. Being a genetically controlled process, apoptosis is susceptible to mutations, and deregulation of the apoptotic machinery is frequently observed in numerous types of cancers. Despite aggressive therapies, resistance of many tumors to established treatment regimens still constitutes a major hurdle in cancer therapy. Thus, attempts to improve the survival of cancer patients largely depend on strategies to target tumor cell resistance. Since induction of apoptosis in target cells is a key mechanism for most antitumor therapies including chemotherapy, γ-irradiation, immunotherapy, or cytokines, defects in apoptosis programs may cause resistance. Although most tumor cells respond to the initial treatment with various chemotherapeutic regimens, relapse due to the emergence and selective outgrowth of drug-resistant variants is frequently observed. Resistant tumor cells evade the action of anticancer agents by increasing their apoptotic threshold.

The emergence of resistance to conventional therapeutic strategies has encouraged the design and/or exploitation of novel compounds with anticancer properties. The key concept is to identify compounds capable of altering the gene expression profile of resistant tumors in such a way to facilitate the action of drugs with minimal toxicity towards the normal healthy tissues. This has spurred the development and/or utilization of novel chemical compounds capable of inducing apoptosis in otherwise chemo-/immune-resistant tumor cells. Because of the close correlation between tumorigenesis and deregulation of apoptosis, any therapeutic strategy aimed at specifically triggering apoptosis in tumor cells might have potential therapeutic applications.

APOPTOSIS AND RESISTANCE TO APOPTOSIS

Apoptosis

Physiological or programmed cell death (PCD), also known as apoptosis, is an innate mechanism by which unwanted, defective, or damaged cells are rapidly and selectively eliminated from the body. It occurs during tissue remodeling, embryonic development, and immune regulation. Apoptosis is a unique type of cell death and phenotypically is associated with cytoskeletal disruption, cellular shrinkage, nuclei and chromosomal DNA condensation and internucleosomal fragmentation into 180–220 bp fragments, membrane blebbing, activation of endonucleases (also known as caspases), as well as the formation of membrane-bound apoptotic bodies [2].

Apoptosis is inherently present in most cells including tumor cells and is activated by the appropriate stimulus. Uncontrolled activation of apoptosis may lead to a variety of diseases such as cancer and autoimmune diseases [3]. The two major apoptotic pathways initiated by activation of cytotoxic T-lymphocytes are the granule-exocytosis pathway, by release of perforin and granzymes, and the death receptor pathways, involving the death receptor signaling pathways such as the tumor necrosis factor-alpha family TNFα, TNF-related apoptosis-inducing ligand (TRAIL), [Apo-2L/TRAIL], and FasL (CD95L) [4]. The granule-exocytosis pathway is initiated by the direct interaction between cytotoxic lymphocytes and tumor cells and involves the recognition of T cell receptor (TCR)/major histocompatibility complex (MHC) leading to the release of cytotoxic granules such as perforin. Upon perforin polymerization, pores similar to those formed by the membrane attack complex of the complement cascade are formed on the target cell membrane, which will allow the unidirectional passage of granzymes from the lymphocyte to the target cell leading to apoptosis. The death receptor signaling pathway is triggered by binding of the death-inducing ligands to their cognate receptors and subsequent induction of apoptosis [5].

The regulation and execution of apoptotic cell death is carried out by a family of cysteine proteases with aspartic acid specificity known as caspases. Caspases are present in the living cells as inactive zymogens, their activation is through autocatalytic processing by caspase cascades, and they are divided into initiators (e.g., [8–10]) and effectors/executioners (e.g., [3,6,7]). Based on the pattern of caspase cascade activation, two types of cells have been characterized so far [6]. In type I cells, caspase cascade is triggered upon the oligomerization of cell surface death receptors and undergoes a sequential activation. Depending on the amount of caspase-8, the apoptotic stimuli either directly induce initiator caspase-8 autocleavage/processing (by yet unknown mechanisms), which leads to activation of the principal mediator of apoptosis caspase-3 and cleavage of death substrates (e.g., poly (ADP) ribose polymerase; PARP) with subsequent induction of apoptosis; or, in the event that cells lack sufficient amounts of caspase-8, the apoptotic stimuli (e.g., most chemotherapeutic drugs) utilize type II apoptotic signaling pathway involving mitochondria and the Bcl-2 family members. In the cytosol, caspase-8 will cleave the proapoptotic Bid. The caspase-cleaved fragment of Bid (truncated Bid; tBid) will then migrate to and reside in the mitochondrial outer membrane, where in association with other proapoptotic molecules such as Bax, Bad, Bcl-$_{xS}$ induce the formation of mitochondrial permeability transition pore (PTP), which will lead to the mitochondrial collapse and decrease in the mitochondrial transmembrane potential ($\downarrow\Delta\psi$m). The integrity of mitochondria is preserved by the protective effects of antiapoptotic Bcl-2 family members (e.g., Bcl-2, Bcl-$_{xL}$, Bfl-1/A1, Mcl-1), which is antagonized by the proapoptotic Bcl-2 family members (e.g., Bax, Bid, Bad, Bik, Bcl-$_{xS}$).

Mitochondrial destabilization will facilitate the unidirectional release of the apoptogenic molecules (cytochrome c and Smac/DIABLO) into the cytosol. Cytochrome c, in the presence of the adaptor molecule apoptosis protease-activating factor-1 (Apaf-1), dATP/ATP, and caspase-9 will participate in the formation of the apoptosome complex. The formation of the apoptosome complex expedites caspase-9 processing and activation. Simultaneously, Smac/DIABLO will physically associate with the cellular inhibitors of apoptosis (c-IAP) family members (c-IAP-1 and -2, XIAP, and survivin), thereby removing the inhibitors of caspase activation. Through autocatalytic processing, caspase-9 becomes activated and in the absence of c-IAPs utilizes caspases-3, -6, and -7 as substrates leading to the cleavage of death substrates (e.g., PARP) and apoptosis ensues. This is the converging point of type I and II pathways [7].

RESISTANCE TO APOPTOSIS

The vast majority of chemotherapeutic drugs and immunotherapeutic strategies eradicate tumor cells by the induction of apoptosis [8]. Tumor cells, in turn, have adopted various mechanisms to resist apoptosis. Natural inhibitors of apoptosis, such as Bcl-2 (Bcl-2, Bcl-$_{xL}$, Mcl-1) and IAP (c-IAP-1, -2, XIAP, survivin) family members, protect the tumor cells from the drug-/immune-induced apoptosis via different mechanisms [8]. Tumor cell drug resistance is further reinforced by the emergence of the multidrug resistance (MDR) phenotype following initial chemotherapy administration due to the action of membrane-bound drug efflux pumps [9,10], which confers resistance to the cytotoxicity induced by structurally and functionally distinct antineoplastic agents. Overexpression of antiapoptotic Bcl-2 family members also contributes to the MDR phenotype of the tumor cells [11,12]. Thus, as tumor cells develop resistance to the apoptotic effects of antineoplastic agents, they may also develop cross-resistance to the cytotoxic action of the immune system. The development of cross-resistance suggests that drugs and death receptors utilize a common apoptotic pathway, and such a cross-resistance phenotype cannot be solely explained by the MDR mechanism [7,13]. This is probably the main reason for the failure of MDR modulators such as retrovirus- or liposome-mediated transfer of MDR1 ribozyme or MDR reversal agents including verapamil, quinidine, and cyclosporine in the treatment of drug-resistant tumor cells [14]. Utilization of these agents is further limited by the presence of redundant cellular mechanisms of resistance, alterations in the pharmacokinetics of the agents, and clinical toxicities [14]. Thus, nontoxic agents that interfere with the function of drug efflux pumps or adversely modulate the signaling pathways leading to alterations in the expression profile of apoptosis-associated gene products can be effectively used in combination with chemotherapy in the clinical treatment of drug- and/or immune-resistant tumors.

SENSITIZATION

Another aspect regarding resistance to apoptosis, such as drug- and/or immune-mediated cytotoxicity, is often due to the failure of cells to carry out the signal transduction pathways ultimately leading to cell death [15]. This may be due to inadequate expression of signaling molecules, overexpression of protective factors, or mutations in apoptotic proteins such as p53. Drugs have been shown to regulate the expression levels of anti- and proapoptotic proteins [13] illustrating the likelihood that therapeutic compounds may not directly induce cytotoxicity but nonetheless possess the ability to modify protein expression profiles in a manner that would allow additional agents to induce apoptosis at much lower threshold. Thus, the direct cytotoxicity and sensitizing attributes exerted by drugs are mediated via distinct mechanisms although some overlap may exist. Thus, identification of sensitizing compounds and elucidation of the intracellular signaling pathways regulated by these agents is of great urgency. In this respect, the functional complementation (two-signal) model is proposed. Accordingly, treatment of tumor cells with a nontoxic sensitizing agent alters the expression profile of apoptosis-associated gene products (signal I), removes the inhibitory block in the apoptotic pathway, and by lowering the apoptosis threshold sensitizes the tumor cells to the cytotoxic effects of the second agent (e.g., biological response modifiers, immune-mediated approaches, and/or chemotherapeutic drugs) (signal II) [7].

Numerous studies have validated the functional complementation model and further attest to the contention that therapeutic compounds, in addition to the ability to directly induce apoptosis, are capable of altering the gene expression profile and decreasing the apoptosis threshold of drug-resistant tumor cells [6,13,16], thus, overcoming the acquired or intrinsic apoptosis resistance. In addition, delineation of the signaling pathways modulated by the sensitizing agents will lead to the identification of intracellular targets for therapeutic intervention in the treatment of drug-refractory tumor cells.

In recent years, naturally occurring antioxidant compounds present in diet and beverages consumed by humans, such as resveratrol (RSV), have gained considerable attention because of their beneficial effects on health as cancer chemopreventive or cardioprotective agents [17]. RSV (3′,4′,5′-trihydroxystilbene) is a polyphenol synthesized by a variety of plant species and is present in grapes or in red wine [18]. RSV was first isolated from the roots of the oriental medicinal plant *Polygonum cuspidatum* (Ko-jo-kon in Japanese) [19]. It is a major active ingredient of stilbene phytoalexins. The relatively high concentration of RSV in red wine and its documented cradioprotective effects form the basis of the so-called "French paradox." Initial work centered on RSV cardiovascular biology and more recently on RSV cancer chemopreventive activity in animals [20]. Recent studies are focused on RSV's mechanism of action and potential clinical applications [21,22].

ANTICANCER THERAPEUTIC EFFECTS OF RSV AND RSV ANALOGS

ANTICANCER THERAPEUTIC EFFECTS OF RSV

The direct effects of RSV and in combination with chemotherapeutic drugs have been reported in several studies. The antitumor effect of RSV alone and in combination with 5-fluorouracil (5-FU) on a transplantable murine hepatoma model was evaluated. RSV could inhibit the growth of murine hepatoma in a dose-dependent manner and arrested the cells at the S phase. Combination of 5-FU and RSV had superior antiproliferative activity compared to each agent alone. The results of this study suggest that RSV can enhance the therapeutic effects of 5-FU [23]. Chen et al. [24] showed that RSV arrests the cells at the S phase of the cell cycle and downregulated p21 whereas cyclin E was upregulated. RSV also increased the apoptosis rate of the neuroblastoma cells and suppressed the growth rate of subcutaneous neuroblastomas, resulting in 70% long-term survival. RSV markedly impaired proliferation of both the TMZ-sensitive M14 and the TMZ-resistant SK-Mel-28 and PR-Mel cell lines mainly through its ability to induce S-phase arrest and apoptosis [25,26]. Laux et al. [27] demonstrated that apoptosis induced by RSV was found to occur only in breast cancer cells expressing wild-type p53 but not in mutant p53-expressing cells suggesting that RSV induces apoptosis via p53-dependent pathways. RSV induced growth inhibition and apoptosis in MDA-MB-231, a highly invasive and metastatic breast cancer cell line, in concomitance with a dramatic endogenous increase of growth inhibitory/proapoptotic ceramide [28]. Ding and Adrian [29] demonstrated the inhibition by RSV of two human pancreatic cancer cell lines, which was accompanied by apoptotic morphologic changes suggestive of a therapeutic value of RSV in the management and prevention of human pancreatic cancer. Analysis of the antiproliferation effect of RSV in Hep G2 and Hep 3B human liver cancer cell lines showed that RSV inhibits cell growth in p53-positive Hep G2 cells only, which was a result of apoptotic cell death coincident with an increase in p21 expression and enhanced Bax expression. In contrast, RSV neither inhibited the proliferation of the p53-negative Hep 3B cells nor showed significant changes in p21 or Fas/APO-1 levels [30].

Roman et al. [31] analyzed the apoptotic and growth inhibitory effects of *trans*-RSV in human B-cell lines derived from chronic B-cell malignancies and in patient-derived B-CLL cells. RSV displayed antiproliferative activity and induced apoptosis in the two cell lines as well as in B-CLL patients' cells. It also inhibited inducible nitric oxide synthase (iNOS) protein expression and *in situ* NO release as well as Bcl-2 expression, which may contribute to the apoptotic effects of RSV in B-CLL. Studies by Kuwajerwala et al. [32] revealed that, in androgen-sensitive LNCaP cells, the effect of RSV on DNA synthesis varied dramatically depending on the concentration and the duration of treatment. Therefore, this unique ability of RSV to exert opposing effects on two important processes in cell cycle progression, induction of S phase and inhibition of DNA synthesis, may be responsible for its apoptotic and antiproliferative effects.

In addition, RSV treatment results in a dose-dependent inhibition of the growth of human prostate carcinoma DU145 cells, which was associated with the inhibition of D-type cyclins and cyclin-dependent kinase (Cdk) 4 expression, and the induction of tumor suppressor p53 and Cdk inhibitor p21. Moreover, the kinase activities, not the protein levels, of cyclin E and Cdk2 were inhibited by RSV. RSV treatment also upregulated the Bax protein and mRNA expression in a dose-dependent manner; however, Bcl-2 and Bcl-$_{xL}$ levels were not significantly affected. These effects correlated with an activation of caspase-3 and caspase-9 [33]. The therapeutic potential of RSV on human medulloblastoma cell lines and its effects on cell growth, differentiation, and death were examined by Wang et al. [34]. The authors showed that RSV could suppress growth, promote differentiation, and commit its target cells to apoptosis. Although the cells constitutively expressed the Fas antigen, anti-Fas antibody neither inhibited growth nor induced apoptosis, thus excluding the Fas-dependent apoptotic pathway. RSV induced caspase-3 and the appearance of the cleaved form of caspase-3 was closely associated with the apoptotic event. These findings suggest that RSV is an effective antimedulloblastoma agent.

Earlier studies have shown that RSV alters the expression of genes involved in cell cycle regulation and apoptosis, including cyclins, Cdks, p53, and Cdk inhibitors. However, most of the p53-controlled effects related to the role of RSV in transcription, either by activation or repression of a sizable number of primary and secondary target genes, are elusive. Narayanan et al. [35] examined whether RSV activates a cascade of p53-directed genes that are involved in apoptosis mechanism(s) or whether it modifies the androgen receptor (AR) and its coactivators directly or indirectly and induces cell growth inhibition. They showed that treatment of androgen-sensitive prostate cancer cells (LNCaP) with RSV (10^{-5} M; 48 h) downregulates prostate-specific antigen (PSA), AR coactivator ARA 24, and NF-κB p65. Altered expression of these genes is associated with an activation of p53-responsive genes such as p53, PIG 7, p21$^{Waf1\text{-}Cip1}$, p300/CBP, and Apaf-1 whereas the effect of RSV on p300/CBP plays a central role in its cancer-preventive mechanisms in LNCaP cells. These results implicate activation of multiple sets of functionally related molecular targets and point to the need for further extensive studies on AR coactivators, such as p300, its central role in posttranslational modifications such as acetylation of p53 and/or AR by RSV at different stages of prostate cancer to elucidate fully the role of RSV as a chemopreventive agent for prostate cancer.

RSV ANALOGS

Widespread interest in this apparently structurally simple molecule and synthetic stilbene analogs has arisen in recent years due to the discovery of its antioxidant, antiinflammatory, and anticarcinogenic activities, among others. Considering the *in vivo* and *in vitro* efficacy of RSV in eradicating tumor cells and its additional beneficial effects, studies were undertaken to evaluate the favorable health benefits of naturally occurring and synthetic RSV analogs [17]. Among these analogs is astringinin (3,3′,4′,5-tetrahydroxystilbene) with considerably higher antioxidative activity and free radical scavenging capacity than RSV, which was evaluated for its cardioprotective effects in ischemia or ischemia-reperfusion rats. The authors concluded that astringinin can be considered as a potent antiarrhythmic agent with cardioprotective activity in ischemic and ischemic-reperfused rat heart. These beneficial effects of astringinin may be attributed to antioxidant activity and upregulation of NO production [36]. Another novel RSV tetramer analog is vaticanol C, isolated from the stem bark of *Vatica rassak*, which markedly suppresses cell growth and induces apoptosis in human colon cancer cell lines [37].

Perhaps the most well-studied naturally occurring analog of RSV is piceatannol, which was previously identified as the active ingredient in herbal preparations in folk medicine and as an inhibitor of p72(Syk). Evaluation of the effects of piceatannol on growth, proliferation, differentiation, and cell cycle distribution profile of the human colon carcinoma cell line Caco-2 revealed a dose- and time-dependent decrease in cell numbers and reduction of the proliferation rate with no obvious effect on differentiation. Analysis of the cell cycle distribution profile revealed an accumulation of cells in the S phase presumably as a result of cyclin D1, cyclin B1, and Cdk-4 downregulation, while cdk-2 and -6, as well as cdc2 were expressed at steady-state levels. Piceatannol also reduced the abundance of $p27^{Kip1}$, whereas the protein level of cyclins A and E was enhanced. Based on its effects on the cell cycle, these investigators concluded that piceatannol can be considered as a promising chemopreventive or anticancer agent [38]. Our group has also reported on the chemosensitizing effects of piceatannol in AIDS-related non-Hodgkin's lymphoma (ARL). Piceatannol exerted its effects via inhibition of the constitutive activity of the signal transducer and activator of transcription 3 (STAT3) culminating in inhibition of the expression of antiapoptotic Bcl-2 protein levels. Consequently, highly drug-resistant ARL cells exhibited a lower apoptosis threshold and were sensitized to the apoptotic effects of various drugs including adriamycin (ADR), cisplatinum (CDDP), etoposide (VP-16), fludarabine, 5-FU, and vinblastine [39].

Recently, a series of styrylquinazoline derivatives (2a-k) were evaluated for their inhibition of prostaglandin E(2) (PGE(2)) production by cyclo-oxygenase-2 (COX-2) upon lipopolysaccharide (LPS) stimulation of macrophage cells RAW264.7. These included 3′,4′-dihydroxylated styryl-quinazolines, 3′-hydroxylated styrylquinazolines, and 3′-acetoxystyrylquinazolines, all of which exhibited good inhibitory effects of PGE(2) production by COX-2 with a range of IC(50) values of 1.19 to approximately 3.56 μM. The potencies were comparable or better than that of RSV (IC(50) = 3.07 μM) suggesting that styrylquinazolines can be considered as potential RSV analogs in the modulation of prostaglandin production by COX-2 [40].

Currently, another synthetic RSV analog has been investigated for its growth inhibitory effects in *in vitro* assays. 3,4,4′,5-Tetramethoxystilbene (DMU 212) exhibits growth-inhibitory and proapoptotic properties in human-derived colon cancer cells HCA-7 and HT-29. The pharmacokinetic properties of DMU 212 were compared with those of RSV in the plasma, liver, kidney, lung, heart, brain, and small intestinal and colonic mucosa of mice. Resveratrol showed significantly higher plasma and liver levels than DMU 212, while DMU 212 exhibited superior availability in the small intestine and colon. In the light of the superior levels achieved in the gastrointestinal tract after DMU 212 administration, compared to RSV, the authors suggest evaluating DMU 212 as a colorectal cancer chemo-preventive agent [41]. Also, the roots of Israeli *Rumex bucephalophorus* (stilbene-*O*-methyl derivatives) contain RSV analogs including 5,4′-dihydroxy-3-methoxystilbene and 3,5-dihydroxy-4′-methoxystilbene, both of which possess antioxidant capabilities [42].

In an attempt to identify new therapeutic compounds, a series of *cis*- and *trans*-stilbene-based RSVs were synthesized and tested *in vitro* for cellular proliferation, inhibition, and apoptosis induction in HL60 promyelocytic leukemia cells. The results of this study show that the *cis*-3, 5-dimethoxy derivatives of rhapontigenin 10a (*trans*-3,3′,5-trihydroxy-4′-methoxystilbene) and its 3′-amino derivative 10b have apoptotic activity at nanomolar concentrations and were more potent than several classic chemotherapeutic compounds in inducing apoptosis in resistant HL60R cells. These investigators suggest that structural alteration of the stilbene motif of RSV can be extremely effective in generating potent apoptosis-inducing agents [43]. Other RSV analogs, *trans*-3,4,5-trimethoxystilbene and *trans*-2′,3,4′,5-tetramethoxystilbene, were found to be more potent than RSV in killing lung and colon cancer cells [44].

To find more efficient antioxidants by structural modification, the RSV analogs *trans*-3,4-dihydroxystilbene (3,4-DHS), *trans*-4,4′-dihydro-xystilbene (4,4′-DHS), *trans*-4-hydroxystilbene (4-HS), and *trans*-3,

5-dihydroxystilbene (3,5-DHS) were synthesized and their antioxidant activity studied for the free radical-induced peroxidation of rat liver microsomes *in vitro*. It was found that all of these *trans*-stilbene derivatives are effective antioxidants against peroxidation of rat liver microsomes with an activity sequence of 3,4-DHS > 4,4′-DHS > RSV > 4-HS > 3,5-DHS [45].

CHEMOPREVENTION

RSV has attracted the attention of several researchers for its role as a cancer chemopreventive agent [21]. Jang et al. [20] reported that RSV may affect either the initiation or promotion and progression of the cancer process. The underlying mechanism, however, of RSV protective effect is unknown, although it has been suggested to include inhibition of cytochrome P450 enzymes, antiinflammatory and antioxidant properties, and effects on cell cycle, cell proliferation, and apoptosis [46]. RSV inhibits CYP1A1, thus inhibiting the metabolic activation of procarcinogens [47,48]. RSV induces oxidative damage-induced H_2O_2 in several cancer cell lines, thus exhibiting its antioxidant activity [49]. Subbaramaiah et al. [50] reported that RSV inhibits the transcription and expression of COX-2 and may protect experimental carcinogenesis. Downstream inhibition of suppression of COX-2 is due in part to RSV-induced inhibition of NF-κB [51,52]. It is possible that the chemopreventive activity can also be accounted for by the immunosensitizing activity of RSV *in vivo* by decreasing the apoptosis threshold for the host cytotoxic immune cells.

SENSITIZING ACTIVITIES OF RSV

IMMUNOSENSITIZATION

Surgical resection, radiotherapy, and chemotherapy remain the dominant weapons in the arsenal for the treatment of cancer. However, these approaches are rarely curative and are often limited by undesirable toxicities and the selective overgrowth of resistant tumor cells. Biological therapies might lead the next front in the battle against cancer, as many hold promise of selectively targeting tumors while minimizing toxic side effects, as well as circumventing or overcoming acquired tumor resistance against conventional treatments. Apo2L/TRAIL is a member of the TNF gene superfamily that induces apoptosis through engagement of death receptors (DRs). *In vitro* studies demonstrate preferential apoptosis-inducing activity of Apo2L/TRAIL against cancer cells when compared with normal cells [7]. *In vivo* studies reveal significant antitumor efficacy of Apo2L/TRAIL in various tumor xenograft models [53]. Furthermore, safety studies in nonhuman primates show that Apo2L/TRAIL administration at relatively high doses is well tolerated, and pharmacokinetic studies suggest that therapeutically relevant doses of this protein may be achievable in humans. Such findings have fueled continued interest in Apo2L/TRAIL as a potential agent for clinical investigation in cancer therapy.

Apo2L/TRAIL was discovered by virtue of its structural homology to members of the TNF gene superfamily. It shares 28 and 23% amino acid sequence identity with Fas ligand and TNF, respectively. Native Apo2L/TRAIL is expressed as a type II transmembrane protein (N-terminal located in the cell interior and C-terminal on the exterior), and its extracellular region can be cleaved proteolytically to release a soluble molecule. Crystallographic studies on soluble Apo2L/TRAIL reveal a homotrimeric molecule, stabilized by an internal zinc atom that is coordinated to three cysteine residues, one at position 230 of each ligand subunit [54]. The complex receptor system for Apo2L/TRAIL comprises four exclusive receptors. Two of these, DR4 (TRAIL receptor-1) and DR5 (TRAIL receptor-2, KILLER, TRICK2), are DRs containing a cytoplasmic death domain (DD) that is able to transduce the apoptotic signal. By contrast, two other receptors, decoy receptor (DcR)1 (TRAIL receptor-3, TRIDD, LIT), which lacks an intracellular death domain, and DcR2 (TRAIL receptor-4, TRUNDD), which contains a truncated nonfunctional death domain, are unable to initiate apoptotic cell death and are believed to act as decoys. Although another TNF family receptor, osteoprotegerin (OPG), binds Apo2L/TRAIL with low affinity, this interaction appears to be of minimal physiological significance. In the mouse, only a single receptor resembling human DR4 and DR5 exists, along with two DcRs that are only distantly related by sequence to human DcR1 and DcR2, and one close homolog of OPG [54].

Apo2L/TRAIL induces apoptosis through activation of DR4 and/or DR5, which engages the "cell-extrinsic" apoptosis pathway. Upon binding of Apo2L/TRAIL, DR4 and DR5 can each recruit and activate the apoptosis-initiating proteases caspase-8 and caspase-10 through the death-domain-containing adaptor molecule Fas-associated death domain (FADD). Caspase-8 and -10 in turn activate effector caspases-3, -6, and -7, which execute the cell's apoptotic demise. In some cancer cell lines, Apo2L/TRAIL-induced activation of caspase-3 is further augmented through engagement of the "cell-intrinsic" apoptosis pathway. Accordingly, caspase-8 or caspase-10 cleaves and activates the proapoptotic Bcl-2 family member Bid, which then interacts with two other proapoptotic family members, Bax and Bak, to induce the release of cytochrome c and Smac/DIABLO from mitochondria. Cytochrome c, together with Apaf-1, activates caspase-9, which contributes to further activation of caspase-3, -6, and -7. By binding to IAPs, Smac/DIABLO prevents IAPs from physically binding caspase-3, hence promoting further caspase-3 activation.

Similar to other apoptosis-inducing systems such as radiation and chemotherapeutic drugs, inherent and/or acquired resistance to Apo2L/TRAIL has been observed [7]. This has spurred the development of strategies to sensitize tumor cells to Apo2L/TRAIL-induced apoptosis. Studies from our laboratory as well as others have demonstrated that RSV can lower the apoptosis threshold and sensitizes non-Hodgkin's lymphoma (NHL) and multiple myeloma (MM) tumor cells to Apo2L/TRAIL [16] through the inhibition of the ERK1/2 pathway and Bcl-$_{xL}$ downregulation. Fulda and Debatin [55] showed that RSV is a potent sensitizer of tumor cells for Apo2L/TRAIL-induced apoptosis through p53-independent induction of p21^{Waf1} and p21^{Waf1}-mediated cell cycle arrest associated with survivin depletion. This effect, however, was not observed in normal human fibroblasts.

CHEMOSENSITIZATION

In addition to immune-sensitizing effects, RSV also has chemosensitizing attributes. Fulda and Debatin [56] showed that RSV pretreatment, in a concentration- and time-dependent manner, cooperated with various drugs including etoposide (VP-16), doxorubicin, cytarabine (AraC), actinomycin D, paclitaxel, and methotrexate to induce apoptosis in SHEP neuroblastoma cells. The molecular mechanisms of chemosensitization involved cell cycle arrest in S phase leading to decreased survivin expression, independent of p53, which subsequently augmented caspase-9 and -3 activation. IAP family members (c-IAP-1, -2, XIAP, survivin) suppress apoptotic pathways initiated by stimuli that release cytochrome c from mitochondria via binding to and ablating of the activation of caspases-9, -3, -6, and -7 but not the initiator caspase-8 [57,58]. Thus, high levels of IAPs confer apoptosis resistance to the tumor cells. Likewise, high expression of survivin confers survival advantage and chemoresistance to tumor cells. Alternatively, decreased levels of IAPs can potentially activate multiple signaling pathways that result in cell cycle delay or block in the G1-S or G2-M transition phase. Because most chemotherapeutic agents use the mitochondrial pathway and the fact that IAPs do not bind to caspase-8, high expression of IAPs represents an additional level of protection of tumor cells against chemotherapy [57,58]. Since inhibition of caspases by IAPs occurs in the effector phase of apoptosis, a central point where multiple apoptosis pathways converge, therapeutic modulation of IAPs could target a key mechanism of cancer resistance. These investigators also showed that RSV cooperates with cytotoxic drugs to inhibit clonogenic tumor cell growth [55]. Kubota et al. [59] studied the *in vitro* biological activity of RSV by examining its effect on proliferation and inducing apoptosis in lung cancer cell lines. RSV inhibited the growth of these cells by 50% (ED50) at concentrations between 5 and 10 μM. Also, RSV (10 μM, 3 days) significantly enhanced the antiproliferative effect of paclitaxel and enhanced the apoptotic activity of paclitaxel. Thus, RSV may be used in the therapy of lung cancer as cells exposed to RSV have a lowered threshold for killing by paclitaxel.

Pretreatment with RSV also sensitizes the drug-resistant NHL and MM cells to CDDP-, etoposide-, ADR-, and paclitaxel-induced apoptosis as well as inhibiting the proliferation rate of the tumor cells, while the combination treatment had minimal toxicity towards freshly derived human peripheral blood mononuclear cells (PBMCs). In our studies RSV inhibited Bcl-$_{xL}$ and Mcl-1 expression [16]. These proteins play an important role in inhibiting apoptosis by protecting mitochondrial collapse. RSV has been shown to upregulate Bax and downregulate Bcl-$_{xL}$ expression concurrent with an increase in sub-G1 fraction [60]. Overexpression of Bcl-$_{xL}$ inhibits paclitaxel-mediated apoptosis [61], which is in agreement with our findings. Thus, RSV-mediated downregulation of Bcl-$_{xL}$ expression may be responsible, in large part, for the observed chemosensitization which was confirmed by functional impairment of Bcl-$_{xL}$ that sensitized the cells to drug-induced apoptosis. The ERK1/2 MAPK pathway regulates Bcl-$_{xL}$ expression via AP-1 [62]. Thus, we extended our studies to show that RSV partially blocks the phosphorylation of ERK1/2 and diminished the DNA binding activity of AP-1 and AP-1-dependent Bcl-$_{xL}$ gene expression. The role of the ERK1/2 pathway in the regulation of Bcl-$_{xL}$ expression was corroborated by using specific pharmacological inhibitors of the ERK1/2 pathway, which decreased Bcl-$_{xL}$ gene expression and induced chemosensitization consistent with previous reports where RSV inhibits the ERK1/2 pathway [63] and inhibition of the ERK1/2 pathway augments the cytotoxic effects of paclitaxel [64,65].

A major survival role of high expression of Mcl-1 in myeloma [66] and lymphoma cell lines [67,68] is proposed. Downregulation of Mcl-1 expression by antisense approach [69] or immunotherapy (anti-CD20 mAb) [70] sensitizes myeloma and leukemia cells to chemotherapy. Decreased Mcl-1 protein stability paralleled the increased sensitivity of paclitaxel-resistant ovarian carcinoma cell lines to paclitaxel [71]. Whereas a major antiapoptotic role of Bcl-2 has been proposed, in Bcl-2-deficient Ramos cells [72], high levels of Bcl-$_{xL}$ and Mcl-1 may indirectly interfere with downstream events that lead to apoptosis, which in turn may contribute to drug resistance. Accordingly, our data suggest that downregulation of Bcl-$_{xL}$ and Mcl-1 by RSV may be partly implicated in the enhanced sensitivity of lymphoma cells to drug-induced apoptosis. Indeed, Mcl-1 expression is also regulated by the ERK1/2 pathway [73,74]. Thus, inhibition of the ERK1/2 signaling pathway and decreased expression of Bcl-$_{xL}$ and Mcl-1 by RSV might, in part, be accountable for chemosensitization.

Although traditional chemotherapeutic compounds eradicate tumor cells mainly through the induction of apoptosis, these agents exert their effects via distinct mechanisms. For instance, the cytotoxic activity of vinca alkaloid paclitaxel is due to its ability to disrupt microtubules, thus inducing metaphase arrest in dividing cells. The antiproliferative effects of the vinca alkaloids at their lowest effective concentrations are due to the inhibition of mitotic spindle function. This results from drug-induced alterations in the dynamics of tubulin addition and loss at the ends of mitotic spindle microtubules rather than by simply depolymerizing the microtubules resulting in the arrest of the cells at the G2/M phase of the cell cycle. Paclitaxel also phosphorylates the antiapoptotic proteins Bcl-2 and Bcl-$_{xL}$ and downregulates their expression [75].

Anthracycline antibiotics (e.g., doxorubicin; ADR) have both cytostasis and cytotoxic actions, through intercalation into DNA with subsequent inhibition of macromolecule biosynthesis, free radical formation with consequent induction of DNA damage or lipid peroxidation, DNA binding and alkylation, DNA cross-linking, interference with DNA unwinding or DNA strand separation and helicase activity, direct membrane effects, and the initiation of DNA damage [76]. Cisplatin (*cis*-diammine-dichloroplatinum, CDDP) is among the most widely used and broadly active anticancer agents. CDDP demonstrates significant activity against tumors of the ovary, bladder, lung, head, and neck, and testicular cancer, and exerts its cytotoxic effects through the formation of DNA adducts in which the chloride ligands of the drug are replaced by specific DNA bases. The DNA adducts of cisplatin mediate cytotoxicity by inhibiting DNA replication and transcription, and by activating the apoptotic pathway [77]. Etoposide (VP-16) induces apoptosis via inhibition of DNA topoisomerase II. Several mechanisms have been implicated in drug resistance, including reduced drug uptake, increased DNA repair, and altered expression of oncogenes that activate the antiapoptotic pathways.

The discovery of an agent that enhances the sensitivity of drug-refractory tumor cells with minimal toxicity towards normal cells is the ultimate goal in cancer research. RSV-mediated augmentation of tumor cell sensitivity to a broad range of structurally and functionally distinct anticancer drugs signifies the important role of this agent, used either alone or combined with other antineoplastic agents, in clinical oncology in the treatment of drug-refractory tumors.

Based on the work from our laboratory and those of others we have proposed a model by which RSV sensitizes the drug- and immune-resistant cells to drug- and immune-induced apoptosis (Figure 15.1). Accordingly, RSV, either via a receptor-mediated pathway or by direct infusion through the plasma membrane, interferes with the type I and type II apoptosis signaling pathways. With respect to type I, depending on the amount of caspase-8, RSV either directly induces caspase-8 autocleavage/processing (via an elusive mechanism), which leads to caspase-3 activation and PARP cleavage with subsequent induction of apoptosis; or, in the event that cells lack sufficient amounts of caspase-8, RSV utilizes type II apoptotic signaling pathway involving mitochondria where the proapoptotic Bcl-2 family member, Bid, will be cleaved. Truncated Bid (tBid) will then migrate to and reside in the mitochondrial outer membrane, where, in collaboration with other proapoptotic molecules, mainly Bax and Bad induces the formation of mitochondrial permeability transition pore (PTP). RSV has also been demonstrated to decrease the expression of antiapoptotic Bcl-$_{xL}$. Decreased levels of Bcl-$_{xL}$ plus the presence of tBid and increased expression of Bax will alter the cellular ratio of pro-/antiapoptosis Bcl-2 family members and favors apoptosis mediated by drugs and/or immune-based strategies. These events will destabilize mitochondria resulting in the release of apoptogenic molecules such as cytochrome c and Smac/DIABLO. RSV also upregulates the expression levels of apoptosis protease activating factor-1 (Apaf-1), which in combination with cytochrome c will facilitate the assembly of the apoptosome complex (dATP/ATP/Apaf-1/cytochrome c/caspase-9). RSV can reduce the expression of certain inhibitors of apoptosis protein (IAP) family members, thereby removing the inhibitors of caspase activation. However, these events by themselves are not sufficient for the full induction of apoptosis. The process is facilitated and expedited by the action of cytotoxic agents, which also alter the expression profile of apoptosis-associated gene products. Through autocatalytic activation, caspase-9 becomes processed/activated and in the absence of natural inhibitors of apoptosis (IAPs) will use caspases-3, -6, -7 as substrate and apoptosis will ensue.

IN VIVO APPLICATION OF RSV

RSV might represent a promising agent to be tested for cancer chemopreventive activity and also as a sensitizing agent for cancer therapeutics in the clinical treatment of resistant tumors. The concentration of RSV with *in vitro* biological activity ranges from 5 to 50 μM. The concentration for *in vivo* effectiveness is not clear. Study in rats shows that RSV is available at 38% after oral administration suggesting that RSV undergoes *in vivo* extensive glucuronidation [78]. Orally administered RSV in mice was shown to remain mostly in its intact form and can penetrate tissues [79]. Clearly, studies in humans must be undertaken to establish the therapeutic clinical dose as well as toxicity.

CONCLUDING REMARKS

In addition to its pleiotropic effects, RSV and new analogs offer a new class of chemo- and immunosensitizing agents with an ability to reverse drug- and immune-resistance phenotypes of cancer cells when used in combination with conventional therapeutics. In addition, these sensitizing effects of RSV on different tumor cells and delineation of the underlying molecular mechanisms involved will identify new targets that regulate resistance and such targets can be used for the development of new therapeutics. The

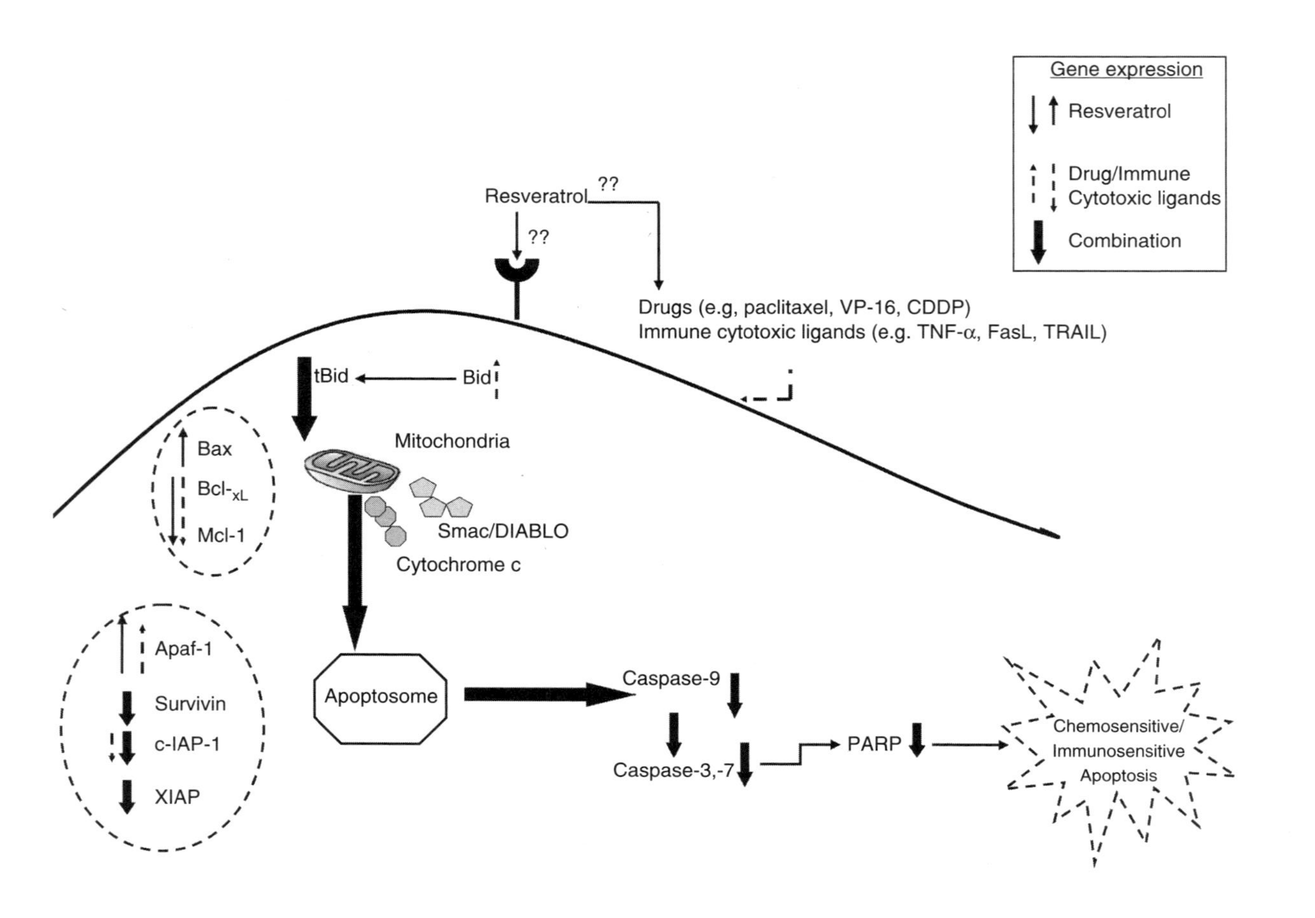
Gene expression
Resveratrol
Drug/Immune
Cytotoxic ligands
Combination
Resveratrol
??
??
Drugs (e.g, paclitaxel, VP-16, CDDP)
Immune cytotoxic ligands (e.g. TNF-α, FasL, TRAIL)
tBid
Bid
Mitochondria
Smac/DIABLO
Cytochrome c
Bax
Bcl-xL
Mcl-1
Apaf-1
Survivin
c-IAP-1
XIAP
Apoptosome
Caspase-9
Caspase-3,-7
PARP
Chemosensitive/
Immunosensitive
Apoptosis

FIGURE 15.1 Proposed model of RSV-mediated chemo-/immunosensitization. Based on the work from our laboratory and those of others we postulate a potential model by which RSV enhances the chemosensitivity of drug-refractory tumor cells, and by lowering the apoptosis threshold, in combination with chemotherapeutic drugs, induces synergistic apoptosis. Although the chemosensitizing effects of RSV have been observed in conjunction with various drugs, paclitaxel is used as a representative drug to elucidate the signaling events. The human NHL cell line Ramos is used as a representative model. According to this model RSV, either via a receptor-mediated pathway or by direct diffusion through the plasma membrane, interferes with apoptosis signal transduction pathways via regulation of gene expression profile. The molecular events triggered by RSV include a decrease in the expression of Bcl-$_{xL}$ (at the transcriptional and translational level, via inhibition of the ERK1/2 MAPK pathway and diminution of AP-1 transcriptional activity) and Mcl-1 antiapoptotic proteins plus significant induction of proapoptotic Bax and apoptosis protease-activating factor-1 (Apaf-1). For example, drugs like paclitaxel downregulate antiapoptotic protein c-IAP-1 and upregulate the expression of proapoptotic proteins Bid and Apaf-1. Also, an immune cytotoxic ligand like TRAIL also modifies antiapoptotic gene products such as Bcl-2 and Bcl-$_{xL}$. However, these various modulatory effects by themselves are insufficient for the full induction of apoptosis. The combination treatment, via functional complementation, results in the formation of proapoptotic truncated Bid (tBid) and the induction of apoptosis. tBid migrates to and resides in the mitochondrial outer membrane. Decreased levels of Mcl-1 and antiapoptotic Bcl-2 family members by resveratrol and drug/immune cytotoxic ligands are more pronounced by the combination treatment. The presence of tBid and high levels of Bax will alter the ratio of proapoptotic/antiapoptotic Bcl-2 family members. Decrease in these ratios, which are key determinants in the cellular fate in response to noxious stimuli, will result in the formation of permeability transition pore and collapse in $\Delta\Psi m$ and will facilitate the cytosolic release of apoptogenic molecules (cytochrome c and Smac/DIABLO). Smac/DIABLO binds to and neutralizes the inhibitory effects of IAPs. Hence, increased levels of Apaf-1 in combination with cytochrome c and dATP/ATP facilitate the assembly of the apoptosome complex. Through autocatalytic processing, caspase-9 becomes activated, and in the presence of diminished levels of natural inhibitors (survivin, XIAP, and c-IAP-1), it activates caspases-3 and -7 to subsequently cleave PARP and induce apoptosis.

apparent low toxicity of RSV also offers advantages such that RSV can be used as adjuvant therapy in cancer patients in remission as it may sensitize arising tumor cells to host-mediated immune destruction. From animal studies, RSV has been shown to be effective when taken orally and its use in humans by this route will be advantageous over other routes. The potential disease-preventive activity mediated by RSV is an area of considerable importance and awaits clinical validation in humans.

ACKNOWLEDGMENT

The authors acknowledge the assistance of Christine Yue in the preparation of the manuscript.

REFERENCES

1. Patel NH and Rothenberg ML, Multidrug resistance in cancer chemotherapy, *Invest New Drugs* 12, 1–13, 1994.
2. Klaus SJ, Sidorenko SP, and Clark EA, CD45 ligation induces programmed cell death in T and B lymphocytes, *J Immunol* 156, 2743–2753, 1996.
3. Thompson I, Feigl P, and Coltman C, Chemoprevention of prostate cancer with finasteride, *Important Adv Oncol* 57–76, 1995.
4. Shresta S, Pham CT, Thomas DA, Graubert TA, and Ley TJ, How do cytotoxic lymphocytes kill their targets? *Curr Opin Immunol* 10, 581–587, 1998.
5. Ashkenazi A and Dixit VM, Death receptors: signaling and modulation, *Science* 281, 1305–1308, 1998.
6. Scaffidi C, Fulda S, Srinivasan A, Friesen C, Li F, Tomaselli KJ, Debatin KM, Krammer PH, and Peter ME, Two CD95 (APO-1/Fas) signaling pathways, *EMBO J* 17, 1675–1687, 1998.
7. Ng CP and Bonavida B, A new challenge for successful immunotherapy by tumors that are resistant to apoptosis: two complementary signals to overcome cross-resistance, *Adv Cancer Res* 85, 145–174, 2002.
8. Ferreria CG, Epping M, Kruyt FAE, and Giaccone G, Apoptosis: target of cancer therapy, *Clin Cancer Res* 8, 2024–2034, 2002.
9. Sandor V, Wilson W, Fojo T, and Bates SE, The role of MDR-1 in refractory lymphoma, *Leuk Lymphoma* 28, 23–31, 1997.
10. Filipits M, Jaeger U, Simonitsch I, Chizzali-Bonfadin C, Heinzl H, and Pirker, R, Clinical relevance of the lung resistance protein in diffuse large B-cell lymphomas, *Clin Cancer Res* 6, 3417–3423, 2000.
11. Reed JC, Bcl-2 family proteins: regulators of chemoresistance in cancer, *Toxicol Lett* 82, 155–158, 1995.
12. Minn AJ, Rudin CM, Boise LH, and Thompson CB, Expression of Bcl-xL can confer a multi-drug resistance phenotype, *Blood* 86, 1903–1910, 1995.
13. Jazirehi AR, Ng C-P, Gan X-H, Schiller G, and Bonavida BB, Adriamycin sensitized the adriamycin-resistant 8226/Dox40 human multiple myeloma cells to Apo2L/TRAIL mediated apoptosis, *Clin Cancer Res* 7, 3874–3883, 2001.

14. Tan B, Piwnica-Worms D, and Ratner L, Multidrug resistance transporters and modulation, *Curr Opin Oncol* 12, 450–458, 2000.
15. Kaufmann SH and Earnshaw WC, Induction of apoptosis by cancer chemotherapy, *Exp Cell Res* 256, 42–49, 2000.
16. Jazirehi AR and Bonavida B, Resveratrol modifies the expression of apoptotic regulatory proteins and sensitizes non-Hodgkin's lymphoma and multiple myeloma cell lines to paclitaxel-induced apoptosis, *Mol Cancer Ther* 3, 71–84, 2004.
17. Park EJ and Pezzuto JM, Botanicals in cancer chemoprevention, *Cancer Metastasis Rev* 21, 231–255, 2002.
18. Cal C, Garban H, Jazirehi A, Yeh C, Mizutani Y, and Bonavida B, Resveratrol and cancer: chemoprevention, apoptosis, and chemo-immunosensitizing activities, *Curr Med Chem Anti-Cancer Agents* 3, 77–93, 2003.
19. Nonomura S, Kanagawa H, and Makimoto A, Chemical constituents of polygonaceous plants: I. Studies on the components of Ko-jo-kon (*Polygonum cuspidatum* SIEB et ZUCC), *Yakugaku Zasshi* 83, 983–988, 1963.
20. Jang M, Cai L, Udeani GO, Slowing KV, Thomas CF, Beecher CW, Fong HH, Farnsworth NR, Kinghorn AD, Mehta RG, Moon RC, and Pezzuto JM, Cancer chemo-preventive activity of resveratrol, a natural product derived from grapes, *Science* 275, 218–220, 1997.
21. Pervaiz S, Resveratrol: from grapevines to mammalian biology, *FASEB J* 17, 1975, 2003.
22. Bianchini F and Vainioh H, Wine and resveratrol: mechanisms of cancer prevention? *Eur J Cancer Prev* 12, 417–425, 2003.
23. Wu SL, Sun ZJ, Yu L, Meng KW, Qin XL, and Pan CE, Effect of resveratrol and in combination with 5-FU on murine liver cancer, *World J Gastroenterol* 10, 3048–3052, 2004.
24. Chen Y, Tseng SH, Lai HS, and Chen WJ, Cellular apoptosis and cell cycle arrest in neuroblastoma cells and antitumor effects on neuroblastoma in mice, *Surgery* 136, 57–66, 2004.
25. Fuggetta MP, D'Atri S, Lanzilli G, Tricarico M, Cannavo E, Zambruno G, Falchetti R, and Ravagnan G, *In vitro* antitumour activity of resveratrol in human melanoma cells sensitive or resistant to temozolomide, *Melanoma Res* 14, 189–196, 2004.
26. Niles RM, McFarland M, Weimer MB, Redkar A, Fu YM, and Meadows GG, Resveratrol is a potent inducer of apoptosis in human melanoma cells, *Cancer Lett* 190, 157–163, 2003.
27. Laux MT, Aregullin M, Berry JP, Flanders JA, and Rodriguez E, Identification of a p53-dependent pathway in the induction of apoptosis of human breast cancer cells by the natural product, resveratrol, *J Alternative Complementary Med* 10, 235–239, 2004.
28. Scarlatti F, Sala G, Somenzi G, Signorelli P, Sacchi N, and Ghidoni R, Resveratrol induces growth inhibition and apoptosis in metastatic breast cancer cells via de novo ceramide signaling, *FASEB J* 17, 2339–2341, 2003.
29. Ding XZ and Adrian TE, Resveratrol inhibits proliferation and induces apoptosis in human pancreatic cancer cells, *Pancreas* 25, e71–76, 2002.
30. Kuo PL, Chiang LC, and Lin CC, Resveratrol-induced apoptosis is mediated by p53-dependent pathway in Hep G2 cells, *Life Sci* 72, 23–34, 2002.

31. Roman V, Billard C, Kern C, Ferry-Dumazet H, Izard JC, Mohammad R, Mossalayi DM, and Kolb JP, Analysis of resveratrol-induced apoptosis in human B-cell chronic leukaemia, *Br J Haematol* 117, 842–851, 2002.
32. Kuwajerwala N, Cifuentes E, Gautam S, Menon M, Barrack ER, and Reddy GP, Resveratrol induces prostate cancer cell entry into s phase and inhibits DNA synthesis, *Cancer Res* 62, 2488–2492, 2002.
33. Kim YA, Rhee SH, Park KY, and Choi YH, Antiproliferative effect of resveratrol in human prostate carcinoma cells, *J Med Food* 6, 273–280, 2003.
34. Wang Q, Li H, Wang XW, Wu DC, Chen XY, and Liu J, Resveratrol promotes differentiation and induces Fas-independent apoptosis of human medulloblastoma cells, *Neurosci Lett* 35, 83–86, 2003.
35. Narayanan BA, Narayanan NK, Re GG, and Nixon DW, Differential expression of genes induced by resveratrol in LNCaP cells: P53-mediated molecular targets, *Int J Cancer* 104, 204–212, 2003.
36. Hung LM, Chen JK, Lee RS, Liang HC, and Su MJ, Beneficial effects of astringinin, a resveratrol analogue, on the ischemia and reperfusion damage in rat heart, *Free Radical Biol Med* 30, 877–883, 2001.
37. Ito T, Akao Y, Tanaka T, Iinuma M, and Nozawa Y, Vaticanol C, a novel resveratrol tetramer, inhibits cell growth through induction of apoptosis in colon cancer cell lines, *Biol Pharm Bull* 25, 147–148, 2002.
38. Wolter F, Clausnitzer A, Akoglu B, and Stein J, Piceatannol, a natural analog of resveratrol, inhibits progression through the S phase of the cell cycle in colorectal cancer cell lines, *J Nutr* 132, 298–302, 2002.
39. Alas S and Bonavida B, Rituximab inactivates STAT3 activity in B-non-Hodgkin's lymphoma through inhibition of the interleukin 10 autocrine/paracrine loop and results in down-regulation of Bcl-2 and sensitization to cytotoxic drugs, *Cancer Res* 61, 5137–5144, 2001.
40. Park JH, Min HY, Kim SS, Lee JY, Lee SK, and Lee YS, Styrylquinazolines: a new class of inhibitors on prostaglandin E2 production in lipopolysaccharide-activated macrophage cells, *Arch Pharm* (Weinheim) 337, 20–24, 2004.
41. Sale S, Verschoyle RD, Boocock D, Jones DJ, Wilsher N, Ruparelia KC, Potter GA, Farmer PB, Steward WP, and Gescher AJ, Pharmacokinetics in mice and growth-inhibitory properties of the putative cancer chemopreventive agent resveratrol and the synthetic analogue trans 3,4,5,4′-tetramethoxystilbene, *Br J Cancer* 90, 736–744, 2004.
42. Kerem Z, Regev-Shoshani G, Flaishman MA, and Sivan L, Resveratrol and two monomethylated stilbenes from Israeli Rumex bucephalophorus and their antioxidant potential, *J Nat Prod* 66, 1270–1272, 2003.
43. Roberti M, Pizzirani D, Simoni D, Rondanin R, Baruchello R, Bonora C, Buscemi F, Grimaudo S, and Tolomeo M, Synthesis and biological evaluation of resveratrol and analogues as apoptosis-inducing agents, *J Med Chem* 46, 3546–3554, 2003.
44. Lee SK, Nam KA, Hoe YH, Min HY, Kim EY, Ko H, Song S, Lee T, and Kim S, Synthesis and evaluation of cytotoxicity of stilbene analogues, *Arch Pharm Res* 26, 253–257, 2003.
45. Cai YJ, Fang JG, Ma LP, Yang L, and Liu ZL, Inhibition of free radical-induced peroxidation of rat liver microsomes by resveratrol and its analogues, *Biochim Biophys Acta* 1637, 31–38, 2003.

46. Gusman J, Malonne H, and Atassi G, A reappraisal of the potential chemopreventive and chemotherapeutic properties of resveratrol, *Carcinogenesis* 22, 1111–1117, 2001.
47. Chun YJ, Kim MY, and Guengerich FP, Resveratrol is a selective human cytochrome P450 1A1 inhibitor, *Biochem Biophys Res Commun* 262, 20–24, 1999.
48. Ciolino HP and Yeh GC, Inhibition of aryl hydrocarbon-induced cytochrome P-450 1A1 enzyme activity and CYP1A1 expression by resveratrol, *Mol Pharmacol* 56, 760–767, 1999.
49. Damianaki A, Bakogeorgou E, Kampa M, Notas G, Hatzoglou A, Panagiotou S, Gemetzi C, Kouroumalis E, Martin PM, and Castanas E, Potent inhibitory action of red wine polyphenols on human breast cancer cells, *J Cell Biochem* 78, 429–441, 2000.
50. Subbaramaiah K, Chung WJ, Michaluart P, Telang N, Tanabe T, Inoue H, Jang M, Pezzuto JM, and Dannenberg AJ, Resveratrol inhibits cyclooxygenase-2 transcription and activity in phorbol ester-treated human mammary epithelial cells, *J Biol Chem* 273, 21875–21882, 1998.
51. Banerjee S, Bueso-Ramos C, and Aggarwal BB, Suppression of 7,12-dimethylbenz(a)anthracene-induced mammary carcinogenesis in rats by resveratrol: role of nuclear factor-kappaB, cyclooxygenase 2, and matrix metalloprotease 9, *Cancer Res* 62, 4945–4954, 2002.
52. Holmes-McNary M, and Baldwin AS Jr, Chemopreventive properties of trans-resveratrol are associated with inhibition of activation of the IkappaB kinase, *Cancer Res* 60, 3477–3483, 2000.
53. Ashkenazi A, Pai RC, Fong S, Leung S, Lawrence DA, Marsters SA, Blackie C, Chang L, McMurtrey AE, Hebert A, DeForge L, Koumenis IL, Lewis D, Harris L, Bussiere J, Koeppen H, Shahrokh Z, and Schwall RH, Safety and antitumor activity of recombinant soluble Apo2 ligand, *J Clin Invest* 104, 155–162, 1999.
54. Kelley SK and Ashkenazi A, Targeting death receptors in cancer with Apo2L/TRAIL, *Curr Opin Pharmacol* 4, 333–339, 2004.
55. Fulda S and Debatin KM, Sensitization for tumor necrosis factor-related apoptosis-inducing ligand-induced apoptosis by the chemopreventive agent resveratrol, *Cancer Res* 64, 337–346, 2004.
56. Fulda S and Debatin KM, Sensitization for anticancer drug-induced apoptosis by the chemopreventive agent resveratrol, *Oncogene* 23, 6702–6711, 2004.
57. Roy N, Deveraux QL, Takahashi R, Salvesen GS, and Reed JC, The c-IAP-1 and c-IAP-2 proteins are direct inhibitors of specific caspases, *EMBO J* 16, 6914–6925, 1997.
58. Deveraux QL, Roy N, Stennicke HR, Arsdale TV, Zhou Q, Srinivasula SM et al, IAPs block apoptotic events induced by caspase-8 and cytochrome c by direct inhibition of distinct caspases, *EMBO J* 17, 2215–2223, 1998.
59. Kubota T, Uemura Y, Kobayashi M, and Taguchi H, Combined effects of resveratrol and paclitaxel on lung cancer cells, *Anticancer Res* 23, 4039–4046, 2003.
60. Nakagawa H, Kiyozuka Y, Uemura Y, Shikata N, Hioki K, and Tsubura A, Resveratrol inhibits human breast cancer cell growth and may mitigate the effect of linoleic acid, a potent breast cancer cell stimulator, *J Cancer Res Clin Oncol* 127, 258–264, 2001.

61. Ibrado AM, Liu L, and Bhalla K, Bcl-xL overexpression inhibits progression of molecular events leading to paclitaxel-induced apoptosis of acute myeloid leukemia HL-60 cells, *Cancer Res* 57, 1109–1116, 1997.
62. Sevilla L, Zaldumbide A, Pognonec P, and Boulukos KE, Transcriptional regulation of the bcl-x gene encoding the anti-apoptotic Bcl-xL protein by Ets, Rel/NF-nB, STAT and AP-1 transcription factor families, *Histol Histopathol* 16, 595–601, 2001.
63. El-Mowafy AM and White RE, Resveratrol inhibits MAPK activity and nuclear translocation in coronary artery smooth muscle: reversal of endothelin-1 stimulatory effects, *FEBS Lett* 451, 63–67, 1999.
64. MacKeigan JP, Collins TS, and Ting JP-Y, MEK inhibition enhances paclitaxel-induced tumor apoptosis, *J Biol Chem* 275, 38953–38956, 2000.
65. Dai Y, Yu C, Sing V, Tang L, Wang Z, Molinstry R et al, Pharmacological inhibitors of the mitogen activated protein kinase (MAPK) kinase/MAPK cascade interact synergistically with UCN-01 to induce mitochondrial dysfunction and apoptosis in human leukemia cells, *Cancer Res* 61, 5106–5115, 2001.
66. Zhang B, Gojo I, and Fenton RG, Myeloid cell factor-1 is a critical survival factor for multiple myeloma, *Blood* 99, 1885–1893, 2002.
67. Pagnano KB, Silva MD, Vassallo J, Aranha FJ, and Saad ST, Apoptosis regulating proteins and prognosis in diffuse large B cell non-Hodgkin's lymphomas, *Acta Haematol* 107, 29–34, 2002.
68. Rassidakis GZ, Lai R, McDonnell TJ, Cabanilla F, Sarris AH, and Mederios LJ, Overexpression of Mcl-1 in anaplastic large cell lymphoma cell lines and tumors, *Am J Pathol* 160, 2309–2310, 2002.
69. Derenne S, Monia B, Dean NM, Taylor JK, Rapp MJ, Harousseau JL et al, Antisense strategy shows that Mcl-1 rather than Bcl-2 or Bcl-x(L) is an essential protein of human myeloma cells, *Blood* 100, 194–199, 2002.
70. Byrd JC, Kitada S, Flinn IW, Aron JL, Pearson M, Lucas D et al, The mechanism of tumor cell clearance by rituximab *in vivo* in patients with B-cell chronic lymphocytic leukemia: evidence of caspase activation and apoptosis induction, *Blood* 99, 1038–1043, 2002.
71. Poruchynsky MS, Giannakakou P, Ward Y, Bulinski JC, Telford WG, Robey RW et al, Accompanying protein alterations in malignant cells with a microtubule-polymerizing drug-resistance phenotype and a primary resistance mechanism, *Biochem Pharmacol* 62, 1469–1480, 2001.
72. Shan D, Ledbetter JA, and Press OW, Apoptosis of malignant human B cells by ligation of CD20 with mAbs, *Blood* 91, 1644–1652, 1998.
73. Huang HM, Huang CJ, and Yen JJ, Mcl-1 is a common target of stem cell factor and IL-5 for apoptosis prevention activity via MEK/MAPK and PI-3K/AKT pathways, *Blood* 96, 1764–1771, 2000.
74. Schubert KM and Duroino V, Distinct roles of extracellular-signal regulated protein kinase (ERK) mitogen activated protein kinases and phosphatidylinositol 3-kinase in the regulation of Mcl-1 synthesis, *Biochem J* 356, 473–480, 2001.
75. Rowinsky ER and Donehower RC, Antimicrotubule agents, in *Cancer Chemotherapy and Biotherapy*, 2nd ed, Lippincott, Williams and Wilkins, Philadelphia, 1996, pp. 263–292.

76. Gewirtz DA, A critical evaluation of the mechanism of action proposed for the antitumor effects of anthracycline antibiotics adriamycin and daunorubicin, *Biochem Pharmacol* 57, 772–741, 1999.
77. Trimmer EE and Essigmann JM, Cisplatin, *Essays Biochem* 34, 191–211, 1999.
78. Marier JF, Vachon P, Gritsas A, Zhang J, Moreau JP, and Ducharme MP, Metabolism and disposition of resveratrol in rats: extent of absorption, glucuronidation, and enterohepatic recirculation evidenced by a linked-rat model, *J Pharmacol Exp Ther* 302, 369–373, 2002.
79. Vitrac X, Desmouliere A, Brouillaud B, Krisa S, Deffieux G, Barthe N, Rosenbaum J, and Merillon JM, Distribution of [14C]-trans-resveratrol, a cancer chemopreventive polyphenol, in mouse tissues after oral administration, *Life Sci* 72, 2219–2233, 2003.

16 Resveratrol as a Radio-Protective Agent

Yogeshwer Shukla, Shannon Reagan-Shaw, and Nihal Ahmad

CONTENTS

Ultraviolet (UV) radiation (particularly its UVB component; 290 to 320 nm) elicits a variety of adverse effects, collectively referred to as the *UV response*. UVB radiation is an established cause of about 90% of skin cancers and is regarded as a causative factor for the most-invasive melanoma as well as the precancerous conditions such as actinic keratosis. UV radiation also elicits a variety of other adverse effects including erythema, sunburn cells, inflammation, hyperplasia, hyperpigmentation, immunosuppression, and premature skin aging. The American Cancer Society and other similar organizations have made considerable efforts in educating the people about (1) the use of sunscreens and (2) protective clothing as preventive strategies for skin cancer. However, despite these efforts, the incidence of skin cancer and other skin-related disorders is on the rise. Therefore, there is an urgent need to develop mechanism-based approaches for prevention/therapy of skin cancer; *chemoprevention* via nontoxic agents could be one such approach.

In this chapter we discuss the preventive effects of resveratrol, an antioxidant found abundantly in grapes and red wine, against UVB exposure-mediated damages *in vitro* as well as *in vivo*. We also discuss studies showing that resveratrol can act as a potential sensitizer to enhance the antiproliferative effects of ionizing radiation against cancer cells. Thus, based on available literature, it could be suggested that resveratrol may be useful (1) for the prevention of UVB-mediated damages including skin cancer and (2) in enhancing the therapeutic response of ionizing radiations against cancer cells.

INTRODUCTION

Radiation can be beneficial or harmful depending on its type. Radiation has a wide range of energies in the electromagnetic spectrum, with two major divisions: nonionizing and ionizing radiation. Radiation that possesses enough energy to move atoms around in a molecule or cause them to vibrate, but not enough to change them chemically, is referred to as nonionizing radiation. Examples of this kind of radiation are sound waves, visible light, and microwaves. The radiation that falls within the ionizing radiation range has enough energy to actually break chemical bonds. This is the type of radiation that is actually perceived as "radiation." Radiations are used to generate electric power, to kill cancer cells, and in many manufacturing processes. In day-to-day life, we take advantage of the properties of nonionizing radiation for common tasks: e.g., microwave radiation in telecommunications and heating food, infrared radiation in infrared lamps used for a variety of purposes, and radio waves in broadcasting. Ionizing radiation such as higher frequency UV radiation begins to have enough energy to break chemical bonds. X-ray and gamma-ray radiation, which are at the higher energy end of the electromagnetic spectrum, have very high frequency. Radiation in this range has extremely high energy that may be harmful for a variety of reasons. Because of its prevalence, ultraviolet (UV) radiation from the sun is regarded as the most important and potentially hazardous radiation and environmental toxicant.

Solar UV radiation (particularly its UVB component; 290 to 320 nm) elicits a variety of adverse effects, collectively referred to as the UV response. UVB radiation is an established cause of about 90% of skin cancers [1]. Further, UV radiation is a causative factor for the most-invasive melanoma as well as the precancerous conditions such as actinic keratosis (AK) [2]. UV radiation also elicits a variety of other adverse effects including erythema, sunburn cells, inflammation, hyperplasia, hyperpigmentation, immunosuppression, and premature skin aging [3]. At present, the two major means of protection from harmful solar radiation are the use of sunscreen and protective clothing. The American Cancer Society and other similar organizations have made considerable efforts in educating people about these two preventive strategies. However, despite these efforts, the incidence of skin cancer and other skin-related disorders is on the rise. Thus, it appears that the available options are inadequate for the management of UV damages including skin cancers. Therefore, there is an urgent need to develop mechanism-based novel approaches for prevention/therapy of skin cancer.

It is known that UV exposure results in an excessive generation of reactive oxygen species (ROS) such as the superoxide anion and hydrogen peroxide [4]. The ROS are short-lived species and can react with DNA, protein, and unsaturated fatty acids thereby causing DNA strand breaks, protein–protein DNA crosslinks, and oxidative damage [5]. ROS have been implicated in many pathological conditions including skin cancer and other adverse effects caused by UV radiation [5]. The skin is known to possess its own "built-in" antioxidant mechanisms to defend itself from the damages caused by the ROS [6]. However, an excessive exposure to ROS may render the defense system incapable of coping with their damaging effects, thereby leading to a variety of pathological conditions including skin cancer. In such a scenario, exogenous supplementation of the antioxidants may be a useful strategy for the management of UV exposure-mediated damages including skin cancer. *Chemoprevention*, a newer dimension in the management of neoplasia (or other diseases), could offer a hope in this direction.

Chemoprevention, by definition, is a means of cancer control in which the occurrence of the disease can be entirely prevented, slowed, or reversed by the administration of one or more naturally occurring and/or synthetic compounds [7]. The expanded definition of cancer chemoprevention also includes the chemotherapy of precancerous lesions. Chemoprevention differs from cancer treatment in that the goal of this approach is to lower the rate of cancer incidence. In recent years, some naturally occurring compounds, especially antioxidants, present in the common diet and beverages consumed by human populations have gained considerable attention as chemopreventive agents for potential human benefit [8,9].

In this chapter we provide information regarding the preventive effects of resveratrol, a strong antioxidant amply present in grapes and red wine, against UV radiation-mediated damages. We also discuss studies showing that resveratrol can act as a potential sensitizer to enhance the antiproliferative effects of ionizing radiation against cancer cells.

SOLAR UV RADIATION AND SKIN DAMAGES

The most dangerous outcome of excessive exposure to solar UV radiation is the development of skin cancer that accounts for considerable worldwide morbidity and mortality [1].

A variety of causative factors, acting over a period of several years, result in the development of cancer. Many specific causes of cancer are now known, the most important being lifestyle, occupational exposure, and environmental toxicants including harmful radiations [10]. The known risk factors for cancer can be broadly divided into two categories: environmental factors and host factors. Environmental factors can be further subdivided into chemical, physical, and biological factors. Physical agents such as UV and the ionizing radiations (x-rays, gamma-rays) are known to be hazardous to nearly all tissues or organs of human or experimental animals depending upon the radiation dose and exposure schedule.

UV radiation from the sun is the major cause of skin cancer. UV radiation is broadly divided into three categories according to wavelength: short-wavelength UVC (200 to 280 nm), mid-wavelength UVB (280 to 320 nm), and long-wavelength UVA (320 to 400 nm). Among the three categories, UVC is effectively blocked by the ozone layer in the Earth's atmosphere and the probability of it having any pathological effects on humans is very low [11]. More than 90% of the total solar radiation that reaches the Earth is UVA radiation, and the relationship between skin cancer and UVA exposure has been clearly established [4]. A number of epidemiological and experimental studies suggested a direct relationship between UVB exposure and skin disorders including cancer. In the United States alone, nonmelanoma skin cancer, which includes both basal and squamous cell carcinoma (SCC), is the most frequently diagnosed cancer and accounts for 50% of all cancer forms [4]. According to recent reports more than 1.3 million new cases of skin cancer are diagnosed every year in the U.S. [1]. It is also important to mention here that among all the cancers, skin cancer is believed to be one of the most preventable and/or curable cancer types.

ANTIOXIDANTS IN CHEMOPREVENTION OF SKIN CANCER AND OTHER SKIN DAMAGES

The development of skin cancer is a multistage process and is best explained by a three-step initiation, promotion, and progression system [12,13]. Initiation is the first step in multistage skin carcinogenesis, which involves carcinogen-induced genetic changes [14,15]. The second step, known as the promotion stage, involves the processes where the initiated cells undergo selective clonal expansion to form visible premalignant lesions [16]. The progression stage involves the conversion of premalignant lesions to malignant conditions.

Studies have suggested that ROS are important in all stages of skin cancer development as well as in a variety of other skin-related disorders. Therefore, it is believed that antioxidants could be useful chemopreventive agents. In recent years, some naturally occurring antioxidant agents, present in the common diet and beverages consumed by human populations, have gained considerable attention as chemopreventive agents against skin cancer [8,9]. Studies have shown that naturally occurring compounds present in human diet/beverages, such as green tea polyphenols, resveratrol, curcumin, silymarin, ginger, and diallyl sulfide, afford protection against the development of skin cancer, both under *in vitro* (in culture systems) and *in vivo* (in animal models) situations. Furthermore, this approach appears to have practical implications in reducing skin cancer risk because unlike the carcinogenic environmental factors that are difficult to control, individuals can easily modify their dietary habits and lifestyle. Further, chemoprevention via nontoxic (often food-based) agents is gaining increasing popularity and attention because it is regarded as safe and is not perceived as "medicine."

"Prevention of a disease is better than its cure" is a well-know phrase. Consistent with this, an approach to prevent cancer or reduce the risk of cancer development in its early stage is a logical and perhaps the most useful strategy for the management of this disease.

RESVERATROL: A STRONG CHEMOPREVENTIVE AGENT

Resveratrol (3,4′,5-trihydroxystilbene), a naturally occurring polyphenolic compound with strong antioxidant properties, is reported to possess a number of beneficial health effects including anticancer property. Resveratrol is abundantly found in grapes, berries, nuts, and red wine. The cancer chemopreventive property of resveratrol was first reported in 1977 by Jang et al. [17], when the authors demonstrated that resveratrol possesses antimutagenic and anticarcinogenic properties and can effectively block all three stages of carcinogenesis (initiation, promotion, and progression). Following this seminal contribution, many studies evaluated and established the cancer chemopreventive and/or therapeutic potential and other health-beneficial and desirable effects of resveratrol [17–20 and references therein]. It is appreciated that an effective and acceptable chemopreventive agent should have certain properties: (1) little or no toxic effects in normal and healthy cells, (2) high efficacy against multiple sites, (3) capable of oral consumption, (4) known mechanism of action, (5) low cost, and (6) acceptance by human populations. Resveratrol possesses most, if not all, of these properties. It is possible that following the extensive ongoing research, this agent could be developed as one of the most ideal chemopreventive agents for a variety of diseases including skin cancer and other skin-related disorders.

EFFECTS OF RESVERATROL ON SHORT-TERM UVB DAMAGES IN MOUSE SKIN

Often, short-term damages caused by UV radiation are regarded as precursors for the development of skin cancers. In a recent study, Afaq and colleagues examined whether resveratrol possesses a potential to ameliorate the damages caused by short-term UVB exposure to the skin in a model having relevance to human situations [20]. In this study, it was found that topical application of resveratrol (25 μmol/0.2 ml acetone/mouse) to SKH-1 hairless mice inhibited UVB (180 mJ/cm^2) exposure-mediated (1) skin edema, (2) inflammation, (3) cyclooxygenase (COX) and ornithine decarboxylase (ODC) enzyme activities, (4) ODC protein expression, (5) lipid peroxidation, and (6) generation of hydrogen peroxide (H_2O_2). A brief description of the findings of this study is given in the following.

Because studies have shown that UV exposure of the skin results in cutaneous edema that is regarded as a marker of UV damage, Afaq et al. evaluated the protective effect of resveratrol on UVB-mediated skin edema in SKH-1 hairless mouse skin [20]. The exposure of mouse skin with UVB radiation (180 mJ/cm^2) resulted in a significant increase in bifold skin thickness and ear punch weight, whereas pretreatment of skin with resveratrol was found to result in a significant inhibition in the increases in bifold skin thickness and ear punch weight 24 hours post-UVB irradiation. Further, in this study UVB irradiation of mouse skin also resulted in an induction of infiltration of leukocytes both in the epidermis and dermis at 24 hours post-UVB; pretreatment of the skin with resveratrol was found to result in a marked reduction in the number of infiltrating leukocytes [20].

Because ODC is regarded as playing an important role in both normal cellular proliferation and the growth and development of tumors [21], Afaq et al. studied the effect of resveratrol on UVB-mediated modulation in ODC activity in SKH-1 hairless mouse skin. Irradiation of mouse skin with UVB was found to result in significant increase in epidermal ODC activity, at 24 hours posttreatment, whereas the preapplication of resveratrol (prior to UVB exposure) resulted in a significant inhibition of ODC activity in the epidermis [20].

COX enzyme is regarded as playing a critical role in inflammation and cancer development [22]. Therefore, Afaq et al. studied the effect of resveratrol on UVB-mediated modulation in COX enzyme activity. The data from this study demonstrated that UVB exposure to the skin resulted in a significant increase in epidermal COX activity at 24 hours following UVB exposure; the preapplication of resveratrol (prior to UVB irradiation) resulted in a significant inhibition of UVB-mediated increase in epidermal COX activity [20].

H_2O_2 generation is a well-accepted marker of oxidative stress [23] and because resveratrol is a strong antioxidant [24,25], Afaq et al. determined whether or not resveratrol inhibits UVB exposure-mediated generation of H_2O_2 in SKH-1 hairless mouse skin. It was found that UVB irradiation of mouse skin resulted in a significant enhanced generation of H_2O_2, whereas application of resveratrol (prior to UVB irradiation) resulted in significant inhibition in the levels of H_2O_2 in the skin [20].

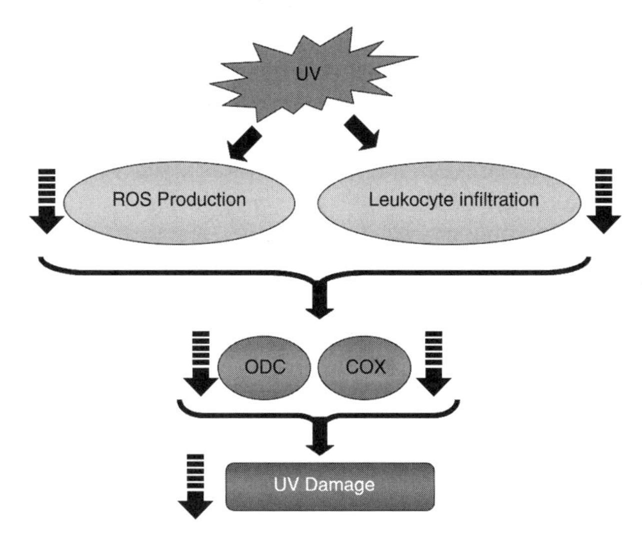

FIGURE 16.1 Schematic of the mechanism of prevention of short-term UVB-mediated damages by resveratrol. The dashed arrows show the changes attributed to resveratrol.

Further, the effects of UVB and/or resveratrol treatments were also assessed on epidermal lipid peroxidation, which is also a well-accepted marker of oxidative stress [26,27]. The data demonstrated that UVB exposure of mouse skin resulted in a significant induction of lipid peroxidation (measured in terms of malondialdehyde production) in the epidermis 24 hours post-UVB irradiation; the pretreatment of skin with resveratrol significantly inhibited the UVB-caused increase in epidermal lipid peroxidation [20].

This study demonstrated that resveratrol affords protection against the damages caused by short-term UVB exposure, and that these protective effects may be mediated via its strong antioxidant properties. A simplified depiction of the possible mechanism of the preventive effects of resveratrol against short-term UVB exposure-mediated damages to mouse skin is shown in Figure 16.1. However, because the markers studied here are regarded as early markers of proliferation, we suggest that resveratrol may be developed as a cancer chemopreventive agent against photo-carcinogenesis and other adverse effects of UVB exposure [20].

EFFECTS OF RESVERATROL ON MULTIPLE UVB EXPOSURE-MEDIATED DAMAGES IN MOUSE SKIN

Because multiple exposures of UV radiation to skin cause an increased risk of the development of skin cancer and other cutaneous disorders which are regarded as precursors to cancer, e.g., actinic keratoses [28,29], in a very recent study, Reagan-Shaw and colleagues studied the preventive effect of resveratrol against multiple UVB exposure-mediated damages to the skin of SKH-1 hairless mice [30]. This animal model of multiple UVB exposures ($180\,mJ/cm^2 \times 7$ exposures on alternate days) was stated to represent human exposure to solar UV radiations on a week-long vacation or outside activities at a sunny time of year. In this model, resveratrol (10 μmol/0.2 ml/mouse) was topically applied to mouse skin prior to each exposure to UVB radiation. A brief description of the findings of this study is provided below.

Reagan-Shaw and colleagues studied the effect of multiple UVB exposures on cutaneous edema in SKH-1 hairless mouse skin and the modulatory effect of resveratrol [30]. In this study, multiple UVB exposures of mouse skin resulted in a significant (1) increase in bifold skin thickness, (2) increase in ear punch weight, (3) hyperplasia, (4) infiltration of leukocytes into the epidermis and dermis, and (5) increase in protein levels of PCNA in the epidermis; pretreatment of skin with resveratrol was found to reverse significantly the UVB-mediated responses [30]. Further, UVB radiations resulted in a significant increase in epidermal thickness (an inflammatory response) that was reversed by topical application of resveratrol prior to each UVB exposure [30].

Reagan-Shaw and colleagues also studied the involvement of cyclin-dependent kinases (cdks) and cyclins, which are critical molecules for the progression of the cell cycle, in chemopreventative effects of resveratrol against UVB exposure-mediated damages in SKH-1 hairless mouse skin. Western blot and immunohistochemical analyses demonstrated that multiple UVB exposures to mouse skin resulted in significant increases in cdk-2, -4, and -6, and cyclin D2 protein levels in the epidermis; resveratrol pretreatment significantly reversed these effects [30].

The authors further evaluated the effect of resveratrol on UVB-mediated modulations in the levels of the cyclin kinase inhibitors WAF1/p21 and KIP1/p27. It was found that multiple UVB exposures caused a significant upregulation in WAF1/p21 and KIP1/p27 protein levels [30]. Interestingly, resveratrol pretreatment was found to result in a further enhancement in UVB exposure-mediated increase in WAF1/p21 [30]. Because studies have shown that the regulation of WAF1/p21 may be dependent on the tumor suppressor p53, the effects of UVB and/or resveratrol treatments was determined on the modulation in p53 protein. Multiple UVB exposures resulted in a significant increase in p53 protein levels; in a fashion similar to the response on WAF1/p21, resveratrol treatment prior to UVB exposures resulted in a further enhancement in p53 protein levels [30]. Finally, the authors also demonstrated that repeated UVB exposures resulted in increased levels in the 42 kDa unphosphorylated isotype of MAPK 1/2 along with mitogen-activated protein kinase kinase (MEK)-1; topical application of resveratrol prior to UVB exposures significantly reversed the UVB-mediated responses in these proteins [30].

Taken together, the results of this study suggested that the antiproliferative effects of resveratrol might be mediated via modulation in the cdk–cyclin–cdk network and MAPK pathway [30]. The authors suggested that resveratrol may be useful in the prevention of UVB-mediated cutaneous damages including skin cancer. However, the need for detailed studies to assess the effectiveness of resveratrol in the prevention of UV exposure-mediated skin tumorigenesis was appreciated by the authors.

In a follow-up study from this group, the authors evaluated the involvement of the IAP protein survivin in the protective effects of resveratrol against multiple UVB exposure-mediated damages in SKH-1 hairless mouse skin. In this study, Aziz et al. demonstrated that multiple UVB irradiation resulted in a significant upregulation of Ki-67 protein levels in both epidermis and dermis at 24 hours following the last UVB exposure [31]. Pretreatment of the skin with resveratrol resulted in a marked reduction in UVB exposure-mediated upregulation of Ki-67 protein levels. Ki-67 is a nuclear protein that is expressed in proliferating cells and is regarded as a marker of proliferation [32,33]. Because Ki-67 is known to be expressed in cells at all proliferative stages of the cell cycle [34,35], this data demonstrated the antiproliferative potential of resveratrol against the hyperproliferative response of UVB. Further, this study demonstrated a significant induction of COX-2 and ODC by multiple UV exposure and pretreatment of skin with resveratrol resulted in a significant downregulation of UVB-mediated increase in COX-2 and ODC levels [31]. The authors next determined the effect of multiple exposures of UVB on protein and mRNA levels of survivin and attenuation by resveratrol. The data demonstrated that multiple UVB exposure resulted in a significant increase in the epidermal survivin protein and mRNA levels; preapplication of the mouse skin with resveratrol resulted in a significant downmodulation of UVB exposure-mediated increases in the levels of survivin [31]. This is an important finding because survivin has been shown to be overexpressed in a variety of human cancers including cancers of prostate, lungs, breast, colon, brain, liver, and pancreas [36–39]. Conversely, the downregulation of survivin by antisense oligonucleotides was shown to induce apoptosis *in vitro* [40–42].

Further, the authors determined the effect of UVB and/or resveratrol treatments on survivin phosphorylation at Thr^{34}. The data from this study demonstrated that multiple UVB exposure of SKH-1 hairless mouse skin resulted in a significant upregulation of survivin phosphorylation at Thr^{34} and pretreatment with resveratrol resulted in a downmodulation of this response of UVB radiation [31]. Another important target in the survivin pathway is the protein Smac/DIABLO that is released from mitochondria and binds to and neutralizes the inhibitory activity of IAPs during apoptosis [43]. Aziz et al. demonstrated that UVB-caused upregulation of survivin was associated with an appreciable downregulation of Smac/DIABLO protein; pretreatment of skin with resveratrol resulted in restoration of UVB-mediated decrease of Smac/DIABLO [31].

Because the survivin pathway is intimately involved with the process of apoptosis, Aziz et al. assessed whether or not the observed modulations in this critical pathway are associated with apoptosis in mouse skin. Thus, the effects of UVB and/or resveratrol treatments were assessed on M30 CytoDEATH protein, which is a unique protein for easy and reliable determination of very early apoptosis [44]. The data demonstrated that multiple UVB exposures resulted in an induction of apoptosis in mouse skin; pretreatment of the skin with resveratrol resulted in a further enhancement of M30 CytoDEATH immunostaining indicating an enhanced apoptosis response [31]. This suggested that resveratrol-caused protection of UVB responses in mouse skin are mediated via an apoptotic elimination of damaged cells via an inhibition of survivin pathway. Based on these findings, the authors suggested that resveratrol may be useful for the prevention of UVB-mediated cutaneous damages including skin cancer.

EFFECTS OF RESVERATROL ON UVB EXPOSURE-MEDIATED DAMAGES IN SKIN KERATINOCYTES *IN VITRO*

Very few *in vitro* studies have assessed the preventive effects of resveratrol against UVB-mediated damages in cell culture systems. Adhami and colleagues investigated the hypothesis that nuclear transcription factor kappa B (NF-κB) plays a critical role in the chemoprevention of UV damages imparted by resveratrol [45]. NF-κB is known to play a critical role in the pathogenesis of cancer and many other hyperproliferative skin conditions [46–48]. NF-κB has been shown to play a particularly central role in epidermal biology and it is believed that perpetually subjected to solar UV radiation, the proliferative cells of the epidermis may rely on NF-κB activation for protection and survival [49–51].

Employing normal human epidermal keratinocyte (NHEK), in dose- as well as time-dependent protocols, Adhami et al. evaluated the effect of resveratrol on UVB-mediated modulation in NF-κB. EMSA and immunoblot analyses of the nuclear cell lysates clearly demonstrated that a 40 mJ/cm^2 dose of UVB radiation resulted in a significant activation of NF-κB/p65; a prior treatment of the keratinocytes with resveratrol resulted is a significant dose-dependent (5, 10, and 25 μM for 24 hours) as well as time-dependent (12, 24, and 48 hour treatment with 5 μM resveratrol) inhibition of NF-κB activation [45]. Following this observation, the authors evaluated the effect of resveratrol on UVB exposure-mediated modulation in IκBα in NHEK cells. The immunoblot analysis demonstrated that UVB exposure (40 mJ/cm^2) of the keratinocytes resulted in a significant decrease in the level of IκBα in the cytosol (postnuclear cell lysate) and pretreatment of keratinocytes with resveratrol resulted in a dose- as well as time-dependent restoration of IκBα degradation [45]. Further, employing a phosphospecific antibody raised against a peptide corresponding to a short amino acid sequence containing phosphorylated serine-32 of IκBα of human origin, Adhami et al. studied the effect of resveratrol on UVB-mediated modulations on serine-32 phosphorylation of IκBα. The data demonstrated that UVB exposure of the keratinocytes results in increased phosphorylation of IκBα at serine-32, whereas resveratrol pretreatment results in a significant dose- and time-dependent inhibition of the phosphorylation [45]. Further, the data from this study also demonstrated that resveratrol inhibited the UVB-mediated upregulation in the levels of IKK, in a dose- as well as time-dependent fashion [45]. Based on the data from this study, the authors suggested that NF-κB pathway plays a critical role in the chemopreventive effects of resveratrol against the adverse effects of UV radiation including photocarcinogenesis.

UV-induced skin pigmentation is widely regarded as a defense mechanism against UV-mediated damages [52,53]. The microphthalmia-associated transcription factor (MITF) is involved in melanocyte development and survival, but its role in UVB-mediated pigment production is not known [54]. Resveratrol (30 μg/ml) has been shown to inhibit UVB-induced (30 mJ/cm^2) MITF promoter activity in B16 murine melanoma cells [55]. Because other strong antioxidants did not affect MITF promoter activity, it could be suggested that the antioxidant property of resveratrol is not involved in MITF transcription. Resveratrol also had an inhibitory effect on tyrosinase and Tyrp1 promoter activity along with tyrosinase activity, which could be rescued by MITF expression [55]. These data suggested that resveratrol, along with MITF, could be used to alter skin pigmentation.

RESVERATROL: AN ENHANCER OF THERAPEUTIC RESPONSE OF IONIZING RADIATION AGAINST CANCER CELLS *IN VITRO*

In a recent study, Zoberi and colleagues [56] tested the hypothesis that resveratrol may modulate the cellular response of cancer cells to ionizing radiation (IR). The effect of pretreatment of resveratrol on IR-mediated modulations in clonogenic cell survival was studies in HeLa and SiHa cells. The data demonstrated that pretreatment of cells with resveratrol resulted in an enhancement of IR exposure-mediated tumor cell killing in a dose-dependent manner [56]. Further, resveratrol pretreatment also resulted in inhibition in cell division as assayed by growth curves and induction in S-phase cell cycle arrest [56]. Based on the data, the authors suggested that that resveratrol alters both cell cycle progression and the cytotoxic response to IR in cervical tumor cell lines [56].

In another recent study, Baatout and colleagues [57] examined whether or not resveratrol could sensitize cancer cells to x-rays. The three different human cancer cell lines used in this study were cervix carcinoma HELA cells, chronic myeloid leukemia K-562 cells, and multiple myeloma IM-9 cells [57]. The effects of resveratrol (0 to 200 μM) and/or x-ray (0 to 8 Gy) treatments were determined on cell viability, cell morphology, cell cycle distribution, and apoptosis [57]. In this study, concomitant treatment of the cells with resveratrol and x-rays was found to result in a synergistic antiproliferative effect at the highest resveratrol concentration (200 μM) [57]. These results suggested that resveratrol can act as a potential radiation sensitizer at high concentrations.

Both these studies suggested that resveratrol may be a useful agent in enhancing the therapeutic response of ionizing radiations against cancer cells. However, detailed studies, especially in preclinical animal models, are required to validate the relevance of *in vitro* observations to the *in vivo* situations.

CONCLUSION

The available *in vitro* and *in vivo* experimental data have suggested that resveratrol could be a useful chemopreventive agent for UV radiation-mediated damages including skin cancer. However, detailed in-depth *in vivo* studies are needed to establish firmly this notion. Based on several studies, as shown in Figure 16.2, it appears that the preventive effects of resveratrol are mediated via its antioxidant potential and via its ability to modulate the

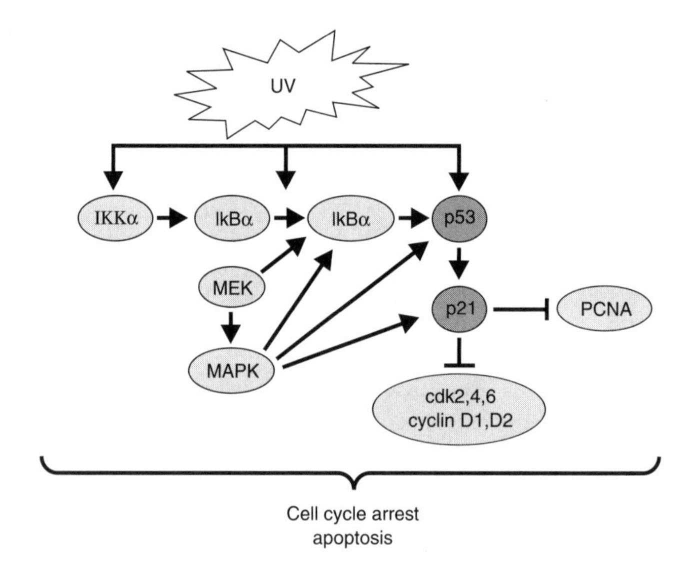

FIGURE 16.2 Simplified depiction of the molecular mechanism of chemoprevention of UVB-mediated damages by resveratrol. The molecules that have been shown to be inhibited by the chemopreventive effects of resveratrol against UV damage are shown in light grey; those that have been shown to be upregulated are depicted in dark grey.

signaling pathways responsible for regulation of cell cycle and apoptosis. Further, based on the available literature, it is conceivable to design resveratrol-containing emollients or patches as well as sunscreens and skin-care products for prevention of skin cancer, premalignant conditions (such as actinic keratoses), as well as other adverse effects of UV radiation such as skin aging. A limited number of studies have also suggested that resveratrol may be useful in enhancing the therapeutic response of ionizing radiations against certain cancer cells; suggesting that this agent may be used as a complementary approach to enhance the therapeutic efficacy of existing anticancer modalities.

REFERENCES

1. Jemal A, Tiwari RC, Murray T, Ghafoor A, Samuels A, Ward E, Feuer EJ, and Thun MJ, Cancer statistics, 2004, *CA Cancer J Clin* 54, 8–29, 2004.
2. Dummer R and Maier T, UV protection and skin cancer, *Recent Results Cancer Res* 160, 7–12, 2002.
3. Matsumura Y and Ananthaswamy HN, Toxic effects of ultraviolet radiation on the skin, *Toxicol Appl Pharmacol* 195, 298–308, 2004.
4. de Gruijl FR, Photocarcinogenesis: UVA vs. UVB radiation, *Skin Pharmacol Appl Skin Physiol* 15, 316–320, 2002.
5. Nishigori C, Hattori Y, and Toyokuni S, Role of reactive oxygen species in skin carcinogenesis, *Antioxid Redox Signal* 6, 561–570, 2004.
6. Sander CS, Chang H, Hamm F, Elsner P, and Thiele JJ, Role of oxidative stress and the antioxidant network in cutaneous carcinogenesis, *Int J Dermatol* 43, 326–335, 2004.
7. Tsao AS, Kim ES, and Hong WK, Chemoprevention of cancer, *CA Cancer J Clin* 54, 150–180, 2004.
8. Manson MM, Gescher A, Hudson EA, Plummer SM, Squires MS, and Prigent SA, Blocking and suppressing mechanisms of chemoprevention by dietary constituents, *Toxicol Lett* 112–113, 499–505, 2000.
9. Tamimi RM, Lagiou P, Adami HO, and Trichopoulos D, Prospects for chemoprevention of cancer, *J Intern Med* 251, 286–300, 2002.
10. Stein CJ and Colditz GA, Modifiable risk factors for cancer, *Br J Cancer* 90, 299–303, 2004.
11. Ohnaka T, Health effects of ultraviolet radiation, *Ann Physiol Anthropol* 12, 1–10, 1993.
12. Hennings H, Glick AB, Greenhalgh DA, Morgan DL, Strickland JE, Tennenbaum T, and Yuspa SH, Critical aspects of initiation, promotion, and progression in multistage epidermal carcinogenesis, *Proc Soc Exp Biol Med* 202, 1–8, 1993.
13. Slaga TJ, Inhibition of skin tumor initiation, promotion, and progression by antioxidants and related compounds, *Crit Rev Food Sci Nutr* 35, 51–57, 1995.
14. DiGiovanni J, Multistage carcinogenesis in mouse skin, *Pharmacol Ther* 54, 63–128, 1992.
15. Yuspa SH, Cutaneous chemical carcinogenesis, *J Am Acad Dermatol* 15 (5 Pt 1), 1031–1044, 1986.

16. Fujiki H, Suganuma M, Okabe S, Sueoka E, Suga K, Imai K et al, A new concept of tumor promotion by tumor necrosis factor-alpha, and cancer preventive agents (–)-epigallocatechin gallate and green tea: a review, *Cancer Detection Prev* 24, 91–99, 2000.
17. Jang M, Cai L, Udeani GO, Slowing KV, Thomas CF, Beecher CW et al, Cancer chemopreventive activity of resveratrol, a natural product derived from grapes, *Science* 275, 218–220, 1997.
18. Granados-Soto V, Pleiotropic effects of resveratrol, *Drug News Perspect* 16, 299–307, 2003.
19. Kapadia GJ, Azuine MA, Tokuda H, Takasaki M, Mukainaka T, Konoshima T et al, Chemopreventive effect of resveratrol, sesamol, sesame oil and sunflower oil in the Epstein-Barr virus early antigen activation assay and the mouse skin two-stage carcinogenesis, *Pharmacol Res* 45, 499–505, 2002.
20. Afaq F, Adhami VM, and Ahmad N, Prevention of short-term ultraviolet B radiation-mediated damages by resveratrol in SKH-1 hairless mice, *Toxicol Appl Pharmacol* 186, 28–37, 2003.
21. Shantz LM and Pegg AE, Translational regulation of ornithine decarboxylase and other enzymes of the polyamine pathway, *Int J Biochem Cell Biol* 31, 107–122, 1999.
22. Zha S, Yegnasubramanian V, Nelson WG, Isaacs WB, and De Marzo AM, Cyclooxygenases in cancer: progress and perspective, *Cancer Lett* 215, 1–20, 2004.
23. Schrader M and Fahimi HD, Mammalian peroxisomes and reactive oxygen species, *Histochem Cell Biol* 122, 383–393, 2004.
24. Murcia MA and Martinez-Tome M, Antioxidant activity of resveratrol compared with common food additives, *J Food Prot* 64, 379–384, 2001.
25. Stivala LA, Savio M, Carafoli F, Perucca P, Bianchi L, Maga G, Forti L, Pagnoni UM, Albini A, Prosperi E, and Vannini V, Specific structural determinants are responsible for the antioxidant activity and the cell cycle effects of resveratrol, *J Biol Chem* 276, 22586–22594, 2001.
26. Requena JR, Fu MX, Ahmed MU, Jenkins AJ, Lyons TJ, and Thorpe SR, Lipoxidation products as biomarkers of oxidative damage to proteins during lipid peroxidation reactions, *Nephrol Dial Transplant* 11 (Suppl 5), 48–53, 1996.
27. Valenzuela A, The biological significance of malondialdehyde determination in the assessment of tissue oxidative stress, *Life Sci* 48, 301–309, 1991.
28. Bowden GT, Prevention of non-melanoma skin cancer by targeting ultraviolet-B-light signalling, *Nat Rev Cancer* 4, 23–35, 2004.
29. Einspahr JG, Bowden GT, and Alberts DS, Skin cancer chemoprevention: strategies to save our skin, *Recent Results Cancer Res* 163, 151–164; discussion 264–266, 2003.
30. Reagan-Shaw S, Afaq F, Aziz MH, and Ahmad N, Modulations of critical cell cycle regulatory events during chemoprevention of ultraviolet B-mediated responses by resveratrol in SKH-1 hairless mouse skin, *Oncogene* 23, 5151–5160, 2004.
31. Aziz MH, Ghotra AS, Shukla Y, and Ahmad N, Ultraviolet (UV) B radiation causes an upregulation of survivin in human keratinocytes and mouse skin, *Photochem Photobiol* 80, 602–608, 2004.

32. Gerdes J, Ki-67 and other proliferation markers useful for immunohistological diagnostic and prognostic evaluations in human malignancies, *Semin Cancer Biol* 1, 199–206, 1990.
33. Rudolph P, Tronnier M, Menzel R, Moller M, and Parwaresch R, Enhanced expression of Ki-67, topoisomerase IIalpha, PCNA, p53 and p21WAF1/Cip1 reflecting proliferation and repair activity in UV-irradiated melanocytic nevi, *Hum Pathol* 29, 1480–1487, 1998.
34. Madewell BR, Cellular proliferation in tumors: a review of methods, interpretation, and clinical applications, *J Vet Intern Med* 15, 334–340, 2001.
35. Scholzen T and Gerdes J, The Ki-67 protein: from the known and the unknown, *J Cell Physiol* 182, 311–322, 2000.
36. Ambrosini G, Adida C, and Altieri DC, A novel anti-apoptosis gene, survivin, expressed in cancer and lymphoma, *Nat Med* 3, 917–921, 1997.
37. Ito T, Shiraki K, Sugimoto K, Yamanaka T, Fujikawa K, Ito M et al, Survivin promotes cell proliferation in human hepatocellular carcinoma, *Hepatology* 31, 1080–1085, 2000.
38. Kawasaki H, Altieri DC, Lu CD, Toyoda M, Tenjo T, and Tanigawa N, Inhibition of apoptosis by survivin predicts shorter survival rates in colorectal cancer, *Cancer Res* 58, 5071–5074, 1998.
39. Tanaka K, Iwamoto S, Gon G, Nohara T, Iwamoto M, and Tanigawa N, Expression of survivin and its relationship to loss of apoptosis in breast carcinomas, *Clin Cancer Res* 6, 127–134, 2000.
40. Chen J, Wu W, Tahir SK, Kroeger PE, Rosenberg SH, Cowsert LM et al, Down-regulation of survivin by antisense oligonucleotides increases apoptosis, inhibits cytokinesis and anchorage-independent growth, *Neoplasia* 2, 235–241, 2000.
41. Olie RA, Simoes-Wust AP, Baumann B, Leech SH, Fabbro D, Stahel RA et al, A novel antisense oligonucleotide targeting survivin expression induces apoptosis and sensitizes lung cancer cells to chemotherapy, *Cancer Res* 60, 2805–2809, 2000.
42. Xia C, Xu Z, Yuan X, Uematsu K, You L, Li K et al, Induction of apoptosis in mesothelioma cells by antisurvivin oligonucleotides, *Mol Cancer Ther* 1, 687–694, 2002.
43. Chai J, Du C, Wu JW, Kyin S, Wang X, and Shi Y, Structural and biochemical basis of apoptotic activation by Smac/DIABLO, *Nature* 406, 855–862, 2000.
44. Chiu PM, Ngan YS, Khoo US, and Cheung AN, Apoptotic activity in gestational trophoblastic disease correlates with clinical outcome: assessment by the caspase-related M30 CytoDeath antibody, *Histopathology* 38, 243–249, 2001.
45. Adhami VM, Afaq F, and Ahmad N, Suppression of ultraviolet B exposure-mediated activation of NF-kappaB in normal human keratinocytes by resveratrol, *Neoplasia* 5, 74–82, 2003.
46. Bell S, Degitz K, Quirling M, Jilg N, Page S, and Brand K, Involvement of NF-kappaB signalling in skin physiology and disease, *Cell Signal* 15, 1–7, 2003.
47. Dhar A, Young MR, and Colburn NH, The role of AP-1, NF-kappaB and ROS/NOS in skin carcinogenesis: the JB6 model is predictive, *Mol Cell Biochem* 234–235, 185–193, 2002.

48. Hsu TC, Young MR, Cmarik J, and Colburn NH, Activator protein 1 (AP-1)- and nuclear factor kappaB (NF-kappaB)-dependent transcriptional events in carcinogenesis, *Free Radical Biol Med* 28, 1338–1348, 2000.
49. Adachi M, Gazel A, Pintucci G, Shuck A, Shifteh S, Ginsburg D et al, Specificity in stress response: epidermal keratinocytes exhibit specialized UV-responsive signal transduction pathways, *DNA Cell Biol* 22, 665–677, 2003.
50. Afaq F, Adhami VM, Ahmad N, and Mukhtar H, Inhibition of ultraviolet B-mediated activation of nuclear factor kappaB in normal human epidermal keratinocytes by green tea constituent (–)-epigallocatechin-3-gallate, *Oncogene* 22, 1035–1044, 2003.
51. Pfundt R, Vlijmen-Willems I, Bergers M, Wingens M, Cloin W, and Schalkwijk J, In situ demonstration of phosphorylated c-jun and p38 MAP kinase in epidermal keratinocytes following ultraviolet B irradiation of human skin, *J Pathol* 193, 248–255, 2001.
52. Ortonne JP, Photoprotective properties of skin melanin, *Br J Dermatol* 146 (Suppl 61), 7–10, 2002.
53. Wulf HC, Sandby-Moller J, Kobayasi T, and Gniadecki R, Skin aging and natural photoprotection, *Micron* 35, 185–191, 2004.
54. Fisher DE, Microphthalmia: a signal responsive transcriptional regulator in development, *Pigment Cell Res* 13, 145–149, 2000.
55. Lin CB, Babiarz L, Liebel F, Roydon PE, Kizoulis M, Gendimenico GJ et al, Modulation of microphthalmia-associated transcription factor gene expression alters skin pigmentation, *J Invest Dermatol* 119, 1330–1340, 2002.
56. Zoberi I, Bradbury CM, Curry HA, Bisht KS, Goswami PC, Roti Roti JL et al, Radiosensitizing and anti-proliferative effects of resveratrol in two human cervical tumor cell lines, *Cancer Lett* 175, 165–173, 2002.
57. Baatout S, Derradji H, Jacquet P, Ooms D, Michaux A, and Mergeay M, Enhanced radiation-induced apoptosis of cancer cell lines after treatment with resveratrol, *Int J Mol Med* 13, 895–902, 2004.

17 Resveratrol as a Phytoestrogen

Barry D. Gehm and Anait S. Levenson

CONTENTS

INTRODUCTION

The existence of phytoestrogens (plant compounds with estrogenic activity) has been known for more than 75 years [1]. In some cases these compounds can impair the fertility of animals that consume them, and it is likely that plants have evolved them as a defense against herbivory [2]. Despite their possible role as endocrine disrupters, in the last few decades phytoestrogens have come to be widely regarded as beneficial to human health, because of observations that populations consuming high-phytoestrogen diets (in particular, soy-rich diets that are common in Asia) have lower rates of some estrogen-modulated conditions such as heart disease, breast cancer, osteoporosis, and menopausal symptoms. However, there are many differences between Asian and Western diets apart from the levels of phytoestrogens, and it is still uncertain to what extent phytoestrogens *per se* are responsible for the health benefits associated with the former.

Chemically, phytoestrogens are typically di- or polyphenolic compounds, with hydroxyl groups that approximate the relative positions of those on the gonadal estrogen 17β-estradiol (E2). Prior to 1997, the major categories of known dietary phytoestrogens were flavonoids (derivatives of flavones, isoflavones, chalcones, and coumestans) and lignans. Representative structures are shown in Figure 17.1. Although these nonsteroidal compounds do not closely mimic the structure of E2, they are capable of binding (with varying degrees of affinity) to estrogen receptors (ERs) and eliciting hormonal effects. Resveratrol's structural similarity to the potent

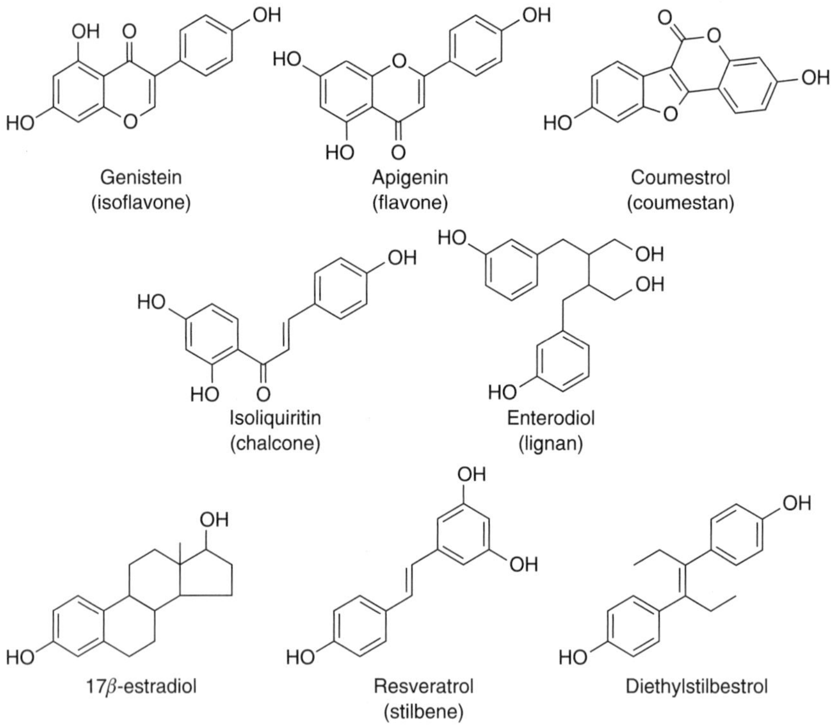

FIGURE 17.1 Chemical structures of representative phytoestrogens from different groups. The structures of E2 and diethylstilbestrol (DES) are also shown. Many of these phytoestrogens occur in plants as hormonally inactive precursors (typically glycosides) that are converted to the estrogenic forms by digestive hydrolysis and/or the actions of intestinal microbes.

synthetic estrogen diethylstilbestrol (DES) suggested that it might be estrogenic, which Gehm et al. confirmed by demonstrating its ability to compete with labeled E2 for receptor binding, activate expression of estrogen-regulated genes, and stimulate the growth of estrogen-dependent breast cancer cells [3]. This discovery revealed a new category of dietary phytoestrogen: as a hydroxystilbene, resveratrol does not have a lignan- or flavone-like structure, although it has sometimes erroneously been called a bioflavonoid [4–8].

The observation that resveratrol has estrogenic activity was all the more provocative because some of the health effects that are (or were) attributed to estrogen are similar to those that have been attributed to red wine, the main dietary source of resveratrol. These include improved HDL/LDL ratios, with decreased risk of coronary heart disease [9], reduced risk of Alzheimer's disease [10], and, according to some reports, increased risk of breast cancer [11,12]. In particular, the discovery prompted speculation that resveratrol's estrogenicity might play a role in the "French paradox" [13]. However, estrogenicity is not always beneficial: resveratrol's ability to stimulate growth of estrogen-dependent breast cancer cells raised concerns about its potential use as an anticancer agent. Accordingly, numerous subsequent studies on the nature and significance of resveratrol's estrogenicity have been published, which have used a variety of experimental approaches and yielded sometimes contradictory results. In this chapter we describe our and others' research in this area and discuss the nature and significance of resveratrol's effects on ERs.

RESVERATROL AS A PHYTOESTROGEN

RESVERATROL AND ER ISOFORMS

Estrogens exert their cellular effects by binding to nuclear receptor proteins that act as ligand-dependent transcription factors. When complexed with E2 or other agonists, the receptors bind to estrogen response elements (EREs) in the promoters of target genes and stimulate transcription [14] through their activating function (AF) domains. There are two distinct isoforms of estrogen receptor, ERα and ERβ, which share 53% sequence homology in their ligand binding domains and 95% homology in their DNA binding domains, but show significant differences in tissue distribution and in their responses to various agonists and antagonists [14,15]. For example, ERβ predominates in the prostate and in ovarian granulosa cells, but ERα is more highly expressed in the uterus. ERα has two AF domains, one (AF-2) closely associated with the ligand binding domain and the other (AF-1) in the N-terminal region of the protein. In ERβ AF-1 is weak or inactive [14].

Phytoestrogens can also bind and activate ERs, thus triggering biological responses. Some phytoestrogens have relatively low binding affinity compared to E2 or DES, but can exert significant effects if consumed in sufficient amounts. Genistein, for example, is about 1000-fold less potent than E2 in its ability to bind to ERs, but its circulating concentration in individuals consuming moderate amounts of soy foods is nearly 1000-fold higher than peak levels of endogenous E2 [16].

Resveratrol was originally recognized as a phytoestrogen based on its ability to bind and activate ERα [3], but has since been found to bind ERβ as well [5,17]. In contrast to some other phytoestrogens, e.g., genistein and coumestrol, which bind preferentially to ERβ [15,18,19], resveratrol shows a small preference for ERα [5,19]. Its binding affinities for the ERs are rather low; in competition studies with labeled E2, Bowers et al. [5] reported IC_{50}s of 58 and 130 μM for binding to ERα and ERβ, respectively, while Mueller et al. [19] reported ~8 and 30 μM. The latter values are more consistent with the 10 μM value reported by Gehm et al. [3] using MCF-7 cell extract (primarily ERα), and with our studies of estrogen-regulated gene expression. Although these results indicate that resveratrol has a low potency, several reports show it to have a high efficacy, equivalent to that of E2 or DES; i.e., it acts as a full agonist [3,5,19–22]. Moreover, in some systems (discussed in more detail below) it shows a higher estrogenic potency than would be expected from the reported binding affinities.

In general, resveratrol appears to be able to activate both ER isoforms with comparable potency, but some studies have found it affects ERα and ERβ differently. For instance, Bowers et al. [5] reported that resveratrol was more agonistic in CHO-K1 cells expressing ERβ but more antagonistic in those expressing ERα, suggesting that the effects of resveratrol on various tissues may depend on which ER isoform is more highly expressed. The same group reported that red wine extracts contain additional phytoestrogens which preferentially activate ERβ [7]. However, our own results (presented and discussed below) indicate that resveratrol may be more efficacious in activating ERα. Much remains to be learned about the roles of the ER isoforms in mediating resveratrol's effects.

Resveratrol's Estrogenic and Antiestrogenic Effects

The initial report of resveratrol's estrogenicity [3] prompted numerous investigations by others, using a variety of experimental systems to measure ER agonism or antagonism. Frequently the results from various laboratories appear contradictory, suggesting that resveratrol's effects on ERs are complex.

In Vitro Studies

Reporter Gene Assays

Artificial estrogen-responsive reporter genes have been widely used in studying ER regulation of gene expression. Gehm et al. [3] used MCF-7 cells

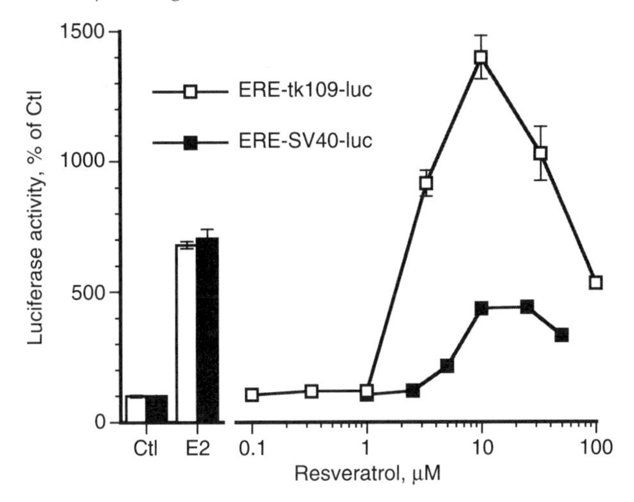

FIGURE 17.2 Activation of reporter gene expression by resveratrol depends on the promoter context of the ERE. MCF-7 cells were transfected with ERE-tk109-luc or ERE-SV40-luc and treated for ~1 day with vehicle (Ctl), 1 nM E2, or the indicated concentrations of resveratrol. Luciferase activities are expressed as percentage of control and plotted as mean ± s.e. of triplicate wells. Construction of the plasmids, conditions for growth and transfection of the cells, and the luciferase assay have been described previously [3,80].

transiently transfected with reporters containing short fragments of the herpes simplex thymidine kinase (tk) promoter, which provide a basal transcriptional activity, attached to one or two *Xenopus* vitellogenin (vit) EREs, which confer estrogen responsiveness. Resveratrol activated these reporters with an EC_{50} of 3 to 10 μM; maximal activity was seen at 10 to 30 μM, with potency depending on the number of EREs. The dose–response curves are relatively sharp; no reporter activation is observed at 1 μM or below, and activity decreases markedly at supraoptimal resveratrol concentrations (Figure 17.2). Strikingly, at optimal concentrations resveratrol induces reporter activity 2 to 3 times as high as the maximal activity induced by E2 or DES. While partial agonists (compounds that produce responses similar to but smaller than those produced by E2) are common, such superagonism is very unusual. It should be noted that treatment with resveratrol alone is sufficient to induce superagonism [3]; these experiments have sometimes been mischaracterized as combining resveratrol and E2 (e.g., [17,23,24]). Similar results, including superagonism although to a lesser degree, were obtained by Basly et al. [25] in MCF-7 cells containing a stably transfected ERE-tk-luciferase reporter. They found that the *cis* isomer of resveratrol also activated the reporter, but was less potent. Superagonism has also been reported in HepG2 cells [26].

Experiments using different cell lines and reporter constructs suggest that the estrogenicity of resveratrol, both in terms of potency and of whether it behaves as a superagonist, full agonist, partial agonist, or antagonist, can depend on the cell type, the specific sequence and promoter context of the ERE, and the ER isoform and subtype. Full agonism and high potency were observed by Wietzke and Welsh [22], using a reporter based on the estrogen-responsive promoter of the vitamin D receptor (VDR) gene. In MCF-7 and T47D breast cancer cells, which express primarily ERα, 4 nM resveratrol produced maximal reporter activity, and concentrations as low as 40 pM produced significant stimulation. No response was seen in the presence of tamoxifen or in an ER-negative cell line, indicating that resveratrol's effects were ER-mediated.

Although the experiments cited above show resveratrol as a full agonist or superagonist, under other conditions it acts as a partial agonist. Gehm et al. observed only partial agonism in ovarian BG-1 cells transfected with the same reporters that displayed superagonism in MCF-7 cells [3]. Even in MCF-7 cells, resveratrol produces only partial agonism on a reporter in which the vit ERE is linked to the SV40 early promoter rather than the tk promoter (Figure 17.2). The SV40 promoter has been widely used in the construction of estrogen-responsive reporters in various laboratories.

Using such reporters, Ashby et al. [27] found that in Cos-1 cells transfected with hERα, resveratrol (~50 μM) displayed full agonism on the vit ERE, but only partial agonism on the ERE from the luteinizing hormone β gene. In contrast, in cells expressing rat ERβ (rERβ), resveratrol acted as a full agonist on both EREs. Klinge's laboratory has obtained similar results in CHO-K1 cells, finding that with a vit ERE reporter, resveratrol is a partial agonist for hERα, but a full agonist for rERβ. Results with other EREs from various genes were more complex, but in general the same pattern was observed [5,7]. More recently, they extended their studies to two subtypes of human ERβ: the 530 aa hERβ1 and a 476 aa variant, hERβ1s. For full activity the human variants required resveratrol concentrations similar to those for rERβ (25 to 100 μM) but significant stimulation was observed even at lower concentrations: 1 μM for hERβ1 and, remarkably, 1 nM for hERβ1s [8].

Since partial agonists can compete with E2 for the ERs' hormone binding sites but induce a less than fully active conformation of the receptor, they can often inhibit E2-stimulated receptor activity. Many phytoestrogens act as mixed agonist/antagonists in this manner [19], and thus it is not surprising that resveratrol is reported to antagonize E2 induction of reporter genes under some circumstances. Klinge's laboratory, for instance, using the system described above, showed resveratrol antagonism (at 10 to 50 μM) on some EREs, including vit, via hERα, but not rERβ [5]. Mueller et al. [19] found that, in Ishikawa endometrial cancer cells stably expressing hERα or hERβ, reporters containing 3 vit EREs or the estrogen-responsive complement C3 promoter were fully activated by 20 to 50 μM resveratrol, but high concentrations (100 μM) antagonized activation by DES. These effects appeared to be mediated, at least in part, by selective alteration of

ER interactions with the coactivators GRIP1 and SRC-1. Interestingly, two metabolites of the soy phytoestrogen daidzein were superagonistic in this system, although resveratrol was not.

In contrast, Pezzuto's laboratory [17] reported that resveratrol had no agonist activity in Ishikawa cells, and antagonized the effect of E2 on an ERE-tk-luc reporter with an IC_{50} between 1 and 10 μM. In four breast cancer cell lines they found that resveratrol showed partial or no agonism when administered alone, and antagonism in the presence of E2 [23]. These results are surprising because the reporter is similar, and two of the cell lines (MCF-7 and S30) identical, to those used by Gehm et al. to observe superagonism [3,20]. Serrero and Lu [28] also observed resveratrol inhibition of ERE-luc reporter activity in E2-treated MCF-7 cells, but did not test its effects in the absence of E2. Despite multiple attempts, we have not yet been able to demonstrate resveratrol antagonism of E2-mediated reporter induction in these cells.

Several reports have examined the roles of hERα's AF-1 and AF-2 domains in its activation by resveratrol. Safe's laboratory, using HepG2 (hepatic), U2 (osteogenic), and MDA-MB-231 (mammary) cell lines transiently transfected with ERα expression vectors and a reporter based on the C3 promoter, found that resveratrol had little or no agonist activity on the wt receptor. However, deleting AF-1 increased resveratrol agonism (relative to E2) in MDA-MB-231 and U2 cells, and deleting AF-2 increased it in MDA-MB-231 cells, although in all cases it was still a partial agonist [29–31]. These results contrast with our own studies, using adenoviral ERE-tk reporter vectors in MDA-MB-231 cells stably expressing wt or mutant hERαs [20]. The wt ER responded superagonistically to resveratrol, with maximal reporter activity at 50 μM. Deletion of AF-1 ablated activation by resveratrol more sharply than that by E2. Deletion of AF-2 eliminated agonism by both ligands, while a triple point mutation in the AF-2 domain enhanced the effect of E2 but inhibited the effect of resveratrol. Consistent with the partial agonism of resveratrol in the latter cells, it acted as an antagonist when administered (at $\geq$10 μM) with E2. No antagonism was observed in cells expressing wt ER. From our results it appears that activation of hERα by resveratrol, compared with E2, is more dependent on both AF domains being present and intact. Perhaps superagonism results from an unusual synergy between AF-1 and AF-2 that is possible for the resveratrol-bound receptor in certain cell and promoter contexts. Computational modeling of resveratrol binding to hERα suggests that it hydrogen-bonds with more binding-site amino acid residues than do E2 or DES [32], which could result in a different protein conformation. Unfortunately the structure of the ERα–resveratrol complex has not been experimentally determined, nor has the conformation of the AF-1 region in any form of the receptor, liganded or unliganded. Elucidating the mechanism of superagonism will likely provide deeper insight into the workings of the ERs.

In addition to mammalian cells, yeast cells stably transfected with the gene for hERα and an estrogen-responsive reporter have been used for *in vitro* assays of estrogenicity. Resveratrol is reported to have very weak [27] or no [7] estrogenic activity in this system, indicating that yeast-based screening may not detect some compounds that are active in mammalian cells.

It is difficult to summarize the complex and sometimes contradictory data that have come from studies with various cell lines and reporters, but some generalities emerge. Most studies find that resveratrol is at least partially agonistic in the absence of E2 or DES. Antagonism, when it appears, typically requires higher concentrations than agonism. The ER-mediated effects of resveratrol can vary greatly from one cell type to another and even between different promoters within a given cell. In this respect it resembles a selective estrogen receptor modulator (SERM) [33].

Endogenous Gene Expression

The expression of endogenous genes in cultured cells has also been used as an endpoint to study the ER-mediated effects of resveratrol. As with the reporter gene assays, there have been conflicting results. Gehm et al. [3] reported that resveratrol acts as a full estrogen agonist in upregulating expression of the pS2 and progesterone receptor (PR) mRNAs in MCF-7 and T47D cells. In contrast, Lu and Serrero [4] reported that it is only a partial agonist as assayed by PR mRNA expression in MCF-7 cells and is antagonistic when combined with E2. All these effects were observed with resveratrol concentrations in the 10 μM range, but a much higher potency was observed by Wietzke and Welsh [22], who found, consistent with the reporter assays described above, that 0.4 to 4 nM resveratrol increased expression of VDR protein in T47D cells. E2 and the soy phytoestrogen genistein also boosted VDR levels.

Transforming growth factor alpha (TGFα) is a much-studied target of estrogen regulation in breast cancer and other cell types, and the ability of wt and mutant (D351Y and D351G) hERαs to modulate its expression in stably transfected MDA-MB-231 cells has been used to categorize ER-active compounds, since residue 351 is crucial in mediating the estrogen-like effects of SERMs [34–36]. We have shown that resveratrol dose-dependently increases TGFα mRNA expression in this system ($EC_{50} \sim 5\,\mu M$); it acts as a full agonist on the wt hERα and both mutants, which is characteristic of type I estrogens such as E2 and DES [21]. No antagonism was observed when resveratrol was administered concurrently with a range of doses of E2 [20,21]. In contrast, Lu and Serrero found that in MCF-7 cells, resveratrol antagonizes E2-stimulated expression of TGFα and other growth factors [4].

We have also examined the role of AF-1 and AF-2 in mediating resveratrol regulation of TGFα expression, using the same stably transfected

MDA-MB-231 cell lines we used in the reporter assays described above [20]. We found that deletion of the hERα AF-1 domain attenuated the stimulatory effect of both E2 and resveratrol, but more severely for E2. Interestingly, deletion of AF-2 eliminated the ability of E2, but not resveratrol, to stimulate TGFα expression. Mutation of the AF-2 domain did not eliminate the responses to either ligand but reduced them both to a similar extent. This pattern of activity is different from that seen in the reporter genes, indicating that the roles of AF-1 and AF-2 vary depending on the target gene.

We also used the TGFα mRNA expression assay to compare the responses of ERα and ERβ to resveratrol in MDA-MB-231 cells stably expressing each isoform (MDA-ERα and MDA-ERβ). We found that resveratrol acts as an estrogen agonist in both cell lines (Figure 17.3) but shows greater potency and produces a larger induction with ERα. However, when a low concentration of E2 was administered concurrently with resveratrol, the dose–response curves of the two cell lines were fairly similar, and the induction of TGFα mRNA was even slightly higher in MDA-ERβ than MDA-ERα. No antagonism was detected. Together with our previously published data [20,21], these results show that in the presence of low doses of E2, resveratrol acts as a full agonist on TGFα mRNA expression in these cells via both ERα and ERβ.

Studies of resveratrol's effects on endogenous estrogen-regulated genes have not been restricted to breast cancer cells. In osteoblastic MC3T3-E1 cells it increases alkaline phosphatase expression [37] and in the pituitary cell line PR1 it potently ($EC_{50} \sim 30\,nM$) stimulates prolactin secretion [38]. These agonistic effects were blocked by antiestrogens.

In addition to its effects on estrogen-modulated gene expression, resveratrol has been shown by Pozo-Guisado et al. [39] to mimic E2's nongenomic activation of the phosphoinositide-3-kinase signaling pathway in MCF-7 cells, with maximal activation at 10 μM. Higher (50 μM) concentrations were inhibitory, consistent with the results of Brownson et al. [40].

Gene Expression Profiling

Microarray technology has enabled measurement of the expression of thousands of genes in a single experiment. We have used this technique to evaluate resveratrol-induced gene expression profiles in MDA-ERα breast cancer cells, using Atlas cDNA expression arrays representing 588 known cancer-related genes [21,41,42]. It should be noted that both E2 and resveratrol are growth-inhibitory in these cells [21]. The bivariate scatterplots of genes from the microarray show that both the number of modulated genes and the fold changes in expression produced by resveratrol treatment were higher than produced by E2 treatment (Figure 17.4). Among 15 genes coregulated by both resveratrol and E2, the largest resveratrol-induced increases were seen in $p21^{cip1}$/WAF1 (23-fold) and IL-7R-alpha

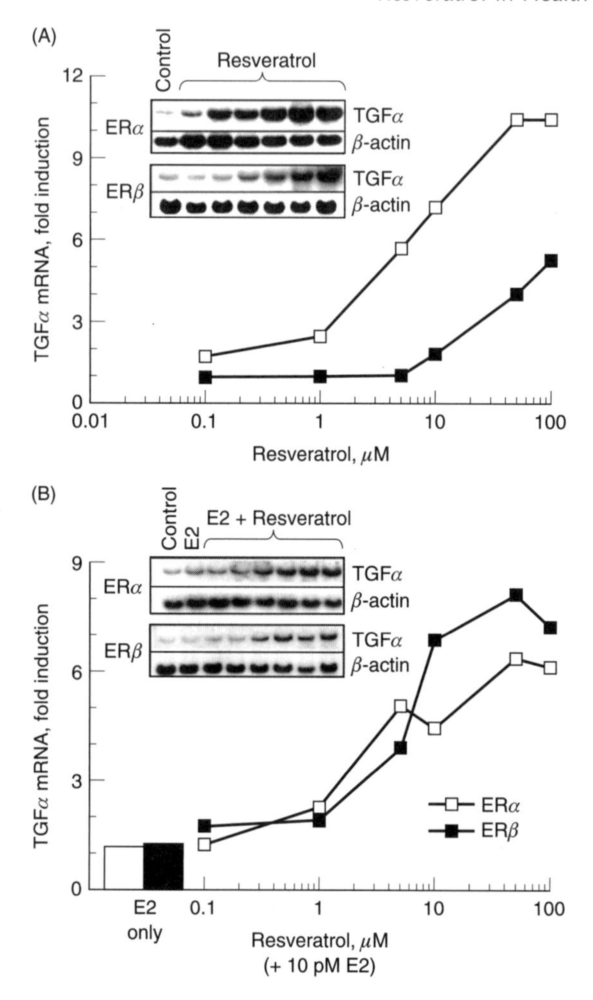

FIGURE 17.3 Resveratrol increases TGFα mRNA via ERα or ERβ in breast cancer cells. MDA-MB-231 cells stably transfected with ERα (MDA-ERα) or ERβ (MDA-ERβ) were estrogen-depleted and treated for 24 hours with the indicated concentrations of resveratrol alone (A) or resveratrol plus 10 pM E2 (B). TGFα mRNA was quantified by Northern blotting using β-actin mRNA for normalization. Results are presented as fold induction compared to control cells (treated with ethanol vehicle only). The techniques for growth of the cells, total RNA isolation, and Northern blotting have been described previously [21,81]. MDA-ERα cells have formerly been known as S30 cells [48], and MDA-ERβ as MDA-MB-231/ERβ41 [81].

(14-fold). Other studies have also shown that resveratrol upregulates p21^{cip1}/WAF1 [43–45] and this has been proposed as a mechanism for resveratrol-induced cell cycle arrest [43]. Resveratrol and E2 also significantly upregulated IGFBP 1 and 3 in our experiments.

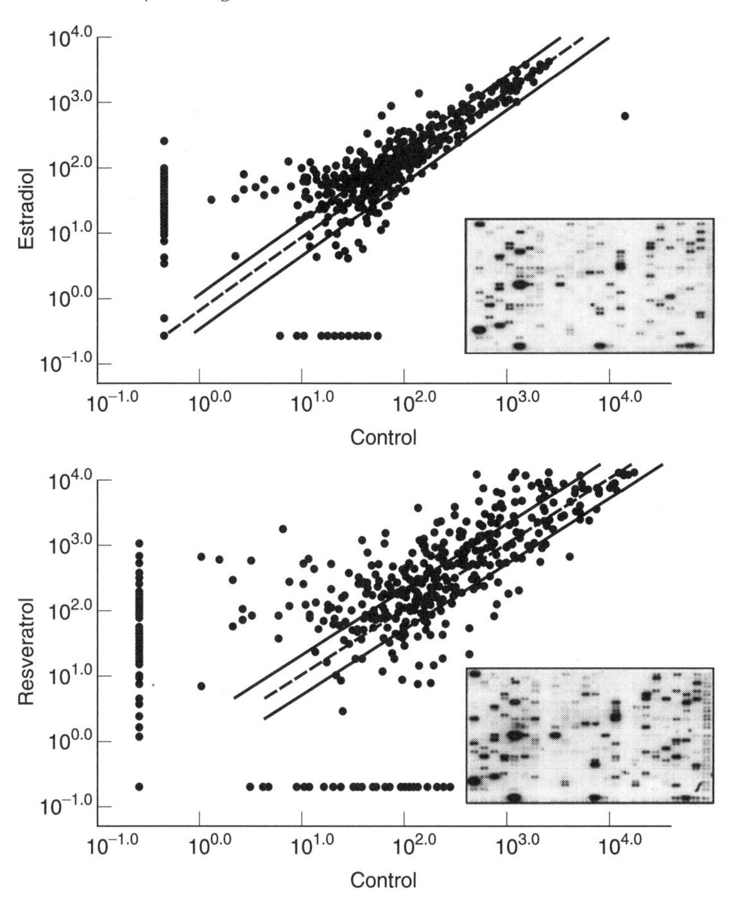

FIGURE 17.4 Estradiol and resveratrol differentially regulate gene expression through ERα. Gene expression profiles of vehicle-treated MDA- ERα cells (control; X-axes) and cells treated with E2 or resveratrol (Y-axes) are shown as bivariate scatterplots of 588 genes from the microarray. The values are normalized adjusted intensities representing levels of expression. Genes with equal expression values lined up on the identity line (central diagonal), whereas outliers correspond to up- and downregulated genes. The scale on each axis is logarithmic. The insets are phosphorimages of the original hybridization membranes showing profiles of the treated cells. (Derived and modified from Levenson AS, Gehm BD, Timm Pearce S, Horiguchi J, Simons LA, Ward JE III, Jameson JL, and Jordan VC, *Int J Cancer* 104, 587–596, 2003.)

There was a unique set of approximately 60 genes, however, whose expression was altered by treatment with resveratrol but not with E2. Of these genes, 24 were increased 4-fold or higher, 17 were increased 2.7- to 3.9-fold, and 19 increased 2- to 2.4-fold. Transcriptional activators, intracellular transducers, effectors, and modulators made up the largest fraction (29%) of the genes that were increased 4-fold or more, while basic transcription factors, growth factors, cytokines, and chemokines made up 26%. The highest induction in this group was observed for integrin beta 1 (ITGB1), which was increased 19.9-fold. Curiously, the level of ITGB1 also increased dramatically in cells treated with raloxifene (13.1-fold), ICI 182780 (fulvestrant) (17.7-fold) and GW7604 (18.6-fold) [41]. Further analysis of overlapping responses to resveratrol and antiestrogens revealed a few more genes showing a greater than 8-fold increase, among them RANTES and superoxide dismutase 1(SOD-1), which have been implicated in the induction of apoptosis and antioxidant activities, respectively.

Compared to E2 and SERMs such as 4-OHT (active metabolite for tamoxifen), raloxifene, GW7604 (active metabolite for GW5638), EM 652 (active metabolite for EM800), idoxifene, and ICI 182780 (fulvestrant), resveratrol had a profound effect on gene expression profiles in our experiments [41,42]. It caused more and larger changes in gene expression than any of these other ligands, presumably because it has both ER-mediated and ER-independent effects. Curiously, cluster analysis and "expression signature" analysis of a subset of upregulated genes grouped resveratrol most closely with idoxifene in the resulting dendrogram [41]. Although idoxifene resembles tamoxifen in chemical structure, its activity in rodent uterus is more similar to raloxifene's [46,47]. Altogether, resveratrol-induced expression profiles reveal the activation of multiple signaling pathways that reflect the mixed agonist/antagonist nature of resveratrol. Accumulation and in-depth analysis of data from large-scale microarray experiments and further functional studies will be needed to fully correlate gene expression patterns with resveratrol's biological effects.

Cultured Cell Proliferation

Resveratrol has been shown to inhibit the proliferation of a wide variety of cell lines, both ER-positive and ER-negative. Inhibition of E2-stimulated cell growth is sometimes adduced as evidence of ER antagonism, but may be due to ER-independent antiproliferative effects. If growth inhibition is due to estrogen antagonism, it should be possible to overcome it with higher concentrations of E2; however, we are not aware of any studies that have attempted to show this. Comparison of ER-positive and ER-negative cell lines provides another possible way of determining whether antiproliferative effects are ER-mediated, but it is difficult if not impossible to be sure that ER expression is the only difference between two cell lines. Transfection of ER-negative cells with ER expression vectors would seem to offer a solution to this problem, but, paradoxically, ER agonists are typically growth-inhibitory in such cells [48,49], which makes interpretation difficult. For instance, we found that resveratrol and E2 each inhibited the growth of MDA-MB-231 cells stably transfected with wtERα or the mutant D351Y.

However, inhibition by resveratrol, unlike that by E2, was not blocked by fulvestrant (ICI 182780); moreover resveratrol inhibited the growth of untransfected ER-negative MDA-MB-231 cells. We were unable to demonstrate any specific ER-mediated growth inhibition against the background of a powerful ER-independent inhibition [21].

However, many cells that express ER endogenously are stimulated to grow by E2, and proliferation of these cells in response to resveratrol can be ascribed to ER agonism with some confidence if the stimulation can be blocked by antiestrogens. For example, Gehm et al. showed that 10 μM resveratrol stimulates the growth of T47D cells to the same extent as a saturating (0.1 nM) dose of E2 [3], but 100 μM is inhibitory (Gehm et al., unpublished results). Wietzke and Welsh [22] obtained similar results, finding that 4 μM resveratrol stimulates T47D growth but concentrations above 20 μM inhibit it. Both groups showed that antiestrogens block the stimulation. Results of these and other studies in breast cancer cell lines are summarized in Table 17.1. It should be noted that some of these studies focused on resveratrol's effects on cell growth and not on estrogenicity *per se*, and so did not employ estrogen-depleted media (phenol red-free formulations and charcoal-stripped serum) to reduce the estrogenic "background." The cells were thus exposed to basal estrogenic stimulation that may have masked the agonistic effects of resveratrol.

Overall, the results shown in Table 17.1 suggest that low (typically 1 to ~10 μM) concentrations of resveratrol have an ER-mediated growth-promoting (i.e., agonistic) effect on estrogen-dependent cell lines, at least in the absence of other estrogens. Two other studies in nonbreast cancer cell lines show agonistic stimulation of growth at even lower concentrations. Mizutani et al. [37] observed that resveratrol was remarkably potent at stimulating the growth of osteoblastic MC3T3-E1 cells (EC_{50} of 10 to 30 nM); this effect was blocked by tamoxifen. Stahl et al. [38] found that resveratrol increased proliferation of PR1 cells: 10 nM produced a modest increase in cell number and 1 μM stimulated growth almost as much as E2.

As Table 17.1 shows, ER-positive cells appear to be less sensitive than ER-negative cells to resveratrol's antiproliferative effects, probably because its agonism offsets its ER-independent growth inhibition to some extent. High concentrations of resveratrol inhibit proliferation even in cell types that are stimulated by lower concentrations, suggesting that at high doses resveratrol is antagonistic and/or that its ER-independent antiproliferative effect becomes overwhelming. The concentrations needed to produce inhibition are variable, but typically in the 0.1 to 10 μM range for ER-negative cells and 20 to 50 μM or higher for ER-positive cells. However, Damianaki et al. [54] have reported growth inhibition by concentrations as low as 1 pM, several orders of magnitude lower than those at which other workers have found any effect. The disparity between these results and the findings of other laboratories remains puzzling.

TABLE 17.1
Effects of Resveratrol on Breast Cancer Cell Proliferation *In Vitro*

Cell line	ER	[Resv.], μM	Effect on growth	Blocked by antiestrogen?	Comments	Ref.
T47D	+	10	↑	Yes (fulvest.)	Full stimulation (≈ 0.1 nM E2)	3
T47D	+	100	↓			Gehm et al. (unpublished data)
T47D	+	4	↑	Yes (tamox.)		22
		20	↓			
MCF-7	+	4	↑	Yes (tamox.)		
MCF-7	+	10–25	↑	N.D.	*Cis* isomer also effective	25
		50	↓		*Cis* isomer ineffective	
		25–50	↓ of E2-stim.		*Cis* isomer less effective	
MCF-7	+	0.1–5	↑	N.D.		4
		10	↓			
		1–10	↓ of E2-stim.			
MCF-7	+	1–10	↑	N.D.		50
		20–60	↓			
MDA-MB-231	–	0.01–60	↓			

MCF-7	+	22–180	↓		Not estrogen-depleted	51
MCF-10	–	22–180	↓			
MDA-MB-231	–	22–180	↓			
MCF-7	+	25	↓		Not estrogen-depleted. Inhibition weaker for MCF-7	52
MDA-MB-435	–	25	↓			
MCF-7	+	4	↑	N.D.	Phenol red-free media but serum not stripped	53
		44–200	↓			
KPL-1	+	4–20	↑	N.D.		
		44–200	↓			
MKL-F	–	0.4–200	↓			
MCF-7	+	0.000001–1	↓		Not estrogen-depleted. IC_{50}s in pM range	54
T47D	+	0.000001–1	↓			
MDA-MB-231	–	0.000001–1	↓			
MDA-ERα[a]	+	3–100	↓	No (fulvest.)	Agonists inhibit growth	21
MDA-MB-231	–	5–100	↓			

Note: ER, estrogen receptor status; N.D., not determined; ↓, inhibition; ↑, stimulation; "↓ of E2-stim.," inhibition of E2-stimulated proliferation when cells were treated with resveratrol and E2 concurrently.

[a]Derived from ER-negative cells stably transfected with ERα.

Resveratrol: Agonist, Antagonist, or Superagonist?

E2 and DES are "pure" ER agonists, and the transcriptional activity induced by these compounds has traditionally been regarded as maximal. The apparent ability of resveratrol to induce even greater activity is surprising and mysterious. This property appears to be specific to resveratrol; tests of half a dozen hydroxystilbenes revealed others showing full or partial agonism, but no other superagonists [20]. Thus far, this phenomenon has been observed only with artificial reporter genes and not with endogenous estrogen-regulated genes (except in cells expressing mutated ERαs that have a depressed response to E2 [20]). However, superagonism is not an artifact of transient transfection; we [20] and others [25] have observed it with reporter constructs stably integrated into the cellular genome. Since natural endogenous genes are subjected to a multiple transcriptional controls mediated by a variety of regulatory sequences in their promoters, it is possible that hyperactivation of one transcription factor (e.g., ER) may be damped or "diluted out" by other control mechanisms. The different complements of transcriptional cofactors present in different cells and assembled on different promoters may also account for the variable degree of agonism/antagonism observed in different experimental systems. In our own studies, we have found resveratrol to be primarily an ER agonist or superagonist. Indeed, we have not observed antagonism except in cells expressing mutant ERα [20]. It is possible, however, that the sharp decline in reporter activity when resveratrol concentrations exceed the optimal level (as seen in Figure 17.2) indicates antagonism at high concentrations; further study is needed.

In Vivo Studies

While cultured cells are powerful tools for screening estrogenic compounds and studying their mechanisms of action, *in vivo* studies on whole organisms are more likely to yield physiologically relevant results. The classic biological test for estrogenic activity has been the rodent uterotrophic assay. E2 greatly increases the wet and dry weight of the uterus in prepubertal or ovariectomized (OVX) rats and mice, and produces other reproductive tract changes typical of estrus. After the initial report of resveratrol's estrogenicity *in vitro*, several papers appeared that examined its effects on the uterus and other estrogen targets in rats, with mostly negative results.

Ashby's laboratory administered resveratrol (0.03 or 0.3 mg/kg/d × 3 d, orally or subcutaneously) to immature female rats. (As a point of comparison, the resveratrol content of many red wines is a few mg/l, or of the order of 10 μM, though higher and lower concentrations are found frequently [55].) Modest increases in uterine weight were seen at the higher dose but these could not be replicated in later experiments, even at much greater doses [27]. In a subsequent paper, a similar range of doses [56] in OVX rats produced no effects on uterine wet weight, plasma HDL cholesterol, or femoral bone density. Similarly, Turner et al. [57] found that resveratrol orally administered to immature rats (1 to 100 μg/d × 6 d; average body weight ~65 g) produced small increases in uterine wet weight but the effect was statistically significant at only one dose (10 μg/d). Resveratrol had no significant effect on body weight, serum cholesterol, or radial bone growth, nor on uterine epithelial cell height or IGF-I mRNA levels, all of which were increased by E2. A high dose (1 mg/d) had no effect on any of the endpoints when administered alone, but antagonized the hypocholesterolemic effect of concurrently administered E2. Freyberger et al. reported that subcutaneous resveratrol (19–575 mg/kg/d × 3d) had no effect on immature rat uteri [81].

Although these results suggest that resveratrol has little if any estrogenicity in rats, Henry and Witt [6] observed that resveratrol in drinking water (100 μM) disrupted estrus cyclicity and increased ovarian weight in gonadally intact rats, which they interpreted as evidence of ERβ agonism. In addition, Mizutani et al. [59] found that resveratrol (5 mg/kg/d orally for several weeks) had estrogen-like effects in a rat model of postmenopausal hypertension and osteoporosis, blocking OVX-induced increases in blood pressure and decreases in femoral breaking energy. These findings suggest that resveratrol may act as a SERM in rats.

The studies described above were performed on adults or weanlings, but Kubo et al. [60] studied the effects of prenatal and perinatal exposure. Pregnant rats were given resveratrol in drinking water (5 mg/l) from the start of gestation until weaning. In female offspring this treatment delayed vaginal opening (a marker of sexual development), decreased sexual receptivity, and disrupted estrous cycling. Male offspring showed feminization of a sexually dimorphic brain region, and their testosterone levels were reduced by ~40%, although the latter result was not statistically significant. Most of these effects were qualitatively similar to, but smaller than, those produced by DES. Thus, resveratrol appears to have estrogenic effects on developing rats.

In vivo results naturally depend on the choice of organism. A paper by Liu et al. [61] suggests that mice may be more sensitive than rats to the estrogenic effects of resveratrol. They reported that in immature mice, subcutaneous resveratrol (2 mg/kg/d × 4 d) increased uterine wet weight and epithelial cell height, and accelerated vaginal opening and keratinization, all consistent with an estrogen agonist effect. Oral administration produced similar but smaller effects.

The range of species examined was further broadened by Ratna and Simonelli [62], who found that roosters injected with resveratrol (2 doses of ~2 to 200 mg/kg i.m.) showed enhanced stabilization of hepatic apolipoprotein II (apo II) mRNA via the estrogen-regulated mRNA stabilizing factor (ERmRNA-SF). Remarkably, resveratrol was about half as potent as E2 in this system.

To summarize, there is little evidence that resveratrol has ER-agonistic effects on classic estrogen targets such as the uterus, at least in the rat. However, some results do point to estrogenic or SERM-like effects on specific targets, and sensitivity to resveratrol's ER-mediated effects may differ between species.

RESVERATROL AND BREAST CANCER

Breast cancer is the main diet-related cancer in women. Numerous epidemiological studies have shown that phytoestrogen-rich diets are associated with a low incidence of breast cancer [63]. Unfortunately, these studies have not dealt with resveratrol, due to its relatively low levels in ordinary foods. However, studies of wine consumption and cancer risk do offer a source of data on possible breast cancer-preventive effects of resveratrol.

High alcohol consumption is associated with increased risk of many kinds of cancer, but for low to moderate intake the picture is less clear, and specific beverages appear to have different effects (reviewed by Bianchini and Vainio [64]). Some studies indicate that wine decreases the risk of several types of cancer, including basal cell carcinoma, non-Hodgkin's lymphoma, esophageal adenocarcinoma, and lung cancer. In contrast, no protective effect was seen for endometrial and colorectal cancers.

Regarding breast cancer, conclusions vary. Viel et al. [12] showed increased risk in premenopausal women consuming more than four liters of red wine per month, i.e., more than one drink (~12 g of alcohol) per day. The effects of white wine, beer, and fortified wine were not significant. In a case-control study, Lê et al. [11] found that postmenopausal women consuming 1 to 2 glasses of wine per day were at higher risk than those who drank less or more. Increased risk was also seen for drinkers of beer but not of distilled sprits. However, other studies have found no significant association between consumption of wine and the risk of breast cancer [65–68]. It should be noted that many studies do not differentiate red and white wines, and there are marked geographical differences in the resveratrol content even of red wines [55]. These factors could play a role in the inconsistent results of epidemiological studies.

Although most of the available data relate to wine's association with breast cancer incidence, recently data became available on its relation to breast cancer mortality. Jain et al. reported that total alcohol intake of 10 to 20 g/day was associated with a slight increase in mortality from breast cancer, and that wine contributed most of the increased risk [69].

Altogether, although studies have shown that there are some beneficial effects of wine that may be attributable to resveratrol's antioxidant, anti-inflammatory, and antitumor activity, available epidemiological data do not support a protective effect of red wine against breast cancer; indeed, some data suggest the reverse. It is possible that the impact of wine consumption on breast cancer is especially complex because of the mixed estrogenic/antiestrogenic properties of resveratrol. Our *in vitro* studies have repeatedly shown agonistic or superagonistic, but not antagonistic, effect on ERs in breast cancer cells [3,20,21 and results presented here]. Further investigation into the estrogenic/antiestrogenic nature of resveratrol is needed before any recommendations can be made regarding its intake for breast cancer prevention or management of postmenopausal symptoms.

ARE RESVERATROL'S ER-MEDIATED ACTIONS PHYSIOLOGICALLY RELEVANT?

The study of resveratrol's effects on ERs, and especially the phenomenon of superagonism, may yield new insights into the mechanisms of ER action, but a more immediate question is: what is the physiological significance of these effects? How relevant are they to human health?

The metabolism and pharmacokinetics of resveratrol are described in detail elsewhere in this volume and will be treated only briefly here. Animal and human studies have shown that ingested resveratrol is rapidly glucuronidated, and consumption of doses corresponding to those in red wine is likely to produce peak plasma concentrations of free resveratrol in the nanomolar or subnanomolar range [70–72]. Even with much larger oral doses (20 mg/kg body weight), plasma concentrations in rats were found to peak at $\sim 1\,\mu M$ and decline rapidly ($\sim 0.1\,\mu M$ after 30 min) [73]. However, most *in vitro* studies indicate that concentrations in the range 3 to 30 μM or more are required for ER-mediated effects. On this basis it would seem quite unlikely that resveratrol's estrogenicity or antiestrogenicity play a role in the health effects of wine.

However, these considerations do not apply only to resveratrol's ER-mediated effects. As Gescher and Steward [74] have pointed out, resveratrol poses a conundrum: nearly all of the mechanisms proposed for resveratrol's cancer chemopreventive activity require concentrations $>1\,\mu M$ *in vitro*, and most require 10 to 100 μM, but animal studies using oral doses of less than 1 mg/kg/d have shown suppression of chemical carcinogenesis [45,75]. Similarly, Pace-Asciak et al. showed that resveratrol inhibited thrombin-induced platelet aggregation *in vitro* with an $EC_{50} > 100\,\mu M$, but observed similar *in vivo* inhibition in human volunteers consuming 2 mg/d [76,77]. For reasons yet to be explained, resveratrol is sometimes efficacious *in vivo* at doses that should produce plasma concentrations well below those required for effectiveness *in vitro*. This may also apply to its ER-mediated effects.

CONCLUSIONS

Like other phytoestrogens, resveratrol appears to act as a SERM, whose effects on ER-mediated gene expression vary depending on cell type, the identity of the target gene, the presence or absence of additional estrogens, and other factors. In many systems it shows a biphasic dose–response curve, acting as an agonist at low concentrations but as an antagonist at high, a property shared by the SERM tamoxifen [78,79]. Given the apparently low bioavailability of ingested resveratrol [70–73], it seems likely that the effects, if any, of dietary exposure would more likely be estrogenic than antiestrogenic. Although resveratrol appears to be less potent than many phytoestrogens, the concentrations required to modulate ER activity *in vitro* are commensurate with those needed to affect other cell signaling pathways. Resveratrol's SERM-like effects are therefore likely to be significant if it is administered in pharmacological doses, and they may contribute to beneficial or adverse side effects if it is used as a therapeutic or preventive agent for cancer or heart disease. Moreover, the reports of agonistic or SERM-like effects at low concentrations *in vitro* [8,22,37,38], and at moderate doses *in vivo* [6,60], suggest that it would be premature to dismiss the possibility that resveratrol's phytoestrogenic nature is relevant to the health effects of red wine.

In some *in vitro* assays, resveratrol acts as an ERα superagonist. This effect is highly dependent on the cell type and the promoter context of the ERE, and on the presence of fully functional AF-1 and AF-2 domains in the receptor. One possible explanation is that the resveratrol-bound receptor adopts a unique conformation that interacts particularly well with certain sets of transcriptional cofactors. Comparative studies with ERβ, in which AF-1 is weak or inactive, may help to explain this puzzling phenomenon. A better understanding of the mechanisms underlying superagonism could lead to the development of more efficacious and selective SERMs.

As the other chapters of this book attest, resveratrol affects an amazing diversity of cellular processes, which can complicate interpretation of its ER-mediated effects. Cell proliferation studies provide one illustration of this, as ER-independent antiproliferative effects compete with ER-agonistic growth promotion. In estrogen-dependent cells, the latter effect appears to predominate at lower concentrations, which may have important implications for the potential use of resveratrol *in vivo* against breast cancer and other estrogen-responsive malignancies. Our gene-array data (Figure 17.4) also illustrate the pleiotropic actions of resveratrol, showing that it alters the expression of many genes that are unaffected by E2 or other SERMs. It is interesting to note that, although resveratrol was categorized as a type

I estrogen (like the pure agonists E2 and DES) based on regulation of TGFα expression via wtERα and D351 point mutants, its gene-array profile more closely resembles that of the SERM idoxifene. Clearly resveratrol's interactions with the ERs are complex, and much work remains to be done before we fully understand them.

ACKNOWLEDGMENTS

The authors thank Dr. Hong Liu of Northwestern University Medical School for assistance in interpreting Chinese studies, Dr. Debra Tonetti for providing MDA-ERβ cells, and Dr. Joanne McAndrews for her work on cell growth assays. Drs. V. Craig Jordan and J. Larry Jameson, as well as past and present members of their laboratories, provided invaluable assistance and advice for our experiments. A.S.L. is grateful to Dr. Michael Schafer for his continuous support. The Department of Defense Breast Cancer Research Program Idea Award DAMD17-99-1-9334 (B.D.G.), the IDPH, Penny Severns Breast and Cervical Cancer Research Fund (A.S.L.), and the Avon Foundation (A.S.L., B.D.G.) provided support for the experiments described herein. The writing of this chapter was supported in part by NIH Grant Number P20 RR-16460-03 (Subcontract 07011-03) from the National Center for Research Resources (B.D.G.) and SPORE No. CA89018-01 (V.C.J.), pilot project (A.S.L.). Its contents are solely the responsibility of the authors and do not necessarily represent the views of the granting agencies.

REFERENCES

1. Bradbury RB and White DE, Oestrogens and related substances in plants, *Vitam Horm* 12, 207–233, 1954.
2. Wynne-Edwards KE, Evolutionary biology of plant defenses against herbivory and their predictive implications for endocrine disruptor susceptibility in vertebrates, *Environ Health Perspect* 109, 443–448, 2001.
3. Gehm BD, McAndrews JM, Chien P-Y, and Jameson JL, Resveratrol, a polyphenolic compound found in grapes and wine, is an agonist for the estrogen receptor, *Proc Natl Acad Sci USA* 94, 14138–14143, 1997.
4. Lu R and Serrero G, Resveratrol, a natural product derived from grape, exhibits antiestrogenic activity and inhibits the growth of human breast cancer cells, *J Cell Physiol* 179, 297–304, 1999.
5. Bowers JL, Tyulmenkov VV, Jernigan SC, and Klinge CM, Resveratrol acts as a mixed agonist/antagonist for estrogen receptors alpha and beta, *Endocrinology* 141, 3657–3667, 2000.
6. Henry LA and Witt DM, Resveratrol: phytoestrogen effects on reproductive physiology and behavior in female rats, *Horm Behav* 41, 220–228, 2002.
7. Klinge CM, Risinger KE, Watts MB, Beck V, Eder R, and Jungbauer A, Estrogenic activity in white and red wine extracts, *J Agric Food Chem* 51, 1850–1857, 2003.

8. Ramsey TL, Risinger KE, Jernigan SC, Mattingly KA, and Klinge CM, Estrogen receptor beta isoforms exhibit differences in ligand-activated transcriptional activity in an estrogen response element sequence-dependent manner, *Endocrinology* 145, 149–160, 2004.
9. Goldberg DM, Hahn SE, and Parkes JG, Beyond alcohol: beverage consumption and cardiovascular mortality, *Clin Chim Acta* 237, 155–187, 1995.
10. Orgogozo JM, Dartigues JF, Lafont S, Letenneur L, Commenges D, Salamon R, Renaud S, and Breteler MB, Wine consumption and dementia in the elderly: a prospective community study in the Bordeaux area, *Rev Neurol*; 153, 185–192, 1997.
11. Lê MG, Hill C, Kramar A, and Flamanti R, Alcoholic beverage consumption and breast cancer in a French case-control study, *Am J Epidemiol* 120, 350–357, 1984.
12. Viel JF, Perarnau JM, Challier B, and Faivre-Nappez I, Alcoholic calories, red wine consumption and breast cancer among premenopausal women, *J Epidemiol* 13, 639–643, 1997.
13. Kopp P, Resveratrol, a phytoestrogen found in red wine. A possible explanation of the conundrum of the "French paradox"? *Eur J Endocrinol* 138, 619–620, 1998.
14. Nilsson S, Makela S, Treuter E, Tujague M, Thomsen J, Andersson G, Enmark E, Pettersson K, Warner M, and Gustafsson J-A, Mechanisms of estrogen action, *Physiol Rev* 81, 1535–1565, 2001.
15. Kuiper GGJM, Carlsson B, Grandien K, Enmark E, Häggblad J, Nilsson S, and Gustafsson J-Å, Comparison of the ligand binding specificity and transcript tissue distribution of estrogen receptors alpha and beta, *Endocrinology* 138, 863–870, 1997.
16. Adlercreutz CH, Goldin BR, Gorbach SL, Hockerstedt KA, Watanabe S, Hamalainen EK, Markkanen MH, Makela TH, Wahala KT, and Adlercreutz T, Soybean phytoestrogen intake and cancer risk, *J Nutr* 125, 757S–770S, 1995.
17. Bhat KP and Pezzuto JM, Resveratrol exhibits cytostatic and antiestrogenic properties with human endometrial adenocarcinoma (Ishikawa) cells, *Cancer Res* 61, 6137–6144, 2001.
18. An J, Tzagarakis-Foster C, Scharschmidt TC, Lomri N, and Leitman DC, Estrogen receptor beta-selective transcriptional activity and recruitment of coregulators by phytoestrogens, *J Biol Chem* 276, 17808–17814, 2001.
19. Mueller SO, Simon S, Chae K, Metzler M, and Korach KS, Phytoestrogens and their human metabolites show distinct agonistic and antagonistic properties on estrogen receptor alpha (ERα) and ERβ in human cells, *Toxicol Sci* 80, 14–25, 2004.
20. Gehm BD, Levenson AS, Liu H, Lee E-J, Amundsen BM, Cushman M, Jordan VC, and Jameson JL, Estrogenic effects of resveratrol in breast cancer cells expressing mutant and wild-type estrogen receptors: role of AF-1 and AF-2, *J Steroid Biochem Mol Biol* 88, 223–234, 2004.
21. Levenson AS, Gehm BD, Timm Pearce S, Horiguchi J, Simons LA, Ward JE III, Jameson JL, and Jordan VC, Resveratrol acts as an estrogen receptor (ER) agonist in breast cancer cells stably transfected with ER alpha, *Int J Cancer* 104, 587–596, 2003.

22. Wietzke JA and Welsh J, Phytoestrogen regulation of a Vitamin D3 receptor promoter and 1,25-dihydroxyvitamin D3 actions in human breast cancer cells, *J Steroid Biochem Mol Biol* 84, 149–157, 2003.
23. Bhat KP, Lantvit D, Christov K, Mehta RG, Moon RC, and Pezzuto JM, Estrogenic and antiestrogenic properties of resveratrol in mammary tumor models, *Cancer Res* 61, 7456–7463, 2001.
24. Pervaiz S, Resveratrol: from grapevines to mammalian biology, *FASEB J* 17, 1975–1985, 2003.
25. Basly JP, Marre-Fournier F, Le Bail JC, Habrioux G, and Chulia AJ, Estrogenic/antiestrogenic and scavenging properties of (E)- and (Z)-resveratrol, *Life Sci* 66, 769–777, 2000.
26. Sun Z, Lu QJ, Wen LQ, Guo SM, Chen YY, and Liu WJ, Establishment and its application of a reporter-based screening model for discovering new ligands of estrogen receptor alpha subtypes [in Chinese], *Yao Xue Xue Bao* [*Acta Pharm Sin*] 35, 747–751, 2000.
27. Ashby J, Tinwell H, Pennie W, Brooks AN, Lefevre PA, Beresford N, and Sumpter JP, Partial and weak oestrogenicity of the red wine constituent resveratrol: consideration of its superagonist activity in MCF-7 cells and its suggested cardiovascular protective effects, *J Appl Toxicol* 19, 39–45, 1999.
28. Serrero G and Lu R, Effect of resveratrol on the expression of autocrine growth modulators in human breast cancer cells, *Antioxid Redox Signal* 3, 969–979, 2001.
29. Yoon K, Pellaroni L, Ramamoorthy K, Gaido K, and Safe S, Ligand structure-dependent differences in activation of estrogen receptor alpha in human HepG2 liver and U2 osteogenic cancer cell lines, *Mol Cell Endocrinol* 162, 211–220, 2000.
30. Yoon K, Pallaroni L, Stoner M, Gaido K, and Safe S, Differential activation of wild-type and variant forms of estrogen receptor alpha by synthetic and natural estrogenic compounds using a promoter containing three estrogen-responsive elements, *J Steroid Biochem Mol Biol* 78, 25–32, 2001.
31. Safe SH, Pallaroni L, Yoon K, Gaido K, Ross S, and McDonnell D, Problems for risk assessment of endocrine-active estrogenic compounds, *Environ Health Perspect* 110 (Suppl 6), 925–929, 2002.
32. El-Mowafy AM, Abou-Zeid LA, and Edafiogho I, Recognition of resveratrol by the human estrogen receptor-alpha: a molecular modeling approach to understand its biological actions, *Med Princ Pract* 11, 86–92, 2002.
33. Levenson AS and Jordan VC, Selective oestrogen receptor modulation: molecular pharmacology for the millennium, *Eur J Cancer* 35, 1974–1985, 1999.
34. Levenson AS and Jordan VC, The key to the antiestrogenic mechanism of raloxifene is amino acid 351 (aspartate) in the estrogen receptor, *Cancer Res* 58, 1872–1875, 1998.
35. Levenson AS, MacGregor Schafer JI, Bentrem DJ, Pease KM, and Jordan VC, Control of the estrogen-like actions of the tamoxifen–estrogen receptor complex by the surface amino acid at position 351, *J Steroid Biochem Mol Biol* 76, 61–70, 2001.
36. Jordan VC, Schafer JM, Levenson AS, Liu H, Pease KM, Simons LA, and Zapf JW, Molecular classification of estrogens, *Cancer Res* 61, 6619–6623, 2001.

37. Mizutani K, Ikeda K, Kawai Y, and Yamori Y, Resveratrol stimulates the proliferation and differentiation of osteoblastic MC3T3-E1 cells, *Biochem Biophys Res Commun* 253, 859–863, 1998.
38. Stahl S, Chun TY, and Gray WG, Phytoestrogens act as estrogen agonists in an estrogen-responsive pituitary cell line, *Toxicol Appl Pharmacol* 152, 41–48, 1998.
39. Pozo-Guisado E, Lorenzo-Benayas MJ, and Fernandez-Salguero PM, Resveratrol modulates the phosphoinositide 3-kinase pathway through an estrogen receptor alpha-dependent mechanism: relevance in cell proliferation, *Int J Cancer* 109, 167–173, 2004.
40. Brownson DM, Azios NG, Fuqua BK, Dharmawardhane SF, and Mabry TJ, Flavonoid effects relevant to cancer, *J Nutr* 132, 3482S–3489S, 2002.
41. Levenson AS, Kliakhandler IL, Svoboda KM, Pease KM, Kaiser SA, Ward JE III, and Jordan VC, Molecular classification of selective oestrogen receptor modulators on the basis of gene expression profiles of breast cancer cells expressing oestrogen receptor alpha, *Br J Cancer* 87, 449–456, 2002.
42. Levenson AS, Svoboda KM, Pease KM, Kaiser SA, Chen B, Simons LA, Jovanovic BD, Dyck PA, and Jordan VC, Gene expression profiles with activation of the estrogen receptor α-selective estrogen receptor modulator complex in breast cancer cells expressing wild-type estrogen receptor, *Cancer Res* 62, 4419–4426, 2002.
43. Ahmad N, Adhami VM, Afaq F, Feyes DK, and Mukhtar H, Resveratrol causes WAF-1/p21-mediated G(1)-phase arrest of cell cycle and induction of apoptosis in human epidermoid carcinoma A431 cells, *Clin Cancer Res* 7, 1466–1473, 2001.
44. Mitchell SH, Zhu W, and Young CY, Resveratrol inhibits the expression and function of the androgen receptor in LNCaP prostate cancer cells, *Cancer Res* 59, 5892–5895, 1999.
45. Tessitore L, Davit A, Sarotto I, and Caderni G, Resveratrol depresses the growth of colorectal aberrant crypt foci by affecting bax and p21(CIP) expression, *Carcinogenesis* 21, 1619–1622, 2000.
46. Nuttall ME, Bradbeer JN, Stroup GB, Nadeau DP, Hoffman SJ, Zhao H, Rehm S, and Gowen M, Idoxifene: a novel selective estrogen receptor modulator prevents bone loss and lowers cholesterol levels in ovariectomized rats and decreases uterine weight in intact rats, *Endocrinology* 139, 5224–5234, 1998.
47. Chander SK, McCague R, Luqmani Y, Newton C, Dowsett M, Jarman M, and Coombes RC, Pyrrolidino-4-iodotamoxifen and 4-iodotamoxifen, new analogues of the antiestrogen tamoxifen for the treatment of breast cancer, *Cancer Res* 51, 5851–5858, 1991.
48. Jiang SY and Jordan VC, Growth regulation of estrogen receptor-negative breast cancer cells transfected with complementary DNAs for estrogen receptor, *J Natl Cancer Inst* 84, 580–591, 1992.
49. Levenson AS and Jordan VC, Transfection of human estrogen receptor (ER) cDNA into ER-negative mammalian cell lines, *J Steroid Biochem Mol Biol* 51, 229–239, 1994.
50. Schmitt E, Lehmann L, Metzler M, and Stopper H, Hormonal and genotoxic activity of resveratrol, *Toxicol Lett* 136, 133–142, 2002.

51. Mgbonyebi OP, Russo J, and Russo IH, Antiproliferative effect of synthetic resveratrol on human breast epithelial cells, *Int J Oncol* 12, 865–869, 1998.
52. Hsieh TC, Burfeind P, Laud K, Backer JM, Traganos F, Darzynkiewicz Z, and Wu JM, Cell cycle effects and control of gene expression by resveratrol in human breast carcinoma cell lines with different metastatic potentials, *Int J Oncol* 15, 245–252, 1999.
53. Nakagawa H, Kiyozuka Y, Uemura Y, Senzaki H, Shikata N, Hioki K, and Tsubura A, Resveratrol inhibits human breast cancer cell growth and may mitigate the effect of linoleic acid, a potent breast cancer cell stimulator, *J Cancer Res Clin Oncol* 127, 258–264, 2001.
54. Damianaki A, Bakogeorgou E, Kampa M, Notas G, Hatzoglou A, Panagiotou S, Gemetzi C, Kouroumalis E, Martin PM, and Castanas E, Potent inhibitory action of red wine polyphenols on human breast cancer cells, *J Cell Biochem* 78, 429–441, 2000.
55. Goldberg DM, Yan J, Ng E, Diamandis EP, Karumanchiri A, Soleas G, and Waterhouse AL, A global survey of trans-resveratrol concentrations in commercial wines, *Am J Enol Vitic* 46, 159–165, 1995.
56. Slater I, Odum J, and Ashby J, Resveratrol and red wine consumption, *Hum Exp Toxicol* 18, 625–626, 1999.
57. Turner RT, Evans GL, Zhang M, Maran A, and Sibonga JD, Is resveratrol an estrogen agonist in growing rats? *Endocrinology* 140, 50–54, 1999.
58. Freyberger A, Hildebrand H, and Krötlinger F, Differential response of immature rat uterine tissue to ethinylestradiol and the red wine constituent resveratrol, *Arch Toxicol* 74, 709–715, 2001.
59. Mizutani K, Ikeda K, Kawai Y, and Yamori Y, Resveratrol attenuates ovariectomy-induced hypertension and bone loss in stroke-prone spontaneously hypertensive rats, *J Nutr Sci Vitaminol* (Tokyo) 46, 78–83, 2000.
60. Kubo K, Arai O, Omura M, Watanabe R, Ogata R, and Aou S, Low dose effects of bisphenol A on sexual differentiation of the brain and behavior in rats, *Neurosci Res* 45, 345–356, 2003.
61. Liu Z, Yu B, Li W, Sun J, Huang J, Huo J, and Liu C, Estrogenicity of trans-resveratrol in immature mice *in vivo* [in Chinese], *Wei Sheng Yan Jiu* [*J Hygiene Res*] 31, 188–190, 2002.
62. Ratna WN and Simonelli JA, The action of dietary phytochemicals quercetin, catechin, resveratrol and naringenin on estrogen-mediated gene expression, *Life Sci* 70, 1577–1589, 2002.
63. Adlercreutz H, Mazur W, Heinonen S-M, and Stumpf K, Phytoestrogens and breast cancer, in *Breast Cancer: Prognosis, Treatment, and Prevention*, Pasqualini JR, Ed, Marcel Dekker, New York, 2002, pp. 527–554.
64. Bianchini F and Vainio H, Wine and resveratrol: mechanisms of cancer prevention? *Eur J Cancer Prev* 12, 417–425, 2003.
65. Zhang Y, Kreger BE, Dorgan JF, Splansky GL, Cupples LA, and Ellison RC, Alcohol consumption and risk of breast cancer: the Framingham Study revisited, *Am J Epidemiol* 149, 93–101, 1999.
66. Smith-Warner SA, Spiegelman D, Yaun SS, van den Brandt PA, Folsom AR, Goldbohm RA, Graham S, Holmberg L, Howe GR, Marshall JR, Miller AB, Potter JD, Speizer FE, Willett WC, Wolk A, and Hunter DJ, Alcohol and breast cancer in women: a pooled analysis of cohort studies, *JAMA* 279, 535–540, 1998.

67. Freudenheim JL, Marshall JR, Graham S, Laughlin R, Vena JE, Swanson M, Ambrosone C, and Nemoto T, Lifetime alcohol consumption and risk of breast cancer, *Nutr Cancer* 23, 1–11, 1995.
68. Martin-Moreno JM, Boyle P, Gorgojo L, Willett WC, Gonzalez J, Villar F, and Maisonneuve P, Alcoholic beverage consumption and risk of breast cancer in Spain, *Cancer Causes Control* 4, 345–353, 1993.
69. Jain MG, Ferrenc RG, Rehm JT, Bondy SJ, Rohan TE, Ashley MJ, Cohe JE, and Miller AB, Alcohol and breast cancer mortality in a cohort study, *Breast Cancer Res Treat* 64, 201–209, 2000.
70. Bertelli AA, Giovannini L, Stradi R, Bertelli A, and Tillement JP, Plasma, urine and tissue levels of trans- and cis-resveratrol (3,4′,5-trihydroxystilbene) after short-term or prolonged administration of red wine to rats, *Int J Tissue React* 18, 67–71, 1996.
71. Goldberg DM, Yan J, and Soleas GJ, Absorption of three wine-related polyphenols in three different matrices by healthy subjects, *Clin Biochem* 36, 79–87, 2003.
72. Meng X, Maliakal P, Lu H, Lee MJ, and Yang CS, Urinary and plasma levels of resveratrol and quercetin in humans, mice, and rats after ingestion of pure compounds and grape juice, *J Agric Food Chem* 52, 935–942, 2004.
73. Asensi M, Medina I, Ortega A, Carretero J, Bano MC, Obrador E, and Estrela JM, Inhibition of cancer growth by resveratrol is related to its low bioavailability, *Free Radical Biol Med* 33, 387–398, 2002.
74. Gescher AJ and Steward WP, Relationship between mechanisms, bioavailibility, and preclinical chemopreventive efficacy of resveratrol: a conundrum, *Cancer Epidemiol Biomarkers Prev* 12, 953–957, 2003.
75. Banerjee S, Bueso-Ramos C, and Aggarwal BB, Suppression of 7,12-dimethylbenz(a)anthracene-induced mammary carcinogenesis in rats by resveratrol: role of nuclear factor-kappaB, cyclooxygenase 2, and matrix metalloprotease 9, *Cancer Res* 62, 4945–4954, 2002.
76. Pace-Asciak CR, Hahn S, Diamandis EP, Soleas G, and Goldberg DM, The red wine phenolics trans-resveratrol and quercetin block human platelet aggregation and eicosanoid synthesis: implications for protection against coronary heart disease, *Clin Chim Acta* 235, 207–219, 1995.
77. Pace-Asciak CR, Rounova O, Hahn SE, Diamandis EP, and Goldberg DM, Wines and grape juices as modulators of platelet aggregation in healthy human subjects, *Clin Chim Acta* 246, 163–182, 1996.
78. Martinez-Campos A, Amara JF, and Dannies PS, Antiestrogens are partial estrogen agonists for prolactin production in primary pituitary cultures, *Mol Cell Endocrinol* 48, 127–133, 1986.
79. Horwitz KB, Koseki Y, and McGuire WL, Estrogen control of progesterone receptor in human breast cancer: role of estradiol and antiestrogen, *Endocrinology* 103, 1742–1751, 1978.
80. Gehm BD, McAndrews JM, Jordan VC, and Jameson JL, EGF activates highly selective estrogen-responsive reporter plasmids by an ER-independent pathway, *Mol Cell Endocrinol* 159, 53–62, 2000.
81. Tonetti DA, Rubenstein R, DeLeon M, Zhao H, Pappas SG, Bentrem DJ, Chen B, Constantinou A, and Jordan VC, Stable transfection of an estrogen receptor beta cDNA isoform into MDA-MB-231 breast cancer cells, *J Steroid Biochem Mol Biol* 87, 47–55, 2003.

18 Resveratrol as an Antibacterial Agent

Gail B. Mahady

CONTENTS

INTRODUCTION

In their natural environment, plants are constantly challenged by numerous potentially pathogenic microorganisms, primarily fungi, bacteria, and viruses. The factors determining the resistance of plants against these pathogens belong to a large cache of constitutive and inducible (active) defense mechanisms [1]. These defense mechanisms are not only activated upon infection by microorganisms, but are also induced by stressors such as induction with ultraviolet light, or by chemicals (respiratory inhibitors, surfactants, antibiotics, plant regulators, or the salts of heavy metals, as well as elicitors released by the pathogens or products resulting from the activity of fungal-degrading enzymes on host cell walls) [2]. Constitutive plant defenses include structural barriers such as waxes, cutin, suberin, lignin, phenolics, cellulose, callose, and cell wall proteins, which are often rapidly

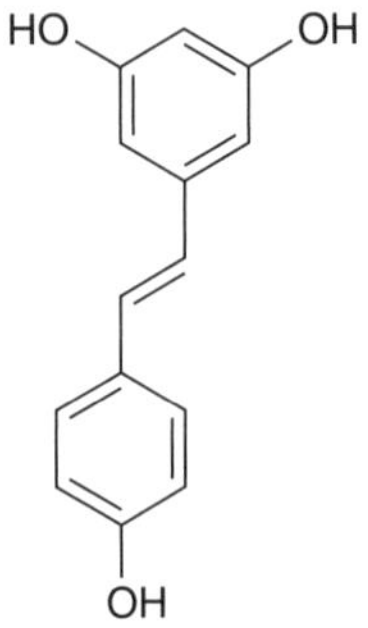

HO OH

OH

FIGURE 18.1 Structure of resveratrol.

reinforced upon the infection process. Active defense mechanisms include the oxidative burst, rapid and localized cell death (hypersensitive response), the synthesis of pathogenesis-related proteins, and accumulation of phytoalexins [1].

Phytoalexins are low-molecular-weight antimicrobial secondary metabolites of wide interest, as they have been shown to possess biological activity against a wide range of plant and human pathogens [2]. Stilbene phytoalexins, from the Vitaceae and other plant families, have been the subject of intense investigation during the past decade, because these compounds are thought to have broad implications in human health [2]. Within the Vitaceae, *Vitis vinifera* L. is one of the most important species, grown worldwide for grape and wine production. Stilbenes (resveratrol and viniferins) are present in grapevine as constitutive compounds of the woody organs (roots, canes, stems) and as induced substances (in leaves and fruit) acting as phytoalexins in the defense response of grape resistance against specific phytopathogens. Resveratrol (3,4′,5-trihydroxystilbene, Figure 18.1), a compound present in grape skins and wine, is thought to be the active principle with antibacterial activity [3]. The compound was first identified in 1963 as the active constituent of the dried roots of *Polygonum cuspidatum*, used in Japanese folk medicine to treat infections [4].

CARDIOVASCULAR AND RESPIRATORY BACTERIA

CHLAMYDIA PNEUMONIAE

Chlamydia pneumoniae (CP), an intracellular Gram-negative bacterium, is known as a leading cause of human acute respiratory tract infections worldwide, accounting for 5 to 10% of all cases of pneumonia [5]. However, while CP infections begin in the respiratory tract, the bacterium is disseminated systemically in the bloodstream within alveolar macrophages, leading to the development of chronic infection. In numerous investigations, chronic CP infections have also been implicated as a causative factor in the development of atherosclerosis and coronary artery disease [6]. This association has been demonstrated in seroepidemiological investigations, even if several prospective studies failed to demonstrate an association between the presence of IgG antibody to CP and incidence of myocardial infarction [7]. Furthermore, CP has been identified in coronary, carotid, and aortic atheroma samples, by immunocytochemical staining, polymerase chain reaction (PCR), and electron microscopy [7]. Thus, there is substantial evidence to suggest that CP is an etiological agent of atherosclerosis or significantly increases the development of atherosclerotic plaque and coronary heart disease (CHD).

France has the lowest incidence of CHD in the world, despite a diet that is relatively high in fat. This phenomenon is commonly referred to as the "French paradox" [8]. Data from epidemiological investigations have indicated that moderate consumption of red wine, rich in resveratrol, may be the most likely explanation of the paradox [8]. Resveratrol is a polyphenolic compound that has a wide range of biological activities, some of which may explain its protective effects on CHD [3]. In a recent report, Schriever et al. [9] have shown that red wine and resveratrol inhibit the growth of CP *in vitro*. A concentrated red wine extract inhibited the growth of two CP strains *in vitro*, with a minimum inhibitory concentration (MIC) of 125 to 250 μg/ml. Resveratrol inhibited the growth of the same two strains, with an MIC of 12.5 μg/ml (Table 18.1). Thus, along with other mechanisms, red wine and resveratrol may exert their beneficial effects on development and progression of CHD through antibacterial effects on CP [9].

TABLE 18.1
Minimum Inhibitory Concentrations (MIC) and Minimum Chlamydicidal Concentrations (MCC) for Resveratrol, Pinot Noir Extracts, and Control Antibiotic (Azithromycin)

Chlamydial strain	Extract	MIC_{TP} (μg/ml)	MIC (μg/ml)	MCC (μg/ml)
AR-39	Pinot Noir (methanol soluble)	> 500	> 500	> 500
TW-183		> 500	> 500	> 500
AR-39	Pinot Noir (concentrated)	125	250	250
TW-39		125	250	250
AR-39	Resveratrol	6.25	12.5	12.5
TW-183		6.25	12.5	12.5
AR-39	Azithromycin (control)	0.0625	0.125	0.125
TW-183		0.0625	0.125	0.125

GASTROINTESTINAL BACTERIA

HELICOBACTER PYLORI

Helicobacter pylori (HP) is a curved or spiral shaped bacteria located on the gastric epithelium of patients with chronic active gastritis. Its discovery as the main etiologic organism of chronic gastritis and peptic ulcer disease was one the most significant discoveries in the field of gastroenterology of the twentieth century [10]. In 1994 HP was classified as a group I carcinogen and a definite cause of gastric cancer in humans by the International Agency for Research on Cancer [11]. Since then, HP has been epidemiologically linked to adenocarcinoma of the distal stomach, and other investigations have also found a positive association between HP infection and colorectal adenomas [12–14].

Data from numerous epidemiological studies show an inverse relationship between alcohol consumption and HP infection [15–17]. In a recent study from Germany, the inverse relationship with HP infection was stronger for alcohol consumed in the form of wine than for alcohol from beer [17]. The results suggested that alcohol consumption, particularly wine, may reduce the odds of active infection with HP [17]. A cross-sectional population study conducted as part of a randomized controlled trial of HP infection eradication in southwest England was performed [18]. A total of 10,537 subjects, recruited from seven general practices, underwent ^{13}C-urea breath testing for active infection with HP and provided data on smoking, usual weekly consumption of alcohol, and daily intake of coffee. After adjustments for age, sex, ethnic status, childhood and adult social class, smoking, coffee consumption, and intake of alcoholic beverages other than wine, subjects drinking 3 to 6 units of wine per week had an 11% lower risk of HP infection as compared with those who did not drink wine (OR = 0.89, 95% CI = 0.80 to 0.99). Higher wine consumption was associated with a further 6% reduction in the risk of infection (OR = 0.83, 95% CI = 0.64 to 1.07). The study suggested that modest consumption of wine (approximately 7 units/week) protects against HP infection, presumably by facilitating eradication of the organism [18].

In 1998 Marimom and Bujanda [19] reported that red wine has strong bactericidal effects on one clinical isolate of HP; however, no MICs were reported and the active constituents were not identified. In 2000 Mahady and Pendland [20] demonstrated that red wine extracts and resveratrol inhibited the growth of HP *in vitro*. The MIC_{50} and MIC_{90} for the red wine extract were 25 and 50 μg/ml, respectively, with a range of 12.5 to 50 μg/ml. The MIC_{50} and MIC_{90} for resveratrol were 12.5 and 25 μg/ml, respectively, with a range of 6.25 to 25 μg/ml. The positive control drug, amoxicillin, had an MIC range of 0.002 to 0.06 μg/ml. These *in vitro* data demonstrated that the antibacterial activity of red wine against HP was due, at least in part, to the presence of resveratrol [20]. Further investigations demonstrated that resveratrol inhibited the growth of *cagA* strains of HP [21]. *CagA* is the strain-specific HP gene that has been linked to the development of premalignant and malignant histological lesions [22]. Thus, susceptibility of *cagA*+ HP strains is of note because as compared with *cagA*– strains, infections caused by *cagA*+ strains significantly increase the risk for developing severe gastric inflammation, atrophic gastritis, and noncardia gastric adenocarcinoma [22]. The antibacterial activities of two red wine extracts (both Pinot Noir) and resveratrol were assessed against 5 *cagA*+ HP strains, accession numbers M23-3, GTD7-13, G1-1, SS1 (Sydney Strain *cagA*+), and the ATCC 43504 possessing the *cagA*+ gene and expressing vacuolating cytotoxin (VacA). Two concentrated red wine extracts were active against all *cagA*+ HP strains, with MICs of 25 and 50 μg/ml, respectively (range of 25 to 50 μg/ml). Resveratrol was also active against all five strains with an MIC of 12.5 μg/ml (range of 6.25 to 25 μg/ml). The control drug, amoxicillin, had an MIC range of 0.0039 to 0.25 μg/ml. These data demonstrate that both red wine and resveratrol inhibit the growth of HP *cagA*+ strains *in vitro*, and further support their role as chemopreventative agents [21].

In an interesting 2003 study, Tombola et al. [23] assessed the effect of a variety of compounds, including compounds from red wine (resveratrol) on the activity of VacA. VacA is a major virulence factor of HP, and causes cell vacuolation and tissue damage by forming anion-selective, urea-permeable channels in plasma and endosomal membranes. These compounds inhibited ion and urea conduction and cell vacuolation by reducing VacA activity. This suggests that red wine and resveratrol may be useful in the prevention or cure of HP-associated gastric diseases [23].

DERMATOLOGICAL BACTERIA

According to traditional European and Chinese medicine, *Vitis vinifera* was a popular remedy for the treatment of skin diseases [24,25]. Very few studies, however, have assessed the effects of red wine and resveratrol on the Gram-positive and Gram-negative bacteria that are responsible for skin and wound infections. Gram-positive organisms such as *Staphylococcus aureus* and *Streptococcus* group D cause a variety of skin conditions, including folliculitis, impetigo, furuncles (boils), and cellulites; *Pseudomonas aeruginosa*, a Gram-negative organism, commonly infects burn wounds. Treatments that combine antimicrobial and antiinflammatory actions are desirable for alleviating many skin conditions that vary in severity. Thus resveratrol has been investigated against bacteria known to be major etiologic agents of human skin infections. In one *in vitro* study, the growth of the bacterial species *S. aureus, Enterococcus faecalis*, and *P. aeruginosa* was inhibited by resveratrol in concentrations of 171 to 342 μg/ml [26].

In another investigation, bioassay-guided fractionation of an ethyl acetate extract of the fruit of *Ficus barteri* Sprague led to the isolation and characterization of 3,4′,5-trihydroxystilbene (*trans*-resveratrol), 3,3′,4′, 5-tetrahydroxystilbene, and catechin as the primary antibacterial constituents. The main antibacterial compound was 3,3′,4′,5-tetrahydroxystilbene, which had an MIC of 25 mg/ml for *S. aureus*, 50 mg/ml for *Bacillus subtilis*, and >400 mg/ml for *Escherichia coli* and *P. aeruginosa* [27].

THERMOACIDOPHILIC BACTERIA

Aticyclobacillus Acidoterrestris

The growth of *Aticyclobacillus acidoterrestris* was inhibited by grape polyphenols, resveratrol (50 mg/ml), ferulic acid (150 mg/ml), *p*-coumalic acid (200 mg/ml), *p*-hydroxybenzoic acid, or "Kyoho" seed proanthocyanidin (900 mg/ml). Resveratrol and *p*-hydroxybenzoic acid enhanced the antibacterial activities of catechin-gallate and the proanthocyanidins [28].

UROGENITAL BACTERIA

Neisseria Gonorrhoeae

Neisseria is the genus of Gram-negative bacteria responsible for the development of the sexually transmitted disease gonorrhea. However, along with the genitals, the bacteria may be found in the blood, heart, joints, eyes, urine, and oral mucosa. *In vitro* studies with resveratrol have shown that resveratrol inhibited the growth of *Neisseria gonorrhoeae* and *N. meningitidis* with an MIC of 25 and 100 μg/ml, respectively, and completely inhibited of growth at concentrations of 75 and 125 μg/ml, after 24 hours [29].

RESVERATROL DERIVATES

Calligonum leucocladum (Schrenk) Bunge, a plant belonging to the Polygonaceae, has been used in Uzbekistan folk medicine for the treatment of syphilis [30]. Two new stilbene derivatives of resveratrol, (*E*)-resveratrol 3-(6″-galloyl)-*O*-β-D-glucopyranoside (1) and (*E*)-resveratrol 3-(4″-acetyl)-*O*-β-D-xylopyranoside (2), and five known stilbene derivatives (3–7) were isolated from the dried aerial parts of *Calligonum leucocladum* (Figure 18.2). When the seven compounds were tested against methicillin-resistant *Staphylococcus aureus* (MRSA), compound 5 showed an MIC of 125 μg/ml. Although not effective alone, compound 1 restored the

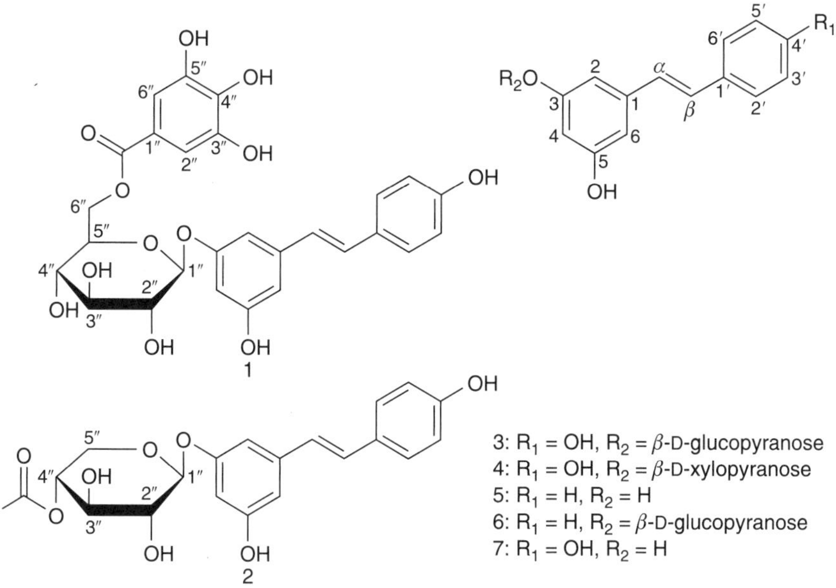

3: R_1 = OH, R_2 = β-D-glucopyranose
4: R_1 = OH, R_2 = β-D-xylopyranose
5: R_1 = H, R_2 = H
6: R_1 = H, R_2 = β-D-glucopyranose
7: R_1 = OH, R_2 = H

FIGURE 18.2 Two stilbene derivatives of resveratrol, (*E*)-resveratrol 3-(6″-galloyl)-*O*-β-D-glucopyranoside (1) and (*E*)-resveratrol 3-(4″-acetyl)-*O*-β-D-xylopyranoside (2), and five known stilbene derivatives (3–7) were isolated from the dried aerial parts of *Calligonum leucocladum*.

efficacy of oxacillin against MRSA when the two compounds were used in combination. Structure–activity relationship data showed that when compared with compound 3, the galloyl group of compound 1 was partially responsible for the activity of this compound against MRSA [30].

Vatica oblongifolia subsp. *oblongifolia* Hook. (Dipterocarpaceae) is a plant native to Sarawak, Malaysia, distributed primarily in Kalimantan and the Malay Peninsula [31]. Over the last two decades, several resveratrol oligomers have been isolated from various *Vatica* species and plants belonging to the Dipterocarpaceae. An ethylacetate-soluble extract of the stem bark of *V. oblongifolia* subsp. *oblongifolia* was found to exhibit moderate activity against MRSA and *Mycobacterium smegmatis* (a model for pathogenic mycobacteria). Two new resveratrol tetramers, hopeaphenol A (1) and isohopeaphenol A (2), along with the known vaticaphenol A (3) were isolated from the stem bark of *V. oblongifolia* subsp. *oblongifolia* through bioassay-guided fractionation (Figure 18.3). The structures and their relative stereochemistry were determined by spectroscopic techniques. Compounds 1 and 3 demonstrated moderate activity against MRSA (100 and 50 μg/ml) and *Mycobacterium smegmatis* (50 and 25 μg/ml); compound 2 was not active [31].

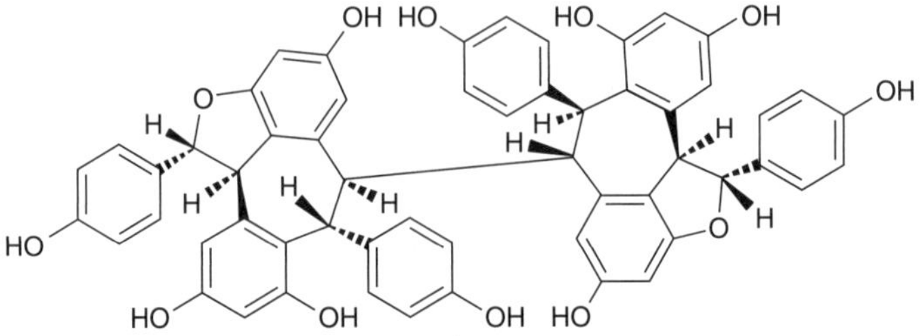

FIGURE 18.3 Structures of three resveratrol tetramers, hopeaphenol A (1), isohopeaphenol A (2), and vaticaphenol A (3), isolated from stem bark of *Vatica oblongifolia* subsp. oblongifolia.

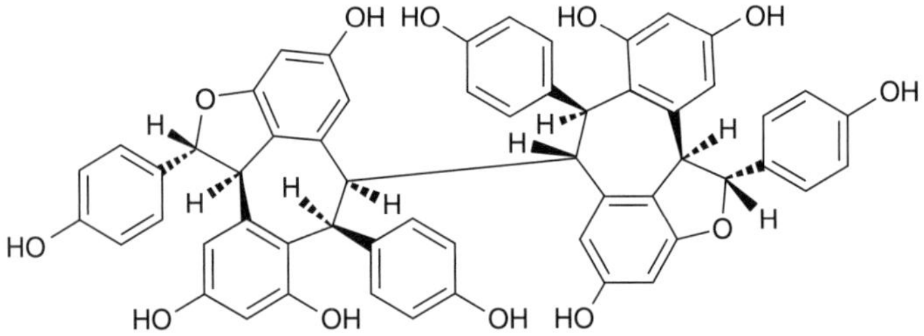

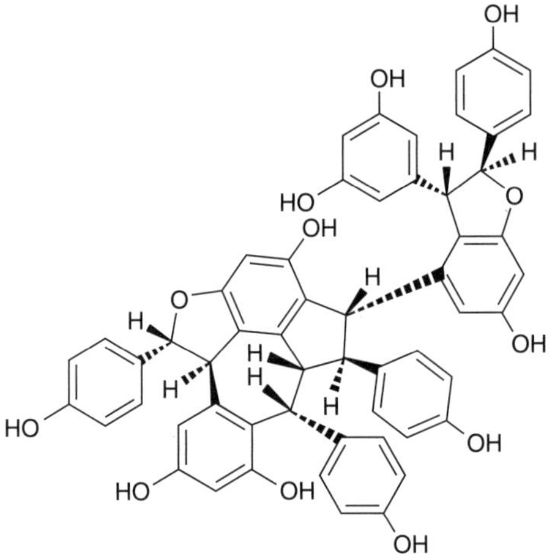

REFERENCES

1. Heil M and Bostock RM, Induced systemic resistance (ISR) against pathogens in the context of induced plant defences, *Ann Bot* (Lond.) 89, 503–512, 2002.
2. Jeandet P, Douillet-Breuil AC, Bessis R, Debord S, Sbaghi M, and Adrian M, Phytoalexins from the Vitaceae: biosynthesis, phytoalexin gene expression in

transgenic plants, antifungal activity, and metabolism, *J Agric Food Chem* 50, 2731–2741, 2002.
3. Kopp P, Resveratrol, a phytoestrogen found in red wine. A possible explanation for the conundrum of the French Paradox?, *Eur J Endocrinol* 138, 619–620, 1998.
4. Nonomura S, Kanagawa H, and Makimoto A, Chemical constituents of polygonaceous plants: I. Studies on the components of Ko-jo-kon (*Polygonum cuspidatum* SIEB et ZUCC), *Yakugaku Zasshi* 83, 983–988, 1963.
5. Halm EA and Teirstein AS, Management of community-acquired pneumonia, *N Eng J Med* 347, 2039–2045, 2002.
6. Grayston JT, Background and current knowledge of *Chlamydia pneumoniae* and atherosclerosis, *J Infect Dis* 181 (Suppl 3), S402–S410, 2000.
7. Noll G, Pathogenesis of atherosclerosis: a possible relation to infection, *Atherosclerosis* 140, S3–S9, 1998.
8. Renaud S and de Lorgeril M, The French paradox: dietary factors and cigarette smoking-related health risks, *Ann NY Acad Sci* 686, 299–309, 1993.
9. Schriever C, Pendland SL, and Mahady GB, Red wine, resveratrol, *Chlamydia pneumoniae* and the French connection, *Atherosclerosis* 171, 379–380, 2003.
10. Graham DY, Evolution of concepts regarding *Helicobacter pylori*: from a cause of gastritis to a public health problem, *Am J Gastroenterol* 89, 469–472, 1989.
11. IARC Working Group on the Evaluation of Carcinogenic Risks to Humans, Schistosomes, Liver Flukes and *Helicobacter pylori*, Infections with Helicobacter pylori, IARC Monographs on the Evaluation of Carcinogenic Risks to Humans, International Agency for Research on Cancer, Lyon, France, 1994, pp. 177–201.
12. Scheiman JM and Cutler AF, *Helicobacter pylori* and gastric cancer, *Am J Med* 106, 222–226, 1999.
13. Figueiredo C, Machado JC, and Pharoah P, *Helicobacter pylori* and interleukin 1 genotyping: an opportunity to identify high-risk individuals for gastric carcinoma, *J Natl Cancer Inst* 94, 1680–1687, 2002.
14. Breuer-Katschinski B, Nemes K, and Marr A, *Helicobacter pylori* and the risk of colonic adenomas, Colorectal Adenoma Study Group, *Digestion* 60, 210–215, 1999.
15. Graham DY, Malaty HM, and Evans DG, Epidemiology of *Helicobacter pylori* in an asymptomatic population in the United States, *Gastroenterology* 100, 1495–1501, 1991.
16. Brenner H, Rothenbacher D, and Bode G, Relation of smoking, alcohol and coffee consumption to active infection with *Helicobacter pylori*, *Br Med J* 315, 1489–1492, 1997.
17. Brenner H, Rothenbacher D, and Bode G, Inverse graded relation between alcohol consumption and active infection with *Helicobacter pylori*, *Am J Epidemiol* 149, 571–576, 1999.
18. Murray LJ et al, Inverse relationship between alcohol consumption and active Helicobacter pylori infection: the Bristol Helicobacter Project, *Am J Gastroenterol* 97, 2750–2751, 2002.
19. Marimom JM and Bujanda I, *In vitro* bactericidal effect of wine against *Helicobacter pylori*, *Am J Gastroenterol* 93, 1392, 1998.

20. Mahady GB and Pendland SL, Resveratrol inhibits the growth of *Helicobacter pylori in vitro*, *Am J Gastroenterol* 95, 1849, 2000.
21. Mahady GB, Pendland SL, and Chadwick LR, Resveratrol and red wine extracts inhibit the growth of CagA+ strains of *Helicobacter pylori in vitro*, *Am J Gastroenterol* 98, 1440–1441, 2003.
22. Censini S, Lange C, and Xiang Z, Cag, a pathogenicity island of *Helicobacter pylori*, encodes type I-specific and disease associated virulence factors, *Proc Natl Acad Sci USA* 93, 14648–14653, 1996.
23. Tombola F, Campello S, De Luca L, Ruggiero P, and Zoratti G, Plant polyphenols inhibit VacA, a toxin secreted by the gastric pathogen *Helicobacter pylori*, *FEBS Lett* 543, 184–189, 2003.
24. Bombardelli E and Morazzoni P, *Vitis vinifera* L., *Fitoterapia* 66, 291–317, 1995.
25. Anon, in *Medicinal Plants in East and Southeast Asia*, Perry L, Ed, MIT Press, London, 1989, p. 436.
26. Chan MM, Antimicrobial effect of resveratrol on dermatophytes and bacterial pathogens of the skin, *Biochem Pharmacol* 63, 99–104, 2002.
27. Ogungbamila FO, Onawunmi GO, Ibewuike JC, and Funmilayo KA, Antibacterial constituents of *Ficus barteri* fruits, *Int J Pharmacog* 35, 185–189, 1997.
28. Oita S and Kohyama N, Antibacterial effect of grape polyphenols against thermoacidophilic bacteria *Alicyclobacillus acidoterrestris*, *Nippon Shokuhin Kagaku Kogaku Kaishi* 49, 555–558, 2002.
29. Docherty JJ, Fu MM, and Tsai M, Resveratrol selectively inhibits *Neisseria gonorrhoeae* and *Neisseria meningitidis*, *J Antimicrob Chemother* 47, 243–244, 2001.
30. Okasaka M, Takaishi Y, Kogure K, Fukuzawa K, Shibata H, Higuti T, Honda G, Ito M, Kodzhimatov OK, and Ashurmetov O, New stilbene derivatives from *Calligonum leucocladum*, *J Nat Prod* 67, 1044–1046, 2004.
31. Zgoda-Pols JR, Freyer AJ, Killmer LB, and Porter JR, Antimicrobial resveratrol tetramers from the stem bark of *Vatica oblongifolia* ssp. oblongifolia, *J Nat Prod* 65, 1554–1559, 2002.

19 Resveratrol as an Antifungal Agent

Marielle Adrian and Philippe Jeandet

CONTENTS

INTRODUCTION

IMPORTANCE OF RESVERATROL IN THE PLANT KINGDOM

Stilbenes are natural compounds occurring in a number of plant families, including Vitaceae, Dipterocarpaceae, Gnetaceae, Leguminoseae, and Cyperaceae [1]. Among stilbenes, the polyphenolic product resveratrol (3,4′,5-trihydroxystilbene) and derived compounds are present in *Vitis* spp. and several other genera including veratrum and arachis [3–6], vaccinium

[7,8], trifolium [9], gnetum [1], especially in *Gnetum parvifolium* [10], *G. hainanense* [11], *G. gnemon* [12], *G. pendulum* [13], *G. klossii* [14], and *G. montanum* [15]. The occurrence of these compounds has also been reported in *Vatica rassak* [16], *Cissus sicyoides* [17], *Polygonum cuspidatum* [18], Itadori tea [19], green tea [20], *Calligonum leucocladum* [21], *Yucca periculosa* [22] and *Y. schidigera* [23,24], *Elephantorrhiza goetzei* [25], *Schoenocaulon officinale* [26], *Cyphostemma crotalarioides* [27], *Pleuropterus ciliinervis* [28], and *Rumex bucephalophorus* [29].

All these studies show that resveratrol and derived compounds are widely distributed among the plant kingdom.

Exciting Potential of This Fascinating Compound in Phytopathology

Resveratrol is present in plant extracts used in Japanese folk medicine for treatment of many ailments including heart diseases, hyperlipidemia, allergic reactions, etc., and considerable evidence based on *in vitro* and animal experiments has accumulated to suggest that this compound possesses antioxidative, anticarcinogenic, and antitumor properties. All these effects are given detailed consideration in the other chapters of this book. Moreover, as is detailed later in this chapter, the secondary metabolite resveratrol and its derivatives show antifungal properties that can be applied for crop protection and in the fruit storage process. This is particularly true with grapevine, which is an agriculturally and economically important crop plant susceptible to attack by numerous fungal phytopathogens.

The antifungal properties of resveratrol and related compounds, together with their interest as regards human health are powerful arguments to justify research on the potential use of stilbenes on an industrial scale.

Specifically, the antifungal activity of resveratrol has led to controversial studies, and the question as to whether resveratrol is (or is not) a phytoalexin according to the definition of Müller and Börger [30], is still debated. Much has been said about the biological activity of resveratrol. Many studies established, using concentrations far exceeding its solubility in water (more than 200 mg/l), that this compound has no antifungal activity and that it should be ruled out as a factor contributing to the resistance of grapevine to disease. However, as is described further in this chapter, we have clearly shown that resveratrol has real inhibitory effects on conidial germination of *Botrytis cinerea* at concentrations comparable to the activity of other phytoalexins, which generally falls within one order of magnitude (10^{-4} to 10^{-5} M). These data, along with the fact that resveratrol is consistently present in high amounts during the phytoalexin response, lead us to conclude that this compound is much more important for the regulation of fungal pathogen–grapevine interactions than was previously thought.

RESVERATROL SYNTHESIS

BIOSYNTHETIC PATHWAY

Resveratrol is synthesized via the phenylalanine/polymalonate pathway [31]. Stilbene synthase (STS) (EC 2.3.1.95) is a key enzyme in the resveratrol biosynthesis pathway since it converts one molecule of *p*-coumaroyl-CoA and three molecules of malonyl-CoA into 3,4′,5-trihydroxystilbene or resveratrol [32]. Phenylalanine ammonia lyase (PAL) and STS genes belong to large multigene families [33–35]. They have a coordinated expression, as has been shown in *V. vinifera* cv. optima cells treated with a fungal cell wall preparation [36]. A coordinated expression of PAL, STS, and cinnamate-4-hydroxylase (C4H) has also been described in ultraviolet (UV)-induced leaves of *Vitis* spp. [37].

Resveratrol (Figure 19.1) is a secondary metabolite, a precursor of glucosylated and polymerized derivatives including, respectively, piceid and viniferins. ε-Viniferin is a cyclic resveratrol dehydrodimer (Figure 19.1) that has been characterized in UVC-irradiated grapevine leaves or leaves infected by *B. cinerea* [38]. This compound is formed through the peroxidase-mediated oxidative dimerization of resveratrol [39]. Douillet-Breuil et al. [40] observed the same profile, but delayed, of the curves of the kinetic of accumulation of resveratrol and ε-viniferin. Three peroxidase isoenzymes implied in the dimerization process of resveratrol and linked to fungus-induced defenses have been identified in grapevines [41,42]. The occurrence of δ-viniferin, an isomer of ε-viniferin resulting from the oxidative dimerization of resveratrol by plant peroxidases or fungal laccases, has also been reported [43,44].

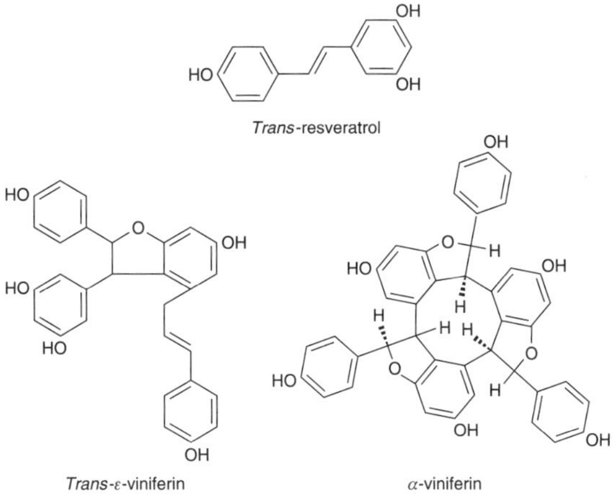

FIGURE 19.1 Chemical structures of *trans*-resveratrol, *trans*-ε-viniferin, and α-viniferin.

α-Viniferin, a cyclic resveratrol trimer (Figure 19.1), has also been well characterized [31]. The occurrence of a β-viniferin, a resveratrol tetramer, and γ-vinferin, a resveratrol oligomer of high molecular weight, was also reported [31].

Following the formation of resveratrol, conjugation to sugars can occur, leading to glucosides such as *trans*- and *cis*-piceid, the 3-*O*-β-D-resveratrol glucoside [45,46]. Krasnow and Murphy [47] have identified a resveratrol glucosyltransferase in cellular extracts of *V. vinifera* (cv. Gamay Fréaux). This enzyme is distinct from the glucosyltransferases that are active on other phenolic-type compounds. Other resveratrol glucosides have been identified in *V. vinifera* cell suspensions but their biosynthesis pathway remains, as far as we know, unclear: astringin (*trans*-3-hydroxypiceid), *cis*-resveratrol-3, 4′-*O*-β-D-diglucoside, and two other stilbene diglucosides [48,49].

The dimethylated resveratrol derivative pterostilbene (3,5-dimethoxy-4′-hydroxystilbene) has been identified in Cabernet Sauvignon leaves [50,51]. The occurrence of piceatannol (*trans*-3,3′,4,5′-tetrahydroxystilbene) has also been reported [52].

The cellular sites of resveratrol synthesis remain unknown, although Blaich and Bachmann [53] have conducted cytological studies that showed the localization of resveratrol within small areas of the cytoplasm or in the periplasm, probably near plasmodesmata, mostly in and below epidermal cells. In elicited grape leaves, stilbenes accumulated at the abaxial side (Figure 19.2), and in grape berries they are especially present in the skin [2,54]. In induced *V. vinifera* cell suspensions, resveratrol is the major stilbene detected in the extracellular medium whereas ε-viniferin is dominant in the cell [55].

All stilbenes are in the *trans* form, but they can be photochemically [53,56] or enzymatically isomerized into the *cis* form.

FIGURE 19.2 (See color insert following page 546.) Observation of the abaxial side of leaves under long-wavelength UV light (365 nm). UV: 24 hours following UV irradiation of the leaf. Note the bright blue purple fluorescence emitted by resveratrol and some of its derivatives. C: untreated leaf (control).

Stilbene Synthase (STS)

STS (EC 2.3.1.95) is a cytosolic polyketide synthase catalyzing the formation of resveratrol [32,57]. STS was first purified from cell suspension cultures of *A. hypogea* [3] and the native form is a homodimer of 90 kDa (with 43 kDa subunits) [57]. In pine, STS converts dihydrocinnamoyl-CoA and three malonyl-CoA units to form dihydropinosylvin [32]. STS is encoded by a multigene family. Schröder et al. [33] were the first to isolate and characterize cDNA and genomic clones for STS in *A. hypogea*. Melchior and Kindl [59] have prepared a full-length STS cDNA from grapevine mRNA and sequence expression in *Escherichia coli* resulted in a catalytically active enzyme. Later, three resveratrol-forming STS genes from grapevines were characterized by Wiese et al. [58]. More recently, a new STS gene has been isolated from *V. riparia* cv. Gloire de Montpellier [60]. Rupprich and Kindl [32] have reported the presence of a noninducible STS form in *Rheum rhaponticum*. In grapevines, STS is generally rapidly and transiently induced [57], except in seedlings where it is constitutively expressed [35]. STS gene expression leads to two waves in the accumulation of the corresponding mRNAs, as was shown in grapevine cell suspensions treated with cell walls of *Phytophthora cambivora* [58] or *B. cinerea* [57] or in UV-irradiated *in vitro*-grown leaves [61]. Similarly, the accumulation of resveratrol in several *Vitis* spp. also shows two peaks [40,61]. The same profile of STS transcripts was also observed in Scots pine (*P. sylvestris*) [62] after ozone exposure. For Wiese et al. [58] the fact that resveratrol accumulation occurs in two waves could correspond to the expression of, at least, two groups of STS genes: those expressed early but with labile mRNAs and those expressed later and giving more stable mRNAs. Lanz et al. [34] have shown a differential regulation of resveratrol synthase (RS) genes in cell cultures of *A. hypogea* in response to various elicitors, i.e., elicitors from *Phytophthora megasperma*, yeast extracts, and dilutions of the cultures. They concluded that RS genes in *A. hypogea* represent a gene family whose members are regulated by different signals. Moreover, Wiese et al. [58], using *V. vinifera* cv. optima cell suspensions, showed that the expression of STS genes differs in magnitude in response to various elicitors.

STS and CHS catalyze common condensation reactions of *p*-coumaroyl-CoA and three units from malonyl-CoA but different cyclization reactions to produce resveratrol and naringenin-chalcone, respectively. CHS is a key enzyme in the flavonoid biosynthesis pathway [63]. STS and CHS share significant homology at the DNA and protein levels. This extends throughout the coding region and the single intron detected in an STS gene is at the same position as a conserved intron in CHS, suggesting a common evolutionary origin for these two enzymes [33]. Schröder et al. [33] have reported 70 to 75% identity at the protein level between STS and the consensus sequence of CHS from *A. hypogea* and the two sequences contain a single essential cystein residue (cys 164) at the same position. Both proteins possess a common scaffold for substrate recognition and condensing region [63] and a single change of His to Glu close to the active site of STS alters the substrate preferences [64]. There are therefore possible cross-reactions between CHS and STS, as shown by Yamaguchi et al. [65].

INDUCTION OF SYNTHESIS

Stilbenes including resveratrol are constitutive compounds of the woody organs, roots, cane, and stems [2,66], and of seeds [56,67]. In green parts of Vitaceae, they are produced and accumulated in response to biotic or abiotic elicitors at concentrations that can reach 600 μg/g FW in some species. The production of resveratrol and its derivatives has been particularly well studied using peanut and grapevine as models.

Langcake and Pryce [2,38] were the first to link the bright blue fluorescence visible in UV-irradiated *Vitis* leaves or leaves infected by *B. cinerea* to the presence of resveratrol. Resveratrol indeed emits a bright blue purple fluorescence when exposed to long-wavelength UV light (365 nm) (Figure 19.2).

UVC irradiation is a well-known elicitor of stilbene production [2,3,37,38,40]. Efficient wavelength for irradiation corresponds to 254 nm (6 mW/cm^2) for a duration depending on the plant material; e.g., for grapevine, irradiation durations are 4, 6, 8, and 10 minutes for, respectively, cell suspensions, *in vitro*-grown plants, plants from a greenhouse, and plants in the vineyard (M. Adrian, unpublished data).

Some chemicals such as aluminum-containing products may also act as potent resveratrol elicitors. Fosetyl-aluminum, for example, is a fungicide that can be associated with the "priming concept," i.e., it enhances the resveratrol accumulation provoked by an attack by *Plasmopara viticola* [68]. The metallic salt aluminum chloride has a different mode of action, since it directly induces resveratrol synthesis [69,70–73].

The synthesis of phytoalexins is one of the most important responses to ozone exposure in pine and grapevine, initiated at the transcription level [62, 74,75]. Specifically, treatments using ozone enable one to maintain quality of cv. Napoleon table grapes by increasing the total content in stilbenoids [76].

Poinssot et al. [55] have shown that the endopolygalacturonase 1 (T4BcPG1) from *B. cinerea* and oligogalacturonides (OGA) with a degree of polymerization of 9 to 20 can activate defense reactions, including resveratrol production, in *V. vinifera* cv. Gamay cell suspensions.

Other stimuli of stilbene production have also been described such as wounding, the β-1,3-glucan laminarin derived from the brown algae *Laminaria digitata* [77], mucic acid [53], methyljasmonate [78], SAR activators [79], or the plant activator benzothiadiazole (BTH, 0.3 mM) which enhanced the *trans*-resveratrol content in grape berries by about 40% [80].

Resveratrol and derived compounds accumulate in plants in response to various fungi: *Plasmopara viticola*, the agent for downy mildew; *Uncinula necator*, the agent for powdery mildew; and *B. cinerea*, the agent for gray mold (2,50,51,53,54,56,57,81–88). Resveratrol can also accumulate during postharvest in *B. cinerea*-infected berries/grapes, as shown by Montero et al. [89] using high-resolution laser-based techniques. Other fungi are known to induce resveratrol production: *Rhizopus stolonifer*, the agent for berry rot [90], and ochratoxin A-producing aspergilli [91].

All the elicitors described above do not induce the same type of response. For example, contrary to what happens when leaves are induced by UV irradiation, resveratrol concentrations do not rapidly decline in the case of infections caused by *B. cinerea* [92,93]. This is linked to the fact that the presence of the fungus within lesions on leaves constantly provokes stilbene elicitation in grapevine. Borie et al. [61] also reported differences in the expression of STS genes from *V. rupestris in vitro* leaves, depending on the nature of the elicitor: UV irradiation, *B. cinerea* infection, or aluminum chloride treatment. Moreover, a promoter deletion analysis of the grapevine resveratrol gene *Vst1* showed an ozone-responsive region different from the basal pathogen responsive sequence [74,94]. Grimmig et al. [94], examining the role of ozone and ethylene in the transcriptional regulation of several genes including *Vst1*, have demonstrated that ethylene is involved in the signaling ozone-induced regulation of this gene and the existence of at least two independent transduction pathways.

Whatever the nature of the elicitor, the level of resveratrol in berries, leaves, canes, and roots is influenced by genetic and viticultural factors such as grape variety and environmental and cultural practices. Higher resveratrol levels are found in red berry grapes than in white berry grapes. There is a positive correlation between vineyard elevation and grape stilbene concentration [52]. Roldan et al. [95] and Jeandet et al. [56] showed that the resveratrol content of Palomino fino or Pinot noir grapes is highly influenced by climatic conditions and the gray mold pressure in the vineyard. Nutritional factors may also intervene. Grapevine needs an optimum nitrogen supply to synthesize large amounts of resveratrol [96], this capacity being decreased at higher doses. Potassium fertilizers also seem to play a role [97].

ANTIFUNGAL PROPERTIES

The question of whether resveratrol may be considered as a true phytoalexin has long remained unclear. According to the definition of Kuc [98], phytoalexins are antimicrobial compounds synthesized by, and accumulated in, plants in response to abiotic or biotic elicitors. They belong to the active defense mechanisms of plants. Specifically, stilbene phytoalexins can act directly on fungal cells and also take part in a protection mechanism as a constituent of lignin-like cell wall incrusts produced by host peroxidases. As will be further described, resveratrol should be considered truly as a grapevine phytoalexin, both for its antifungal properties [2,31,38,43,51, 68,82,92,99,100] and its role as precursor of more toxic compounds [43,82]. The following discussion of the antifungal properties of resveratrol is based on a review of the three main approaches documented in the literature: (1) development of biotests, (2) establishment of correlation between phytoalexin production and resistance of plants to fungal infections, and (3) use of transgenic plants.

Biotests

In the heartwood of some *Vitis* spp., resveratrol and its derivatives can accumulate up to 700 μg/g FW and have been implicated in preventing wood decay [101,102] and in the disease resistance of plants to pathogens [102,103]. The antifungal properties of resveratrol have been demonstrated using biotests. Langcake [51] reported a low fungitoxicity for resveratrol with an ED (effective dose required for 50% mortality) of 71 to 200 μg/ml upon spores of *P. viticola* and conidia of *B. cinerea*. The radial growth of fungal mycelia of *B. cinerea* and *P. viticola* on agar substrates was found to be decreased in the presence of resveratrol quantities of, respectively, 50 and 100 ppm [104]. However, these experiments were carried out with concentrations beyond the limits of solubility of stilbene compounds. Adrian et al. [100] observed an inhibition of the germination on *B. cinerea* conidia in the presence of 7×10^{-4} M resveratrol. This value is comparable to the activity of other phytoalexins, which generally falls within one order of magnitude (10^{-4} to 10^{-5} M). Pterostilbene is far more active since it completely inhibits *B. cinerea* conidia germination at a concentration of 2.3×10^{-4} M [51,100,105]. According to Langcake [51], the fungitoxicity of ε-viniferin seems to be intermediate between that of resveratrol and pterostilbene. Despite being active at high concentrations (as compared to pterostilbene or viniferins), resveratrol is quantitatively the major component of the grapevine phytoalexin response, whereas its more toxic derivatives are less abundant or not detectable (pterostilbene) [40,50,54,69].

Resveratrol and pterostilbene used at sublethal or lethal doses (60 to 140 μg/ml and 20 to 40 μg/ml, respectively) also alter fungal morphogenesis and cause cytological abnormalities in *B. cinerea* conidia. This includes the formation of curved germ-tubes, cessation of growth of germ-tubes that enhances the emission of other ones, cytoplasmic granulations, and disruption of the plasma membrane [100] (Figure 19.3). Pezet and Pont [105] reported a rapid cessation of respiration and strong modifications of the endocellular membrane system of *B. cinerea* conidia treated with 5×10^{-4} M. Such effects have been previously described (for a review, see [106]). The formation of curved germ tubes results from an asymmetric growth that may correspond to the ability of stilbenes, namely resveratrol, to interact

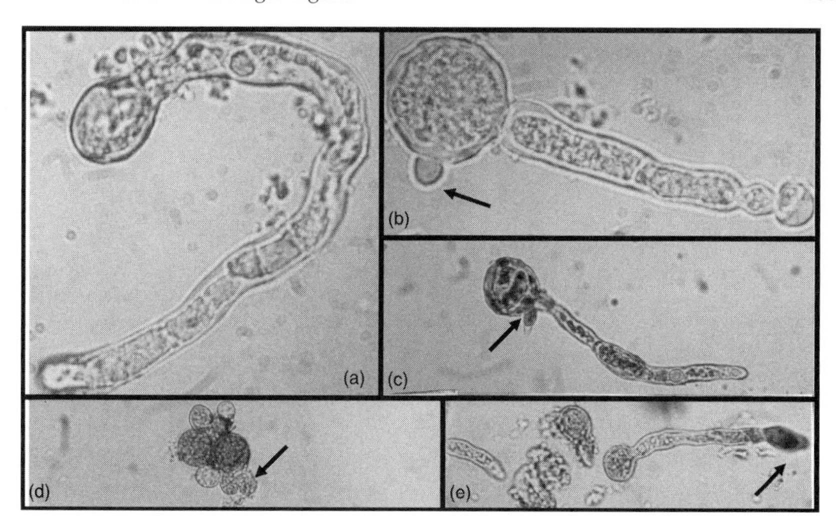

FIGURE 19.3 Alterations of *Botrytis cinerea* morphogenesis following treatment by sublethal doses of resveratrol. (A) Formation of curved germ-tubes. (B, C) Cessation of the growth of germ-tubes that enhances the emission of other ones (arrows). (D) Disruption of the plasma membrane (arrow) and leakage of cell constituents. Note the drakening due to the laccase-mediated oxidation of resveratrol inside conidia. (E) Death of the apical cells, as indicated by staining with Trypan blue (1% w/v).

with tubulin with resultant disruption of microtubule assembly at this level. This phenomena was described by Woods et al. [107] working with various stilbenes based on combretastatin A-4 and used in cancer therapy. This mode of action is typical for many other fungicides, such as benomyl, a fungicide used for the control of gray mold in vineyards [108].

Cytoplasmic granulations, plasma membrane disruptions, and the leakage of cellular contents have been reported in zoospores of *Phytophthora infestans*, *P. porriet*, and *P. cactorum* treated by terpenoid-type phytoalexins (rishitin, phytuberin, anhydro-β-rotunol, and solavetivone) [109,110] or in fungal cells treated with isoflavonoid-type phytoalexins (phaseollin and kievitone) [106,111–113]. Otherwise, cessation of growth of germ-tubes and protoplasmic retraction in the hyphal tip cells is due to the death of these cells, as has been shown using phaseollin and the phenanthrenes orchinol and dehydroorchinol [103,106,111–113]. Apical cells of hyphae indeed possess a weak wall that facilitates the entry of phytoalexins, contrary to subapical and interstitial hyphal cells (which have mature walls) [112,114–117]. Finally, surviving conidia can attempt to escape from the action of phytoalexins by emitting a secondary or tertiary germ-tube [116,117].

The mode of action of hydroxystilbenes on fungal cells has been studied extensively by the group of Pezet. It appears that 4′-hydroxystilbenes, especially those containing electron-attracting substituents, participate in the formation of charge transfer complexes favoring contact and affinity with proteins and acting as uncoupling agents of electron transport and photophosphorylation [105]. Moreover, hydroxystilbenes, such as resveratrol and piceatannol, are capable of inhibiting some fungal ATPases and inducing the dissociation of chaperones and cochaperones, two proteins frequently associated with the cytoskeleton [118].

Resveratrol accumulates locally to high levels at sites of infection in diseased leaves and the quantities found correspond to effective doses that can contribute effectively to cessation of parasite growth in the host tissue. *Trans*-resveratrol has also been shown to enhance resistance of vine plants to pathogens such as *Phomopsis viticola* [104] or *Rhizopus stolonifer* [90]. Treatment of grapevine (*Vitis vinifera* L.) plants with laminarin (an elicitor of plant defense mechanisms including resveratrol production) reduces the infection by *B. cinerea* and *P. viticola* by approximately 55 and 75%, respectively [77].

Correlations between Phytoalexin Production and Disease Resistance

There is a positive correlation between the production of resveratrol and its derivatives and the resistance of *Vitis* spp. to pathogens such as *P. viticola* and *B. cinerea*. The pioneering work of Pool et al. [92] underlined the importance of both the speed and the intensity of resveratrol production in such correlations. Resistance of grapevine to *P. viticola* has been correlated with the ability of *Vitis* spp. to synthesize resveratrol and ε-viniferin following UVC irradiation [99]. Based on the analysis of 95 *Vitis* spp., Stein and Hoos [93] have established a close positive correlation between stilbene production potential and resistance to *B. cinerea*. Their results were confirmed by Jeandet et al. [119] and Sbaghi et al. [120] who have studied resveratrol production in *in vitro*-grown plantlets from 13 Vitis cultivars following UVC treatment. The rapid accumulation of resveratrol and ε-viniferin has been associated with the resistance of grapevine against *B. cinerea*. Such studies have demonstrated that American species and interspecific hybrids generally show a higher capacity to synthesize stilbenes than do *Vitis vinifera* cultivars [40,93,96,97,120] and that, within *V. vinifera* varieties, not all cultivars have the same capacity for resveratrol production. Many muscadine grapes accumulate resveratrol in nonelicited berries [67] while Okuda and Yokotsuka [121] found very low levels of resveratrol in ripe and apparently healthy berries of *V. vinifera* and interspecific varieties. Thus, resveratrol assessment has been used as a tool for the selection of grapevine varieties resistant to *B cinerea* [92,96,120] and has been considered as a disease resistance index in grape breeding programs [83,92].

Transgenic Plants

Traditional breeding to develop plants with a higher degree of tolerance or resistance against pathogens is an extremely slow process. As an alternative, a strategy of genetic engineering offers the possibility of introducing new characteristics into existing commercial cultivars. STS is a key enzyme of resveratrol synthesis, using as substrates precursor molecules that are present throughout the plant kingdom. Therefore, the introduction of a single gene is sufficient to synthesize resveratrol in heterologous plant species. Transformations were then operated to investigate the potential of stilbene biosynthetic genes to confer resistance to pathogens. Transformations that have been conducted with the introduction of STS genes are listed in Table 19.1.

Tobacco plants transformed by an STS gene from *Arachis hypogea* are capable of producing resveratrol [122] and are found to be more resistant to *B. cinerea* infections [122]. In the same way, Hipskind and Paiva [123] have transformed alfalfa (*Medicago sativa*) with a peanut cDNA encoding resveratrol synthase transcriptionally regulated by an enhanced cauliflower mosaic virus (CaMV) 35S promoter. The transgenic plants so obtained accumulated *trans*-resveratrol 3-*O*-β-D-glucopyranoside (piceid) and showed an increased resistance to *Phoma medicaginis*. 41-B plants overexpressing stilbene synthase produce high stilbene levels and exhibit a reduction of the symptoms in response to an infection by *B. cinerea* [124]. Similar transformations have improved the resistance of rice to *Pyricularia oryzae* [125], tomato to *P. infestans* [126], barley and wheat to *B. cinerea* [127,128], wheat to *Oïdium tuckerei* [127,128], alfalfa to *Phoma medicaginis*

TABLE 19.1
Effect of the Overexpression of Resveratrol Synthase Gene in Transformed Grapevines or Its Expression in Foreign Plants

Plant	Pathogen	Results	Ref.
Alfalfa	*Phoma medicaginis*	Enhanced resistance	123
Tobacco	*Botrytis cinerea*	Enhanced resistance	122
Rice	*Pyricularia oryzae*	Enhanced resistance	125
Kiwi	*Botrytis cinerea*	No increased resistance	131
Barley	*Botrytis cinerea*	Enhanced resistance	127,128
Wheat	*Botrytis cinerea*	Enhanced resistance	127,128
Grapevine rootstock 41B	*Botrytis cinerea*	Enhanced resistance	124
White poplar	*Melampsora pulcherrima*	No increased resistance	130
Papaya	*Phytophthora palmivora*	Enhanced resistance	129
Tomato	*Phytophthora infestans*	Enhanced resistance	126

[123], and papaya to *P. palmivora* [129]. All these results show that resveratrol is a determinant factor in the expression of resistance of plants to phytopathogens and corresponds to the definition of phytoalexins given by Müller and Börger [30]. However, there are some cases for which no resistance was observed after transforming the plants with STS genes. For example, transformation of white poplar (*Populus alba*) with a cDNA insert coding STS that leads to the biosynthesis of both *cis* and *trans* isomers of piceid does not confer any increased resistance to *Melaspora pulcherrima* (rust disease) [130]. Similarly, no increased resistance against *B. cinerea* has been observed in STS transgenic kiwi plants [131].

These contradictory results show that pathogen control by transgenic STS plants can be significant but may be considered as empiric and not predictable. In tobacco and petunia, overexpression of a transgenic STS gene leads to substrate competition between STS and CHS and causes male sterility [132]. It is thus preferable to transform plants with a construct having a pathogen-inducible promoter.

DETOXIFICATION

It now clearly appears that resveratrol, as a fungitoxic compound and as a precursor of other active derivatives, plays an important role in the resistance of grapevines to fungal pathogens. However, the numerous infections that occur in greenhouses and in the field clearly show a lack of efficiency of the plant defense mechanisms. In the following discussion, some examples are given to explain why resveratrol failed, in some instances, to help grapevine to withstand fungal infections.

The resveratrol production potential of grape berries changes during the season with the highest potential reached at veraison but declining thereafter until complete maturity [54,86]. This phenomenon corresponds to the decline in inducible STS gene expression during maturation [133]. Based on the results obtained from assaying resveratrol and anthocyanins, Jeandet et al. [54] hypothesized that STS may compete with CHS for substrates. This situation may explain the increase of susceptibility of mature berries to gray mold [56] whereas *B. cinerea* is maintained in a latent stage in unripe berries [135]. Peristomatal cracks or openings naturally present on the surface of berries are also important for infections by *B. cinerea*.

Stilbenes are inhibitory to fungi but, at the same time, botrytis is able to inactivate the resveratrol defense line of the host. This is a reason why the defense strategy of the plant may depend both on the speed and the intensity of stilbene production [93].

B. cinerea, like other fungi, produces laccase [134] whose function may be an oxidative detoxification of stilbenes [43]. The stilbene oxidase laccase is an enzyme capable of oxidizing resveratrol and pterostilbene [43,104,105] (Figure 19.3D). Stilbene oxidase activity can be strongly inhibited by grape phenolic compounds, such as catechin, epicatechin 3-*O*-gallate, *trans*-caftaric, *trans*- and *cis*-coutaric, *trans*-coumaric acids, taxifoline 3-*O*-rhamnoside, and quercetine 3-*O*-glucuronide. Some of these compounds are largely involved in the quiescent stage of *B. cinerea* in grape berries [135].

Sbaghi et al. [136] have established a relation between the ability of *B. cinerea* strains to degrade phytoalexins (laccase-mediated degradation) and their pathogenicity to grapevine. The laccase-deficient strains were nonpathogenic. Stilbenes in the hard wood of conifers are metabolized only by white rot laccase-producing fungi. Brown rot fungi lacking laccase are far more sensitive to stilbenes [137,138]. Schouten et al. [139] have characterized a laccase gene from *B. cinerea* (*Bclcc2*) that plays an active role in the oxidation process of resveratrol since this stilbene is converted into compounds that are more toxic. The major compound resulting from the oxidation of resveratrol by *B. cinerea* laccase is a resveratrol dehydrodimer different from ε-viniferin [43], later called δ-viniferin by Pezet et al. [44]. The findings of Breuil et al. [43] have been confirmed by Cichewicz et al. [140]. Similarly, the laccase-mediated oxidation of pterostilbene, the dimethylated derivative of resveratrol, leads to the formation of a dehydrodimer very similar to that obtained from resveratrol [140]. Resveratrol can also be oxidized by soybean lipoxygenase, leading to a complex mixture of products similar to that obtained in the case of the oxidation of this stilbene by H_2O_2.

Fungal transporters may extrude plant defense products, as well as fungicides, and thereby play an important role in pathogenicity [142–144]. Recent studies have shown that ATP-binding cassette (ABC) transporters might provide protection against plant defense compounds and fungicides by ATP-driven efflux mechanisms. An ABC transporter gene, *BcatrB*, affecting the sensitivity of *B. cinerea* to resveratrol has been detected and characterized in *B. cinerea* [144]. Disruptant mutants for the gene *BcatrB* of *B. cinerea* were reported to display increased sensitivity to resveratrol and the phenylpyrrole fungicides and to show slightly reduced virulence on grapevine leaves [145]. In *Aspergillus nidulans*, the transcription of the ABC transporter gene *atrB* is induced by resveratrol [146]. *MgAtr* (ABC transporter gene) deletion mutants of *Mycosphaerella graminicola* showed an increase in the sensitivity to resveratrol, suggesting a role for this transporter in protecting the fungus against plant defense compounds [147]. Resveratrol also induces specifically the transcription of an ABC transporter gene, *PMR5*. This gene has a role in multidrug resistance and has been cloned from the phytopathogenic fungus *Penicillium digitatum* [148]. To elucidate the function of *PMR5*, investigations on the susceptibility of the Dpmr5 mutant to resveratrol and other antifungal agents have been conducted. An increased sensitivity of the mutants has been demonstrated.

It appears clearly evident that, in the phytopathogenetic fungi, ABC transporters act as virulence factors, providing protection against defense compounds produced by the host. In several plant–fungus interactions, it has become evident that the ability to weaken or neutralize the effects of phytoalexins is one of the essential determinants of fungal host range [149,150].

CONCLUSION

The antifungal properties of resveratrol, its interest as regards human health, together with its key role as precursor of other active derivatives are powerful arguments to justify further research to develop its potential on an industrial scale. The use of resveratrol to improve fruit storage has begun to be developed and looks promising. Other possible developments concern the research of elicitors capable of inducing its production in crops to limit the use of pesticides and in cosmetics to preserve their life duration.

REFERENCES

1. Sotheeswaran S and Pasupathy V, Distribution of resveratrol oligomers in plants, *Phytochemistry* 32, 1083–1092, 1993.
2. Langcake P and Pryce RJ, The production of resveratrol by *Vitis vinifera* and other members of the Vitaceae as a response to infection or injury, *Physiol Plant Pathol* 9, 77–86, 1976.
3. Schoeppner A and Kindl H, Purification and properties of a stilbene synthase from induced cell suspension of peanut, *J Biol Chem* 259, 6806–6811, 1984.
4. Sobolev VS and Cole RJ, *Trans*-resveratrol content in commercial peanuts and peanut products, *J Agric Food Chem* 47, 1435–1439, 1999.
5. Sanders TH, McMichael RW Jr, and Hendrix KW, Occurrence of resveratrol in edible peanuts, *J Agric Food Chem* 48, 1243–1246, 2000.
6. Chen RS, Wu PL, and Chiou RY, Peanut roots as a source of resveratrol, *J Agric Food Chem* 50, 1665–1667, 2002.
7. Lyons MM, Yu C, Toma RB, Cho SY, Reiboldt W, Lee J, and van Breemen RB, Resveratrol in raw and baked blueberries and bilberries, *J Agric Food Chem* 51, 5867–5870, 2003.
8. Rimando AM, Kalt W, Magee JB, Dewey J, and Ballington JR, Resveratrol, pterostilbene, and piceatannol in *Vaccinium* berries, *J Agric Food Chem* 52, 4713–4719, 2004.
9. Ingham JL, Isoflavonoid and stilbene phytoalexins of the genus *Trifolium*, *Biochem Systematics Ecol* 6, 217–223, 1978.
10. Li JB, Lin M, Li SZ, and Song WZ, Studies on the structure of gnetifolin A of *Gnetum parvifolium* (Warb.) CY, *Cheng Yao Xue Xue Bao* 26, 437–441, 1991.
11. Huang KS, Wang YH, Li RL, and Lin M, Five new stilbene dimers from the lianas of *Gnetum hainanense*, *J Nat Prod* 63, 86–89, 2000.
12. Iliya I, Ali Z, Tanaka T, Iinuma M, Furusawa M, Nakaya K, Murata J, Darnaedi D, Matsuura N, and Ubukata M, Stilbene derivatives from *Gnetum gnemon* Linn, *Phytochemistry* 62, 601–606, 2003.

13. Li XM, Wang YH, and Lin M, Stilbenoids from the lianas of *Gnetum pendulum*, *Phytochemistry* 58, 591–594, 2001.
14. Ali Z, Tanaka T, Iliya I, Iinuma M, Furusawa M, Ito T, Nakaya K, Murata J, and Darnaedi D, Phenolic constituents of *Gnetum klossii*, *J Nat Prod* 66, 558–560, 2003.
15. Xiang W, Jiang B, Li XM, Zhang HJ, Zhao QS, Li SH, and Sun HD, Constituents of *Gnetum montanum*, *Fitoterapia* 73, 40–42, 2002.
16. Tanaka T, Ito T, Nakaya K, Iinuma M, and Riswan S, Oligostilbenoids in stem bark *of Vatica rassak*, *Phytochemistry* 54, 63–69, 2000.
17. Quilez AM, Saenz MT, Garcia MD, and de la Puerta R, Phytochemical analysis and anti-allergic study of *Agave intermixta* Trel. and *Cissus sicyoides* L., *J Pharm Pharmacol* 56, 1185–1189, 2004.
18. Vastano BC, Chen Y, Zhu N, Ho CT, Zhou Z, and Rosen RT, Isolation and identification of stilbenes in two varieties of *Polygonum cuspidatum*, *J Agric Food Chem* 48, 253–256, 2000.
19. Burns J, Yokota T, Ashihara H, Lean ME, and Crozier A, Plant foods and herbal sources of resveratrol, *J Agric Food Chem* 50, 3337–3340, 2002.
20. Pillai SP, Mitscher LA, Menon SR, Pillai CA, and Shankel DM, Antimutagenic/antioxidant activity of green tea components and related compounds, *J Environ Pathol Toxicol Oncol* 18, 147–158, 1999.
21. Okasaka M, Takaishi Y, Kogure K, Fukuzawa K, Shibata H, Higuti T, Honda G, Ito M, Kodzhimatov OK, and Ashurmetov O, New stilbene derivatives from *Calligonum leucocladum*, *J Nat Prod* 67, 1044–1046, 2004.
22. Torres P, Avila JG, Romo de Vivar A, Garcia AM, Marin JC, Aranda E, and Cespedes CL, Antioxidant and insect growth regulatory activities of stilbenes and extracts from *Yucca periculosa*, *Phytochemistry* 64, 463–473, 2003.
23. Oleszek W, Sitek M, Stochmal A, Piacente S, Pizza C, and Cheeke P, Resveratrol and other phenolics from the bark of *Yucca schidigera* Roezl, *J Agric Food Chem* 49, 747–752, 2001.
24. Piacente S, Montoro P, Oleszek W, and Pizza C, *Yucca schidigera* bark: phenolic constituents and antioxidant activity, *J Nat Prod* 67, 882–885, 2004.
25. Wanjala CC and Majinda RR, A new stilbene glycoside from *Elephantorrhiza goetzei*, *Fitoterapia* 72, 649–655, 2001.
26. Kanchanapoom T, Suga K, Kasai R, Yamasaki K, Kamel MS, and Mohamed MH, Stilbene and 2-arylbenzofuran glucosides from the rhizomes of *Schoenocaulon officinale*, *Chem Pharm Bull* (Tokyo) 50, 863–865, 2002.
27. Bala AEA, Kollmann A, Ducrot PH, Majira A, Kerhoas L, Leroux P, Delorme R, and Einhorn J, *Cis*-viniferin: a new antifungal resveratrol dehydrodimer from *Cyphostemma crotalarioides* roots, *J Phytopath* 148, 29–32, 2000.
28. Lee JP, Min BS, An RB, Na MK, Lee SM, Lee HK, Kim JG, Bae KH, and Kang SS, Stilbenes from the roots of *Pleuropterus ciliinervis* and their antioxidant activities, *Phytochemistry* 64, 759–763, 2003.
29. Kerem Z, Regev-Shoshani G, Flaishman MA, and Sivan L, Resveratrol and two monomethylated stilbenes from Israeli *Rumex bucephalophorus* and their antioxidant potential, *Nat Prod* 66, 1270–1272, 2003.
30. Müller KO and Börger H, Experimentelle Untersuchungen über die *Phytophthora* Resistenz der Kartoffel, *Arb Biol Reichsant* 23, 189–231, 1940.

31. Langcake P and Pryce RJ, A new class of phytoalexins from grapevines, *Experientia* 33, 151–152, 1977.
32. Rupprich N and Kindl H, Stilbene synthases and stilbenecarboxylate synthases: I. Enzymatic synthesis of 3,5′,4-trihydroxystilbene from p-coumaroyl coenzyme A and malonyl coenzyme A, *Hoppe Seylers Z Physiol Chem* 359, 165–172, 1978.
33. Schröder G, Brown JWS, and Schröder J, Molecular analysis of resveratrol synthase cDNA, genomic clones and relationship with chalcone synthase, *Eur J Biochem* 197, 161–169, 1988.
34. Lanz T, Schröder G, and Schröder J, Differential regulation of genes for resveratrol synthase in cell cultures of *Arachis hypogea*, *Planta* 181, 169–175, 1990.
35. Sparvoli F, Martin C, Scienza A, Gavazzi G, and Tonelli C, Cloning and molecular analysis of structural genes involved in flavonoid and stilbene biosynthesis in grape (*Vitis vinifera* L.), *Plant Mol Biol* 24, 743–755, 1994.
36. Melchior F and Kindl H, Coordinate- and elicitor-dependant expression of stilbene synthase and phenylalanine ammonia lyase genes in *Vitis* cv Optima, *Arch Biochem Biophys* 288, 552–557, 1991.
37. Fritzemeier KH and Kindl H, Coordinate induction by UV light of stilbene synthase, phenylalanine ammonia lyase and cinnamate 4-hydroxylase in leaves of *Vitaceae*, *Planta* 151, 48–52, 1981.
38. Langcake P and Pryce RJ, The production of resveratrol and the viniferins by grapevines in response to ultraviolet irradiations, *Phytochemistry* 16, 1193–1196, 1977.
39. Langcake P and Pryce RJ, Oxidative dimerization of 4-hydroxystilbenes *in vitro*: production of a grapevine phytoalexin mimic, *J Chem Soc Chem Commun* 208–210, 1977.
40. Douillet-Breuil AC, Jeandet P, Adrian M, and Bessis R, Changes in the phytoalexin content of various *Vitis* spp. in response to ultraviolet C elicitation, *J Agric Food Chem* 47, 4456–4461, 1999.
41. Calderon AA, Pedreno MA, Ros Barcelo A, and Munoz R, Zymographic screening of plant peroxidase isoenzymes oxidizing 4-hydroxystilbenes, *Electrophoresis* 11, 507–508, 1990.
42. Calderon AA, Zapata JM, Pedreno MA, Munoz R, and Ros-Barcelo A, Levels of 4-hydroxystilbene oxidizing isoperoxidases related to constitutive disease resistance in *in vitro*-cultured grapevine, *Plant Cell Tissue Org Cult* 29, 63–70, 1992.
43. Breuil AC, Adrian M, Pirio N, Weston LA, Meunier P, Bessis R, and Jeandet P, Metabolism of stilbene phytoalexins by *Botrytis cinerea*: characterization of a resveratrol dehydrodimer, *Tetrahedron Lett* 39, 537–540, 1998.
44. Pezet R, Perret C, Jean-Denis JB, Tabacchi R, Gindro K, and Viret O, Delta-viniferin, a resveratrol dehydrodimer: one of the major stilbenes synthesized by stressed grapevine leaves, *J Agric Food Chem* 51, 5488–5492, 2003.
45. Waterhouse AL and Lamuela-Raventos RM, The occurrence of piceid, a stilbene glucoside in grape berries, *Phytochemistry* 37, 571–573, 1994.
46. Jeandet P, Breuil AC, Adrian M, Weston LA, Debord S, Meunier P, Maume G, and Bessis R, HPLC analysis of grapevine phytoalexins coupling photodiode array detection and fluorometry, *Anal Chem* 69, 5172–5177, 1997.

47. Krasnow MN and Murphy TM, Polyphenol glucosylating activity in cell suspensions of grape (*Vitis vinifera*), *J Agric Food Chem* 52, 3467–3472, 2004.
48. Waffo-Teguo P, Fauconneau B, Deffieux G, Huguet F, Vercauteren J, and Mérillon JM, Isolation, identification, and antioxidant activity of three stilbene glucosides newly extracted from *Vitis vinifera* cell cultures, *J Nat Prod* 61, 655–657, 1998.
49. Decendit A, Waffo-Teguo P, Richard T, Krisa S, Vercauteren J, Monti JP, Deffieux G, and Merillon JM, Galloylated catechins and stilbene diglucosides in *Vitis vinifera* cell suspension cultures, *Phytochemistry* 60, 795–798, 2002.
50. Langcake P, Cornford CA, and Pryce RJ, Identification of pterostilbene as a phytoalexin from *Vitis vinifera* leaves, *Phytochemistry* 18, 1025–1027, 1979.
51. Langcake P, Disease resistance of *Vitis* spp. and the production of the stress metabolites resveratrol, ε-viniferin, α-viniferin and pterostilbene, *Physiol Plant Pathol* 18, 213–226, 1981.
52. Bavaresco L, Role of viticultural factors on stilbene concentrations of grapes and wine, *Drugs Exp Clin Res* 29, 181–187, 2003.
53. Blaich R and Bachmann O, Die Resveratrolsynthese bei Vitaceen Induktion und zytologische Beobachtungen, *Vitis* 19, 230–240, 1980.
54. Jeandet P, Bessis R, and Gautheron B, The production of resveratrol (3,5, 4′-trihydroxystilbene) by grape berries in different developmental stages, *Am J Enol Vitic* 42, 41–46, 1991.
55. Poinssot B, Vandelle E, Bentejac M, Adrian M, Levis C, Brygoo Y, Garin J, Sicilia F, Coutos-Thevenot P, and Pugin A, The endopolygalacturonase 1 from *Botrytis cinerea* activates grapevine defence reactions unrelated to its enzymatic activity, *Mol Plant Microbe Interact* 16, 553–564, 2003.
56. Jeandet P, Bessis R, Sbaghi M, and Meunier P, Production of the phytoalexin resveratrol by grapes as a response to *Botrytis* attacks in the vineyard, *J Phytopathology* 143, 135–139, 1995.
57. Liswidowati F, Melchior F, Hohmann F, Schwer B, and Kindl H, Induction of stilbene synthase by *Botrytis cinerea* in cultured grapevine cells, *Planta* 183, 307–314, 1991.
58. Wiese W, Vornam B, Krause E, and Kindl H, Structural organization and differential expression of three stilbene synthase genes located on a 13 kb grapevine DNA fragment, *Plant Mol Biol* 26, 667–677, 1994.
59. Melchior F and Kindl H, Grapevine stilbene synthase cDNA only slightly differing from chalcone synthase cDNA is expressed in *Escherichia coli* into a catalytically active enzyme, *FEBS Lett* 268, 17–20, 1990.
60. Goodwin PH, Hsiang T, and Erickson L, A comparison of stilbene and chalcone synthases including a new stilbene synthase gene from *Vitis riparia* cv. Gloire de Montpellier, *Plant Sci* 151, 1–8, 2000.
61. Borie B, Jeandet P, Bessis R, and Adrian M, Comparison of resveratrol and stilbene synthase mRNA production from grapevine leaves treated with biotic and abiotic phytoalexin elicitors, *Am J Enol Vitic* 55, 60–64, 2004.
62. Zinser C, Jungblut T, Heller W, Seidlitz HK, Schnitzler JP, Ernst D, and Sandermann H, The effect of ozone in Scots pine (*Pinus sylvestris* L.): gene expression, biochemical changes and interactions with UV-B radiation, *Plant Cell Environ* 23, 975–982, 2000.
63. Schröder J and Schröder G, Stilbene and chalcone synthases: related enzymes with key functions in plant-specific pathways, *Z Naturforsch* (C) 45, 1–8, 1990.

64. Suh DY, Kagami J, Fukuma K, and Sankawa U, Evidence for catalytic cysteine-histidine dyad in chalcone synthase, *Biochem Biophys Res Commun* 275, 725–730, 2000.
65. Yamaguchi T, Kurosaki F, Suh D.Y, Sankawa U, Nishioka M, Shibuya M, and Ebizuka Y, Cross-reaction of chalcone synthase and stilbene synthase overexpressed in *Escherichia coli*, *FEBS Lett* 460, 457–461, 1999.
66. Bavaresco L, Fregoni C, Cantu E, and Trevisan M, Stilbene compounds: from the grapevine to wine, *Drugs Exp Clin Res* 25, 57–63, 1999.
67. Ector BJ, Magee JB, Hegwood CP, and Coign MJ, Resveratrol concentration in muscadine berries, juice, pomace, purees, seeds and wine, *Am J Enol Vitic* 47, 57–62, 1996.
68. Dercks W and Creasy LL, Influence of fosetyl-Al on phytoalexin accumulation in the *Plasmopara viticola*–grapevine interaction, *Physiol Mol Plant Pathol* 34, 203–213, 1989.
69. Adrian M, Jeandet P, Bessis R, and Joubert JM, Induction of phytoalexin (resveratrol) synthesis in grapevine leaves treated with aluminum chloride ($AlCl_3$), *J Agric Food Chem* 44, 1979–1981, 1996.
70. Jeandet P, Bessis R, Adrian M, Joubert JM, and Yvin JC, Use of Aluminum Chloride as a Resveratrol Formation Elicitor in Plants, French Patent 95 13462, PCT Int Appl WO 97 18,715.
71. Jeandet P, Adrian M, Breuil AC, Sbaghi M, Joubert JM, Weston LA, Harmon R, and Bessis R, Chemical stimulation of phytoalexin synthesis in plants as an approach to crop protection, in *Recent Research Developments in Agricultural and Food Chemistry*, Vol. 2, Pandalai SG, Ed, Research Signpost, Trivandrum, 1998, pp. 501–511.
72. Jeandet P, Adrian M, Breuil AC, Debord S, Sbaghi M, Joubert JM, Weston LA, Harmon R, and Bessis R, Potential use of phytoalexin induction in plants as a basis for crop protection, in *Modern Fungicides and Antifungal Compounds*, Vol. 2, Lyr H, Russell PE, Dehne HW, and Sisler HD, Eds, Intercept, Andover, 1999, pp. 349–356.
73. Jeandet P, Adrian M, Breuil AC, Sbaghi M, Debord S, Bessis R, Weston LA, and Harmon R, Chemical induction of phytoalexin synthesis in grapevines: application to the control of grey mould (*Botrytis cinerea* Pers.) in the vineyard, *Acta Hortic* 528, 591–596, 2000.
74. Schubert R, Fischer R, Hain R, Schreier PH, Bahnweg G, Ernst D, and Sandermann H Jr, An ozone-responsive region of the grapevine resveratrol synthase promoter differs from the basal pathogen-responsive sequence, *Plant Mol Biol* 34, 417–426, 1997.
75. Chiron H, Drouet A, Lieutier F, Payer H.D, Ernst D, and Sandermann H, Gene induction of stilbene biosynthesis in Scots pine in response to ozone treatment, wounding, and fungal infection, *Plant Physiol* 124, 865–872, 2000.
76. Artes-Hernandez F, Artes F, and Tomas-Barberan FA, Quality and enhancement of bioactive phenolics in cv. Napoleon table grapes exposed to different postharvest gaseous treatments, *J Agric Food Chem* 51, 5290–5295, 2003.
77. Aziz A, Poinssot B, Daire X, Adrian M, Bezier A, Lambert B, Joubert JM, and Pugin A, Laminarin elicits defense responses in grapevine and induces protection against *Botrytis cinerea* and *Plasmopara viticola*, *Mol Plant Microbe Interact* 16, 1118–1128, 2003.

78. Krisa S, Larronde F, Budzinski H, Descendit A, Deffieux G, and Mérillon JM, Stilbene production by *Vitis vinifera* cell suspension cultures: methyljasmonate induction and ^{13}C biolabelling, *J Nat Prod* 62, 1688–1690, 1999.
79. Busam G, Junghanns KT, Kneusel RE, Kassemeyer HH, and Matern U, Characterization and expression of caffeoyl-coenzyme A 3-O-methyltransferase proposed for the induced resistance response of *Vitis vinifera* L., *Plant Physiol* 115, 1039–1048, 1997.
80. Iriti M, Rossoni M, Borgo M, and Faoro F, Benzothiadiazole enhances resveratrol and anthocyanin biosynthesis in grapevine, meanwhile improving resistance to *Botrytis cinerea*, *J Agric Food Chem* 52, 4406–4413, 2004.
81. Pryce RJ and Langcake P, α-Viniferin: an antifungal resveratrol trimer from grapevines, *Phytochemistry* 16, 1452–1454, 1977.
82. Langcake P and MacCarthy WV, The relationship of resveratrol production to infection of grapevine leaves by *Botrytis cinerea*, *Vitis* 18, 244–253, 1979.
83. Barlass M, Miller RM, and Douglas TJ, Development of methods for screening grapevines for resistance to downy mildew: II. Resveratrol production, *Am J Enol Vitic* 38, 65–68, 1987.
84. Dercks W, Creasy LL, and Luczka-Bayles CJ, Stilbene phytoalexins and disease resistance in *Vitis*, in *Handbook of Phytoalexin Metabolism and Action*, Daniel M and Purkayashta RP, Eds, Marcel Dekker, New York, 1995, pp. 287–315.
85. Dai GH, Andary C, Mondolot-Cosson L, and Boubals D, Histochemical studies on the interaction between three species of grapevine, *Vitis vinifera*, *V. rupestris* and *V. rotundifolia* and the downy mildew fungus, *Plasmopara viticola*, *Physiol Mol Plant Pathol* 46, 177–188, 1995.
86. Bavaresco L, Petegolli D, Cantu E, Fregoni M, Chiusa G, and Trevisan M, Elicitation and accumulation of stilbene phytoalexins in grapevine berries infected by *Botrytis cinerea*, *Vitis* 36, 77–83, 1997.
87. Adrian M, Jeandet P, Douillet-Breuil AC, Tesson L, and Bessis R, Stilbene content of mature *Vitis vinifera* berries in response to UV-C elicitation, *J Agric Food Chem* 48, 6103–6105, 2000.
88. Romero-Perez AI, Lamuela-Raventos RM, Andres-Lacueva C, and de La Torre-Boronat MC, Method for the quantitative extraction of resveratrol and piceid isomers in grape berry skins. Effect of powdery mildew on the stilbene content, *J Agric Food Chem* 49, 210–215, 2001.
89. Montero C, Cristescu SM, Jimenez JB, Orea JM, te Lintel Hekkert S, Harren FJM, and Gonzalez Urena A, *Trans*-resveratrol and grape disease resistance. A dynamical study by high-resolution laser-based techniques, *Plant Physiol* 131, 129–138, 2003.
90. Sarig P, Zutkhi Y, Monjauze A, lisker N, and Ben-Arie R, Phytoalexin elicitation in grape berries and their susceptibility to *Rhizopus stolonifer*, *Physiol Mol Plant Pathol* 50, 337, 1997.
91. Bavaresco L, Vezzulli S, Battilani P, Giorni P, Pietri A, and Bertuzzi T, Effect of ochratoxin A-producing Aspergilli on stilbenic phytoalexin synthesis in grapes, *J Agric Food Chem* 51, 6151–6157, 2003.
92. Pool R.M, Creasy LL, and Frackelton AS, Resveratrol and the viniferins, their application to screening for disease resistance in grape breeding programs, *Vitis* 20, 136–145, 1981.

93. Stein U and Hoos G, Induktions und Nachweismethoden für Stilbene bei Vitaceen, *Vitis* 23, 179–194, 1984.
94. Grimmig B, Gonzalez-Perez MN, Leubner-Metzger G, Vogeli-Lange R, Meins F Jr, Hain R, Penuelas J, Heidenreich B, Langebartels C, Ernst D, and Sandermann H Jr, Ozone-induced gene expression occurs via ethylene-dependent and -independent signalling, *Plant Mol Biol* 51, 599–607, 2003.
95. Roldan A, Palacios V, Caro I, and Perez L, Resveratrol content of Palomino fino grapes: influence of vintage and fungal infection, *J Agric Food Chem* 51, 1464–1468, 2003.
96. Bavaresco L and Eibach R, Investigations on the influence of N fertilizer on resistance to powdery mildew (*Oidium tuckeri*), downy mildew (*Plasmopara viticola*) and on phytoalexin synthesis in different grapevine varieties, *Vitis* 26, 192–200, 1987.
97. Bavaresco L, Fregoni M, and Petegolli D, Effect of nitrogen and potassium fertilizer on induced resveratrol synthesis in two grapevine genotypes, *Vitis* 33, 175–176, 1994.
98. Kuc J, Phytoalexins, stress metabolism, and disease resistance in plants, *Annu Rev Phytopathol* 33, 275–297, 1995.
99. Dercks W and Creasy LL, The significance of stilbene phytoalexins in the *Plasmopara viticola*–grapevine interaction, *Physiol Mol Plant Pathol* 34, 189–202, 1989.
100. Adrian M, Jeandet P, Veneau J, Weston LA, and Bessis R, Biological activity of resveratrol, a stilbenic compound from grapevines, against *Botrytis cinerea*, the causal agent for gray mold, *J Chem Ecol* 23, 1689–1702, 1997.
101. Hart JH and Shrimpton DM, Role of stilbenes in resistance of wood to decay, *Phytopathology* 69, 1138–1143, 1979.
102. Hart JH, Role of phytostilbenes in decay and disease resistance, *Annu Rev Phytopathol* 19, 437–458, 1981.
103. Ward E, Unwin WB, and Stoessl A, Postinfectional inhibitors from plants: XV. Antifungal activity of the phytoalexin orchinol and related phenanthrenes and stilbenes, *Can J Bot* 53, 964–971, 1975.
104. Hoos G and Blaich R, Influence of resveratrol on germination of conidia and mycelial growth of *Botrytis cinerea* and *Phomopsis viticola*, *J Phytopathol* 129, 102–110, 1990.
105. Pezet R and Pont V, Ultrastructural observations of pterostilbene fungitoxicity in dormant conidia of *Botrytis cinerea* Pers., *J Phytopathology* 129, 19–30, 1990.
106. VanEtten HD and Pueppke SG, Isoflavonoid phytoalexins, in *Biochemical Aspects of Plant–Parasite Relationships*, Friend J and Threlfall DR, Eds, Academic Press, New York, 1976, pp. 239–289.
107. Woods JA, Hadfield JA, Pettit GR, Fox BW, and McGown AT, The interaction with tubulin of a series of stilbenes based on combretastatin A-4, *Br J Cancer* 71, 705–711, 1995.
108. Hoang-Van K, Rossier C, Baija F, and Turian G, Characterization of tubulin isotypes and of β-tubulin mRNA of *Neurospora crassa* and effects of benomyl on their developmental time courses, *Eur J Cell Biol* 49, 42–47, 1989.
109. Harris JE and Dennis C, Antifungal activity of post-infectional metabolites from potato tubers, *Physiol Plant Pathol* 9, 155–165, 1976.

110. Harris JE and Dennis C, The effect of post-infectional potato tuber metabolites and surfactants on zoospores of oomycetes, *Physiol Plant Pathol* 11, 163–169, 1977.
111. VanEtten HD and Bateman DF, Studies on the mode of action of the phytoalexin phaseollin, *Phytopathology* 61, 1363–1372, 1971.
112. Skipp RA and Bailey JA, The effect of phaseollin on the growth of *Colletotrichum lindemuthianum* in bioassays designed to measure fungitoxicity, *Physiol Plant Pathol* 9, 45–55, 1976.
113. Smith DA, Some effects of the phytoalexin, kievitone on the vegetative growth of *Aphanomyces euteiches*, *Rhizoctonia solani*, and *Fusarium solani* f. sp. *phaseoli*, *Physiol Plant Pathol* 9, 253–263, 1976.
114. Bartnicki-Garcia S and Lippman E, The bursting tendency of hyphal tips of fungi: presumptive evidence for a delicate balance between wall synthesis and wall lysis in apical growth, *J Gen Microbiol* 73, 487–500, 1972.
115. Grisebach H and Ebel J, Phytoalexins, chemical defence substances of higher plants, *Angew Chem Int Ed Engl* 17, 635–647, 1978.
116. Higgins VJ, The effect of some pterocarpanoid phytoalexins on germ tube elongation of *Stemphylium botryosum*, *Phytopathology* 68, 338–345, 1978.
117. Rossall S, Mansfield JW, and Huston RA, Death of *Botrytis cinerea* and *B. fabae* following exposure to wyerone derivatives *in vitro* and during infection development in broad bean leaves, *Physiol Plant Pathol* 16, 135–146, 1980.
118. Kindl H, Interplay *Botrytis*–Plant: Plant Stilbene Synthase Gene Promoters Responsive to *Botrytis*-Made Compounds and *Botrytis* Chaperones Sensitive to Plant Stilbene Phytoalexins, 12th International Botrytis Symposium, Reims, France, July 3–7, 2000.
119. Jeandet P, Sbaghi M, and Bessis R, The production of resveratrol (3,5, 4′-trihydroxystilbene) by grapevine *in vitro* cultures, and its application to screening for grey mould resistance, *J Wine Res* 3, 47–57, 1992.
120. Sbaghi M, Jeandet P, Faivre B, Bessis R, and Fournioux JC, Development of methods using phytoalexin (resveratrol) assessment as a selection criterion to screen grapevine *in vitro* cultures for resistance to grey mould (*Botrytis cinerea*), *Euphytica* 86, 41–47, 1995.
121. Okuda T and Yokotsuka K, *Trans*-resveratrol concentrations in berry skins and wines from grapes grown in Japan, *Am J Enol Vitic* 47, 93–99, 1996.
122. Hain R, Reif HJ, Krause E, Langebartels R, Kindl H, Vornam B, Wiese W, Schmelzer E, Schreier P, Stöcker R, and Stenzel K, Disease resistance results from foreign phytoalexin expression in a novel plant, *Nature* 361, 153–156, 1993.
123. Hipskind JD and Paiva NL, Constitutive accumulation of a resveratrol glucoside in transgenic alfalfa increases resistance to *Phoma medicaginis*, *Mol Plant Microbe Interact* 13, 551–562, 2000.
124. Coutos-Thévenot P, Poinssot B, Bonomelli A, Yean H, Breda C, Buffard D, Esnault R, Hain R, and Boulay M, *In vitro* tolerance to *Botrytis cinerea* of grapevine 41B rootstock in transgenic plants expressing the stilbene synthase *Vst 1* gene under the control of a pathogen-inducible PR 10 promoter, *J Exp Bot* 52, 901–910, 2001.

125. Stark-Lorenzen P, Nelke B, Hänbler G, Mühlbach HP, and Thomzik JE, Transfer of a grapevine stilbene synthase gene to rice (*Oryza sativa* L.), *Plant Cell Rep* 16, 668–673, 1997.
126. Thomzik JE, Stenzel K, Stöcker R, Schreier PH, Hain R, and Stahl DJ, Synthesis of a grapevine phytoalexin in transgenic tomato (*Lycopersicon esculentum* Mill.) conditions resistance against *Phytophthora infestans*, *Physiol Mol Plant Pathol* 51, 265–278, 1997.
127. Leckband G and Lörz H, Transformation and expression of a stilbene synthase gene of *Vitis vinifera* L, in barley and wheat for increased fungal resistance, *Theor Appl Genet* 96, 1001–1012, 1998.
128. Fettig S and Hess D, Expression of a chimeric stilbene synthase gene in transgenic wheat lines, *Transgenic Res* 8, 179–189, 1999.
129. Zhu YJ, Agbayani R, Jackson MC, Tang CS, and Moore PH, Expression of the grapevine stilbene synthase gene *VST1* in papaya provides increased resistance against diseases caused by *Phytophthora palmivora*, *Planta* 12, 807–812, 2004.
130. Giorcelli A, Sparvoli F, Mattivi F, Tava A, Balestrazzi A, Vrhovsek U, Calligari P, Bollini R, and Confalonieri M, Expression of the stilbene synthase (StSy) gene from grapevine in transgenic white poplar results in high accumulation of the antioxidant resveratrol glucosides, *Transgenic Res* 13, 203–214, 2004.
131. Kobayashi S, Ding CK, Nakamura Y, Nakajima I, and Matsumoto R, Kiwifruits (*Actinidia deliciosa*) transformed with a *Vitis* stilbene synthase gene produce piceid (resveratrol-glucoside), *Plant Cell Rep* 19, 904–910, 2000.
132. Fischer R, Budde I, and Hain R, Stilbene syntahse gene expression causes changes in flower colour and male sterility in tobacco, *Plant J* 11, 489–498, 1997.
133. Bais AJ, Murphy PJ, and Dry IB, The molecular regulation of stilbene phytoalexin biosynthesis in *Vitis vinifera* during grape berry development, *Aust J Plant Physiol* 27, 425–433, 2000.
134. Marbach I, Harel E, and Mayer AM, Molecular properties of the extracellular *Botrytis cinerea* laccase, *Phytochemistry* 23, 2713–2717, 1984.
135. Goetz G, Fkyerat A, Métais N, Kunz M, Tabacchi R, Pezet R, and Pont V, Resistance factors to grey mould in grape berries: identification of some phenolics inhibitors of *Botrytis cinerea* stilbene oxidase, *Phytochemistry* 52, 759–767, 1999.
136. Sbaghi M, Jeandet P, Bessis R, and Leroux P, Metabolism of stilbene-type phytoalexins in relation to the pathogenicity of *Botrytis cinerea* to grapevines, *Plant Pathol* 45, 139–144, 1996.
137. Lyr H, Enzymatische Detoxifikation der Kernholztoxine, *Flora* 152, 289–290, 1962.
138. Loman AA, Bioassays of fungi isolated from *Pinus concorta* var. *latifolia* with pinosylvin, pinosylvinmonomethyl ether, pinobanksin and pinocembrin, *Can J Bot* 48, 1303–1308, 1970.
139. Schouten A, Wagemakers L, Stefanato FL, van der Kaaij RM, and van Kan JA, Resveratrol acts as a natural profungicide and induces self-intoxication by a specific laccase, *Mol Microbiol* 43, 883–894, 2002.

140. Cichewicz RH, Kouzi SA, and Hamann MT, Dimerization of resveratrol by the grapevine pathogen *Botrytis cinerea*, *J Nat Prod* 63, 29–33, 2001.
141. Breuil AC, Jeandet P, Adrian M, Chopin F, Pirio N, Meunier P, and Bessis R, Characterization of a pterostilbene dehydrodimer produced by laccase of *Botrytis cinerea*, *Phytopathology* 89, 298–302, 1999.
142. Urban M, Bhargava T, and Hamer JE, An ATP-driven efflux pump is a novel pathogenicity factor in rice blast disease, *EMBO J* 18, 512–521, 1999.
143. Del Sorbo G, Schoonbeek H, and De Waard MA, Fungal transporters involved in efflux of natural toxic compounds and fungicides, *Fungal Genet Biol* 30, 1–15, 2000.
144. Schoonbeek H, Del Sorbo G, and De Waard MA, The ABC transporter BcatrB affects the sensitivity of *Botrytis cinerea* to the phytoalexin resveratrol and the fungicide fenpiclonil, *Mol Plant Microbe Interact* 14, 562–571, 2001.
145. Vermeulen T, Schoonbeek H, and De Waard MA, The ABC transporter BcatrB from *Botrytis cinerea* is a determinant of the activity of the phenylpyrrole fungicide fludioxonil, *Pest Manag Sci* 57, 393–402, 2001.
146. Del Sorbo G, Andrade AC, Van Nistelrooy JGM, Van Kan JA, Balzi E, and De Waard MA, Multidrug resistance in *Aspergillus nidulans* involves novel ATP-binding cassette transporters, *Mol Gen Genet* 254, 417–426, 1997.
147. Zwiers LH and De Waard MA, Characterization of the ABC transporter genes *MgAtr1* and *MgAtr2* from the wheat pathogen *Mycosphaerella graminicola*, *Fungal Genet Biol* 30, 115–125, 2000.
148. Nakaune R, Hamamoto H, Imada J, Akutsu K, and Hibi T, A novel ABC transporter gene, *PMR5*, is involved in multidrug resistance in the phytopathogenic fungus *Penicillium digitatum*, *Mol Genet Genomics* 267, 179–185, 2002.
149. VanEtten HD, Matthews DE, and Matthews PS, Phytoalexin detoxification: importance for pathogenicity and practical implications, *Annu Rev Phytopathol* 27, 143–164, 1989.
150. Osbourn AE, Preformed antimicrobial compounds and plant defense against fungal attack, *Plant Cell* 8, 1821–1831, 1996.

20 Protective Effects of Resveratrol in Age-Related Neurodegenerative Diseases and Gene Regulatory Action

Sofiyan Saleem, Abdullah Shafique Ahmad, and Sylvain Doré

CONTENTS

INTRODUCTION

Polyphenols found in red wine are believed to have beneficial health effects stemming from their ability to act as antioxidants. Polyphenols are divided into two main classes: flavonoids and nonflavonoids. Flavonoids, which have been measured in several natural extracts and edible plants and fruits, are the most abundant polyphenols, whereas stilbenes are only a minor class. The natural stilbene resveratrol (*trans*-3,4′,5-trihydroxystilbene) is a nonflavonoid that appears to be one of the most biologically active polyphenols. The reported antioxidant properties of red wine that are thought to be associated with resveratrol are likely due to a unique cascade of events leading to activation of an antioxidant pathway. This observation becomes especially relevant when comparing the molar ratio of free radicals to the very modest amount of resveratrol. In other words, it appears that antioxidant effective concentrations are unlikely to be reached in plasma *in vivo*. Therefore, resveratrol is likely to stimulate an intracellular pathway leading to cytoprotection. As discussed here, several genes and proteins have been proposed as targeted pathways. Heme oxygenase (HO) is a good candidate, as we have discovered that resveratrol can significantly induce HO in cells, and also since HO activity and its metabolites have been demonstrated to be neuroprotective *in vivo* and *in vitro*, to be antiapoptotic and antiinflammatory, and to have vasodilatory effects. Several other enzymatic systems have been described to be either directly or indirectly modulated by members of the stilbene family. We propose that HO is neuroprotective and is a prototypical candidate for providing brain resistance against pharmacologic or physiologic stress. Resveratrol's effect on cerebral blood flow, cell death, and inflammatory processes can contribute therapeutic actions in either acute or chronic neurodegenerative conditions. These actions are especially important when there is a reduction in cerebral blood flow, a lack of oxygen, and/or free radical damage, affecting especially vulnerable classes of neurons, leading to neuroinflammation and cell death. Polyphenol stilbenes can precondition the neurons and brain against such damage. More support and research are required to test resveratrol's potential in preventing a number of neurodegenerative conditions such as stroke, amyotrophic lateral sclerosis, Parkinson's and Alzheimer's diseases, and a variety of age-related vascular disorders.

It has been reported that moderate wine consumption is linked to a lower incidence of cardiovascular disease, the so-called "French paradox" [1]. This paradox (i.e., low incidence of cardiovascular events in spite of a diet relatively high in saturated fat) has been attributed to the drinking of red wine in southern France. Red wine, as well as grape juice and other natural extracts, contains several phenolic compounds that are mainly divided into two categories: flavonoids and nonflavonoids. The major classes of flavanoids are the flavanols, flavonols, and anthocyanins. The flavanoids constitute the majority of phenols in red wine, whereas the nonflavanoids are less abundant. The latter are mainly found as hydroxycinnamic acids, benzoic acids, and stilbenes. Of these, the stilbenes are only a minor class. The principal stilbene present in grapes and wine is resveratrol. During the production of red wine, the sugar from the grapes is fermented into alcohol, which acts as a good solvent for polyphenol extraction in the presence of the skins and seeds. The fermentation process provides ample opportunity for the extraction of many of the polyphenols.

Unlike red wines, white wines contain very low concentrations of phenols [2–4]. Resveratrol is found mainly in the skin of the grapes. Therefore, more resveratrol is found in red wine than in white, because the skin is not used in the production of white wine. It has been reported that the resveratrol level in most red wines is approximately 7 mg/l, but it is only 0.5 mg/l in white wines [5,6]. In comparison, flavonoid concentrations in red wines range between 1300 and 1500 mg/l.

Resveratrol is virtually absent from commonly eaten fruits and vegetables [7,8]. It is found in a soluble form in wine, potentially making it the predominant bioavailable dietary source [2]. The very simple chemical structure of resveratrol may interact with a variety of cellular and molecular targets. Resveratrol exists as *cis* and *trans* isomers (Figure 20.1), both of which are found in wine, although only the *trans* isomer is found in grapes. Light is known to cause isomerization of *trans*- to *cis*-resveratrol, and our preliminary observations in primary cultured neurons indicate that the active isomer is *trans*-resveratrol.

Since the initial reports in *Science* and *Nature* implicating a biological role for resveratrol in 1997 [9] and 2001 [10], over 900 original research articles have been published on the subject. It is somewhat counterintuitive that a "simple" ingredient such as resveratrol could be the active ingredient responsible for the classic "French paradox," considering the small amount present in either wine or juice. The theory proposed to explain resveratrol's bioactivity is mainly based on its direct antioxidant capability. If one considers the molar ratio necessary for resveratrol to be active in such conditions, taking into account its absorption rate, modifications/conjugations, etc., one would have to consume liters of such drinks to achieve its health effects. In other words, it appears that concentrations required for antioxidant effects are unlikely to be reached in plasma *in vivo*. In this chapter we discuss other pathways that could potentially be activated, and ultimately lead to the neurological protective effects (Table 20.1).

Polyphenols in red wine may also reduce the likelihood of atherosclerosis [27–29]. Resveratrol has been shown to dose-dependently decrease platelet aggregation, which contributes to the process of atherosclerosis [30,31]. Platelets stick to the endothelial surface of blood vessels where they can activate the process of thrombus formation, and their aggregation could set

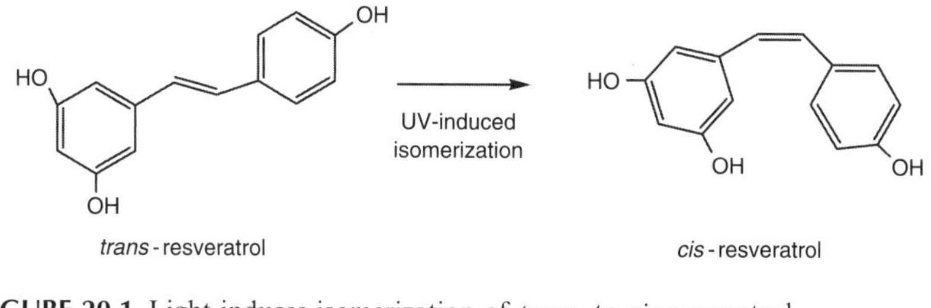

HO OH OH

UV-induced isomerization

HO OH OH

trans - resveratrol

cis - resveratrol

FIGURE 20.1 Light induces isomerization of *trans*- to *cis*-resveratrol.

TABLE 20.1
Experimental Neurologic Benefits of Resveratrol

Neurologic disease	Neurologic benefits	Species
Cerebral ischemia	Reduces infarct size	Rat [11]
	Reduces focal ischemia damage	Rat [12]
Amyotrophic lateral sclerosis	Directly stimulates BK(Ca) channel activity in vascular endothelial cells	Human cells [13]
Parkinson's disease	Protects embryonic mesencephalic cells	Rat [14]
Spinal cord lesion	Protects spinal cord from ischemia-reperfusion injury	Rabbit [15]
Brain edema	Inhibits expression of NFκB p65	Rat [16]
Brain tumors	Reduces paclitaxel-induced apoptosis in neuroblastoma SH-SY5Y cells	Human cells [17]
Chemoprevention	Inhibits critical cell cycle regulatory proteins	Mice [18]
	Converts into an anticancer agent by cytochrome P_{450} enzymes	Human cells [19]
Seizure	Reduces generalized tonic-clonic convulsions	Rat [20]
	Delays $FeCl_3$-induced epileptiform EEG changes	Rat [21]
	Reduces kainate-induced lesions in hippocampus	Rat [22]
Pain	Decreases carrageenan-induced hyperalgesia	Rat [23]
Cognitive impairment	Prevents ICV streptozotocin-induced cognitive impairment	Rat [24]
Aging	Mimics beneficial effects of caloric restriction	Yeast [25,26]

into motion the process of vascular occlusion. Eicosanoids synthesized from arachidonic acid have been shown to modulate platelet adhesion [31]. The ability of resveratrol to inhibit platelet aggregation could be linked to its ability to inhibit eicosanoid synthesis [8]. In addition to lowering the risk of cardiovascular disease, red wine consumption has been shown by some studies to correlate with diminished risk of age-related macular degeneration, vascular dementia, Alzheimer's disease, and cognitive deficits [32–35]. However, the mechanism by which resveratrol is neuroprotective continues to be unknown. Table 20.2 lists several genes that have been shown to be modulated by resveratrol.

Living organisms are continuously threatened by the damage caused by reactive oxygen species that are produced during normal oxygen metabolism and mitochondrial function or generated by exogenous damage. Most of the polyphenol studies have necessitated pretreatment to observe their biologic

TABLE 20.2
Example of Resveratrol-Regulated Genes/Proteins Affected and in Cells (<25 μM)

Gene/protein	Effect (major function)
HO1	Induces protein expression (role as antioxidant system and vasodilation) [36]
COX-2	Inhibits COX-2 transcription (role in inflammation) [37]
Cytochrome P_{450} enzymes	Reduces mRNA expression (role in metabolism of drugs) [38]
eNOS	Stabilizes mRNA and increases protein expression (role in NO production) [6]
iNOS	Induces protein expression (role in NO production) [39]
ET-1	Reduces protein expression (role in vasoconstriction) [10,40]
GADD45	Increases mRNA (role in cell growth) [41]
Retinoic acid-binding protein II	Increases mRNA (role in transduction) [41]
TNFα-inducible protein	Increases mRNA (role in cell death) [41]
ATF3	Increases mRNA (role in cell growth) [41]
Adenylyl cyclase-associated protein	Decreases mRNA (role in structure regulation) [41]
Hemopoietic cell protein-tyrosine kinase	Decreases mRNA (role in transduction) [41]
IGFBP-5	Decreases mRNA (role in cell growth) [41]

Note: HO1, heme oxygenase 1; COX-2, cyclooxygenase 2; eNOS, endothelial nitric oxide synthase; iNOS, inducible nitric oxide synthase; ET-1, endothelin-1; GADD45, growth arrest and DNA damage-induced protein 45; TNFα, tumor necrosis factor alpha; ATF3, activating transcription factor 3; IGFBP, insulin-like growth factor binding protein.

functions. The effect of a pretreatment is required either to increase the intracellular concentration of polyphenolic compound antioxidant constituents or to increase a cascade involving an endogenous antioxidant system. Activation of antioxidant pathways is particularly important as cytoprotection for tissue with relatively weak endogenous antioxidant defenses, such as the heart and the brain. Considering the proven antioxidant properties of polyphenolic compounds from red wine and their possible therapeutic efficacy, we investigated whether neuroprotective effects of resveratrol could be mediated by induction of the HO antioxidant system in neurons. Induction of HO could be a protective system in acute or chronic neurodegenerative conditions [42].

HEME OXYGENASE

HO catalyzes the cleavage of heme to form iron, carbon monoxide, and biliverdin, which is immediately reduced to bilirubin (Figure 20.2). Two

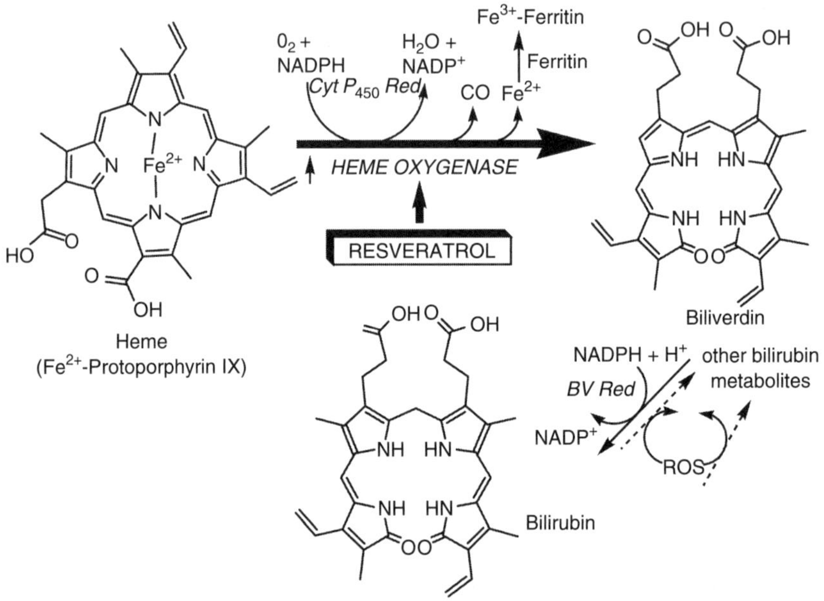

FIGURE 20.2 Stimulation of heme degradation by resveratrol-mediated induction of heme oxygenase levels. Results demonstrate that low concentrations of resveratrol are sufficient to induce significantly heme oxygenase 1 protein levels in primary cultures of cortical neurons.

isoforms of HO have been distinguished and well studied [42–45]. A third isoform has been reported [46], although our results and those of others indicate that it is not present in mice [47]. HO1, the first to be isolated [44], is an inducible enzyme and concentrated in tissues such as the spleen and liver. It is barely detectable in the brain under normal conditions, although many reports have shown that it is significantly increased with stress. In contrast, HO2 is constitutively expressed and highly concentrated in the brain and testis [48]. Several lines of evidence suggest that HO1 plays a role in neuromodulatory activities. It is considered to be a stress protein that can be induced in a cell-specific manner by multiple factors [49], including hyperthermia [50], Alzheimer's disease [51,52], global ischemia [53], and subarachnoid hemorrhage [54]. In a reperfusion model of focal ischemia, we and others have shown that HO1 mRNA and proteins are induced [55–58]. Upregulation of HO1 in the substantia nigra has been demonstrated in Parkinson's disease patients. In these patients, intense HO1 immunoreactivity was found in the proximity of nigral neurons containing cytoplasmic Lewy bodies [59]. Several reports have shown HO1 induction in neurons [60], and our preliminary data indicated that HO1 can be significantly induced within neurons after resveratrol treatment. We believe that modulation of HO1 activity could be a pathway by which wine can be protective in the brain. The abundant heme-containing enzymes that are located in the mitochondria and endoplasmic reticulum presumably undergo turnover during oxidative stress. HO1 is likely to play an important role in ensuring that prooxidant-free heme does not increase to toxic levels. Tissue injury after an ischemic event appears to be due to enhanced oxidative stress by free radicals generated during the reperfusion phase [61]. We showed nearly double the infarct volume in HO2 knockout mice versus wildtype after stroke. Interestingly, Maines' group has limited the infarct damage after stroke by increasing HO1 activity in neurons (using a transgenic mouse overexpressing HO1 with a neuron-specific promoter) [62]. Taken together, these results suggest that if HO1 activity within the neurons can be increased, neuroprotection can be provided. We believe that modulation of neuronal HO1 activity is important to cell survival.

HEME

Heme metabolism is an important metabolic process (Figure 20.2 and Table 20.3). Free heme in the cell can be rapidly generated from heme-containing proteins/enzymes (such as catalase, glutathione peroxidase, cytochrome, guanylate cyclase, superoxide dismutase, nitric oxide synthase,

TABLE 20.3
Example of Potential Outcomes from Gene/Protein Regulation (i.e., HO1) and Potential Enzymatic Role in the Brain

Heme oxygenase reaction	Biological actions
Degradation of heme	Free heme is prooxidant (regulation of intracellular heme synthesis) (heme from hemoprotein degradation) (heme from hemoglobin)
Production of iron	Participates in the classical Fenton reaction and generates free radicals Regulates ferritin, transferring, and iron regulatory protein levels
Production of biliverdin/bilirubin	Scavenge free radicals Bilirubin metabolites may have vasoactive actions
Production of carbon monoxide	Regulates soluble guanylate cyclase (sGC) activity and cGMP synthesis Regulation of several hemoproteins/enzymes Plays potential role in learning and memory processes

Note: This list of potential actions would occur at physiologic levels, while abnormal or pharmacologic levels of these compounds could have deleterious effects.

TABLE 20.4
Nonexhaustive List of Potential Neurologic Benefits of Resveratrol-Regulating HO1

Heme oxygenase activity	Neurologic disorders
Antioxidant properties	Most neurologic disorders [36]
Inflammation reduction	Most neurologic disorders [88,89]
Restoration of normal blood flow	Cerebral ischemia [90]
	Alzheimer's (age-related vascular dementia) [42]
Reduction of neuronal cell death	In stroke models [55]
	In induced apoptosis [91]
	Cholinergic neuron in AD models [92]
	Dopaminergic neurons in PD models [59]
Regulation of iron homeostasis	Parkinsonism [93]
	Ataxia (Friedrich) [94]
	Movement disorders, tremors [95]
	Restless leg syndrome [96]
	Hallervorden-Spatz syndrome [97]
	Aceruloplasminemia [98]
	Neuroferritinopathy [99]
Others	Sleep pattern [100]
	Circadian rhythm [101,102]
	Amyotrophic lateral sclerosis [103]
	Spinal cord lesion [104]
	Head trauma [105]
	Brain edema [62]
	Huntington's [106]
	Kernicterus [107]
	Brain tumors [108]
	Chemoprevention [109]
	Pain [110]
	Seizure [111]
	Atherosclerosis [112]
	Age-related cognitive impairment [113]

etc.). For example, hypoxia during ischemic injury could trigger micromolar concentrations of heme to be released in the intracellular pool. When these hemoproteins are degraded, the heme is not salvaged and should be degraded. HO is the enzyme that, by rapidly cleaving prooxidant heme, can limit its capacity to enter into the generation of a free radical cycle. Thus, HO could be considered an antioxidant enzyme. Our observation that HO1 is specifically induced in neurons following resveratrol treatment (Figure 20.2), associated with HO1 biologic properties (Table 20.4), makes it an interesting target for regulating the redox state of cells and may potentially afford neuroprotection.

IRON

The degradation of heme by HO1 generates iron (Figure 20.2). We have accumulated evidence that suggests that modulation of HO1 protein levels could be a limiting factor in controlling the rate by which iron is eliminated from cells [63]. Controlling the iron homeostasis within the cell is critical. For example, free iron is a key element in the Fenton reaction, which, by reaction with H_2O_2, generates free radicals. The cellular homeostasis of iron is a complex system regulated by many proteins, some of which are still being identified and characterized. The rapid upregulation of HO1 in neurons by resveratrol is likely to affect directly the intracellular iron levels. Conversely, one would expect that a decrease in HO1 activity would be sufficient to change the iron levels within cells. Indeed, experiments with HO1 knockout mice showed accumulation of iron in several organs [64]. Numerous iron-binding proteins have been found, and their expression can modulate the iron intracellular free pool. To cite only one example, ferritin in the cell can sequester free iron, and its intracellular levels can be very rapidly induced by free iron [65]. The therapeutic implications of controlling iron levels are numerous (Table 20.3). For example, one study revealed that the administration of desferoxamine, a trivalent ion chelator, over a two-year period slowed the clinical progression of symptoms in Alzheimer's disease [66]. Through studying the regulation of HO1 levels, one could also expect the enzyme to control iron homeostasis. Very little is known about the possible role of iron in mediating the neuroprotective role of resveratrol.

CARBON MONOXIDE (CO)

CO is a soluble gas that is almost exclusively generated in cells by the degradation of heme by HO (Figure 20.2) [67,68]. CO can travel freely throughout intracellular and extracellular compartments. The literature regarding the role of CO is complex, with many controversies that have yet to be resolved (Table 20.3) [69–74]. CO is known to be toxic, and the inhalation of high concentrations can cause death, whereas low concentrations can be protective, as reviewed before [75]. The affinity of CO for several proteins is generally lower than that of nitric oxide, although its longer half-life could be a key factor in modifying several key enzymes containing a heme moiety [76]. At a cellular level, physiologic levels of CO generated from degradation of intracellular heme are likely to have biologic actions on several heme-containing proteins. For example, CO may act as a vasodilator by binding to soluble guanylate cyclase (sGC) and modulating its activity. CO can also act by opening calcium-activated potassium channels (K_{Ca} channels) [77,78]. It has also been reported to have specific antiinflammatory and antiapoptotic effects [79,80]. Rapid induction of HO1 would be a means by which resveratrol could increase CO within physiologic levels, allowing cells and tissues to benefit from many of CO's biologic actions, especially in a scenario in which blood flow is reduced and cell survival is triggered.

BILIRUBIN (BR)

BR is better known in the central nervous system for its toxicity at high micromolar concentrations in neonates; however, under physiologic concentration, BR could be a significant endogenous antioxidant [81,82]. In an extensive series of antioxidants, BR displayed significant superoxide and peroxyl radical scavenger activity [83]. An animal model of hyperbilirubinemia is also associated with protection against cerebral ischemia. Stroke damage is reported to be diminished in these animals. We have observed that BR can be protective at low concentrations in primary hippocampal and cortical neuronal cultures [84,85]. BR is a product of biliverdin reductase, which is an abundant and ubiquitous enzyme with a high turnover rate. Along with previous observations of HO contribution in Alzheimer's disease pathology [51,86], levels of bilirubin derivatives in the cerebral spinal fluid have been reported to be increased significantly in the brains of Alzheimer's disease patients as compared with those of controls [87]. Increased HO1 levels are likely to decrease the heme level and significantly increase the bilirubin level, within physiologic ranges. Such a unique pathway could also explain some of the antioxidant properties often associated with resveratrol.

All together, regulation of genes by resveratrol could present a new understanding of the "potential" prophylactic use of resveratrol against either acute or chronic neurologic disorders (Table 20.4). For example, in human ischemic stroke, recirculation occurs frequently after focal ischemia, especially in cases of cerebral embolism and transient ischemic attack, which are warning signs of stroke. Recurrence is a very prevalent phenomenon in patients who have suffered from one episode of stroke. Therefore, the availability of a prophylactic approach in such patients would be a desired goal in preventive medicine. In that the incidence of coronary heart disease is reduced in chronic consumers of red wine, a full understanding of the mechanism by which polyphenols and red wine can protect against ischemic or neurotoxic insults is optimal.

SUMMARY

Wine has been shown to have intrinsic medicinal properties. It has been reported that resveratrol, a component particularly of red wine, can protect against heart and brain ischemia [22,114]. It is possible that these protective effects could be extended to the treatment of impairments and symptoms in traumatic brain injury, stroke, multiinfarct dementia, cerebral atherosclerosis, cerebral edema, and inflammation, as well as Alzheimer's disease and age-/vascular-associated dementia. The determinants of neuronal cell death in acute and chronic neurodegenerative conditions have been postulated to be mediated by free radical damage. Resveratrol and moderate red wine consumption have been suggested as potential preventive medicines [31,115,116], but the underlying cellular mechanisms are still unclear. Knowing that HO also plays various roles in oxidative stress and inflammation [117], additional work and funding are necessary to assess its unique properties and whether it participates in resveratrol neuroprotectivity. Together, these results will address possible mechanisms by which the consumption of red wine could be beneficial and propose new pathways by which to explain this neuroprotective effect and how it could provide the brain's resistance in acute (such as in ischemia) and chronic (such as age-/vascular-related dementia and Alzheimer's disease) neurodegenerative debilitating conditions.

ACKNOWLEDGMENTS

This work was supported in part by the NIAAA, NCCAM, the Wine Institute, and the ABMR Foundation.

REFERENCES

1. Renaud S and de Lorgeril M, Wine, alcohol, platelets, and the French paradox for coronary heart disease, *Lancet* 339, 1523–1526, 1992.
2. Soleas GJ, Diamandis EP, and Goldberg DM, Resveratrol: a molecule whose time has come? And gone?, *Clin Biochem* 30, 91–113, 1997.
3. Goldberg D, Tsang E, Karumanchiri A, Diamandis E, Soleas G, and Ng E, Method to assay the concentrations of phenolic constituents of biological interest in wines, *Anal Chem* 68, 1688–1694, 1996.
4. Celotti E, Ferrarini R, Zironi R, and Conte LS, Resveratrol content of some wines obtained from dried Valpolicella grapes: Recioto and Amarone, *J Chromatogr A* 730, 47–52, 1996.
5. Ribeiro de Lima MT, Waffo-Teguo P, Teissedre PL, Pujolas A, Vercauteren J, Cabanis JC, and Merillon JM, Determination of stilbenes (trans-astringin, cis- and trans-piceid, and cis- and trans-resveratrol) in Portuguese wines, *J Agric Food Chem* 47, 2666–2670, 1999.
6. Wallerath T, Deckert G, Ternes T, Anderson H, Li H, Witte K, and Forstermann U, Resveratrol, a polyphenolic phytoalexin present in red wine, enhances expression and activity of endothelial nitric oxide synthase, *Circulation* 106, 1652–1658, 2002.
7. Goldberg DM, Hahn SE, and Parkes JG, Beyond alcohol: beverage consumption and cardiovascular mortality, *Clin Chim Acta* 237, 155–187, 1995.
8. Soleas GJ, Diamandis EP, and Goldberg DM, Wine as a biological fluid: history, production, and role in disease prevention, *J Clin Lab Anal* 11, 287–313, 1997.

9. Jang M, Cai L, Udeani GO, Slowing KV, Thomas CF, Beecher CW, Fong HH, Farnsworth NR, Kinghorn AD, Mehta RG, Moon RC, and Pezzuto JM, Cancer chemopreventive activity of resveratrol, a natural product derived from grapes, *Science* 275, 218–220, 1997.
10. Corder R, Douthwaite JA, Lees DM, Khan NQ, Viseu Dos Santos AC, Wood EG, and Carrier MJ, Endothelin-1 synthesis reduced by red wine, *Nature* 414, 863–864, 2001.
11. Huang SS, Tsai MC, Chih CL, Hung LM, and Tsai SK, Resveratrol reduction of infarct size in Long-Evans rats subjected to focal cerebral ischemia, *Life Sci* 69, 1057–1065, 2001.
12. Sinha K, Chaudhary G, and Gupta YK, Protective effect of resveratrol against oxidative stress in middle cerebral artery occlusion model of stroke in rats, *Life Sci* 71, 655–665, 2002.
13. Wu SN, Large-conductance Ca^{2+}-activated K^+ channels: physiological role and pharmacology, *Curr Med Chem* 10, 649–661, 2003.
14. Karlsson J, Emgard M, Brundin P, and Burkitt MJ, Trans-resveratrol protects embryonic mesencephalic cells from tert-butyl hydroperoxide: electron paramagnetic resonance spin trapping evidence for a radical scavenging mechanism, *J Neurochem* 75, 141–150, 2000.
15. Kiziltepe U, Turan NN, Han U, Ulus AT, and Akar F, Resveratrol, a red wine polyphenol, protects spinal cord from ischemia-reperfusion injury, *J Vasc Surg* 40, 138–145, 2004.
16. Wang YJ, He F, and Li XL, The neuroprotection of resveratrol in the experimental cerebral ischemia (in Chinese), *Zhonghua Yi Xue Za Zhi* 83, 534–536, 2003.
17. Nicolini G, Rigolio R, Scuteri A, Miloso M, Saccomanno D, Cavaletti G, and Tredici G, Effect of trans-resveratrol on signal transduction pathways involved in paclitaxel-induced apoptosis in human neuroblastoma SH-SY5Y cells, *Neurochem Int* 42, 419–429, 2003.
18. Reagan-Shaw S, Afaq F, Aziz MH, and Ahmad N, Modulations of critical cell cycle regulatory events during chemoprevention of ultraviolet B-mediated responses by resveratrol in SKH-1 hairless mouse skin, *Oncogene* 23, 5151–5160, 2004.
19. Potter GA, Patterson LH, Wanogho E, Perry PJ, Butler PC, Ijaz T, Ruparelia KC, Lamb JH, Farmer PB, Stanley LA, and Burke MD, The cancer preventative agent resveratrol is converted to the anticancer agent piceatannol by the cytochrome P450 enzyme CYP1B1, *Br J Cancer* 86, 774–778, 2002.
20. Gupta YK, Chaudhary G, and Srivastava AK, Protective effect of resveratrol against pentylenetetrazole-induced seizures and its modulation by an adenosinergic system, *Pharmacology* 65, 170–174, 2002.
21. Gupta YK, Chaudhary G, Sinha K, and Srivastava AK, Protective effect of resveratrol against intracortical FeCl3-induced model of posttraumatic seizures in rats, *Methods Find Exp Clin Pharmacol* 23, 241–244, 2001.
22. Virgili M and Contestabile A, Partial neuroprotection of *in vivo* excitotoxic brain damage by chronic administration of the red wine antioxidant agent, trans-resveratrol in rats, *Neurosci Lett* 281, 123–126, 2000.
23. Gentilli M, Mazoit JX, Bouaziz H, Fletcher D, Casper RF, Benhamou D, and Savouret JF, Resveratrol decreases hyperalgesia induced by carrageenan in the rat hind paw, *Life Sci* 68, 1317–1321, 2001.

24. Sharma M and Gupta YK, Chronic treatment with trans resveratrol prevents intracerebroventricular streptozotocin induced cognitive impairment and oxidative stress in rats, *Life Sci* 71, 2489–2498, 2002.
25. Howitz KT, Bitterman KJ, Cohen HY, Lamming DW, Lavu S, Wood JG, Zipkin RE, Chung P, Kisielewski A, Zhang LL, Scherer B, and Sinclair DA, Small molecule activators of sirtuins extend Saccharomyces cerevisiae lifespan, *Nature* 425, 191–196, 2003.
26. Wood JG, Rogina B, Lavu S, Howitz K, Helfand SL, Tatar M, and Sinclair D, Sirtuin activators mimic caloric restriction and delay ageing in metazoans, *Nature* 430, 686–689, 2004.
27. Bors W and Saran M, Radical scavenging by flavonoid antioxidants, *Free Radical Res Commun* 2, 289–294, 1987.
28. Afanas'ev IB, Dorozhko AI, Brodskii AV, Kostyuk VA, and Potapovitch AI, Chelating and free radical scavenging mechanisms of inhibitory action of rutin and quercetin in lipid peroxidation, *Biochem Pharmacol* 38, 1763–1769, 1989.
29. Slater TF, Cheeseman KH, Davies MJ, Hayashi M, Sharma OP, Nigam S, and Benedetto C, Free radical scavenging properties of modulators of eicosanoid metabolism, *Adv Prostaglandin Thromboxane Leukot Res* 17B, 1098–1102, 1987.
30. Orsini F, Pelizzoni F, Verotta L, Aburjai T, and Rogers CB, Isolation, synthesis, and antiplatelet aggregation activity of resveratrol 3-O-beta-D-glucopyranoside and related compounds, *J Nat Prod* 60, 1082–1087, 1997.
31. Olas B and Wachowicz B, Resveratrol and vitamin C as antioxidants in blood platelets, *Thromb Res* 106, 143, 2002.
32. Leibovici D, Ritchie K, Ledesert B, and Touchon J, The effects of wine and tobacco consumption on cognitive performance in the elderly: a longitudinal study of relative risk, *Int J Epidemiol* 28, 77–81, 1999.
33. Mukamal KJ, Longstreth WT, Jr., Mittleman MA, Crum RM, and Siscovick DS, Alcohol consumption and subclinical findings on magnetic resonance imaging of the brain in older adults: the cardiovascular health study, *Stroke* 32, 1939–1946, 2001.
34. Obisesan TO, Hirsch R, Kosoko O, Carlson L, and Parrott M, Moderate wine consumption is associated with decreased odds of developing age-related macular degeneration in NHANES-1, *J Am Geriatr Soc* 46, 1–7, 1998.
35. Orgogozo JM, Dartigues JF, Lafont S, Letenneur L, Commenges D, Salamon R, Renaud S, and Breteler MB, Wine consumption and dementia in the elderly: a prospective community study in the Bordeaux area, *Rev Neurol* (Paris) 153, 185–192, 1997.
36. Zhuang H, Kim YS, Koehler RC, and Doré S, Potential mechanism by which resveratrol, a red wine constituent, protects neurons, *Ann NY Acad Sci* 993, 276–286; discussion 287–288, 2003.
37. Subbaramaiah K, Chung WJ, Michaluart P, Telang N, Tanabe T, Inoue H, Jang M, Pezzuto JM, and Dannenberg AJ, Resveratrol inhibits cyclooxygenase-2 transcription and activity in phorbol ester-treated human mammary epithelial cells, *J Biol Chem* 273, 21875–21882, 1998.
38. Cheung CY, Chen J, and Chang TK, Evaluation of a real-time polymerase chain reaction method for the quantification of CYP1B1 gene expression

in MCF-7 human breast carcinoma cells, *J Pharmacol Toxicol Methods* 49, 97–104, 2004.

39. Imamura G, Bertelli AA, Bertelli A, Otani H, Maulik N, and Das DK, Pharmacological preconditioning with resveratrol: an insight with iNOS knockout mice, *Am J Physiol Heart Circ Physiol* 282, H1996–2003, 2002.
40. Liu JC, Chen JJ, Chan P, Cheng CF, and Cheng TH, Inhibition of cyclic strain-induced endothelin-1 gene expression by resveratrol, *Hypertension* 42, 1198–1205, 2003.
41. Shi T, Liou LS, Sadhukhan P, Duan ZH, Novick AC, Hissong JG, Almasan A, and DiDonato JA, Effects of resveratrol on gene expression in renal cell carcinoma, *Cancer Biol Ther* 3, 1538–4047, 2004.
42. Doré S, Decreased activity of the antioxidant heme oxygenase enzyme: implications in ischemia and in Alzheimer's disease, *Free Radical Biol Med* 32, 1276–1282, 2002.
43. Ewing JF and Maines MD, Histochemical localization of heme oxygenase-2 protein and mRNA expression in rat brain, *Brain Res Brain Res Protoc* 1, 165–174, 1997.
44. Shibahara S, Muller R, Taguchi H, and Yoshida T, Cloning and expression of cDNA for rat heme oxygenase, *Proc Natl Acad Sci USA* 82, 7865–7869, 1985.
45. Maines MD, The heme oxygenase system: a regulator of second messenger gases, *Annu Rev Pharmacol Toxicol* 37, 517–554, 1997.
46. McCoubrey WK Jr, Huang TJ, and Maines MD, Isolation and characterization of a cDNA from the rat brain that encodes hemoprotein heme oxygenase-3, *Eur J Biochem* 247, 725–732, 1997.
47. Zhuang H, Pin S, Li X, and Doré S, Regulation of heme oxygenase expression by cyclopentenone prostaglandins, *Exp Biol Med* 228, 499–505, 2003.
48. Doré S, Law A, Blackshaw S, Gauthier S, and Quirion R, Alteration of expression levels of neuronal nitric oxide synthase and haem oxygenase-2 messenger RNA in the hippocampi and cortices of young adult and aged cognitively unimpaired and impaired Long-Evans rats, *Neuroscience* 100, 769–775, 2000.
49. Ewing JF and Maines MD, Rapid induction of heme oxygenase 1 mRNA and protein by hyperthermia in rat brain: heme oxygenase 2 is not a heat shock protein, *Proc Natl Acad Sci USA* 88, 5364–5368, 1991.
50. Ewing JF, Haber SN, and Maines MD, Normal and heat-induced patterns of expression of heme oxygenase-1 (HSP32) in rat brain: hyperthermia causes rapid induction of mRNA and protein, *J Neurochem* 58, 1140–1149, 1992.
51. Schipper HM, Cisse S, and Stopa EG, Expression of heme oxygenase-1 in the senescent and Alzheimer-diseased brain, *Ann Neurol* 37, 758–768, 1995.
52. Smith MA, Kutty RK, Richey PL, Yan SD, Stern D, Chader GJ, Wiggert B, Petersen RB, and Perry G, Heme oxygenase-1 is associated with the neurofibrillary pathology of Alzheimer's disease, *Am J Pathol* 145, 42–47, 1994.
53. Takeda A, Kimpara T, Onodera H, Itoyama Y, Shibahara S, and Kogure K, Regional difference in induction of heme oxygenase-1 protein following rat transient forebrain ischemia, *Neurosci Lett* 205, 169–172, 1996.

54. Kuroki M, Kanamaru K, Suzuki H, Waga S, and Semba R, Effect of vasospasm on heme oxygenases in a rat model of subarachnoid hemorrhage, *Stroke* 29, 683–688; discussion 688–689, 1998.
55. Doré S, Sampei K, Goto S, Alkayed NJ, Guastella D, Blackshaw S, Gallagher M, Traystman RJ, Hurn PD, Koehler RC, and Snyder SH, Heme oxygenase-2 Is neuroprotective in cerebral ischemia, *Mol Med* 5, 656–663, 1999.
56. Koistinaho J, Miettinen S, Keinanen R, Vartiainen N, Roivainen R, and Laitinen JT, Long-term induction of haem oxygenase-1 (HSP-32) in astrocytes and microglia following transient focal brain ischaemia in the rat, *Eur J Neurosci* 8, 2265–2272, 1996.
57. Nimura T, Weinstein PR, Massa SM, Panter S, and Sharp FR, Heme oxygenase-1 (HO-1) protein induction in rat brain following focal ischemia, *Brain Res Mol Brain Res* 37, 201–208, 1996.
58. Matz P, Weinstein P, States B, Honkaniemi J, and Sharp FR, Subarachnoid injections of lysed blood induce the hsp70 stress gene and produce DNA fragmentation in focal areas of the rat brain, *Stroke* 27, 504–512; discussion 513, 1996.
59. Yoo MS, Chun HS, Son JJ, DeGiorgio LA, Kim DJ, Peng C, and Son JH, Oxidative stress regulated genes in nigral dopaminergic neuronal cells: correlation with the known pathology in Parkinson's disease, *Brain Res Mol Brain Res* 110, 76–84, 2003.
60. Takizawa S, Hirabayashi H, Matsushima K, Tokuoka K, and Shinohara Y, Induction of heme oxygenase protein protects neurons in cortex and striatum, but not in hippocampus, against transient forebrain ischemia, *J Cereb Blood Flow Metab* 18, 559–569, 1998.
61. McCord JM, Oxygen-derived free radicals in postischemic tissue injury, *N Engl J Med* 312, 159–163, 1985.
62. Panahian N, Yoshiura M, and Maines MD, Overexpression of heme oxygenase-1 is neuroprotective in a model of permanent middle cerebral artery occlusion in transgenic mice, *J Neurochem* 72, 1187–1203, 1999.
63. Ferris CD, Jaffrey SR, Sawa A, Takahashi M, Brady SD, Barrow RK, Tysoe SA, Wolosker H, Baranano DE, Doré S, Poss KD, and Snyder SH, Haem oxygenase-1 prevents cell death by regulating cellular iron, *Nat Cell Biol* 1, 152–157, 1999.
64. Poss KD and Tonegawa S, Heme oxygenase 1 is required for mammalian iron reutilization, *Proc Natl Acad Sci USA* 94, 10919–10924, 1997.
65. Quinlan GJ, Chen Y, Evans TW, and Gutteridge JM, Iron signalling regulated directly and through oxygen: implications for sepsis and the acute respiratory distress syndrome, *Clin Sci* (Lond) 100, 169–182, 2001.
66. Crapper McLachlan DR, Dalton AJ, Kruck TP, Bell MY, Smith WL, Kalow W, and Andrews DF, Intramuscular desferrioxamine in patients with Alzheimer's disease, *Lancet* 337, 1304–1308, 1991.
67. Vreman HJ, Wong RJ, Sanesi CA, Dennery PA, and Stevenson DK, Simultaneous production of carbon monoxide and thiobarbituric acid reactive substances in rat tissue preparations by an iron-ascorbate system, *Can J Physiol Pharmacol* 76, 1057–1065, 1998.
68. Yoshida T, Noguchi M, and Kikuchi G, The step of carbon monoxide liberation in the sequence of heme degradation catalyzed by the reconstituted microsomal heme oxygenase system, *J Biol Chem* 257, 9345–9348, 1982.

69. Alkadhi KA, Al-Hijailan RS, Malik K, and Hogan YH, Retrograde carbon monoxide is required for induction of long-term potentiation in rat superior cervical ganglion, *J Neurosci* 21, 3515–3520, 2001.
70. Verma A, Hirsch DJ, Glatt CE, Ronnett GV, and Snyder SH, Carbon monoxide: a putative neural messenger, *Science* 259, 381–384, 1993.
71. Maines M, Carbon monoxide and nitric oxide homology: differential modulation of heme oxygenases in brain and detection of protein and activity, *Methods Enzymol* 268, 473–488, 1996.
72. Meffert MK, Haley JE, Schuman EM, Schulman H, and Madison DV, Inhibition of hippocampal heme oxygenase, nitric oxide synthase, and long-term potentiation by metalloporphyrins, *Neuron* 13, 1225–1233, 1994.
73. Poss KD, Thomas MJ, Ebralidze AK, O'Dell TJ, and Tonegawa S, Hippocampal long-term potentiation is normal in heme oxygenase-2 mutant mice, *Neuron* 15, 867–873, 1995.
74. Linden DJ, Narasimhan K, and Gurfel D, Protoporphyrins modulate voltage-gated Ca current in AtT-20 pituitary cells, *J Neurophysiol* 70, 2673–2677, 1993.
75. Cowan RL and Doré S, Toxicity and neuroprotective effects of carbon monoxide: consequences to suicide and survival, *Encycl Psychol Behav Sci* 3, 994–997, 2004.
76. Hartsfield CL, Cross talk between carbon monoxide and nitric oxide, *Antioxid Redox Signal* 4, 301–307, 2002.
77. Wang R and Wu L, The chemical modification of KCa channels by carbon monoxide in vascular smooth muscle cells, *J Biol Chem* 272, 8222–8226, 1997.
78. Wu L, Cao K, Lu Y, and Wang R, Different mechanisms underlying the stimulation of K(Ca) channels by nitric oxide and carbon monoxide, *J Clin Invest* 110, 691–700, 2002.
79. Otterbein LE, Carbon monoxide: innovative anti-inflammatory properties of an age-old gas molecule, *Antioxid Redox Signal* 4, 309–319, 2002.
80. Brouard S, Berberat PO, Tobiasch E, Seldon MP, Bach FH, and Soares MP, Heme oxygenase-1-derived carbon monoxide requires the activation of transcription factor NF-kappa B to protect endothelial cells from tumor necrosis factor-alpha-mediated apoptosis, *J Biol Chem* 277, 17950–17961, 2002.
81. Gopinathan V, Miller NJ, Milner AD, and Rice-Evans CA, Bilirubin and ascorbate antioxidant activity in neonatal plasma, *FEBS Lett* 349, 197–200, 1994.
82. Stocker R, Glazer AN, and Ames BN, Antioxidant activity of albumin-bound bilirubin, *Proc Natl Acad Sci USA* 84, 5918–5922, 1987.
83. Farrera JA, Jauma A, Ribo JM, Peire MA, Parellada PP, Roques-Choua S, Bienvenue E, and Seta P, The antioxidant role of bile pigments evaluated by chemical tests, *Bioorg Med Chem* 2, 181–185, 1994.
84. Doré S, Takahashi M, Ferris CD, Zakhary R, Hester LD, Guastella D, and Snyder SH, Bilirubin, formed by activation of heme oxygenase-2, protects neurons against oxidative stress injury, *Proc Natl Acad Sci USA* 96, 2445–2450, 1999.
85. Doré S and Snyder SH, Neuroprotective action of bilirubin against oxidative stress in primary hippocampal cultures, *Ann NY Acad Sci* 890, 167–172, 1999.
86. Takahashi M, Doré S, Ferris CD, Tomita T, Sawa A, Wolosker H, Borchelt DR, Iwatsubo T, Kim SH, Thinakaran G, Sisodia SS, and Snyder SH,

Amyloid precursor proteins inhibit heme oxygenase activity and augment neurotoxicity in Alzheimer's disease, *Neuron* 28, 461–473, 2000.

87. Kimpara T, Takeda A, Yamaguchi T, Arai H, Okita N, Takase S, Sasaki H, and Itoyama Y, Increased bilirubins and their derivatives in cerebrospinal fluid in Alzheimer's disease, *Neurobiol Aging* 21, 551–554, 2000.
88. Zhuang H, Kim YS, Namiranian K, and Doré S, Prostaglandins of J series control heme oxygenase expression: potential significance in modulating neuroinflammation, *Ann NY Acad Sci* 993, 208–216; discussion 287–288, 2003.
89. Bishop A and Cashman NR, Induced adaptive resistance to oxidative stress in the CNS: a discussion on possible mechanisms and their therapeutic potential, *Curr Drug Metab* 4, 171–184, 2003.
90. Goto S, Sampei K, Alkayed NJ, Doré S, and Koehler RC, Characterization of a new double-filament model of focal cerebral ischemia in heme oxygenase-2-deficient mice, *Am J Physiol Regul Integr Comp Physiol* 285, R222–230, 2003.
91. Doré S, Goto S, Sampei K, Blackshaw S, Hester LD, Ingi T, Sawa A, Traystman RJ, Koehler RC, and Snyder SH, Heme oxygenase-2 acts to prevent neuronal cell death in brain cultures and following transient cerebral ischemia, *Neuroscience* 99, 587–592, 2000.
92. Calabrese V, Butterfield DA, and Stella AM, Nutritional antioxidants and the heme oxygenase pathway of stress tolerance: novel targets for neuroprotection in Alzheimer's disease, *Ital J Biochem* 52, 177–181, 2003.
93. Schipper HM, Heme oxygenase-1: role in brain aging and neurodegeneration, *Exp Gerontol* 35, 821–830, 2000.
94. Wagner KR, Sharp FR, Ardizzone TD, Lu A, and Clark JF, Heme and iron metabolism: role in cerebral hemorrhage, *J Cereb Blood Flow Metab* 23, 629–652, 2003.
95. Thompson K, Menzies S, Muckenthaler M, Torti FM, Wood T, Torti SV, Hentze MW, Beard J, and Connor J, Mouse brains deficient in H-ferritin have normal iron concentration but a protein profile of iron deficiency and increased evidence of oxidative stress, *J Neurosci Res* 71, 46–63, 2003.
96. Krieger J and Schroeder C, Iron, brain and restless legs syndrome, *Sleep Med Rev* 5, 277–286, 2001.
97. Ponka P, Hereditary causes of disturbed iron homeostasis in the central nervous system, *Ann NY Acad Sci* 1012, 267–281, 2004.
98. Xu X, Pin S, Gathinji M, Fuchs R, and Harris ZL, Aceruloplasminemia: an inherited neurodegenerative disease with impairment of iron homeostasis, *Ann NY Acad Sci* 1012, 299–305, 2004.
99. Crompton DE, Chinnery PF, Fey C, Curtis AR, Morris CM, Kierstan J, Burt A, Young F, Coulthard A, Curtis A, Ince PG, Bates D, Jackson MJ, and Burn J, Neuroferritinopathy: a window on the role of iron in neurodegeneration, *Blood Cells Mol Dis* 29, 522–531, 2002.
100. Reid G, Association of sudden infant death syndrome with grossly deranged iron metabolism and nitric oxide overload, *Med Hypotheses* 54, 137–139, 2000.
101. Artinian LR, Ding JM, and Gillette MU, Carbon monoxide and nitric oxide: interacting messengers in muscarinic signaling to the brain's circadian clock, *Exp Neurol* 171, 293–300, 2001.

102. Dioum EM, Rutter J, Tuckerman JR, Gonzalez G, Gilles-Gonzalez MA, and McKnight SL, NPAS2: a gas-responsive transcription factor, *Science* 298, 2385–2387, 2002.
103. Ilzecka J and Stelmasiak Z, Serum bilirubin concentration in patients with amyotrophic lateral sclerosis, *Clin Neurol Neurosurg* 105, 237–240, 2003.
104. Gordh T, Sharma HS, Azizi M, Alm P, and Westman J, Spinal nerve lesion induces upregulation of constitutive isoform of heme oxygenase in the spinal cord. An immunohistochemical investigation in the rat, *Amino Acids* 19, 373–381, 2000.
105. Beschorner R, Adjodah D, Schwab JM, Mittelbronn M, Pedal I, Mattern R, Schluesener HJ, and Meyermann R, Long-term expression of heme oxygenase-1 (HO-1, HSP-32) following focal cerebral infarctions and traumatic brain injury in humans, *Acta Neuropathol* (Berl) 100, 377–384, 2000.
106. Browne SE, Ferrante RJ, and Beal MF, Oxidative stress in Huntington's disease, *Brain Pathol* 9, 147–163, 1999.
107. Cooke RW, New approach to prevention of kernicterus, *Lancet* 353, 1814–1815, 1999.
108. Hara E, Takahashi K, Tominaga T, Kumabe T, Kayama T, Suzuki H, Fujita H, Yoshimoto T, Shirato K, and Shibahara S, Expression of heme oxygenase and inducible nitric oxide synthase mRNA in human brain tumors, *Biochem Biophys Res Commun* 224, 153–158, 1996.
109. Solowiej E, Kasprzycka-Guttman T, Fiedor P, and Rowinski W, Chemoprevention of cancerogenesis: the role of sulforaphane, *Acta Pol Pharm* 60, 97–100, 2003.
110. Liang DY, Li X, and Clark JD, Formalin-induced spinal cord calcium/calmodulin-dependent protein kinase II alpha expression is modulated by heme oxygenase in mice, *Neurosci Lett* 360, 61–64, 2004.
111. Carratu P, Pourcyrous M, Fedinec A, Leffler CW, and Parfenova H, Endogenous heme oxygenase prevents impairment of cerebral vascular functions caused by seizures, *Am J Physiol Heart Circ Physiol* 285, H1148–1157, 2003.
112. Zhuang H, Littleton-Kearney MT, and Doré S, Characterization of heme oxygenase in adult rodent platelets, *Curr Neurovascular Res* 2, 163–168, 2005.
113. Law A, Doré S, Blackshaw S, Gauthier S, and Quirion R, Alteration of expression levels of neuronal nitric oxide synthase and haem oxygenase-2 messenger RNA in the hippocampi and cortices of young adult and aged cognitively unimpaired and impaired Long-Evans rats, *Neuroscience* 100, 769–775, 2000.
114. Sato M, Ray PS, Maulik G, Maulik N, Engelman RM, Bertelli AA, Bertelli A, and Das DK, Myocardial protection with red wine extract, *J Cardiovasc Pharmacol* 35, 263–268, 2000.
115. Fang JG, Lu M, Chen ZH, Zhu HH, Li Y, Yang L, Wu LM, and Liu ZL, Antioxidant effects of resveratrol and its analogues against the free-radical-induced peroxidation of linoleic acid in micelles, *Chemistry* 8, 4191–4198, 2002.

116. Belguendouz L, Fremont L, and Linard A, Resveratrol inhibits metal ion-dependent and independent peroxidation of porcine low-density lipoproteins, *Biochem Pharmacol* 53, 1347–1355, 1997.
117. Ryter SW and Tyrrell RM, The heme synthesis and degradation pathways: role in oxidant sensitivity. Heme oxygenase has both pro- and antioxidant properties, *Free Radical Biol Med* 28, 289–309, 2000.

21 Protective Effects of Resveratrol against Ischemia-Reperfusion

Dipak K. Das

CONTENTS

INTRODUCTION

A number of studies have been devoted to understanding the cause of the so-called "French paradox", the anomaly whereby in several parts of France and other Mediterranean countries the morbidity and mortality related to coronary heart diseases in absolute value and in consideration of its rate to other manners of death is significantly lower than that in other developed countries, despite the high consumption of fat and saturated fatty acids [1,2]. The cause of this cardioprotective effect is believed to be, among others, regular consumption of wine. Wines, especially red wines, consist of about 1800 to 3000 mg/l of polyphenolic compounds of which only about 0.5 to 8 mg/l is resveratrol [3]. Many polyphenolic compounds are potent antioxidants capable of scavenging free radicals and inhibiting lipid peroxidation both *in vitro* and *in vivo* [4].

Resveratrol, a polyphenol phytoalexin (*trans*-3,4′,5-trihydroxystilbene) abundantly found in grape skins and in wines, possesses diverse biochemical and physiological actions, which include estrogenic, antiplatelet, and anti-inflammatory properties [7,8]. *Trans*-resveratrol was originally identified as the active ingredient of an oriental herb ('Ko-jo-kon') used for treatment of a wide variety of diseases including dermatitis, gonorrhea, fever, hyperlipidemia, atherosclerosis, and inflammation. Resveratrol can be found in a variety of plants including grapevines, peanuts, and berries.

Recently, resveratrol was found to protect kidney, heart, and brain from ischemia-reperfusion injury [9–11]. The cardioprotective mechanism of resveratrol included its role as an intracellular antioxidant and anti-inflammatory agent and the involvement of nitric oxide [12]. In kidney cells, resveratrol was also found to exert its protective action through upregulation of nitric oxide (NO) [13,14]. This chapter discusses the possible mechanisms of resveratrol-mediated cardioprotection.

RESVERATROL

RESVERATROL: A GRAPE-DERIVED NATURAL ANTIOXIDANT

Resveratrol, a polyphenolic antioxidant, possesses a spectrum of physiological activities including its ability to protect tissues such as brain, kidney, and heart from ischemic injury, and its role as a cancer chemoprotective agent. Its biological activities include its role as a neuroprotective, anti-inflammatory, and antiviral compound [14,15]. This compound also exerts diverse biochemical and physiological actions, which include estrogenic, antiplatelet, and antiinflammatory properties [16,17]. Recently, resveratrol obtained from grape skins and wines has been found to protect the heart from ischemia-reperfusion injury [18,19]. In kidney cells, resveratrol was found to exert its protective action through upregulation of NO [20].

Resveratrol can scavenge some reactive oxygen species (ROS). Although it possesses antioxidant properties, it does not function as a strong antioxidant *in vitro* [21]. While resveratrol behaves as a poor *in vitro* scavenger of ROS, it functions as a potent antioxidant *in vivo*. The *in vivo* antioxidant property of resveratrol is probably due to its ability to upregulate NO synthesis, which in turn functions as an *in vivo* antioxidant by its ability to scavenge superoxide radicals. In the ischemic-reperfused heart, brain, and kidney, resveratrol has been found to induce NO synthesis and lower oxidative stress [22,23].

Most of the *in vivo* antioxidant properties of resveratrol are believed to be achieved through its ability to upregulate NO. NO, with an unpaired electron, behaves as a potent antioxidant *in vivo*. NO can rapidly react at or near the diffusion-limited rate (6.7×10^9 l/mol/sec) with the superoxide anion (O_2^-), which is presumably formed in the ischemic reperfused myocardium [24]. The affinity of NO for O_2^- is far greater than superoxide dismutase (SOD) for O_2^-. In fact, NO may compete with SOD for O_2^-, thereby removing O_2^- and sparing SOD, further supporting its antioxidant role. Thus, resveratrol functions as an excellent *in vivo* antioxidant through NO.

In cellular systems, resveratrol can scavenge ROS. For example, resveratrol was found to inhibit 12-*O*-tetradecanoylphosbol-13-acetate (TPA)-induced free radical formation in cultured HL-60 cells [25]. In DU 145 prostate cancer cell line, resveratrol inhibited growth accompanied by a reduction in NO production and inhibition of inducible nitric oxide synthase (iNOS) [26]. Resveratrol also inhibited the formation of O_2^- and H_2O_2 produced by macrophages stimulated by lipopolysaccharide (LPS) or TPA [27]. In a related study, resveratrol inhibited reactive oxygen intermediates and lipid peroxidation induced by tumor necrosis factor (TNF) in a wide variety of cells [28].

RESVERATROL: A PHYTOESTROGEN

Resveratrol has been recognized as a phytoestrogen based on its structural similarities to diethylstilbestrol. Resveratrol can bind to estrogen receptors (ERs) thereby activating transcription of estrogen-responsive reporter genes transfected into cells [29,30]. Resveratrol was shown to function as a superagonist when combined with estradiol (E_2), and induce the expression of estrogen-regulated genes [30]. However, several other studies showed conflicting results. In another related study using the same cell line, resveratrol showed antiestrogen activity, because it suppressed progesterone receptor expression induced by E_2. Another recent study showed that both isomers of resveratrol possessed superestrogenic activity only at moderate concentration ($>10\,\mu M$), while at lower concentration ($<1\,\mu M$) antiestrogenic effects prevailed [31].

Most of the *in vivo* studies failed to confirm the estrogenic potential of resveratrol. At physiologic concentration, resveratrol could not induce any changes in uterine weight, uterine epithelial cell height, or serum cholesterol [32]. Only at a very high concentration did resveratrol modulate the serum cholesterol lowering activity of E_2 [32]. In another study, resveratrol given orally as well as subcutaneously did not affect uterus weight at any concentration ranging from the lowest to highest doses (0.03 to 120 mg/kg/day) [33]. In another related study, resveratrol reduced uterine weight and decreased the expression of ERα mRNA and protein and PR mRNA [34].

In contrast, resveratrol was found to possess estrogenic properties in stroke-prone spontaneously hypertensive rats [35]. When ovariectomized rats were fed resveratrol at a concentration of 5 mg/kg/day, it attenuated an increase in systolic blood pressure. In concert, resveratrol enhanced endothelin-dependent vascular relaxation in response to acetylcholine and prevented ovariectomy-induced decreases in femoral bone strength in a manner similar to estradiol. Recently, resveratrol was found to act as an ER agonist in breast cancer cells stably transfected with ERα [36]. While more data accumulate on the estrogenic behavior of resveratrol, the controversy continues.

Resveratrol: Effective Against a Variety of Degenerative Diseases

As mentioned earlier, several recent studies determined the cardioprotective abilities of resveratrol. Both in acute experiments and chronic models, resveratrol was found to reduce myocardial ischemia-reperfusion injury [37,38]. In addition to heart cells, resveratrol has also been found to protect kidney and brain cells from ischemia-reperfusion injury. Similar to the heart, the ability of resveratrol to stimulate NO production during ischemia-reperfusion is believed to play a crucial role in its ability to protect kidney cells from ischemia-reperfusion injury [39]. The maintenance of constitutive NO release is a critical factor in the recovery of function after an ischemic injury. Release of constitutive NO is significantly reduced after ischemia-reperfusion, and maintenance of NO by any means such as induction of NO production with L-arginine can restore the postischemic myocardial function [40].

While resveratrol protects brain, kidney, and heart cells, it preferentially kills cancer cells. For example, intraperitoneal administration of resveratrol caused an increase in the G2/M phase of the cell cycle and apoptosis and reduced tumor growth [41]. In oral squamous carcinoma cells, resveratrol caused growth inhibition, both alone and in combination with quercetin [42]. In a recent study, resveratrol inhibited the growth of highly metastatic B16-BL6 melanoma cells [43]. In a rat colon carcinogenesis model, resveratrol induced proapoptotic bax expression in colon aberrant cryptic foci [44]. In fact, resveratrol was found to affect three major stages of carcinogenesis and inhibit the formation of preneoplastic lesions in a mouse mammary organ culture model [45].

A recent study demonstrated inhibition of the growth of CagA+ strains of *Helicobacter pylori in vitro* [46]. In another study, resveratrol inhibited the growth of 15 clinical strains of *H. pylori in vitro* and it was suggested that the anti-*H. pylori* activity of resveratrol may play a role in its chemopreventive effects [47]. Resveratrol was also found to increase the activity of antiaging gene SIRT1 13-fold [48]. Resveratrol-mediated activation of life-extending genes in human cells may open a new horizon of resveratrol research.

RESVERATROL AND THE HEART

The role of NO in the cardioprotective abilities of grape products first became apparent in 1995, when certain wines, grape juices, grape skins, or their components were found to relax precontracted smooth muscle of intact rat aortic rings but had no effect on aortas in which the endothelium had been removed [49]. The extracts of wine or grape increased cGMP levels in intact vascular tissue, and both relaxation and the increase in cGMP were reversed by NG-monomethyl-L-arginine and NG-nitro-L-arginine, competitive inhibitors of the synthesis of the endothelium-derived relaxing factor, NO, suggesting that vasorelaxation induced by grape products was mediated by the NO–cGMP pathway. A direct role of NO was shown from a study that found resveratrol-mediated increase in NOS activity in cultured pulmonary artery endothelial cells, suggesting that resveratrol could afford cardioprotection by affecting the expression of NOS [50]. A related study demonstrated beneficial effects of astringinin, a resveratrol analog, on ischemia and reperfusion damage in rat heart in an NO-dependent manner [51]. In this study, pretreatment with astringinin dramatically reduced the incidence and duration of ventricular tachycardia (VT) and ventricular fibrillation (VF) during either ischemia or ischemia-reperfusion period. Astringinin at 2.5×10^{-5} and 2.5×10^{-4} g/kg completely prevented the mortality of animals during ischemia or ischemia-reperfusion. During the same period, astringinin pretreatment also increased NO and decreased LDH levels in the carotid blood. In animals subjected to 4 h coronary occlusion, the myocardial infarct size was reduced significantly by astringinin. Consistent with these results, resveratrol was found to protect isolated perfused working rat heart through the upregulation of iNOS. The cardioprotective ability of resveratrol was abolished with an iNOS inhibitor, aminoguanidine. Resveratrol failed to provide cardioprotection in iNOS knockout mice devoid of any copy of iNOS gene [52]. In a recent study, resveratrol reduced myocardial ischemia-reperfusion injury in both an iNOS-dependent and iNOS-independent manner [53]. In this study, resveratrol modulated iNOS, eNOS, and nNOS and provided cardioprotection. L-NAME, an NO inhibitor, abolished cardioprotective effects of resveratrol including infarction size-lowering ability, but failed to alter resveratrol's ability to reduce the incidence of ventricular arrhythmias.

The cardioprotective ability of resveratrol has become apparent from its antiinflammatory function in the ischemic heart. In a recent study, resveratrol significantly improved postischemic ventricular function and reduced myocardial infarct size compared to a nontreated control group [54].

The amounts of proadhesive molecules including soluble intracellular adhesion molecule-1 (sICAM-1), endothelial leukocyte adhesion molecule-1 (sE-selectin), and vascular cell adhesion molecule-1 (sVCAM-1) were each significantly decreased during reperfusion in the resveratrol group. L-NAME completely abolished such beneficial effects of resveratrol. The results support an antiinflammatory action of resveratrol through an NO-dependent mechanism.

The antiinflammatory role of resveratrol is also evident from a recent study where resveratrol effectively suppressed the aberrant expression of tissue factor (TF) and cytokines in vascular cells [55]. Resveratrol, in a dose-dependent manner, inhibited the expression of TF in endothelial cells stimulated with a variety of agonists, including interleukin (IL)-1β, TNFα, and LPS. Nuclear run-on analysis in endothelial cells showed that resveratrol inhibited TF expression at the level of transcription. In another recent study, resveratrol reduced the agonist-induced increase in TF mRNA in endothelial and mononuclear cells [56]. The inhibition of TF mRNA originated from a reduction in nuclear binding activity of the transacting factor c-Rel/p65. Western blot analysis revealed that the diminished c-Rel/p65 activity was dependent upon inhibition of degradation of the c-Rel/p65 inhibitory protein IκBα. In another related study, resveratrol inhibited platelet aggregation induced by collagen, thrombin, or ADP [57]. A study also found that *trans*-resveratrol interfered with the release of inflammatory mediators by activated PMN and downregulated adhesion-dependent thrombogenic PMN functions [58]. Hypercholesterolemic rabbits showed enhanced ADP-induced platelet aggregation, which was blocked by resveratrol. The antiinflammatory action of resveratrol is also apparent from its ability to prevent the generation of eicosanoids involved in pathological processes. For example, resveratrol could inhibit dioxygenase activity of lipooxygenase, and oxidized resveratrol was as efficient a lipoxygenase inhibitor as in its reduced form [59]. In another study, *trans*-resveratrol inhibited the synthesis of TxB2, HHT, and to a lesser extent 12-HETE, from arachidonate in a dose-dependent manner [60].

Recently, resveratrol was found to protect the heart against ROS-mediated menadione toxicity through the induction of NAD(P)H:quinone oxidoreductase, also known as DT-diaphorase, a detoxifying enzyme for quinone-containing substances, due to its ability to prevent their one-electron reduction and consequent generation of ROS [61]. The cardio-protective effect of resveratrol was also attributed to its ability to upregulate catalase activity in the myocardium. Resveratrol functions as an *in vivo* antioxidant and can scavenge peroxyl radicals in the heart [62,63]. A commercial preparation of resveratrol (Protykin) also scavenged peroxyl radicals and protected the heart from ischemia-reperfusion injury [64].

Resveratrol appears to induce an antiapoptotic signal for the protection of the heart. In porcine coronary arteries, short-term treatment with resveratrol significantly inhibited mitogen-activated protein kinase (MAPK) activities and immunoblot analyses revealed consistent reduction in the phosphorylation of extracellular signal-regulated protein kinase (ERK) 1/2, c-Jun NH2-terminal kinase (JNK)-1, and p38 MAPK [65]. The same study found that resveratrol attenuated basal and ET-1-evoked protein tyrosine phosphorylation. The antiapoptotic function of resveratrol is further supported by several other studies, which have demonstrated reduction of apoptotic cardiomyocytes in the ischemic-reperfused heart that had been pretreated with resveratrol.

PRECONDITIONING WITH RESVERATROL

MYOCARDIAL PROTECTION WITH PRECONDITIONING

Preconditioning (PC) is the most powerful technique for cardioprotection ever known [66–70]. The most generalized method of classic PC is mediated by cyclic episodes of several short durations of reversible ischemia, each followed by another short duration of reperfusion. In most laboratories including our own, PC is achieved by four cycles of 5 min of ischemia each followed by 10 min of reperfusion [71,72]. Such PC makes the heart resistant to subsequent lethal ischemic injury.

The mechanisms underlying cardiac PC have been studied extensively. Several regulatory pathways have been identified in different systems. Three important factors, adenosine A1 receptor, multiple kinases including protein kinase C (PKC), MAPKs, and tyrosine kinases, and mitochondrial ATP-sensitive potassium (K_{ATP}) channels are known to play a crucial role in PC-mediated cardioprotection. For example, cardioprotection achieved by PC can be abolished by adenosine A1 receptor antagonists [73]. Adenosine A1 receptor agonists can limit myocardial infarct size [74]. While there is a general agreement regarding the beneficial role of adenosine on ischemic tissue, the adenosine hypothesis remains controversial. To reconcile the adenosine hypothesis, an argument has been made that adenosine could trigger a secondary mechanism such as activation of Gi protein, which in turn could open the K_{ATP} channels. This hypothesis is inconsistent with several findings that inhibition of K_{ATP} channels blocks the effects of PC and a K_{ATP} channel opener can simulate ischemic PC [75,76].

Another intriguing hypothesis stemmed from the concept of stimulation of an endogenous protective mechanism by myocardial adaptation to ischemic stress. PC has been found to induce the expression of endogenous antioxidant enzymes such as SOD and glutathione peroxidase-1 (GSHPx-1) [77,78] and heat shock proteins (HSPs) such as HSP 27, 32, and 70 [79,80]. Additionally, PC potentiates a signal transduction cascade by inhibiting death signal and activating survival signal. Thus, several pro- and anti-apoptotic genes and transcription factors including Jnk-1, c-Jun, NFκB, and AP-1 are likely to play a crucial role in PC [81–83]. Recently, NO has been found to act as the mediator of PC [84,85].

It appears that ROS play a crucial role in PC, which is realized probably by the activation of adenosine A1 receptor, stimulating PKC and MAPKs and by opening mitochondrial K_{ATP} channels. Recent studies from our laboratory determined that resveratrol may function as a pharmacological PC agent as it fulfills several criteria for PC including activation of adenosine A1 and A3 receptors, PKC and MAPKs, and K_{ATP} channels.

Adenosine accumulates in tissues under metabolic stress. In myocardial cells, the nucleoside interacts with various receptor subtypes [A(1), A(3), and probably A(2A) and A(2B)] that are coupled, via G proteins, to multiple effectors, including enzymes, channels, transporters, and cytoskeletal components. Studies using adenosine receptor agonists and antagonists, as well as animals overexpressing the A(1) receptor indicate that adenosine exerts antiischemic action. Adenosine released during PC by short periods of ischemia followed by reperfusion induces cardioprotection to a subsequent sustained ischemia. This protective action is mediated by A(1) and A(3) receptor subtypes and involves the activation and translocation of PKC to sarcolemmal and to mitochondrial membranes. PKC activation leads to an increased opening of K_{ATP} channels. Other effectors possibly contributing to cardioprotection by adenosine or PC, and which seem particularly involved in the delayed (second window of) protection, include MAPKs, HSPs, and iNOS and eNOS. Because of its antiischemic effects, adenosine has been tested as a protective agent in clinical interventions such as PTCA, CABG, and tissue preservation, and was found in most cases to enhance the postischemic recovery of function. The mechanisms underlying the role of adenosine and of mitochondrial function in PC are not completely clear, and uncertainties remain concerning the role played by newly identified potential effectors such as free radicals, the sarcoplasmic reticulum, etc. In addition, more studies are needed to clarify the signaling mechanisms by which A(3) receptor activation or overexpression may promote apoptosis and cellular injury.

Ischemic PC is a receptor-mediated process, and is realized via signal transduction pathways. Several investigators have proposed a unifying hypothesis that activation of PKC represents a link between cell surface receptor activation and a putative end-effector sarcolemmal or mitochondrial K_{ATP} channel [86,87]. The possible involvement of protein tyrosine kinases in PC was proposed for the first time by Maulik et al. [88] and subsequently confirmed by Ping et al. [89]. It is now increasingly clear that protein tyrosine kinases play a crucial role in mediating PC in some animal species. Protein tyrosine kinases may act in parallel to [90,91], downstream of [92,93], or upstream of [94] PKC in eliciting PC. However, the identity of exact members of protein tyrosine kinases involved in PC remains unclear. Among a large number of tyrosine kinases, Src family tyrosine kinase has received much attention [95]. Src tyrosine kinase has been implicated in the mechanism of cell survival and death, which is regulated by complex signal transduction processes [96]. Rapid activation of Src family tyrosine kinases after ischemia has also been documented in the isolated guinea pig heart [97]. Src family tyrosine kinases are activated by stimulation of G-protein-coupled receptors [98], an increase in intracellular Ca^{2+} [99], oxidative stress [100], and enhanced NO synthesis [101], all of which can be elicited by PC challenges.

The intracellular signaling mechanisms that mediate preconditioning require one or more members of MAPK cascades. Among the three distinct MAPK families, stress-activated protein kinase (SAPK), also known as JNK, and p38 MAPK are known to be regulated by extracellular stresses including environmental stress, oxidative stress, heat shock, and ultraviolet radiation [102]. JNKs and p38 MAPK appear to be involved in distinct cellular function, because they possess different cellular targets and are located on different signaling pathways. Thus, JNKs activate c-Jun while p38 MAPK stimulates MAPKAP kinase 2 [103]. A recent study demonstrated that PC triggered a tyrosine kinase-regulated signaling pathway leading to the translocation and activation of p38 MAPK and MAPKAP kinase 2 [104].

Activation of K_{ATP} channels appears to be an adaptive mechanism that protects the myocardium against ischemia-reperfusion injury [105]. The activation of this ion channel is at least partially responsible for the increase in outward K^+ currents, shortening of action potential duration (APD), and an increase in extracellular K^+ concentration during anoxic and globally ischemic condition [106]. It appears that delayed PC, irrespective of PC stimulus, is always mediated by K_{ATP} channels [107]. It was shown that a mitochondrial K_{ATP} channel opener, diazoxide, significantly reduced the rate of cell death following simulated ischemia in adult ventricular cardiac myocytes [108]. The protective effects of diazoxide were blocked by 5-HD, a selective blocker of the mitochondrial K_{ATP} channels. Intravenous injection of diazoxide 10 min before ischemia greatly reduced infarct size in the rabbit heart [109].

Pharmacological Preconditioning with Resveratrol

Unlike pharmacological therapeutic interventions, PC protects the heart by upregulating its endogenous defense mechanisms [110]. Unfortunately, ischemic PC-mediated cardioprotection has a limited life, classic or early PC lasting for several hours, and delayed PC lasting for several days [111]. There is a definite need to identify a pharmacological PC agent to render the PC stimulus everlasting. Recently, our researchers and others found that monophosphoryl lipid A (MLA) induces dose-dependent cardioprotection against myocardial infarction [112]. Such cardioprotection was achieved through the ability of MLA to upregulate endogenous NO formation [113], which also plays an essential role as mediator of ischemic PC. Similar to this, resveratrol was also found to protect the ischemic myocardium through NO, because inhibition of NO with L-NAME abolished the cardioprotective effects of resveratrol [16]. In this study, PC of the heart with resveratrol provided cardioprotection as evidenced by improved postischemic ventricular functional recovery (developed pressure and aortic flow) and reduced myocardial infarct size and cardiomyocyte apoptosis. Resveratrol-mediated cardioprotection was completely abolished by both L-NAME and AG (iNOS inhibitor). Resveratrol caused an induction of the expression of iNOS mRNA beginning at 30 min after reperfusion, increasing steadily up to 60 min of reperfusion and then decreasing progressively up to 2 hours after reperfusion. Preperfusion of the heart with AG almost completely blocked the induction of iNOS. In this study, resveratrol also diminished the amount of ROS activity as evidenced by reduced malonaldehyde formation. The results of this study demonstrated that resveratrol could pharmacologically precondition the heart in an NO-dependent manner.

Another related study showed that iNOS knockout mice could not be preconditioned with resveratrol, further indicating that this polyphenol provides cardioprotection through NO, and specifically through the induction of iNOS [52]. In this study, control experiments were performed with wild-type and iNOS knockout hearts, which were not treated with resveratrol. Resveratrol-treated wild-type mouse hearts displayed significant improvement in postischemic ventricular functional recovery compared to nontreated hearts. Both resveratrol-treated and nontreated iNOS knockout mouse hearts resulted in relatively poor recovery in ventricular function compared to wild-type resveratrol-treated hearts. Myocardial infarct size was lower in the resveratrol-treated wild-type mouse hearts as compared to the other group of hearts. In concert, the number of apoptotic cardiomyocytes was lower in the wild-type mouse hearts treated with resveratrol. Cardioprotective effects of resveratrol were abolished when the wild-type mouse hearts were simultaneously perfused with aminoguanidine, an iNOS inhibitor. Resveratrol induced the expression of iNOS in the wild-type mouse hearts, but not in the iNOS knockout hearts, after only 30 min of reperfusion. Expression of iNOS remained high even after 2 hours of reperfusion. Resveratrol-treated wild-type mouse hearts were subjected to lower amount of oxidative stress as evidenced by reduced amount of malonaldehyde content in these hearts compared to iNOS knockout and untreated hearts. These results demonstrated that resveratrol was unable to precondition iNOS knockout mouse hearts while it could successfully precondition the wild-type mouse hearts indicating an essential role of iNOS in resveratrol PC of the heart.

The above results strongly support the notion that resveratrol-mediated cardioprotection is achieved by PC. In normal tissue, resveratrol downregulates iNOS. However, as seen from the above studies, in ischemic heart, resveratrol induces iNOS and with time it also induces eNOS [18,23,53], an observation that is consistent with the ischemic PC [84,85]. Recently resveratrol was found to protect the ischemic heart through the upregulation of adenosine A1 and A3 receptors [114], a property shared by ischemic PC. The results of this study demonstrated significant cardioprotection with resveratrol as evidenced by improved ventricular recovery, and reduced infarct size and cardiomyocyte apoptosis. An adenosine A_1 receptor blocker, CPT, and an adenosine A_3 receptor blocker, MRS 1191, but not adenosine A_{2a} receptor blocker CSC, abrogated the cardioprotective abilities of resveratrol suggesting a role of adenosine A_1 and A_3 receptors in resveratrol PC. Resveratrol induced the expression of Bcl_2, and caused its phosphorylation along with phosphorylation of CREB, Akt, and Bad. CPT blocked the phosphorylation of Akt and Bad without affecting CREB, while MRS 1191 blocked phosphorylation of all the compounds including CREB. A phosphoinositide 3-kinase (PI3K) inhibitor, LY 294002, partially blocked the cardioprotective abilities of resveratrol. The results indicate that resveratrol preconditions the heart through the activation of adenosine A_1 and A_3 receptors, the former transmitting a survival signal through PI3K-Akt-Bcl_2 signaling pathway, while the latter protects the heart through a CREB-dependent Bcl_2 pathway in addition to Akt-Bcl_2 pathway (Figure 21.1).

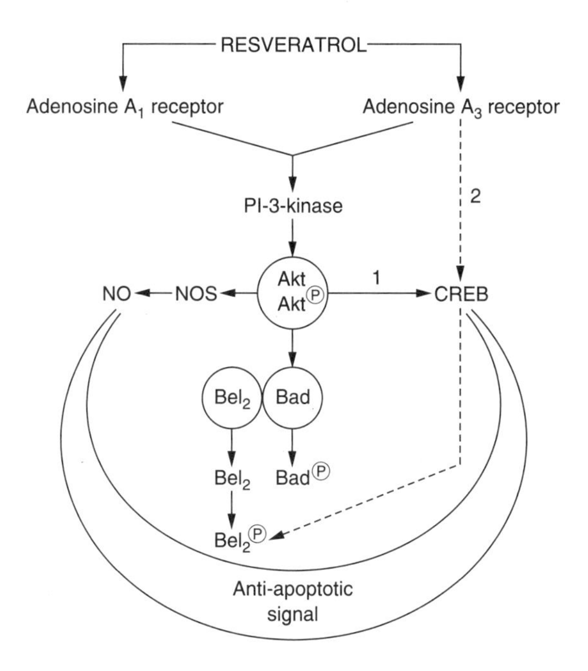

FIGURE 21.1 Possible mechanisms of action for resveratrol preconditioning of the heart.

SUMMARY AND CONCLUSION

It should be clear from the above discussion that resveratrol has two faces: it protects cells through augmentation of NO in the ischemic tissue by functioning as a pharmacological PC agent and it selectively kills cancer cells. It is tempting to speculate that NO may play a crucial role in such dual behavior of resveratrol. Resveratrol shares many properties of NO and, similar to resveratrol, NO also has two entirely opposite faces. While constitutive expression of NO is protective, it is equally destructive to cells. Evidence is rapidly accumulating in support of the cardioprotective role of resveratrol. It appears that resveratrol-mediated cardioprotection is achieved through the PC effect, rather than direct protection. As mentioned earlier, PC, the state-of-the-art method of cardioprotection, is achieved through stress conditioning by subjecting the heart to a therapeutic amount of stress, thereby disturbing normal cardiovascular homeostasis and reestablishing a modified system with upregulated cardiac defense that can withstand subsequent stress insult. The most common and well-studied method of PC is ischemic PC. Such a method, however, cannot be extrapolated to clinics, because of practical problems. There has been a desperate search for pharmaceutical PC agents. Resveratrol appears to fulfill the definition of a pharmaceutical PC compound. Future studies will reveal the mystery of the two faces of resveratrol and pinpoint its precise mechanism of action in cardioprotection.

ACKNOWLEDGMENTS

This study was supported in part by NIH HL 22559, HL 33889, HL 56803, and HL 56322.

REFERENCES

1. Gaziano JM, Buring JE, Breslow JL, Goldhaber SZ, Rosner B, VanDenburgh M, et al., Moderate alcohol intake, increased levels of high-density lipoprotein and its sub-fractions and decreased risk of myocardial infarction, *N Engl J Med* 329, 1829–1834, 1993.
2. Renaud S and De Lorgeril M, Wine, alcohol, platelets and the French Paradox for coronary heart disease, *Lancet* 339, 1523–1526, 1992.
3. Soleas GJ, Diamandis EP, and Goldberg DM, Wine as a biological fluid: history, production and role in disease prevention, *J Clin Lab Anal* 11, 287–313, 1997.
4. Saija A, Scalese M, and Laiza M, Flavonoids as antioxidant agents: importance of their interaction with biomembranes, *Free Radical Biol Med* 19, 481–486, 1995.
5. Das DK and Maulik N, Evaluation of antioxidant effectiveness in ischemia reperfusion tissue injury, *Methods Enzymol* 233, 601–610, 1994.

6. Cordis GA, Maulik G, Bagchi D, Riedel W, and Das DK, Detection of oxidative DNA damage to ischemic reperfused rat hearts by 8-hydroxy-deoxyguanosine formation, *Mol Cell Cardiol* 30, 1939–1944, 1998.
7. Bertelli AAE, Giovannini L, De Caterina R, Bernini W, Migliori M, Fregoni M, Bavaresco L, and Bertelli A, Antiplatelet activity of cis-resveratrol, *Drugs Exp Clin Res* 22, 61–63, 1996.
8. Ferrero ME, Bertelli AE, Fulgenzi A, Pellegatta F, Corsi MM, Bonfrate M, Ferrara F, DeCaterina R, Giovannini L, and Bertelli A, Activity *in vitro* of resveratrol on granulocyte and monocyte adhesion to endothelium, *Am J Clin Nutr* 68, 1208–1214, 1998.
9. Bastianetto S, Zheng WH, and Quirion R, Neuroprotective abilities of resveratrol and other red wine constituents against nitric-oxide-related toxicity in cultured hippocampal neurons, *Br J Pharmacol* 131, 711–720, 2000.
10. Das DK, Sato M, Ray PS, Maulik G, Engelman RM, Bertelli AAE, and Bertelli A, Cardioprotection with red wine: role of polyphenolic antioxidants, *Drugs Exp Clin Res* 25, 115–120, 1999.
11. Giovannini L, Migliori M, Longoni BM, Das DK, Bertelli AAE, Panichi V, Filippi, and Bertelli A, Resveratrol, a polyphenol found in wine, reduces ischemia reperfusion injury in rat kidneys, *J Cardiovasc Pharmacol* 37, 262–270, 2001.
12. Ray PS, Maulik G, Cordis GA, Bertelli AAE, Bertelli A, and Das DK, The red wine antioxidant resveratrol protects isolated rat hearts from ischemia reperfusion injury, *Free Radical Biol Med* 27, 160–169, 1999.
13. Guo Y, Jones WK, Xuan YT, Tang XL, Bao W, Wu WJ, Han H, Laubach VE, Ping P, Yang Z, Qiu Y, and Bolli R, The late phase of ischemic preconditioning is abrogated by targeted disruption of the inducible NO synthase gene, *Proc Natl Acad Sci USA* 96, 11507–11512, 1999.
14. Ferrero ME, Bertelli AE, Fulgenzi A, Pellegatta F, Corsi MM, Bonfrate M, Ferrara F, DeCaterina R, Giovannini L, and Bertelli A, Activity *in vitro* of resveratrol on granulocyte and monocyte adhesion to endothelium, *Am J Clin Nutr* 68, 1208–1214, 1998.
15. Chen CK and Pace-Asciak CR, Vasorelaxing activity of resveratrol and quercetin in isolated rat aorta, *Gen Pharmacol* 27, 363–366, 1996.
16. Hao HD and He LR, Mechanisms of cardiovascular protection by resveratrol. *J. Med. Food* 7, 290–298, 2005.
17. Gehm BD, McAndrews JM, Chien PY, and Jameson JL, Resveratrol, a polyphenolic compound found in grapes and wine, is an agonist for the estrogen receptor, *Proc Natl Acad Sci USA* 94, 14138–14143, 1997.
18. Hung L, Chen J, Hunag S, Lee R, and Su M, Cardioprotective effect of resveratrol, a natural antioxidant derived from grapes, *Cardiovasc Res* 47, 549–555, 2000.
19. Ignatowicz E and Baer-Dubowska W, Resveratrol, a natural chemo-preventive agent against degenerative diseases, *Polish J Pharmacol* 53, 557–69, 2001.
20. Bertelli AA, Migliori M, Panichi V, Origlia N, Filippi C, Das DK, and Giovannini L, Resveratrol, a component of wine and grapes, in the prevention of kidney disease, *Ann NY Acad Sci* 957, 230–238, 2002.
21. Bhat KPL, Kosmeder JW II, and Pezzuto JM, Biological effects of resveratrol, *Antioxidant Redox Signal* 3, 1041–1064, 2001.

22. Cadenas S and Barja G, Resveratrol, melatonin, vitamin E, and PBN protect against renal oxidative DNA damage induced by the kidney carcinogen $KBrO_3$, *Free Radical Biol Med* 26, 1531–1537, 1999.
23. Hattori R, Otani H, Maulik N, and Das DK, Pharmacological preconditioning with resveratrol: a role of nitric oxide, *Am J Physiol* 282, H1988–H1995, 2002.
24. Engelman DT, Watanabe M, Maulik N, Cordis GA, Engelman RM, Rousou JA, Flack JE, Deaton DW, and Das DK, L-arginine reduces endothelial inflammation and myocardial stunning during ischemia/reperfusion, *Ann Thoracic Surg* 60, 1275–1281, 1995.
25. Lee SK, MbWambo ZH, and Chung H, Evaluation of the antioxidant potential of natural products, *Comb Chem High Throughput Screen* 1, 35–46, 1998.
26. Kampa MA, Hatzoglou A, and Notas G, Wine antioxidant polyphenols inhibit the proliferation of human prostate cancer cell lines, *Nutr Cancer* 37, 223–233, 2000.
27. Martinez J and Moreno JJ, Effect of resveratrol, a natural polyphenolic compound, on reactive oxygen species and prostaglandin production, *Biochem Pharmacol* 59, 865–870, 2000.
28. Manna SK, Mukhopadhyay A, and Aggarwal BB, Resveratrol suppresses TNF-induced activation of nuclear transcription factors NFkB, activator protein-1, and apoptosis: potential role of reactive oxygen intermediates and lipid peroxidation, *J Immunol* 164, 6509–6519, 2000.
29. Wiseman H, The therapeutic potential of phytoestrogens, *Expert Opin Invest Drugs* 9, 1829–1840, 2000.
30. Turner RT, Evans GL, Zhang M, Maran A, and Sibonga JD, Is resveratrol an estrogen agonist in growing rats? *Endocrinology* 140, 50–54, 1999.
31. Basly JP, Marre-Fournier F, and LeBail JC, Estrogenic/antiestrogenic and scavenging properties of (E)- and (Z)-resveratrol, *Life Sci* 66, 769–777, 2000.
32. Turner RT, Evans GL, and Zhang M, Is resveratrol an estrogen agonist in growing rats?, *Endocrinology* 140, 50–54, 1999.
33. Ashby JH, Tinwell W, and Pennie W, Partial and weak oestrogenecity of the red wine constituent resveratrol: consideration of its superagonist activity in MCF-7 cells and its suggested cardiovascular protective effects, *J Appl Toxicol* 19, 39–45, 1999.
34. Freyberger A, Hartmann E, Hildebrand H, and Krotlinger F, Differential response of immature rat uterine tissue to ethinyltradiol and the red wine constituent resveratrol, *Arch Toxicol* 11, 709–715, 2001.
35. Mizutani K, Ikeda K, Kawai Y, and Yamori Y, Resveratrol attenuates ovarectomy-induced hypertension and bone loss in stroke-prone spontaneously hypertensive rats, *J Nutr Sci Vitaminol* 46, 78–83, 2000.
36. Levenson AS, Gehm BD, Pearce ST, HoriguchiJ, Simons LA, Ward JE, Jameson JL, and Jordan VC, Resveratrol acts as an estrogen receptor (ER) agonist in breast cancer cells stably tranfected with Erα, *Int J Cancer* 104, 587–596, 2003.
37. Bagchi D, Das DK, Tosaki A, Bagchi M, and Kothari SC, Benefits of resveratrol in women's health, *Drugs Exp Clin Res* 27, 233–248, 2001.
38. Bradamante S, Barenghi L, Piccinini F, Bertelli AA, De Jonge R, Beemster P, and De Jong JW, Resveratrol provides late-phase cardioprotection by means of

a nitric oxide- and adenosine-mediated mechanism, *Eur J Pharmacol* 465, 115–123, 2003.
39. Nihei T, Miura Y, and Yagasaki K, Inhibitory effect of resveratrol on proteinuria, hypoalbuminemia and hyperlipidemia in nephritic rats, *Life Sci* 68, 2845–2452, 2001.
40. Zhao T, Xi L, Chelliah J, Levasseur MS, and Kukreja RC, Inducible nitric oxide synthase mediates delayed myocardial protection induced by activation of adenosine A_1 receptors, *Circulation* 102, 902–908, 2001.
41. Carbo N, Costelli P, Baccino MF, Lopez-Soriano FJ, and Argiles JM, Resveratrol, a natural product present in wine, decreases tumor growth in a rat tumor model, *Biochem Biophys Res Commun* 254, 739–743, 1999.
42. ElAttar TMA and Virji AS, Modulating effect of resveratrol and quercetin on oral cancer cell growth and proliferation, *Anticancer Drugs* 10, 187–193, 1999.
43. Caltagirone S, Rossi C, Poggi A, Ranelletti FO, Natali PG, Brunetti M, Aiello FB, and Piantelli M, Alavonoids apigenin and quercetin inhibit melanoma growth and metastatic potential, *Int J Cancer* 87, 595–600, 2000.
44. Tessitore L, Davit A, Sarotto I, and Caderni G, Resveratrol depresses the growth of colorectal aberrant crypt foci by affecting bax and p21 expression, *Carcinogenesis* 21, 1619–1622, 2000.
45. Jang M, Cai L, Udeani GO, Slowing KV, Thomas CF, Beecher CWW, Fong HHS, Farnsworth NR, Kinghorn AD, Mehta RG, Moon RC, and Pezzuto JM, Cancer chemopreventive activity of resveratrol, a natural product derived from grapes, *Science* 275, 218–220, 1997.
46. Mahady GB, Pendland SL, and Chadwick LR, Resveratrol and red wine extracts inhibit the growth of CagA+ strains of *Helicobacter pylori in vitro*, *Am J Gastroenterol* 98, 1440–1441, 2003.
47. Mahady GB and Pendland SL, Resveratrol inhibits the growth of *Helicobacter pylori in vitro Am J Gasteroenterol* 95, 1849, 2000.
48. Hall SS, Longevity research, in vino vitalis? Compounds activate life-extending genes, *Science* 301, 1165, 2003.
49. Fitzpatrick DF, Hirschfield SL, and Coffey RG, Endothelium-dependent vasorelaxing activity of wine or other grape products, *Am J Physiol* 265, H774–748, 1993.
50. Hsieh TC, Juan G, Darzynkiewicz Z, and Wu JM, Resveratrol increases nitric oxide synthase, induces accumulation of p53 and p21 (WAF1/CIP1), and suppresses cultured bovine pulmonary artery endothelial cell proliferation by perturbing progression through S and G2, *Cancer Res* 59, 2596–2601, 1999.
51. Hung LM, Chen JK, Lee RS, Liang HC, and Su MJ, Beneficial effects of astringinin, a resveratrol analogue, on the ischemia and reperfusion damage in rat heart, *Free Radical Biol Med* 30, 877–883, 2001.
52. Imamura G, Bertelli AA, Bertelli A, Otani H, Maulik N, and Das DK, Pharmacologic preconditioning with resveratrol: an insight with iNOS knock-out mice, *Am J Physiol* 282, H1996–2003, 2002.
53. Hung LM, Su MJ, and Chen JK, Resveratrol protects myocardial ischemia-reperfusion injury through both NO-dependent and NO-independent mechanisms, *Free Radical Biol Med* 36, 774–781, 2004.
54. Das S, Bertelli AA, Bertelli A, Maulik N, and Das DK, Antiinflammatory action of resveratrol: a novel mechanism of action, *Drugs Exp Clin Res*, in press.

55. Pendurthi UR, Williams JT, and Rao LV, Resveratrol, a polyphenolic compound found in wine, inhibits tissue factor expression in vascular cells: a possible mechanism for the cardiovascular benefits associated with moderate consumption of wine, *Arterioscleros Throm Vasc Biol* 19, 419–426, 1999.
56. Di Santo A, Mezzetti A, Napoleone E, Di Tommaso R, Donati MB, De Gaetano G, and Lorenzet R, Resveratrol and quercetin down-regulate tissue factor expression by human stimulated vascular cells, *J Thromb Haemostasis* 1, 1089–1095, 2003.
57. Wang Z, Huang Y, Zou J, Cao K, Xu Y, and Wu JM, Effects of red wine and wine polyphenol resveratrol on platelet aggregation *in vivo* and *in vitro*, *Int J Mol Med* 9, 77–79, 2002.
58. Rotondo S, Rajtar G, Manarini S, Celardo A, Rotillo D, de Gaetano G, Evangelista V, and Cerletti C, Effects of trans-resveratrol, a natural polyphenolic compound, on human polymorphonuclear leukocyte function, *Br J Pharmacol* 123, 1691–1699, 1998.
59. Pinto MC, Garcia-Barrado JA, and Macias P, Resveratrol is a potent inhibitor of the dioxygenase activity of lipoxygenase, *J Agric Food Chem* 47, 4842–4846, 1999.
60. Pace-Asciak CR, Hahn S, Diamandis EP, Soleas G, and Goldberg DM, The red wine phenolics trans-resveratrol and quercetin block human platelet aggregation and eicosanoid synthesis: implications for protection against coronary heart disease, *Clin Chim Acta* 235, 207–219, 1995.
61. Floreani M, Napoli E, Quintieri L, and Palatini P, Oral administration of trans-resveratrol to guinea pigs increases cardiac DT-diaphorase and catalase activities, and protects isolated atria from menadione toxicity, *Life Sci* 72, 2741–2750, 2003.
62. Sato M, Ray PS, Maulik G, Maulik N, Engelman RM, Bertelli AA, Bertelli A, and Das DK, Myocardial protection with red wine extract, *J Cardiovasc Pharmacol* 35, 263–268, 2000.
63. Shigematsu S, Ishida S, Hara M, Takahashi N, Yoshimatsu H, Sakata T, and Korthuis RJ, Resveratrol, a red wine constituent polyphenol, prevents superoxide-dependent inflammatory responses induced by ischemia/reperfusion, platelet-activating factor, or oxidants, *Free Radical Biol Med* 34, 810–817, 2003.
64. Sato M, Maulik G, Bagchi D, and Das DK, Myocardial protection by protykin, a novel extract of trans-resveratrol and emodin, *Free Radical Res* 32, 135–144, 2000.
65. El-Mowafy AM and White RE, Resveratrol inhibits MAPK activity and nuclear translocation in coronary artery smooth muscle: reversal of endothelin-1 stimulatory effects, *FEBS Lett* 45, 63–67, 1999.
66. Mitchell MB, Meng X, AO Lihua, Brown JM, Harken AH, and Banerjee A, Preconditioning of isolated rat heart mediated by protein kinase C, *Circulation Res* 76, 73–81, 1995.
67. Flack J, Kimura Y, Engelman RM, and Das DK, Preconditioning the heart by repeated stunning improves myocardial salvage, *Circulation* 84 (Suppl II), 369–374, 1991.
68. Sato M, Cordis GA, Maulik N, and Das DK, SAPKs regulation of ischemic preconditioning, *Am J Physiol* 279, H901–H907, 2000.

69. Maulik N, Wei ZJ, Engelman RM, Lu D, Moraru II, Rousou JA, and Das DK, Interleukin-1a preconditioning reduces myocardial ischemic reperfusion injury, *Circulation* 88, 387–394, 1993.
70. Das DK, Engelman RM, and Kimura Y, Molecular adaptation of cellular defenses following preconditioning of the heart by repeated ischemia, *Cardiovasc Res* 27, 578–584, 1993.
71. Hattori R, Maulik N, Otani H, Zhu L, Cordis G, Engelman RM, Siddiqui MAQ, and Das DK, Role of Stat 3 in ischemic preconditioning, *J Mol Cell Cardiol* 33, 1929–1936, 2001.
72. Flack J, Kimura Y, Engelman RM, and Das DK, Preconditioning the heart by repeated stunning improves myocardial salvage, *Circulation* 84 (Suppl III), 369–374, 1991.
73. Tanno M, Tsuchida A, Nozawa Y, Matsumoto T, Hasegawa T, Miura T, and Shimamoto K, Roles of tyrosine kinase and protein kinase C in infarct size limitation by repetitive ischemic preconditioning in the rat, *J Cardiovasc Pharmacol* 35, 345–352, 2000.
74. Fryer RM, Schultz JEJ, Hsu AK, and Gross GJ, Importance of PKC and tyrosine kinase in single or multiple cycles of preconditioning in rat hearts, *Am J Physiol* 276, H1229–H1235, 1999.
75. Liu Y, Satyo T, O'Rourke B, and Marban E, Mitochondrial ATP-dependent potassium channels-novel effectors of cardioprotection?, *Circulation* 97, 2463–2469, 1998.
76. Parrat JR and Kane KA, KATP channels in ischemic preconditioning, *Cardiovasc Res* 28, 783–787, 1994.
77. Das DK, Prasad MR, Lu D, and Jones RM, Preconditioning of heart by repeated stunning. Adaptive modification of antioxidant defense system, *Cell Mol Biol* 38, 739–749, 1992.
78. Das DK, Maulik N, Engelman RM, Yoshida T, and Zu YL, Preconditioning potentiates molecular signaling for myocardial adaptation to ischemia, *Ann NY Acad Sci* 793, 191–209, 1996.
79. Liu X, Engelman RM, Moraru II, Rousou JA, Flack JE, Deaton DW, Maulik N, and Das DK, Heat shock. A new approach for myocardial preservation in cardiac surgery, *Circulation* 86, II358–II363, 1992.
80. Maulik N, Engelman RM, Wei Z, Liu X, Rousou JA, Flack J, Deaton D, and Das DK, Drug induced heat shock improves post-ischemic ventricular recovery after cardiopulmonary bypass, *Circulation* 92, II381–II388, 1995.
81. Maulik N, Yoshida T, Zu Y L, Banerjee A, and Das DK, Ischemic stress adaptation of heart triggers a tyrosine kinase regulated signaling pathway: a potential role for MAPKAP kinase 2, *Am J Physiol* 275, H1857–H1864, 1998.
82. Maulik N, Engelman RM, Rousou JA, Flack JE, Deaton D, and Das DK, Ischemic preconditioning reduces apoptosis by upregulating anti-death gene Bcl-2, *Circulation* 100, II369–II375, 1999.
83. Maulik N, Sasaki H, Addya S, and Das DK, Regulation of cardiomyocyte apoptosis by redox-sensitive transcription factors, *FEBS Lett* 485, 7–12, 2000.
84. Guo Y, Jones WK, Xuan YT, Tang XL, Bao W, Wu WJ, Han H, Laubacvh VE, Ping P, Yang Z, Qiu Y, and Bolli R, The late phase of ischemic preconditioning is abrogated by targeted disruption of the inducible NO synthase gene, *Proc Natl Acad Sci USA* 96, 11507–11512, 1999.

85. Tosaki A, Maulik N, Elliott GT, Blasig IE, Engelman RM, and Das DK, Preconditioning of rat heart with monophosphoryl lipid A: a role of nitric oxide, *J Pharmacol Exp Ther* 285, 1274–1279, 1998.
86. Yao Z and Gross GJ, Role of nitric oxide, muscarinic receptors, and the ATP-sensitive K+ channel in mediating the effects of acetylcholine to mimic preconditioning in dogs, *Circulation Res* 73, 1193–1201, 1993.
87. Gross GJ and Auchampach JA, Blockade of ATP-sensitive potassium channel prevents myocardial preconditioning in dogs, *Circulation Res* 70, 223–233, 1992.
88. Maulik N, Watanabe M, Zu YL, Huang C-K, Cordis GA, Schley JA, and Das DK, Ischemic preconditioning triggers the activation of MAP kinases and MAPKAP kinase 2 in rat hearts, *FEBS Lett* 396, 233–237, 1996.
89. Ping P, Zhang J, Zheng Y-T, Li RCX, Dawn B, Tang X-L, Takano H, Balafanova Z, and Bolli R, Demonstration of selective protein kinase C-dependent activation of Src and Lck tyrosine kinases during ischemic preconditioning in conscious rabbits, *Circulation Res* 85, 542–550, 1999.
90. Baines CP, Wang L, Cohen WV, and Downey JM, Protein tyrosine kinase is downstream of protein kinase C for ischemic preconditioning's anti-infarct effect in the rabbit heart, *J Mol Cell Cardiol* 30, 383–392, 1998.
91. Gniadecki R, Nongenomic signaling by vitamin D: a new face of Src, *Biochem Pharmacol* 56, 1273–1277, 1997.
92. Vahlhaus C, Schulz R, Post H, Rose J, and Heusch G, Prevention of ischemic preconditioning only by combined inhibition of protein kinase C and protein tyrosine kinase in pigs, *J Mol Cell Cardiol* 30, 197–209, 1998.
93. Schlessinger J, New roles for Src kinases in control of cell survival and angiogenesis, *Cell* 100, 293–296, 2000.
94. Takeishi Y, Abe J, Lee JD, Kawakatsu H, Walsh RA, and Berk BC, Differential regulation of p90 ribosomal S6 kinase and Big mitogen-activated protein kinase 1 by ischemia/reperfusion and oxidative stress in perfused guinea pig heart, *Circulation Res* 85, 1164–1172, 1999.
95. Abe J, Takahashi M, Ishida M, Lee JD, and Berk BC, c-Src is required for oxidative stress-mediated activation of Big mitogen-activated protein kinase 1 (BMK1), *J Biol Chem* 272, 20389–20394, 1997.
96. Akhand AA, Pu M, Senga T, Kato M, Suzuki H, Miyata T, Hamaguchi M, and Nakashima I, Nitric oxide controls Src kinase activity through a sulfhydryl group modification-mediated Tyr-527-independent and Tyr-416-lonked mechanism, *J Biol Chem* 274, 25821–25826, 1999.
97. Kyriakis JM, Banerjee, P, Nikolakaki E, Dai T, Rubie EA, Ahmad MF, Avruch J, and Woodgett JD, The stress-activated protein kinase subfamily of c-Jun kinases, *Nature* 369, 156–160, 1994.
98. Hu Y, Metzler B, and Xu Q, Discordant activation of stress-activated protein kinase or c-jun-NH2-terminal protein kinases in tissues of heat stressed mice, *J Biol Chem* 272, 9113–9119, 1997.
99. Eguchi S, Numaguchi K, Iwasaki H, Mitsumoto T, Yamakawa T, Utsunomiya H, Motley ED, Kawakatsu H, Owada KM, Hirata Y, Marumo F, and Inagami T, Calcium-dependent epidermal growth factor receptor transactivation mediates the angiotensin II-induced mitogen activated protein kinase activation in vascular smooth muscle cells, *J Biol Chem* 273, 8890–8896, 1998.

100. Maulik N, Watanabe M, Engelman D, Engelman RM, and Das DK, Oxidative stress adaptation improves postischemic ventricular recovery, *Mol Cell Biochem* 144, 67–74, 1995.
101. Tosaki A, Maulik N, Elliott GT, Blasig IE, Engelman RM, and Das DK, Preconditioning of rat heart with monophosphoryl lipid A: a role for nitric oxide, *J Pharmacol Exp Ther* 285, 1274–1279, 1998.
102. Abe J, Bains CP, and Berk BC, Role of mitogen-activated protein kinases in ischemia and reperfusion injury, *Circulation Res* 86, 607–609, 2000.
103. Hattori R, Otani H, Uchiyama T, Imamura H, Cui J, Maulik N, Cordis GA, Zhu L, and Das DK, Src tyrosine kinase is the trigger but not the mediator of ischemic preconditioning, *Am J Physiol* 281, H1066–H1074, 2001.
104. Maulik N, Watanabe M, Zu YL, Huang CK, Cordis GA, Schley JA, and Das DK, Ischemic preconditioning triggers the activation of MAP kinases and MAPKAP kinase 2 in rat hearts, *FEBS Lett.* 396, 233–237, 1996.
105. Cohen MV, Baines CP, and Downey JM, Ischemic preconditioning: from adenosine receptor to K_{ATP} channel, *Ann Rev Physiol* 62, 79–109, 2000.
106. Cole WC, McPherson CD, and Sontag D, ATP-regulated K^+ channels protect the myocardium against ischemic/reperfusion damage, *Circulation Res* 69, 571–581, 1991.
107. Bedheit SS, Restivo M, Boutjdir M, Henkin P, Gooyandeh K, Assadi M, Khatib S, Gough WB, and El-Sherif N, Effects of glyburide on ischemia-induced changes in extracellular potassium and local myocardial activation: a potential new approach to the management of ischemia-induced malignant ventricular arrhythmias, *Am Heart J* 119, 1025–1033, 1990.
108. Liu Y, Satyo T, O'Rourke B, and Marban E, Mitochondrial ATP-dependent potassium channels: novel effectors of cardioprotection?, *Circulation* 97, 2463–2469, 1998.
109. Baines CP, Liu GS, Birincioglu M, Critz SD, Cohen MV, and Downey JM, Ischemic preconditioning depends on interaction between mitochondrial KA-TP channels and actin cytoskeleton, *Am J Physiol* 276, H1361–H1368, 1999.
110. Das DK, Engelman RM, and Cherian KM, Eds, Myocardial Preservation, Preconditioning, and Cellular Adaptation, *Ann NY Acad Sci*, Vol. 793, 1996.
111. Meerson FZ, Pshennikova MG, and Malyshev IY, Adaptive defense of the organism, *Ann NY Acad Sci* 793, 371–385, 1996.
112. Maulik N, Tosaki A, Elliott GT, Maulik G, and Das DK, Induction of iNOS expression by monophosphoryl lipid A: a pharmacological approach for myocardial adaptation to ischemia, *Drugs Exp Clin Res* 24, 117–124, 1998.
113. Elliott GT, Pharmacologic myocardial preconditioning with monophosphoryl lipid A (MLA) reduces infarct size and stunning in dogs and rabbits. *Ann NY Acad Sci* 793, 386–399, 1996.
114. Das S, Cordis GA, Maulik N, and Das DK, Pharmacological preconditioning with resveratrol: a role of CREB-dependent Bcl-2 signaling via adenosine A3 receptor activation, *Am J Physiol* 288, H328–H335, 2005.

22 Resveratrol as Cardioprotective Agent: Evidence from Bench and Bedside

Sukesh Burjonroppa and Ken Fujise

CONTENTS

INTRODUCTION

Atherosclerosis and coronary artery disease (CAD), its most feared complication, is the major cause of morbidity and mortality in the modern world. In the United States, atherosclerosis is a major killer [1].

Data from the West of Scotland Coronary Prevention Study and from the Air Force/Texas Coronary Atherosclerosis Prevention Study have provided strong evidence that the lowering of cholesterol by HMG-CoA inhibitors (also known as statins) reduces the occurrence of and death from CAD [2,3]. Secondary prevention trials such as the Scandinavian Simvastatin Survival Study (SSSS) [4] and Long Term Intervention with Pravastatin in Ischemic Disease (LIPID) Study [5] have revealed that lowering low-density lipoprotein (LDL) cholesterol levels can retard the progression of coronary atherosclerosis and reduce morbidity and mortality due to CAD and stroke. For individuals with serum cholesterol levels between 250 and 300 mg/dl, each 1% reduction in serum cholesterol would bring about a 2% reduction in CAD-related morbidity and mortality rates. The absolute magnitude of these benefits would be even greater in those individuals having other risk factors for CAD, such as cigarette smoking, diabetes, and hypertension. In a recent and highly publicized clinical study, a very aggressive lipid-lowering treatment achieved by high-dose atorvastatin blocked, but a moderate lipid-lowering strategy by pravastatin failed to prevent, the progression of coronary atherosclerosis [6]. In addition, an intensive statin therapy provided CAD patients with a greater protection against death or other adverse events, following hospital discharge [7]. However, a pitfall of the cholesterol-lowering strategy is that only about 50% of patients with CAD have elevated cholesterols, suggesting that even a perfect cholesterol level would not guarantee the freedom from atherosclerosis and its complications such as CAD [1].

To identify patients whose cholesterol and LDL are normal but are at risk of developing CAD, cardiologists have extensively studied risk factors of CAD other than cholesterol, including elevated C-reactive protein (CRP) level [8,9], *Chlamydia pneumoniae* infection [10], and elevated homocystine level [11,12].

In addition, a number of nutritional approaches have been evaluated, including the administration of vitamin B complex [13] and the Mediterranean diet, which consists of an abundance of plant food, moderate consumption of wine, olive oil as the principal source of fat, modest consumption of fish and poultry, and restricted consumption of red meat. The first clinical trail that proved the cardioprotective nature of the

Mediterranean diet was the Lyon Diet Heart Study, showing a startling 70% mortality reduction with the Mediterranean diet [14–17]. Red wine has attracted significant attention, partly because of its association with the Mediterranean diet, and because of a body of epidemiological literature known as the French paradox that suggests red wine's beneficial effects on the cardiovascular system in the face of high fat consumption [18–21]. Resveratrol has caught much public attention because it is abundantly present in red wine [22].

In this chapter we evaluate epidemiological, clinical, translational, and basic scientific evidence that supports resveratrol as a cardioprotective agent, reviewing selected key references.

EPIDEMIOLOGICAL EVIDENCE TO SUPPORT RESVERATROL AS A CARDIOPROTECTIVE AGENT

Red Wine as a Cardioprotective Agent: French Paradox

Despite their preference for high-fat and high-cholesterol diet, the French have a much lower prevalence of CAD. This paradoxical phenomenon has given rise to the term "French paradox" referring to the vasoprotective effect of red wine [19,21,23]. Samuel Black (1762–1832), an Irish physician, was probably the first to describe the French paradox [24].

St. Leger and others evaluated the mortality data from 18 countries and found that the death rates from coronary heart disease were higher with larger fat consumption but that they were lower with larger consumption of wine and other alcohol beverages [18]. Again, France has the highest per capita wine consumption [21]. In similar studies, Balkau [25] and others [20,26,27] found that with a large consumption of red wine, the French, who eat meals high in saturated fats, did not necessarily suffer from CAD as much as people in other countries (Figure 22.1).

Examining highly variable cardiovascular mortality rates among several countries with different dietary habits, Renaud and de Lorgeril found that the difference in fat consumption alone could statistically explain 53% of such variations and that differences in both fat and wine consumption could explain as much as 76% of them [28]. They concluded that a larger consumption of ethanol, and wine in particular, was associated with lower cardiovascular mortality [28].

Ulbricht and others showed that the French consume 3.8 times as much butter and 2.8 times as much lard as Americans, have higher serum cholesterol levels, and suffer more from hypertension than Americans, and that the French cardiovascular death rate was only one fourth of that of Americans [29]. Knowing that the French consumed wine 7 times as much as the Americans and 3 to 13 times the northern Europeans and Britons [21], the authors speculated that wine was protective against

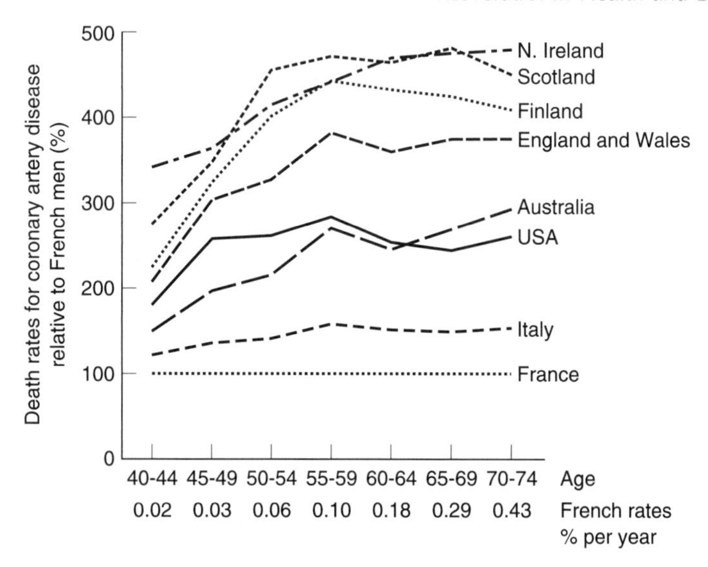

FIGURE 22.1 French paradox: red wine and CAD. Age-specific male death rates in 1990 for CAD, in comparison to France (100%). The French death rates due to CAD (% per year) are given by age group. In 1990 the death rate at all ages was the lowest in France. Taken together with the fact that the French consume a far larger amount of red wine than people from other countries, these data support the beneficial effect of red wine in the prevention of atherosclerosis and its complications. (From Balkau B, Eschwege F, and Eschwege E, *Ann Epidemiol* 7, 490–497, 1997. With permission.)

cardiovascular disorders in the presence of hypercholesterolemia and hypertension [29,30].

The MONICA study is a 10-year study that closely monitored CAD-related deaths and a variety of potential risk factors in men and women of many countries and regions [30]. In this study, France again enjoyed a low incidence rate for fatal and nonfatal cardiovascular events: 297 and 508 per 100,000 men in France and the United States, respectively, suffered the events. Intriguingly, there was a significant discrepancy in cardiovascular events in different regions of France: 240 and 336 per 100,000 men in Toulouse, a southern province of France in which wine consumption is highest, and Strasbourg, a northern province of France where wine consumption is lower, respectively [31].

In a similar study, Ducimetiere and others [32] examined mortality rates and dietary habits among European nations, including Belgium, West Germany, Italy, and Spain. Again, despite the consumption of dairy products that was higher than most of the countries studied, France enjoyed a lower than expected cardiovascular mortality than these countries. Taken together with the fact that the French consumed at least twice as much wine as the average, while consuming significantly less beer than did peoples of other countries, the authors concluded that the lower cardiovascular death rate in France was most likely due to their large wine consumption [32].

In summary, multiple epidemiological studies described above in detail collectively suggest that red wine is protective against cardiovascular diseases.

Resveratrol as a Major Bioactive Agent in Red Wine

Although red wine itself has been well recognized to be cardioprotective, there are many molecules in red wine that can potentially be responsible for the effect. Red wines contain polyphenols/flavonoids, quercetin, catechin, astringinin and rutin, all of which are known to have biological activities that may lead to the prevention of atherosclerosis. Among the polyphenol compounds is resveratrol, which was originally isolated from the roots of the oriental medicinal plant *Polygonum cuspidatum*. Resveratrol is one of the phytoalexins, plant molecules produced in response to a variety of noxious stimuli. In grapevines, resveratrol is synthesized in leaf tissue in response to various types of infections [33]. The average concentration of resveratrol in red wine is 10 to 20 μM while its concentration in white wine is normally less than 1 μM [34]. It has been speculated that chronic and moderate consumption of red wine will keep the serum resveratrol concentration high enough for it to be biologically active [34].

Evidence in Clinical Trials to Support Resveratrol as a Cardioprotective Agent

It has been noted that red wine can reduce LDL lipid peroxidation in humans whereas white wine does not have such an effect [35]. Hung and others showed that resveratrol blocked LDL peroxidation through its antioxidant and free radical scavenging activities in an *ex vivo* rat heart model [36]. Although there has been no clinical study that directly tested the role of resveratrol in atherosclerogenesis, LDL peroxidation is known to facilitate the progression of atherosclerosis, through its facilitative roles in the recruitment of circulating monocytes into the intima and the formation and apoptosis of foam cells. In addition, lipid peroxidation has been shown to accelerate the progression atherosclerosis of human carotid arteries [37]. Furthermore, the Antioxidant Supplementation in Atherosclerosis Prevention (ASAP) study tested the effect of vitamins E and C supplementation on the progression of common carotid atherosclerosis and showed that these antioxidant vitamin supplementations slowed down the progression of carotid atherosclerosis in patients with high plasma cholesterol levels [38].

Based on these observations, it is likely that resveratrol will prove to be protective against atherosclerosis, at least partly through its antioxidant activities and the prevention of LDL oxidation and peroxidation.

ANIMAL STUDIES TO SUPPORT RESVERATROL AS A CARDIOPROTECTIVE AGENT

Here we review several animal studies that evaluated the protective roles of resveratrol in restenosis, platelet aggregation, and endothelial dysfunction. We then briefly discuss studies that demonstrated the beneficial roles of red wine in animal models of atherosclerosis.

RESVERATROL INHIBITS NEOINTIMAL HYPERPLASIA AFTER VASCULAR INJURY

Postangioplasty and in-stent restenoses are the renarrowing of the balloon-dilated and stent-expanded coronary artery lumens, respectively, and represent serious complications of percutaneous coronary intervention (PCI). The predominant pathological picture of restenosis is the proliferation of the vascular smooth muscle cells and the production of extracellular matrix by them [39,40]. Zou and others demonstrated that resveratrol blocked the proliferation of vascular smooth muscle cells and inhibited neointimal hyperplasia in a rabbit model of restenosis where endothelial cells were denudated by overinflated balloons [41] (Figure 22.2). At a tissue culture level, resveratrol has been shown to inhibit strongly vascular smooth muscle cell proliferation. In one study, resveratrol's inhibitory effect on vascular smooth muscle cells was evident at 1 μM [42]. Since Clowes and others have shown that the proliferation of vascular smooth muscle cells occurs within the first 2 weeks after vascular injury [43] and since negative regulators of cell cycle progression have been shown to block neointimal hyperplasia after vascular injury [44–46], it is tempting to postulate that resveratrol administration in the peri-PCI period might reduce the degree of neointimal hyperplasia and restenosis. Furthermore, it is possible that a long-term resveratrol administration would inhibit vascular smooth muscle cell migration into intima and the progression of primary atherosclerosis.

RESVERATROL INHIBITS PLATELET ACTIVATION AND AGGREGATION

Wang and others demonstrated that platelets from rabbits fed a high-cholesterol diet aggregated more readily than those fed a normal chow *in vitro* and that oral resveratrol treatment blocked the aggregation of

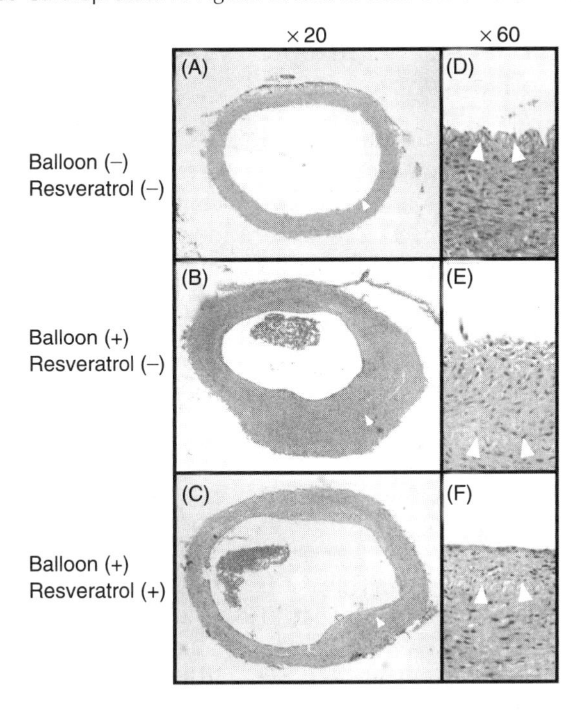

FIGURE 22.2 (See color insert following page 546.) Inhibitory effect of resveratrol on neointimal proliferation after vascular injury. White arrows: internal elastic laminae; Balloon: balloon-induced endothelial denudation; Resveratrol: intragastrical administration of resveratrol (4 mg/kg/d). The ability of resveratrol to inhibit neointimal proliferation was tested in an experimental model in which endothelial denudation was performed in the normal rabbit iliac artery. Control arteries without balloon injury had no cells in the intima (Balloon (–), Resveratrol (–); A and D). Balloon injury resulted in robust neointimal proliferation (Balloon (+), Resveratrol (–); B and E). Resveratrol (4 mg/kg/d), administered intragastrically for 5 weeks beginning 1 week before balloon injury, effectively blocked the neointimal hyperplasia (Balloon (+), Resveratrol (+); C and F). (From Zou J, Huang Y, Cao K, Yang G, Yin H, Len J, Hsieh TC, and Wu JM, *Life Sci* 68, 153–163, 2000. With permission.)

platelets from high-cholesterol-fed rabbits. Importantly, resveratrol did not change total plasma LDL levels in this study [47]. Wang's study is consistent with the findings by other investigators that resveratrol inhibited the activation and aggregation of platelets in response to thrombin, ADP, and collagen, probably through the blockade of thromboxane B2 synthesis [48,49]. Platelet activation has been shown to play a critical role in the pathogenesis of CAD-related complications. Wang's study suggests that resveratrol administration may positively modify the outcomes of these clinical conditions.

Resveratrol Dilates Arteries through Nitric Oxide-Mediated Pathways

Orallo and others showed that resveratrol (1 to 10 μmol/l) reversed the phenylephrine-induced contraction of rat aortas with intact endothelium. In addition, the relaxation of the endothelium-containing rat aorta by resveratrol was blocked by the administration of inhibitors of nitric oxide (NO) synthase [50]. Using inducible NO synthase (iNOS) knockout mice, Imamura and others showed that the protection by resveratrol of myocardial ischemia-reperfusion injury required the presence of iNOS [51]. Consistent with these finding, Giovannini and others demonstrated the upregulation of NO by resveratrol as a principal factor for the prevention of ischemia-reperfusion injury [52].

Meanwhile, Kawashima and others showed that endothelial NO (eNOS) was protective against atherosclerosis by demonstrating that eNOS deficiency accelerated atherosclerosis in hyperlipidemic environment [53]. Endothelial dysfunction, the earliest sign of atherosclerotic changes in the vasculature, manifests itself in the lack of the endothelium-dependent, NO-mediated dilation of arteries [1]. Taken together with the fact that resveratrol upregulates NO production and potentiates NO-induced vasodiatory effects, it is possible that resveratrol's protective effects against the initiation and progression of atherosclerosis may be mediated through NO pathways [53].

Resveratrol and Atherosclerosis

Although there is no animal study that directly tested the role of resveratrol in atherosclerogenesis, several studies have shown the positive effects of red wine against atherosclerosis. First, Klurfield and Kritchevsky showed that red wine was more effective in preventing atherosclerotic changes in rabbits than were other alcoholic beverages such as beer, white wine, and whiskey [54]. In addition, Hayek and others showed that resveratrol decreased LDL oxidation and aggregation and that Apo-E-deficient mice consuming red wine diluted to contain 1.1% ethanol had smaller atherosclerotic lesion areas than the same mice consuming 1.1% ethanol alone [55]. Finally, Vinson and others showed in hamster models of atherosclerosis that polyphenols within red wine and grape juice slowed the development of atherosclerosis [56]. These animal studies, taken together, further support the French paradox and suggest that resveratrol, a major polyphenol within red wine, may at least partly explain the antiatherosclerotic effects of red wine. An animal study using resveratrol and Apo-E-deficient mice has not been performed but will be important to test definitively the role of resveratrol in atherosclerogenesis.

FIGURE 1.1 Plant sources of resveratrol.

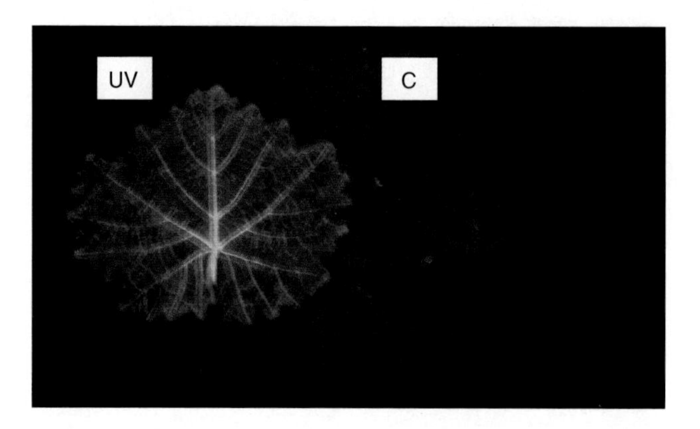

FIGURE 19.2 Observation of the abaxial side of leaves under long-wavelength UV light (365 nm). UV: 24 hours following UV irradiation of the leaf. Note the bright blue purple fluorescence emitted by resveratrol and some of its derivatives. C: untreated leaf (control).

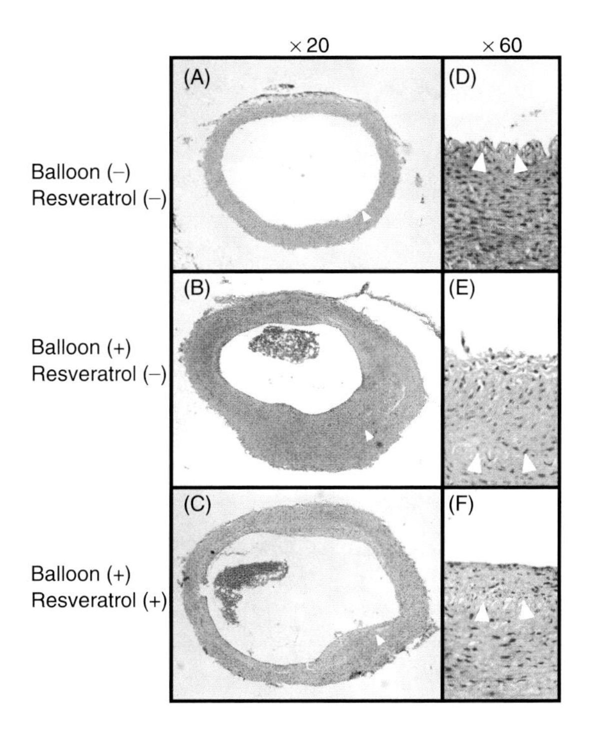

FIGURE 22.2 Inhibitory effect of resveratrol on neointimal proliferation after vascular injury. White arrows: internal elastic laminae; Balloon: balloon-induced endothelial denudation; Resveratrol: intragastrical administration of resveratrol (4 mg/kg/d). The ability of resveratrol to inhibit neointimal proliferation was tested in an experimental model in which endothelial denudation was performed in the normal rabbit iliac artery. Control arteries without balloon injury had no cells in the intima (Balloon (−), Resveratrol (−); A and D). Balloon injury resulted in robust neointimal proliferation (Balloon (+), Resveratrol (−); B and E). Resveratrol (4 mg/kg/d), administered intragastrically for 5 weeks beginning 1 week before balloon injury, effectively blocked the neointimal hyperplasia (Balloon (+), Resveratrol (+); C and F). (From Zou J, Huang Y, Cao K, Yang G, Yin H, Len J, Hsieh TC, and Wu JM, *Life Sci* 68, 153–163, 2000. With permission.)

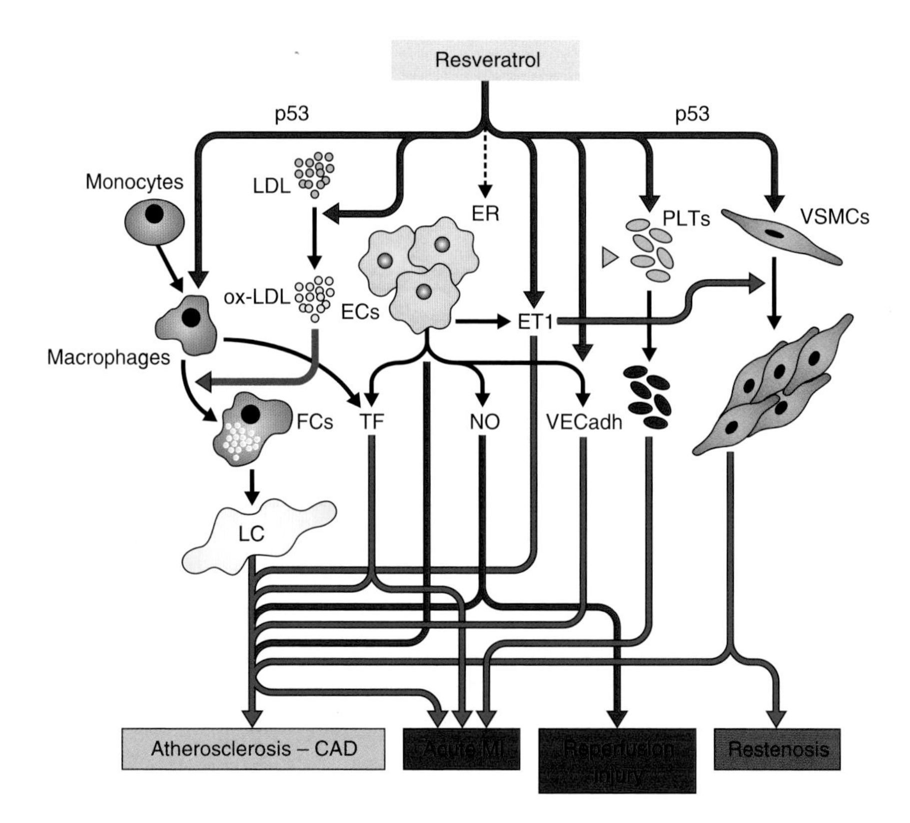

FIGURE 22.3 The link between the biological activities of resveratrol and the prevention of cardiovascular disorders. FCs: foam cells; LC: lipid core within atherosclerotic lesions; CAD: coronary artery disease; LDL: low-density lipoprotein; ox-LDL: oxidized LDL; ER: estrogen receptor; ECs: endothelial cells; TF: tissue factor; NO: nitric oxide; ET1: endothelin-1; VECadh: vascular endothelial cadherin; PLTs: platelets; VSMCs: vascular smooth muscle cells. Red arrows represent inhibitory effects while blue arrows represent facilitative effects. Black arrows denote either progression or secretion. A dotted arrow suggests a possible link. Resveratrol positively modulates all vascular cells, including monocytes and their derivatives, endothelial cells, vascular smooth muscle cells, and platelets.

BENCH RESEARCH TO SUPPORT RESVERATROL AS A CARDIOPROTECTIVE AGENT

RESVERATROL IS AN ANTIENDOTHELIN AGENT IN VASCULAR CELLS

Endothelin-1 (ET-1), a 21-amino-acid peptide, is a potent vasoconstrictor as well as a survival and mitogenic factor of vascular cells [57,58].

Resveratrol inhibits the transcription of ET-1 gene and the expression and secretion of ET-1 protein in endothelial cells that are subjected to strain, shear stress, and reactive oxygen species [59]. Consistently, red wine also inhibits the synthesis of ET-1 [60].

Plasma ET-1 levels are elevated in patients with severe atherosclerosis and CADs [61]. In addition, Ihling and others demonstrated that ET-1 was overexpressed in atherosclerotic tissue, predominantly localizing in the plaque exhibiting a picture of chronic inflammation [62]. Furthermore, Zeiher and others observed that coronary arteries from unstable angina patients exhibited a robust ET-1-like immunoreactivity [63]. Finally, ET-1 has been shown to function as the mitogen of vascular smooth muscle cells that play a critical role in the pathogenesis of atherosclerosis [64]. It is possible that resveratrol, through the inhibition of ET-1 production, protects the cardiovascular system against atherosclerotic changes and their complications.

RESVERATROL IS A TISSUE FACTOR INHIBITOR IN VASCULAR CELLS

Tissue factor (TF) is a cell surface glycoprotein and the primary initiator of the coagulation cascade in both physical and pathological conditions [65,66].

Pendurthi and others demonstrated that resveratrol suppresses the induction of TF by interleukin (IL)-1β, tumor necrosis factor alpha (TNFα), lipopolysaccharide (LPS), and phorbol myristate acetate (PMA) [67] in endothelial cells. In monocytes, resveratrol blocked LPS-induced macrophage activation and TF expression.

While TF is constitutively expressed in several extravascular cells such as fibroblasts and pericyte [68,69], little TF is found in cells of the intima or media of normal arteries [68,69]. Strikingly, TF is abundantly expressed in atherosclerotic plaques [70]. TF antigenicity in human atherosclerotic plaques is localized within macrophages, smooth muscle cells, and endothelial cells near the acellular lipid-rich core [70,71]. Activation of TF pathway not only enhances the plaque thrombogenicity but also promotes vascular smooth cell proliferation [65]. These data suggest that TF pathway plays an important role in the development and progression of atherosclerosis and that resveratrol, through its inhibitory effects on TF pathway, passivates the thrombogenic environment of atherosclerotic arteries and helps prevent thrombosis-related complications of atherosclerosis.

Resveratrol Is an Estrogen Receptor Agonist in Vascular Cells

Estrogen binds to its nuclear receptor that initiates transcription of estrogen-responsive target genes. Resveratrol is structurally similar to diethylstilbestrol [46].

Gehm and others showed that resveratrol could specifically activate the estrogen receptor transcriptional element using luciferase reporter assays in MCF-7 cells [72,73]. Although this study used a cancer cell line, the findings suggest that resveratrol possesses estrogen receptor stimulatory effects (i.e., phytoestrogen) also in the vascular system. A question is: Are resveratrol's possible cardioprotective effects mediated through its phytoestrogen activity? It is true that women develop atherosclerotic complications 10 years later than men. This is presumably because of the vasoprotective effects of estrogen. At the same time, estrogen supplementation either with or without progestin failed to reduce the risk of coronary heart diseases in postmenopausal women in the Women's Health Initiative Study [74]. It is to be experimentally addressed whether resveratrol's phytoestrogen activity is important for the prevention of human atherosclerosis.

Resveratrol Upregulates and Activates p53 Tumor Suppressor Protein in Vascular Cells

Tumor suppressor protein p53 has been called the "guardian of the genome" because it enforces a variety of anticancer functions by encouraging cells to arrest or die in the face of DNA damage and other noxious stimuli. P53 knockout ($p53^{-/-}$) mice develop a variety of neoplasms, multiple and multifocal in most cases, at early ages.

Resveratrol has been shown to activate p53 pathway in cancerous cells and in vascular smooth muscle cells. The transcriptional activation of p53 target genes by resveratrol appears to involve the accumulation of p53 within the cell and the activation of p53 through serine phosphorylation [75]. Intriguingly, at lower concentrations (6.25 to 12.5 μM), resveratrol effectively blocks cell cycle progression of serum-stimulated vascular smooth muscle cells without inducing apoptosis, while a higher concentration of resveratrol (25 μM) selectively induced apoptosis in the same vascular smooth muscle cells [42]. George and others showed, using adenoviral p53 gene transfer to vascular smooth muscle cells, that p53 overexpression promoted the apoptosis and inhibited the migration of vascular smooth muscle cells, blocking neointimal hyperplasia after vascular injury, suggesting that functional p53 is preventive of postangioplasty restenosis [76].

It has been also established that p53 is protective against atherosclerosis. Guevara and others compared with $p53^{+/+}apoE^{-/-}$ mice and

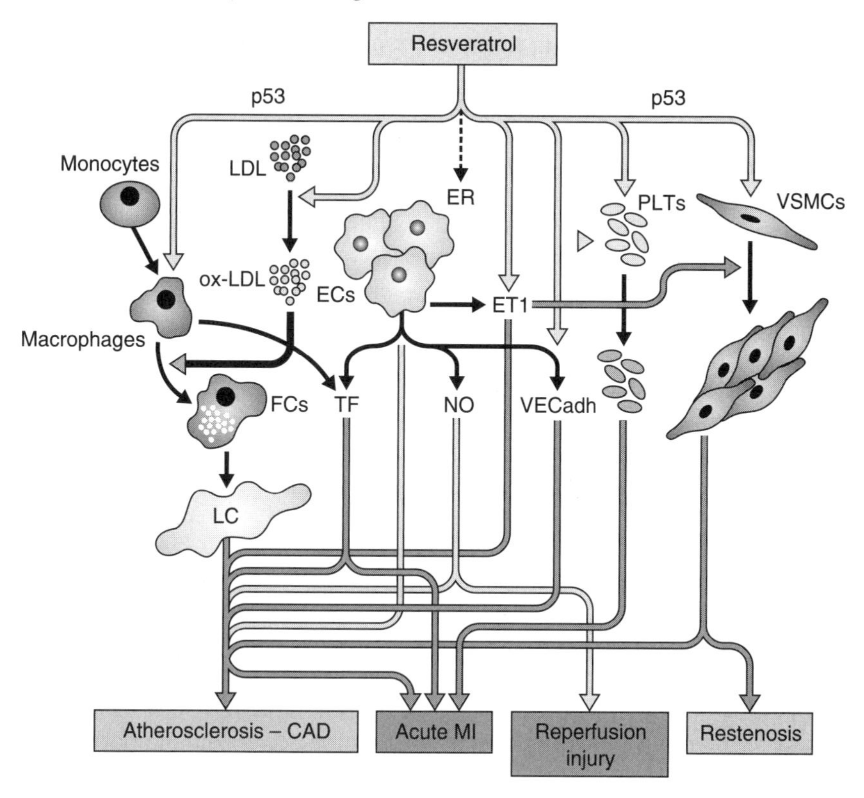

FIGURE 22.3 (See color insert following page 546.) The link between the biological activities of resveratrol and the prevention of cardiovascular disorders. FCs: foam cells; LC: lipid core within atherosclerotic lesions; CAD: coronary artery disease; LDL: low-density lipoprotein; ox-LDL: oxidized LDL; ER: estrogen receptor; ECs: endothelial cells; TF: tissue factor; NO: nitric oxide; ET1: endothelin-1; VECadh: vascular endothelial cadherin; PLTs: platelets; VSMCs: vascular smooth muscle cells. Red arrows represent inhibitory effects while blue arrows represent facilitative effects. Black arrows denote either progression or secretion. A dotted arrow suggests a possible link. Resveratrol positively modulates all vascular cells, including monocytes and their derivatives, endothelial cells, vascular smooth muscle cells, and platelets.

$p53^{-/-}apoE^{-/-}$ mice and found that $p53^{-/-}apoE^{-/-}$ mice developed far accelerated atherosclerotic lesions despite similar serum cholesterol levels [77]. Merched and others showed that macrophage-specific p53 deficiency stimulates cellular proliferation leading to a vulnerable-appearing phenotype of lesions in $Ldlr^{-/-}$ mice fed a high-cholesterol diet [78]. These data, taken together, suggest that resveratrol's vasoprotective effects may be at least partly mediated by its induction of p53 in vascular smooth muscle cells.

EPILOGUE – KEN FUJISE

Over ten years ago, I was an internal medicine resident at the Bronx VA Medical Center. One day I encountered a 95-year-old gentleman who was going home after being treated for a minor pneumonia. His electrocardiogram was normal and his heart function was excellent. He looked well and in good spirit, pushing around a friend of his in a wheelchair. When I asked what he did for his longevity and health, his answer was a glass of red wine every night. Despite the rapid progress in percutaneous and surgical revascularization procedures, the fact remains that atherosclerosis represents the process that is diffuse not focal. The chemoprevention of atherosclerosis by resveratrol is intriguing when one considers that resveratrol positively influences so many biological processes relevant to atherosclerogenesis (Figure 22.3). Someday in the future when more basic and clinical investigations have been completed, clinicians may be willing to suggest a resveratrol tablet instead of a glass of red wine for the longevity and health of their patients.

ACKNOWLEDGMENT

The authors thank Ms. Estella C. Wheatley, University of Texas Health Science Center at Houston, for her outstanding secretarial support in the preparation of the manuscript. Dr. Fujise is an Established Investigator of the American Heart Association (0540054N). This work is supported in part by grants from the National Institutes of Health (HL068024) and the Roderick D. MacDonald General Research Fund at St. Luke's Episcopal Hospital, Houston, Texas.

REFERENCES

1. Braunwald E, Shattuck lecture. Cardiovascular medicine at the turn of the millennium: triumphs, concerns, and opportunities, *N Engl J Med* 337, 1360–1369, 1997.
2. WOSCOPS Study Group, Screening experience and baseline characteristics in the West of Scotland Coronary Prevention Study, *Am J Cardiol* 76, 485–491, 1995.
3. Gotto AM Jr, Whitney E, Stein EA, Shapiro DR, Clearfield M, Weis S, Jou JY, Langendorfer A, Beere PA, Watson DJ, Downs JR, and de Cani JS, Relation between baseline and on-treatment lipid parameters and first acute major coronary events in the Air Force/Texas Coronary Atherosclerosis Prevention Study (AFCAPS/TexCAPS), *Circulation* 101, 477–484, 2000.
4. Randomised trial of cholesterol lowering in 4444 patients with coronary heart disease: the Scandinavian Simvastatin Survival Study (4S), *Lancet* 344, 1383–1389, 1994.

5. Tonkin AM, Management of the Long-Term Intervention with Pravastatin in Ischaemic Disease (LIPID) study after the Scandinavian Simvastatin Survival Study (4S), *Am J Cardiol* 76, 107C–112C, 1995.
6. Nissen SE, Tuzcu EM, Schoenhagen P, Brown BG, Ganz P, Vogel RA, Crowe T, Howard G, Cooper CJ, Brodie B, Grines CL, and DeMaria AN, Effect of intensive compared with moderate lipid-lowering therapy on progression of coronary atherosclerosis: a randomized controlled trial, *JAMA* 291, 1071–1080, 2004.
7. Cannon CP, Braunwald E, McCabe CH, Rader DJ, Rouleau JL, Belder R, Joyal SV, Hill KA, Pfeffer MA, and Skene AM, Intensive versus moderate lipid lowering with statins after acute coronary syndromes, *N Engl J Med* 350, 1495–1504, 2004.
8. Blake GJ and Ridker PM, Novel clinical markers of vascular wall inflammation, *Circulation Res* 89, 763–771, 2001.
9. Ridker PM, Clinical application of C-reactive protein for cardiovascular disease detection and prevention, *Circulation* 107, 363–369, 2003.
10. Sessa R, Di Pietro M, Santino I, del Piano M, Varveri A, Dagianti A, and Penco M, *Chlamydia pneumoniae* infection and atherosclerotic coronary disease, *Am Heart J* 137, 1116–1119, 1999.
11. McCully KS, Vascular pathology of homocysteinemia: implications for the pathogenesis of arteriosclerosis, *Am J Pathol* 56, 111–128, 1969.
12. Clarke R, Daly L, Robinson K, Naughten E, Cahalane S, Fowler B, and Graham I, Hyperhomocysteinemia: an independent risk factor for vascular disease, *N Engl J Med* 324, 1149–1155, 1991.
13. Lenhart SE and Nappi JM, Vitamins for the management of cardiovascular disease: a simple solution to a complex problem?, *Pharmacotherapy* 19, 1400–1414, 1999.
14. Kris-Etherton P, Eckel RH, Howard BV, St Jeor S, and Bazzarre TL, AHA Science Advisory: Lyon Diet Heart Study benefits of a Mediterranean-style, National Cholesterol Education Program/American Heart Association Step I dietary pattern on cardiovascular disease, *Circulation* 103, 1823–1825, 2001.
15. de Lorgeril M, Salen P, Martin JL, Monjaud I, Delaye J, and Mamelle N, Mediterranean diet, traditional risk factors, and the rate of cardiovascular complications after myocardial infarction: final report of the Lyon Diet Heart Study, *Circulation* 99, 779–785, 1999.
16. Leaf A, Dietary prevention of coronary heart disease: the Lyon Diet Heart Study, *Circulation* 99, 733–735, 1999.
17. de Lorgeril M, Salen P, Caillat-Vallet E, Hanauer MT, Barthelemy JC, and Mamelle N, Control of bias in dietary trial to prevent coronary recurrences: the Lyon Diet Heart Study, *Eur J Clin Nutr* 51, 116–122, 1997.
18. St Leger AS, Cochrane AL, and Moore F, Factors associated with cardiac mortality in developed countries with particular reference to the consumption of wine, *Lancet* 1, 1017–1020, 1979.
19. Renaud S and de Lorgeril M, Wine, alcohol, platelets, and the French paradox for coronary heart disease, *Lancet* 339, 1523–1526, 1992.
20. Hegsted DM and Ausman LM, Diet, alcohol and coronary heart disease in men, *J Nutr* 118, 1184–1189, 1988.
21. Criqui MH and Ringel BL, Does diet or alcohol explain the French paradox?, *Lancet* 344, 1719–1723, 1994.

22. Bertelli AA, Giovannini L, Giannessi D, Migliori M, Bernini W, Fregoni M, and Bertelli A, Antiplatelet activity of synthetic and natural resveratrol in red wine, *Int J Tissue React* 17, 1–3, 1995.
23. Mukamal KJ, Conigrave KM, Mittleman MA, Camargo CA, Jr., Stampfer MJ, Willett WC, and Rimm EB, Roles of drinking pattern and type of alcohol consumed in coronary heart disease in men, *N Engl J Med* 348, 109–118, 2003.
24. Sandler M and Pinder R, *Wine: A Scientific Exploration*, Taylor and Francis, New York, 2002.
25. Balkau B, Eschwege F, and Eschwege E, Ischemic heart disease and alcohol-related causes of death: a view of the French paradox, *Ann Epidemiol* 7, 490–497, 1997.
26. Nanji AA, Alcohol and ischemic heart disease: wine, beer or both?, *Int J Cardiol* 8, 487–489, 1985.
27. Nanji AA and French SW, Alcoholic beverages and coronary heart disease, *Atherosclerosis* 60, 197–198, 1986.
28. Renaud S and de Lorgeril M, The French paradox: dietary factors and cigarette smoking-related health risks, *Ann NY Acad Sci* 686, 299–309, 1993.
29. Ulbricht TL and Southgate DA, Coronary heart disease: seven dietary factors, *Lancet* 338, 985–992, 1991.
30. Tuomilehto J and Kuulasmaa K, WHO MONICA Project: assessing CHD mortality and morbidity, *Int J Epidemiol* 18, S38–S45, 1989.
31. Tunstall-Pedoe H, Kuulasmaa K, Amouyel P, Arveiler D, Rajakangas AM, and Pajak A, Myocardial infarction and coronary deaths in the World Health Organization MONICA Project. Registration procedures, event rates, and case-fatality rates in 38 populations from 21 countries in four continents, *Circulation* 90, 583–612, 1994.
32. Ducimetiere R and Richard J, Dietary lipids and coronary heart disease. Is there a French paradox?, *Nutr Metab Cardioivasc Dis* 2, 195–201, 1992.
33. Jang M, Cai L, Udeani GO, Slowing KV, Thomas CF, Beecher CW, Fong HH, Farnsworth NR, Kinghorn AD, Mehta RG, Moon RC, and Pezzuto JM, Cancer chemopreventive activity of resveratrol, a natural product derived from grapes, *Science* 275, 218–220, 1997.
34. Bertelli A, Bertelli AA, Gozzini A, and Giovannini L, Plasma and tissue resveratrol concentrations and pharmacological activity, *Drugs Exp Clin Res* 24, 133–138, 1998.
35. Fuhrman B, Lavy A, and Aviram M, Consumption of red wine with meals reduces the susceptibility of human plasma and low-density lipoprotein to lipid peroxidation, *Am J Clin Nutr* 61, 549–554, 1995.
36. Hung LM, Su MJ, Chu WK, Chiao CW, Chan WF, and Chen JK, The protective effect of resveratrols on ischaemia-reperfusion injuries of rat hearts is correlated with antioxidant efficacy, *Br J Pharmacol* 135, 1627–1633, 2002.
37. Salonen JT, Yla-Herttuala S, Yamamoto R, Butler S, Korpela H, Salonen R, Nyyssonen K, Palinski W, and Witztum JL, Autoantibody against oxidised LDL and progression of carotid atherosclerosis, *Lancet* 339, 883–887, 1992.
38. Salonen RM, Nyyssonen K, Kaikkonen J, Porkkala-Sarataho E, Voutilainen S, Rissanen TH, Tuomainen TP, Valkonen VP, Ristonmaa U, Lakka HM,

Vanharanta M, Salonen JT, and Poulsen HE, Six-year effect of combined vitamin C and E supplementation on atherosclerotic progression: the Antioxidant Supplementation in Atherosclerosis Prevention (ASAP) Study, *Circulation* 107, 947–953, 2003.
39. Liu MW, Roubin GS, and King SB III, Restenosis after coronary angioplasty. Potential biologic determinants and role of intimal hyperplasia, *Circulation* 79, 1374–1387, 1989.
40. Gravanis MB and Roubin GS, Histopathologic phenomena at the site of percutaneous transluminal coronary angioplasty: the problem of restenosis, *Hum Pathol* 20, 477–485, 1989.
41. Zou J, Huang Y, Cao K, Yang G, Yin H, Len J, Hsieh TC, and Wu JM, Effect of resveratrol on intimal hyperplasia after endothelial denudation in an experimental rabbit model, *Life Sci* 68, 153–163, 2000.
42. Mnjoyan ZH and Fujise K, Profound negative regulatory effects by resveratrol on vascular smooth muscle cells: a role of p53-p21(WAF1/CIP1) pathway, *Biochem Biophys Res Commun* 311, 546–552, 2003.
43. Clowes AW, Reidy MA, and Clowes MM, Mechanisms of stenosis after arterial injury, *Lab Invest* 49, 208–215, 1983.
44. Morishita R, Gibbons GH, Ellison KE, Nakajima M, von der Leyen H, Zhang L, Kaneda Y, Ogihara T, and Dzau VJ, Antisense oligonucleotides directed at cell cycle regulatory genes as strategy for restenosis therapy, *Trans Assoc Am Physicians* 106, 54–61, 1993.
45. Jackson CL and Schwartz SM, Pharmacology of smooth muscle cell replication, *Hypertension* 20, 713–736, 1992.
46. Braun-Dullaeus RC, Mann MJ, and Dzau VJ, Cell cycle progression: new therapeutic target for vascular proliferative disease, *Circulation* 98, 82–89, 1998.
47. Wang Z, Huang Y, Zou J, Cao K, Xu Y, and Wu JM, Effects of red wine and wine polyphenol resveratrol on platelet aggregation *in vivo* and *in vitro*, *Int J Mol Med* 9, 77–79, 2002.
48. Pace-Asciak CR, Hahn S, Diamandis EP, Soleas G, and Goldberg DM, The red wine phenolics trans-resveratrol and quercetin block human platelet aggregation and eicosanoid synthesis: implications for protection against coronary heart disease, *Clin Chim Acta* 235, 207–219, 1995.
49. Kirk RI, Deitch JA, Wu JM, and Lerea KM, Resveratrol decreases early signaling events in washed platelets but has little effect on platalet in whole food, *Blood Cells Mol Dis* 26, 144–150, 2000.
50. Orallo F, Alvarez E, Camina M, Leiro JM, Gomez E, and Fernandez P, The possible implication of trans-resveratrol in the cardioprotective effects of long-term moderate wine consumption, *Mol Pharmacol* 61, 294–302, 2002.
51. Imamura G, Bertelli AA, Bertelli A, Otani H, Maulik N, and Das DK, Pharmacological preconditioning with resveratrol: an insight with iNOS knockout mice, *Am J Physiol Heart Circ Physiol* 282, H1996–H2003, 2002.
52. Giovannini L, Migliori M, Longoni BM, Das DK, Bertelli AA, Panichi V, Filippi C, and Bertelli A, Resveratrol, a polyphenol found in wine, reduces ischemia reperfusion injury in rat kidneys, *J Cardiovasc Pharmacol* 37, 262–270, 2001.
53. Kawashima S and Yokoyama M, Dysfunction of endothelial nitric oxide synthase and atherosclerosis, *Arterioscler Thromb Vasc Biol* 24, 998–1005, 2004.

54. Klurfeld DM and Kritchevsky D, Differential effects of alcoholic beverages on experimental atherosclerosis in rabbits, *Exp Mol Pathol* 34, 62–71, 1981.
55. Hayek T, Fuhrman B, Vaya J, Rosenblat M, Belinky P, Coleman R, Elis A, and Aviram M, Reduced progression of atherosclerosis in apolipoprotein E-deficient mice following consumption of red wine, or its polyphenols quercetin or catechin, is associated with reduced susceptibility of LDL to oxidation and aggregation, *Arterioscler Thromb Vasc Biol* 17, 2744–2752, 1997.
56. Vinson JA, Teufel K, and Wu N, Red wine, dealcoholized red wine, and especially grape juice, inhibit atherosclerosis in a hamster model, *Atherosclerosis* 156, 67–72, 2001.
57. Sargent CA, Liu EC, Chao CC, Monshizadegan H, Webb ML, and Grover GJ, Role of endothelin receptor subtype B (ET-B) in myocardial ischemia, *Life Sci* 55, 1833–1844, 1994.
58. Tonnessen T, Giaid A, Saleh D, Naess PA, Yanagisawa M, and Christensen G, Increased *in vivo* expression and production of endothelin-1 by porcine cardiomyocytes subjected to ischemia, *Circulation Res* 76, 767–772, 1995.
59. Liu JC, Chen JJ, Chan P, Cheng CF, and Cheng TH, Inhibition of cyclic strain-induced endothelin-1 gene expression by resveratrol, *Hypertension* 42, 1198–1205, 2003.
60. Corder R, Douthwaite JA, Lees DM, Khan NQ, Viseu Dos Santos AC, Wood EG, and Carrier MJ, Endothelin-1 synthesis reduced by red wine, *Nature* 414, 863–864, 2001.
61. Lerman A, Edwards BS, Hallett JW, Heublein DM, Sandberg SM, and Burnett JC Jr, Circulating and tissue endothelin immunoreactivity in advanced atherosclerosis, *N Engl J Med* 325, 997–1001, 1991.
62. Ihling C, Szombathy T, Bohrmann B, Brockhaus M, Schaefer HE, and Loeffler BM, Coexpression of endothelin-converting enzyme-1 and endothelin-1 in different stages of human atherosclerosis, *Circulation* 104, 864–869, 2001.
63. Zeiher AM, Goebel H, Schachinger V, and Ihling C, Tissue endothelin-1 immunoreactivity in the active coronary atherosclerotic plaque. A clue to the mechanism of increased vasoreactivity of the culprit lesion in unstable angina, *Circulation* 91, 941–947, 1995.
64. Komuro I, Kurihara H, Sugiyama T, Yoshizumi M, Takaku F, and Yazaki Y, Endothelin stimulates c-fos and c-myc expression and proliferation of vascular smooth muscle cells, *FEBS Lett* 238, 249–252, 1988.
65. Taubman MB, Fallon JT, Schecter AD, Giesen P, Mendlowitz M, Fyfe BS, Marmur JD, and Nemerson Y, Tissue factor in the pathogenesis of atherosclerosis, *Thromb Haemost* 78, 200–204, 1997.
66. Rapaport SI and Rao LV, The tissue factor pathway: how it has become a "prima ballerina", *Thromb Haemost* 74, 7–17, 1995.
67. Pendurthi UR, Williams JT, and Rao LV, Resveratrol, a polyphenolic compound found in wine, inhibits tissue factor expression in vascular cells: a possible mechanism for the cardiovascular benefits associated with moderate consumption of wine, *Arterioscler Thromb Vasc Biol* 19, 419–426, 1999.
68. Fleck RA, Rao LV, Rapaport SI, and Varki N, Localization of human tissue factor antigen by immunostaining with monospecific, polyclonal anti-human tissue factor antibody, *Thromb Res* 59, 421–437, 1990.

69. Drake TA, Ruf W, Morrissey JH, and Edgington TS, Functional tissue factor is entirely cell surface expressed on lipopolysaccharide-stimulated human blood monocytes and a constitutively tissue factor-producing neoplastic cell line, *J Cell Biol* 109, 389–395, 1989.
70. Wilcox JN, Smith KM, Schwartz SM, and Gordon D, Localization of tissue factor in the normal vessel wall and in the atherosclerotic plaque, *Proc Natl Acad Sci USA* 86, 2839–2843, 1989.
71. Thiruvikraman SV, Guha A, Roboz J, Taubman MB, Nemerson Y, and Fallon JT, In situ localization of tissue factor in human atherosclerotic plaques by binding of digoxigenin-labeled factors VIIa and X, *Lab Invest* 75, 451–461, 1996.
72. Gehm BD, McAndrews JM, Chien PY, and Jameson JL, Resveratrol, a polyphenolic compound found in grapes and wine, is an agonist for the estrogen receptor, *Proc Natl Acad Sci USA* 94, 14138–14143, 1997.
73. Gehm BD, Levenson AS, Liu H, Lee EJ, Amundsen BM, Cushman M, Jordan VC, and Jameson JL, Estrogenic effects of resveratrol in breast cancer cells expressing mutant and wild-type estrogen receptors: role of AF-1 and AF-2, *J Steroid Biochem Mol Biol* 88, 223–234, 2004.
74. Manson JE, Hsia J, Johnson KC, Rossouw JE, Assaf AR, Lasser NL, Trevisan M, Black HR, Heckbert SR, Detrano R, Strickland OL, Wong ND, Crouse JR, Stein E, and Cushman M, Estrogen plus progestin and the risk of coronary heart disease, *N Engl J Med* 349, 523–534, 2003.
75. Haider UG, Sorescu D, Griendling KK, Vollmar AM, and Dirsch VM, Resveratrol increases serine15-phosphorylated but transcriptionally impaired p53 and induces a reversible DNA replication block in serum-activated vascular smooth muscle cells, *Mol Pharmacol* 63, 925–932, 2003.
76. George SJ, Angelini GD, Capogrossi MC, and Baker AH, Wild-type p53 gene transfer inhibits neointima formation in human saphenous vein by modulation of smooth muscle cell migration and induction of apoptosis, *Gene Ther* 8, 668–676, 2001.
77. Guevara NV, Kim HS, Antonova EI, and Chan L, The absence of p53 accelerates atherosclerosis by increasing cell proliferation *in vivo*, *Nat Med* 5, 335–339, 1999.
78. Merched AJ, Williams E, and Chan L, Macrophage-specific p53 expression plays a crucial role in atherosclerosis development and plaque remodeling, *Arterioscler Thromb Vasc Biol* 23, 1608–1614, 2003.

23 Immunomodulation by Resveratrol

Subhash C. Gautam, Xiaohua Gao, and Scott A. Dulchavsky

CONTENTS

INTRODUCTION

Resveratrol (3,4′,5-trihydroxystilbene) is a nonflavonoid polyphenolic compound synthesized mainly by spermatophytes in response to injury, environmental stress (ultraviolet irradiation), or fungal infections. It was first detected in grapevines (*Vitis vinifera*) in 1976 by Langcake and Pryce [1] and then in wine in 1992 [2]. In addition to grapes, resveratrol is also found in lesser amounts in peanuts and mulberries. In grapes, resveratrol is concentrated in grape skin (50 to 100 mg/g), and to lesser extent in seeds [3]. Resveratrol has been identified as the active ingredient in the root powder of *Polygonum cuspidatum*, which has been used for centuries in traditional Japanese and Chinese medicine to treat inflammation and liver and heart diseases [4]. Resveratrol exists in *cis* and *trans* isomeric forms; it is *trans*-resveratrol that is predominantly present in grapes, wines, and grape juice [1,5]. The amount and isoforms of resveratrol in wines are affected by a number of factors including grape cultivar, vinification techniques, and aging of the wines [6,7]. In general, red wines contain higher amounts of *trans*-resveratrol (up to 8 mg/l) than white wines (< 0.1 mg/l) [8].

Several epidemiological studies have shown an inverse correlation between moderate red wine consumption and the incidence of cardiovascular disease: a phenomenon known as the "French paradox." Despite the consumption of a high-fat diet, little exercise, and widespread smoking in southern France and other Mediterranean populations, the incidence of coronary heart disease is very low [9,10]. The health benefits of red wine are attributed to the presence of polyphenolic substances such as resveratrol and flavonoids with strong antioxidant activity. The antioxidant activity of these polyphenolics derives from the ability of the aromatic hydroxyl group to donate H^+ to free radicals [11] which has been suggested to reduce the oxidation of low-density lipoprotein and prevent atherogenicity [12,13]. In addition, the ability of these polyphenolic compounds to inhibit platelet aggregation and to induce relaxation of blood vessels through production of nitric oxide has also been implicated in the cardioprotective effect of red wines [14,15]. The beneficial effects of red wine have sparked an intense interest in the pharmacological and biological activities of individual polyphenolic compounds present in red wine. Studies with synthetic resveratrol have shown that it also exhibits strong antioxidant activity, inhibits oxidation of low-density lipoprotein, prevents platelet aggregation, and decreases production of eicosanoids by platelets and leukotrienes by neutrophils [16–18]. These effects may contribute to the prevention of coronary heart disease.

Besides its cardioprotective effects, resveratrol has been shown to inhibit cellular events associated with tumor initiation, promotion, and progression [19,20]. Resveratrol has also shown antiproliferative and growth inhibitory activity against tumor cell lines of breast, prostate, liver, colorectal/intestinal, lung, and blood cancers [21,22]. The growth inhibitory and antimetastatic effects of resveratrol are also observed *in vivo* [23,24]. Although the mechanism of antitumor effects of resveratrol is not fully known, it may be related to its ability to inhibit ribonucleotide reductase and DNA polymerase, protein kinase C, cyclooxygenase-2, and cell cycle progression [25–28]. The antitumor activity of resveratrol may also result from its activation of Fas–Fas ligand interaction [29] and through inhibition of angiogenesis [30]. Thus there are numerous molecular targets of resveratrol which interact with cellular processes (inflammation, cell cycle, angiogenesis, apoptosis, etc.) to inhibit tumor growth.

IMMUNOMODULATION BY RESVERATROL

The cells and molecules responsible for protection against infectious and noninfectious agents collectively constitute the immune system, which is organized into the incredibly intricate arrangement of central (bone marrow, thymus) and peripheral (lymph nodes, spleen) lymphoid organs and tissues. A healthy individual has two levels of defense against foreign agents: innate (natural) immunity and adaptive or acquired immunity. The mechanisms of innate immunity exist before an encounter with microbes and are rapidly activated before the development of adaptive immune responses. The principal effector cells of innate immunity are phagocytes (neutrophils, monocytes, macrophages) and natural killer (NK) cells. These cells engage and destroy microbes that have breached the epithelial barriers. Neutrophils and phagocytes produce toxic mediators, such as hydrogen peroxide, superoxide anion (O_2^-), and nitric oxide which are directly toxic to microbes. Free radicals are generated by lysosomal NADPH oxidases in a process known as respiratory burst.

EFFECT ON INFLAMMATORY PROCESSES

Activated macrophages release proinflammatory cytokines (tumor necrosis factor (TNF), interleukin (IL)-1, etc.) to induce inflammation in tissue. The inflammatory process is further aided by the production by macrophages and neutrophils of prostaglandins, leukotrienes, and thromboxanes, collectively known as eicosanoids. These mediators are synthesized through enzymatic degradation of arachidonic acid (AA) contained in the phospholipids of cell membranes. Prostaglandin synthesis is catalyzed by cylcooxygenase (COX), also known as prostaglandin G/H synthase. Two distinct isoforms of COX, COX-1 and COX-2, have been discovered [31]. COX-1 is constitutively expressed and is important in mucosal integrity, maintenance of renal function, and platelet stabilization and activity. COX-2 is induced by inflammatory mediators, endotoxins, cytokines, growth factors, tumor promoters, and stress. COX-2 catalyzes the synthesis of prostaglandins (PGE2, PGF2-α, PGI2, PGD2), thromboxanes, and leukotrienes (LT) by mononuclear phagocytes, endothelial cells, polymorphonuclear leukocytes, and platelets [32–34]. Eicosanoid byproducts of AA are also produced via lipooxygenase (LOX) and cytochrome P450 (cyp450) pathways. The LOX group of enzymes metabolizes AA to produce hydroperoxyeicosatetraenoic acids (HpETEs), which are converted into series-4 leukotrienes (e.g., 5-HETE, 12-HETE, and 15-HETE); cyp450 can directly catalyze the formation of 12-HETE and 16-HETE. Together, these AA-derived compounds contribute to pain, inflammation, swelling, vasoconstriction, and thrombosis. Eicosanoids are also elevated in most human cancers and potentially play a significant role in tumor cell proliferation, angiogenesis, spread of cancers, and suppression of immune response [35].

Resveratrol has been shown to suppress inflammatory responses through its antioxidative activity and by interfering with many of the pathways previously described that are involved in the metabolism of AA and production of eicosanoids. Resveratrol inhibits TPA-induced free radical formation in HL-60 leukemic cells in a dose-dependent manner [36]. It also inhibited the formation of superoxide radicals (O_2^-) and H_2O_2 by macrophages stimulated by lipopolysaccharide (LPS) and phorbol esters (TPA) [37], and generation of reactive oxygen intermediates by TNF in wide variety of cell lines [38]. In addition, resveratrol is reported to suppress unopsonized zymosan-induced oxygen radical production in mouse and human phagocytic cells [39]. In another report, resveratrol was shown to inhibit nitric oxide production in macrophages by suppressing nitric oxide synthase [40]. In rats, pretreatment with resveratrol completely prevents oxidative damage to kidneys by the kidney-specific carcinogen $KBrO_3$ [41].

Early studies by Kimura et al. [42] showed that resveratrol, isolated from the roots of *Polygonum* spp., inhibited the activity of 5-LOX and COX products in rat peritoneal leukocytes. Resveratrol inhibited the cyclo-oxygenase and hydroperoxidase activity of COX-1 isolated from the seminal vesicles of sheep in a concentration-dependent manner [43]. In contrast, enzymatic activity of COX-2 was not significantly affected by resveratrol [44]. The *in vivo* antiinflammatory activity of resveratrol in a rat model of carrageenan-induced paw edema was attributed to the impairment of prostaglandin synthesis via selective inhibition of COX-1 [44]. In contrast to the inhibition of COX-1 activity *in vivo*, resveratrol inhibited phorbol myristate acetate (PMA)-induced PGE2 production *in vitro* by suppressing the transcription and activity of COX-2 [45]. The inhibition of protein kinase C signaling pathway was found to be the mechanism by which resveratrol suppressed COX-2 gene expression and activity [45]. In another study, resveratrol inhibited the FCS- and PDGF-stimulated production of reactive oxygen species (ROS), phospholipase A2 activity, AA release, and PGE2 synthesis in 3T6 fibroblasts [46]. The generation of ROS, COX-2 induction, AA release, and prostaglandin synthesis were also markedly inhibited in murine peritoneal macrophages stimulated with LPS, PMA, O_2^-, or H_2O_2 [38]. In human PMNL, resveratrol decreased the 5-LOX proinflammatory products (5-HETE, 5,12-dHETE, and leukotriene C4), inhibited the release of lysozyme and β-glucuronidase from lysosomes, and decreased ROS generation upon exposure to calcium ionophore [17,47]. Resveratrol also inhibits collagen and thrombin+ADP-induced aggregation of platelets and synthesis of TxB_2 and HHT, the stable products formed from thromboxane A_2 [14]. However, production of 12-HETE from AA via 12-lipooxygenase is not affected. Together, these published data indicate that the antioxidant and antiinflammatory effects of resveratrol are mediated by its ability to inhibit strongly the generation of ROS (superoxide anion, hydrogen peroxide, and hydroxyl radicals), and production of eicosanoids from AA through COX and LOX pathways.

MODULATION OF CELL-MEDIATED IMMUNITY BY RESVERATROL

The effect of resveratrol on various lymphoid cell populations and their functional activity has not been vigorously investigated. The immunomodulatory effects of resveratrol, primarily on the generation of cell-mediated cytotoxicity and cytokine production, have only recently been published in scattered reports. An extensive database search has failed to discover any publication dealing with the effect of resveratrol on humoral (antibody-mediated) immune responses; therefore, data published on the effect of resvertrol on production of cell-mediated immune responses and cytokines are reviewed.

EFFECT ON LYMPHOID CELLS

Based on the observations that resveratrol inhibits the clonal growth and induces apoptosis of mouse and human leukemia cell lines, Gautam et al. [48] investigated the effect of resveratrol on growth of mouse hematopoietic progenitor cells *in vitro*. The incorporation of resveratrol into cultures during development of colony-forming units-total (CFU-total) inhibited the proliferation and expansion of bone marrow-derived hematopoietic progenitor cells in a dose-dependent fashion. Unlike the irreversible growth inhibitory effect of resveratrol on leukemic progenitor cells, the suppressive effect of resveratrol on growth of normal hematopoietic progenitor cells was reversible; removal of resveratrol from the cultures significantly restored clonal expansion of these cells. These investigators examined the effect of resveratrol on CFU-total only; therefore, the effect of resveratrol on the proliferation of lineage-specific progenitors (e.g., CFU-GEMM, CFU-GM, CFU-M, CFU-E) is not known. Furthermore, treatment with resveratrol does not appear to interfere with the engraftment of bone marrow cells as nearly 80% of the lethally irradiated mice transplanted with resveratrol-treated bone marrow cells survived, whereas 100% of the irradiated mice without bone marrow transplant died. The CFU-total in marrow of mice transplanted with normal or resveratrol-treated bone marrow cells were identical six weeks after transplantation. Hematologic recovery was also comparable in mice transplanted with untreated or treated bone marrow cells. Based on these findings, the authors suggested that resveratrol could potentially be used for *ex vivo* purging of bone marrow transplants to destroy contaminating tumor cells.

The effect of oral administration of resveratrol on various lymphoid populations in different lymphoid tissues has been evaluated by two independent investigative teams. In a study by Juan et al. [49], oral treatment of Sprague-Dawley rats with resveratrol did not cause adverse affects or mortality during the treatment period. There was no significant change in body weight, red blood cells (RBC), white blood cells (WBC; leukocytes, neutrophils, lymphocytes, monocytes, eosinophils, basophils), or platelets in the blood samples of animals treated with resveratrol. Repeated administration of resveratrol caused no pathological changes in spleen or other major organs. The author concluded that long-term administration of *trans*-resveraro1 at a dose significantly higher than the amount ingested (8 mg) by consuming three glasses of red wine daily is not harmful. In another study by Gao et al. [50], administration of higher doses of resveratrol to mice was nontoxic. Administration of resveratrol to mice orally at 2 mg/day, 5 days a week, for 2 or 4 weeks resulted in insignificant changes in body weight. In addition, there were no significant changes in the hematopoietic progenitors (CFU-total) in the bone marrow or WBC, RBC, or platelets in the peripheral blood. Furthermore, there was no change in the number of nucleated lymphoid cells in bone marrow or spleen after feeding mice with resveratrol for 2 or 4 weeks. Flow cytometric analysis revealed no change in $CD4^+$ or $CD8^+$ subsets of T lymphocytes in the spleen. These investigators also concluded that oral intake of resveratrol, even in large quantities, does not induce hematopoietic or hematologic deficiencies.

Effect on Antigen Presenting Cells

The production of an acquired immune response involves capturing, processing, and presentation of antigens by antigen presenting cells (APCs), macrophages and dendritic cells, to T or B lymphocytes. Resveratrol may modulate immune responses by interfering with antigen presentation by APCs. Resveratrol was studied for its effect on phagocytosis and intracellular killing of *Candida albicans* by the human promonocytic U937 cell line [51]. Treatment with resveratrol enhanced phagocytosis of *C. albicans* and decreased intracellular killing of these microorganisms at 10 μM resveratrol. However, phagocytosis was enhanced when the concentration of resveratrol was reduced to 1 μM without an increase in intracellular killing. In a recent publication, Kim et al. [52] described an inhibitory effect of resveratrol on the phenotypic and functional maturation of mouse bone marrow-derived dendritic cells (BM-DC). Resveratrol inhibited the expression of costimulatory molecules CD80 and CD86, and major histocompatibility complex class I and II molecules on BM-DC. Resveratrol-treated DC were very efficient in antigen capture via mannose-mediated endocytosis; however, they were poor inducers of an allogeneic mixed leukocyte reaction. Although the effect of resveratrol on antigen processing and presentation was not addressed in both of these studies, the inhibition of antigen capture by resveratrol indicates that resveratrol may downregulate production of immune responses by interfering with antigen handling by antigen presenting cells.

Effect on T Cell Proliferation

The presentation of antigen to responsive T lymphocytes by antigen presenting cells or the binding of antigen directly to B lymphocytes leads to the activation, proliferation, and clonal expansion of the antigen reactive immune cells. The inhibition or augmentation of the proliferative response of lymphocytes is frequently used to evaluate potential therapeutic agents for their immunomodulatory effects. Resveratrol has been investigated for its effect on mitogen- or alloantigen-induced proliferation of T cells *in vitro*. Using incorporation of ^{3}H-thymidide in duplicating DNA to measure cell proliferation, Gao et al. [53] showed that resveratrol (10 μM) suppresses the proliferation of murine splenic T cells stimulated with conconavalin A (Con A), IL-2, or allogeneic cells in a dose-dependent manner (Figure 23.1). At a concentration of 50 μM, resveratrol completely inhibited the proliferative response of these cells. It was also reported that at concentrations lower than 10 μM, resveratrol had no effect on cell proliferation or slightly enhanced the proliferation of T cells responding to Con A or allogeneic cells. Feng et al. [54] also found that low doses of resveratrol (6 μM or less) enhanced Con A-induced proliferation of mouse splenic cells; however, these investigators did not examine the effect of higher concentrations of resveratrol. In two studies, resveratrol was also shown to inhibit the proliferation of human peripheral mononuclear cells (PMNC). Resveratrol inhibited the proliferation of PMNC stimulated with phytohemagglutinin A (PHA) at 10^{-4} M but not at concentrations of 10^{-5} or 10^{-6} M [55]. Resveratrol was not found to affect the spontaneous proliferation of PMNC at these concentrations. In another study, resveratrol at 5 or 10 μg/ml was shown to inhibit completely the proliferation of PMNC stimulated with anti-CD3 or anti-CD28 antibodies whereas at lower concentrations the agent had no significant effect [56]. The concentrations of resveratrol used in these experiments had no significant effect on the viability of PMNC, indicating that suppression of PMNC proliferation at higher concentrations is not due to a toxic effect of resveratrol. In order to examine the *in vivo* effect of resveratrol on the proliferation of lymphocytes, Gao et al. [50] administered resveratrol orally (2 mg per day) to mice for 2 to 4 weeks before evaluating mitogen/IL-2-induced proliferation of splenic T cells *in vitro*. There was no significant change in the Con A-induced proliferation of splenic T cells whether animals were treated with resveratrol for 2 or 4 weeks. Although feeding resveratrol for 2 weeks did not affect the proliferative response to IL-2, it was significantly increased after feeding for 4 weeks. To rule out the

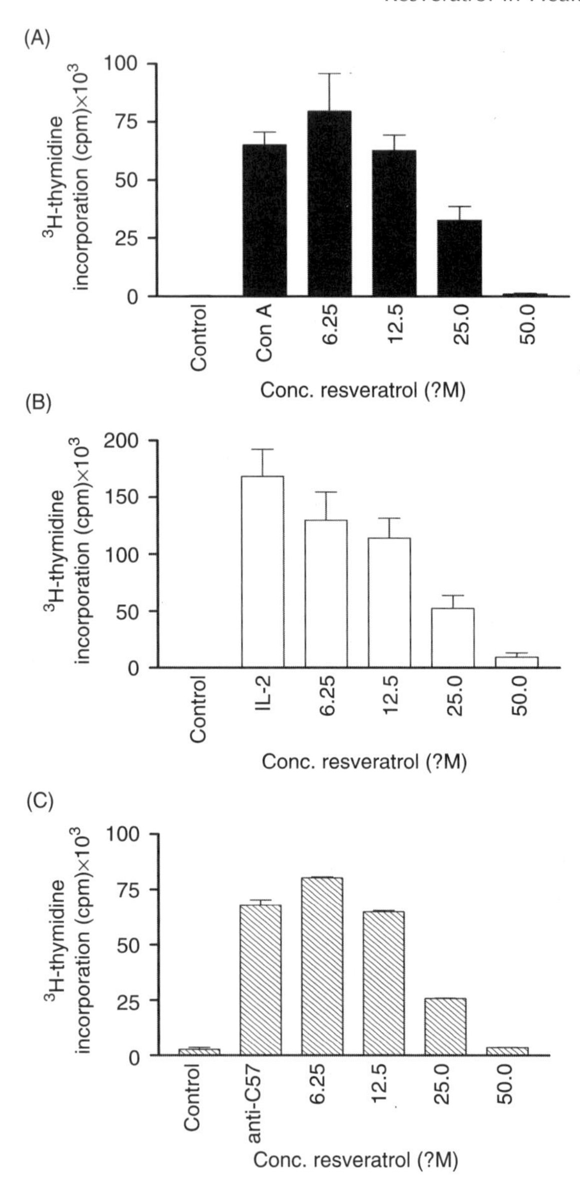

FIGURE 23.1 Effect of resveratrol on proliferation of spleen cells. C3H spleen cells (10^6 cells/ml) were stimulated with Con A (1.5 μg/ml) (A) or IL-2 (150 ng/ml) (B), or irradiated C57BL/6 spleen cells (1:1) (C) for 4 days in the absence or in the presence of resveratrol (6.25 to 50 μM). Cells (2×10^5) from each culture were transferred to the wells of a 96-well microtiter tissue culture plate in triplicate. Cultures were pulsed with [^{3}H]-thymidine (0.25 μCi/well) for 8 h. [^{3}H]-thymidine incorporation was determined by liquid scintillation spectrometry. (Reproduced from Gao X, Xu YX, Janakiraman N, Chapman RA, and Gautam SC, *Biochem Pharmacol* 62, 1299–1308, 2001. With permission. Copyright 2001, Elsevier Publishers.)

possibility that cells might have recovered from immunosuppression during the culture period to induce proliferation with Con A or IL-2, the authors examined the proliferative response of lymphocytes in the popliteal draining lymph nodes of mice injected with allogeneic cells in the paws. Only a slight decrease (10 to 15%) in proliferation of lymphocytes was observed whether mice were treated with resveratrol for 2 or 4 weeks. The authors concluded that unlike the suppressive effect of resveratrol *in vitro*, feeding mice resveratrol for 2 to 4 weeks does not induce significant antiproliferative effect *in vivo*.

Effect on Cell-Mediated Cytotoxicity

The production of antigen-specific $CD8^+$ cytotoxic T lymphocytes (CTLs) plays a crucial role in the killing of virus-infected cells and neoplastic cells, and rejection of allogeneic transplants. In addition to antigen-specific CTLs, broadly nonspecific effector mechanisms such as activated macrophages and NK cells also play an important role in defense against infectious agents and tumors. The effect of resveratrol on the generation of antigen-specific CTLs and other nonspecific mediators of cell-mediated cytotoxicity was addressed in recent publications. Falchetti et al. [56] investigated the effect of resveratrol on the generation of alloreactive CTLs from PBMC *in vitro* by analyzing the interferon-gamma (IFN-γ)-secreting $CD8^+$ T lymphocytes and measuring target specific cytolysis by LDH release. Resveratrol was observed to increase significantly the frequency of IFN-γ-producing $CD8^+$ T cells at a concentration of 0.6 μg/ml; however, the development of this T cell subset was completely blocked at a concentration of 5 μg/ml. In addition, exposure to resveratrol also resulted in significant enhancement of CTL activity at a concentration of 0.6 μg/ml. In contrast, exposure to resveratrol at higher concentrations (2.5 and 5 μg/ml) significantly inhibited the induction of CTL activity. These findings indicate that resveratrol increases the production of alloantigen-specific CTLs at lower concentrations, whereas CTL induction is suppressed at higher concentration of resveratrol. The concentration-dependent biphasic effect of resveratrol on the generation of allo-specific CTLs was not observed by Gao et al. [53]. They tested the effect of resveratrol, over a wide range of concentrations, on the production of alloantigen-specific CTLs. There was no effect on the generation of CTL activity up to a concentration of 12.5 μM. The response was partially suppressed at 25 μM but was completely abolished at 50 μM resveratrol (Figure 23.2). The reason for the discrepancy between the results of these two reports is not clear; however, it may be related to the difference in species employed in these studies. In a subsequent publication, Gao et al. [50] demonstrated that unlike the inhibitory effect of resveratrol on generation of CTLs *in vitro*, CTL induction *in vivo* was only minimally affected after treating mice with resveratrol (2 mg/daily) orally for up to 4 weeks (Figure 23.3). These findings stress the importance of being cautious

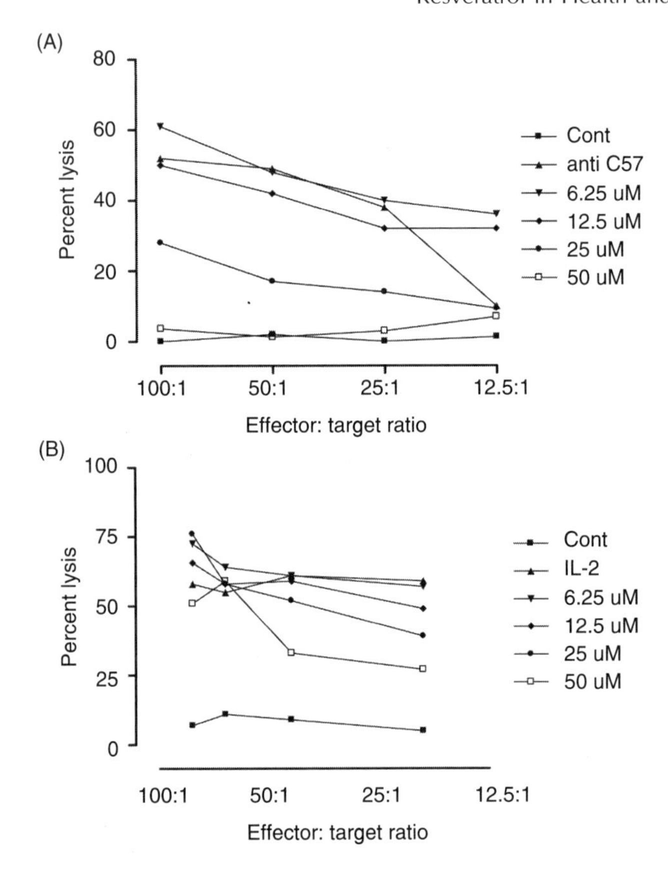

FIGURE 23.2 Effect of resveratrol on the development of cell-mediated cytotoxicity. For the effect of resveratrol on the generation of CTLs, 10^7 C3H spleen cells were cocultured with an equal number of irradiated C57BL/6 spleen cells for 5 days in the absence or in the presence of resveratrol (6.25 to 50 μM). Cytotoxicity of the effector against C57BL/6 splenic blast cells was determined in a 4 h ^{51}Cr-release assay (A). Effect of resveratrol on LAK cell generation was examined by incubating C3H spleen cells (5×10^5 cells/ml) with IL-2 (200 ng/ml) for 3 days in the absence or the presence of resveratrol (6.25 to 50 μM). Cytotoxicity of effector cells against 32Dp210 mouse leukemia cells was measured in a 4 h ^{51}Cr-release assay (B). In each panel, the results are presented from a representative experiment as percent lysis at different effector-to-target ratios. (Reproduced from Gao X, Xu YX, Janakiraman N, Chapman RA, and Gautam SC, *Biochem Pharmacol* 62, 1299–1308, 2001. With permission. Copyright 2001, Elsevier Publishers.)

in interpreting the results of *in vitro* studies, as they may not always be reproducible *in vivo*.

The effect of resveratrol on the activity of nonspecific cytotoxic effector cells such as NK cells, activated macrophages, and lymphokine activated killer (LAK) cells has also been examined. Incubation of human PBMC

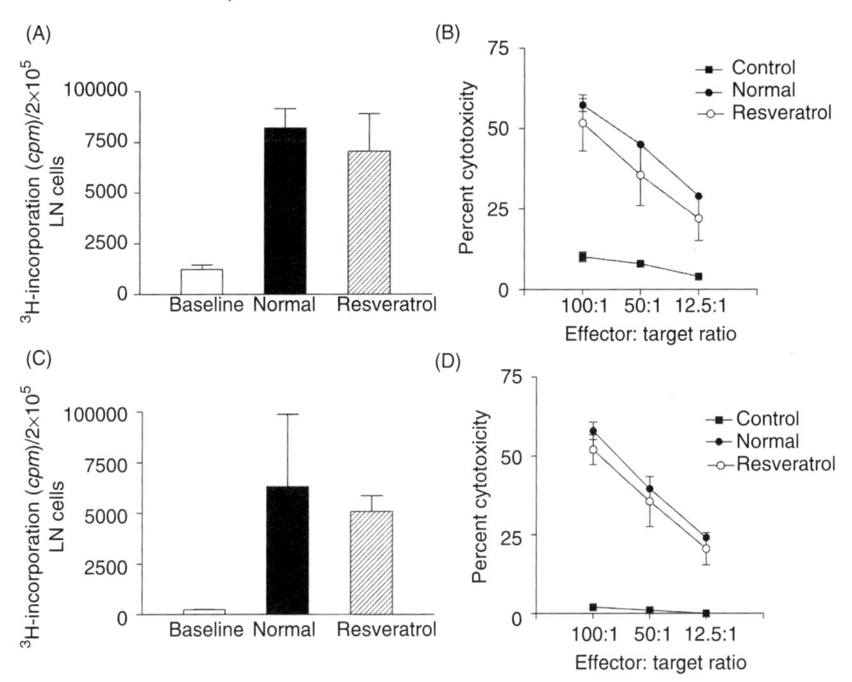

FIGURE 23.3 Effect of resveratrol on *in vivo* development of T cell-mediated immune responses. Normal C3H mice or mice treated with resveratrol orally for 2 or 4 weeks were injected in the hind paws with 2.5×10^7 irradiated (20 Gy) C57BL/6 spleen cells. Five days later, draining popliteal lymph nodes were removed and tested for T cell proliferation by incorporation of [^{3}H]-thymidine and cell-mediated cytotoxicity against EL-4 cells of C57BL/6 origin in a 4 h ^{51}Cr-release assay. (A) T cell proliferation and (B) T cell-mediated cytotoxicity after treatment with resveratrol for 2 weeks. (C) Proliferative and (D) cytotoxic response after treatment with resveratrol for 4 weeks. (Reproduced from Gao X, Dorrah D, Media J, Divine G, Jiang H, Janakiraman N, Chapman RA, and Gautam SC, *Biochem Pharmacol* 66, 2427–2435, 2003. With permission. Copyright 2003, Elsevier Publishers.)

with resveratrol at concentrations between 0.075 and 1.25 μg/ml for 18 hours enhanced their cytotoxic activity against NK-sensitive K562 target cells, with maximum enhancement occurring at 0.3 μg/ml [56]. In contrast, exposure of PBMC to resveratrol at 5 and 20 μg/ml resulted in dose-related inhibition of NK activity. Resveratrol at a concentration of 2.5 mg/l reduced the cytotoxicity of peritoneal macrophages against mouse hepatocellular carcinoma H22 target cells, but it was significantly increased by resveratrol at a concentration of 20 mg/l [57]. The concentration-related effect of resveratrol on cytotoxicity of macrophages contrasts with those on the production of CTLs and NK activity as noted above. Gao et al. [53] examined the effect of resveratrol on IL-2 activated LAK cells at concentrations ranging from 6.25 to 50.0 μM. Unlike the suppression of CTL generation at concentrations of 25 and 50 μM, the effect of resveratrol on production of LAK cells was minimal at concentrations ranging from 6.25 to 50 μM (Figure 23.2). In addition, treatment of mice with resveratrol at a dose of 80 mg/kg body weight for 2 to 4 weeks also did not have an impact on the production of LAK cells [50], suggesting that, unlike CTL production, the generation of NK cells, tumoricidal macrophages, and LAK cells is relatively resistant to resveratrol. Overall, resveratrol seems to have a concentration-dependent biphasic effect on the generation of CTL and NK activity from human PBMC, and it inhibits the generation of CTLs from mouse spleen cells at relatively high concentrations that also inhibit tumoricidal activity of macrophages without affecting the generation of LAK cells.

Effect on Production of Cytokines

Cytokines are polypeptides produced by lymphocytes, monocytes, and a variety of tissue cells, including endothelial and epithelial cells. Cytokines are usually produced in response to microbes, noninfectious antigens, and physiological stress. Cytokines stimulate the response of cells involved in immunity and inflammation. The generation of T cell-mediated immune responses involves a complex network of intracellular and intercellular cytokine signals generated by antigen presenting cells and T helper1 (Th1) and Th2 cells. Activated monocytes/macrophages release several proinflammatory cytokines (monokines) such as TNFα, IL-1, and IL-6, which play a prominent role in immune and inflammatory responses. Th1 cells secrete IL-2 and IFN-γ, and promote predominantly cell-mediated immunity. Th2 cells secrete IL-4, IL-5, IL-6, and TGF-β that upregulate humoral immunity and negatively regulate cell-mediated immunity. Because resveratrol exerts immunoregulatory effects on production of immune responses, it has been investigated for effects on the production of cytokines. In a study by Nair et al. [58] grape seed extract was found to enhance gene transcription and secretion of IFN-γ, a Th1 cytokine, by human PBMC, without any effect on Th2-derived cytokine IL-6. In another study, *in vitro* exposure of PBMC to resveratrol produced biphasic effect on the anti-CD3- or anti-CD28-induced development of IFN-γ, IL-2, and IL-4 producing $CD8^+$ and $CD4^+$ T cells [56]. The production of these cytokines was stimulated at low concentrations and inhibited at high concentrations of resveratrol. Boscolo et al. [55] demonstrated suppression of IFN-γ and TNFα production by PHA-stimulated PBMC by resveratrol at a concentration of 10^{-4} M but not 10^{-5} or 10^{-7} M without a clear concentration-dependent biphasic effect as reported by Falchetti et al. [56]. The discrepancy between the results of these two reports may be related to different stimuli applied to the cells to induce cytokine production. In a study by Gao et al. [53] resveratrol was found to impair irreversibly IFN-γ and IL-2 production by mouse spleen cells stimulated with Con A. Removal of resveratrol from the cultures following treatment of spleen cells for 20 hours at 50 μM failed to restore production of these cytokines upon stimulation of cells with Con A. These investigators also showed that although *in vitro* treatment of spleen cells with resveratrol readily inhibited the production of cytokines, oral administration of resveratrol (80 mg/kg body weight) for 4 weeks had no effect on production of these cytokines [50]. The authors suggested that discrepant *in vitro* and *in vivo* effects of resveratrol might be due to poor bioavailability or rapid metabolism and failure to achieve immunosuppressive concentration of resveratrol *in vivo*.

Resveratrol also modulates the production of cytokines by monocytes and macrophages. Treatment of the macrophage cell line RAW 264.7 with resveratrol at concentrations of 0.05 and 0.1 mM increased the basal and LPS-induced levels of TNFα mRNA and protein secretion 2- and 8-fold, respectively [59]. The compound also inhibited IL-6 release from peritoneal macrophages stimulated with calcium ionophore A23187 and fMLP [60], and glial cells subjected to hypoxia/hypoglycemia followed by reoxygenation [61]. Since resveratrol also inhibited calcium ion influx into macrophages [60], blocking calcium ion influx into cells was proposed as a possible mechanism by which resveratrol may inhibit biosynthesis of IL-6. Gao et al. [53] also showed partial but irreversible suppression of LPS-induced release of TNFα and IL-12 (p40) from murine peritoneal macrophages by resveratrol. Secretion of TNFα from macrophages was partially reduced after treatment of mice with resveratrol orally (80 mg/kg body weight) for 4 weeks; however, production of IL-12 (p40) remained unaffected [50]. In another study, Liu et al. [57] showed that i.p. administration of resveratrol in doses of 500 to 1500 mg/kg body weight modestly increased LPS-induced TNFα levels in tumor-bearing mice. Resveratrol was also found to suppress LPS-induced intracellular IL-12 p40/p70 and secretory IL-12 p70 in bone marrow-derived dendritic cells [52]. Cytokine production by human THP-1 macrophages was differentially regulated by resveratrol. It enhanced TNFα, IL-12, and IL-1β induction by LPS, but in cells primed with IFN-γ, resveratrol inhibited the production of TNFα, IL-12, and IL-1β by LPS [62]. In human monocytic cells, resveratrol inhibited accumulation of mRNA and production of PMA-induced IL-8 protein at concentrations of 10, 1, and 0.1 μM [63]. In another study, resveratrol was shown to inhibit both basal and induced production of IL-8 and granulocyte-macrophage colony stimulating factor by alveolar macrophages from patients with chronic obstructive pulmonary disease [64].

Thus, it is quite evident from the results of various published reports that resveratrol exerts strong immunoregulatory effects on the production of cytokines by lymphocytes and monocytes/macrophages. In general, cytokine production is enhanced at low concentrations and suppressed at high concentrations of resveratrol; however, additional studies are needed to establish the *in vivo* effect of resveratrol on cytokine production.

EFFECT ON NF-κB ACTIVATION

Nuclear factor kappaB (NF-κB) is a member of the Rel family of pleiotropic transcription factors that regulate transcription of genes involved in immune and inflammatory responses, cell proliferation, apoptosis, oncogenesis, and atherosclerosis [65,66]. Under normal conditions, NF-κB is sequestered in the cytoplasm, as a heterodimer of Rel proteins p50 and p65, by an inhibitory protein IκBα [67]. In response to a variety of stimuli, including cytokines, oxygen radicals, bacterial and viral products, ultraviolet light, and chemotherapeutic agents, IκBα is rapidly phosphorylated and degraded by cytosolic proteosome [68]. Active NF-κB translocates to the nucleus, where it regulates the expression of target genes containing κB regulatory elements. Phosphorylation of IκBα is catalyzed by a stimulus-responsive IκB kinase (IKK) complex [69]. Since NF-κB plays a central role in many inflammatory disease processes and oncogenesis, it provides an attractive molecular target for the development of novel therapeutic agents. Holmes-McNary and Baldwin [70] showed that resveratrol is a potent inhibitor of both NF-κB activation and NF-κB-dependent gene expression in human monocyte (THP-1) and macrophage (U937) cell lines. Treatment with resveratrol at 30 μM inhibited TNF- or LPS-induced activation of NF-κB by inhibiting IκB kinase activity. It also blocked the transcription of monocyte chemoattractant protein-1, a NF-κB-regulated gene.

Recent findings suggest that chemoprevention by resveratrol also involves suppression of NF-κB and NF-κB-dependent COX-2 and MMP-2 expression [71]. Although inhibition of proinflammatory molecules produced by activated macrophages by resveratrol has been well documented, the role of suppression of NF-κB remains controversial. Tsai et al. [40] demonstrated that suppression of LPS-inducible nitric oxide synthase in RAW 264.7 cells, both at the mRNA and protein levels, is associated with inhibition of activation of NF-κB. However, Wadsworth and Koop [59] showed that resveratrol decreased LPS-induced NO release without affecting NF-κB activation. In another report, resveratrol was demonstrated in a number of cell types to inhibit NF-κB activation by TNF and several other inducers, such as PMA, LPS, IL-1β, H_2O_2, okadaic acid, and ceramide [38,72]. The suppression of NF-κB by resveratrol coincided with the inhibition of AP-1, MAP kinase kinase, cJNK, ROS generation, lipid peroxidation, and apoptosis. The authors concluded that anticarcinogenic, antiinflammatory, and growth inhibitory effects of resveratrol may be ascribed to the inhibition of the activation of NF-κB, AP-1, and associated kinases. Suppression of IL-8 release by resveratrol from PMA-stimulated U937 cells was shown to coincide with inhibition of AP-1 but not NF-κB activation [63]. This result contrasts with the results of

Manna et al. [38] showing inhibition of PMA-induced activation of NF-κB in U937 cells. Gao et al. [53] demonstrated that inhibition of IL-2 release by spleen cells activated with Con A was associated with inhibition of NF-κB activation. Collectively, these data indicate that resveratrol modulates NF-κB activation in cell-specific and stimulus-dependent manners. Furthermore, inhibition of NF-κB and NF-κB-dependent gene products provides part of the molecular basis for the immunomodulatory effects of resveratrol.

CONCLUSION

There has been a steady increase in the integration of natural remedies and conventional medicine in recent years. Resveratrol has attracted considerable attention due to the wide range of biological effects associated with this molecule, including, but not limited to, the prevention of chronic heart disease and neoplastic transformation, inhibition of eicosanoid synthesis, cell cycle inhibition, modulation of lipid metabolism, and antioxidant activity. Many of the health benefits of resveratrol are attributed to its antioxidant and antiinflammatory activity. While the cardioprotective and chemopreventive activity of resveratrol against cancers has been well documented in several epidemiological and experimental studies, the immunoregulatory effects of resveratrol have not yet been fully explored. To date, whether or not resveratrol modulates the induction of humoral immune responses has not been demonstrated. There are few studies, mainly *in vitro* in cell culture, that have investigated the effect of resveratrol on the development of cellular immune responses and cytokine production. The results of these studies have been inconsistent and even contradictory concerning the selectivity and concentration-dependent biphasic effect of resveratrol on production of cell-mediated immune responses and their soluble mediators. The mechanism of immunosuppression or immunoenhancement by resveratrol remains to be elucidated. Clearly, more research is required to establish thoroughly the immunomodulatory properties of resveratrol and their mechanism of action.

REFERENCES

1. Langcake P and Pryce RJ, The production of resveratrol by *Vitis vinifera* and other members of the Vitaceae as a response to infection and injury, *Physiol Plant Pathol* 9, 77–86, 1976.
2. Siemann EH and Creasy LL, Concentration of the phytoalexin resveratrol in wine, *Am J Enol Vitic* 43, 49–52, 1992.
3. Creasy LL and Coffee M, Phytoalexin production potential of grape berries, *J Am Soc Hortic Sci* 113, 230–234, 1988.

4. Nonomura S, Kanagawa H, and Makimoto A, Chemical constituents of polygonaceous plants: 1. Studies on the components of Ko-jo-kon (*Polygonum cuspidatum* SIEB et ZUCC), *Yakugaku Zasshi* 83, 983–988, 1963.
5. Jeandet P, Bessis R, and Gautherson B, The production of resveratrol (3,5,4′-trihydroxystilbene) by grapes berries in different developmental stages, *Am J Enol Vitic* 42, 41–46, 1991.
6. Price SF, Breen PJ, Valladao, and Watson BT, Cluster sun exposure and quercetin in Pinot noir grapes and wines, *Am J Enol Vitic* 46, 26–29, 1995.
7. McDonald MS, Hughes M, Burns J, Lean MEJ, Matthews D, and Crozier A, A survey of the free and conjugated myricetin and quercetin content of red wines of different geographical origins, *J Agric Food Chem* 46, 368–375, 1998.
8. Goldberg DM, Yan J, Ng E, Diamandis EP, Karumnchiri A, Soleas G, and Waterhouse AL, A global survey of trans-resveratrol concentration in commercial wines, *Am J Enol Vitic* 46, 155–187, 1995.
9. Renaud S and Lorgeril M, Wine, alcohol, platelets, and the French paradox for coronary heart disease, *Lancet* 339, 1523–1526, 1992.
10. Constant J, Alcohol, ischemic heart disease, and the French paradox, *Coron Artery Dis* 8, 645–649, 1997.
11. Kanner J, Frankel E, Grant R, German B, and Kinsella E, Natural antioxidants in grapes and wines, *J Agric Food Chem* 42, 64–69, 1994.
12. Frankel EN, Waterhouse AL, and Teissedre PL, Principal phenolic phytochemicals in selected Californian wines and their antioxidant activity in inhibiting oxidation of human low-density lipoproteins, *J Agric Food Chem* 43, 890–894, 1995.
13. Nigdikar SV, Williams NR, Griffin BA, and Howard AN, Consumption of red wine polyphenols reduces the susceptibility of low-density lipoproteins to oxidation *in vivo*, *Am J Clin Nutr* 68, 258–265, 1998.
14. Pace-Asciak CR, Hahn S, Diamandis EP, Soleas G, and Goldberg DM, The red wine phenolics trans-reversatrol and quercetin block human platelet aggregation and eicosanoid synthesis: implication for protection against coronary heart disease, *Clin Chim Acta* 235, 207–219, 1995.
15. Fitzpatrick DF, Hisschfield SL, and Coffey RG, Endothelium-dependent vasorelaxing activity of wine and other grape products, *Am J Physiol* 265, H774–H778, 1993.
16. Soleas GJ, Diamndis EO, and Goldberg DM, Resveratrol: a molecule whose time has come? And gone?, *Clin Chem* 30, 91–113, 1997.
17. Rotondo S, Rajtar G, Manarini S, Celardo A, Rotilio D, de Gaetano G, Evangelista V, and Cerletti C, Effects of trans-resveratrol, a natural polyphenolic compound, on human polymorphonuclear leukocyte function, *Br J Pharmacol* 123, 1691–1699, 1998.
18. Fremont L, Biological properties of resveratrol, *Life Sci* 66, 663–673, 2000.
19. Jang M and Pezzuto JM, Cancer chemopreventive activity of resveratrol, *Drugs Exp Clin Res* 25, 65–77, 1999.
20. Gusman J, Malonne H, and Atassi G, A reappraisal of potential chemopreventive and chemotherapeutic properties of resveratrol, *Carcinogenesis* 22, 1111–1117, 2001.
21. Aziz MH, Kumar R, and Ahmad N, Cancer chemoprevention by resveratrol: *in vitro* and *in vivo* studies and underlying mechanisms (review), *Int J Oncol* 23, 17–28, 2003.

22. Gautam SC, Xu YX, Dumaguin M, Janakiraman N, and Chapman RA, Resveratrol selectively inhibits leukemia cells: a prospective agent for ex vivo bone marrow purging, *Bone Marrow Transplant* 25, 639–645, 2000.
23. Kimura Y and Okuda H, Resveratrol isolated from *Polygonum cuspidatum* root prevents tumor growth and metastasis to lung and tumor-induced neovascularization in Lewis lung carcinoma-bearing mice, *J Nutr* 131, 1844–1849, 2001.
24. Carbo N, Costelli P, Baccino FM, Lopez-Soriano FJ, and Argiles JM, Resveratrol, a natural product present in wine, decreases tumor growth in a rat model, *Biochem Biophys Res Commun* 254, 739–743, 1999.
25. Fontcave M, Lepoivre M, Elleingand E, Gerez C, and Guittet O, Resveratrol, a remarkable inhibitor of ribonucleotide reductase, *FEBS Lett* 421, 277–279, 1997.
26. Jayatilake GS, Jayasuria H, Lee HS, Koonchanok NM, Geahlen RL, Ashendel CL, McLaughlin JL, and Chang CJ, Kinase inhibitors from *Polygonum cuspidatum*, *J Nat Prod* 56, 1805, 1993.
27. Ragione FD, Cucciolla V, Borriello A, Pietra VD, Racioppi L, Soldati G, Manna C, Galletti P, and Zappia V, Resveratrol arrests the cell division cycle at S/G2 phase transition, *Biochem Biophys Res Commun* 250, 53–58, 1998.
28. Kuwajerwala N, Cifuentes E, Gautam S, Menon M, Barrack ER, and Reddy GVP, Resverarol induces prostate cancer cell entry into S phase and inhibits DNA synthesis, *Cancer Res* 62, 2488–2492, 2002.
29. Clement MV, Hirpara JL, Chawdhury SH, and Pervaiz S, Chemopreventive agent resveratrol, a natural product derived from grapes, triggers CD95 signaling-dependent apoptosis in human tumor cells, *Blood* 92, 996–1002, 1998.
30. Tseng SH, Lin SM, Chen JC, Su YH, Huang HY, Chen CK, Lin PY, and Chen Y, Resveratrol suppresses the angiogenesis and tumor growth of gliomas in rats, *Clin Cancer Res* 10, 2190–2202, 2004.
31. Sharma S and Sharma SC, An update on eicosanoids and inhibitors of cyclooxygenase enzyme system, *Indian J Exp Biol* 35, 1025–1031, 1997.
32. Zimmermann GA, Whatley RE, McIntyre TM, McIntyre TM, Benson DM, and Prescott SM, Endothelial cells for studies of platelet activating factor and arachidonic metabolites, *Methods Enzymol* 187, 520–535, 1990.
33. Goldstein IM, Malmsten CM, Kindhal H et al., Thromboxane generation by human peripheral blood polymorphonuclear leukocytes, *J Exp Med* 48, 787–792, 1978.
34. Henderson WR and Klebanoff SJ, Leukotriene production and inactivation by normal, chronic granulomatous disease and myeloperoxidase-deficient neutrophils, *J Biol Chem* 258, 13522–13527, 1983.
35. Wallace JM, Nutritional and botanical modulation of the inflammatory cascade — eicosanoids, cyclooxygenase, and lipooxygenase — as an adjunct to cancer therapy, *Integ Cancer Ther* 1, 7–37, 2002.
36. Lee SK, Mbwabo ZH, Chung H et al, Evaluation of the antioxidant potential of natural products, *Comb Chem High Throughput Screen* 1, 35–46, 1998.
37. Martinez J and Moreno JJ, Effect of resveratrol, a natural polyphenolic compound, on reactive oxygen species and prostaglandin production, *Biochem Pharmacol* 59, 865–870, 2000.
38. Manna SK, Mukhopadhyaya A, and Aggarwal BB, Resveratrol suppresses TNF-induced activation of nuclear transcription factor NF-kB, activation

protein-1, and apoptosis: potential role of reactive oxygen intermediates and lipid peroxidation, *J Immunol* 164, 6509–6519, 2000.
39. Jang DS, Kang BS, Ryu SY et al., Inhibitory effect of resveratrol analogs on unopsonized zymosan-induced oxygen radical production, *Biochem Pharmacol* 57, 705–712, 1999.
40. Tsai SH, Lin-Shiau SY, and Lin JK, Suppression of nitric oxide synthase and down-regulation of the activation of NFkB in macrophages by resveratrol, *Br J Pharmacol* 126, 673–680, 1999.
41. Cadenas S and Barja G, Resveratrol, melatonin, vitamin E and PBN protect against renal oxidative damage by the kidney carcinogen $KbrO_3$, *Free Radical Biol Med* 26, 1531–1537, 1999.
42. Kimura Y, Okuda H, and Arichi S, Effects of stilbenes on arachidonate metabolism in leukocytes, *Biochim Biophys Acta* 834, 275–278, 1985.
43. Shin NH, Ryu SY, Lee H, Min KR, and Kim Y, Inhibitory effects of hydroxystilbenes on cyclooxygenase from sheep seminal vesicles, *Planta Med* 64, 283–284, 1998.
44. Jang M, Cai L, Udeani GO, Slowing KV, Thomas CF, Beecher CWW, Fong HHS, Farnsworth NR, Kinghorn AD, Mehta RG, Moon RC, and Pezzuto JM, Cancer chemopreventive activity of resveratrol, a natural product derived from grapes, *Science* 275, 218–220, 1997.
45. Subbaramaiah K, Chung WJ, Michaluart P, Telang N, Tanabe T, Inoue H, Jang M, Pezzuto JM, and Dannenberg AJ, Resveratrol inhibits cyclooxygenase-2 transcription and activity in phorbol ester-treated human mammary epithelial cells, *J Biol Chem* 273, 21875–21882, 1998.
46. Moreno JJ, Resveratrol modulates arachidonic acid release, prostaglandin synthesis, and 3T6 fibroblast growth factor, *J Pharm Exp Ther* 294, 333–338, 2000.
47. Kimura Y, Okuda H, and Kubo M, Effects of stilbenes isolated from medicinal plants on arachidonate metabolism, *J Ethnopharmacol* 45, 131–139, 1995.
48. Gautam SC, Xu YX, Dumaguin M, Janakiraman N, and Chapman RA, Resveratrol selectively inhibits leukemia cells: a prospective agent for ex vivo bone marrow purging, *Bone Marrow Transplant* 25, 639–645, 2000.
49. Juan ME, Vinardell MP, and Planas JM, The daily administration of high doses of trans-resveratrol to rats for 28 days is not harmful, *J Nutr* 132, 257–260, 2002.
50. Gao X, Dorrah D, Media J, Divine G, Jiang H, Janakiraman N, Chapman RA, and Gautam SC, Immunomodulatory activity of resveratrol: discrepant *in vitro* and *in vivo* immunological effects, *Biochem Pharmacol* 66, 2427–2435, 2003.
51. Bertelli AA, Ferrara F, Diana G, Fulgenzi A, Corsi M, Ponti W, Ferrero ME, and Bertelli A, Resveratrol, a natural stilbene in grapes and wine, enhances intraphagocytosis in human promonocytes: a co-factor in anti-inflammatory and anticancer chemopreventive activity, *Int J Tissue React* 21, 93–104, 1999.
52. Kim GY, Cho H, Ahn SC, Oh YH, Lee CM, and Park YM, Resveratrol inhibits phenotypic and functional maturation of murine bone marrow-derived dendritic cells, *Int Immunopharmacol* 4, 145–153, 2004.
53. Gao X, Xu YX, Janakiraman N, Chapman RA, and Gautam SC, Immunomodulatory activity of resveratrol: suppression of lymphocyte

proliferation, development of cell-mediated cytotoxicity, and cytokine production, *Biochem Pharmacol* 62, 1299–1308, 2001.
54. Feng Y-H, Zhou W-L, Wu QL, Li X-Y, Zhoa W-M, and Zou J-P, Low dose of resveratrol enhanced immune response of mice, *Acta Pharmacol Sin* 23, 893–897, 2002.
55. Boscolo P, del Signore A, Sabbioni E, Di Gioacchino M, Di Giampaola L, Reale M, Conti P, Paganelli R, and Giaccio M, Effects of resveratrol on lymphocyte proliferation and cytokine release, *Ann Clin Lab Sci* 33, 226–231, 2003.
56. Falchetti R, Pia Fuggetta M, Lanzilli G, Tricarico M, and Ravagnam G, Effects of resveratrol on human immune cell function, *Life Sci* 70, 81–96, 2001.
57. Liu HS, Pan CE, Yang W, and Liu XM, Antitumor and immunomodulatory activity of resveratrol on experimentally implanted tumor of H22 in Balb/c mice, *World J Gastroenterol* 9, 1474–1476, 2003.
58. Nair N, Mahajan S, Chawda R, Kandaswami C, Shanahan TN, and Schwartz SA, Grape seed extract activates Th1 cells *in vitro*, *Clin Diag Lab Immunol* 9, 470–476, 2002.
59. Wadsworth TL and Koop DR, Effects of the wine polyphenolics quecetin and resveratrol on pro-inflammatory cytokine expression in RAW264.6 macrophages, *Biochem Pharmacol* 57, 941–949, 1999.
60. Zhong M, Cheng GF, Wang WJ, Guo Y, Zhu XY, and Zhang JT, Inhibitory effect of resveratrol on IL-6 release by stimulated peritoneal macrophages in mice, *Phytomedicine* 6, 79–84, 1999.
61. Wang MJ, Huang SJ, Jeng KCG, and Kuo JS, Resveratrol inhibits interleukin-6 production in cortical mixed glial cells under hypoxia/hypoglycemia followed by reoxygenation, *J Neuroimmunol* 112, 28–34, 2001.
62. Feng YH, Zhu YN, Liu J, Ren YX, Xu JY, Yang YF, Li XY, and Zou JP, Differential regulation of resveratrol on lipopolysaccharide-stimulated human macrophages with or without IFN-gamma pre-priming, *Int Immunopharmacol* 4, 713–720, 2004.
63. Shen F, Chen SJ, Dong XJ, Zhong H, Li YT, and Cheng GF, Suppression of IL-8 gene transcription by resveratrol in phorboester treated human monocytic cells, *J Asian Nat Prod Res* 5, 151–157, 2003.
64. Culpitt SV, Gogers DF, Fenwick PS, Shah P, De Matos C, Russell RE, Barnes PJ, and Donnelly LE, Inhibition by red wine extract, resveratrol, of cytokine release by alveolar macrophages in COPD, *Thorax* 58, 942–946, 2003.
65. Baeuerle PA and Henkle T, Function and activation of NF-kB in the immune system, *Annu Rev Immunol* 12, 141, 1994.
66. Karin M and Lin A, NF-kB at the crossroad of life and death, *Nature Immunol* 3, 221–227, 2002.
67. Verma IM and Stevensen J, IkB kinase: beginning not the end, *Proc Natl Acad Sci* 94, 11758–11760, 1997.
68. Brown K, Gerstberger S, Carlson L, Franzoso G, and Siebenlist U, Control of IkB-a proteolysis by site-specific, signal-induced phosphorylation, *Science* 267, 1485–1491, 1995.
69. Karin M, How NF-kB is activated: the role of the IkB kinase (IKK) complex, *Oncogene* 18, 6867–6874, 1999.

70. Holmes-McNary M and Baldwin AS, Chemopreventive properties of trans-resveratrol are associated with inhibition of activation of the IkB kinase, *Cancer Res* 60, 3477–3483, 2000.
71. Banerje S, Bueso-Ramos C, and Aggarwal BB, Suppression of 7,12-dimethylbenz(a)anthracene-induced mammary carcinogenesis in rats by resveratrol: role of nuclear factor-kB, cyclooxygenase 2, and matrix metalloprotease 9, *Cancer Res* 62, 4945–4954, 2002.
72. Estrov Z, Shishodia S, Faderl S, Harris D, Van Q, Kantarjian HM, Talpaz M, and Aggarwal BB, Resveratrol blocks interleukin-1 beta-induced activation of the nuclear transcription factor NF-kappaB, inhibits proliferation, causes S-phase arrest, and induces apoptosis of acute myeloid leukemia cells, *Blood* 102, 987–995, 2003.

24 Biological Effects of *Cis*- Versus *Trans*-Resveratrol

Francisco Orallo

CONTENTS

INTRODUCTION

Resveratrol (3,4′,5-trihydroxystilbene, RESV, Figure 24.1) is a natural phenolic component of *Vitis vinifera* L. (Vitaceae), abundant in the skin of grapes and present in wines, especially red wines. Depending on a number of factors including grape cultivar, geographical origin, and fermentation time, a wide range of RESV concentrations has been reported in the different wines analyzed [1–4].

RESV is not unique to *Vitis* but is also present in at least 72 other plant species (distributed in 12 families and 31 genera; e.g., *Veratrum*, *Arachis*, *Morus*, and *Trifolium*), some of which are components of the human diet,

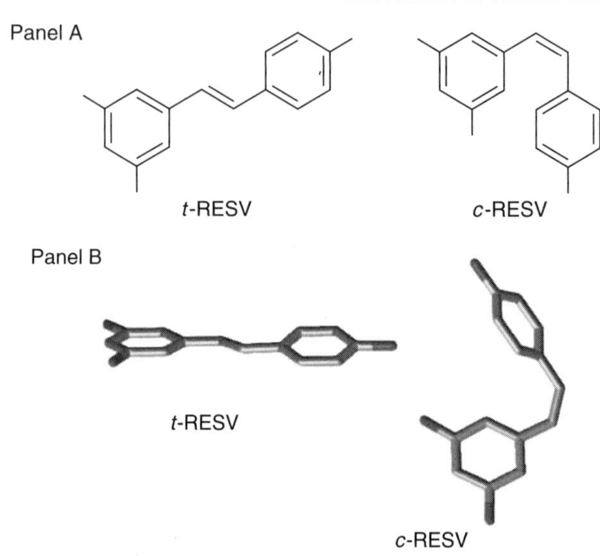

FIGURE 24.1 Chemical structures (panel A) and three-dimensional molecular configurations (panel B) of the *cis* and *trans* isomers of RESV.

such as mulberries and peanuts [2,5,6,7]. The physiological significance of RESV in the plant kingdom is still unclear. However, it is thought to be a phytoalexin, i.e., an antimicrobial and/or antifungal compound produced by the plant in response to injury, environmental stress, pathogenic attack, and other types of stimuli [2,5].

RESV was first identified in 1940 by Michio Takaoka [8] as a constituent of the roots of white hellebore [*Veratrum grandiflorum* (Maxim. ex Baker) Loes. (Liliaceae)], and later (in 1963) by Nonomura et al. [9] (together with its β-glucoside, so-called polydatin or piceid) as an active constituent of the Japanese and Chinese folk medicine Ko-jo-kon, which is used for the treatment of allergic, inflammatory, and fungal diseases, gonorrhea, hyperlipemia, and other afflictions. Ko-jo-kon is the dried powdered root of Japanese knotweed (*Polygonum cuspidatum* Sieb. et Zucc. [Polygonaceae]) (e.g., [2,4,6]). Later, in 1976 RESV was detected in grapevines by Langcake and Pryce [10], who found that it is synthesized by leaf tissues in response to fungal infection (mainly *Botrytis cinerea*) or exposure to ultraviolet light. In 1992, Siemann and Creasy [1] reported the presence of this polyphenol in wines (basically red wine). At the same time, a number of large-scale epidemiological studies were clearly suggesting that prolonged moderate consumption of wine (especially red wine) by the southern French and other Mediterranean populations was associated with a very low incidence of cardiovascular diseases, notably coronary heart disease, despite a high-fat diet, little exercise, and widespread smoking: this is the so-called "French paradox" (e.g., [11]). When it became known that the protective effects of wine consumption were independent of alcohol content, a number of studies were initiated with the aim of identifying the component(s) responsible.

Since then, the pharmacological activity of RESV has been extensively investigated. Almost all of these studies have used the *trans* isomer of RESV (*t*-RESV or (*E*)-RESV; Figure 24.1), which is the isomer that is currently available commercially. *t*-RESV has shown a number of biological activities including antiinflammatory, antioxidant, platelet antiaggregatory, and anticarcinogenic properties, and modulation of lipoprotein metabolism (for reviews, see, e.g., [2,4,7,12–15]). Some of these activities have been implicated in the cardiovascular protective effects attributed to RESV and to red wine [12,16–18].

RESV also exists as a *cis* isomer (*c*-RESV or (*Z*)-RESV; Figure 24.1), which is not currently available commercially [4,6]; as a result, little is known about this isomer's pharmacological activity. In most studies (e.g., [19,20]), the *cis* isomer has not been detected in grapes, unlike *t*-RESV (although see [21]), but is present in wines at variable concentrations, suggesting that it may be produced from *t*-RESV by yeast isomerases during fermentation, or released from RESV polymers called viniferins or from *c*-RESV glucosides (e.g., the 3-*O*-β-D-glucoside, so-called *cis*-piceid or *cis*-polydatin) [2,4]. *c*-RESV can also be obtained from *t*-RESV by exposure to ultraviolet radiation (e.g., [22–24]).

Thus it is important to bear in mind that there are two isomers of RESV, and that these may not have identical pharmacological activities. Nevertheless, most research to date has considered *t*-RESV, and for most of this chapter I focus on this isomer: specifically, I focus on the remarkable vasorelaxant activity of *t*-RESV in rat aorta, which has been a central concern of my group's research in the cardiovascular pharmacology laboratory at the University of Santiago de Compostela. Towards the end of the chapter, I briefly discuss what is known about the biological effects of *c*-RESV, including our preliminary findings on the vasorelaxant activity of *c*-RESV in rat aorta.

VASORELAXANT EFFECTS OF *t*-RESV IN RAT AORTA

INITIAL EXPERIMENTS

When we started our pharmacological studies with *t*-RESV, about five years ago, the *in vitro* vasodilator activity of this natural compound had rather surprisingly received little research attention. Jager and Nguyen-Duong [25] and Naderali et al. [26] had described the vasorelaxant effects of *t*-RESV on porcine coronary arteries, and on mesenteric and uterine arteries from female guinea pigs, respectively. Furthermore, Li et al. [27] had demonstrated that this drug can enhance the activity of Ca^{2+}-activated K^+ channels (K_{Ca}) in endothelial cells derived from human umbilical veins, leading to hyperpolarization of vascular myocytes and dilatation of blood vessels. Later, in 2002 El-Mowafy [28] described the vasorelaxant effects of *t*-RESV on sheep coronary arteries. However, in rat aorta, somewhat controversial results had been obtained. Thus, Fitzpatrick et al. [29] had reported that *t*-RESV (at concentrations of up to 0.1 mM) had no effects on phenylephrine (PHE)-induced contractions in endothelium-intact rat aorta, whereas Chen and Pace-Asciak [30] had reported that *t*-RESV (at $>30\,\mu M$) induces relaxation of PHE-precontracted endothelium-intact rat aorta, and at higher concentrations ($>60\,\mu M$) also induces the relaxation of endothelium-denuded rat aorta. These apparent discrepancies are probably due to the different experimental conditions used by the two groups, such as the absence or presence of light (which degrades *t*-RESV), the use of different anesthetics prior to tissue isolation, and the use of different rat strains and different buffers.

Taking into account these previous findings and with the aim of obtaining new data related to the possible involvement of *t*-RESV in the protective effects of long-term moderate wine consumption against the incidence of cardiovascular diseases, in 2001 we initiated a detailed study of the possible endothelium-dependent vasodilator effects of this compound in rat aorta, and its possible action on the L-arginine–nitric oxide ($NO^{\bullet}$)–guanosine 3′,5′-cyclic monophosphate (cGMP) pathway, using various experimental protocols. The principal results obtained in this project, which have been previously published [31], are briefly outlined in what follows.

The first series of experiments was designed to study the effects of *t*-RESV on the contractions induced by two vasoconstrictor agents (PHE and high extracellular KCl concentration), which are known to act through different mechanisms. In these experiments, *t*-RESV (1 to $10\,\mu M$) had no effect on either PHE-induced ($1\,\mu M$) or high KCl-induced (60 mM) contractions in endothelium-denuded rat aortic rings. However, it concentration-dependently relaxed the contractile response produced by both vasoconstrictor agents in endothelium-intact rat aorta (for more details and numerical data, see [31]). In addition, the vasorelaxant effects of *t*-RESV on high KCl-induced contractions were significantly weaker than those on PHE-induced contractions. Since $NO^{\bullet}$ has been reported to more effectively relax the contractions induced by PHE than those induced by high KCl [32], these preliminary results indicated that the characteristic endothelium-dependent vasorelaxation caused by *t*-RESV may be due to an enhancement of the L-arginine–$NO^{\bullet}$–cGMP pathway.

L-Arginine–$NO^{\bullet}$–cGMP Pathway

The L-arginine–$NO^{\bullet}$–cGMP pathway starts in endothelial cells, in which under physiological conditions low levels of $NO^{\bullet}$ (pico- to nanomolar range) are synthesized from the terminal guanidino nitrogen atom of L-arginine by means of a Ca^{2+}/calmodulin-dependent endothelial constitutive $NO^{\bullet}$

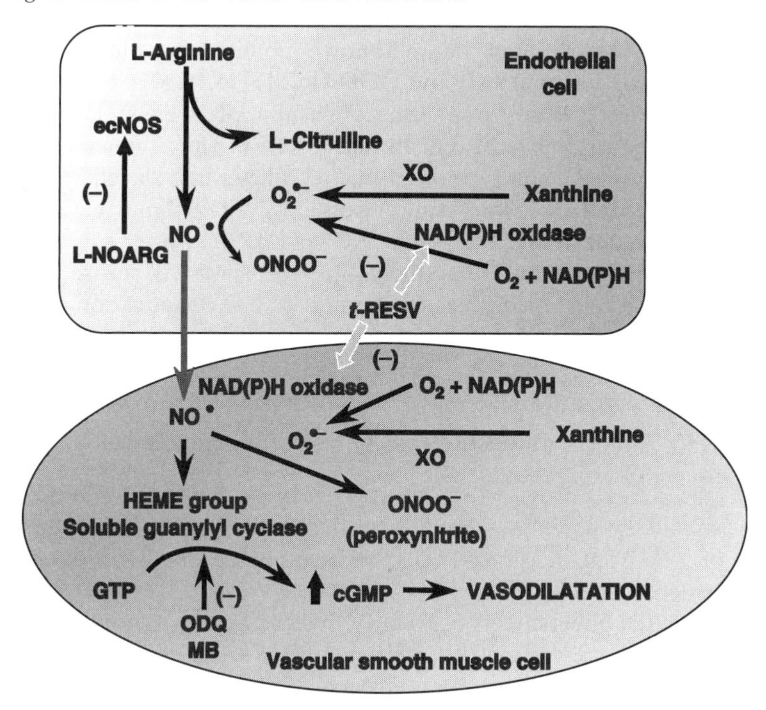

FIGURE 24.2 The L-arginine–NO$^{\bullet}$–cGMP pathway. Steps that may be interfered with by *t*-RESV, and pharmacological tools such as L-NOARG, methylene blue (MB) and ODQ, are indicated by arrows.

synthase (ecNOS or NOS-3), mainly located in the plasmalemma (for reviews, see, e.g., [33–37]). Thereafter, the process involves the diffusion of NO$^{\bullet}$ into the smooth muscle cells of the medial layer and the stimulation of the soluble (cytosolic) guanylyl cyclase, leading to an increase in cytosolic cGMP concentrations, which relaxes vascular smooth muscle via various poorly understood mechanisms [38–42] (Figure 24.2).

NO$^{\bullet}$ can rapidly react with superoxide radicals ($O_2^{\bullet-}$) synthesized by vascular cells [smooth muscle cells (myocytes) or endothelial cells] (see below) to form the powerful oxidant peroxynitrite ($ONOO^-$) [43], which seems to take part in protein oxidation under physiological conditions (e.g., [34,36]). This is the principal pathway for inactivation of NO$^{\bullet}$ in the human body. $ONOO^-$ also reacts with and damages many important biological molecules including thiols, lipids, proteins, and nucleic acids by a number of mechanisms [44,45]. Three different superoxide dismutase (SOD) isoforms may dismute $O_2^{\bullet-}$ into hydrogen peroxide (H_2O_2) and molecular oxygen (O_2) under physiological conditions in vascular cells [46].

The L-arginine–NO$^{\bullet}$–cGMP pathway may be pharmacologically modulated by use of a number of pharmacological tools that act at different levels, such as N^G-nitro-L-arginine (L-NOARG, 0.1 mM, an inhibitor of all

NO$^{\bullet}$ synthase isoforms), and the well-known inhibitors of soluble guanylyl cyclase methylene blue (10 μM) and ODQ (1 μM) [35,36,47].

All the above mentioned pharmacological tools were able to block completely the vasorelaxant effects of *t*-RESV (for numerical data, see [31]), which once again supported the hypothesis that the characteristic endothelium-dependent vasorelaxation caused by *t*-RESV may be mediated by an enhancement of the L-arginine–NO$^{\bullet}$–cGMP pathway. *t*-RESV may enhance this pathway by either increasing the synthesis/release of NO$^{\bullet}$ from endothelial cells, or decreasing the rate of NO$^{\bullet}$ inactivation. In what follows, I summarize the results of experiments designed to distinguish between these two possibilities.

Does *t*-RESV Affect Synthesis/Release of NO$^{\bullet}$ from Endothelial Cells?

To investigate the potential stimulatory effects of *t*-RESV on the synthesis/release of NO$^{\bullet}$ from endothelial cells, we investigated whether it directly or indirectly activates ecNOS.

The potential direct effects of *t*-RESV on ecNOS activity were evaluated by measuring the conversion of L-[^{3}H]arginine to L-[^{3}H]citrulline, using rat aorta homogenates as source of enzyme. The results obtained clearly demonstrate that *t*-RESV (1 to 10 μM) did not increase ecNOS activity in rat aortic homogenates, which suggests that its characteristic endothelium-dependent vasodilator effects are not due to direct activation of this enzyme, and thus not due to increased NO$^{\bullet}$ biosynthesis (for more details and numerical data, see [31]). As far as I know, there has only been one previous study of direct effects of *t*-RESV on ecNOS activity, that of Leikert et al. [48] in human umbilical-vein endothelial cells: as in our study, these authors found that *t*-RESV did not stimulate ecNOS enzymatic activity. However, Hsieh et al. [49] and Wallerath et al. [50] found that chronic treatment with *t*-RESV (3 days) may enhance *expression* of this enzyme in cultured bovine pulmonary artery endothelial cells and human umbilical-vein endothelial cells, although in both studies only at rather high concentrations (50 to 100 μM).

The possibility of indirect stimulation of ecNOS by *t*-RESV was evaluated by means of other experiments. It has been widely reported that, in rat aorta, the activation of endothelial muscarinic receptors or α_2-adrenoceptors by acetylcholine (Ach) or α_2-adrenoceptor agonists (e.g., B-HT 920) produces an increase in the cytosolic free calcium concentration $[Ca^{2+}]_c$, which stimulates the Ca^{2+}/calmodulin-dependent ecNOS (i.e., the ecNOS present in the aortic endothelium) and the subsequent and rapid production/release of NO$^{\bullet}$ [51–53] (for review, see, e.g., [54]). This increase in $[Ca^{2+}]_c$ has been suggested to be due to Ca^{2+} release from intracellular stores, and possibly also to transmembrane Ca^{2+} influx through a receptor-operated ion channel. These effects are specifically blocked by

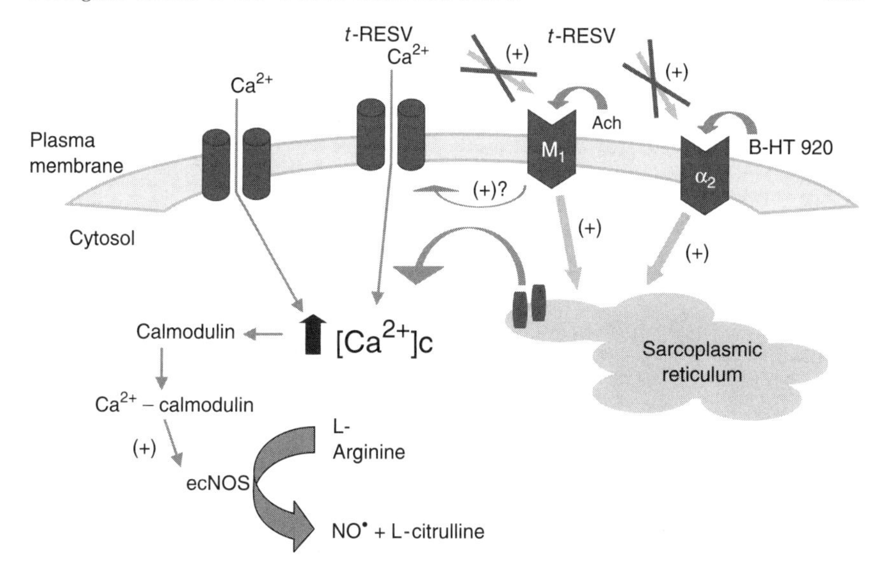

FIGURE 24.3 Pathway of indirect stimulation of ecNOS enzymatic activity in endothelial cells through activation of endothelial muscarinic receptors and α_2-adrenoceptors. As noted in the text, our experiments indicate that *t*-RESV is not an agonist of these receptors.

selective antagonists of the above mentioned receptors (Figure 24.3). To investigate whether *t*-RESV acts as an agonist of these receptors, we therefore investigated the influence of atropine (a known muscarinic receptor blocking agent) and yohimbine (a selective α_2-adrenoceptor antagonist) on the vasorelaxant effects of *t*-RESV. Neither atropine (10 μM) nor yohimbine (1 μM) had significant effects on the vasorelaxant effects of *t*-RESV, which suggests that it does not activate endothelial muscarinic receptors or α_2-adrenoceptors and, therefore, that it does not indirectly stimulate the Ca^{2+}/calmodulin-regulated ecNOS of rat aorta (for numerical data, see [31]).

Does *t*-RESV Affect Inactivation of NO$^•$?

To investigate whether *t*-RESV decreases the rate of inactivation of NO$^•$, we investigated its potential $O_2^{•-}$ scavenging properties, as well as its possible inhibitory effects on the cellular biosynthesis of $O_2^{•-}$ (since NO$^•$ biotransformation requires $O_2^{•-}$; see above and Figure 24.2).

The possible activity of *t*-RESV as a selective scavenger of $O_2^{•-}$ was evaluated using the hypoxanthine (HX)–xanthine oxidase (XO) system. Under standard assay conditions for this system, using the standard commercial form of XO from buttermilk, XO converts HX to xanthine, H_2O_2, and $O_2^{•-}$, and xanthine to uric acid, H_2O_2, and $O_2^{•-}$ (e.g., [55])

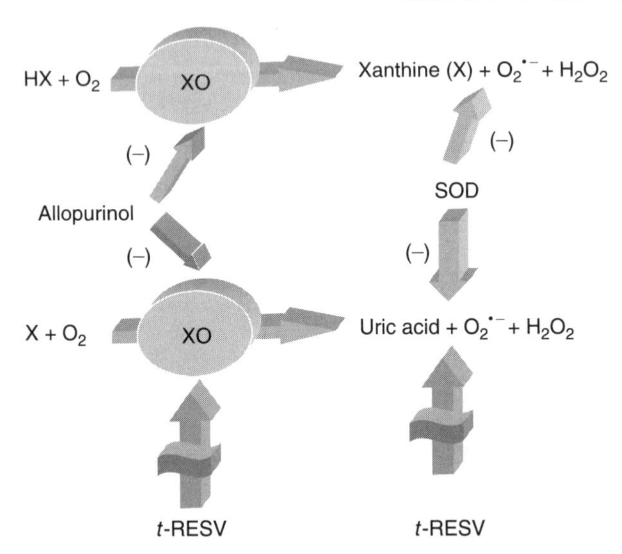

FIGURE 24.4 Summarized reaction steps in the HX–XO system. As noted in the text, our experiments indicate that *t*-RESV is not a scavenger of $O_2^{\bullet -}$ or an inhibitor of commercial (buttermilk) XO.

(Figure 24.4); $O_2^{\bullet -}$ generated in this way then chemically reduces nitroblue tetrazolium (NBT) to produce the colored compound formazan; the amount of $O_2^{\bullet -}$ generated by the assay system is estimated on the basis of spectrophotometric determination of formazan at 560 nm [56]. In parallel assays, the production of uric acid by XO may be estimated on the basis of spectrophotometric determination at 265 nm. When a test substance lowers the amount of $O_2^{\bullet -}$ (i.e., decreases the rate of NBT reduction) and at the same time does not affect the formation of uric acid, it can be considered a selective scavenger of $O_2^{\bullet -}$; on the other hand, reduced levels of both $O_2^{\bullet -}$ and uric acid indicate that the test substance is inhibiting XO activity. In our experiments, and unlike the reference $O_2^{\bullet -}$ scavenger SOD (0.1 to 10 U/ml), *t*-RESV (1 to 10 μM) did not decrease the rate of NBT reduction. In addition, and unlike allopurinol (1 to 10 μM), a well-known reference inhibitor of XO, *t*-RESV (1 to 10 μM) had no effects on uric acid production in this assay system (for numerical data, see [31]). These results clearly indicate that *t*-RESV does not selectively scavenge $O_2^{\bullet -}$ or directly inhibit XO, and we can thus rule out the possibility that these activities are involved in the observed endothelium-dependent vasorelaxant effects of *t*-RESV in rat aorta.

These results agree with those obtained by Hung et al. [57], who found that low concentrations of *t*-RESV (10 μM) do not scavenge $O_2^{\bullet -}$ generated by the HX–XO system. In contrast, our results disagree with those of Zhou et al. [58], who found that *t*-RESV inhibits the enzymatic activity of XO with an IC_{50} of about 11 μM. On the other hand, at first sight, our data do not seem to agree with those obtained by Fauconneau et al. [59], Waffo-Teguo et al. [60], Basly et al. [61], and Stivala et al. [23], who reported that both isomers of RESV (*t*-RESV and *c*-RESV) directly scavenge the stable free radical 1,1-diphenyl-2-picryl-hydrazyl (DPPH). However, it should be noted that the concentrations of RESV used by these authors were rather high (>10 μM). In addition, DPPH tests are not specific and, therefore, of questionable validity for demonstrating the ability of RESV to scavenge selectively $O_2^{\bullet-}$. Similarly, our results do not agree with those obtained by Belguendouz et al. [22], who found that *c*-RESV and *t*-RESV are capable of scavenging free radicals generated by the radical initiator 2,2′-azobis(2-amidinopropane) dihydrochloride (AAPH) (see below).

Although our results indicate that *t*-RESV is neither a direct scavenger of $O_2^{\bullet-}$ nor an inhibitor of commercial buttermilk XO, it is possible that *t*-RESV may inhibit the XO and/or nicotinamide adenine dinucleotide/nicotinamide adenine dinucleotide phosphate (NADH/NADPH) oxidase (NAD(P)H oxidase) present in rat aorta. To investigate this possibility, we performed experiments to assess the possible effects of *t*-RESV on XO and NAD(P)H oxidase activities (as measured by lucigenin-enhanced chemiluminescence) in rat aorta homogenates. In this connection, it is worth noting that $O_2^{\bullet-}$ has been reported to be generated by two main pathways in vascular cells (myocytes and endothelial cells) (Figure 24.2 and Figure 24.5). First, $O_2^{\bullet-}$ generation may occur as a result of activation of an NAD(P)H

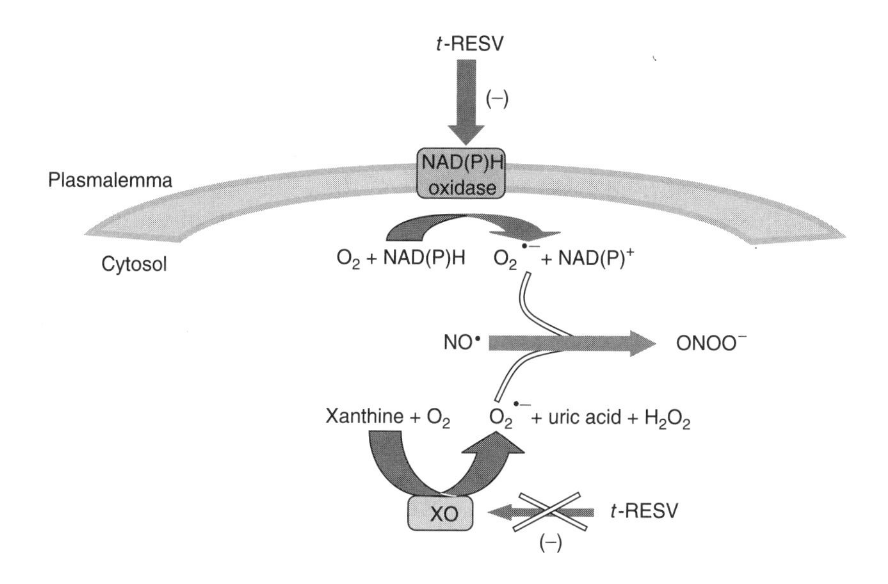

FIGURE 24.5 Summarized pathways of $O_2^{\bullet-}$ biosynthesis in myocytes and endothelial cells, through activation of NAD(P)H oxidase and XO. As detailed in the text, *t*-RESV reduces NAD(P)H oxidase activity but not XO activity.

oxidase that is assembled from different membrane and cytosolic subunits at the plasma membrane, and that catalyzes the vectorial synthesis of $O_2^{\bullet -}$ from O_2 and cellular NADH (the preferred substrate in vascular cells) (stoichiometry: $2O_2 + NADH \rightarrow 2O_2^{\bullet -} + NAD^+ + H^+$). The vascular NAD(P)H oxidase differs from the phagocytic form in a number of subunits, as recently reviewed [62,63]. Second, $O_2^{\bullet -}$ generation may occur through stimulation of xanthine oxidoreductase, an enzyme mainly located in the cytosol, which catalyzes the oxidative hydroxylation of purine substrates (e.g., xanthine or HX) at the molybdenum center (the reductive half-reaction) with production of uric acid, and the subsequent reduction of O_2 at the flavin center with generation of either $O_2^{\bullet -}$ or H_2O_2 (the oxidative half-reaction) (for more details, see, e.g., [64–66]). This enzyme has two interconvertible forms, xanthine oxidase (XO) and xanthine deshydrogenase, which transfer electrons from xanthine to two different preferred acceptors, O_2 and NAD^+, respectively [65].

In these experiments, *t*-RESV (1 to 10 μM) (unlike allopurinol (1 to 10 μM), a reference inhibitor of XO) had no effects on $O_2^{\bullet -}$ generation from xanthine as specifically quantified by lucigenin-enhanced chemiluminescence, using rat aorta homogenates as sources of XO. However, *t*-RESV (1 to 10 μM) [like diphenyleneiodonium (DPI), 0.5 to 3 μM, a reference inhibitor of NAD(P)H oxidase] decreased the specific chemiluminescence signal emitted by the reaction between lucigenin and $O_2^{\bullet -}$ generated from NADH in rat aortic homogenates (for numerical data, see [27]). This suggests that the observed enhancement of the L-arginine–$NO^{\bullet}$–cGMP pathway by *t*-RESV, apparently responsible for its characteristic endothelium-dependent vasodilator effects, may be basically due to inhibition by *t*-RESV of basal $O_2^{\bullet -}$ synthesis via direct inhibition of NAD(P)H oxidase activity, leading to a reduced rate of biotransformation of $NO^{\bullet}$ (Figure 24.2 and Figure 24.5).

We have recently obtained similar results in the laboratory using rat aortic myocytes and human umbilical-vein endothelial cells as sources of enzymes (unpublished data). These data agree with those obtained by Liu et al. [67], who found that *t*-RESV inhibits mechanical strain-induced NAD(P)H oxidase activity in endothelial cells isolated from human umbilical cords.

Implications of the Increased $NO^{\bullet}$ Levels Induced by *t*-RESV

As noted in the preceding section, *t*-RESV appears to reduce the rate of inactivation of $NO^{\bullet}$, and thus increase $NO^{\bullet}$ levels in vascular tissues. In this context, it is relevant that $NO^{\bullet}$ is thought to have significant cardioprotective activity (e.g., [35,68]). The various possible mechanisms suggested for this activity include inhibition of low-density lipoprotein (LDL) oxidation, and inhibition of platelet aggregation and adhesion to damaged blood vessels.

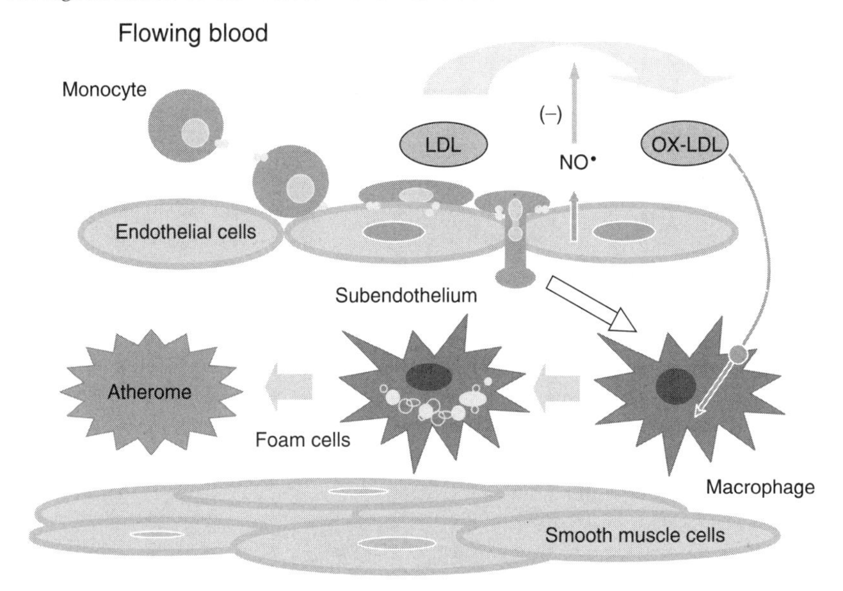

FIGURE 24.6 Schematic representation of the cardioprotective effects of NO$^{\bullet}$ arising from inhibition of LDL oxidation (for details see text). OX-LDL = oxidized LDL.

LDL is generated in the circulation by remodeling of very-low-density lipoprotein (VLDL) through lipolysis, and exchange of lipids and proteins with the high-density lipoprotein (HDL) fraction. During its transit in the circulation, LDL may be oxidized. The endocytosis of normal LDL is tightly regulated, but oxidized LDL is taken up by macrophages via the unregulated macrophage scavenger receptor system. In atherosclerotic patients, over-accumulation of oxidized LDL by macrophages in subendothelial locations (derived from blood monocytes migrated though interendothelial cell junctions; e.g., [69]) leads to the "foam cell" morphology characteristic of the early atherosclerotic lesion known as the "fatty streak" [70,71] (Figure 24.6). NO$^{\bullet}$ has been reported to inhibit the oxidation of LDL during its transit in the circulation, thus reducing the risk of atherosclerosis (e.g., [72]).

As regards the second possible mechanism, NO$^{\bullet}$ is a potent inhibitor of platelet adhesion and aggregation [35,43,73,74]. It may thus prevent the formation of intravascular white thrombi, i.e., thrombi resulting from the adhesion and aggregation of platelets to the atherosclerotic plaque (Figure 24.7). Thrombi of this type are those generally responsible for coronary heart disease.

Certainly, and in view of our findings in rat aorta, it seems reasonable to suppose that the apparent cardioprotective effects of moderate red wine consumption may be at least partially due to the cardioprotective effects of NO$^{\bullet}$, the levels of which are increased in vascular tissues by *t*-RESV.

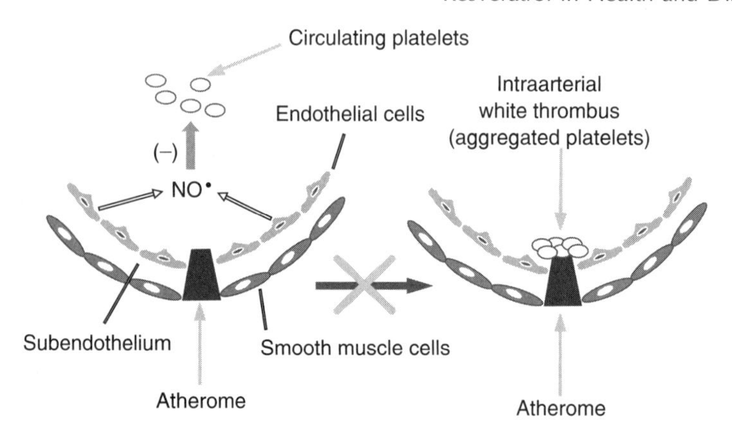

FIGURE 24.7 Inhibitory effects of NO$^{\bullet}$ on the adhesion and aggregation of platelets to the atherosclerotic plaque.

Finally, it should be noted that a defective L-arginine–NO$^{\bullet}$–cGMP pathway (and thus an altered ratio of NO$^{\bullet}$ to $O_2^{\bullet-}$ production) appears to be present in diverse cardiovascular pathologies and conditions, including atherosclerosis and hypertension (e.g., [34,75]), probably as a consequence of NAD(P)H overactivity [62,76,77].

VASORELAXANT AND OTHER BIOLOGICAL EFFECTS OF *c*-RESV

Vasorelaxant Effects of *c*-RESV in the Rat Aorta

As noted, the great majority of studies of the effects of RESV in rat aorta have used *t*-RESV, and much less is known about *c*-RESV. However, we are currently studying the effects of the *cis* isomer, and here briefly outline our preliminary findings. Our experiments are indicating that the vasorelaxant effects of *c*-RESV (unlike those of *t*-RESV) are endothelium-independent.

Notably, *c*-RESV (10 μM to 0.1 mM) concentration-dependently relaxed with almost equal effectiveness the contractions induced by PHE (1 μM) and phorbol 12-myristate 13-acetate [PMA, 1 μM, a known activator of protein kinase C (PKC)], and did not affect the influx of $^{45}Ca^{2+}$ induced by PHE (1 μM) in endothelium-intact or endothelium-denuded rat aortic rings (unpublished preliminary results). These data suggest that the vasorelaxant effects are possibly due, at least in part, to a direct inhibition of PKC activity and/or to inhibition of some of the mechanisms involved in the contractile process triggered by the activation of this kinase. The pathway of activation of PKC in vascular smooth muscle is summarized in Figure 24.8 (for a review, see, e.g., [39]). Binding of a selective agonist (e.g., PHE) to a

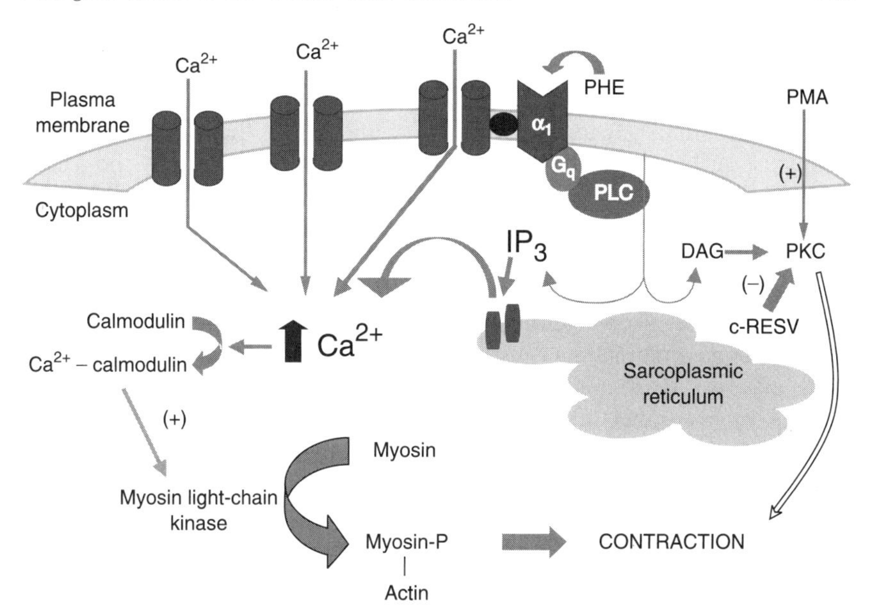

FIGURE 24.8 Hypothesized interference of *c*-RESV with the PKC signal transduction pathway in vascular smooth muscle (for details see text).

PKC-coupled postsynaptic α_1-adrenoceptor (mainly the α_{1D} subtype in rat aorta) activates a G-protein (probably Gq). This sets off a sequence of events that leads to the activation of PKC (probably the Cβ isoform). The activated PKC catalyzes the splitting of the membrane phospholipid phosphatidyl inositol 4,5-bisphosphate (PIP_2) into two messengers: inositol 1,4,5-trisphosphate (IP_3) and diacylglycerol (DAG). IP_3 diffuses from the cell membrane into the cytosol, where it induces release of calcium from the IP_3-sensitive storage compartment, which is followed by a Ca^{2+} influx through the so-called "receptor-operated Ca^{2+} channels." The suddenly abundant cytosolic Ca^{2+} binds to calmodulin, forming a complex containing four calcium ions for every calmodulin molecule, which, in turn, activates the contractile apparatus. In response to these increases in $[Ca^{2+}]_c$ and in phosphoinositide levels, a number of Ca^{2+}- and phospholipid-dependent PKC isoforms migrate from the cytosol to the cell membrane, where they are activated by DAG. Phorbol esters containing the DAG structure, such as PMA, also activate PKCs. Once activated, the DAG-regulated PKC appears to induce slow-developing contraction of vascular smooth muscle via a poorly understood mechanism (for reviews, see, e.g., [38,39,78,79]).

The vasorelaxant effects of *c*-RESV observed by us in rat aorta are qualitatively different to those previously observed for *t*-RESV. This suggests that the different spatial configuration of *c*-RESV (see Figure 24.1) modifies its interaction with the potential cellular targets in rat aorta.

Other Biological Effects of *c*-RESV

This chapter focuses on the vasorelaxant effects of *t*-RESV in rat aorta, and presents in summary form our preliminary findings on the effects of *c*-RESV in this tissue. In this connection, it is of interest to review briefly the results of other studies that have compared the effects of *t*-RESV and *c*-RESV.

Most such comparative studies of *c*-RESV versus *t*-RESV have demonstrated only quantitative differences in the activities of the two forms. Thus, for example, Bertelli et al. [80] found that the *cis* isomer induces a slightly greater decrease in collagen-induced human platelet aggregation than the *trans* isomer, while Varache-Lembège et al. [81] reported that *t*-RESV is more efficient than *c*-RESV for inhibiting human platelet aggregation induced by a number of aggregatory agents (arachidonic acid, ADP, and collagen).

Waffo-Teguo et al. [82] investigated the potential cancer-chemopreventive activities of several wine stilbenoids and flavans extracted from grape (*Vitis vinifera*) cell cultures and found that the *trans* isomer appears to inhibit more effectively cyclooxygenase (COX)-1 activity, COX-2 activity, and the development of 7,12-dimethylbenz[*a*]anthracene-induced preneoplastic lesions in mouse mammary glands. In line with this, Pettit et al. [83], in a preliminary structure-activity relationship study of the potential antineoplastic activity of various stilbene derivatives, have likewise reported that the *cis* isomer exhibits slightly weaker inhibitory effects than the *trans* isomer on a number of cancer cell lines.

In contrast, Roberti et al. [84] have reported that *c*-RESV, unlike *t*-RESV, is scarcely active as an antiproliferative and apoptosis-inducing agent in HL60 cells. Similarly, Stivala et al. [23] have reported that *t*-RESV inhibits the proliferation of normal human fibroblasts and HT1080 cells, whereas *c*-RESV and a number of RESV derivatives had no significant effects on cell growth at the concentrations tested. In addition, Stivala et al. [23] reported that *t*-RESV, unlike *c*-RESV, inhibits DNA biosynthesis in normal human fibroblasts, and inhibits the *in vitro* enzymatic activity of the B-type DNA polymerases α and δ.

De Ruvo et al. [85] investigated the potential effects of a number of nutritional antioxidants as antidegenerative agents and found that, unlike ascorbic acid, *c*-RESV and *t*-RESV (5 to 200 μM) did not have a significant protective effect against apoptosis induced by low K^+ concentrations in cerebellar granule neurons.

Jayatilake et al. [86] studied the effects of both isomers of RESV and various RESV derivatives isolated from *Polygonum cuspidatum* on PKC partially purified from rat brain and on protein-tyrosin kinase ($p56^{lck}$) partially purified from bovine thymus. Both *c*-RESV and *t*-RESV inhibited PKC with a potency comparable to their inhibition of $p56^{lck}$. The inhibitory effects of the *cis* isomer were slightly stronger than those of the *trans* isomer.

Basly et al. [61] investigated the estrogenic/antiestrogenic effects of *t*-RESV and *c*-RESV in the human breast cancer cell lines MCF-7 and MVLN. Both isomers increased the *in vitro* growth of MCF-7 cells at moderate concentrations (10 and 25 μM), whereas the lowest concentrations tested (0.1 and 1 μM) had no effect, and the highest concentration (50 μM) was cytotoxic. *t*-RESV at 25 μM and *c*-RESV at 50 μM reduced the proliferation induced by estradiol. In MVLN cells, *t*-RESV at 10 and 25 μM (but not 1 or 0.1 μM) and *c*-RESV at 25 μM (but not 10 μM, 1 or 0.1 μM) functioned as superagonists of the estrogen receptor (i.e., induced a greater response than estradiol). Recently, Abou-Zeid and El-Mowafy [87] have reported (on the basis of molecular modeling) that *c*-RESV can be expected to bind to the human estrogen receptor α with lower affinity than *t*-RESV.

Basly et al. [61] also compared the free-radical scavenging properties of *t*-RESV and *c*-RESV, using the stable free radical DPPH. These authors found that *c*-RESV was more efficient than *t*-RESV as a scavenger of DPPH. However, this finding contradicts the results of Fauconneau et al. [59] and Waffo-Teguo et al. [60], who reported that *c*-RESV was less efficient than *t*-RESV for scavenging DPPH, the results of Stivala et al. [23], who demonstrated that the two isomers scavenge DPPH with similar efficiency, and the results of Belguendouz et al. [22] who reported that *c*-RESV and *t*-RESV scavenge free radicals generated by the radical generator AAPH with similar potency. (And in any case, note that the above results with DPPH and AAPH are not paralleled by our experiments, which indicate that neither *t*-RESV nor *c*-RESV show $O_2^{\bullet -}$ scavenging activity; see above.)

Fauconneau et al. [59] similarly found that *t*-RESV was more effective than *c*-RESV for preventing Fe^{2+}-induced lipid peroxidation in rat liver microsomes and Cu^{2+}-induced lipid peroxidation in human LDL. Similar results regarding the effects of *t*-RESV and *c*-RESV on lipid peroxidation were obtained by Waffo-Teguo et al. [60] in human LDL, by Belguendouz et al. [22] in porcine LDL (possibly due, at least in part, to the lower Cu^{2+} chelating capacity of the *cis* isomer), and by Stivala et al. [23] in rat liver microsomes (using Fe^{2+}–ascorbate in place of Fe^{2+} alone) and in cultured normal human fibroblasts (using the free radical generator *tert*-butylhydroperoxide (TBHP)). By contrast, Belguendouz et al. [22] found that *c*-RESV and *t*-RESV inhibit AAPH-mediated porcine LDL oxidation with similar potency. In addition, Stivala et al. [23] demonstrated that *t*-RESV inhibits more effectively than *c*-RESV the citroneal thermo-oxidation.

Finally, our group has recently demonstrated that:

1. *c*-RESV (like *t*-RESV) exhibits a typical antioxidant activity, i.e., it blocks (though less effectively than *t*-RESV) reactive oxygen species production by inflammatory murine peritoneal macrophages, apparently through inhibition of NAD(P)H oxidase

activity, not direct scavenging of $O_2^{\bullet -}$ or decreased XO activity. In addition, *c*-RESV (again like *t*-RESV) attenuates the synthesis of the proinflammatory mediators $NO^{\bullet}$ and PGE_2 in inflammatory murine peritoneal macrophages, at least in part by inhibition of inducible $NO^{\bullet}$ synthase (iNOS or NOS-2) and of COX-2 gene and protein expression [24].

2. *c*-RESV has a significant modulatory effect on a number of genes whose expression is controlled by nuclear trancription factor NF-κB [88]. Similar results have been obtained for *t*-RESV by us (unpublished results).
3. Both isomers of RESV increase cytosolic calcium levels in A7r5 cells, *c*-RESV being significantly more effective than the *trans* isomer [89].

CONCLUDING REMARKS

As noted in the Introduction, *t*-RESV has been found to be present at higher concentrations in red than in white wines, possibly because in the preparation of many white and rosé wines, even though red grapes are sometimes used, the grape skins (the principal source of *t*-RESV) are removed before fermentation, allowing very little time for extraction of ethanol-soluble grape-skin components like *t*-RESV [1,29]. Furthermore, it is known that red wines are far more effective vasorelaxants and protectants against cardiovascular diseases than are white wines (e.g., [11]).

On the other hand, it is interesting to note that the *t*-RESV concentrations reached in plasma and tissues after oral administration to rats or humans [90,91] appear to approach the concentrations that are active *in vitro* (usually in the range 1 to 30 μM; e.g., [12,90]).

In addition to free RESV isomers (*t*-RESV and *c*-RESV) present at variable concentrations in red wines (see Introduction), a number of RESV derivatives (mainly β-glucosides) are also present. These may be absorbed directly, as reported for the rat small intestine [92,93], and/or hydrolyzed before absorption by glucosidases present in the human intestinal tract, with subsequent release of free RESV ([94]; for review, see, e.g., [2]). These RESV derivatives may contribute to the biologically available RESV dose.

Moreover, *t*-RESV is a lipophilic substance which is effectively absorbed after oral administration in rats [95–97], mice [97,98], and humans [97,99,100], and which accumulates in rat and mouse tissues including heart, liver, and kidney ([90,98,101]; for review, see, e.g., [102]). For this reason, Bertelli et al. [90] concluded that an average drinker of wine can absorb a sufficient amount of RESV, at least in the long term, to explain the beneficial effects of red wine on health.

Bearing in mind the above reports, and assuming that *t*-RESV shows similar behavior in rat and human blood vessels, our results in rat aorta may at least partially explain the protection offered by prolonged consumption of moderate amounts of wine, especially red wine, against cardiovascular diseases, especially coronary heart disease.

Moreover, taking into account the vascular effects of *t*-RESV described above, it can also be concluded that:

1. *t*-RESV may be of value as a structural template for the development of new drugs capable of inhibiting vascular NAD(P)H oxidase activity, of protecting $NO^{\bullet}$ from inactivation by $O_2^{\bullet -}$ (via inhibition of $O_2^{\bullet -}$ generation catalyzed by NAD(P)H oxidase), and thus for reducing the risk of mortality from ischemic cardiopathy and other cardiovascular disorders, and/or for improving their pharmacological treatment.
2. Since long-term consumption of alcoholic drinks (e.g., wine) may have a number of adverse effects in humans, nonalcoholic drinks and foods prepared from grape skins may constitute a healthier source of *t*-RESV.

Finally, our preliminary studies of *c*-RESV suggest that its vasorelaxant effects in rat aorta are similar but not identical to those of *t*-RESV. Notably, the effects of *c*-RESV (unlike those of *t*-RESV) appear to be endothelium-independent and may be mediated by direct inhibition of PKC activity. *t*-RESV and *c*-RESV seem to have different hepatic metabolisms [103] (specifically, regio- and stereoselective glucuronidation, catalyzed by different UDP-glucuronosyltransferase isoforms); but notwithstanding this, *c*-RESV, like *t*-RESV (see above), has been reported to be effectively absorbed after oral administration in rats, and to accumulate in rat tissues such as the heart, liver, and kidney ([104,105]; for review, see, e.g., [102]).

Bearing in mind the above considerations and assuming that the effects of the two RESV isomers in human blood vessels are similar to those seen in rat aorta, it seems likely that the apparent beneficial effects of moderate red wine consumption may be due to the combined effects of *t*-RESV and *c*-RESV. Certainly, there is a clear need for future research to explore the specific effects and utility of *c*-RESV.

ACKNOWLEDGMENTS

I apologize for failing to cite many relevant primary papers because of space constraints. The work in my laboratory was supported in part by grants from the Spanish Ministerio de Ciencia y Tecnología (SAF2002-0245), Almirall-Prodesfarma Laboratories (Pharmacology Award 2003), and the Xunta de Galicia (PGIDIT02BTF20301PR), Spain. I am especially grateful to Almirall-Prodesfarma Laboratories and the Spanish Pharmacological Society for granting me the 2003 Pharmacology Award.

REFERENCES

1. Siemann EH and Creasy LL, Concentration of the phytoalexin resveratrol in wine, *Am J Enol Vitic* 43, 49–52, 1992.
2. Soleas GJ, Diamandis EP, and Goldberg DM, Resveratrol: a molecule whose time has come? And gone?, *Clin Biochem* 30, 91–113, 1997.
3. Burns J, Gardner PT, O'neil, J, Crawford S, Morecroft I, McPhail DB, Lister C, Matthews D, Mclean MR, Lean MEJ, Duthie GG, and Crozier A, Relationship among antioxidant activity, vasodilatation capacity, and phenolic content of red wines, *J Agric Food Chem* 48, 220–230, 2000.
4. Frémont L, Biological effects of resveratrol, *Life Sci* 66, 663–673, 2000.
5. Soleas GJ, Diamandis EP, and Goldberg DM, The world of resveratrol, *Adv Exp Med Biol* 492, 159–182, 2001.
6. Pervaiz S, Resveratrol: from grapevines to mammalian biology, *FASEB J* 17, 1975–1985, 2003.
7. Alarcón de la Lastra C, and Villegas I, Resveratrol as an anti-inflammatory and anti-aging agent: Mechanisms and clinical implications, *Mol Nutr Food Res*, 49, 405–430, 2005.
8. Takaoka MJ, Of the phenolic substances of white hellebore (*Veratrum grandiflorum* Loes. fil.), *J Fac Sci Hokkaido Imp Univ*, 3, 1–16, 1940.
9. Nonomura S, Kanagawa H, and Makimoto A, Chemical constituents of polygonaceous plants. I. Studies on the components of ko-jo-kon (*Polygonum cuspidatum* Sieb. et Zucc.), *Yakugaku Zasshi* 83, 988–990, 1963.
10. Langcake P and Pryce RJ, The production of resveratrol by *Vitis vinifera* and other members of the Vitaceae as a response to infection or injury, *Physiol Plant Pathol* 9, 77–86, 1976.
11. Renaud S and De Lorgeril M, Wine, alcohol, platelets and the French paradox for coronary heart disease, *Lancet* 339, 1523–1526, 1992.
12. Wu JM, Wang ZR, Hsieh TC, Bruder JL, Zou JG, and Huang YZ, Mechanism of cardioprotection by resveratrol, a phenolic antioxidant present in red wine, *Int J Mol Med* 8, 3–17, 2001.
13. Granados-Soto V, Pleiotropic effects of resveratrol, *Drug News Perspect* 16, 299–307, 2003.
14. Aggarwal BB, Bhardwaj A, Aggarwal RS, Seeram NP, Shishodia S, and Takada Y, Role of resveratrol in prevention and therapy of cancer: preclinical and clinical studies, *Anticancer Res* 24, 2783-2840, 2004.
15. Ulrich S, Wolter F, and Stein JM, Molecular mechanisms of the chemopreventive effects of resveratrol and its analogs in carcinogenesis, *Mol Nutr Food Res* 49, 452–461, 2005.
16. Bradamante S, Barenghi L, and Villa A, Cardiovascular protective effects of resveratrol, *Cardiovasc Drug Rev* 22, 169–188, 2004.
17. Hao HD and He LR, Mechanisms of cardiovascular protection by resveratrol, *J Med Food* 7, 290–298, 2004.
18. Delmas D, Jannin B, and Latruffe N, Resveratrol: preventing properties against vascular alterations and ageing, *Mol Nutr Food Res* 49, 377–395, 2005.
19. Palomino O, Gomez-Serranillos MP, Slowing K, Carretero E, and Villar A, Study of polyphenols in grape berries by reversed-phase high-performance liquid chromatography, *J Chromatogr A* 870, 449–451, 2000.

20. Burns J, Yokota T, Ashihara H, Lean ME, and Crozier A, Plant foods and herbal sources of resveratrol, *J Agric Food Chem* 50, 3337–3340, 2002.
21. Wang Y, Catana F, Yang Y, Roderick R, and van Breemen RB, An LC-MS method for analyzing total resveratrol in grape juice, cranberry juice, and in wine, *J Agric Food Chem* 50, 431–435, 2002.
22. Belguendouz L, Frémont L, and Linard A, Resveratrol inhibits metal ion-dependent and independent peroxidation of porcine low-density lipoproteins, *Biochem Pharmacol* 53, 1347–1355, 1997.
23. Stivala LA, Savio M, Carafoli F, Perucca P, Bianchi L, Maga G, Forti L, Pagnoni UM, Albini A, Prosperi E, and Vannini V, Specific structural determinants are responsible for the antioxidant activity and the cell cycle effects of resveratrol, *J Biol Chem* 276, 22586–22594, 2001.
24. Leiro J, Álvarez E, Arranz JA, Laguna R, Uriarte E, and Orallo F, Effects of *cis*-resveratrol on inflammatory murine macrophages: antioxidant activity and down-regulation of inflammatory genes, *J Leukoc Biol* 75, 1156–1165, 2004.
25. Jager U and Nguyen-Duong H, Relaxant effect of trans-resveratrol on isolated porcine coronary arteries, *Arzneimittelforschung* 49, 207–211, 1999.
26. Naderali EK, Doyle PJ, and Williams G, Resveratrol induces vasorelaxation of mesenteric and uterine arteries from female guinea-pigs, *Clin Sci* 98, 537–543, 2000.
27. Li HF, Chen SA, and Wu SN, Evidence for the stimulatory effect of resveratrol on Ca^{2+}-activated K^{+} current in vascular endothelial cells, *Cardiovasc Res* 45, 1035–1045, 2000.
28. El-Mowafy AM, Resveratrol activates membrane-bound guanylyl cyclase in coronary arterial smooth muscle: a novel signaling mechanism in support of coronary protection, *Biochem Biophys Res Commun* 291, 1218–1224, 2002.
29. Fitzpatrick DF, Hirschfield SL, and Coffey RG, Endothelium-dependent vasorelaxing activity of wine and other grape products, *Am J Physiol* 65, 774–778, 1993.
30. Chen CK and Pace-Asciak CR, Vasorelaxing activity of resveratrol and quercetin in isolated rat aorta, *Gen Pharmacol* 27, 363–366, 1996.
31. Orallo F, Álvarez E, Camiña M, Leiro JM, Gómez E, and Fernández P, The possible implication of trans-resveratrol in the cardioprotective effects of long-term moderate wine consumption, *Mol Pharmacol* 61, 294–302, 2002.
32. Furchgott RF, Role of endothelium in responses of vascular smooth muscle, *Circulation Res* 53, 557–573, 1983.
33. Förstermann U, Gath I, Schwarz P, Closs EI, and Kleinert H, Isoforms of nitric oxide synthase. Properties, cellular distribution and expressional control, *Biochem Pharmacol* 50, 1321–1332, 1995.
34. Marín J and Rodríguez-Martínez MA, Role of vascular nitric oxide in physiological and pathological conditions, *Pharmacol Ther* 75, 111–134, 1997.
35. Hobbs AJ, Higgs A, and Moncada S, Inhibition of nitric oxide synthase as a potential therapeutic target, *Annu Rev Pharmacol Toxicol* 39, 191–220, 1999.
36. Domenico R, Pharmacology of nitric oxide: molecular mechanisms and therapeutic strategies, *Curr Pharm Des* 10, 1667–1676, 2004.
37. Mariotto S, Menegazzi M, and Suzuki H, Biochemical aspects of nitric oxide, *Curr Pharm Des* 10, 1627–1645, 2004.

38. Kuriyama H, Kitamura K, and Nabata H, Pharmacological and physiological significance of ion channels and factors that modulate them in vascular tissues, *Pharmacol Rev* 47, 387–573, 1995.
39. Orallo F, Regulation of cytosolic calcium levels in vascular smooth muscle, *Pharmacol Ther* 69, 153–171, 1996.
40. Lucas KA, Pitari GM, Kazerounian S, Ruiz-Stewart I, Park J, Schulz S, Chepenik KP, and Waldman SA, Guanylyl cyclases and signaling by cyclic GMP, *Pharmacol Rev* 52, 375–414, 2000.
41. Friebe A and Koesling D, Regulation of nitric oxide-sensitive guanylyl cyclase, *Circulation Res* 93, 96–105, 2003.
42. Napoli C and Ignarro LJ, Nitric oxide-releasing drugs, *Annu Rev Pharmacol Toxicol* 43, 97–123, 2003.
43. Tiefenbacher CP and Kreuzer J, Nitric oxide-mediated endothelial dysfunction: is there need to treat?, *Curr Vasc Pharmacol* 1, 1231–1233, 2003.
44. Murphy MP, Packer MA, Scarlett JL, and Martin SW, Peroxynitrite: a biologically significant oxidant, *Gen Pharmacol* 31, 179–186, 1998.
45. Dröge W, Free radicals in the physiological control of cell function, *Physiol Rev* 82, 47–95, 2002.
46. Faraci FM and Didion SP, Vascular protection. Superoxide dismutase isoforms in the vessel wall, *Arterioscler Thromb Vasc Biol* 24, 1367–1373, 2004.
47. Moro MA, Russel RJ, Cellek S, Lizasoain I, Su Y, Darley-Usmar VM, Radomski MW, and Moncada S, cGMP mediates the vascular and platelet actions of nitric oxide: confirmation using an inhibitor of the soluble guanylyl cyclase, *Proc Natl Acad Sci USA* 93, 1480–1485, 1996.
48. Leikert JF, Rathel TR, Wohlfart P, Cheynier V, Vollmar AM, and Dirsch VM, Red wine polyphenols enhance endothelial nitric oxide synthase expression and subsequent nitric oxide release from endothelial cells. *Circulation* 106, 1614–1617, 2002.
49. Hsieh TC, Juan G, Darzynkiewicz Z, and Wu JM, Resveratrol increases nitric oxide synthase, induces accumulation of p53 and p21 (WAF1/CIP1) and suppresses cultured bovine pulmonary artery endothelial cell proliferation by perturbing progression through S and G(2), *Cancer Res* 59, 2596–2601, 1999.
50. Wallerath T, Deckert G, Ternes T, Anderson H, Li H, Witte K, and Förstermann U, Resveratrol, a polyphenolic phytoalexin present in red wine, enhances expression and activity of endothelial nitric oxide synthase, *Circulation* 106, 1652–1658, 2002.
51. Eglème C, Godfraind T, and Miller RC, Enhanced responsiveness of rat isolated aorta to clonidine after removal of the endothelial cells, *Br J Pharmacol* 81, 16–18, 1984.
52. Carrier GO and White RE, Enhancement of α_1 and α_2 adrenergic agonist-induced vasoconstriction by removal of endothelium in rat aorta, *J Pharmacol Exp Ther* 232, 682–687, 1985.
53. Boulanger ChM, Morrison KJ, and Vanhoutte P, Mediation by M_3-muscarinic receptors of both endothelium-dependent contraction and relaxation to acetylcholine in the aorta of the spontaneously hypertensive rat, *Br J Pharmacol* 112, 519–524, 1994.
54. Busse R and Fleming I, Regulation and functional consequences of endothelial nitric oxide formation, *Ann Med* 27, 331–340, 1995.

55. Cos P, Ying L, Calomme M, Hu JP, Cimanga K, Van Poel B, Pieters L, Vlietinck AJ, and Berghe DV, Structure–activity relationship and classification of flavonoids as inhibitors of xanthine oxidase and superoxide scavengers, *J Nat Prod* 61, 71–76, 1998.
56. Robak J and Gryglewski RJ, Flavonoids are scavengers of superoxide anions, *Biochem Pharmacol* 37, 837–841, 1988.
57. Hung LM, Su MJ, Chu WK, Chiao CW, Chan WF, and Chen JK, The protective effect of resveratrols on ischaemia-reperfusion injuries of rat hearts is correlated with antioxidant efficacy, *Br J Pharmacol* 135, 1627–1633, 2002.
58. Zhou CX, Kong LD, Ye WC, Cheng CH, and Tan RX, Inhibition of xanthine and monoamine oxidases by stilbenoids from *Veratrum taliense*, *Planta Med* 67, 158–161, 2001.
59. Fauconneau B, Waffo-Teguo P, Huguet F, Barrier L, Decendit A, and Mérillon JM, Comparative study of radical scavenger and antioxidant properties of phenolic compounds from *Vitis vinifera* cell cultures using *in vitro* tests, *Life Sci* 61, 2103–2110, 1997.
60. Waffo-Teguo P, Fauconneau B, Deffieux G, Huguet F, Vercauteren J, and Mérillon JM, Isolation, identification, and antioxidant activity of three stilbene glucosides newly extracted from *Vitis vinifera* cell cultures, *J Nat Prod* 61, 655–657, 1998.
61. Basly JP, Marre-Fournier F, Le Bail JC, Habrioux G, and Chulia AJ, Estrogenic/antiestrogenic and scavenging properties of (*E*)- and (*Z*)-resveratrol, *Life Sci* 66, 769–777, 2000.
62. Cai H, Griendling KK, and Harrison DG, The vascular NAD(P)H oxidases as therapeutic targets in cardiovascular diseases, *Trends Pharmacol Sci* 24, 471–478, 2003.
63. Griendling KK, Novel NAD(P)H oxidases in the cardiovascular system, *Heart* 90, 491–493, 2004.
64. Harrison R, Structure and function of xanthine oxidoreductase: where are we now?, *Free Radical Biol Med* 33, 774–797, 2002.
65. Berry CE and Hare JM, Xanthine oxidoreductase and cardiovascular disease: molecular mechanisms and pathophysiological implications, *J Physiol* 555, 589–606, 2004.
66. Hille R, Molybdenum-containing hydroxylases, *Arch Biochem Biophys* 433, 107–116, 2005.
67. Liu JC, Chen JJ, Chan P, Cheng CF, and Cheng TH, Inhibition of cyclic strain-induced endothelin-1 gene expression by resveratrol, *Hypertension* 42, 1198–1205, 2003.
68. Landmesser U, Hornig B, and Drexler H, Endothelial function: a critical determinant in atherosclerosis?, *Circulation* 109, II27–33, 2004.
69. Michiels C, Endothelial cell functions, *J Cell Physiol* 196, 430–443, 2003.
70. Steinberg D, Parthasarathy S, Carew TE, Khoo JC, and Witztum JL, Beyond cholesterol. Modifications of low-density lipoprotein that increase its atherogenicity, *N Engl J Med* 320, 915–924, 1989.
71. Stocker R and Keaney JF Jr, Role of oxidative modifications in atherosclerosis, *Physiol Rev* 84, 1381–1478, 2004.
72. Rubbo H, Trostchansky A, Botti H, and Batthyany C, Interactions of nitric oxide and peroxynitrite with low-density lipoprotein, *Biol Chem* 383, 547–552, 2002.

73. Freedman JE and Loscalzo J, Nitric oxide and its relationship to thrombotic disorders, *J Thromb Haemost* 1, 1183–1188, 2003.
74. Barbato JE and Tzeng E, Nitric oxide and arterial disease, *J Vasc Surg* 40, 187–193, 2004.
75. Hamilton CA, Brosnan MJ, McIntyre M, Graham D, and Dominiczak A, Superoxide excess in hypertension and aging. A common cause of endothelial dysfunction, *Hypertension* 37, 529–534, 2001.
76. Meyer JW and Schmitt ME, A central role for the endothelial NADPH oxidase in atherosclerosis, *FEBS Lett* 472, 1–4, 2000.
77. Zalba G, Beaumont FJ, San José G, Fortuño A, Fortuño MA, Etayo JC, and Díez J, Vascular NADH/NADPH oxidase is involved in enhanced superoxide production in spontaneously hypertensive rats, *Hypertension* 35, 1055–1061, 2000.
78. Karaki H, Ozaki H, Hori M, Mitsui-Saito M, Amano K, Harada K, Miyamoto S, Nakazawa H, Won KJ, and Sato K, Calcium movements, distribution, and functions in smooth muscle, *Pharmacol Rev* 49, 157–230, 1997.
79. Wier WG and Morgan KG, α_1-adrenergic signaling mechanisms in contraction of resistance arteries, *Rev Physiol Biochem Pharmacol* 150, 91–139, 2003.
80. Bertelli AA, Giovannini L, Bernini W, Migliori M, Fregoni M, Bavaresco L, and Bertelli A, Antiplatelet activity of *cis*-resveratrol, *Drugs Exp Clin Res* 22, 61–63, 1996.
81. Varache-Lembège M, Waffo-Teguo P, Richard T, Monti JP, Deffieux G, Vercauteren J, Mérillon JM, and Nuhrich A, Structure–activity relationships of polyhydroxystilbene derivatives extracted from *Vitis vinifera* cell cultures as inhibitors of human platelet aggregation, *Med Chem Res* 10, 253–267, 2000.
82. Waffo-Teguo P, Hawthorne ME, Cuendet M, Mérillon JM, Kinghorn AD, Pezzuto JM, and Mehta RG, Potential cancer-chemopreventive activities of wine stilbenoids and flavans extracted from grape (*Vitis vinifera*) cell cultures, *Nutr Cancer* 40, 173–179, 2001.
83. Pettit GR, Grealish MP, Jung MK, Hamel E, Pettit RK, Chapuis JC, and Schmidt JM, Antineoplastic agents. 465. Structural modification of resveratrol: sodium resverastatin phosphate, *J Med Chem* 45, 2534–2542, 2002.
84. Roberti M, Pizzirani D, Simoni D, Rondanin R, Baruchello R, Bonora C, Buscemi F, Grimaudo S, and Tolomeo M, Synthesis and biological evaluation of resveratrol and analogues as apoptosis-inducing agents, *J Med Chem* 46, 3546–3554, 2003.
85. De Ruvo C, Amodio R, Algeri S, Martelli N, Intilangelo A, D'Ancona GM, and Esposito E, Nutritional antioxidants as antidegenerative agents, *Int J Dev Neurosci* 18, 359–366, 2000.
86. Jayatilake GS, Jayasuriya H, Lee ES, Koonchanok NM, Geahlen RL, Ashendel CL, McLaughlin JL, and Chang CJ, Kinase inhibitors from *Polygonum cuspidatum*, *J Nat Prod* 56, 1805–1810, 1993.
87. Abou-Zeid LA and El-Mowafy AM, Differential recognition of resveratrol isomers by the human estrogen receptor-α: molecular dynamics evidence for stereoselective ligand binding, *Chirality* 16, 190–195, 2004.
88. Leiro J, Arranz JA, Fraiz N, Sanmartín ML, Quezada E, and Orallo F, Effect of *cis*-resveratrol on genes involved in nuclear factor kappa B signalling, *Int Immunopharmacol* 5, 393–406, 2005.

89. Campos-Toimil M, Elíes J, and Orallo F, *Trans*- and *cis*-resveratrol increase cytoplasmatic calcium levels in vascular smooth muscle, *Mol Nutr Food Res* 49, 396–404, 2005.
90. Bertelli A, Bertelli AAE, Gozzini A, and Giovannini L, Plasma and tissue resveratrol concentrations and pharmacological activity, *Drugs Exp Clin Res* 24, 133–138, 1998.
91. Soleas GJ, Yan J, and Goldberg DM, Ultrasensitive assay for three polyphenols (catechin, quercetin and resveratrol) and their conjugates in biological fluids utilizing gas chromatography with mass selective detection, *J Chromatogr B* 757, 161–172, 2001.
92. Andlauer W, Kolb J, Siebert K, and Furst P, Assessment of resveratrol bioavailability in the perfused small intestine of the rat, *Drugs Exp Clin Res* 26, 47–55, 2000.
93. Kuhnle G, Spencer JP, Chowrimootoo G, Schroeter H, Debnam ES, Srai SK, Rice-Evans C, and Hahn U, Resveratrol is absorbed in the small intestine as resveratrol glucuronide, *Biochem Biophys Res Commun* 272, 212–217, 2000.
94. Goldberg DM, Ng E, Karumanchiri A, Diamandis EP, and Soleas GJ, Resveratrol glucosides are important components of commercial wines, *Am J Enol Vitic* 47, 415–420, 1996.
95. Soleas GJ, Angelini M, Grass L, Diamandis EP, and Goldberg DM, Absorption of *trans*-resveratrol in rats, *Methods Enzymol* 335, 145–154, 2001.
96. Marier JF, Vachon P, Gritsas A, Zhang J, Moreau JP, and Ducharme MP, Metabolism and disposition of resveratrol in rats: extent of absorption, glucuronidation, and enterohepatic recirculation evidenced by a linked-rat model, *J Pharmacol Exp Ther* 302, 369–373, 2002.
97. Meng X, Maliakal P, Lu H, Lee MJ, and Yang CS, Urinary and plasma levels of resveratrol and quercetin in humans, mice, and rats after ingestion of pure compounds and grape juice, *J Agric Food Chem* 52, 935–942, 2004.
98. Vitrac X, Desmouliére A, Brouillaud B, Krisa S, Deffieux G, Barthe N, Rosenbaum J, and Mèrillon JM, Distribution of [^{14}C]-*trans*-resveratrol, a cancer chemopreventive polyphenol, in mouse tissues after oral administration, *Life Sci* 72, 2219–2233, 2003.
99. Goldberg DM, Yan J, and Soleas GJ, Absorption of three wine-related polyphenols in three different matrices by healthy subjects, *Clin Biochem* 36, 79–87, 2003.
100. Walle T, Hsieh F, DeLegge MH, Oatis JE, and Walle UK, High absorption but very low bioavailability of oral resveratrol in humans, *Drug Metab Dispos* 32, 1377–1382, 2004.
101. Bertelli AAE, Giovannini L, Stradi R, Urien S, Tillement JP, and Bertelli A, Evaluation of kinetic parameters of natural phytoalexin in resveratrol orally administered in wine to rats, *Drugs Exp Clin Res* 24, 51–55, 1998.
102. Wenzel E and Somoza V, Metabolism and bioavailability of *trans*-resveratrol, *Mol Nutr Food Res* 49, 472–481, 2005.
103. Aumont V, Krisa S, Battaglia E, Netter P, Richard T, Mérillon JM, Magdalou J, and Sabolovic N, Regioselective and stereospecific glucuronidation of *trans*- and *cis*-resveratrol in human, *Arch Biochem Biophys* 393, 281–289, 2001.

104. Bertelli AA, Giovannini L, Stradi R, Urien S, Tillement JP, and Bertelli A, Kinetics of *trans*- and *cis*-resveratrol (3,4′,5-trihydroxystilbene) after red wine oral administration in rats, *Int J Clin Pharmacol Res* 16, 77–81, 1996.
105. Bertelli AA, Giovannini L, Stradi R, Bertelli A, and Tillement JP, Plasma, urine and tissue levels of *trans*- and *cis*-resveratrol (3,4′,5-trihydroxystilbene) after short-term or prolonged administration of red wine to rats, *Int J Tissue React* 18, 67–71, 1996.

25 Resveratrol as an Antiinflammatory Agent

Young-Joon Surh and Joydeb Kumar Kundu

CONTENTS

INTRODUCTION

Resveratrol (*trans*-3,4′,5-trihydroxystilbene) is a naturally occurring polyphenol synthesized by around 70 edible plant species including grapes, mulberries, and peanuts in response to injury, ultraviolet (UV) irradiation, and fungal attack [1]. Because of its protective role against external stress in plants, resveratrol is known as a phytoalexin [1]. Resveratrol is also a pharmacologically active component of dried roots of *Polygonum cuspidatum*, and has been used in traditional oriental folk medicine for curing diseases that are contemporarily known as inflammation, allergy, and hyperlipidemia [2]. Epidemiological studies have revealed an inverse relationship between red wine consumption and the risk of cardiovascular disease, the so-called

"French paradox," and resveratrol has been considered as an active cardioprotective component. In addition, laboratory studies have demonstrated that resveratrol possesses a wide spectrum of biological activities including antibacterial, antifungal, antioxidant, free radical scavenging, vasorelaxing, chemopreventive, and antiinflammatory activities [3].

Inflammation is considered as a primary physiologic defense mechanism that helps the body fight infection, burns, toxic chemicals, allergens, or other noxious stimuli. However, uncontrolled and persistent inflammation may cause deleterious effects [4]. Although a handful of synthetic antiinflammatory drugs are readily available, nontoxic antiinflammatory substances of natural origin, especially from dietary sources, are currently being sought. This chapter focuses on the antiinflammatory activity of resveratrol and its underlying mechanisms.

INFLAMMATION AS A PREDISPOSING FACTOR FOR VARIOUS HUMAN DISEASES

Although inflammation is fundamentally a protective mechanism, the harmful sequelae of inflammation underlie various chronic diseases. Inflammation has been recognized as an etiologic factor in a wide array of human diseases including respiratory, cardiovascular, and neurodegenerative disorders, rheumatoid arthritis, cancer, diabetes, etc. (Figure 25.1). Inflammatory responses, both acute and chronic, flare through vascular and cellular reactions that are mediated by chemical factors in response to inflammatory stimuli. The mediators of inflammation constitute a wide variety of chemical substances including vasoactive amines (e.g., histamine and serotonin), plasma proteases (e.g., kinins and complement proteins), arachidonic acid metabolites [e.g., prostaglandins (PGs), leukotrienes, thromboxanes], cytokines [e.g., interleukins (IL), tumor necrosis factor (TNF)], growth factors such as platelet-derived growth factor, fibroblast growth factor (FGF), and vascular endothelial growth factor (VEGF), and nitric oxide (NO). The vascular and cellular reactions of inflammation generally involve vasodilation, increased vascular permeability to proteins, increased infiltration of inflammatory cells such as granulocytes and lymphocytes into the inflamed tissue, release of cytokines and chemokines, activation of leukocytes, etc. [4].

Dysregulation of immune and inflammatory responses, including excessive production of cytokines and acute phase proteins, has been associated with insulin resistance, diabetes, and cardiovascular disease [5–7]. Characteristic features of inflammation, such as neutrophil infiltration and generation of proinflammatory cytokines and eicosanoids, are clearly involved in the pathogenesis of inflammatory bowel diseases including ulcerative colitis and Crohn's disease. Chronic inflammation of the airways, associated with release of inflammatory cytokines from activated

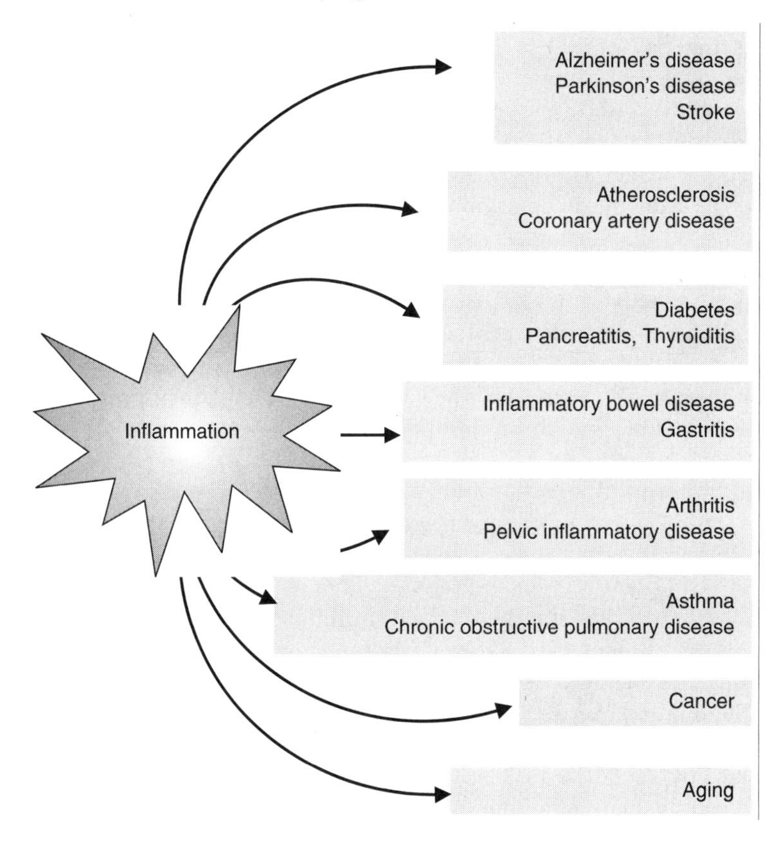

FIGURE 25.1 Inflammation as a cause of various diseases.

bronchoalveolar macrophages, is one of the distinct features of chronic obstructive pulmonary diseases (COPD) and mild intermittent asthma [8]. Data from epidemiological and clinical studies suggest inflammation as a prognostic marker of cardiovascular diseases [9]. Inflammatory processes play critical roles in all stages of atherosclerosis, beginning from its inception to the development of complications [10].

A variety of inflammatory cytokines and growth factors, including transforming growth factor (TGF), TNFα, IL-1, IL-6, and FGF, contribute to tubulointerstitial inflammation and fibrosis, which are major pathologic components of obstructive renal injury [11]. While a chronic inflammatory reaction in the brain is particularly prominent in Alzheimer's disease (AD) [12], a substantial body of data also suggests a possible role of inflammatory response in the progression of Parkinson's disease (PD). A dramatic increase in the levels of inflammatory cytokines, such as IL-1β, IL-6, and TNFα, has been noted in the basal ganglia and cerebrospinal fluid of PD patients [13]. In addition, a recent epidemiological study has demonstrated that regular use of nonsteroidal antiinflammatory drugs reduces the risk of PD [14]. Recently, persistent inflammation has been considered to contribute to multistage carcinogenesis [15,16]. Reactive oxygen species (ROS), typical byproducts of eicosanoid metabolism, produced during the inflammatory tissue damage can also trigger a series of reactions which are responsible for malignant transformation, particularly at the stage of tumor promotion [17]. Clinical and experimental studies have revealed that antiinflammatory agents, particularly COX-2 inhibitors, constitute a noble class of chemopreventive agents [18,19]. Inflammatory processes are actively associated with several other diseases such as pancreatitis, glomerulonephritis, acute pyelonephritis, thyroiditis, pelvic inflammatory diseases, acute and chronic inflammatory dermatoses, arthritis, peritonitis, etc. [4]. Therefore, the control of inflammatory processes, may represent the first-line prevention of various human ailments.

RESVERATROL: A NATURAL ANTIINFLAMMATORY AGENT

Recent interest in searching for antiinflammatory substances from edible sources identified a wide variety of dietary phytochemicals including resveratrol as potent antiinflammatory agents [20]. The antiinflammatory activity of resveratrol has been associated with its potential efficacy in preventing various human ailments including COPD, AD, cardiovascular disorders, and cancer. Resveratrol has been shown to protect against inflammation of rat colon [21] and human airway epithelial cells [22]. The leukocyte recruitment and endothelial barrier disruption induced by superoxide-dependent proinflammatory stimuli such as ischemia/reperfusion, platelet-activating factor, and oxidants generated from the hypoxanthine/xanthine oxidase system were inhibited by resveratrol [23]. The attenuation of immune cell-mediated release of various inflammatory cytokines by resveratrol has been documented. Resveratrol inhibited hypoxia/reoxygenation-induced release of IL-6 from cortical mixed glial cells suggesting that the compound is effective in controlling the ischemia-induced inflammatory process in stroke [24]. In addition, it reduced IL-6 release from mouse peritoneal macrophages stimulated with calcium ionophore A23187 [25]. Resveratrol also inhibited the release of several inflammatory mediators such as histamines, TNFα, leukotrienes, and PGD_2 from bone marrow-derived mouse mast cells stimulated with immunoglobin E (IgE) [26]. Resveratrol inhibited both basal and IL-1β or cigarette smoke media (CSM)-induced release of IL-8 and granulocyte macrophage-colony stimulating factor (GM-CSF) from alveolar macrophages in COPD patients [27]. It has been reported that resveratrol effectively prevents the production of interferon (IFN)-γ and TNFα as well as proliferation of peripheral blood mononuclear cells [28].

The expression of cell adhesion molecules, such as intracellular adhesion molecule-1 (ICAM-1) and vascular cell adhesion molecule-1 (VCAM-1), is a critical step in inflammatory response. Activation of cell adhesion molecules leads to the enhanced adhesion of monocytes to activated endothelium, resulting in increased vascular permeability and infiltration of neutrophils and leukocytes in inflamed tissues. The inhibition of TNFα-induced expression of VCAM-1 in human umbilical vein endothelial cells in culture and a decrease in TNFα-induced vascular permeability *in vivo* by resveratrol have also been reported [29]. Ferrero et al. [30] reported that resveratrol inhibited the expression of ICAM-1 and VCAM-1 in TNFα-treated human umbilical endothelial and human saphenous endothelial cells, respectively. In addition, the adhesion of granulocytes and monocytes to TNFα-stimulated endothelium was diminished by resveratrol [30].

Multiple lines of evidence suggest that resveratrol possesses significant inhibitory effects on the expression of iNOS and COX-2, two representative proinflammatory enzymes [31]. The inhibitory effect of resveratrol on the activation of these critical enzymes and underlying molecular mechanisms are discussed in the subsequent section.

MOLECULAR MECHANISMS OF THE ANTIINFLAMMATORY ACTIVITY OF RESVERATROL

Resveratrol exerts antiinflammatory effects by interfering with various events in the course of inflammation. The inhibition of cytokine release, blocking of expression of proinflammatory genes, downregulation of intracellular signal transduction molecules and transcription factors that regulate transcription of proinflammatory genes, and inhibition of leukocyte activation by blocking the expression of cell adhesion molecules are key molecular mechanisms underlying the antiinflammatory activity of resveratrol (Table 25.1; Figure 25.2).

INHIBITION OF iNOS AND COX-2 BY RESVERATROL

Diverse inflammatory stimuli induce expression of iNOS and COX-2. While the induction of COX-2 contributes to inflammatory processes by generating PGs, prostacyclins, and thromboxanes, iNOS plays a pivotal role in mediating inflammation through the production of another inflammatory mediator, NO, by oxidative deamination of L-arginine [32]. A recent study by Chun et al. [33] has demonstrated that NO induces COX-2 expression in mouse skin *in vivo*, which further suggests the significance of iNOS in the inflammatory process. Resveratrol has been shown to inhibit the induction of both iNOS and COX-2 [21,34–38]. In a recent study, *cis*-resveratrol has also been shown to exert antiinflammatory

TABLE 25.1
Molecular Basis of Antiinflammatory Activity of Resveratrol

Molecular targets	Tissues/cells	Stimuli	Ref.
Release of Inflammatory Mediators			
↓IL-6	Human cortical mixed glial cells	Hypoxia/reoxygenation	24
	Mouse peritoneal macrophage	Calcium ionophore	25
↓IL-8	Alveolar macrophage of COPD patients	IL-1β, CSM	27
↓TNFα	Bone marrow-derived mouse mast cell	IgE	26
	Human peripheral blood mononuclear cells	PHA	28
↓PGE_2	Human mammary and oral epithelial cells	TPA	34, 41
	Mouse peritoneal macrophage	LPS, TPA, or H_2O_2	35
↓NO	Mouse peritoneal macrophage	LPS	37
	Murine macrophage RAW 264.7 and J774 cells	LPS	49
↓GM-CSF	Alveolar macrophage of COPD patients	IL-1β, CSM	27
↓ IFN-γ	Human peripheral blood mononuclear cells	PHA	28
↓ Histamine, PGD_2, leukotrienes	Bone marrow-derived mouse mast cell	IgE/calcium ionophore	26
Expression of Proinflammatory Genes and Gene Products			
↓ COX-2 mRNA expression	Human mammary and oral epithelial cells	TPA	34, 41
	Esophageal tumors in F344 rats	NMBA	43
↓ COX-2 protein expression	Rat colon	Trinitrobenzenesulfonic acid	22
	Mouse peritoneal macrophage	LPS, TPA, or H_2O_2	35
	RAW 264.7 murine macrophage	LPS and IFN-γ	36
	Female ICR mouse skin	TPA	38
	Rat mammary tumor	DMBA	57
↓ *De novo* formation of iNOS	Mouse macrophage	LPS	44
↓ iNOS protein expression	RAW 264.7 murine macrophage	LPS and IFN-γ	36

	Mouse peritoneal macrophage	LPS	37
↓Expression of IL-8 protein and mRNA	Human myeloid (U937) cells	PMA	64
Activation of Transcription Factors Regulating Proinflammatory Genes			
↓NF-κB activation	Rat mammary tumor	DMBA	57
	Human mammary carcinoma (MCF-7) cells	—	57
	Acute myeloid leukemia cell	IL-1β	56
	Myeloid (U937) cells	—	65
↓IκBα phosphorylation and degradation; ↓NF-κB activity	RAW 264.7 cells	LPS	50
↓NF-κB nuclear translocation	Murine macrophage RAW 264.7 and J774 cells	LPS	49
↓IκB kinase activity; ↓NF-κB activation; ↓NF-κB-dependent gene expression	Rat fibroblast	*Ras* transformed	52
↓Phosphorylation and transactivational activity of p65	Monocytic cell THP1	LPS	59
↓Phosphorylation and nuclear translocation of p65; ↓NF-κB activation	Myeloid (U937), lymphoid (Jurkat), epithelial (HeLa), and glioma (H4) cells	TNFα	54
↓AP-1 activation	Human mammary and oral epithelial cells	TPA	34, 41
	Female ICR mouse skin	TPA	38
	Myeloid (U937), lymphoid (Jurkat), epithelial (HeLa), and glioma (H4) cells	TNFα	54
	HeLa cells	PMA and UVC	72
	Myeloid (U937) cells	PMA	64
Activation of Upstream Signaling Molecules			
↓PKC activation	Human mammary and oral epithelial cells	TPA	34, 41
↓Expression of pERK; ↓activity of ERK and p38	Female ICR mouse skin	TPA	38

(*continued*)

TABLE 25.1
Continued

Molecular targets	Tissues/cells	Stimuli	Ref.
↓MAPK activation	Myeloid (U937), lymphoid (Jurkat), epithelial (HeLa), and glioma (H4) cells	TNFα	54
↓Activation of ERK2, JNK1, and p38 MAP kinase	HeLa cells	PMA and UVC	72
Expression of Cell Adhesion Molecules			
↓VCAM-1 expression	Human umbilical vein endothelial cells	TNFα	29
	Human saphenous vein endothelial cells	TNFα	30
↓VCAM-1 mRNA and protein expression	Human umbilical vein endothelial cells	TNFα, LPS, and PMA	73
↓ ICAM-1 expression	Human umbilical vein endothelial cells	TNFα	30

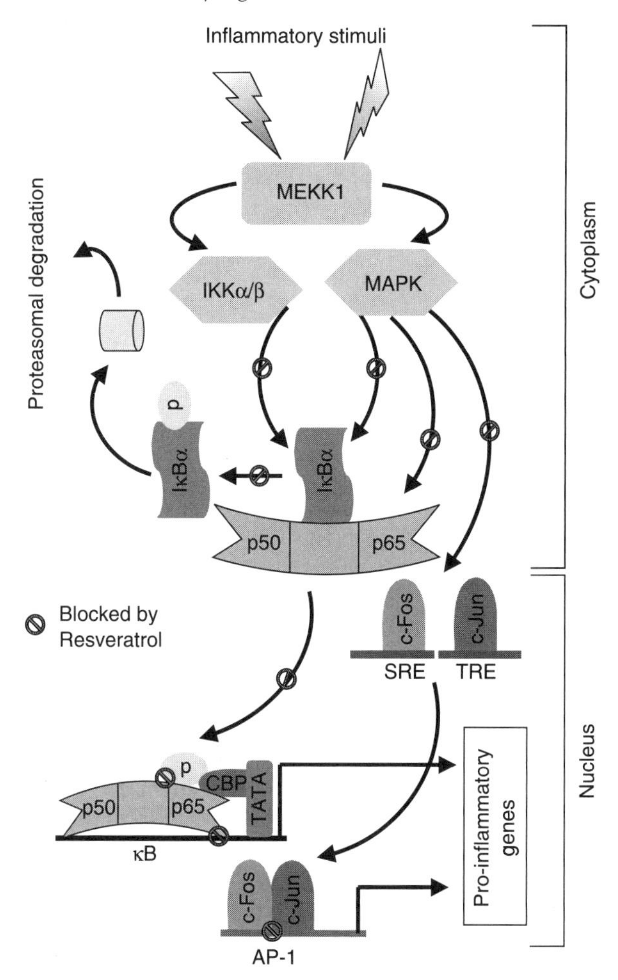

FIGURE 25.2 Molecular basis of antiinflammatory activity of resveratrol.

activity through downregulation of NOS-2 and COX-2, and subsequent production of NO and PGE_2, respectively, in lipopolysaccharide (LPS)- or IFN-γ-stimulated murine macrophages [39]. Resveratrol has been reported to inhibit COX-2, at both transcriptional [40,41] and posttranscriptional levels [42,43]. It has been demonstrated that resveratrol reduces 12-*O*-tetradecanoyl phorbol-13-acetate (TPA)-induced PGE_2 production by downregulating COX-2 gene transcription in human mammary and oral epithelial cells [34,41]. While resveratrol downregulated the expression of both COX-1 and COX-2 mRNA transcripts in *N*-nitrosomethylbenzylamine (NMBA)-induced esophageal tumors in F344 rats [43], TPA-induced expression of both COX-1 and COX-2 in CD-1 mouse skin remained unaffected by resveratrol pretreatment [42]. However, recent studies have demonstrated that resveratrol has a significant inhibitory effect on COX-2 protein expression in rat colonic inflammation induced by trinitrobenzene-sulfonic acid [21] as well as in mouse peritoneal macrophages stimulated with LPS, TPA, or H_2O_2 [35]. The latter study has also demonstrated that resveratrol prevents mobilization of arachidonic acid and decreases PGE_2 production [35]. In addition, resveratrol reduced iNOS expression in LPS-activated macrophages without altering COX-2 expression [44]. However, according to Murakami et al. [36], the expression of both iNOS and COX-2 in LPS- and IFN-γ-treated RAW 264.7 macrophages was strongly inhibited by resveratrol. It also inhibited NO production and iNOS protein expression in LPS-stimulated macrophages [37]. Moreover, topically applied resveratrol resulted in the inhibition of TPA-induced COX-2 expression in female ICR mouse skin [38].

Modulation of Transcription Factors Regulating Proinflammatory Genes

Effects on the Activation of NF-κB

Intracellular signaling pathways responsible for induction of proinflammatory gene products such as TNFα, IL-8, IL-1, iNOS, COX-2, VCAM-1, ICAM-1, IL-6, and GM-CSF involve the activation of eukaryotic transcription factor nuclear factor-kappaB (NF-κB) [45]. The 5′-promoter region of COX-2 contains two putative NF-κB binding sites. The functionally active NF-κB exists mainly as a heterodimer consisting of subunits of Rel family (e.g., p65/Rel A, p50, p52, c-Rel, v-Rel, and Rel B), which is normally present in the cytoplasm by forming an inactive complex with the inhibitory protein IκB. Exposure of cells to mitogens, inflammatory cytokines, UV radiation, ionizing radiation, bacterial toxins, etc., causes rapid phosphorylation and subsequent proteasomal degradation of IκB. The free NF-κB dimer then translocates to the nucleus, where it binds to the *cis*-acting κB element located in the promoter regions of COX-2 [46] and iNOS [47], thereby regulating their expression. NF-κB is considered as the hallmark of inflammation and suggested to make a bridge between inflammation and cancer [48].

Resveratrol inhibited nuclear translocation and DNA binding of NF-κB subunits in LPS-stimulated RAW 264.7 cells by blocking phosphorylation and degradation of IκBα [49,50]. Recently, the inhibition of IκBα degradation by resveratrol in LPS- or IFN-γ-stimulated RAW 264.7 macrophage has also been reported [36]. In contrast, an early study by Wadsworth and Koop [51] demonstrated that resveratrol had no inhibitory effect on LPS- or H_2O_2-induced NF-κB activation in the same cell line. Resveratrol blocked NF-κB-dependent transcription of monocyte chemoattractant protein-1 in TNFα-treated THP-1 cells [52]. Similarly, resveratrol diminished TNFα- and LPS-induced NF-κB DNA binding in human monocyte (THP-1) and myeloid (U-937) cells through inhibition of IκB kinase (IKK) activity [52]. In normal human epidermal keratinocytes, resveratrol inhibited UVB-induced activation of NF-κB by blocking the activation of IKKα as well as phosphorylation and degradation of IκBα [53]. Alternatively, resveratrol blocked TNFα-induced activation of NF-κB in a dose- and time-dependent manner by suppressing phosphorylation and nuclear translocation of p65 without affecting IκBα degradation in U937 cells [54]. While LPS-induced NF-κB transcriptional activity and phosphorylation of IκBα in human colon cancer HT-29 cells were inhibited by resveratrol at a relatively higher concentration, the compound increased the same activity at a lower concentration [55].

According to Estrov et al. [56], resveratrol significantly decreased the production of IL-1β in acute myeloid leukemia OCIM2 cells and suppressed IL-1β-induced activation of NF-κB. In addition, resveratrol abrogated NF-κB DNA binding in rat-1 fibroblasts stimulated with oncogenic H-*ras* [52], TNFα-treated human breast carcinoma (MCF-7) cells, and DMBA-induced rat mammary tumors [57]. Moreover, resveratrol inhibited Cr (VI)-stimulated activation of NF-κB in mouse epidermal JB6 cells as assessed by the luciferase reporter gene assay [58]. It also inhibited NF-κB activation induced by other stimuli such as TPA, LPS, H_2O_2, okadaic acid, and ceramide in lymphoid (Jurkat-T), HeLa and glioma (H4) cells [54]. In another study, resveratrol suppressed LPS-induced expression of tissue factor in THP-1 cells by inhibiting phosphorylation and transactivation potential of p65, but failed to inhibit the activation or translocation of NF-κB/Rel proteins [59].

Effects on the Activation of AP-1

In addition to the suppression of NF-κB, resveratrol also inhibited the activation of activator protein (AP-1). The transcription factor AP-1 is formed by different dimeric combination of proteins of Jun (c-Jun, JunB, and JunD) and Fos (c-Fos, FosB, Fra-1, and Fra-2) family, Jun dimerization partners (JDP1 and JDP2), and the closely related activating transcription factor (ATF2, LRF1/ATF3, and B-ATF) subfamilies, which are basic leucine zipper proteins [60–62]. The activation of AP-1 also induces a battery of proinflammatory genes.

In our recent study, resveratrol pretreatment inhibited DNA binding of AP-1 [38] in TPA-stimulated mouse skin. Similarly, TPA-induced activation of AP-1 was inhibited by resveratrol in U937 cells [63]. While TPA-induced DNA binding activity of AP-1 in U937 cells was significantly inhibited by resveratrol [64], treatment of these cells with resveratrol alone inhibited DNA binding activity of NF-κB [65]. Resveratrol suppressed NF-κB-driven but not AP-1-driven transcriptional activity in LPS-treated THP-1 cells [59]. While resveratrol induced both protein and mRNA expression of c-Jun and c-Fos, components of AP-1 in papillary thyroid cancer cells [66], it enhanced the DNA-binding activity of c-Fos in Caco-2 cells without affecting the c-Jun DNA binding [67]. Resveratrol has also been reported to modulate the transcriptional activity of AP-1 [34,41].

Effects on the Intracellular Upstream Kinases

Although the molecular signaling mechanisms that lead to the induction of COX-2 through activation of NF-κB and/or AP-1 in response to various external stimuli have not been fully clarified [32], the mitogen-activated protein (MAP) kinase pathway, an extensively investigated intracellular signaling cascade, has been implicated in proinflammatory responses. Three distinct groups of well-characterized MAP kinase subfamily members, which are serine/threonine protein kinases, include extracellular signal-regulated protein kinase (ERK), c-Jun NH_2-terminal kinase (JNK)/stress-activated protein kinase (SAPK), and p38 MAP kinase. The activation of any one or all of the above-mentioned MAP kinases results in the activation of a set of transcription factors including NF-κB and AP-1, thereby inducing target genes. The pharmacological inhibition [68,69] or dominant-negative mutation of the corresponding MAP kinases [70,71] results in diminished induction of COX-2 and/or PG production, suggesting the regulatory function of MAP kinase cascades, at least in part, in the upregulation of COX-2. The modulation of the above mentioned MAP kinases by resveratrol partly constitutes the molecular basis of the antiinflammatory activity exhibited by this dietary phytochemical. Pretreatment with resveratrol has been shown to block UVC- and TPA-induced activation of ERK2, JNK1, and p38 MAP kinase and subsequently transcription of AP-1 reporter gene in HeLa cells [72]. Resveratrol also inhibited TPA-induced activation of ERK and p38 MAP kinase in mouse skin *in vivo* [38]. Besides MAP kinases, the activation of other upstream signaling kinases such as protein tyrosine kinase and protein kinase C is also inhibited by resveratrol [34,41].

CONCLUSION

In recent years, there has been growing interest in therapeutic intervention in various disease processes by naturally occurring substances, particularly dietary phytochemicals. Multiple lines of laboratory studies suggest that resveratrol, one of the dietary components, may act as a potent antiinflammatory agent. Considering the role of inflammation in the pathogenesis of various chronic illness, the antiinflammatory property of resveratrol may be attributed to the cardioprotective, antiinfective, immunomodulatory, and chemopreventive activities of this phytochemical. Although substantial progress in exploring the molecular mechanisms underlying the antiinflammatory activity of resveratrol in different experimental systems has already been made, it still remains a subject of future research how resveratrol targets different components of intracellular signaling cascades in a coordinated fashion.

ACKNOWLEDGMENT

This work was supported by the National Research Laboratory (NRL) Grant from the Ministry of Science and Technology, Republic of Korea.

REFERENCES

1. Gusman J, Malonne H, and Atassi G, A reappraisal of the potential chemopreventive and chemotherapeutic properties of resveratrol, *Carcinogenesis* 22, 1111–1117, 2001.
2. Ignatowicz E and Baer-Dubowska W, Resveratrol, a natural chemopreventive agent against degenerative diseases, *Pol J Pharmacol* 53, 557–569, 2001.
3. Dong Z, Molecular mechanism of the chemopreventive effect of resveratrol, *Mutat Res* 523–524, 145–150, 2003.
4. Kumar V, Abbas AK, and Fausto N, Eds, *Robbins and Cotran Pathologic Basis of Disease*, Elsevier Saunders, Philadelphia, 2004, pp. 47–86.
5. Pillarisetti S and Saxena U, Role of oxidative stress and inflammation in the origin of Type 2 diabetes: a paradigm shift, *Expert Opin Ther Targets* 8, 401–408, 2004.
6. Gomes MB and Nogueira VG, Acute-phase proteins and microalbuminuria among patients with type 2 diabetes, *Diabetes Res Clin Pract* 66, 31–39, 2004.
7. Glass CK and Witztum JL, Atherosclerosis. The road ahead, *Cell* 104, 503–516, 2001.
8. van den Toorn LM, Clinical implications of airway inflammation in mild intermittent asthma, *Ann Allergy Asthma Immunol* 92, 589–594, 2004.
9. Willerson JT and Ridker PM, Inflammation as a cardiovascular risk factor, *Circulation* 109, II2–II10, 2004.
10. Kris-Etherton PM, Lefevre M, Beecher GR, Gross MD, Keen CL, and Etherton TD, Bioactive compounds in nutrition and health-research methodologies for establishing biological function: the antioxidant and antiinflammatory effects of flavonoids on atherosclerosis, *Annu Rev Nutr* 24, 511–538, 2004.
11. Misseri R, Rink RC, Meldrum DR, and Meldrum KK, Inflammatory mediators and growth factors in obstructive renal injury, *J Surg Res* 119, 149–159, 2004.
12. McGeer EG and McGeer PL, Inflammatory processes in Alzheimer's disease, *Prog Neuropsychopharmacol Biol Psychiatry* 27, 741–749, 2003.
13. McGeer PL and McGeer EG, Inflammation and neurodegeneration in Parkinson's disease, *Parkinsonism Relat Disord* 10, S3–S7, 2004.
14. Chen H, Zhang SM, Hernan MA, Schwarzschild MA, Willett WC, Colditz GA, Speizer FE, and Ascherio A, Nonsteroidal antiinflammatory drugs and the risk of Parkinson disease, *Arch Neurol* 60, 1059–1064, 2003.

15. Philip M, Rowley DA, and Schreiber H, Inflammation as a tumor promoter in cancer induction, *Semin Cancer Biol.* 14, 433–439, 2004.
16. Marx J, Cancer research. Inflammation and cancer: the link grows stronger, *Science* 306, 966–968, 2004.
17. Surh YJ, Anti-tumor promoting potential of selected spice ingredients with antioxidative and antiinflammatory activities: a short review, *Food Chem Toxicol* 40, 1091–1097, 2002.
18. Chun K-S and Surh Y-J, Signal transduction pathways regulating cyclooxygenase-2 expression: potential molecular targets for chemoprevention, *Biochem Pharmacol* 68, 1089–1100, 2004.
19. Sinicrope FA, Half E, Morris JS, Lynch PM, Morrow JD, Levin B, Hawk ET, Cohen DS, Ayers GD, and Stephens LC, Cell proliferation and apoptotic indices predict adenoma regression in a placebo-controlled trial of celecoxib in familial adenomatous polyposis patients, *Cancer Epidemiol Biomarkers Prev* 13, 920–927, 2004.
20. Surh Y-J, Cancer chemoprevention with dietary phytochemicals, *Nature Rev Cancer* 3, 768–780, 2003.
21. Martin AR, Villegas I, La Casa C, and de la Lastra CA, Resveratrol, a polyphenol found in grapes, suppresses oxidative damage and stimulates apoptosis during early colonic inflammation in rats, *Biochem Pharmacol* 67, 1399–1410, 2004.
22. Donnelly LE, Newton R, Kennedy GE, Fenwick PS, Leung RH, Ito K, Russell RE, and Barnes PJ, Antiinflammatory effects of resveratrol in lung epithelial cells: molecular mechanisms, *Am J Physiol Lung Cell Mol Physiol* 287, L774–L783, 2004.
23. Shigematsu S, Ishida S, Hara M, Takahashi N, Yoshimatsu H, Sakata T, and Korthuis RJ, Resveratrol, a red wine constituent polyphenol, prevents superoxide-dependent inflammatory responses induced by ischemia/reperfusion, platelet-activating factor, or oxidants, *Free Radical Biol Med* 34, 810–817, 2003.
24. Wang MJ, Huang HM, Hsieh SJ, Jeng KC, and Kuo JS, Resveratrol inhibits interleukin-6 production in cortical mixed glial cells under hypoxia/hypoglycemia followed by reoxygenation, *J Neuroimmunol* 112, 28–34, 2001.
25. Zhong M, Cheng GF, Wang WJ, Guo Y, Zhu XY, and Zhang JT, Inhibitory effect of resveratrol on interleukin 6 release by stimulated peritoneal macrophages of mice, *Phytomedicine* 6, 79–84, 1999.
26. Baolin L, Inami Y, Tanaka H, Inagaki N, Iinuma M, and Nagai H, Resveratrol inhibits the release of mediators from bone marrow-derived mouse mast cells *in vitro*, *Planta Med* 70, 305–309, 2004.
27. Culpitt SV, Rogers DF, Fenwick PS, Shah P, De Matos C, Russell RE, Barnes PJ, and Donnelly LE, Inhibition by red wine extract, resveratrol, of cytokine release by alveolar macrophages in COPD, *Thorax* 58, 942–946, 2003.
28. Boscolo P, del Signore A, Sabbioni E, Di Gioacchino M, Di Giampaolo L, Reale M, Conti P, Paganelli R, and Giaccio M, Effects of resveratrol on lymphocyte proliferation and cytokine release, *Ann Clin Lab Sci* 33, 226–231, 2003.
29. Bertelli AA, Baccalini R, Battaglia E, Falchi M, and Ferrero ME, Resveratrol inhibits TNF alpha-induced endothelial cell activation, *Therapie* 56, 613–616, 2001.

30. Ferrero ME, Bertelli AE, Fulgenzi A, Pellegatta F, Corsi MM, Bonfrate M, Ferrara F, De Caterina R, Giovannini L, and Bertelli A, Activity *in vitro* of resveratrol on granulocyte and monocyte adhesion to endothelium, *Am J Clin Nutr* 68, 1208–1214, 1998.
31. Kundu JK and Surh Y-J, Molecular basis of chemoprevention by resveratrol: NF-kappaB and AP-1 as potential targets, *Mutat Res* 555, 65–80, 2004.
32. Surh Y-J, Chun K-S, Cha H-H, Han S-S, Keum Y-S, Park K-K, and Lee S-S, Molecular mechanisms underlying chemopreventive activities of anti-inflammatory phytochemicals: down-regulation of COX-2 and iNOS through suppression of NF-kappa B activation, *Mutat Res* 480–481, 243–268, 2001.
33. Chun K-S, Cha H-H, Shin J-W, Na H-K, Park K-K, Chung W-Y, and Surh Y-J, Nitric oxide induces expression of cyclooxygenase-2 in mouse skin through activation of NF-kappaB, *Carcinogenesis* 25, 445–454, 2004.
34. Subbaramaiah K, Michaluart P, Chung WJ, Tanabe T, Telang N, and Dannenberg AJ, Resveratrol inhibits cyclooxygenase-2 transcription in human mammary epithelial cells, *Ann NY Acad Sci* 889, 214–223, 1999.
35. Martinez J and Moreno JJ, Effect of resveratrol, a natural polyphenolic compound, on reactive oxygen species and prostaglandin production, *Biochem Pharmacol* 59, 865–870, 2000.
36. Murakami A, Matsumoto K, Koshimizu K, and Ohigashi H, Effects of selected food factors with chemopreventive properties on combined lipopolysaccharide- and interferon-gamma-induced IkappaB degradation in RAW264.7 macrophages, *Cancer Lett* 195, 17–25, 2003.
37. Matsuda H, Kageura T, Morikawa T, Toguchida I, Harima S, and Yoshikawa M, Effects of stilbene constituents from rhubarb on nitric oxide production in lipopolysaccharide-activated macrophages, *Bioorg Med Chem Lett* 10, 323–327, 2000.
38. Kundu JK, Chun K-S, Kim SO, and Surh Y-J, Resveratrol inhibits phorbol ester-induced cyclooxygenase-2 expression in mouse skin: MAPKs and AP-1 as potential molecular targets, *Biofactors* 21, 33–39, 2004.
39. Leiro J, Alvarez E, Arranz JA, Laguna R, Uriarte E, and Orallo F, Effects of cis-resveratrol on inflammatory murine macrophages: antioxidant activity and down-regulation of inflammatory genes, *J Leukoc Biol* 75, 1156–1165, 2004.
40. Afaq F, Adhami VM, and Ahmad N, Prevention of short-term ultraviolet B radiation-mediated damages by resveratrol in SKH-1 hairless mice, *Toxicol Appl Pharmacol* 186, 28–37, 2003.
41. Subbaramaiah K, Chung WJ, Michaluart P, Telang N, Tanabe T, Inoue H, Jang M, Pezzuto JM, and Dannenberg AJ, Resveratrol inhibits cyclooxygenase-2 transcription and activity in phorbol ester-treated human mammary epithelial cells, *J Biol Chem* 273, 21875–21882, 1998.
42. Jang M and Pezzuto JM, Effects of resveratrol on 12-O-tetradecanoylphorbol-13-acetate-induced oxidative events and gene expression in mouse skin, *Cancer Lett* 134, 81–89, 1998.
43. Li ZG, Hong T, Shimada Y, Komoto I, Kawabe A, Ding Y, Kaganoi J, Hashimoto Y, and Imamura M, Suppression of N-nitrosomethylbenzylamine (NMBA)-induced esophageal tumorigenesis in F344 rats by resveratrol, *Carcinogenesis* 23, 1531–1536, 2002.

44. Jang M and Pezzuto JM, Cancer chemopreventive activity of resveratrol, *Drugs Exp Clin Res* 25, 65–77, 1999.
45. Rahman I, Marwick J, and Kirkham P, Redox modulation of chromatin remodeling: impact on histone acetylation and deacetylation, NF-kappaB and pro-inflammatory gene expression, *Biochem Pharmacol* 68, 1255–1267, 2004.
46. Ramsay RG, Ciznadija D, Vanevski M, and Mantamadiotis T, Transcriptional regulation of cyclo-oxygenase expression: three pillars of control, *Int J Immunopathol Pharmacol* 16, 59–67, 2003.
47. Kleinert H, Pautz A, Linker K, and Schwarz PM, Regulation of the expression of inducible nitric oxide synthase, *Eur J Pharmacol* 500, 255–266, 2004.
48. Greten FR, Eckmann L, Greten TF, Park JM, Li ZW, Egan LJ, Kagnoff MF, and Karin M, IKKbeta links inflammation and tumorigenesis in a mouse model of colitis-associated cancer, *Cell* 118, 285–296, 2004.
49. Cho DI, Koo NY, Chung WJ, Kim TS, Ryu SY, Im SY, and Kim KM, Effects of resveratrol-related hydroxystilbenes on the nitric oxide production in macrophage cells: structural requirements and mechanism of action, *Life Sci* 71, 2071–2082, 2002.
50. Tsai SH, Lin-Shiau SY, and Lin JK, Suppression of nitric oxide synthase and the down-regulation of the activation of NFkappaB in macrophages by resveratrol, *Br J Pharmacol* 126, 673–680, 1999.
51. Wadsworth TL and Koop DR, Effects of the wine polyphenolics quercetin and resveratrol on pro-inflammatory cytokine expression in RAW 264.7 macrophages, *Biochem Pharmacol* 57, 941–949, 1999.
52. Holmes-McNary M and Baldwin AS Jr, Chemopreventive properties of trans-resveratrol are associated with inhibition of activation of the IkappaB kinase, *Cancer Res* 60, 3477–3483, 2000.
53. Adhami VM, Afaq F, and Ahmad N, Suppression of ultraviolet B exposure-mediated activation of NF-kappaB in normal human keratinocytes by resveratrol, *Neoplasia* 5, 74–82, 2003.
54. Manna SK, Mukhopadhyay A, and Aggarwal BB, Resveratrol suppresses TNF-induced activation of nuclear transcription factors NF-kappa B, activator protein-1, and apoptosis: potential role of reactive oxygen intermediates and lipid peroxidation, *J Immunol* 164, 6509–6519, 2000.
55. Jeong WS, Kim IW, Hu R, and Kong AN, Modulatory properties of various natural chemopreventive agents on the activation of NF-kappaB signaling pathway, *Pharm Res* 21, 661–670, 2004.
56. Estrov Z, Shishodia S, Faderl S, Harris D, Van Q, Kantarjian HM, Talpaz M, and Aggarwal BB, Resveratrol blocks interleukin-1beta-induced activation of the nuclear transcription factor NF-kappaB, inhibits proliferation, causes S-phase arrest, and induces apoptosis of acute myeloid leukemia cells, *Blood* 102, 987–995, 2003.
57. Banerjee S, Bueso-Ramos C, and Aggarwal BB, Suppression of 7,12-dimethylbenz(a)anthracene-induced mammary carcinogenesis in rats by resveratrol: role of nuclear factor-kappaB, cyclooxygenase 2, and matrix metalloprotease 9, *Cancer Res* 62, 4945–4954, 2002.
58. Leonard SS, Xia C, Jiang BH, Stinefelt B, Klandorf H, Harris GK, and Shi X, Resveratrol scavenges reactive oxygen species and effects radical-induced cellular responses, *Biochem Biophys Res Commun* 309, 1017–1026, 2003.

59. Pendurthi UR, Meng F, Mackman N, and Rao LV, Mechanism of resveratrol-mediated suppression of tissue factor gene expression, *Thromb Haemost* 87, 155–162, 2002.
60. Angel P and Karin M, The role of Jun, Fos and the AP-1 complex in cell-proliferation and transformation, *Biochim Biophys Acta* 1072, 129–157, 1991.
61. Wisdom R, AP-1: one switch for many signals, *Exp Cell Res* 253, 180–185, 1999.
62. Young MR, Yang HS, and Colburn NH, Promising molecular targets for cancer prevention: AP-1, NF-kappa B and Pdcd4, *Trends Mol Med* 9, 36–41, 2003.
63. Li YT, Shen F, Liu BH, and Cheng GF, Resveratrol inhibits matrix metalloproteinase-9 transcription in U937 cells, *Acta Pharmacol Sin* 24, 1167–1171, 2003.
64. Shen F, Chen SJ, Dong XJ, Zhong H, Li YT, and Cheng GF, Suppression of IL-8 gene transcription by resveratrol in phorbol ester treated human monocytic cells, *J Asian Nat Prod Res* 5, 151–157, 2003.
65. Asou H, Koshizuka K, Kyo T, Takata N, Kamada N, and Koeffier HP, Resveratrol, a natural product derived from grapes, is a new inducer of differentiation in human myeloid leukemias, *Int J Hematol* 75, 528–533, 2002.
66. Shih A, Davis FB, Lin HY, and Davis PJ, Resveratrol induces apoptosis in thyroid cancer cell lines via a MAPK- and p53-dependent mechanism, *J Clin Endocrinol Metab* 87, 1223–1232, 2002.
67. Wolter F, Turchanowa L, and Stein J, Resveratrol-induced modification of polyamine metabolism is accompanied by induction of c-Fos, *Carcinogenesis* 24, 469–474, 2003.
68. Jaffee BD, Manos EJ, Collins RJ, Czerniak PM, Favata MF, Magolda RL, Scherle PA, and Trzaskos JM, Inhibition of MAP kinase kinase (MEK) results in an antiinflammatory response *in vivo*, *Biochem Biophys Res Commun* 268, 647–651, 2000.
69. Niiro H, Otsuka T, Ogami E, Yamaoka K, Nagano S, Akahoshi M, Nakashima H, Arinobu Y, Izuhara K, and Niho Y, MAP kinase pathways as a route for regulatory mechanisms of IL-10 and IL-4 which inhibit COX-2 expression in human monocytes, *Biochem Biophys Res Commun* 250, 200–205, 1998.
70. Guan Z, Buckman SY, Miller BW, Springer LD, and Morrison AR, Interleukin-1beta-induced cyclooxygenase-2 expression requires activation of both c-Jun NH2-terminal kinase and p38 MAPK signal pathways in rat renal mesangial cells, *J Biol Chem* 273, 28670–28676, 1998.
71. Xie W and Herschman HR, v-src induces prostaglandin synthase 2 gene expression by activation of the c-Jun N-terminal kinase and the c-Jun transcription factor, *J Biol Chem* 270, 27622–27628, 1995.
72. Yu R, Hebbar V, Kim DW, Mandlekar S, Pezzuto JM, and Kong AN, Resveratrol inhibits phorbol ester and UV-induced activator protein 1 activation by interfering with mitogen-activated protein kinase pathways, *Mol Pharmacol* 60, 217–224, 2001.
73. Carluccio MA, Siculella L, Ancora MA, Massaro M, Scoditti E, Storelli C, Visioli F, Distante A, and De Caterina R, Olive oil and red wine antioxidant polyphenols inhibit endothelial activation: antiatherogenic properties of Mediterranean diet phytochemicals, *Arterioscler Thromb Vasc Biol* 23, 622–629, 2003.

26 Neuroprotective Effects of Resveratrol

Ying-Shan Han, Stéphane Bastianetto, and Rémi Quirion

CONTENTS

INTRODUCTION

Resveratrol (*trans*-3,4′,5-trihydroxystilbene; Figure 26.1) is a naturally occurring stilbene found in more than 70 plants, particularly in polygonum roots, grape skin, and peanut seeds [1]. In plants, resveratrol functions as a phytoalexin produced in response to injury or fungal attack. Resveratrol has been shown to have a broad range of pharmacological properties, including antioxidative, anticarcinogenic, antiinflammatory, neuroprotective, and antiviral effects [2]. Epidemiological studies indicated that a moderate consumption of alcohol, particularly red wine, reduced the risk of coronary diseases [3] and neurodegenerative diseases such as Alzheimer's disease (AD) [4–6]. Recent *in vitro* and a limited number of *in vivo* studies have demonstrated that resveratrol exerts neuroprotective effects against various models of toxicity induced by oxidative stress [7–12], beta-amyloid peptides (Aβ) [13–16], axotomy [17], and hypoxia/ischemia [18–21], suggesting that this molecule could contribute to the purported beneficial effects of red wine [15,22–25].

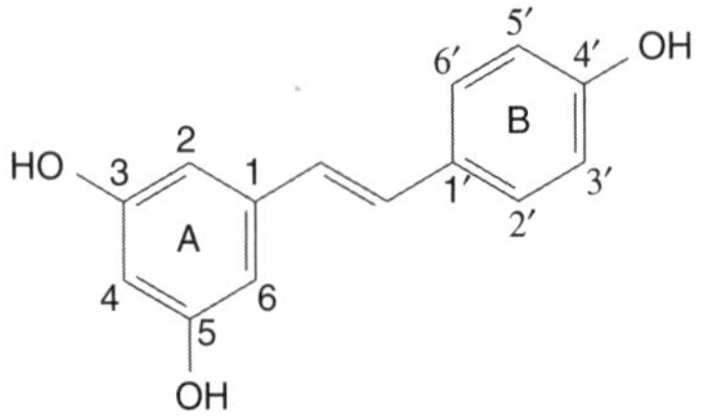

FIGURE 26.1 Structure and numbering scheme of *trans*-resveratrol.

Recent studies raised the hypothesis that inhibition of free radicals is not the sole mechanism by which resveratrol exerts its neuroprotective effects [23,24] and may involve its purported modulatory effects on various intracellular enzymes, such as protein kinase C (PKC) [16], extracellular signal-regulated kinases (ERK1) [26], mitogen-activated protein kinase (MAPK) [27], and heme oxygenase 1 (HEME 1) [28], as well as on the expression of various genes and transcription factors, including nuclear factor-kappaB (NF-κB) [23,24]. These data are of particular interest in the clinical context, given that these intracellular enzymes likely play roles in the neurodegenerative process occurring in diseases such as AD and stroke. This chapter attempts to summarize some of the most recent *in vitro* and animal studies reporting on the neuroprotective effects of resveratrol and its structure–activity relationship. We finally discuss the potential mechanisms of the neuroprotective actions of resveratrol, and the possible relevance of these studies for the treatment of neurodegenerative diseases.

NEUROPROTECTIVE EFFECTS OF RESVERATROL AND ITS ANALOGS

In Vitro and Animal Studies

Numerous *in vitro* and animal studies have shown that resveratrol possesses neuroprotective abilities (Table 26.1). The effects of resveratrol have been widely reported in various types of cultured cells against toxicities induced by oxidative stress, Aβ, and glutamate. For example, resveratrol (25 or 50 μM) effectively protected PC12 cells from cell death induced by Fe^{2+}, *t*-butylhydroperoxide, and ethanol [7,29], as well as by oxidized lipoproteins [8]. Using rat primary hippocampal cell cultures, our laboratory [10,11] found that both co- and posttreatments with resveratrol attenuated hippocampal cell death induced by sodium nitroprusside (SNP) for the same range of concentrations known to block intracellular reactive oxygen species (ROS) accumulation generated by SNP, suggesting that the inhibition of ROS likely contributes to the neuroprotective action of resveratrol, whereas it failed to protect against 3-morpholinosydnonimine

TABLE 26.1
Examples of the Neuroprotective Effects of Resveratrol in Cell Culture and Animal Model Systems

Model system	Insult	Effect	Effective dose	Ref.
PC12	Fe^{2+}, *tert*-butylhydroperoxide, ethanol	Protective effects against lipid peroxidation and cell death	25 μM	7, 29
PC12	Oxidized lipoprotein	Protected the cells from the activation of NF-κB/DNA binding activity and apoptotic cell death		8
PC12	Dopamine + H_2O_2	Decreased cell death	50 μM	9
HT22	Glutamate toxicity	Neuroprotective	20 μM	31
Rat embryonic mesencephalic tissue	*Tert*-butylhydroperoxide	Protected cells from *tert*-butylhydroperoxide-induced toxicity	50, 500 μM	33
Primary cerebellar granule neurons	Low K^+, glutamate	Has no protective effects on apoptosis	5–200 μM	34
Rat primary hippocampal cultured cells	Sodium nitroprusside (SNP), 3-morpholinosydnonimine (SIN-1)	Attenuated hippocampal cell death and intracellular reactive oxygen species accumulation produced by SNP, but no protection against SIN-1-induced toxicity	5–25 μM	10, 11
PC12	H_2O_2	Attenuated H_2O_2-induced cytotoxicity, DNA fragmentation, and ROS	25, 50 μM	12
SH-SY5Y	Paclitaxel	Reduced paclitaxel-induced cellular death	50 μM	32
PC12	1-Methyl-4-phenyl pyridium (MPP+)	Reverted MPP+-induced cellular death	0.1 μM	30
PC12	$A\beta_{1-41}$	Reduced $A\beta_{1-41}$-induced cytotoxicity	25, 40 μM	13
PC12	$A\beta_{25-35}$, $A\beta_{1-42}$	Attenuated β-amyloid-induced cytotoxicity	10 or 25 μM	14

(*continued*)

TABLE 26.1
Continued

Model system	Insult	Effect	Effective dose	Ref.
SH-SY5Y	β-amyloid or H_2O_2	Attenuated H_2O_2 or β-amyloid-induced cytotoxicity	15 μM	15
Rat primary hippocampal cultured cells	$A\beta_{25\text{-}35}$, $A\beta_{1\text{-}42}$	Reduced Aβ-induced cytotoxicity	10-40 μM	16
Primary dorsal root ganglion explant neurons	Mechanical transection	Decreased axonal degradation	10–100 μM	17
Rats	Kainic acid	Partial neuroprotection of *in vivo* excitotoxic brain damage	8 mg/kg	35
Long-Evans rats	Middle cerebral artery (MCA) occlusion	Reduced infarct size	10^{-6}, 10^{-7} g/kg	18
Mongolian gerbils	Common carotid arteries occlusion	Decreased neuronal cell death	30 mg/kg	19
Rats	MCA occlusion	Decreased volume of infarct	20 mg/kg i.p. for 21 days	20
Rats	Intracerebroventricular streptozotocin	Prevented cognitive impairment	10, 20 mg/kg, 21 days	36
Mice, primary cortical cultures, and vascular endothelial cells	Middle cerebral artery occlusion	Reduced infarct size	20 mg/kg, 3 days	21
Rabbits	Spinal cord ischemia	Protected spinal cord from ischemia-reperfusion injury	10 mg/kg	37

(SIN-1)-induced toxicity. Moreover, resveratrol, at concentrations ranging from 0.1 to 500 μM, protected against excitotoxicity in PC12 [9,12,30], HT22 [31], and SH-SY5Y [32] cells as well as rat embryonic mesencephalic tissue [33] and cerebellar granule neurons [34]. Resveratrol (10 to 40 μM) was also shown to promote significantly cell survival in PC12 [12–14], SH-SY5Y [15], and primary hippocampal [16] cells exposed to Aβ peptides. Araki et al. [17] reported that a pretreatment with resveratrol (10 to 100 μM) resulted in a decrease in axonal degradation in primary dorsal root ganglion explant neurons exposed to axotomy. In these studies, neuronal protection was assessed using morphological markers, dye exclusion techniques, lactate dehydrogenase release, and formazan dye conversion.

Although purported effective concentrations of resveratrol occur in a broad range of concentrations varying from 0.1 [30] to 500 μM [33], maximal effects are usually seen at 20 μM [11,14–16,31]. Interestingly, De Ruvo et al. [34] reported that both *trans*- and *cis*-resveratrol (5 to 200 μM) failed to have any protective effects on apoptosis induced by low K^+ concentration in rat primary cultures of cerebellar granule neurons. According to the authors, the lack of effect may be explained by the inability of resveratrol to reach intracellular sites involved in generating ROS. It would be important to confirm that cerebellar granule neurons are particularly resistant to the neuroprotective effects of resveratrol.

Also reported were potent neuroprotective effects of resveratrol in animal models of toxicity induced by ischemia and excitotoxicity (Table 26.1). It has been shown that chronic administration of resveratrol (8 mg/kg) to young-adult rats significantly protected olfactory cortex and hippocampus from neuronal damage caused by systemic injection of the excitotoxin kainic acid [35]. However, resveratrol (8 mg/kg) failed to have significant neuroprotective effects in hippocampal slices in an *ex vivo* model of ischemia [35]. Sinha et al. [20] reported that a pretreatment with *trans*-resveratrol (20 mg/kg, i.p. for 21 days) prevented motor impairment and decreased infarct volume induced by middle cerebral artery. These protective effects were accompanied by increased levels of malondialdehyde, a molecule that reflects polyunsaturated fatty acid peroxidation, and reduced glutathione [20]. It was also reported that pre- and cotreatments with resveratrol (0.1 to 1 μg/kg) significantly reduced infarction size in rats from the damaging effects of focal cerebral ischemia [18]. Moreover, an intraperitoneal treatment with resveratrol (30 mg/kg) significantly decreased neuronal cell death induced by common carotid artery occlusion in Mongolian gerbils [19]. Furthermore, using streptozotocin (STZ) model of sporadic dementia in rats, Sharma and Gupta [36] reported that a treatment with *trans*-resveratrol significantly prevented cognitive impairments induced by intracerebroventricular injection of STZ. Taken together, these results suggest that resveratrol could be effective in the treatment of cognitive decline occurring in neurodegenerative diseases. More recently, it has been shown that resveratrol (10 mg/kg) decreased oxidative stress, increased nitric oxide release, and protected spinal cord from ischemia-reperfusion injury in rabbits [37].

Structure–Activity Relationships of Resveratrol

Since naturally occurring resveratrol analogs such as piceatannol are also present in grape [38], natural or synthetic resveratrol analogs with enhanced neuroprotective activities may be more useful than resveratrol as neuroprotective agents. Unfortunately, structure–activities studies are still very limited and rare. It has been shown that oxyresveratrol (*trans*-2,3′,4,5′-tetrahydroxystilbene) promotes neuronal survival in an *in vivo* model of stroke [39]. In addition, diethylstilbestrol, a stilbene derivative, has also been shown to increase survival in serum-deprived human SK-N-SH neuroblastoma cells [40]. In that study it was shown that a phenolic A ring and at least one phenolic hydroxyl group are necessary for the neuroprotective action of diethylstilbestrol. Using rat primary hippocampal cells and Aβ-induced toxicity as a model system, we recently investigated the neuroprotective potency of 13 natural and synthetic resveratrol analogs with different numbers and positions of hydroxyl and/or methoxy groups on rings A and B. Among the compounds tested, piceatannol (*trans*-3,3′,4′,5-tetrahydroxystilbene) and *trans*-3,5-dimethoxy-4′-hydroxystilbene, a synthetic analog, exerted concentration-dependent neuroprotective effects similar to those seen with resveratrol. In contrast, other compounds such as *trans*-stilbene failed to protect against Aβ-induced toxicity or some compounds such as diethylstilbestrol, 4-hydroxystilbene, and 4,4′-dihydroxystilbene were even cytotoxic. In brief, our results indicated that the presence of 3,5-hydroxy substituents on ring A and 4′-hydroxy or methoxy groups on ring B are essential for the neuroprotective effects of the resveratrol [41].

POTENTIAL MECHANISMS OF NEUROPROTECTIVE ACTIONS OF RESVERATROL

Recent studies focused on the underlying mechanisms of action by which resveratrol exerts its neuroprotective effects. Data obtained from these studies indicated that resveratrol possesses a broad range of actions (i.e., antioxidant/metal-chelating activities, modulation of intracellular signaling pathways, and activation of transcription factors and gene expression) that are responsible for its effects on neuronal survival [23,24]. However, the exact role of each of these mechanisms in resveratrol-mediated neuroprotection remains to be fully elucidated.

Antioxidant/Metal-Chelating Activity

Oxidative stress is likely to mediate neuronal losses and memory deficits occurring in age-related neurodegenerative diseases such as AD and

Parkinson's disease [42]. Antioxidants able to block oxidative damage are thus considered as promising agents to prevent age-related neurodegeneration. Resveratrol is known to have potent antioxidant properties that are likely responsible, at least in part, for its neuroprotective action [43]. Karlsson et al. [33] investigated the ability of resveratrol to protect rat embryonic dopaminergic mesencephalic tissue exposed to the prooxidant *t*-butylhydroperoxide. Using a spin-trapping technique, the authors demonstrated that cytoprotection promoted by resveratrol was due to its ability to scavenge free radicals. Our laboratory previously found that resveratrol inhibited ROS production (mainly peroxides) in the same range of concentration (5 to 25 μM) that is required to protect hippocampal cells against toxicity induced by the free radical NO donor SNP [10]. In addition, other studies have shown that resveratrol has the ability to protect cells from oxidative stress induced by a number of toxic agents including hydroperoxide, ethanol, oxidized lipoprotein, and *tert*-butylhydroperoxide [7–9,20,29,33]. It has also been shown that resveratrol protected low-density lipoprotein against peroxidative degradation via its iron chelating property [44].

Effects on Intracellular Signaling Pathways

In addition to its antioxidant activity, resveratrol may act by modulating intracellular signaling pathways. Miloso et al. [26] have shown that resveratrol can induce the activation of the MAP kinases ERK1 and ERK2 in SH-SY5Y neuroblastoma cells. It was also shown that PC12 cells treated with Aβ exhibited increased accumulation of intracellular ROS and induced apoptosis [14]. Resveratrol attenuated Aβ-induced cytotoxicity and activation of NF-κB in this model [14]. Conte et al. [13] also showed that resveratrol had protective action against $A\beta_{1\text{-}41}$-induced toxicity in PC12 cells that was accompanied by the inhibition of tyrosine kinase activity. Resveratrol is also able to significantly reduce paclitaxel-induced apoptosis in human neuroblastoma SH-SY5Y cells, by modulating several cellular signaling pathways [32]. For example, these authors demonstrated that resveratrol reversed the phosphorylation of the antiapoptotic molecule Bcl-2 and blocked Raf-1 phosphorylation induced by paclitaxel. It also reversed the sustained phosphorylation of JNK/SAPK, known to occur after exposure to paclitaxel [32]. In HL-60 cells, resveratrol, but not the antioxidants hydroxytyrosol and pyrrolidine dithiocarbamate, increased early growth response 1 gene (*egr1*) that encodes for an immediate to early response transcription factor, suggesting that it can modulate selective signal transduction pathways [45]. Recently, we have shown that resveratrol activated the phosphorylation of PKC in rat primary hippocampal cell cultures exposed to Aβ; MAP and Akt kinases were not affected, suggesting that PKC was involved in the neuroprotective action of resveratrol against Aβ-induced toxicity [16].

Effects on Transcription Factors and Gene Expression

There is increasing interest in investigating the effects of resveratrol on transcription factors. Oxidative stress in the central nervous system can cause the oxidation of lipoproteins, which may in turn damage cellular and subcellular membranes and other biomolecules, leading to tissue injury and cell death [42]. NF-κB, a redox-regulated transcription factor, has been recognized as an important molecule that regulates the expression of early response genes involved in cell survival [46]. It has been shown that oxidized low-density lipoprotein (LDL) and very low-density lipoprotein (VLDL) induced cell death in PC12 cells and activated the binding of NF-κB to consensus sequence in the promoter region of target genes, leading to apoptotic cell death. Resveratrol has been reported to protect cells from both the activation of NF-κB and DNA binding activity [8]. In addition, a pretreatment with resveratrol suppressed Aβ-induced activation of NF-κB in PC12 cells [14].

Peroxisome proliferator-activated receptors (PPARs) are members of the nuclear hormone receptor family of ligand-dependent transcription factors that comprise three isotypes: PPARα, PPARβ, and PPARγ. It was reported that PPARα and PPARβ were activated by resveratrol (10 μM) in primary cortical cultures and vascular endothelial cells [21]. Moreover, it was demonstrated that resveratrol requires PPARα expression to exert its neuroprotective effects in focal cerebral ischemia, as resveratrol failed to protect against ischemia in PPARα gene knockout mice [21]. Resveratrol also modulates the transcription and activity of cyclooxygenase-2 (COX-2), an enzyme that is involved in inflammation triggered by prostaglandin E2 [47]. Interestingly, using primary cortical neuronal cultures, resveratrol (5 to 100 μM) was shown to induce significantly heme oxygenase 1, a redox-sensitive inducible protein that acts as a key defense mechanism in neurons exposed to oxidative stress [28]. These authors suggested that increase in heme oxygenase activity induced by resveratrol is a novel pathway by which resveratrol can exert its neuroprotective actions [28]. More recently, the activation of SIRT1, a nicotinamide adenine dinucleotide-dependent deacetylase of histones and other proteins, by resveratrol has been suggested to contribute to its promoting effect on axonal degradation in dorsal root ganglion neurons [17].

CONCLUSION AND FUTURE DIRECTIONS

The studies that are summarized in this chapter support a neuroprotective role of resveratrol through several different mechanisms, including the inhibition of free radicals, metal chelating, the modulation of various intracellular signal pathways, and the regulation of the expression of various genes and transcription factors. Further studies are necessary to clarify the specific role of each pathway modulated by resveratrol in a given pharmacological effect. Ultimately, this may lead to the identification of key molecular targets through which resveratrol can exert its neuroprotective actions. Efforts are also necessary to establish the role of resveratrol metabolites in its neuroprotective effects as well as key elements of structure–activity relationships. Understanding the molecular mechanisms underlying the neuroprotective actions of resveratrol and its derivatives should ultimately prove to be helpful toward the designing of new and more effective neuroprotective agents useful in the treatment of neurodegenerative diseases.

REFERENCES

1. Bavaresco L, Fregoni C, Cantu E, and Trevisan M, Stilbene compounds: from the grapevine to wine, *Drugs Exp Clin Res* 25, 57–63, 1999.
2. Bhat KPL, Kosmeder JW, and Pezzuto JM, Biological effects of resveratrol, *Antioxid Redox Signal* 3, 1041–1064, 2001.
3. Goldberg DM, Hahn SE, and Parkes JG, Beyond alcohol: beverage consumption and cardiovascular mortality, *Clin Chim Acta* 237, 155–187, 1995.
4. Leibovici D, Ritchie K, Ledesert B, and Touchon J, The effects of wine and tobacco consumption on cognitive performance in the elderly: a longitudinal study of relative risk, *Int J Epidemiol* 28, 77–81, 1999.
5. Lemeshow S, Letenneur L, Dartigues JF, Lafont S, Orgogozo JM, and Commenges D, Illustration of analysis taking into account complex survey considerations: the association between wine consumption and dementia in the PAQUID study. Personnes Ages Quid, *Am J Epidemiol* 148, 298–306, 1998.
6. Orgogozo JM, Dartigues JF, Lafont S, Letenneur L, Commenges D, Salamon R, Renaud S, and Breteler MB, Wine consumption and dementia in the elderly: a prospective community study in the Bordeaux area, *Rev Neurol* 153, 185–192, 1997.
7. Chanvitayapongs S, Draczynska-Lusiak B, and Sun AY, Amelioration of oxidative stress by antioxidants and resveratrol in PC12 cells, *Neuroreport* 8, 1499–1502, 1997.
8. Draczynska-Lusiak B, Chen YM, and Sun AY, Oxidized lipoproteins activate NF-kappaB binding activity and apoptosis in PC12 cells, *Neuroreport* 9, 527–532, 1998.
9. Frankel D and Schipper HM, Cysteamine pretreatment of the astroglial substratum (mitochondrial iron sequestration) enhances PC12 cell vulnerability to oxidative injury, *Exp Neurol* 160, 376–385, 1999.
10. Bastianetto S, Zheng WH, and Quirion R, Neuroprotective abilities of resveratrol and other red wine constituents against nitric oxide-related toxicity in cultured hippocampal neurons, *Br J Pharmacol* 131, 711–720, 2000.
11. Bastianetto S and Quirion R, Natural extracts as possible protective agents of brain aging, *Neurobiol Aging* 23, 891–897, 2002.
12. Jang JH and Surh YJ, Protective effects of resveratrol on hydrogen peroxide-induced apoptosis in rat pheochromocytoma (PC12) cells, *Mutation Res* 496, 181–190, 2001.

13. Conte A, Pellegrini S, and Tagliazucchi D, Effect of resveratrol and catechin on PC12 tyrosine kinase activities and their synergistic protection from beta-amyloid toxicity, *Drugs Exp Clin Res* 29, 243–255, 2003.
14. Jang JH and Surh YJ, Protective effect of resveratrol on beta-amyloid-induced oxidative PC12 cell death, *Free Radical Biol Med* 34, 1100–1110, 2003.
15. Savaskan E, Olivieri G, Meier F, Seifritz E, Wirz-Justice A, and Muller-Spahn F, Red wine ingredient resveratrol protects from beta-amyloid neurotoxicity, *Gerontology* 49, 380–383, 2003.
16. Han YS, Zheng WH, Bastianetto S, Chabot JG, and Quirion R, Neuroprotective effects of resveratrol against beta-amyloid-induced neurotoxicity in rat hippocampal neurons: involvement of protein kinase C, *Br J Pharmacol* 141, 997–1005, 2004.
17. Araki T, Sasaki Y, and Milbrandt J, Increased nuclear NAD biosynthesis and SIRT1 activation prevent axonal degeneration, *Science* 305, 1010–1013, 2004.
18. Huang SS, Tsai MC, Chih CL, Hung LM, and Tsai SK, Resveratrol reduction of infarct size in Long-Evans rats subjected to focal cerebral ischemia, *Life Sci* 69, 1057–1065, 2001.
19. Wang Q, Xu J, Rottinghaus GE, Simonyi A, Lubahn D, Sun GY, and Sun AY, Resveratrol protects against global cerebral ischemic injury in gerbils, *Brain Res* 958, 439–447, 2002.
20. Sinha K, Chaudhary G, and Gupta YK, Protective effect of resveratrol against oxidative stress in middle cerebral artery occlusion model of stroke in rats, *Life Sci* 71, 655–665, 2002.
21. Inoue H, Jiang XF, Katayama T, Osada S, Umesono K, and Namura S, Brain protection by resveratrol and fenofibrate against stroke requires peroxisome proliferator-activated receptor alpha in mice, *Neurosci Lett* 352, 203–206, 2003.
22. Pace-Asciak CR, Rounova O, Hahn SE, Diamandis EP, and Goldberg DM, Wines and grape juices as modulators of platelet aggregation in healthy human subjects, *Clin Chim Acta* 246, 163–182, 1996.
23. Bastianetto S and Quirion R, Resveratrol and red wine constituents: evaluation of their neuroprotective properties, *Pharm. News* 8, 33–38, 2001.
24. Sun AY, Simonyi A, and Sun GY, The "French Paradox" and beyond: neuroprotective effects of polyphenols, *Free Radical Biol Med* 32, 314–318, 2002.
25. Russo A, Palumbo M, Aliano C, Lempereur L, Scoto G, and Renis M, Red wine micronutrients as protective agents in Alzheimer-like induced insult, *Life Sci* 72, 2369–2379, 2003.
26. Miloso M, Bertelli AA, Nicolini G, and Tredici G, Resveratrol-induced activation of the mitogen-activated protein kinases, ERK1 and ERK2, in human neuroblastoma SH-SY5Y cells, *Neurosci Lett* 264, 141–144, 1999.
27. Tredici G, Miloso M, Nicolini G, Galbiati S, Cavaletti G, and Bertelli A, Resveratrol, map kinases and neuronal cells: might wine be a neuroprotectant?, *Drugs Exp Clin Res* 25, 99–103, 1999.
28. Zhuang H, Kim YS, Koehler RC, and Dore S, Potential mechanism by which resveratrol, a red wine constituent, protects neurons, *Ann NY Acad Sci* 993, 276–286, 2003.

29. Sun AY, Chen YM, James-Kracke M, Wixom P, and Cheng Y, Ethanol-induced cell death by lipid peroxidation in PC12 cells, *Neurochem Res* 22, 1187–1192, 1997.
30. Gelinas S and Martinoli MG, Neuroprotective effect of estradiol and phytoestrogens on MPP+-induced cytotoxicity in neuronal PC12 cells, *J Neurosci Res* 70, 90–96, 2002.
31. Moosmann B and Behl C, The antioxidant neuroprotective effects of estrogens and phenolic compounds are independent from their estrogenic properties, *Proc Natl Acad Sci USA* 96, 8867–8872, 1999.
32. Nicolini G, Rigolio R, Scuteri A, Miloso M, Saccomanno D, Cavaletti G, and Tredici G, Effect of *trans*-resveratrol on signal transduction pathways involved in paclitaxel-induced apoptosis in human neuroblastoma SH-SY5Y cells, *Neurochem Int* 42, 419–429, 2003.
33. Karlsson J, Emgard M, Brundin P, and Burkitt MJ, *Trans*-resveratrol protects embryonic mesencephalic cells from tert-butyl hydroperoxide: electron paramagnetic resonance spin trapping evidence for a radical scavenging mechanism, *J Neurochem* 75, 141–150, 2000.
34. De Ruvo C, Amodio R, Algeri S, Martelli N, Intilangelo A, D'Ancona GM, and Esposito E, Nutritional antioxidants as antidegenerative agents, *Int J Dev Neurosci* 18, 359–366, 2000.
35. Virgili M and Contestabile A, Partial neuroprotection of *in vivo* excitotoxic brain damage by chronic administration of the red wine antioxidant agent, *trans*-resveratrol in rats, *Neurosci Lett* 281, 123–126, 2000.
36. Sharma M and Gupta YK, Chronic treatment with *trans* resveratrol prevents intracerebroventricular streptozotocin induced cognitive impairment and oxidative stress in rats, *Life Sci* 71, 2489–2498, 2002.
37. Kiziltepe U, Turan NN, Han U, Ulus AT, and Akar F, Resveratrol, a red wine polyphenol, protects spinal cord from ischemia-reperfusion injury, *J Vasc Surg* 40, 138–145, 2004.
38. Sovak M, Grape extract, resveratrol, and its analogs: a review, *J Med Food* 4, 93–105, 2001.
39. Andrabi SA, Spina MG, Lorenz P, Ebmeyer U, Wolf G, and Horn TF, Oxyresveratrol (*trans*-2,3′,4,5′-tetrahydroxystilbene) is neuroprotective and inhibits the apoptotic cell death in transient cerebral ischemia, *Brain Res* 1017, 98–107, 2004.
40. Green PS, Gordon K, and Simpkins JW, Phenolic A ring requirement for the neuroprotective effects of steroids, *J Steroid Biochem Mol Biol* 63, 229–235, 1997.
41. Han Y, Dumont Y, and Quirion R, On a Possible Novel Mechanism of Action for Resveratrol, Program No. 219.9, abstract, Society for Neuroscience's 34th Annual Meeting, San Diego, Oct. 23–27, 2004.
42. Smith MA, Perry G, Richey PL, Sayre LM, Anderson VE, Beal MF, and Kowall N, Oxidative damage in Alzheimer's, *Nature* 382, 120–121, 1996.
43. Miller NJ and Rice-Evans CA, Antioxidant activity of resveratrol in red wine, *Clin Chem* 41, 1789, 1995.
44. Belguendouz L, Fremont L, and Linard A, Resveratrol inhibits metal ion-dependent and independent peroxidation of porcine low-density lipoproteins, *Biochem Pharmacol* 53, 1347–1355, 1997.

45. Della Ragione F, Cucciolla V, Criniti V, Indaco S, Borriello A, and Zappia V, Antioxidants induce different phenotypes by a distinct modulation of signal transduction, *FEBS Lett* 532, 289–294, 2002.
46. Martindale JL and Holbrook NJ, Cellular response to oxidative stress: signaling for suicide and survival, *J Cell Physiol* 192, 1–15, 2002.
47. Subbaramaiah K, Chung WJ, Michaluart P, Telang N, Tanabe T, Inoue H, Jang M, Pezzuto JM, and Dannenberg AJ, Resveratrol inhibits cyclooxygenase-2 transcription and activity in phorbol ester-treated human mammary epithelial cells, *J Biol Chem* 273, 21875–21882, 1998.

27 Pharmacokinetics and Metabolism of Resveratrol

Alberto A. E. Bertelli

CONTENTS

INTRODUCTION

Resveratrol is a naturally occurring compound that is found in a small number of plants. Some of them have been used throughout the ages in Chinese and Japanese medicine for the treatment of dyslipidemia, arterial hypertension, arteriosclerosis, and a variety of inflammatory and allergic conditions. Indeed, one of them, Huzang (*Rhizoma polygoni cuspidati*), is still included in the Chinese pharmacopoeia.

In the last few decades Chinese and Japanese researchers have made efforts to identify the active ingredients of their traditional herbal remedies. One of the active ingredients they found was resveratrol and they established a number of its important properties: inhibition of lipid peroxidation [1], 5-lipoxygenase and cyclooxygenase products in rat peritoneal polymorphonuclear (PMN) leukocytes [2], platelet aggregation in rabbits [3], and lysosomal enzyme release from human PMN leukocytes [4].

In 1992 French researchers drew attention to an epidemiological study that showed that the consumption of saturated fat was high, but coronary artery disease mortality was low in France: a phenomenon that was called the "French paradox". The discrepancy was attributed to extensive, but moderate, alcohol consumption associated with a reduction in platelet aggregation [5]. Subsequently, resveratrol was detected in wine, and Siemann and Creasy [6] speculated that it could be responsible for the French paradox. Studies on the antiplatelet properties of wine supported this hypothesis, as this property was found to be related to the concentration of resveratrol in wine [7]. However, two questions needed to be answered before such a hypothesis could be accepted:

1. Is resveratrol absorbed from the gut?
2. If it is, are plasma levels sufficiently high to achieve biological effects *in vivo*?

RESVERATROL ABSORPTION

A study was performed with 84 male Wistar rats with a mean body weight of 300 g [8]. The rats were given food and drink *ad libitum* and were kept according to the policy set down by National Research Council guidelines.

The rats were randomly subdivided into two groups, each consisting of 42 animals. In the first group 36 rats were given a single dose of 4 ml red wine (Cabernet Sauvignon from central Italy containing 6.5 mg/l of *cis*- and *trans*-resveratrol), equivalent to 80 μg resveratrol/kg body weight by gavage; the other 6 rats were used as controls. The rats were sacrificed in groups of 6 immediately before administration (control group), and after 0.5, 1, 2, 4, 8, and 12 h. Blood samples were collected and the heart, liver, and kidneys were removed from each rat to measure resveratrol levels.

In the second group 36 rats were given 2 ml of the same wine daily for 15 days and the remaining 6 rats were used as controls. The animals were sacrificed, and blood samples and the main organs were collected for the measurement of resveratrol concentrations as above; also urine was collected during the last 24 h before termination.

Resveratrol was extracted from the tissues, purified and analyzed with HPLC-UVI visible-HPLC/MS equipment. The detection limit was 1 ng/ml in serum and 1 ng/g in tissue.

The concentrations of *trans*- and *cis*-resveratrol measured in plasma and the organs at the various time points are shown in Figure 27.1. Resveratrol was absorbed rapidly, being detectable in both plasma and the liver already 30 min after the single administration. Fairly similar peak concentrations around 20 ng/ml were achieved in the liver already after 30 min, and in the plasma and the kidneys after 1 h. Concentrations in the heart were much lower and achieved later, after 2 h. After 4 h the compound was no longer detectable in the plasma, but tissue levels were still measurable. No resveratrol was detectable 12 h after a single administration.

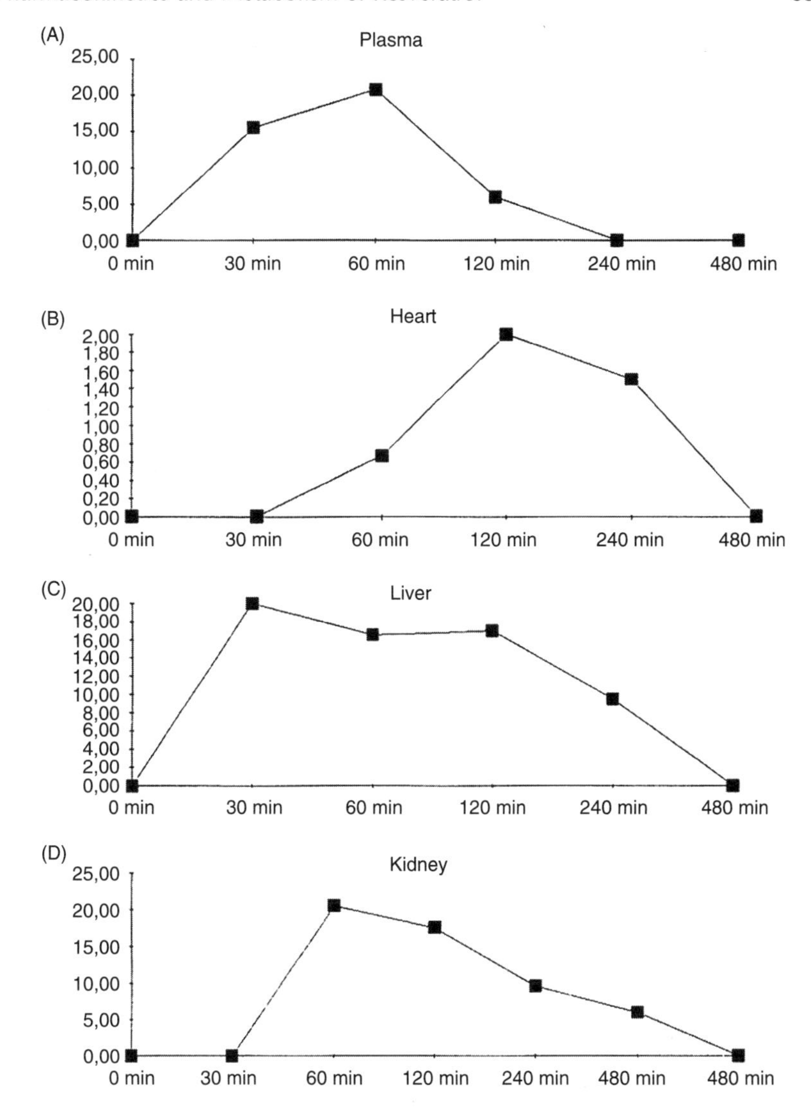

FIGURE 27.1 Total resveratrol (*trans* and *cis*) concentrations in rats. Ordinates are: ng/ml for (A) plasma and ng/g for (B) heart, (C) liver, and (D) kidney. (From Bertelli AAE, Giovannini L, Stradi R, Bertelli A, and Tillement JP, *Int J Tissue React* 18, 67–71, 1996.)

After chronic administration for 15 days the mean plasma concentration was 7.6 ± 0.55 ng/ml. The concentrations in the liver, urine, and kidney were much higher (53.5 ± 1.5, 66.2 ± 1.7, and 44 ± 1.3 ng/ml, respectively). Resveratrol was also detectable in the heart, but at much lower concentrations as was the case after single administrations (3 ± 0.3 ng/ml).

The data show that resveratrol is readily absorbed from the gut and that measurable concentrations are achieved in various tissues. The high concentrations in the urine and the kidney suggest that the compound is mainly excreted by the urinary tract, whereas the high concentrations in the liver suggest that it accumulates there. Tissue concentrations increased after chronic treatment, suggesting that regular wine consumption could be associated with higher concentrations than those measured in this study.

However, subsequent animal work showed that small quantities of resveratrol are absorbed and that bioavailability following oral administration is low. Meng et al [9] established that not more than 2.3% of the dose of resveratrol given in concentrated grape juice to mice is excreted in the urine. Another research group [10] found low tissue concentrations (< 1 nmol/g fresh tissue) in the brain, lung, liver, and kidney, which paralleled those in plasma, after oral administration of 20 mg *trans*-resveratrol/kg, given to mice in the absence of grape juice or wine.

RESVERATROL KINETICS IN ANIMALS

The data of the study described in the previous section on absorption were used to assess the kinetics of resveratrol, which were modeled by the MicroPharm-K software, ascribing time–concentration data to one or two open compartment models [11]. When the two-compartment model was used, the plasma concentrations were ascribed to the central compartment and the tissue concentrations were ascribed to the peripheral compartment. A model overview is provided in Figure 27.2.

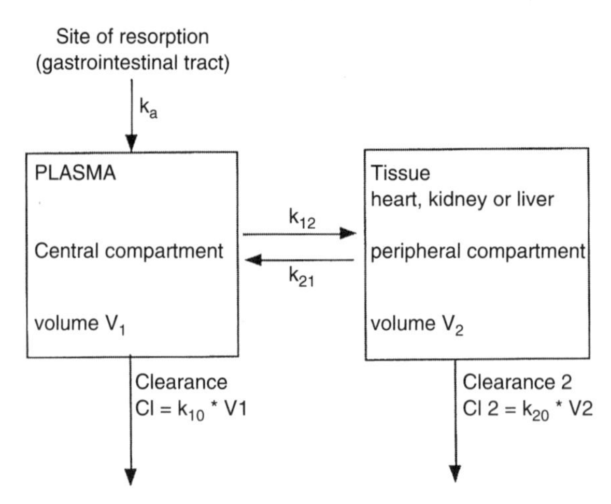

FIGURE 27.2 Schematic representation of the model used to fit the resveratrol plasma and tissue time–concentration data simultaneously. (From Bertelli AAE, Giovannini L, Stradi R, Urien S, Tillement JP, and Bertelli A, *Int J Clin Pharm Res* 16, 77–81, 1996.)

Resveratrol was detected in the plasma at only three time points (30 min, 1 h, and 2 h), so the time–concentration data could be ascribed only to a one-compartment model; moreover, the last concentration could not be adequately fitted (Figure 27.3). However, tissue concentrations were available at later time points. The delay in tissue concentration peaks and the fact that the concentrations in the liver and in the kidneys were higher than in the plasma suggested that resveratrol distributes into a peripheral compartment, so the time–concentration tissue data were fitted taking the additional compartment into account according to the scheme in Figure 27.2. This approach is illustrated in Figure 27.3 and the results are shown in Table 27.1. Plasma and tissue data were analyzed simultaneously for each organ, so that information on the slow elimination phase in the plasma could be obtained from the tissue data.

Interestingly, the main pharmacokinetic parameters (clearance, central volume $V1$ and absorption rate Ka) obtained by this approach were similar to the parameters obtained using the one-compartment model. When the comparison was refined by combining the analyses of plasma data with heart data, and plasma data with kidney data, the estimates of the main pharmacokinetic data, including those obtained from one-compartment modeling, were still similar. In the kidney, the model with an additional elimination pathway from the tissue itself (clearance 2) was better than the model with a single elimination pathway from the central compartment. This indicates that the kidney is an important route for the elimination of resveratrol.

The following kinetic parameters were derived from plasma + kidney modeling and best-curve fitting:

- Resorption half-life $(\log 2/Ka) = 0.46$ h
- Elimination half-life $(\log 2 \times V1/C1) = 0.50$ h
- Distribution alpha half-life $(\log 2/K1) = 0.48$ h
- Terminal beta plasma half-life $(\log 2/K2) = 25$ h
- Elimination plasma half-life $(\log 2/K10) = 0.5$ h

The results obtained combining plasma data with liver data differed considerably from the others. This suggests that the liver acts more as a filter than as a compartment. Thus, the plasma pharmacokinetics of resveratrol appear to be described satisfactorily by an open two-compartment model.

The distribution of resveratrol in the three tissues was also estimated via a noncompartmental approach. After the areas under the time–concentration curves (AUCs) had been calculated following the trapezoidal rule from $t = 0$ to the last sampling time point at which resveratrol had still been detected, tissue to plasma AUC ratios were computed as the expression of relative tissue bioavailability. The plasma AUC resulted to be 26.3 ng h/ml. The AUC in the liver was much higher (57.35 ng h/ml) and the AUC in the kidneys was even higher (77.75 ng h/ml), corresponding to a relative tissue

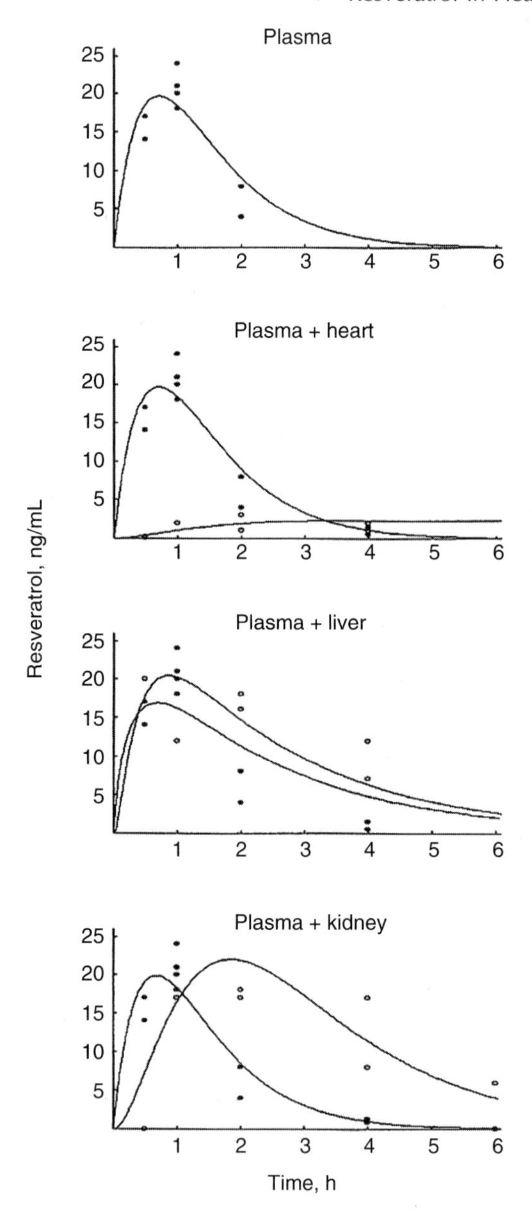

FIGURE 27.3 Curve-fitting of the resveratrol time–concentration data. From top to bottom: plasma alone (one compartment) and combined analyses from scheme 1 (two compartment), plasma + heart, plasma + liver, and plasma + kidney (•, plasma; ○, tissue). The plasma data after 3 h are not observed values (not actually detectable), but low and high values around the predicted value from the one-compartment model for comparison. (From Bertelli AAE, Giovannini L, Stradi R, Urien S, Tillement JP, and Bertelli A, *Int J Clin Pharm Res* 16, 77–81, 1996.)

TABLE 27.1
Mean Total and Free Serum *Trans*-resveratrol Concentrations in Four Healthy Volunteers Given 25 mg *Trans*-resveratrol Orally in Three Different Matrices

Time point (h)	Vegetable juice	White wine	Grape juice
Total (Free and Conjugated) Resveratrol (μg/l)			
0	3	4	2
0.5	471	416	424
1	250	191	344
2	112	154	220
4	56	100	106
Free Resveratrol (μg/l)			
0	0.9	2	1.1
0.5	8.5	7.1	8.0
1	5.3	5.1	4.8
2	1.3	1.8	2.0
4	0.7	1.4	1.8

Adapted from Goldberg DM, Yan J, and Soleas GJ, *Clin Biochem* 36, 79–87, 2003.

bioavailability versus plasma of 218 and 295%, respectively. The AUC in the heart was 6.5 ng h/ml, equivalent to a tissue bioavailability of 24%. Thus, relative bioavailability was high not only in the liver and in the kidneys as expected, but also, albeit to a lesser degree, in the heart. The last heart concentrations were not clearly in the decreasing phase, so the computed bioavailability may actually underestimate the true value.

The contribution of the plasma terminal phase to the total plasma AUC amounted to less than 1%. This is consistent with the fact that plasma concentrations in the terminal phase were so low that they were not detectable.

Thus, the kinetics of resveratrol in plasma following oral administration can be described by an open one- or two-compartment model, as both models produced similar values of the main kinetic parameters. High tissue concentrations are reached in the liver, which appears to act as a filter, and in the kidneys, which appear to be the main route of excretion. Important concentrations are reached also in the heart.

Subsequent experiments [12,13] were performed in isolated preparations of rat small intestine perfused with solutions spiked with resveratrol in physiological, nutritionally relevant concentrations (28, 34, and 57 μmol/l). Resveratrol concentrations were measured by liquid chromatography coupled with mass spectrometry. About 20% of resveratrol was absorbed by the jejunum and ileum, mostly already in metabolized form. More than 80% of the compound was conjugated yielding resveratrol glucuronide and the rest was transformed into resveratrol sulfate. Free resveratrol intestinal tissue concentrations were negligible.

In vitro studies [14] in which *trans*-resveratrol was incubated with rat and human hepatocytes, confirmed that the main metabolites of resveratrol are *trans*-resveratrol-3-*O*-glucuronide and *trans*-resveratrol-3-sulfate and showed that the compound is not subject to oxidation, reduction, or hydrolysis in the liver. Only trace amounts of *cis*-resveratrol were found, a result that indicates that isomerization does not play an important role in resveratrol metabolism. The same research group performed also *in vivo* studies in rats and mice and detected practically no unconjugated resveratrol in urine or serum samples.

Trans-resveratrol (50 mg/kg) was administered orally to intact rats in a solution of beta-cyclodextrin, and the aglycone and glucuronide forms were measured using an electrospray ionization/liquid chromatography/tandem mass spectrometry method in another animal study [15]. Enterohepatic recirculation was also studied in a linked rat model. Enterohepatic recirculation of resveratrol proved to be important in the kinetic profile of the compound, as it accounts for 24% of the exposure to the compound following oral administration. This study also documented extensive first-pass glucuronidation and negligible urinary excretion of the parent compound during the first 12 hours after dosing.

In conclusion, animal studies indicate that free resveratrol is absorbed from the gut, but that it is rapidly conjugated.

RESVERATROL KINETICS IN HUMANS: ARE DATA FROM ANIMAL MODELS PREDICTIVE?

The absorption, bioavailability, and metabolism of resveratrol (25 mg ^{14}C-resveratrol given both orally and intravenously) have been studied in humans [16]. At least 70% of the administered dose was absorbed, peak plasma levels of resveratrol and metabolites amounting to 491 ± 90 ng/ml (about 2 μM). However, as was the case in animals, plasma levels of unmodified resveratrol were negligible (< 5 ng/ml). Liquid chromatography/mass spectrometry analysis established that resveratrol metabolites were the result not only of sulfate and glucuronic acid conjugation, but also of hydrogenation of the aliphatic double bond, possibly produced by human intestinal microflora.

The most in-depth investigation on the kinetics of resveratrol in humans is the study by Goldberg et al. [17], who assessed the extent of the absorption of three polyphenols, including resveratrol, in 12 healthy male volunteers aged 25 to 45 years, who were nonsmokers and moderate drinkers, following oral administration in three different matrices: white grape juice, white wine (1998 Chardonnay 11.5% v/v ethanol), and vegetable juice. Resveratrol (25 mg/70 kg body weight) was dissolved in 100 ml of each of the three beverages, which were given on different occasions after an overnight fast to four of the healthy volunteers.

Blood samples were drawn before administration and 0.5, 1, 2, and 4 h later, and urine was collected for 24 h after administration. Resveratrol concentrations were measured by gas chromatography.

Resveratrol was readily absorbed from the gut independently of the matrix used and concentrations were significantly higher than at baseline after all three administrations ($p < 0.001$; Table 27.1). Total absorption was similar in all three matrices, but plasma levels declined at different rates, the rate being faster when resveratrol had been given in vegetable juice and slower when it had been given in grape juice ($p < 0.04$). Free resveratrol accounted for a negligible fraction of total concentrations (1.7 to 1.9%; Table 27.1).

Also 24 h urinary excretion of resveratrol was independent of the matrix used: it amounted to 17% of the dose administered in the vegetable juice, 16.8% of the dose administered in wine, and to 16% of the dose administered in grape juice.

Thus, the data in humans are consistent with the experimental data in rodents, showing that resveratrol is absorbed in the gut, but that it is rapidly metabolized, so that the concentrations of free compound in the plasma are very low.

Rapid glucuronidation of resveratrol following incubation with human liver microsomes has been documented [18,19]. Both *trans*- and *cis*-resveratrol were subject to glucuronidation, which was regio- and stereoselective and resulted in the formation of 3-*O*- and 4′-*O*-glucuronides. Also Wang et al. have found 3-*O*- and 4′-*O*-glucuronides in humans [20].

Resveratrol sulfation has been documented also in human liver and human duodenum [21]. The mean ± standard deviation resveratrol sulfation rate was found to be 90 ± 21 pmol/min/mg cytosolic protein in the liver and 74 ± 60 pmol/min/mg cytosolic protein in the duodenum.

No differences in free resveratrol levels were found in humans according to the matrix used [17]. However, it should be pointed out that the wine used was a white wine. Red wine contains higher concentrations of the flavonoid quercetin than white wine [22]. Quercetin has proved to be able to inhibit both resveratrol sulfation and glucuronidation [19,21]. The IC50 for sulfation is 12 ± 2 pm in the liver and 15 ± 2 pm in the duodenum, and the IC50 for glucuronidation is 10 ± 1 μM. This phenomenon could permit unmetabolized quantities of resveratrol to enter the bloodstream, thus improving the bioavailability of resveratrol in red wine.

RESVERATROL BIOAVAILABILITY AND BIOLOGICAL EFFECTS

The second question, i.e., whether such low plasma concentrations of resveratrol are able to exert biological effects has not been answered.

The statement by Goldberg et al. [17] that "The voluminous literature reporting powerful *in vitro* anticancer and antiinflammatory effects of the free polyphenols is irrelevant, given that they are absorbed as conjugates" should be taken into serious consideration. In view of the very low resveratrol concentrations that were found, only studies in which very low doses were given (100 to 500 nM) should be taken into account.

Resveratrol has proved to exert a number of biological effects even at such low doses. For instance, red wine containing 1.2 mg/l of natural *trans*-resveratrol was still able to inhibit platelet aggregation by approximately 42% even when it was diluted 1000-fold (final resveratrol concentration 1.2 μg/l) [23].

At concentrations as low as 1 μmol/l and 100 nmol/l resveratrol was able to inhibit significantly the expression of intracellular adhesion molecule 1 (ICAM-1) by tumor necrosis factor alpha (TNFα)-stimulated human umbilical vein endothelial cells and vascular cell adhesion molecule 1 (VCAM-1) by lipopolysaccharide-stimulated human saphenous vein endothelial cells, thus blocking the adhesion of monocytes and granulocytes to endothelial cells — an important step in the pathogenesis of atherosclerosis [24].

Overnight incubation with resveratrol at physiologic concentrations (≤1 μmol/l) modulated the NF-κB intracellular signaling pathway after TNFα stimulation of human umbilical vein endothelial cells [25].

Resveratrol is able to exert biological effects also on other kinds of cells, such as human neuroblastoma cells, in which concentrations as low as 1 nM were able to induce phosphorylation of the mitogen-activated protein kinases (MAPK) extracellular signal-regulated kinase (ERK) 1 and ERK2, whereas higher concentrations (50 to 100 μM) inhibited MAP kinase phosphorylation [26].

Resveratrol has also proved to be able to protect purified rat brain mitochondria submitted to anoxia and reoxygenation from dysfunction at very low concentrations. Oxygen consumption by mitochondria in the presence of NADH and cytochrome c was inhibited with a low EC50 of 18.34 pM. At a concentration of 1 nM the compound significantly decreased superoxide anion production. At concentrations < 0.1 μM the compound reversed the respiratory control ratio decrease by 23.3% and at a concentration of 0.1 μM it protected about 70% of mitochondrial membranes [27].

Finally, low-dose resveratrol has also proved to be able to influence sirtuin enzymes, which have been shown to mediate the slowing down of the aging process and the extension of lifespan following calorie restriction in various organisms. In yeast, resveratrol administration actually mimics the effects of calorie restriction by stimulating SIR2 and is associated with a 70% lengthening of lifespan [28].

In conclusion, in the absence of information on the biological effects of resveratrol metabolites, future research on the antiinflammatory and anticancer properties of this compound should take into account its low levels in the human body following oral administration and use them as a reference for designing new studies.

REFERENCES

1. Kimura Y, Ohminami H, Okuda H, Baba K. Kozawa M, and Arichi S, Effects of stilbene components of roots of Polygonum ssp. on liver injury in peroxidized oil-fed rats, *J Med Plant Res* 49, 51–54, 1983.
2. Kimura Y, Okuda H, and Arichi S, Effects of stilbenes on arachidonate metabolism in leukocytes, *Biochim Biophys Acta* 834, 275–278, 1985.
3. Chung M, Teng C, Cheng K, Ko F, and Lin C, An antiplatelet principle of Veratrum formosanum, *Planta Med* 58, 274–276, 1992.
4. Kimura Y, Okuda H, and Kubo M, Effects of stilbenes isolated from medicinal plants on arachidonate metabolism and degranulation in human polymorphonuclear leukocytes, *J Ethnopharmacol* 45, 131–139, 1995.
5. Renaud S and De Lorgeril M, Wine, alcohol, platelets, and the French paradox for coronary heart disease, *Lancet* 339, 1523–1526, 1992.
6. Siemann EH and Creasy LL, Concentration of the phytoalexin resveratrol in wine, *Am J Enol Vitic* 43, 49–52, 1992.
7. Bertelli AAE, Giovannini L, Giannessi D, Migliori M, Bernini W, Fregoni M, and Bertelli A, Antiplatelet activity of synthetic and natural resveratrol in red wine, *Int J Tissue React* XVII, 1–3, 1995.
8. Bertelli AAE, Giovannini L, Stradi R, Bertelli A, and Tillement JP, Plasma, urine and tissue levels of trans- and cis-resveratrol (3,4′-trihydroxystilbene) after short-term or prolonged administration of red wine to rats, *Int J Tissue React* 18, 67–71, 1996.
9. Meng X, Maliakal P, Lu H, Lee MJ, and Yang CS, Urinary and plasma levels of resveratrol and quercetin in humans, mice and rats after ingestion of pure compounds and grape juice, *J Agric Food Chem* 52, 935–942, 2004.
10. Asensi M, Medina I, Ortega A, Carretero J, Baño MC, Obrador E, and Estrema JM, Inhibition of cancer growth by resveratrol is related to its low bioavailability, *Free Radical Biol Med* 33, 387–398, 2002.
11. Bertelli AAE, Giovannini L, Stradi R, Urien S, Tillement JP, and Bertelli A, Kinetics of trans- and cis-resveratrol (3,4′, 5-trihydroxystilbene) after red wine oral administration in rats, *Int J Clin Pharm Res* 16, 77–81, 1996.
12. Andlauer W, Kolb J, Siebert K, and Fuerst P, Assessment of resveratrol bioavailability in the perfused small intestine of the rat, *Drugs Exp Clin Res* 226, 47–55, 2000.
13. Kuhnle G, Spencer JPE, Chowrimootoo G, Schroeter H, Debnam ES, Srai SKS, Rice-Evans C, and Hahn U, Resveratrol is absorbed in the small intestine as resveratrol glucuronide, *Biochem Biophys Res Commun* 272, 212–217, 2000.
14. Yu C, Shin YG, Chow A, Li Y, Kosmeder JW, Lee YS, Hirschelman WH, Pezzuto JM, Mehta RG, and van Breemen RB, Human, rat and mouse metabolism of resveratrol, *Pharm Res* 19, 1907–1914, 2002.
15. Marier JF, Vachon P, Gritsas A, Zhang J, Moreau JP, and Ducharme MP, Metabolism and disposition of resveratrol in rats: extent of absorption,

glucuronidation, and enteropatic recirculation evidenced by a linked-rat model, *J Pharmacol Exp Ther* 302, 369–373, 2002.

16. Walle T, Hsieh F, Delegge MH, Oatis JE Jr, and Walle UK, High absorption but very low bioavailability of oral resveratrol in humans, *Drug Metab Disposition* 32, 1377–1382, 2004.
17. Goldberg DM, Yan J, and Soleas GJ, Absorption of three wine-related polyphenols in three different matrices by healthy subjects, *Clin Biochem* 36, 79–87, 2003.
18. Aumont V, Krisa S, Battaglia E, Netter P, Richard T, Merillon JM, Magdalou J, and Sabolovic N, Regioselective and stereospecific glucuronidation of trans- and cis-resveratrol in human, *Arch Biochem Biophysics* 393, 281–289, 2001.
19. De Santi C, Pietrabissa A, Mosca F, and Pacifici GM, Glucuronidation of resveratrol, a natural product present in grape and wine, in the human liver. *Xenobiotica* 30, 1047–1054, 2000.
20. Wang LX, Heredia A, Song H, Zhang Z, Yu B, Davis C, and Redfield R, Resveratrol glucuronides as the metabolites of resveratrol in humans: characterization, synthesis and anti-HIV activity, *J Pharm Sci* 93, 2448–2457, 2004.
21. De Santi C, Pietrabissa A, Spisni R, Mosca F, and Pacifici GM, Sulphation of resveratrol, a natural product present in grapes and wine, in the human liver and duodenum, *Xenobiotica* 30, 609–617, 2000.
22. Flesch M, Schwarz A, and Bohm M, Effects of red and white wine on endothelium-dependent vasorelaxation of rat aorta and human coronary arteries, *Am J Physiol* 275, H1183–H1190, 1998.
23. Bertelli AAE, Giovannini L, Giannessi D, Migliori M, Bernini W, Fregoni M, and Bertelli A, Antiplatelet activity of synthetic and natural resveratrol in red wine, *Int J Tissue React* 17, 1–3, 1995.
24. Ferrero ME, Bertelli AAE, Fulgenzi A, Pellegatta F, Corsi MM, Bonfrate M, Ferrara F, De Caterina R, Giovannini L, and Bertelli A, Activity *in vitro* of resveratrol on granulocyte and monocyte adhesion to endothelium, *Am J Clin Nutr* 68, 1208–1214, 1998.
25. Pellegatta F, Bertelli AAE, Staels B, Duhem C, Fulgenzi A, and Ferrero ME, Different short- and long-term effects of resveratrol on nuclear factor-KB phosphorylation and nuclear appearance in human endothelial cells, *Am J Clin Nutr* 77, 1220–1228, 2003.
26. Miloso M, Bertelli AAE, Nicolini G, and Tredici G, Resveratrol-induced activation of the mitogen-activated protein kinases, ERK1 and ERK2, in human neuroblastoma SH-SY5Y cells, *Neuroscience Lett* 264, 141–144, 1999.
27. Zini R, Morin C, Bertelli A, Bertelli AAE, and Tillement JP, Resveratrol-induced limitation of dysfunction of mitochondria isolated from rat brain in an anoxia-reoxygenation model, *Life Sci* 71, 3091–3108, 2002.
28. Howitz KT, Bitterman KJ, Cohen HY, Lamming DW, Lavu S, Wood JG, Zipkin RE, Chung P, Kisielewski A, Zhang LL, Scherer B, and Sinclair DA, Small molecule activators of sirtuins extend Saccharomyces cerevisiae lifespan, *Nature* 425, 191–196, 2003.

Index

C

D

F

G

H

I

J

K

M

N

Q

R

T

U

V